INVASIVE AQUATIC SPECIES OF EUROPE. DISTRIBUTION, IMPACTS AND MANAGEMENT

INVASIVE AQUATIC SPECIES OF EUROPE. DISTRIBUTION, IMPACTS AND MANAGEMENT

Edited by

Erkki Leppäkoski

Department of Biology,
Åbo Akademi University,
Turku, Finland

Stephan Gollasch

GoConsult,
Hamburg, Germany

and

Sergej Olenin

Coastal Research and Planning Institute,
Klaipeda University,
Klaipeda, Lithuania

KLUWER ACADEMIC PUBLISHERS
DORDRECHT / BOSTON / LONDON

A C.I.P. Catalogue record for this book is available from the Library of Congress.

ISBN 1-4020-0837-6

Published by Kluwer Academic Publishers,
P.O. Box 17, 3300 AA Dordrecht, The Netherlands.

Sold and distributed in North, Central and South America
by Kluwer Academic Publishers,
101 Philip Drive, Norwell, MA 02061, U.S.A.

In all other countries, sold and distributed
by Kluwer Academic Publishers,
P.O. Box 322, 3300 AH Dordrecht, The Netherlands.

Printed on acid-free paper

Photo:Stephan Gollasch, 1998
Juvenile Chinese Mitten Crabs migrating upstream in the river Elbe near Hamburg. The crab was first recorded in German waters in 1912. Mass developments occurred in 1930s, 1960s and 1990s.

Printed in the Netherlands.

Contents

Introduction

Who is Who Among Nonindigenous Species

Vectors

Regional Overviews

Impacts

Risk Assessment

Treatment Measures

Databases

Epilogue

Preface

The global scale of alien species invasions is becoming more and more evident in the beginning of the new millennium. Though the problem of biological invasions became a rapidly growing research area, there are large gaps still, both geographically and thematically, to be filled in the near future.

This book is the first attempt to provide an overall picture of aquatic species invasions in Europe. Its geographical scope stretches from Irish waters in the west to Volga River and the Caspian Sea in the east, and from Mediterranean in the south up to the Arctic coast of Europe. Not all parts of the continent could be equally covered, as in some countries species invasions are not studied yet. The book tends to represent the array of all major European aquatic systems on the broadest geographical and ecological scope possible from fully saline seas, semi-enclosed brackish water bodies and coastal lagoons to freshwater lakes, major river systems and waterways. The key objectives include the present status and impacts caused by non-native aquatic species in European waters. Please note that lengthy species lists submitted for publication and additional information were put on the Internet, as the electronical version of these tables benefits from computer assisted search for data (http://www.ku.lt/nemo/EuroAquaInvaders.htm).

Altogether more than 100 scientists from 24 countries have joined to synthesize the available information on bioinvasions. However, the book does not claim to be fully comprehensive. To complete the picture, we asked non-European authors to contribute with facts not yet being dealt with in the European research arena.

The idea of this volume originates from activities of the Baltic Marine Biologists (BMB), a non-governmental scientific organisation that established a working group on nonindigenous estuarine and marine species in 1994. The book benefited a lot from co-operation with similar working groups of the International Maritime Organization (IMO), the International Council for the Exploration of the Sea (ICES), and the Nordic Council of Ministers. We are indebted to these organisations as well as to the Academy of Finland for funding major parts of the editorial work. We thank all our colleagues for their individual efforts in gathering information in many hardly accessible and scattered local sources and making this available to an international reader. We also gratefully acknowledge Camilla Roos, MSc, for her patient editing of miscellaneous versions of "continental" English and technical assistance in shaping up of the book.

Turku, Hamburg and Klaipeda, June 2002

The Editors

Introduction

ALIEN SPECIES IN EUROPEAN WATERS

ERKKI LEPPÄKOSKI[1*], STEPHAN GOLLASCH[2] &
SERGEJ OLENIN[3]
[1]*Department of Biology, Åbo Akademi University, Turku, Finland*
[2]*GoConsult, Hamburg, Germany*
[3]*Coastal Research and Planning Institute, Klaipeda University, Klaipeda, Lithuania*
[*]Corresponding author *eleppakoski@abo.fi*

Abstract
It is evident that the history of introduction of nonindigenous aquatic species into Europe dates back to pre-historic times. In this paper, we review from a historical perspective some of the human activities that have facilitated the potential of alien species from all continents but Antarctica to become established in coastal and inland waters of Europe.

1 Biological invasions: global challenge for new age

Biological invasions associated with human activities, resulting in global mixing of previously isolated biota, are recognised as one of the major elements of the global change. Many natural dispersal barriers for both terrestrial and aquatic species have been weakened, whereby the number of both potential invaders and remote source areas has increased. Geographic isolation of seas and continents as a creator and conservator of global biodiversity was breached several centuries ago and continues to melt down at increasing rate in recent decades (Bright 1998; Dukes & Mooney 1999; Leppäkoski & Olenin 2001).

Results of this global exchange of species are evident in most seas and inland waters of Europe. Much of its present structural and functional diversity is of foreign origin. We have defined this human-mediated addition of nonindigenous species as xenodiversity (Gr. *xenos* - strange) to indicate diversity caused by nonindigenous (non-native, alien, exotic, introduced) species (Leppäkoski & Olenin 2000). In some heavily invaded aquatic ecosystems, xenodiversity tends to reach and even exceed native biodiversity in terms of the number of species and life forms, and strongly affects the rate of ecosystem functions. Even if the receiving ecosystems can become more diverse in terms of species number they also become more similar to the rest of the world.

Biological invasions, one of the top research areas in current days, have a long history. Ancient tribes and their primeval agriculture, nomadism and lifestyle as hunter-

E. Leppäkoski et al. (eds.), Invasive Aquatic Species of Europe, 1–6.

gatherers contributed to the spread of plants and animals when people moved across the European continent, carrying ecto- and endoparasites, seeds and other propagules over long distances in similar ways as animals do. The ability of the first settlers to survive and spread was based on multi-source use of resources available. The early tribes most probably carried not only terrestrial plants and animals but also fish and crayfish from one lake to another. Those actions in old historical times are naturally not documented; this is true even for a great number of introductions dating back to the 18th and 19th centuries only.

The early intentional transfers at the dawn of agriculture and stocking of waters with edible species remain unknown and nameless. The speed at which transfers of species happen has increased dramatically. The world's growing human population in combination with the inherent mobility and entrepreneurship of the human race is revealed in accelerating mixing of aboriginal plant and animal life continent and world wide created in the course of millions of years of evolution. For the aquatic life, the "discovery" of the Americas was a turning point. In the year 2000 it was 1,000 years since the Viking Leif Eirikson set foot on North American soil. The early explorers opened an invasion corridor across the Atlantic Ocean. This pathway was broadened by Columbus and his successors 500 years later. The introduction of nonindigenous species became a global issue. Nearly half a century ago, Elton (1958) wrote in his classical work on invasions: "We must make no mistake: we are seeing one of the great historical convulsions in the world's fauna and flora."

As each introduced species poses an impact to the environment and, in many cases, economies and human health, biological invasions are of high concern for a plethora of human activities from nature conservation to agriculture, shipping industry and biosecurity policy. Non-native species are living organisms that tend to reproduce and spread in contrast to chemical pollutants and most other forms of environmental degradation. Harmful chemical substances can be forbidden or substituted by less harmful ones, and discharges of them can be reduced or even stopped. The nature will recover from chemical or physical impacts within shorter or longer time, whereas a single fertilized female individual can initiate an expanding basin-wide invasion of a potentially harmful species. In addition, a newly established species can also adapt, getting better at exploiting available resources and strengthening its position in relation to competitors and predators over time.

Tens of invaders have become a nuisance in lakes, rivers or coastal seas of Europe. On the other hand, several alien species have appeared to be beneficial, in some way or another, in man's economy as hunting and fishery objects or as food items supporting stocks of economically valuable species. In all types of European waters biopollution by several highly invasive alien species has caused structural and functional changes in the invaded ecosystems the spatial scales of which extend from local ones (in, e.g., receiving waters for cooling water from power plants) to basin-wide ecosystem-level impacts, like those caused by the invasion of the comb jelly *Mnemiopsis leidyi* in the Black Sea. Thus, the previous, mainly biogeographical, interest in non-native species introductions, focusing on the patterns and processes of invasions, has been changed to cover a broad spectrum of disciplines involved in the total science of invasion biology. In recent years, it stretches from molecular genetics and experimental ecology to water treatment technology and management options and makes a scientific basement for development of international conventions and national legislation, aiming at preventing the establishment of invasive plants, animals and disease agents and restricting their spread within the country.

The introduction of nonindigenous aquatic organisms to Europe has resulted in populations of many species being established in self-sustaining populations outside their native ranges. Whilst some species are introduced deliberately to new areas for aquaculture and stocking purposes, shipping is often regarded as one of the main vectors for the accidental transfer of aquatic organisms. While intentional introductions are controlled more effectively than before through guidelines, national legislation and international conventions, the unintentional ones seem to be in steady increase as a result of the progressive globalisation of trade. It is most likely that the global nature of the shipping industry, with faster ships transporting large quantities of ballast between geographically separate areas, has an increasing potential to transport nonindigenous species to new areas. It has been estimated that major cargo vessels annually transport 3 to 12 billion tonnes of ballast water worldwide, indicating the global dimension of the problem. On average, 3,000-4,000 species may be transported in ships' ballast water each day (Carlton & Geller 1993; Gollasch 1996). Recent revised estimates yielded a still higher number of species being in transport. There are more than 35,000 merchant vessels at sea at any given time, carrying more than 7,000 species during any given 24 hour period; if one adds all kinds of vessels, the number exceeds 10,000 (Carlton 1999).

2 Europe - a source and receiving area for bioinvasions

Europe is known as the source continent for many invaders in all continents, as well as a recipient continent for thousands of nonindigenous species from all over the world. Several introduced species are known to take roles that the native species do not have. New life forms and thus a "foreign style" of aquatic animal and plant life have been introduced into European waters. In many cases an invader represents a new functional group in the invaded community and differs substantially from natives in life form and efficiency of resource utilisation. The most successful invaders in European waters have been capable of altering fundamental ecosystem level processes and thus control the functioning of whole ecosystems. There are several modern-time analogies to the post-Columbus and post-Cook colonisation of remote oceanic islands by both herbivores and predators that have changed or even destroyed much of their native habitats. The newcomers affect also the ecosystem services available for man, such as primary production, fish production, degradation capacity, recreational services and amenities.

Since prehistoric times, European inland and coastal waters have been exposed to invasions of aquatic plant and animal species. Non-native plants can be divided into two categories: archaeophytes, imported before the end of the Middle Ages, and neophytes, introduced after the year 1500 (di Castri 1989; Sykora 1990). This time limit fits roughly also for aquatic nonindigenous species. The term 'native' is used for organisms, which have been native to a particular geographical area in historical times. However, the concept of "historical times" itself is rarely defined in the literature. For most of nonindigenous species, there is evidence (archaeological, historical, ecological, biogeographical or genetic) of their origin. However, it is often difficult to determine whether a species is native or introduced; such species with an unknown origin (either being native or not) are termed cryptogenic (Carlton 1996). Perhaps thousands of introductions date far back in time and their invasion history will be unclear forever. Unrecognised historical transport may have led to false conclusions of natural cosmopolitanism and many introduced species erroneously taken as native (Carlton & Geller 1993). For example, > 30% of 780 species from the Chesapeake Bay, USA, also occur in Europe. A few (5%) of them are known invaders in the bay or Europe; the invasion status of the rest has never been evaluated, suggesting that 30% should be considered cryptogenic (Ruiz et al. 2000).

Expansion of the European culture to other continents began at the time of the Great Discoveries: Europe became a major source of intentional and unintentional introductions to all parts of the world. The success of emigrating Europeans and the colonies established under the rule of them largely depended on their ability to "Europeanise" the fauna and flora of the New World (Crosby 1972). Generally, European plants and animals seem to have been more successful at invading other regions that vice versa (Crosby 1986); their opportunity to reach other continents was greater, so the same success rate may have produced a larger number of successes (Simberloff 1989). In fact, 80% of alien plants in the other continents have a Eurasian origin (di Castri 1989).

The European seas and inland waters have been invaded by nonindigenous species belonging to several major zoogeographic groups (Kinzelbach 1995; Gollasch 1996, Leppäkoski & Olenin 2001). The principal sources are (i) the Mediterranean fauna and flora expanding to the north, (ii) the Ponto-Caspian element, and (iii) transoceanic or transcontinental invasions originating from remote areas (e.g. South-East Asia, New Zealand, North America, Lake Baikal). There are examples of both deliberate and unintentional introductions among all these groups.

3 What opened European waters for bioinvasions?

There is great number of factors that facilitated the potential of alien aquatic organisms to invade Europe, get naturalized and spread within the continent. Some of them are consequences or reasons for others, some act independently, not all are equal in importance. However, even if the list is far from complete and not presented in chronological order, it allows an impression why the European continent was so heavily subjected to biological invasions for thousands of years.

(i) Compared with other continents, Europe had an early history of aquatic introductions, which facilitated their secondary within-continent spread over time; in the 19th century, 22% of known introductions into Europe took place (Welcomme 1992)

(ii) Building up of coastal colonies and large empires and later early market and trade economies covering the known world up to the Far East, with expansion of transportation systems (di Castri 1989). The Roman Empire had lively contacts with Northern Africa and Middle East and in this or early medieval period, e.g. the carp (*Cyprinus carpio*) is supposed to have been introduced north of the Alps (Kinzelbach 1995).

(iii) Early seafarers (e.g. Carthaginians, Arabs, Vikings) expanded sea trade far beyond the European coast, thus increasing potential for biological invasions.

(iv) Intense trade contacts with the Americas, established in the 15th century, connected European coasts with source areas of similar climate regions. The natural barriers, in this case the oceans, preventing coast-bound and freshwater species to spread, were broken down by ship traffic across the oceans.

(v) The history of Euro-American contacts is reflected in the high proportion of Neo-Europeans of American origin in the semi-enclosed seas of Europe. In the Baltic and Black Seas alone, > 35 species of American origin have been recorded, i.e. one third of all nonindigenous species known from these two seas (Leppäkoski & Olenin 2000).

(vi) There are no strict environmental barriers between the European and Asian continents; moreover there is great matching in climatic factors.

(vii) The major rivers of Europe provide cross-continent invasion corridors for non-native species to spread either by natural means or assisted by ship traffic (Kinzelbach 1995). Almost half of all neophytes naturalized in Europe occur along inland waterways, canals and rivers; two thirds of the Polish neophytes can be found in river valleys (Sykora 1990). Intracontinental canals connect most river systems since the 18th century, providing access from the Mediterranean and the Ponto-Caspian basin to Central, Western and Northern Europe.

(viii) Opening of the interoceanic Suez Canal in 1869 exposed the Eastern Mediterranean Sea to Lessepsian migration by hundreds of Indo-Pacific species (Por 1978; Galil 1993).

(ix) Intentional introductions from other continents, especially of North American animal species into European lakes in the 19th and 20th centuries were preceded by oyster culture. The Japanese oyster *Crassostrea gigas* appeared in Europe soon after Portuguese explorers returned from Asia in the 16th century (Carlton 1985, 1999). Intended imports of animals and plants showed a dramatic increase in the 19th and 20th centuries, in the wake of the Industrial Revolution (Kinzelbach 1995).

(x) In the former Soviet Union, Ponto-Caspian crustaceans were transplanted as prey items to stimulate fish production in lakes and water reservoirs in the 1950s-1970s. More than 30 amphipod species and several mysids from the Caspian complex were used for acclimatization purposes. Some of these species have expanded their range into central and western European waters (e.g. Tittizer 1996; Leppäkoski & Olenin 2001).

(xi) Zoos and botanical gardens (e.g., the Royal Botanic Gardens in Kew, Great Britain, founded in 1759), heated greenhouses, public aquaria and other ornamental use of exotic species became popular and opened a gateway for species from tropical and subtropical species to get established in Europe mainly due to accidental releases.

(xii) There were several vacant Eltonian niches available in aquatic communities (e.g., those of large semiaquatic rodents such as the muskrat *Ondatra zibethicus* and the coypu *Myocastor coypus*, large migratory decapods in brackish waters, such as the Chinese mitten crab *Eriocheir sinensis*, and big predacious gastropods (the whelk *Rapana venosa* in the Black Sea). A common hypothesis is that homoclimatic conditions between the source territory and the receiving area facilitate introductions (e.g. Simberloff 1989; di Castri 1990).

(xiii) With intensified agriculture, increasing human population and industrialisation, the European inland waters and later even coastal waters become altered through pollution, eutrophication and other forms of human disturbance. A common hypothesis postulates that disturbed ecosystems and communities are or *seem* to be more open to biological invasions than pristine systems (Elton 1958; Simberloff 1989). It can be presumed that environmental changes result in considerable re-sharing of space and energy resources available for native and non-native species and even open new vacant niches for the most adaptable and tolerant invaders. In polluted or eutrophicated waters, where the native fauna are already severely decimated, newcomers may play a conservative role. They assist ecosystems to cope with stresses in order to maintain basic functions and ecosystem services, such as nutrient cycling, production and degradation of organic material.

(xiv) Increasing shipping activities resulted in the development of busy ports in many European countries connecting Europe to ports all around the world and shipping is believed to be one, if not the most important vector for species invasions (i.e. ballast water, sediments in ballast tanks and hull fouling).

4 Conclusions

Much of the biogeographical peculiarities of Europe that were created over millions of years of evolutionary separation and specialisation of animal and plant life has been lost forever during the last 500 hundred years because of different human activities. In this evolutionary perspective, the alterations related to the introduction and invasion of non-native species can be seen as a part of the global change (Leppäkoski & Olenin 2001) - evolution in reverse? In fact this trend towards increasing homogeneity is one of the most important aspects of the geography of life since the retreat of the continental glaciers (Crosby 1972). Humans have started a revolution in the structure of marine and other aquatic ecosystems through the worldwide introduction in ecological time of non-native species that had been isolated for millions of years over evolutionary time (Carlton 1999). In the beginning of the 3rd millennium, the complete globalisation of human activities will further facilitate species exchange between and within continents and aggravate problems, both environmental and economic, related to nonindigenous species.

There is currently no way to exactly predict how a nonindigenous species will behave in a new habitat with an area-specific set of abiotic and biotic factors that regulate its population growth and modify the species′ impact on the invaded ecosystem. Therefore, treating all invasive species as "guilty until proven innocent" is the only environmentally sound approach (Gollasch & Leppäkoski 1999). Once an aquatic invader is present in high enough numbers to be recorded, it is too late to attempt to eliminate it, especially in marine habitats. Prevention of further arrivals is the only way to go and thus a crucial issue for the European biosecurity strategy (Leppäkoski 2002).

BIOINVASION ECOLOGY: ASSESSING INVASION IMPACT AND SCALE

JAMES T. CARLTON
Maritime Studies Program, Williams College - Mystic Seaport, Mystic, Connecticut, U.S.A.
jcarlton@williams.edu

Abstract
Two aspects of assessing the impact and scale of invasions are explored here. First, the argument is made that it is not possible to place introduced species into one of two bins - those with, and those without, "impact" or "harm". Species invasions lead to a broad and deep sliding scale of alterations in invaded communities, as well as multiple ranges of societal (industrial, economic, social, recreational, health) impact. Second, invasion scale (and thus impact) have been underestimated by historical biases and size biases. A further bias arises out of the selective study of habitats presumed to be more invaded than other habitats. Examples of invasions in many different marine environments are presented.

1 Introduction

"*An investigation of the effects of the disturbance of what may be termed the faunal equilibrium that exists in Nature, by the intentional, incidental, or accidental interference of man, would prove to be not only an interesting and curious study in itself, but would show how intimately interwoven are the various forms of animal life, as well as animal and vegetable life, and further demonstrates the necessity of looking ahead to detect if possible what advantage or disadvantage may follow in the train of causes and effects, through the modification of the fauna of any region by the introduction and permanent addition of a new species, an exotic, strange, and not to the region born.*" - Robert E. C. Stearns (1891)

The science of studying invasions has expanded rapidly in the past quarter century. In the aquatic environment, a series of introductions in the 1980s in both freshwater (for example, the arrival of the zebra mussel in North America, and numerous new invasions throughout western and northern Europe) and marine environments (Asian dinoflagellates in Australia, American comb jellyfish in the Black Sea, American razor clams and polychaete worms in western Europe, among scores of others) precipitated new global awareness and concern in the 1990s about both the vectors that move species around the world and the impact of successful introductions. Modern interest has burgeoned in terrestrial ecosystems as well (Drake et al. 1989). With a torrent of interest, study, and literature on this flood of invasions has come a concomitant - and not unexpected - surge of terminology, interpretations, and analysis.

Carlton (2000) discussed perceptions of invasions relative to (i) the underestimation of the historical scale of invasions in the ocean over time and space, calling attention to the "Missing 1,000" - a minimum of 1,000 marine and estuarine species introduced around the world from the 16th to the 18th centuries that remain largely unrecognized (also discussed by Carlton 1999a). Carlton (2000) further discussed (ii) the ambiguous assumptions behind the concept of "ecological integration" of introduced species into

E. Leppäkoski et al. (eds.), Invasive Aquatic Species of Europe, 7–19.

communities, (iii) the misperception that if dispersal corridors and vectors have been in place for a long time that all or most species able to be introduced via those corridors and vectors would have been introduced by now, (iv) the myth that humans are simply speeding up the phenomenon of natural invasions, and (v) species size biases as another factor in the underestimation of the diversity and scale of introductions. Carlton (2000) also introduced (vi) the concept of ILTA (invasion lag time analysis), and the need to make progress on understanding why species invade when they do.

I explore here two additional aspects of perceptions of the impacts and scale of invasions: (vii) the challenge of categorizing invasion impact as a dichotomy of either "harmful" or "not harmful", and (viii) whether biases introduced by the study of invasions in selected marine habitats (and concomitantly the lack of such study in other habitats) have lead to further underestimations of global invasions.

2 Assessing invasion impact: dualism and "invasive" species

The concepts and thus words used to describe non-native species vary among countries and among scientists, and in the near future show no clear indication of achieving either intranational or international uniformity (or understanding). In addition, terminology has shifted over the 1990s. Phrases such as "introduced species" (Boudouresque et al. 1994) and "nonindigenous species" (OTA 1993) were largely replaced by "invasive" species" (Sandlund et al., 1999; Bossard et al. 2000; Preston et al. 2000; Sherley 2000; Mooney and Hobbs 2000) by the end of the 1990s and early 2000s.

"Invasive species" have in turn been defined as follows:

"Invasive species means an alien species whose introduction does or is likely to cause economic or environmental harm or harm to human health" (Clinton 1999).

"Alien invasive species means an alien species which becomes established in natural or semi-natural ecosystems or habitat, is an agent of change, and threatens native biological diversity" (IUCN 2000).

"Invasive taxa (are those that are) spreading where they are not native, represent(ing) a subset of naturalized taxa (non-native, forming sustainable populations without direct human help but not necessarily spreading)" (Rejmanek 2000).

"(Certain colonizers) are novel and have a large impact, usually undesirable, on the new environment. These colonizers are the .... invasive colonizers, the true invaders" (Davis & Thompson 2000).

In an insightful essay, Bachrach (1972) distinguished three "levels" of definitions of words: *daily*, *poetic*, and *scientific*. Bachrach noted that a "daily definition is one that is universally accepted and for which there is a general understanding", a "poetic" definition "is treated within the realm of individual license and creativity", and a "scientific" definition is one that "is restricted to a limited group for which the definition must have specific meaning". Bachrach observed that "a major error in scientific method is the use of daily definitions", further quoting Quine (1946), who noted that "the less a science has advanced the more its terminology tends to rest upon an uncritical assumption of mutual understanding".

In the definitions of "invasive" given above, and in most other similar definitions, "harm", "threaten", "impact", and "spread" are not defined quantitatively, nor are concepts such as invasive species being a subset of introductions that cause changes in species composition, community structure, or ecosystem function, as invoked by Cronk & Fuller (1995). While, as Carlton (2001) has noted, "invasive is a powerful word conveying a sense of impact and urgency", in the absence of a quantitative, objective definition, most current explanations of the meaning of "invasive" fall within Bachrach's concept of a "daily" definition.

Of interest in this regard is Richardson et al.'s (2000) suggestion that introduced terrestrial plants (specifically for those taxa that spread by seeds and other propagules) be designated "invasive" if they have spread more than 100 meters from parent plants in less than 50 years. As Jenkins (2001) noted, however, relative to a somewhat different subject, "the number is suspiciously round" - even for plants - and would have little bearing, at any rate, on many other taxa or in other habitats. Nevertheless, Richardson et al. (2000) have begun a thread of thinking and inquiry into potential quantification methods for the scale of invasiveness.

Invasions do not divide up conveniently into those taxa that have an impact and those that do not - invasions are not either "harmless" or "harmful". They are thus not a category of species (evolutionarily, biogeographically, ecologically, or socioeconomically) that lend themselves to such a dichotomous approach. It may be naturally tempting to downplay the potential impact of species in systems where they did not evolve, if for no other reasons than to focus limited time, money, and other resources - and the limited attention span of the public, press, and politicians - on a subset of species that are perceived to require greater attention than others, for whatever reasons. However, the natural world is a continuum of kaleidoscopic interactions, direct and indirect linkages, and intimately interwoven (Stearns, 1891; see quotation at the beginning of this paper) species interrelationships. The impact (see also Parker et al. 1999) of introduced species is a long sliding scale of modifications, alterations, and adjustments from a pre-existing state of the community prior to a introduction. These changes range from small deviations to demonstrably very great ones, and may impact one or more pre-existing species (native and introduced). Similarly, societal (industrial, economic, social, recreational, health, and so on) impact may range vastly from low to medium to high costs, with every nuance in between. To arbitrarily and subjectively draw an imaginary line somewhere along these scales - based upon an unquantified perception of "harm" - for dualistic purposes (harmful/harmless, impact/no impact, threat/no threat, and so forth) recalls the thoughts of Quine (1946) above.

Nevertheless, extraordinarily common in invasions literature, particularly in the last 25 years, have been statements such as,

"(The species) has negligible effects on the environment" (Eno et al. 1997), or,

"(The species) did not have any harmful effect" (Boalch 1994), or,

"(Two species of) extremely common plants in England (are having no) discernible impact on their environment" (Davis & Thompson 2000), or,

"... this introduction may be relatively benign" (Southward et al. 1998).

There are hundreds if not thousands of similar statements in the literature. In virtually all such cases the lack of impact is assumed by the absence of evidence - but if such evidence has never been sought, clearly no such conclusions can be made. It cannot be overemphasized that for most invasions *experimental* work is required to determine whether or not there is a statistically significant alteration in one or more parameters of the populations or communities of those species that existed at a given site prior to the introduction. In contrast, however, most conclusions about impact, or the lack thereof, are based upon anecdotes and correlations, or, even less, on conjecture, suppositions, and presumptions.

Daehler (2001) noted that "If we consider all kinds of ecological impacts, the fraction of invaders defined by the spread criterion that also have an ecological impact ... probably approaches 100%". Davis & Thompson (2001) took issue with Daehler's attempts to remove the concept of impact from the definition of "invasive", in part because, they argued, "invasive species are usually explicitly defined on the basis of their impact", citing Clinton's (1999) definition (quoted above) and another similar and closely related U.S. federal definition. However, such definitions of "invasive" are by their very construction precisely *not* explicit (as defined by *Webster's New World Dictionary*: "explicit is applied to that which is so clearly stated or distinctly set forth that there should be no doubt as to the meaning"). Most definitions of invasive (including that of Clinton 1999) invoke concepts, as noted above, that are unclear and indistinct. "Impact" need not necessarily be removed from all definitions of invasions, but it would require defendable, logical, objective and rigorous quantification.

3 Assessing invasion impact: habitat diversity of marine invasions

The perception of the global scale of invasions in the ocean over time and space has been compromised by historical biases (Carlton 1999a, 2000) and by size biases (Carlton 2000). I discuss here an additional bias: the underestimation of marine global invasions due to the selective study of certain habitats and environments. Briefly considered here, relative to marine bioinvasions, are 11 primary marine habitats: two oceanic, seven neritic, and two supratidal. These are as follows:

Oceanic Systems
- Open ocean (mid ocean, high seas): epipelagic habitat
- Deep sea: benthic habitat

Neritic Systems
Open (Higher Energy) Marine Waters
- Continental shelf: pelagic habitat
- Continental shelf (sublittoral): benthic habitat
* Kelp beds
- Intertidal zone: sandy beaches
* Surf zone
- Intertidal zone: rocky shore
- Intertidal zone: soft-sediment shores
- Intertidal to sublittoral: coral reefs

Protected (Lower Energy) Brackish to Marine Waters
- Estuaries (including hard and soft bottoms, intertidal and sublittoral, and pelagic habitat)
* Salt marshes
* Mangroves

Supratidal Systems
- Supralittoral fringe (maritime zone, strand zone)
- Coastal dunes

Marine systems can be classified using many different approaches: degree of exposure (low to high energy), productivity (low to high), salinity (brackish to fully marine), substrate (hard or soft bottom, and in the latter, gradations from mud to sand to coarser fractions), depth and tidal range (supratidal to deep sea), and so on. Each resulting classification produces a different view of the comparative diversity and ecological biogeography of the oceans. For the present exercise, habitats are chosen to represent a diverse array of marine habitats with no further implication that the habitats as divided here are a suggested or fundamental arrangement of the diversity or range of marine communities. Thus many intertidal ecologists tend to work either in (and become specialists on) rocky intertidal systems, salt marshes, mudflats, or exposed sandy beaches, although all of these habitats could be captured under one category of "the intertidal zone", regardless of wave exposure. There are endless transitions, systems within systems, and diffuse boundaries. Marine mudflats may grade landward into estuarine mudflats. Seagrasses form distinct communities both in estuaries and on some rocky shores. Many habitats are adjacent to each other and thus may have some portion of their biota overlapping in an ecotonal gradation.

Not surprisingly, therefore, marine habitats, communities and ecosystems have been divided and subdivided numerous ways. Some workers take broader strokes (for example, Bertness et al. (2001), who treat eight "community types": rocky intertidal, soft-sediment, salt marsh, seagrass, rocky subtidal, deep-sea, coral reefs, and mangroves), while others take finer strokes (for example, Britton & Morton (1989) who under three broad categories (hard shores, soft shores, and "subtidal sands, banks and coral reefs") focus on more narrow subdivisions ("man-made hard shores", "the oyster community", "unvegetated subtidal bayfloors") with mangroves included under "soft shores"). Kozloff (1983) divides coastal cold-water temperate habitats into five systems: floating docks and pilings, rocky shores of enclosed waters, rocky shores of the open coast, sandy beaches, and bays and salt marshes, with some of these divided into finer categories. There are nearly as many categorizations of marine habitats as there are treatises, texts, and monographs on marine biota.

The purpose of the exercise here is not to examine how introduced species would be resorted (and thus be viewed differently) based upon such different classifications of ocean habitats, although such an exercise would be of interest. Rather, our goal here is to ask whether invasions have occurred in a diverse array of habitats.

Note that in the framework presented here, "harbors" and "ports" would fall under different habitat types. While harbors and ports are often constructed in estuaries, a port created in sheltered water on the open coast (in the lee of a point of land, or in a natural but relatively open bight) may be fully marine in salinity. Such open marine ports tend to be colonized on intertidal pilings by species from adjacent rocky shores, and on floats and subtidal fixed structures by species from adjacent sublittoral hard bottoms. Marine ports with little or no estuarine influence may thus also provide a habitat for species from nearby estuaries, bays, and lagoons, that can survive in higher salinities but require lower energy environments.

Table 1 lists examples of invasions in this diverse array of marine habitats, and some examples of places in the world where they have invaded.

3.1 OCEANIC SYSTEMS

No invasions are known to have occurred in the mid-ocean (the open high seas). Carlton et al. (1995) commented on the potential for the introduction between and within ocean basins of open ocean species via ships' ballast water. They noted that water may be taken up in one portion of the high seas and discharged in another. It is safe to say that when the open ocean is sampled it is rarely with invasions in mind, which may have lead to biases in the taxonomic working up of plankton samples. Cheng (1989) speculated that the open ocean pelagic sea strider (insect) *Halobates micans* might owe its presence in the Atlantic Ocean to ship-mediated dispersal (as eggs on vessel hulls) from the Pacific Ocean. However, Damgaard et al. (2000) and Damgaard (2001) have found that divergence of Atlantic from Pacific populations, based upon molecular genetics, predates the appearance of the Panamanian Isthmus.

In a rare deep sea example of a successful human-mediated invasion, the North Pacific king crab *Paralithodes camtschatica* was intentionally transported and successfully released by Russian fisheries biologists into the Barents Sea in the 1960s. Large males are now common in depths to at least 300 meters (Kuzmin et al. 1996). Whether accidental means of transport - such as deep-sea trawling operations - could effectively move benthic species to new regions appears to have been largely uninvestigated, although Uriz (1991) has explored this concept, in part, for deep-water sponges off the coast of South Africa and Namibia.

3.2 NERITIC SYSTEMS

3.2.1 Open (higher energy) marine waters

While the preponderance of work on invasions in the ocean has been in estuaries, invasions are also known in most other neritic waters (Table 1). What is not known is the relative extent of invasions in many non-estuarine environments. Pelagic and benthic habitats on open continental shelves have been invaded, as have most common intertidal environments (Table 1). Limited study on invasions leaves the appearance of few invasions in certain specific habitats, such as kelp beds or open sandy beach surf zones. Particularly noteworthy relative to the latter habitat is the apparently ballast water-mediated appearance about 1950 of an Australasian diatom, *Attheya armatus* (= *Chaetoceros armatum*, = *Gonioceros armatus*) in the American Pacific Northwest (Table 1). The species is now common in winter and spring from Oregon to British Columbia, turning the surf a yellow-brown color. It seems improbable that this is the sole species in the world that has invaded surf zones.

No invasions are yet clearly reported from the intertidal zone of open sandy beaches, a habitat characterized in many parts of the world by such well-known indigenous taxa as anomuran sandcrab (mole crabs) and haustoriid amphipods. This said, centuries of moving beach sand around the world as "solid" ballast (Carlton 1985) may be a more compelling historical explanation for the apparent cosmopolitan distributions of many meiofaunal taxa, rather than invoking plate tectonics and the inexplicable lack of allopatric speciation after tens of millions of years of separation if continental drift had played a role.

Eldredge (1987, 1994) has provided a valuable introduction to the concept that *coral reefs* are not invasion free. While the Hawaiian Islands are better known relative to coral reef invasions than many other regions, it is reasonable to assume that a large number of species have been transported among, for example, Pacific coral reef systems, especially during and after World War II, when fouled barges and landing craft were towed in and among islands.

Invasions have occurred on rocky shores around the world, including the most exposed, high-energy environments (Table 1). For example, the Mediterranean mussel *Mytilus galloprovincialis* occurs on the open wave-swept coast of South Africa (Griffiths et al. 1992; Carlton, pers. obs.). The Pacific Ocean red alga *Pikea californica* has invaded the high-energy wave exposed habitats in Isles of Scilly in the British Isles (Maggs & Ward 1996). The periwinkle *Littorina littorea* (whatever its early historic or prehistoric history in eastern Canada (Steneck & Carlton 2001), and otherwise prehistorically and historically also known from Europe) invaded the rocky intertidal environment of New England, from the Bay of Fundy south, commencing in the 1860s, as did a number of other species earlier and later (Table 1).

An example of the potential underestimation of invasions on rocky shores in general is provided by examining these northwest Atlantic Ocean rocky shore systems invaded by *Littorina littorea*. We recognize no clear invasions in these systems prior to about 1800, although vessels from Europe had been arriving on these shores for centuries. Rather, other than a relatively few species recognized as post-1800 introductions, most other rocky shore taxa that are found both in North America and in Europe are regarded as naturally amphiatlantic, either as a result of preglacial colonization episodes (and thus surviving through various glaciations as refugial relicts) or as post-Wisconsin glaciation colonists. An unknown number of these species are, however, cryptogenic (Carlton 1996), raising the possibility that species perhaps as important in community regulation as *Littorina littorea* may be historical, post-1500 introductions to North America. Carlton (1999a) suggested that a minimum of 3 to 5 species per year may have been transported by ships around the world in the 300 year period between 1500 and 1800. If only 1 species every 10 years arrived from Europe alone to New England shores in this same period, a minimum of 30 additional introduced species are now masquerading as native species. Conversely, an unknown number of cryptogens with possible American origins, the forgotten invasions of centuries past, now reside on European shores due to historical shipping.

The difficulty in recognizing such introductions in even ostensibly well-known rocky shore systems, such as those of New England with relatively long histories of biological investigation, is admirably illustrated by the cryptic history of one of the most common barnacles now occurring in southern New England, *Chthamalus fragilis. C. fragilis* today ranges from the north shore of Cape Cod (Cape Cod Bay) to Texas (Dando & Southward 1980; Carlton, field obs.). All literature on this American Atlantic species since 1913 fails to mention that this species was apparently - and remarkably - not known in New England until 1898 (Sumner et al. 1913, p. 191). As Sumner (1909) and Sumner et al. (1913) argued, it is inconceivable that this barnacle was overlooked by several generations of New England invertebrate zoologists. *Chthamalus* is not mentioned in the monographs of Gould (1841), Binney & Gould (1870) or Verrill & Smith (1873), nor in scores of other publications, on the invertebrates of the intertidal zone from Massachusetts to New Jersey in the 19th century.

As Sumner (1909) stated, "It is surely difficult to explain how this barnacle has been so long overlooked upon our own Atlantic shores. It is hard to believe that the present species has been habitually confused with *Balanus balanoides* by the long succession of field naturalists and systematic zoologists who have [explored] the shores of New England for over a century". Sumner noted that *Chthamalus* and *Balanus* were so "plainly distinguishable" by color and external morphology that the "confusion of the two... seems incredible". Sumner suggested that "An alternative explanation is that *Chthamalus* has only recently invaded New England waters...", further noting the then still-recent arrivals in New England of the snail *Littorina littorea* and the Asian sea anemone *"Sagartia luciae"* (= *Diadumene lineata*) as "doubtless the most striking local examples of this phenomenon". Since his remarks (repeated by Sumner et al. 1913), no further comments appear to have been made regarding this barnacle's history in the past century.

Two scenarios suggest themselves:

(i) *Chthamalus fragilis* was first known from collections from South Carolina (the type locality of the species, described by Darwin in 1854). *Chthamalus* may have been transported north naturally and successfully colonized Cape Cod and southern New England in the 1890s, when it was unable to do so previously.

(ii) *Chthamalus* may have been carried north into southern New England on ship fouling; as with (i) it was able to successfully establish when it had not done so before. Chthamalid barnacles, including members of the genus *Chthamalus*, are known from ship fouling communities (Southward et al. 1998).

The timing of the apparent northward movement of *C. fragilis* is apparently coincident with the movement north of Cape Cod by the European green crab *Carcinus maenas*, from southern New England (where it had remained isolated for almost a century). Coastal warming commenced after 1850 with the cessation of the so-called "Little Ice Age". Alternatively, the snail *Littorina littorea* was moving south down the Atlantic coast in the last half of the 19th century, and became common to abundant in southern Cape Cod and Long Island Sound waters by the 1880s (Steneck & Carlton 2001). It is thus tempting to also consider whether or not the new arrival of this relatively large and extremely abundant snail may have in some fashion altered the uppermost rocky shores in some manner such as to facilitate the colonization of *Chthamalus. L. littorea* may have depressed high intertidal algae (Vadas 1992), or depressed the populations of grazing *Littorina saxatilis*, a native periwinkle. In turn, both may have direct or indirect impacts on the settlement potential of higher intertidal barnacles. While *L. littorea* is capable of consuming large numbers of the cyprids and juveniles of newly settled *Semibalanus balanoides* (Brenchley & Carlton 1983), which competes with *Chthamalus* in the lower intertidal zones, *Chthamalus* occurs in an upper intertidal refugium where *Semibalanus* does not occur, such that indirect interaction whereby *Littorina* may have depressed *Semibalanus* leading to enhanced *Chthamalus* success would appear less likely.

Regardless of why it was able to colonize southern New England in the 1890s, and pending evidence of its aboriginal presence, *Chthamalus* appears to be a 19th century invasion in New England waters. If the invasion of an abundant barnacle in the rocky intertidal shore of New England as late as the 1890s has been long overlooked, recognizing invasions of the 1500s,1600s or 1700s is that much more difficult.

Invasions on rocky shores around the world remain similarly poorly understood, although such invasions are known in many different regions (Table 1), including both highly diverse systems (such as Pacific North America) and low diversity systems (such as Atlantic North America).

3.2.2 Protected (lower energy) brackish to marine waters

Estuarine, lagoon, and inland sea invasions have been reviewed extensively in recent years (Eno et al. 1997 [British Isles]; Reise et al. 1999 [North Sea]; Ruiz et al. 2001 [United States]; Cohen & Carlton 1995 [San Francisco Bay]; Hewitt et al. 1999 [Port Phillip Bay], among others). However, invertebrate and algae invasions in estuarine salt marshes and mangroves are not well known.

3.2.3 Supratidal systems

The *supralittoral fringe*, or the "maritime" or "strand" zone, has been invaded by a number of crustaceans, insects, and plants, many likely derived from the days of the dumping of "solid" or "dry" ballast. Coastal marine dunes have been similarly invaded, but no general review of plant or animal invasions in dune systems appears to be available.

4 Discussion

Despite the existence of invasions in a wide variety of marine communities, a common perception is that ocean habitats other than estuaries may be largely or entirely invasion-free. Thus Zevina &Kuznetsova (1965) remarked that, "It should be stated that an open coast is not suited to the introduction of new organisms. Only closed inlets and bays, from which the larvae will not be carried into the open sea, and in which they will be able to set in close proximity to each other, are suitable". Clearly, we now know that open coasts can be invaded.

For certain habitats - such as the deep sea or high seas - invasions may in fact be rare (but see discussion above relative to open ocean invasions). For other habitats, invasions have been without doubt underestimated. One pathway to improve predictive ability in invasion science is a greater understanding of the differential resistance, or susceptibility, of different habitats to invasions. Clearly, however, the same habitat type is not equally resistant or susceptible to invasions around the world, and thus this understanding must be modified by the unique evolutionary and ecological histories of environments at a given location. The rocky shores of the Pacific coast of North America, the Atlantic coast of North America, and of Europe, all have different late Tertiary histories, leaving each not only with unique endemic elements but also different capacities to rebuff or absorb invasions. Examining a wide diversity of marine habitats, and the invasions that have occurred in them, as tempered by those regional histories that naturally shaped the biota, may provide further foundations for understanding invasion potential.

Table 1. Examples of invasions in a diversity of marine habitats.

Environment / Examples of invasions	**Introduced locations**	**References**
(Species / (Native to)) (* = see footnote)	(examples only; the species may occur in more places than shown, and in other habitats)	
OCEANIC SYSTEMS		
Open ocean (mid ocean, high seas): epipelagic habitat	see text discussion	
Deep sea: benthic habitat		
King Crab *Paralithodes camtschatica* (N Pacific Ocean)	Barents Sea	Kuzmin et al. 1996
NERITIC SYSTEMS		
Open (Higher Energy) Marine Waters		
Continental shelf: pelagic habitat		
Jellyfish *Phyllorhiza punctata* (Pacific Ocean)	Gulf of Mexico	Carlton 2001
Jellyfish *Aurelia labiata* (Pacific N America)	Hawaiian Islands	Gershwin 2001
Jellyfish *Rhopilema nomadica* (Red Sea)	Mediterranean Sea	Galil et al. 1990
Diatom *Coscinodiscus wailesii* (Pacific/Indian Oceans)	Europe	Eno et al. 1997; Laing1999
Diatom *Odontella sinensis* (Pacific Ocean)	Europe	Eno et al. 1997
Diatom *Gyrodinium aureolum* (Atlantic N America)	Europe	Macdonald1999
Striped Bass *Morone saxatilis* (Atlantic N America)	Pacific N America	Fuller et al. 1999; Cohen & Carlton 1995
Continental shelf (sublittoral): benthic habitat		
Coral *Oculina patagonica* (S America)	Mediterranean	Zibrowius 1992
Flatworm *Convoluta convoluta* (Europe)	Atlantic N America	Rivest et al. 1999
Seaslug (nudibranch) *Philine auriformis* (New Zealand)	California	Gosliner 1995; D. Cadien, pers. comm.
Snail *Rapana venosa* (Japan)	Mediterranean	Mann & Harding 2000
Snail *Crepidula fornicata* (Atlantic N America)	Europe	de Montaudouin & Sauriau 1999; Minchin 1999
Snail *Maoricolpus roseus* (New Zealand)	Australia	Edgar 1997
Razor clam *Ensis directus* (Atlantic N America)	Europe	Armonies 2001
Crab *Charybdis longicollis* (Red Sea)	Mediterranean Sea	Galil 1994; Galil & Lützen 1998
Seasquirt (ascidian) *Styela clava* (Japan)	Atlantic N America	Harris & Tyrrell 2001
Goatfish *Upeneus moluccensis, U. pori* (Red Sea)	Mediterranean Sea	Golani 1994
Algae *Undaria pinnatifida* (Japan)	New Zealand, Europe, California	Wallentinus 1999a; Carlton 2001
Algae *Caulerpa taxifolia* (Australia)	Mediterranean	Jousson et al.1998
Algae *Codium fragile tomentosoides* (Japan)	Atlantic N America, Europe, Australia	Harris & Tyrrell 2001; Trowbridge 1998; Campbell 1999
*** Kelp beds**		
Bryozoan *Membranipora membranacea* (Europe)	Atlantic N America	Harris & Tyrrell 2001
Intertidal zone: rocky shores		
Polychaete worm *Terebrasabella heterouncincata* (S Africa) (*)	California	Kuris & Culver 1999
Sea anemone *Diadumene lineata* (Japan)	Atlantic N America	Steneck & Carlton 2001
Periwinkle snail *Littorina littorea* (Europe)	N America	Steneck & Carlton 2001

Environment / Examples of invasions	Introduced locations	References
Mussel *Mytilus galloprovincialis* (Mediterranean)	S Africa	Griffiths et al. 1992
Barnacle *Chthamalus proteus* (Caribbean)	Hawaiian Islands	Southward et al. 1998
Barnacle *Chthamalus fragilis* (southern U.S.)	New England	This chapter
Barnacle *Elminius modestus* (New Zealand)	W Europe	Eno et al. 1997
Barnacle *Balanus glandula* (Pacific N America)	Argentina	Vallarino & Elias 1997
Shore crab *Carcinus maenas* (Europe)	S Africa, Australia	Griffiths et al. 1992
Shore crab *Hemigrapsus sanguineus* (Japan)	Atlantic N America	McDermott 2000
Shore crab *Hemigrapsus penicillatus* (Japan)	W Europe	Noël et al. 1997
Seasquirt (ascidians) *Botryllus schlosseri* (Europe) and *Botrylloides violaceous* (Japan)	Atlantic N America	Steneck & Carlton 2001
Seasquirt (ascidian) *Styela clava* (Japan)	Atlantic N America	Steneck & Carlton 2001
Algae *Sargassum muticum* (Japan)	Pacific N America, Europe	De Wreede 1983; Wallentinus 1999b
Algae *Pikea californica* (California)	British Isles	Maggs & Ward 1996
Algae *Codium fragile atlanticum*(Japan)	W Europe	Trowbridge & Todd 1999a
Algae *Codium fragile tomentosoides* (Japan)	Atlantic N America; W Europe; Australia	Carlton & Scanlon 1985; Trowbridge & Todd 1999b; Campbell 1999
Algae *Undaria pinnatifida* (Japan)	France	Castric-Fey 1999a,b
Intertidal zone: sandy beaches (see Supralittoral zone)		
*** Surf Zone**		
Diatom *Attheya armatus* (Australasia)	Pacific NW: British Columbia to Oregon	Lewis & Norris 1970
Intertidal zone: soft-sediment shores		
Razor clam *Ensis directus* (Atlantic N America) (*)	Europe	Armonies 2001
Amphipod *Corophium volutator* (Europe)	N America (Bay of Fundy)	Chapman 2000
Intertidal to sublittoral: coral reefs		
Vermetid snail *Vermetus alii* (W Atlantic, E Pacific)	Hawaiian Islands	Carlton 1999b
Top snail *Trochus niloticus* (Fiji Islands)	Pacific Islands (Micronesia and Polynesia)	Bour 1990; Eldredge1994
Mantis shrimp *Gonodactylaceus falcatus* (Pacific Ocean)	Hawaiian Islands	Kinzie 1984
Bluestripe snapper *Lutjanus kasmira* (Marquesas Islds.)	Hawaiian Islands	Randall 1987
Blacktail snapper *Lutjanus fulvus* (Marquesas Islds.)	Hawaiian Islands	Randall 1987
Blackchin Tilapia *Sarotherodon melanotheron* (Africa)	Hawaiian Islands	Randall 1987; Fuller et al. 1999
Bluespotted Grouper *Cephalopholis argus*	Hawaiian Islands	Randall 1987
Algae *Kappaphycus alvarezii* (Philippines)	Hawaiian Islands	Rodgers & Cox 1999
Algae *Kappaphycus striatum* (Philippines)	Hawaiian Islands	Woo et al. 2000
Algae *Acanthophora spicifera* (Guam)	Hawaiian Islands	Russell 1992
Algae *Hypnea musciformis* (Florida)	Hawaiian Islands	Russell 1992
Protected (Lower Energy) Brackish to Marine Waters		
Estuaries (including hard and soft bottoms, intertidal and sublittoral, and pelagic habitat)		
Sea anemone *Diadumene lineata* (Japan)	Atlantic N America, Europe	Eno et al. 1997; Steneck & Carlton 2001
Comb jellyfish *Mnemiopsis leidyi* (Atlantic Americas)	Black Sea, Caspian Sea	Ivanov et al. 2000; Volovik 2000
Polychaete worm *Ficopomatus enigmaticus* (southern hemisphere)	Europe, N America, Japan, Hawaiian Islands, S America	Eno et al. 1997; Cohen & Carlton 1995
Polychaete worm *Marenzelleria viridis* (N America)	Europe	Daunys et al. 1999
Polychaete worm *Sabella spallanzanii* (Europe)	Australia	Currie et al. 2000
Slipper snail *Crepidula fornicata* (Atlantic N America)	Europe	Eno et al. 1997
Snail *Rapana venosa* (Japan)	Chesapeake Bay	Mann & Harding 2000
Clam *Mya arenaria* (Atlantic N America)	Pacific N America, Europe	Carlton 1992; Eno et al. 1997
Clam *Venerupis philippinarum* (Japan) (*)	N America, Europe, Hawaiian Islands	Carlton 1992, 1999b; Flassch & Leborgne 1992; Laruelle et al. 1994
Clam *Potamocorbula amurensis* (China)	California	Carlton 1992, 1999b
Clam *Nuttallia obscurata* (Japan)	Pacific NW: British Columbia to Oregon	Carlton 1999b
Mussel *Mytilus galloprovincialis* (Mediterranean)	Pacific N America	Carlton 1992, 1999b

Environment / Examples of invasions	Introduced locations	References
Mussel *Mytilopsis leucophaeata* (Southern U.S.)	Long Island Sound	Carlton 1992
Mussel *Musculista senhousia* (Japan)	Pacific N America, New Zealand, Australia, Europe	Carlton 1992; Creese et al.1997; Zibrowius 1992
Barnacle *Balanus improvisus* (Atlantic N America)	Europe, Australia-New Zealand	Leppäkoski 1999; Jones 1992
Copepod *Pseudodiaptomus marinus* (Japan)	Pacific N America	Fleminger & Kramer 1988
Copepod *Pseudodiaptomus inopinus* (Japan)	Pacific N America	Cordell & Morrison 1996
Amphipod *Grandidierella japonica* (Japan)	Pacific N America	Chapman & Dorman 1975
Shore crab *Carcinus maenas* (Europe)	Pacific N American coast, S America, Japan, Australia	Cohen et al. 1995
Mitten crab *Eriocheir sinensis* (China)	Europe; California	Rudnick et al. 2000; Gollasch 1999
Mud crab *Rhithropanopeus harrisii* (Atlantic N America)	Europe, Pacific N America	Eno et al.1997; Cohen & Carlton 1995
Seastar *Asterias amurensis* (Japan)	Australia	Morrice 1995; Grannum et al. 1996
Seasquirt (ascidian) *Styela clava* (Japan)	N America, Europe, Australia	Lambert & lambert 1998
Seasquirt (ascidians) *Botryllus schlosseri* (Europe) and *Botrylloides violaceus* (Japan)	Atlantic N America	Steneck & Carlton 2001
Goby *Acanthogobius flavimanus* (Japan)	Pacific N America, Australia	Fuller et al. 1999
Goby *Neogobius melanostomus* (Black/Caspian Seas)	Baltic Sea, Great Lakes	Skora et al., 1999; Fuller et al. 1999
Mozambique Tilapia *Oreochromis mossambicus* (Africa)	Hawaiian Islands	Fuller et al. 1999
Algae *Codium fragile tomentosoides* (Japan)	Atlantic N America, Europe	Carlton & Scanlon 1985; Eno et al.1997
Algae *Caulerpa taxifolia* (Australia)	California	Anderson & Keppner 2001; Jousson et al. 2000
Eelgrass *Zostera japonica* (Japan)	Pacific NW: British Columbia to Oregon	Harrison & Bigley 1982
*** Mangroves**		
Isopod *Sphaeroma terebrans* (Indian Ocean)	W Atlantic	Carlton & Ruckelshaus 1997
Mayan cichlid *Cichlasoma urophthalmus* (Mexico and Central America)	Florida	Faunce & Lorenz 2000; Fuller et al. 1999
Blackchin Tilapia *Sarotherodon melanotheron* (Africa)	Florida	Jennings & Williams 1992; Fuller et al. 1999
Mozambique Tilapia *Oreochromis mossambicus* (Africa)	Hawaiian Islands; Singapore	Fuller et al. 1999; Ng & Sivasothi 2001
Green chromide fish *Etroplus suratensis*	Singapore	Ng & Sivasothi 2001
Mangrove *Rhizophora mangle* (tropical Pacific-Atlantic)	Hawaiian Islands; California	Cox & Allen 1999
Mangrove *Rhizophora stylosa* (Australia)	Taiwan, French Polynesia	Smith 1996; Lu et al. 1999
*** Salt Marshes**		
Snail *Myosotella myosotis* (Europe)	N America	Carlton 1992
Isopod *Sphaeroma quoyanum*(New Zealand)	California	Talley et al. 2001
Cordgrass *Spartina alterniflora, S. patens* (Atlantic N America)	Pacific N America	Daehler & Strong 1996
Cordgrass *Spartina angelica* (*S. alterniflora* (America) x *S. maritima* (Europe))	Europe, Pacific N America, Australia	Daehler & Strong 1996
Cordgrass *Spartina densiflora* (Chile)	Pacific N America	Daehler & Strong 1996
Brass Buttons *Cotula coronopifolia* (S Africa)	Pacific N America; Atlantic N America (Massachusetts)	Cohen & Carlton 1995
Sand Spurrey *Spergularia media* (Europe)	Atlantic N America, Pacific N America, S America	Cohen & Carlton 1995
SUPRATIDAL SYSTEMS		
Supralittoral fringe (maritma zone, strand zone)		
Amphipod *Transorchestia chiliensis* (Chile/New Zealand)	Pacific N America (California)	Cohen & Carlton 1995

Environment / Examples of invasions	Introduced locations	References
Amphipod *Orchestia gammarella* (Europe)	Atlantic N America	Chapman 2000
Isopod *Halophiloscia couchii* (Europe)	Hawaiian Islands, Bermuda, Argentina, Atlantic, N America, Argentina	Taiti 1999
Isopod *Porcellio lamellatus* (Europe)	Hawaiian Islands, Bermuda, Cuba, Argentina, Australia	Taiti & Howarth 1996
Maritime earwig *Anisolabis maritima* (Atlantic)	Pacific N America, Japan, New Zealand	Cohen & Carlton 1995
Searockets *Cakile maritima* (Europe), *C. edentula* (Atlantic N America)	Pacific N America, Australia	Barbour & Rodman 1970; Rodman 1986
Sand Spurrey *Spergularia media* (Europe)	Atlantic N America, Pacific N America, S America	Cohen & Carlton 1995
Coastal dunes		
Dune grass *Ammophila arenaria*	Pacific N America, S Africa	Huiskes 1979; Wiedemann & Pickart 1996
Asiatic sand sedge *Carex kobomugi* (Japan)	Atlantic N America	Standley 1983

Notes:

* *Terebrasabella heterouncincata*: a polychaete that became established in rocky intertidal snails at a single location on the open California coast, but believed to have been subsequently eradicated (Culver & Kuris 2000).

* *Ensis directus:* Species name follows usage in Turgeon et al. (1998).

* *Venerupis philippinarum* is widely known in the literature by other names, including *Tapes philippinarum, Ruditapes philippinarum, Tapes japonica,* and *Venerupis philippinarum*; the taxonomy of Coan et al. (2000) is followed here.

Who is Who Among Nonindigenous Species

PROTISTS - A DOMINANT COMPONENT OF THE BALLAST-TRANSPORTED BIOTA

NORBERT HÜLSMANN[1*] & BELLA S. GALIL[2]
[1]*Institute of Zoology/Protozoology, Free University Berlin, Berlin, Germany*
[2]*Israel Oceanographic and Limnological Research, National Institute of Oceanography, Haifa, Israel*
[*]Corresponding author *hulsmann@zedat.fu-berlin.de*

Abstract
Ship's ballast water and sediments serve as a main vector in the transportation and spreading of protists: toxic dinoflagellates, parasitic labyrinthulids and other potentially harmful and harmless unicellular beings. The omnipresence of protists in the sea means that photo- and heterotrophic protists invariably occur in ballast water and sediments. Ballast tanks - especially in container vessels with a high turnover of ballast water - serve as mesocosms supporting rich assemblages of heterotrophic protists. We estimate that more than 250 protist taxa are commonly present in a ballast tank, since their size and diet favour survival in there. As protists form an important part of the marine food web, otherwise harmless protists may be indirectly involved in the successful transport of bioinvaders by increasing the chances of survival of entrained filter-feeding and biofilm-grazing metazoans. Furthermore, we suggest that ballast-transported protists, harmless in their native habitats, may modify or trigger changes in native assemblages, or may affect them by wholly outcompeting or partially displacing native taxa. Of momentous concern is whether protists, that are harmless in their native habitats, may become toxic after ballast-mediated spreading into the microbial associations of a recipient area, or vice versa, that the introduction of bacteria may start toxin production in formerly harmless eukaryotes.

1 Introduction

Protists occur in all marine and freshwater habitats, and play a more important role in the marine food webs than previously considered (Caron & Finlay 1994). Many species, especially the smaller ones, have high metabolism rates and input of organic matter is promptly followed by growth and proliferation. As consumers of dissolved organic matter, bacteriovores, detritovores, algal grazers, predators and parasites, as well as prey, protists form the most important link between bacterial degradation processes and the higher trophic levels of the marine food web.

The omnipresence of protists in the marine environment means that photo- and heterotrophic protists invariably occur in ballast water and sediments (Galil & Hülsmann

E. Leppäkoski et al. (eds.), Invasive Aquatic Species of Europe, 20–26.

1997; Pierce et al. 1997). Studies have shown that many protists were viable on arrival in destination port, and may be successfully dispersed along shipping routes. The global spread of these microorganisms may alter communities and ecosystems, and therefore is of major concern. However, the invasion potential of marine heterotrophic protists is unknown due to ignorance of their biogeographical distribution patterns and lack of port-specific studies. To date only two harbours were investigated: Ostend, Belgium (Persoone 1968), and Port Jackson, Sydney, Australia (Tong et al. 1998). We assume that since the introduction of water as ballast in the middle of the 19th century, protists may have spread globally, unheeded by protozoologists.

2 How many protists occur in ballast tanks?

Heterotrophic protists, pelagic as well as benthic, have attracted considerable attention following the development of sampling techniques and procedures. Sampling ballast-entrained microbiota is complicated as only the last step is hazard-free, from the investigator's point of view:

(i) ballasting (water and accidentally sediment uptake) follows each vessel's stability and safety constraints;
(ii) the necessity of ballasting and deballasting follows economic and safety reasons, as the import and export of goods changes from port to port and with time;
(iii) deballasting – near port or during open-ocean-exchange - is constrained by route, weather and vessel's safety;
(iv) sampling is restricted mostly to berthed vessels and therefore time-constrained, and taken at the courtesy of ship-owners or crew;
(v) sampling methods depend on the accessibility of ballast tanks and therefore on spatio-temporal constraints due to the ship architecture and load.

From these arguments follows that each ballast sample - irrespective of the sampling methods - will contain a unique and unpredictable collection of taxa. The singularity of each sample is due the unrepeated chain of events leading to permanently changing mixtures of tank-internal biota. The singular character of ballast samples sets them apart from samples taken from natural habitats under more controlled and even reproducible conditions. The irreproducible nature of ballast samples sets a premium on samples obtained on identical shipping routes and on vessels with a comparable design. Investigation of ballast-entrained protists in 63 dedicated ballast tanks (and 6 ballasted cargo-holds) discloses a remarkable uniformity (Fig. 1): Hauls taken with 55 µm plankton nets through open manholes (Fig. 1, 2) show a low diversity of protist taxa, comparable with samples obtained through the pumping system (Fig. 1, 1). Twenty to forty taxa were recorded from the tanks' water column, even when the tanks were partially emptied (Fig. 1, 4), or in an experimental tank used as "plankton tower" (Fig. 1, 9, cf. Gollasch et al. in press). Many more taxa (> 120) were recorded from ballast sediments, with the highest numbers noted where the sediment was flooded with few centimetres of seawater (Fig. 1, 3), and somewhat lower numbers in tanks nearly emptied for inspection (Fig. 1, 5). The lowest diversity was found in remnants of ballast water from emptied cargo holds (Fig. 1, 6). In the rare opportunities where samples from the tank's water column were compared with samples of tank's sediment, there were two and even three times more taxa in the sediment samples, thus confirming our observations. Since the volume of our samples (60 ml to 20 l) is miniscule compared with that of the ballast tanks sampled (about 1,000 MT), it is clear the number of taxa transported in ballast tanks is underestimated. We estimate that more than 250 protist taxa may occur in a ballast tank under normal conditions, higher by one order of magnitude than entrained metazoans (Galil & Hülsmann 1997).

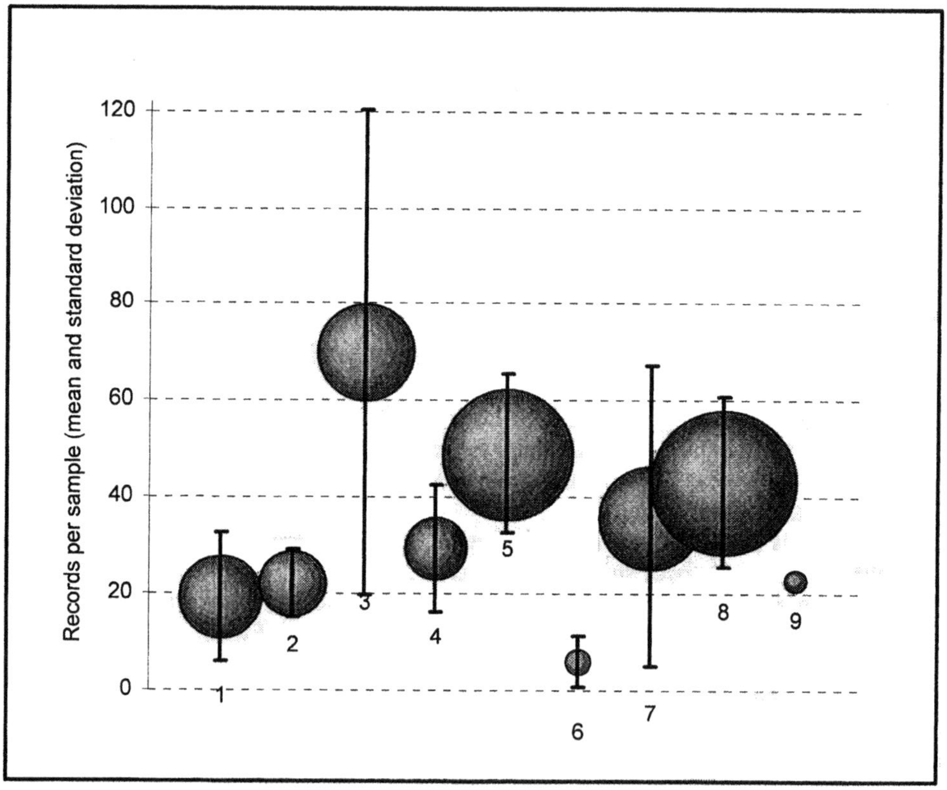

Figure 1. Classification of sampling methods: comparison of cumulative ballast water samples collected by plankton net (mesh size 55 μm (1, 2, 4, 7) and sediment samples by scoop (3, 5, 8), and a single item from an experimental plankton tower (9). The diversity is indicated by the centre of the circles (mean value) and by the lines (standard deviation). The plane of the circles is representative for the total number of singular registrations of taxa. Results obtained for normal tanks with water remnants (2, 3) and tanks more or less emptied for inspection purposes (4, 5) are summarized in 7 and 8.

3 What makes protists successful survivors in ballast tanks?

There are approximately 38,000 named protist species, but most known heterotrophic protists are parasites or occur in freshwater or terrestrial habitats. Since 5% of the taxa recorded from ballast tanks were previously undescribed (mainly small amoebae and net slimemolds), it seems that the free-living marine protist biota is underinvestigated.

Heterotrophic protists' size and diet favour survival in ballast tanks. Galil & Hülsmann (1997) have shown that ballast tanks - especially in container vessels with a higher turnover of ballast water - serve as mesocosms supporting rich assemblages of heterotrophic protists. The ballast-entrained protists may feed on the remains of metazoans destroyed by the mechanical stress of the ballasting process, or on phototrophic organisms that perish for lack of light. The mass mortalities (discernible in fresh isolates from ballast tanks) precipitate as organic-rich deposits on the tank bottom, where the dissolved organic components (DOC) are consumed by bacteria – a food source for small flagellates, ciliates and amoebae (Fig. 2).

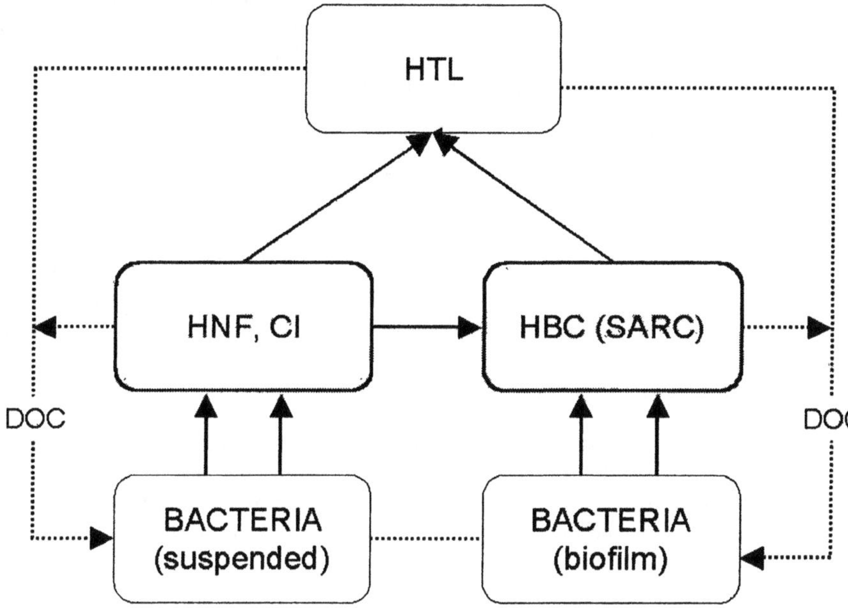

Figure 2. The interaction of Bacteria and protists with different nutritional strategies (filter-feeding versus grazing on biofilms). The filter-feeders are mainly represented by heterotrophic nanoflagellates (HNF) and ciliates (CI), the biofilm grazers by heterotrophic benthic community (HBC) represented by sarcodins (SARC = amoebae, labyrinthulids and foraminifers). DOC = dissolved organic matter, HTL = higher trophic levels.

The tank-enclosed food web (Table 1) favours protists and metazoans feeding on bacteria and protists, they in turn may be consumed at higher trophic levels. Water intake - during ballasting or open ocean exchange - renews this cycle. Ballast tanks that are filled, partially drained and refilled in the course of routine voyages accumulate a diverse assemblage. This phenomenon may be termed the "ballast paradox" - interspecific competition under variable medium-sized conditions allows the co-existence of more species than one can expect from the number of limiting resources (intermediate disturbance hypothesis, Sommer 1995).

Dilution of in-tank ballast with oceanic or coastal waters may reduce potentially harmful metabolic residuals, increase levels of oxygen and organic and inorganic nutrients (DOC, DIC), admit organisms destroyed by the mechanical stress of the ballasting, and import new live taxa into the tank. These consequences have to be considered in the discussion on the effects of flushing procedures and ballast water management.

Table 1. The development of food chains of heterotrophic protists in sub-cultivation experiments (A – E) of a ballast sample. (F) experimental food source, 1- 5 trophic levels developing in dependence of randomly inoculated protists. The foraminifers *Allogromia* and *Nemogullmia*, several ciliates inclusive a cyst-forming *Chilodonella* species, and the net slimemold *Labyrinthula* can grow successfully on an experimental biofilm consisting of living bakery yeast (and bacteria). As predators of taxa listed on level 1 occur *Labyrinthula* and benthic amoebae, mainly *Mayorella* and *Nuclearia* species in the next trophic level, which may serve also as food for *Allogromia* and *Labyrinthula* in the higher levels. The development of the most complex succession shown in C was apparent after two months in culture.

	A	B	C	D	E
5			*Allogromia*		
4			*Labyrinthula*		
3		*Allogromia*	*Nemogullmia*	*Allogromia*	
2		*Labyrinthula*	*Mayorella* *Nuclearia*	*Labyrinthula*	
1	*Allogromia*	*Nemogullmia*	small ciliates	*Chilodonella*	*Labyrinthula*
F	yeast	yeast	yeast (bacteria)	yeast	yeast

4 Do ballast-mediated protist invasions pose a risk?

Since protists form an important part of the marine food web, increasing the chances of survival of entrained filter-feeding and biofilm-grazing metazoans, even harmless protists may be indirectly involved in the successful transport of bioinvaders. In vitro cultivation experiments with ballast biota have demonstrated possible trophic pathways (summarized in Table 2). We suggest that these trophic interactions may account for many successful ballast-mediated bioinvasions.

Much attention has been paid to toxic algal blooms resulting from international shipping (Hallegraeff & Bolch 1992; Hallegraeff 1993, 1998; Hamer et al. 2000). We suggest that invasive ballast-transported protists, harmless in their native habitats, may modify or trigger changes in native assemblages, or may affect them by wholly outcompeting or partially displacing native taxa. The diatoms *Odontella (Biddulphia) sinensis* and *Coscinodiscus wailesii* are examples of successful invasions into the North Sea and the Baltic Sea, as well as other seas, that have precipitated drastic changes in the local plankton assemblages, though (yet) without noticeable negative effects on the ecosystem as a whole.

A remarkably large number of protists pose a risk by themselves, as being organisms releasing toxins to the surroundings or when acting as parasites or opportunistic predators. All of them have at least one benthic stage (as cysts) or at least one planktonic stage (as zygotes, gametes or zoospores). Therefore they can be isolated from ballast water column as well as from ballast sediments. It is as yet unclear whether protists are responsible for autochthonous production of toxic substances, or whether the released toxins may be traced to the synthetic activity of endobiotic bacteria (Mitretek Systems 2001).

Table 2. Trophic interrelationships between ballast-entrained phototrophic phytoplankton (PPP = INPUT), pre-existing communities of heterotrophic protists and deballasted metazoans representing higher trophic levels (OUTPUT). Particulate organic matter (POM) or dissolved organic matter (DOM) stem from degradation processes inside the tank (--->) as well as from INPUT of newly ballasted seawater and serve as carbon source for bacteria. These (and phototrophic protists) are consumed by small and medium-sized heterotrophic protists and (to a lower degree) by some metazoans (→). Dotted arrows refer to protists serving as food for larger metazoans or their larvae. All trophic groups are ordered according to the respective size.

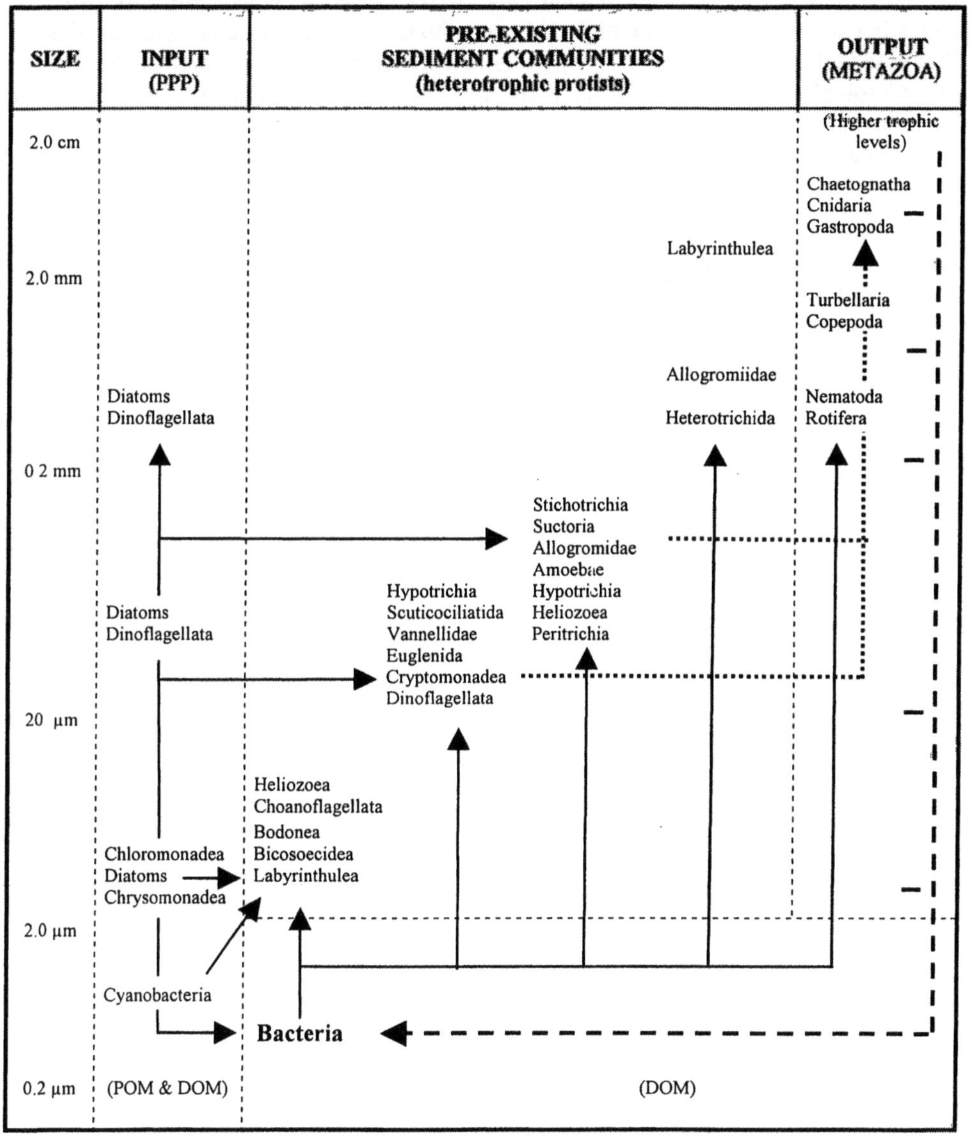

It is well known that some organisms may harbour intracellular symbionts which serve as pathogenic agents after coming in contact with humans: *Legionella pneumophila*, a Gram-negative bacterium, is responsible for lethal disorders once its pathological activity is triggered and enhanced by passage through the cytoplasm of amoebae (Mandell et al. 1995; Robert-Koch-Institute 2001). *Harmannella vermiformis* may be infected by *L. pneumophila,* as well as other amoebae (*Acanthamoeba*, *Naegleria*), and the ciliate *Tetrahymena.* Only freshwater protists were shown to harbour *Legionella*, but these exact taxa are frequently found in ballast assemblages. Given the number of freshwater harbours, and the fact that the Panama Canal is a freshwater lake, the probability of ballasting such vectors is quite high. Of momentous concern is whether protists, that are harmless in their native habitats, may become toxic after ballast-mediated spreading into the microbial associations of a recipient area, or vice versa, that the introduction of bacteria may start toxin production in formerly harmless eukaryotes. Such circumstances may affect "toxic" dinoflagellates and diatoms, as well as metazoans of higher trophic levels, benthic amoebae and small ciliates.

Since the first outbreak in the 1930s, species of *Labyrinthula* (Heterokonta) - net or marine slime molds - have been associated with the wasting disease of marine sea grass communities, as well as with the natural decomposition and the opportunistic degradation of biological material. It is now known that *Labyrinthula zosterae* is the causative agent of the wasting disease in *Zostera* (Muehlstein et al. 1988; 1991). The high metabolic activity and nutrient availability in ballast sediments favours the survival of labyrinthulids: half of the ballast sediment samples we examined contained from one to four *Labyrinthula* species. Given the relative small size of the samples (60 ml water-sediment-mixture), we believe labyrinthulids are omnipresent in ballast tanks. Recently *L. zosterae* was recorded as invader in Wismar Bight, Baltic Sea (Hülsmann unpubl.).

Is there a "silver lining" in the form of ballast-mediated transfer of biocontrol agents? Recently, species-specific amoebae and flagellates that may serve as biological control agents of the aforementioned diatoms Odontella and Coscinodiscus were detected in the North Sea (Kühn et al. 1996; Kühn 1996/97, 1997a,b). It is assumed that these parasites have arrived from the diatoms' native environment by the same route.

INTRODUCED MARINE ALGAE AND VASCULAR PLANTS IN EUROPEAN AQUATIC ENVIRONMENTS

INGER WALLENTINUS
Department of Marine Botany, Göteborg University, Göteborg, Sweden
Inger.Wallentinus@marbot.gu.se

Abstract
Literature compilations for all Europe revealed totally 529 introduced vascular plants (established or occasional) in aquatic environments, 214 being true aquatic or semiaquatic species. Species are listed for all countries and around 75% came from other continents. Totally, 72% occurred in freshwater surroundings, but only 37% in marine-brackish environments, some in both habitats (thus the sum is > 100%). For the aquatic and semiaquatic vascular plants corresponding figures were 95% and 17%. Around 50% are weeds, slightly more among the aquatic and semiaquatic species, and totally 20% have a potential to cause serious problems, of which 65% are aquatic. Of the 113 likely introduced marine macroalgae 27 are invasive. Tentatively, 40 marine microalgae were introductions, nearly 50% bloom-forming and/or toxic. Half of the introduced vascular plants are ornamentals or cultivated, and 43 species used in aquaria, but also occur as rice-field weeds. Other vectors have also been important. Macroalgae have mainly been spread accidentally as epibionts in aquaculture, with ship-fouling also being significant. Ballast is most likely the main vector for microalgae, but also mollusc movements contribute. For some algae and vascular plants different vectors have been involved in different areas. Effects, survival and dispersal strategies, management and utilization are exemplified.

1 Introduction

During the three last decades we have seen an increasing number of new, introduced seaweeds being established in all parts of the world, but especially along the Atlantic and Mediterranean coasts. Some macroalgal introductions also date back more than a century ago (e.g. *Codium fragile* ssp. *atlanticum*, *Polysiphonia harveyi* and *Bonnemaissonia hamifera* in Britain; for references see Farnham 1997; Maggs & Stegenga 1999; Wallentinus 1999a). For vascular plants quite many introductions are more than hundred years old, but some are also quite recent. Plants are generally thought of in terms of primary producers, but an as important role of the macrophytes in temperate waters is their architectural structure, providing shelter for the mobile fauna and increased surface for epibionts. The impact of the introduced seaweeds occurs on many levels of the ecosystem through e.g. competition with native plants for space, light and nutrients; competition for space with sessile animals; reduced water movements - especially when growing on previously barren shores; accumulation of sedimenting particles - which can reduce food abundance for benthic filter-feeders - and in some cases by producing deterrents or toxic compounds affecting grazers. On fisheries and hence economics, their impact mainly occurs through clogging of fish nets and/or by changing the character of the benthic communities, by being a nuisance in aquaculture through fouling on supporting structure and aquaculture target species (e.g. oysters and mussels), clogging of cages and in extreme cases by lifting and sailing away with free-living oysters and mussels. Fouling on ship hulls is both an economic and a recreational problem and for small boats some large introduced seaweeds can hinder navigation.

E. Leppäkoski et al. (eds.), Invasive Aquatic Species of Europe, 27–52.

In freshwater the introduced benthic plants can clog small streams and rivers and a biomass of up to 45 kg fw m^{-2} has been found for the Australian-New Zealand weed *Crassula helmsii* in Britain (Dawson 1996). Plants forming dense mats floating on the surface may be detrimental for the benthos by drastically decreasing the light penetration for the benthic plants and by reducing oxygen for animals through decreased exchange through the water surface. The problems with almost total clogging of tropical waterways caused by the water hyacinth and the fern *Salvinia* are well-known, as are the problems when such huge biomasses are decomposing. Even in Britain the much smaller floating fern *Azolla* can reach a biomass of almost 10 kg fw m^{-2} (Janes 1998a). Both macroalgae and vascular plants may exceptionally develop into such biomasses as to monopolize the substrate, but it is not unusual that they decrease with time.

This review covers vascular plants in all aquatic surroundings and algae in marine and brackish waters. I have not found any compilation of introduced vascular plants covering all European aquatic environment, although national documents exist which cover such species (e.g. Jansson 1994; Tømmerås 1994; Eno et al. 1997; Josefsson 1999). For macroalgae only a few broad publications on introductions were published before the 1990s (e.g. Por 1978; Farnham 1980; Verlaque 1981, 1989; Anon. 1982; Rueness 1989). Thus when asked by the International Council for Exploration of the Sea (ICES) to compile the second status report on introduced marine plants and algae during the decade 1981-1991 in its member states (i.e. countries bordering the North Atlantic and the Baltic Sea, including also the Canadian and US Pacific coasts, the Spanish and French Mediterranean coasts), I decided not only to list new introductions, but for each country list all known introduced marine plants and algae as well as times of their first records with a substantial list of references included. The details and all references given there are not repeated in this review.

The compilation was presented as a poster with handouts at the 14th International Seaweed Symposium in France in 1992, and the manuscript was finished in 1993. Unfortunately, the printing of the report was delayed for several years and it did not appear until 1999 (Wallentinus 1999a). In the meantime several review papers on introduced species had appeared (e.g. Boudouresque 1994; Farnham 1994; Ribera 1994; Verlaque 1994; Boudouresque & Ribera 1995; Farnham 1997; Fletcher & Farell 1999; Maggs & Stegenga 1999). In 2001 Verlaque published a new paper describing new and the status of previously introduced macroalgae in the French Thau lagoon.

2 Compilation of the data presented

The material presented in Appendices 1-2 was achieved in several ways (Appendix 1 only available at http://www.ku.lt/nemo/EuroAquaInvaders.htm, Appendix 2 at the end of this contribution). For all material presented, I am certain many readers will find gaps of information and also have different opinions of whether a species is native or not, or occasional or established. This may also be the case when reading other chapters in this book. That a species is missing in a country thus can be due to either that it is considered native there, or that there is no information of its occurrence. Thus the material should not be seen as a final, fully comprehensive status report for introduced vascular plants and marine algae in European aquatic environments. It is more a starting point to compile information, which can be useful when discussing biodiversity and other ecological aspects not only in a national perspective but also on a larger geographical scale. Hopefully, the lists could also be useful for countries where little such information is compiled. The final goal of course would be to have a continuously updated European database for introduced species in aquatic environments, including also other taxonomical groups not treated here.

2.1 VASCULAR PLANTS

For vascular plants, in the first hand I have consulted the Flora Europaea (Tutin et al. 1968, 1972, 1976, 1980, 1993), the Atlas Florae Europeae (data used from Jalas & Suominen 1980, 1994; Jalas et al. 1999) and "Süsswasserflora von Mitteleuropa" (Casper & Krause 1980, 1981). The former two series cover all of Europe and the latter also has information for more than central Europe. They give information of which species are introduced or native in European countries (most volumes produced before all new European states were created, see comments to geographical areas in the head of the Tables, Figures and Appendices). However, they do not have specific lists of introductions, meaning one has to search for each species separately, which is very time-consuming if not knowing what to look for. That is the reason for the list of species in Appendix 1 (which can also be obtained from the author as a searchable Excel-file with references). I have also consulted several regional floras and lists (see references in App. 1), where much information on introduced species can be found. I am aware of that this may bias the results, since for logistic and linguistic reasons not all parts of Europe were equally well covered by available modern floras, which was especially evident for eastern and southeastern Europe. For many species also digitized literature databases were consulted, where information in more recent scientific journals could be collected, as well as information available on the Internet, but this was impossible to do for all species. A world-wide presentation of aquatic plant genera have been published by Cook (1990).

Taxonomically, I have tried to use the names of species and genera used in the most recent works (synonyms not listed), but this has not been possible to check throughout. The species have been sorted by families to ensure that even if new names exist, that way it should be easier to find a species than if just listed alphabetically. However, within the families species are sorted alphabetically and do not follow the different tribes and sections as in true floristic papers. The listing of families and their names follow the modern British flora works (Stace 1991; Sell & Murell 1996), because many taxonomic changes have occurred since most volumes of Flora Europeae were published. However, the order of families may have changed also after the British floras were published, depending on new phylogenetic results. Cross-references are generally not given when the same information was available in several floras. Information on introduced ornamental plants mostly originates from the floras, but if a species is used as ornamental or not has also been achieved from other sources, which are not given. Anyhow, such usage varies much, both locally and regionally, as well as changes with time.

2.2 ALGAE

For macroalgae the taxonomical affinities used follow the database "AlgaeBase" (Guiry & Nic Dhonncha 2001), although other sources have different opinions, especially for the affinities of green algae (e.g. van den Hoek et al. 1995). In this database also all synonyms as well as higher taxonomical affinities can be found. The order of families follows Guiry (2001). The material is compiled mainly from my own and other review papers, as well as from a variety of published articles by using digitized literature databases complemented with information in the "AlgaeBase". No attempt could be made to scan the literature on introduced limnic macroalgae. As pointed out above, there are different opinions of whether a species is introduced or not, and thus in some cases they should be seen as tentative introductions. Also for macroalgae the amount of information varies between countries.

For microscopic algae a tentative list is presented, based mainly on compilations by Nehring (1998), Elbrächter (1999) and Wallentinus (1999a), with some additional input from other sources. The classes follow van den Hoek et al. (1995). With no time to go through all the literature, the list is probably very incomplete, especially for phytoplankton not belonging to the potentially harmful species, and it does not contain any freshwater species.

3 Results

3.1 VASCULAR PLANTS

3.1.1 Numbers, taxonomic affinities and distribution

Totally, 529 introduced vascular plant species (including occasional species as well as some hybrids; App. 1) were found to occur in or very close to aquatic environments in Europe. Certainly, this figure is not a fixed one, but very much depends on how close to the water one chooses to draw the limit when including an introduced species or not. The purpose of producing the list with a definite number was twofold. Firstly, to give a species list which encompassed the whole of Europe and also included species, which often only are discussed by terrestrial biologists. Secondly, to enable some quantitative comparisons and to show how many different introduced plants that may turn up in the European aquatic surroundings.

The total numbers of species were distributed among six classes, 101 families and 286 genera. Of these 214 species (40% of the total) could be called aquatic or semiaquatic species, i.e. with their main occurrence in aquatic habitats within their original distribution areas (cf. Casper & Krausch 1980, 1981; Cook 1990). However, some of the latter species, especially those used as ornamentals or cultivated for crops can also grow in other habitats, which quite often is the case for many of these species when established in other countries. The aquatic and semiaquatic plants were distributed among three classes, 48 families and represented 110 genera which is around 40-50% of the total. Most examples given below are from those two groups. The other plants were included when listed as growing on shores, river-banks, by streams or on beaches, sand-dunes or maritime rocky shores, either in their native area (references not given) or as introduced. Even though they often are as, or perhaps even more, important in other habitats, their capacity to grow around waterways or lakes may influence the total watershed, especially if occurring as weeds. Furthermore, tall plants may affect the light climate for small rivers and streams. The majority (44%) of these 315 non-aquatic plants are tall (> 0.5 m) herbs, and 24% tall (> 1 m) shrubs or trees, 28% small herbs and a minority of small bushes.

Britain had by far the highest number of introductions with almost 300 species (Table 1, Fig. 1), nearly twice as many as France and almost three times or more as many as most other countries. Of all introductions listed here, 368 species (70%) came from areas outside Europe as did 167 of the aquatic and semiaquatic species (78%), while the rest were introductions within Europe. The latter is especially true for the British Isles where many species native elsewhere in Europe have been introduced and 23% (34 spp.) of the total within-Europe dispersal were listed only for that area. However, also the intercontinental introductions were often listed only for the British Isles and constituted 19% (69 spp.) of the total intercontinental ones. It is difficult to judge whether that reflects the British people's well-known interest in gardening, many contacts with much imported goods (wool, grains, oil- and bird-seeds etc.) from the former colonies, mild winters in the southern parts making it easier for exotic plants to survive, or simply better floristic documentation. However, ornamental plants (incl. crop plants etc., see 3.1.3) dominate and make up as much as 2/3 of all species introduced into Britain (cf. 3.1.3). Overall, many species were established only in some few areas (see App. 1 for areas) and of all plants coming from other continents only 99 species (27% of the intercontinentally dispersed ones) and 32 aquatic and semiaquatic ones (19%) where listed for five countries or more. Especially plants growing as weeds in rice-fields were often found only in a few countries since the warm climate needed by most of these plants is not provided further north. Also the number of introduced aquatic and semiaquatic species were highest for Britain (82 species), while the percentage of those species of the total were highest in countries having rice-fields (Table 1).

3.1.2 Habitats, survival and dispersal strategies

Most plants were associated with inland waters and 380 species of the total can establish in or around inland waters (incl. swamps, rice-fields etc; see App. 1), while only 195 species can be expected to be associated with the sea (incl. brackish areas and sand dunes). Some species could grow in both types of habitat (making the sum exceed 100%). Of the aquatic and semiaquatic species 203 can be found in inland areas while only 37 occur in the sea or on seashores. Some of the introduced aquatic plants have become so established as neophytes that communities even in Europe have been named after them such as Azollo filiculoidis-Lemnetum minusculae (=*L. minuta*), Azolletum carolineanae (=*A. mexicana*), Spartinetum townsendii (Schaminée et al. 1995) and Acoretum calami and the *Veronica beccabunga-Mimulus guttatus* community (Casper & Krausch 1980, 1981).

Once established a perennial strategy enhances the chances of the plants' survival in the new environments, if the habitat is stable enough. However, annuals which can manage to mature numerous seeds, often saved in seed-banks, may be as successful, as can be witnessed among many ordinary weeds, especially on disturbed soils. Almost 80% of the total number of introductions as well as of the aquatic and semiaquatic plants (410 and 164 spp., respectively) have a perennial life strategy, however, in some cases also an annual strategy may be exhibited.

Table 1. Number of introduced vascular plants in the different European areas, including also occasional species. (Aq) aquatic spp., (Aq+Saq) aquatic+semiaquatic spp. (Spp) total number of listed introductions in the aquatic environments, (%) percentage of Aq+SAq of total. For areas the international ISO codes are used (capital letters) except for (Balt) Estonia, Latvia, Lithuania and Kaliningrad (Russia), (Ire) Ireland+Northern Ireland, (CzS) the Czech Republic and Slovakia, (Yug) former Yugoslavia. For the European part of the former Soviet Union the areas follow Flora Europeae.

	IS	FO	NO	SE	FI	Balt	PL	DK	DE	NL	BE	GB	Ire	FR	PT	ES
Aq	1	0	4	8	3	2	1	7	14	12	9	29	12	21	13	17
Aq+SAq	2	0	25	31	17	9	9	25	37	30	23	82	30	54	47	43
Spp	7	2	105	115	80	24	50	95	93	74	64	298	92	163	115	108
%	29	0	24	27	21	38	18	26	40	41	36	28	33	33	41	40

	IT	MT	CH	AT	CzS	Yug	AL	GR	HU	RO	BG	Ru|w|	Ru|s|	Ru|c|	Ru|e|	Ru|n|
Aq	23	1	10	9	4	6	3	5	11	10	5	6	4	2	5	1
Aq+SAq	68	2	26	20	20	12	6	8	24	19	14	12	14	8	7	2
Spp	124	7	69	67	71	42	11	22	58	53	27	36	25	32	18	10
%	55	29	38	30	28	29	55	36	41	36	52	33	56	25	39	20

Many of the aquatic species listed (e.g. *Elodea* spp., *Egeria densa*, *Myriophyllum* spp. and several other plants used in aquaria; for references see App. 1) can multiply by fragmentation, vegetative buds or cuttings, which especially for submerged or floating plants is an efficient way of dispersing into new areas, on the condition that they can survive the winters. In several cases introduced plants are seldom flowering (e.g. *Elodea nuttallii, Lagarosiphon major, Murdannia blumei, Myriophyllum aquatica*; for references see App. 1) or dioecious plants miss one sex as introduced (e.g. only females found in much of Europe for *Elodea candensis*, *Acorus calamus*; for references see App. 1), or are sterile triploids (e.g. *Acorus calamus, Spartina townsendii*; for references see App. 1). The hexaploid *Spartina anglica* on the other hand is fertile and a serious weed. Many of the aquatic and semiaquatic species, as well as many of the others, have tubers or rhizomes (e.g. *Hydrilla verticillata, Acorus calamus, Spartina* spp., *Halophila stipulacea*) or suckers which help them rejuvenate, thus getting an advantage over other species by having an early start in growth. Overall, a vegetative dispersal is highly likely to enhance the risks that the species behaves like a weed, especially if it can establish all year through from a propagule bank (cf. Cellot et al. 1998, and macroalgae 3.2.2).

In many cases the colder climate in the north is the reason that casual, tropical-subtropical species have not survived (e.g. *Salvinia* spp. and *Azolla* spp. in Scandinavia). However, it has been shown (Janes 1998a,b) that *Azolla filiculoides* can survive in ice and air temperatures even below -5°C, and sporocarps in sediments survived winters with below-zero temperatures. It also may have acclimatized to the cooler British climate and could grow even at 5°C, although optimally at 15°C for sporelings. The subtropical nymphoid *Cabomba caroliniana* has survived in southern Sweden for more than five years (Jonsell 2001). When the tropical weed, water lettuce (*Pistia stratiotes*) was forming dense floating mats by vegetative propagation in The Netherlands in the hot summer of 1976, there were speculations that it could become a serious pest there. Although not happening, *Pistia* still could set ripen seeds even in colder summers and Pieterse et al. (1981) showed that the viability of the seeds were not affected by two months at 4°C, and that the seeds even could stand ice for a few weeks, but needed temperatures above 20°C to germinate. Thus alternative survival strategies may apply depending on climate. Water hyacinth, however, seems less hardy, but can stand near-freezing for a limited period (Owens & Madsen 1995).

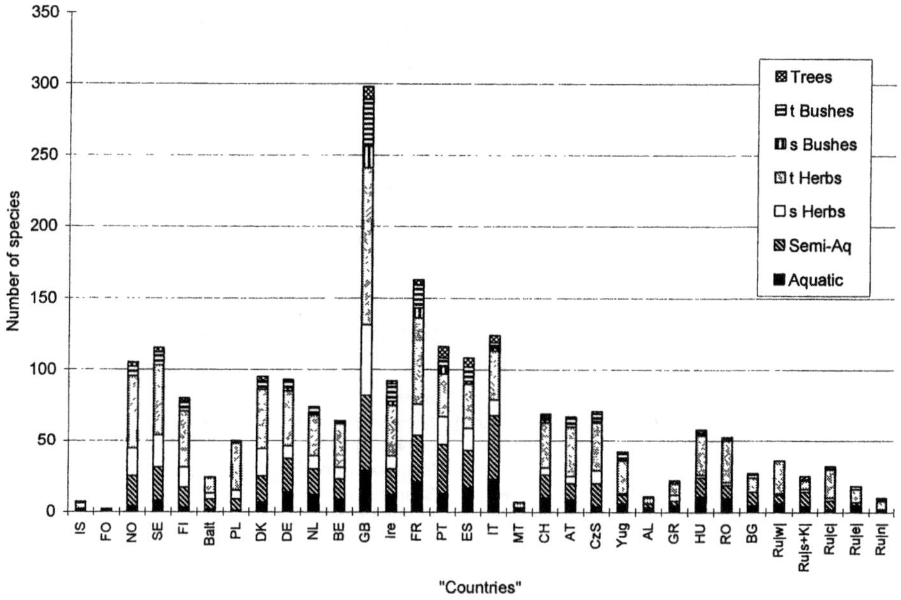

Figure 1. Introduced (established or occasional) vascular plants for the different European areas. For areas the ISO codes are used except for (Balt) Estonia, Latvia, Lithuania and Kaliningrad (Russia), (Ire) Ireland+Northern Ireland, (CzS) the Czech Republic and Slovakia, (Yug) former Yugoslavia. For the European part of the former Soviet Union the areas follow Flora Europeae. (s Bushes) small bushes (< 1 m), (t) tall, (s Herbs) small herbs (< 0.5 m). (Semi-Aq) semiaquatic species according to text.

3.1.3 Vectors

Of the total number of plants, 50% (262 spp.) are ornamentals or cultivated as crop plants (marked by ° or ^{c} in App. 1), in some cases, however, not used any longer. Thus even totally, (cf. the even higher percentage for Britain in 3.1.2) this is by far the most dominant vector of the total number of introduced plants, including also crop plants (e.g. *Oryza sativa*, *Cholocasia esculenta*), fodder and oil plants (e.g. *Oenothera* spp.) as well as some trees planted for timber (e.g. *Eucalyptus* trees), plants used to bind mud on tidal flats (*Spartina* spp., but note that *Spartina pectinata* also can grow in freshwater), to stabilize sand-dunes (e.g. the *Casuarina* and *Acia* trees and the grasses *Ammophila* spp., *Lagurus ovatus*) or as protecting hedges close to the seashore (e.g. the shrubs *Pittosporum* spp., *Hebe* spp.).

Table 2. Genera found on sale within the aquaria trade[a] (some also for ponds or as potted plants[o]) but with no species recorded as introductions in Europe (those are listed in App. 1-2). [a] or [o] denotes one species, [aa] several species on sale. () denote only European species found on sale. Genera are noted in **bold** when species on sale are considered as weeds, in **bold and underlined** when being principal or serious weeds and when marked in *italic* other species than those on sale are weeds. (SW) saltwater, (FW) freshwater. The list is most certainly not complete.

CLASS and Genus	Family =X-ceae	References (see below)
CHLOROPHYCEAE		
(**Enteromorpha**)[aa] SW	Ulva-	5, 8, w
(**Ulva**)[aa] SW	Ulva-	5, 8, w
Valonia[a] SW	Valonia-	8
Avrainvillea[a] SW	Udotea-	8
Halimeda[aa] SW	Udotea-	6, 8
Penicillus[a] SW	Udotea-	6, 8
Udotea[aa] SW	Udotea-	6, 8
CHAROPHYCEAE		
(***Chara***)[aa] FW	Chara-	1, 2
(***Nitella***)[a] FW	Chara-	1, 2
RHODOPHYCEAE		
Galaxaura[a] SW	Galaxaura-	8
Mesophyllum[a] SW	Corallina-	w
Amphiroa[a] SW	Corallina-	8
Euchema[a] SW	Solieria-	8
MIXED		
"Living stones" / "Live rocks" SW		8, w (cf. 3.2.3)
HEPATICE		
(Riccia)[a] FW	Riccia-	1, 5, 6, w
(**Ricciocarpus**)[a] FW	Riccia-	9
MUSCI		
(Leptodictyum)[a] FW	Amblystegia-	9
(Fontinalis)[a] FW	Fontinala-	6
Vesicularia[a] FW	Hypna-	5, 6, 9
Glosadelphus[a] FW	Sematophylla-	5, 9
LYCOPODIOPSIDA		
(Isoëtes)[aa] FW	Isoëta-	6
PTEROPSIDA		
(Pilularia)[a] FW	Marsilea-	6, 9
Regnelidium [a] FW	Marsilea-	1, 9
Bolbitis[aa] FW	Lomariopsida-	3, 5, 7, 9
Microsorium[a] FW	Polypodia-	1, 5, 6, 7, 9
MAGNOLIOPSIDA		
Barclaya[a,o] FW	Nympha-	4, 5, 6, 7, 9, w
(**Ceratophyllum**)[aa] FW	Ceratophylla-	1, 5, 6, 9
Houttuynia[o] FW	Saurura-	7, w
(Aldrovanda)[a] FW	Drosera-	9
Dionaea[a] FW	Drosera-	w
Samolus[a] FW	Primula-	9
Neptunia[aa] FW	Faba-	1, 4
Proserpinaca[a] FW	Haloraga-	4, w
Didiplis[a] FW	Lythra-	6, 7, 9
Hydrolythrum[a] FW	Lythra-	9
Nesaea[a,o] FW	Lythra-	7, w
Phyllanthus[a] FW	Euphorbia-	1, 4, 9, w
(Hippuris)[a] FW	Hippurida-	5
Glossostigma[a] FW	Scrophularia-	7, w
Hydrotriche[a,o] FW	Scrophularia-	4, 7, 9, w
Limnophila [aa] FW	Scrophularia-	1, 4, 5, 6, 7, 9 w
Micranthemum[a] FW	Scrophularia-	4, 7, 9, w
Hemigraphis[a] FW	Acantha-	1, w
Hygrophila [aa] FW	Acantha-	1, 4, 5, 6, 7, 9, w
Nomaphila[a] FW	Acantha-	6, 9, w
Synema[a] FW	Acantha-	6
(Pinguicola)[aa] FW	Lentibularia-	6
Lobelia[a] FW	Lobelia-	1, 4, 5, w
Gymnocoronis[a,o] FW	Astera-	1, 7, w
Shinnersia[a,o] FW	Astera-	7, w
LILIOPSIDA		
Hydrocleis[a,o] FW	Butoma-	3, 7, 9
Echinodorus[aa,o] FW	Alismata-	1, 3, 5, 6, 7, 9, w
Limnobium[aa] FW	Hydrocharita-	1, 6, 9, w
Amauriella[a] FW	Ara-	9
Anubias[aa,o] FW	Ara-	5, 6, 7, 9, w
Cryptocoryne[aa] FW	Ara-	1, 5, 7, 9, w
Lagenandra[aa,o] FW	Ara-	6, 7, 9
Spathiphyllum[a] FW	Ara-	5, 9
Syngonium[a] FW	Ara-	1, w
Mayaca[a,o] FW	Mayaca-	3, 7
Xyris[a] FW	Xyrida-	1, 9
Eleocharis[aa] FW	Cypera-	3, 5, 9
Hygrorhyza[a] FW	Poa-	1, 9, w
Zosterella[a] FW	Pontederia-	3, 7, 9
Crinum[aa] FW	Lilia-	1, 5, 7, 9, w
Ophiopogum[a] FW	Lilia-	w

References: 1) Holm et al. 1979; 2) Krause 1997; 3) Casper & Krausch 1980; 4) Casper & Krausch 1981; 5) Anon. 2000; 6) Laursen 1971; 7) Cook 1990; 8) Fosså & Nilsen 1996; 9) Jacobsen 1979; (w) has been found advertised on one or several sites on the Internet (references not given).

Of the introduced plants 43 species are used in the freshwater aquaria trade (20% of the aquatic and semiaquatic plants). Among the most well known examples of aquaria plants are the submerged or floating species such as *Elodea* spp., *Egeria densa, Laga-*

rosiphon major, Hydrilla verticillata, Najas graminea, Myriophyllum aquaticum, Crassula helmsii and the aquatic ferns. However, aquaria release may not be the true reason for their introduction since many of these are also weeds in rice-fields (App. 1). Furthermore, some aquaria plants are also used in out-door ponds and thus may easily disperse to nearby water courses. Also a large number of semiaquatic plants living on freshwater shores are often used in aquaria (App. 1). A large number of other genera are also sold in the aquaria trade, but so far have not turned up in the lists of species found in the wild in Europe (Table 2). Thus it is difficult to estimate the importance of the aquaria trade as a vector for vascular plants, but some of the aquatic plants most certainly originate from the aquaria trade (App. 1).

Other important vectors have been imports of bird- or oil-seed, as weed among imported grain, or plants having contaminated imported plants or soil or been entangled in imported wool. For example, wool is listed as a likely vector for the Australian aquatic *Myriophyllum verrucosum* as well as for a large number of the semiaquatic *Cyperus* species from all continents. Dry ballast is a likely reason for some plants introduced before the 20th century, the plants often being found around old docks (e.g. *Spartina alterniflora, Sueda altissima;* cf. also *Chara connivens* 3.2.3).

Some few species are old medical plants (e.g. *Acorus calamus*). The seagrass *Halophila stipulacea* is an often mentioned Lessepsian immigrant from the Red Sea (but cf. also discussion by Por 1978, Verlaque 1994, that it may have arrived earlier), and it is the only Lessepsian immigrant among the vascular plants. It has recently also been found in the western part of the Mediterranean (Rindi et al. 1999). For several species different vectors are responsible in different parts of Europe (App. 1).

For some species transport vehicles such as small boats, military vehicles or even tires of tourist cars are possible vectors, especially for secondary regional dispersal, but may be of marginal importance (e.g. Lonsdale & Lane 1994). There is no mentioning of the fishing industry or angling as a vector for introductions of vascular plants into Europe (cf. e.g. *Spartina japonica* used for packing of molluscs from Japan to US, Wallentinus 1999a and references therein). However, Casper and Krausch 1981 stated that *Myriophyllum heterophyllum* has been secondarily spread within central Europe by anglers (cf. 3.2.3) and fish movements were suggested as one possible explanation forthe establishment of *Elodea canadensis* in Lake Baikal in the 1970s (Kozova & Izhboldina 1993). It is difficult to judge how many vascular plants that have been introduced by scientific experiments, since this has not been stated as a vector in the literature available. Most likely some species have been introduced that way, at least locally. However, the large number of ornamental and crop plants would probably dominate anyhow, but one should of course never use introduced plants when carrying out experiments in environments where they do not already occur.

3.1.4 Weed species, control attempts and competition

An attempt has been made to see how many of the introduced plants are considered to be weeds (marked in bold in App. 1), some of which are serious or principal weeds (underlined and marked in bold in App. 1) either in their native countries or as introduced or both. The data are mainly taken from Holm et al. (1978), or in some cases from other published data (e.g. Williamson 1994) as well as from the many lists of noxious or illegal weeds on the Internet. The maximum potential for weediness has been used, irrespectively if this may not be realized when introduced in many European countries. As stated by Holm et al. (1978) it is a somewhat subjective measure if a species is a serious weed or not depending on the persons in the different countries sending in the information. However, the concern is serious and in e.g. Australia and the US many species are forbidden to possess or sale (see e.g. references in App. 1).

Of the total number of introduced vascular plants in aquatic environments 44% can be considered as weeds with 20% (107 spp.) being serious or principal weeds somewhere in the world. Of the aquatic or semiaquatic plants 57% are weeds, and of the total number of serious weeds they make up 65% (70 species), i.e. the introduced non-aquatic plants higher up on the shores have a much lower percentage of serious weeds. This of course does not hold for other terrestrial environments. Of these serious weeds 57 species are aquatic or semiaquatic plants coming from other continents, but only 13 of these are listed for five countries or more, since many of the serious pest species grow in rice-fields and thus do not have suitable conditions further north.

A very good basic review of management of aquatic weeds was presented by Nichols (1991) where he discussed several techniques such as mechanical control (incl. the risk of spreading more propagules), physical (e.g. changes in water levels, light availability, sediment structure), herbicides and biocontrol. Since then the field of biocontrol of introduced aquatic weeds has developed vastly. Organisms used are insects, mostly imported weevils but also native insects, and with as many as one new agent developed each 2.9 years (Cofrancesco 1998), but not always successful due to predation, parasitism or inbreeding (e.g. Grodowitz et al. 1997); fungi (e.g. Barreto et al. 2000) and bacteria (e.g. Morris et al. 1999), as well as combinations of several techniques (e.g. Center et al. 1999). Generally, the dominant management techniques used in Europe are mechanical harvesting (e.g. Moreira et al. 1999) and release of introduced grass carps, the latter, however, also used to control native aquatic weeds. Herbicides are used in some European areas to fight e.g. *Crassula helmsii*, but restrictions often apply (Dawson 1996). Getting rid of one weed by management, however, may not solve the problem, since it has also been shown that this can make way for other introduced weeds (e.g. Chikwenhere 1994).

There are examples in Europe that a second introduced species has been successfully competing with a previous introduction. *Elodea nuttaillii* has replaced *Elodea canadensis* in parts of France and The Netherlands (Casper & Krausch 1980, Thiebaut et al. 1997), probably due to its higher capacity of nutrient uptake. Also *Egeria densa* has replaced *Elodea canadensis* in some warmer parts of Europe (Casper & Krausch 1980). This seems to be an even more common phenomenon in other continents, where series of introductions have followed each other with different dominants e.g. in New Zealand lakes (the European *Ceratophyllum demersum* and the South American *Egeria densa* > the South African *Lagarosiphon major* > the North American *Eldoea canadensis* > native species; Wells et al. 1997). Many more such examples exist. The South American *Carpobrotus chilensis*, which has been established in western US since long time ago, is now threatened by the much later introduced *Carpobrotus edulis,* which also forms vigorous hybrids with the former making the threat worse (e.g. Weber & D'Antonio 1999). Likewise, within the cordgrass genus *Spartina* hybridization is said to constitute a threat to the shores in California since both the introduced species and the hybrid may occupy other zones than the native species (Anttila et al. 1998).

There are several examples that introduced species have been outcompeting native ones by having better growth rates versus light climate, faster nutrient absorption rates or are resistant to grazing or growing in inhostile environments e.g. *Acorus calamus* tolerating very strongly anoxic sediments (Weber & Brandle 1996), which besides competitive advantages also may be useful for bioremediation. Surprisingly, Casper and Krausch (1980) reported that *Hydrilla verticillata* may be a weak competitor in Europe and that it has been replaced by *Elodea canadensis* in some areas. The North-American *Bidens frondosa* has in many areas outcompeted the native *Bidens tripartita* (Casper & Krausch 1981) and *Elodea nuttallii* replaced natural vegetation of charophytes in experimental ditches (Portielje & Roijackers 1995), especially at higher nutrient levels. There are also many reports on how the floating introduced species of the genera *Lemna* and *Azolla* have an effect on benthic plants, both native and introduced. On the contrary, the South African invasive species *Aponogeton distachyos* has declined in is native area, but since it is used as an ethnic food item, it has now been domesticated (Pemberton 2000). This might increase the risk of spreading that way into new countries, since it can proliferate by the rhizome.

3.1.5 Utilization

For plants introduced to be harvested as crop or timber etc. (see 3.1.3), the benefit is obvious, but these plants may also become serious pests in the wild. Perhaps it is less well known for readers in colder countries that the fern *Azolla* has been widely spread intentionally among tropical rice-fields. The reason is its symbiotic nitrogen-fixing cyanobacteria (*Anabaena azollae*), which leaks nitrogen to plants growing in sediments, which otherwise are prone to denitrification, and hence give better harvests (e.g. Lejeune et al. 1999). However, this may not be enough in nutrient-poor sediments (cf. Cary & Weerts 1992).

Introduced vascular plants may even be used as management tools in turbid waters in fishery (cf. *Sargassum muticum* 3.2.5), where the introduced species can grow but not the native ones (e.g. Engel 1995). Even serious introduced weeds are often used later on and thus also may have some positive effect, although often impossible to stop from taking over in the tropics. Several species such as *Eichhornia crassipes*, *Azolla filiculoides* have been tested or used as green manure and to feed cattle (e.g. Costa et al. 1999; Woomer et al. 2000), to cultivate mushrooms (Murugesan et al. 1995) and for bioremediation of both urban and industrial wastes (e.g. Zakova et al. 1994; Redding et al. 1997; Costa et al. 1999; incl. also petrochemical disharges e.g. De Casabianca et al. 1995). However, in most cases these species are a huge environmental threat due to their rapid growth in tropical climates. Water hyacinth as well as many other weed species such as members of the genera *Cyperus*, *Hydrilla*, *Nymphaea*, *Utricularia* and aquatic ferns also have a potential as methane producers by fermentation (e.g. Abbasi et al. 1990).

The nymphoid *Cabamba caroliniana,* introduced into Europe and Japan, and the in many continents established Euroasian watermilfoil (*Myriophyllum spicatum)* have been shown to develop allelopathic polyphenols, which retard the growth of some harmful cyanobacteria by e.g. inactivating some of their extracellular enzymes (Gross et al. 1996; Nakai et al. 1999). What importance that has in practice, however, needs to be seen. Perhaps it could be a parallel to the usage of straw to prevent algal blooms in ponds, but it should also be evaluated for areas where those species are native.

3.2 MACROALGAE IN MARINE AND BRACKISH WATERS

For several reasons it is more difficult to list introduced macroalgae than vascular plants. Firstly, the taxonomy of especially small ceramoid red algae, but also other groups of algae, is continuously changing due to small differences in criteria used. Thus many species described as new for an area, later turn up as synonyms to earlier described species, to a much higher extent than for vascular plants. This makes it quite difficult to judge whether they have been introduced or not. Furthermore, the small sizes and sterility of many species recorded as new for an area may also mean that they have not been detected or properly identified before.

These taxonomical changes are likely to shift even more in the future, when molecular tools will be used more frequently than today. Molecular tools have already proved to be valuable, when determining the origin of some species turning up in the Atlantic or in the Mediterranean (e.g. Peters & Breeman 1992; Maggs & Ward 1996; Marston & Villalard-Bohnsack 2000; Rueness & Rueness 2000), as well as to elucidate the aquarium origin of *Caulerpa taxifolia* in the Mediterranean (Wiedenmann et al. 2001). By using such techniques van Oppen et al. (1995) showed that recent introductions can occur also within the native range of a species as they discussed for the common red alga *Phycodrys rubens*. Maggs et al. (1992) concluded, based on molecular and morphological evidences, that a species within the *Ahnfeltiopsis devoniensis* complex, quite likely had crossed the North Atlantic by shipping in the early 19th century, now occurring along the western North Atlantic as well as in Northern Ireland, Scotland, Denmark and France. However, they did not resolve which was the original site and thus the species is not included in the numbers given nor in Appendix 2.

Secondly, many macroalgae grow quite deep and are thus less well studied in many areas, especially if they are small. Thus when a disjunct geographical pattern occurs, it is difficult to judge if they are introduced recently, or if they may occur in areas in between, but without having been recorded. There are several such papers, which have not been included here, since the authors did not believe the species in question had been introduced. Thirdly, relicts from previous geological times may also occur and some Mediterranean species may have existed in the former Tethys Sea, instead of being recent immigrants from the Indian Ocean through the Suez Canal or by other means (e.g. Por 1978; Athanasiadis 1990; Perrone and Cecere 1994; Verlaque 1994 and references therein).

3.2.1 Numbers, taxonomic affinities and distribution

A total of 113 likely introduced macroalgal species were listed for marine and brackish waters in Europe (App. 2). This also includes some *in situ* experimental tests and some species with disjunct distribution that perhaps are native. Because of time limits it was not possible to search for introduced limnic macroalgae. As stated above this figure is approximate and much depending on development in taxonomy. As for vascular plants the purpose of producing a list with a definite number was twofold, although more compilations occur for macroalgae than for the vascular plants (see 1). Totally, the macroalgal species were distributed among five classes, 48 families and 79 genera. Although the number of species is only about half of the aquatic and semiaquatic vascular plants, the number of classes is higher, of families the same and the number of genera around 3/4. Of the introductions there are 78 red, 26 brown and 9 green algae.

As stated by several others (e.g. Verlaque 1994, 2001; Boudourcsque & Ribera 1995; Wallentinus 1999a) the French Mediterranean area has a large number of introductions with 61 species recorded, followed by Italy with 41 species, while most other areas showed around 20 introduced species or fewer (Table 3, Fig. 2). Many of these introductions, as well as many of those spread secondarily further within Europe, are primarily due to movements of oysters, mainly from Japan and Korea, but also within Europe (App. 2). In contrast to animals only one introduced macroalga occurs in the Baltic proper, the estuarine green alga *Chara connivens*. Also the northernmost countries, although comparatively well studied, so far have few introduced species (e.g. *Bonnemaisonia hamifera, Sargassum muticum* and *Codium fragile*). For some other countries, lack of investigations probably partly contribute to the low numbers. I have only been able to trace one introduced macroalgae along the Black Sea coasts, the Atlantic brown alga *Desmarestia viridis* (Zaitsev & Mamaev 1997), also being introduced in several Mediterranean countries (App. 2). There was no available information on introduced macroalgae from the northern shores of European Russia.

3.2.2 Habitats, survival and dispersal strategies

Many of the introduced macroalgae have first become established at aquaculture sites or in harbour areas. In many cases they have later dispersed by currents to other shores, mainly rocky, but they also occupy small stones and shells on sediments. Eutrophication has favoured some species (e.g. *Fucus evanescens, Sargassum muticum,* ulvoid algae). *Undaria pinnatifida* is a frequent fouling organisms on ships hulls and growth on artificial substrates can be a competitive advantage as discussed by several authors (e.g. Fletcher & Farrell 1999; Maggs & Stegenga 1999; Wallentinus 1999a,b,c).

Both survival and dispersal is enhanced by the possibility to grow after fragmentation and many introduced red and some green algae lack fertility in one or several stages relying on vegetative dispersal (see also Perrone & Cecere 1994; Chapman 1999; Maggs & Stegenga 1999; Wallentinus 1999a and references therein). Especially for the *Caulerpa* species fragmentation has been a very potent dispersal mechanism (e.g. Davis et al. 1997; Ceccherelli & Cinelli 1999; Smith & Walters 1999; Wallentinus 1999a; Ceccherelli & Piazzi 2001; Zuljevic et al. 2001 and references therein). On the other hand the most common, introduced brown algae have a high fecundity and cannot use vegetative reproduction. Being monoecious and carrying young sporelings as well as having air-bladders has been a great advantage for the successful dispersal of *Sargassum muticum* to new areas (Wallentinus 1999c and references therein). Farnham (1997) listed several more characters, which have made *Sargassum muticum* so successful, amongst them the combination of r- and K-strategies. For *Undaria pinnatifida* the microscopic, very desiccation tolerant, gametophytic stage has enhanced its dispersal capacity, since it can survive dried out for a month and tolerate darkness for even longer

(Wallentinus 1999b and references therein). The to Helgoland introduced North Atlantic red alga *Mastocarpus stellatus* has been found to gain in competition with the native *Chondrus crispus*, when growing at the shoreline, by tolerating more UV-radiation through production of mycosporine-like amino acids (Bischof et al. 2000). Generally, variations in life cycle strategies are important features for the establishment and survival of the introduced macroalgae, especially when there are asexual short cuts.

Table 3. Number of introduced marine and brackish-water macroalgal species in the different European areas, including also occasional and drift specimens as well as *in situ* experiments. For areas the international ISO codes are used (capital letters) except for (Balt) Estonia, Latvia, Lithuania and Kaliningrad (Russia), (Ire) Ireland+Northern Ireland, (Yug) former Yugoslavia. For countries bordering two different seas, figures have been split. Canary Islands and the Azores are listed under Atlantic Spain and Portugal, respectively. (R) red, (B) brown and (G) green macroalgae.

	IS	FO	NO	w SE	e SE	FI	Balt	PL	n DE	DK	w DE	NL	BE	GB	Ire
%R	50	100	50	33						43	60	67	50	68	70
%B			38	50					50	43	30	28	33	26	20
%G	50		13	17	100		100	100	50	14	10	6	17	5	10
Spp	2	1	8	6	1	0	1	1	2	7	10	18	6	19	10

	FR(Atl)	ES(Atl)	PT	ES(Med)	FR(Med)	IT	MT	Yug	AL	GR	Black S
%R	81	79	71	73	66	66	100	33	100	75	
%B	14	17	21	14	23	24		33		13	100
%G	5	4	7	14	11	10		33		13	
Spp	21	24	14	22	61	41	6	3	1	16	1

3.2.3 Vectors

The most dominant vector is movements of oysters and other molluscs which may be responsible for 44% of the introductions, both intercontinentally and within Europe, with the northwest Pacific as the major donor area. Tentatively, 27% have arrived via shipping mostly on hulls or in ballast, with one example of a solid ballast arrival, *Chara connivens.* It was introduced in the mid and late 19th century in several port areas around the Baltic proper and in the southwestern part (Luther 1979). Shipping has also been suggested as a vector for some macroalgae being new to the Antarctica (Clayton et al. 1997). Fourteen species (12%) are probably Lessepsian immigrants (or due to older communications between the Mediterranean and the Red Sea; cf. 3.1.3 with only one species). Six listed species are experimental introductions (including farming, see below) of which one, *Mastocarpus stellatus* has been very successful in establishing on Helgoland (Wallentinus 1999a; Bartsch & Kuhlenkamp 2000), where it previously did not grow. However, this can be considered to be within its large-scale native distribution area. The number of experimental introductions could increase, if tank experiments are included (see Wallentinus 1999a).

There are few examples among macroalgae of intentional introduction for farming in Europe, the best known being the transfer of the Japanese brown kelp *Undaria pinnatifida* to northern France from the Mediterranean, where it had been accidentally introduced by oyster imports (Wallentinus 1999a,b and references therein). A few Atlantic macroalgae (*Polysiphonia fucoides, Fucus spiralis*; see Verlaque 1994; Wallentinus

1999a for references) are believed to have been brought to the Mediterranean by anglers discarding bait packings. Dumping of by-catches are listed as possible reasons to have brought *Fucus evanescens* to the Swedish west coast and *Bonnemaisonia hamifera* to Helgoland (see Wallentinus 1999a for references). Some macroalgae have multiple responsible vectors. *Undaria pinnatifida* has been spread to new countries by oyster or mussel movements, by ships and been brought deliberately for farming (see references in e.g. Fletcher & Farell 1999; Wallentinus 1999a,b).

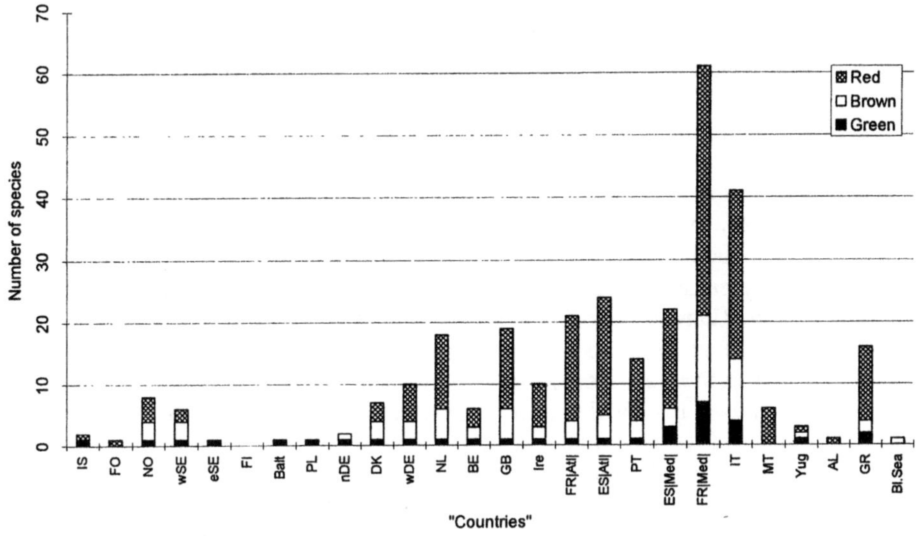

Figure 2. Introduced marine and brackish-water macroalgal species in the different European areas, including also drift specimens and *in situ* experiments. For areas the international ISO codes are used except for (Balt) Estonia, Latvia, Lithuania and Kaliningrad (Russia), (Ire) Ireland+Northern Ireland, (Yug) former Yugoslavia. Countries bordering two different seas are split. Canary Islands and the Azores are listed under Atlantic Spain and Portugal, respectively.

The aquaria trade so far is only known to have released *Caulerpa taxifolia* in Europe (Wiedenmann et al. 2001), although also other *Caulerpa* species are used and sold and can even be ordered through the Internet. However, its further dispersal has included currents, boats and fishing nets (e.g. Sant et al. 1996). Also other macroalgae are sold for salt-water aquaria (Table 2) including also e.g. *Sargassum* spp. It is difficult to judge to what extent "Living stones/live rocks" (Table 2) may have been released in European waters with their mixture of epibionts. Fosså & Nilsen (1996) gave many examples of marine tropical macroalgae growing out in aquaria from such pieces (e.g. species of the red algal genera *Acanthophora, Halymenia, Hydrolithon Mesophyllum, Peyssonnelia, Sporolithon*; the brown algal genera *Dictyota, Lobophora, Padina*; the green algal genera *Bryopsis, Cladophoropsis, Chlorodesmis, Derbesia, Dictyosphaeria, Neomeris, Valonia, Ventricaria* as well as the cosmopolitan green algae *Enteromorpha* and *Ulva* and red turf algae).

Some epiphytic algae (e.g. the red algae *Polysiphonia harveyi* and *Acrochaetium codiolum* and the small brown alga *Corynophlaea umbellatum*) are suggested to have arrived with their larger host (for references see App. 2). I found no reports of marine or brackish macroalgae being imported with wool, but Krause (1997) mentioned an adventive growth in Britain of the freshwater green alga *Chara braunii* in discharges from a cotton mill. For quite a large number of macroalgae it is difficult to prove which vector has been involved, and shipping cannot be excluded as a vector for some of those species. As mentioned above (3.2), recent introductions may also occur within the native distribution range of a species, making the real number of macroalgal introductions higher. Neither are there any figures of how many of the so called cosmopolitan macroalgae (e.g. *Enteromorpha* spp., *Ulva* spp.) having been carried around the world, and the number of cryptogenic species can be high in many areas (cf. Carlton 1996).

3.2.4 Invasive species, control attempts and competition

The impact on the environment does not correspond to the numbers of introduced species within the groups. Although quite few introduced species are green algae, *Codium fragile* (e.g. Garbary et al. 1997; Chapman 1999 and references therein), and the *Caulerpa* species (Hénocque 1997; Meinesz 1997; Verlaque et al. 2000; Piazzi et al. 2001) are known to have caused large problems in the western North Atlantic and Mediterranean, respectively. There are two *Ulva* and one *Monostroma* species listed as introduced by Verlaque (2001) in southern France. Whether they create the same environmental problems as their native relatives do in eutrophicated areas is not stated. On the other hand *Chara connivens* is considered to be an endangered species in the Baltic Sea due to eutrophication and destroyed habitats. Also among the brown algae, species such as *Sargassum muticum* (e.g. Farnham 1997; Fletcher & Farell 1999; Wallentinus 1999a,c and references therein) and *Undaria pinnatifida* (e.g. Fletcher & Farell 1999; Wallentinus 1999a,b and references therein) are well known for their environmental impact in many coastal areas, not the least through their sizes. Also *Colpomenia peregrina* can cause some trouble as an oyster thief. Had the experiment with the giant kelp *Macrocystis* in northern France gone wrong in the early 1970s, many phycologists believed this could have caused a real ecological problem (for references see Wallentinus 1999a). Although being by far the largest introduced group, the red algae only have some few species considered as real weeds in Europe such as the filamentous turf algae *Womersleyella setacea* and *Acrothamnion preissii* (e.g. Piazzi & Cinelli 2000 and references therein), as well as the large folious alga *Grateloupia doryphora* (Cabioch et al. 1997; Marston & Villalard-Bohnsack 2000; Simon et al. 2001). Also the "old" introductions such as *Asparagopsis armata* and *Bonnemaisonia hamifera* can be considered troublesome in some areas, where the small tetrasporophytes can be dominant epiphytes. Many of the other (at least tentatively) introduced species are small red algae, but can also constitute a dominant species as e.g. *Antithamnion pectinatum* in Venice (Curiel et al. 1998). Totally, 27 macroalgae can be considered as invasive and nine species as highly invasive.

There have been fewer attempts to eradicate accidentally introduced marine macroalgae than vascular plants, certainly partly due to the problem of reaching the established populations, but also to that many of the introduced species are quite small. When *Sargassum muticum* was first recorded in Britain an eradication campaign was initiated to manually clear the shores but without any success (Critchley et al 1986). Neither has it been possible to remove the populations of *S. muticum* establishing in Northern Ireland in the mid 1990s. In many countries great efforts have been made to clear *Undaria* to protect the biodiversity or because it hinders mussel harvest (e.g. Sinner et al. 2000; Stuart & Chadderton 2001; Wetsteyn 2001), but often with limited success. Also other introduced species cause problems by fouling on aquaculture structures, on mussels or entangling in fishing tools and have to be cleared away. However, mechanical eradication may even worsen the case if the alga can reproduce vegetatively by fragments such as the *Caulerpa* species and most red algae (cf. 3.1.4). In the Mediterranean several mechanical and chemical techniques have been tried to destroy *Caulerpa taxifolia* (e.g. Hénocque 1997) and a proposal to use biological control with imported Ascoglossan snails (e.g. Meinesz 1997; Thibaut et al. 2001) was much debated and turned down (e.g. Thellier et al. 1997). Also in Canada different methods have recently been tested to eradicate *Codium fragile* (Kieser & Campell 2000). For the newly recorded populations of *Caulerpa taxifolia* in the US (e.g. Jousson et al. 2000), which grew only in limited areas, cover with black plastic was tried as a method to kill them, as is practiced for some terrestrial weeds.

Although many of the large brown algae cannot propagate vegetatively, drifting fertile parts can drift away and establish new populations. For the experimental test with the giant kelp *Macrocystis pyrifera* it was possible, however, to remove the plants before fertility commenced and there have been no further sightings of the species in the eastern North Atlantic. However, one should always be concerned of the risk of loosing whole or fertile parts of plants or whole aquaculture structures, which can be moved by storms and currents. The establishment of *Undaria pinnatifida* in the wild, after farming in northern France, also showed that the total physiological ranges of the species must be considered (for references see Wallentinus 1999a,b).

There are many reports on *Caulerpa taxifolia* smothering native Mediterranean seaweeds and seagrasses but this may not occur everywhere (e.g. Jaubert et al. 1999; Molenaar et al. 2001; Thibaut & Meinesz 2001). Also *C. racemosa* has been seen to overgrow the native flora, which can happen in less than half a year (e.g. Piazzi et al. 2001 and references therein). *Sargassum muticum* has in many countries been found to affect the native flora, and in several areas especially slow-growing seaweeds such as fucoids and kelps, but also the introduced green alga *Codium fragile*, have decreased (for references see e.g. Viejo 1997; Wallentinus 1999a,c; Staehr et al. 2000; Occhipinti Ambrogi 2001). It has been discussed if also seagrass communities may be altered (den Hartog 1997). During the 1990s *S. muticum* has been found to drastically reduce the kelp *Laminaria digitata* in northern France (Cosson 1999). However, in some areas in Scandinavia the establishment of *S. muticum* has mainly been in areas almost devoid of other macrophytes (e.g. Karlsson & Loo 1999), where its main impact is by modifying the physical environment as well as the microphytobentic communities by reducing water movements and light penetration. *Codium fragile* has been reported to reduce populations of the native *C. tomentosum* in Britain (Farnham 1980) and *C. vermilara* in Norway (Silva 1957). On the other hand, *Chara connivens* as many other charophytes seems little competitive and is noted on the Swedish list of endangered species, but at least is surviving on the Swedish east coast (P. Snoeijs pers. comm.).

Even red turf algae such as *Acrothamnion preisii* and *Womersleyella setacea* have been found to change the structure of epiphytic algal communities with a decreased biodiversity, partly due to accumulation of sediment in the dense turfs (Piazzi & Cinelli 2000). Many of the introduced red algae, however, are quite small without much impact. A striking exception is *Grateloupia doryphora*, which is considered a nuisance species due to its size. For *Mastocarpus stellatus* see 3.2.2. As pointed out by e.g. Maggs & Stegenga (1999) halogenated or sulphated antigrazing compounds may be an important advantage for a successful competition from introduced red algae. Phlorotannins may have the same role for brown algae as have different terpenoids among the *Caulerpa* species (Wallentinus 1999a and references therein). However, such substances are also common among many native macroalgae.

3.2.5 Utilization

Established wild populations of *Undaria pinnatifida* have been harvested for wakami products in northern Spain (Cremades 1993), as well as in New Zealand (e.g. Sinner et al. 2000), besides the farming sites on the Atlantic French coast (see 3.2.3). In the mid 1990s two patents were taken out in France to use the biochemical compounds in *Asparagopsis armata*, a reason for it is now being grown in cultures (see references in Haslin & Pellegrini 2001). *Codium fragile* also contains chemicals, which may be of medical use (e.g. Jurd et al. 1995). Species within red algal genera such as *Gracilaria, Grateloupia, Hypnea, Soliera* as well as kelps such as *Laminaria* have in theory a potential as phycocolloid producers but probably with small economic feasibility for most introduced wild populations. If *Macrocystis* had established, it most probably could have been harvested, but probably without stopping its ecological impact (Wallentinus 1999a and references therein). If taken out of the water macroalgae can act as bioremediators and be used as fertilizers, but this has seldom been used in practice (cf. Wallentinus 1999b,c) and also fermentation to methane is possible. If establishing in areas previously devoid of macrophytes large seaweeds such as *Sargassum muticum* can also attract wild fish (Wallentinus 1999c) seeking shelter and food.

3.3 MICROALGAE

Generally, the main concern with phytoplankton is not introduced species as such but toxic algae and harmful algal blooms, whether native or introduced. As emphasized and exemplified by several scientists (e.g. Nehring 1998; Elbrächter 1999), it is often very difficult to assess if a planktonic species is introduced or not, although it has been proved for some cases by dated sediment probes for cyst-producing species (McMinn et al. 1997). Molecular tools have been tried, but so far the question of origin still remains in several cases (e.g. Bolch et al. 1999; Blackburn et al. 2001). The many synonyms occurring in literature makes the task even worse and a species may well be first described from a place where it has been introduced (Elbrächter 1999). Many cosmopolitan microalgae may also be cryptogenic (cf. Carlton 1996).

3.3.1 Numbers, taxonomic affinities and distribution

Species believed to have been introduced in European marine and brackish waters are listed in Table 4. The list is still very incomplete and some species may be native or brought by currents. For the Black Sea seven introduced phytoplankton species have been included in Table 4. Altogether 40 species are listed for the different sea areas (generally, distribution per country is not given), belonging to 4 classes, and 21 genera. It should be pointed out that the true *Gymnodinium catenatum* so far has not been encountered in recent sediments in northern Europe, although referred to in several papers from the 1990s (see also e.g. Elbrächter 1999; Bravo & Ramilo 1999 and references therein). The species is mostly considered as a likely introduction in southern Europe (Spain, Portugal, the Mediterranean; Table 4), but recent studies (Bolch 2000) indicate that the origin instead may be Europe. *Heterosigma akashiwo* is a species often listed as introduced into Europe (for areas see Nehring 1998), but according to Elbrächter (1999) it is a native species. The species of *Chattonella* forming blooms in Scandinavian waters since 1998 may in fact not have arrived that recently, since specimens have been found retrospectively in older preserved samples (M. Kuylenstierna pers. comm.). However, it could still be an introduced species. Some species were included in the list on the basis of having a disjunct distribution. In some cases this might be caused by the difficulty to identify them, being morphologically very similar to other species (e.g. many species of *Alexandrium*), while others are quite distinct and ought to have been recognized (e.g. *Dicroerisma psiloneriella* and *Gyrodinium corallinum*, Karlson & Kuylenstierna 2001; M. Kuylenstierna pers. comm.).

3.3.2 Vectors and survival strategies

The main vector most certainly would be ballast discharges (e.g. Hallegraeff 1998; many chapters in this volume), but this is mostly difficult to prove, once the species has become established. The classical example is the introduction of the large diatom *Odontella sinensis* into European waters in the beginning of the 20th century (e.g. Eno et al. 1997; Nehring 1998; Wallentinus 1999a and references therein). The potential of ballast introductions can be seen in that nearly 500 phytoplankton taxa (incl. heterotrophic flagellates and resting stages) have been encountered in ballast tanks by European scientists during the last decade, although the majority of these species are native or cosmopolitan (Gollasch et al. in press). However, even among such species, discharged ballast may add a new strain, and often not all algal strains have the potential to become toxic.

Dispersal of phytoplankton also has occurred through movements of live shellfish (e.g. Dijkema 1992; O'Mahony 1993; Hallegraeff 1998), with the main concern being the risk of moving toxic species. Shellfish movements have been suggested as a vector for *Coscinodiscus wailesii* and *Thalassiosira punctigera* and *T. tealata* to Europe (Nehring 1998), and for movements of *Alexandrium tamarense* from Japan to Pacific North America (Hallegraeff 1998). One benthic diatom (*Pleurosira laevis*) probably has arrived by eel-stocking in cooling water discharges (Jansson 1994). Halim (1990) presented a tentative list of 17 dinoflagellate species, which may be Lessepsian immigrants in the Mediterranean, but he did not state in which part of the Mediterranean they occur. Thus they have not been included in Table 4, but may increase the total number in Europe. Two diatoms were listed by Por (1978) as highly probable Lessepsian immigrants, but only recorded from Israel and thus excluded here.

Secondary dispersal by currents and winds is almost compulsory, if populations (incl. resting stages) are dense enough. The capacity of many phytoplankton species within several classes to produce resting stages (cysts or spores, Table 4) - very often extremely tolerant to anoxic conditions in the sediment - adds to the chances of their survival during adverse transport conditions. Persson (in press) showed that, if a sediment slurry culture was set up, many more phytoplankton species could be recorded from the sediments than found by immediately counting resting stages, revealing a higher potential of transported species than often is stated.

Table 4. Tentative list of introduced microalgae listed for countries (according to ISO code) or sea areas. Species in **bold** are bloomforming, underlined harmful. Notes: (PSP) Paralythic Shellfish Poisoning, (NSP) Neurotic Shellfish Poisoning. Sea areas: (Atl) the Atlantic, (Med) the Mediterranean, (IndO) the Indian Ocean, (Pac) the Pacific, (EngCha) the English Channel, (N.Sea) the North Sea, (Ska) the Skagerrak, (Kat) the Kattegat. Vectors: (Bal) ballast, (Curr) currents, (Mol) mollusc movements.

CLASS, Genus, species & Author	Notes	Resting stages	Established in:	Origin	Vector	References
DINOPHYCEAE						
Alexandrium andersoni Balech	PSP	Cysts	seIT	w N.Atl?	Bal?	20, 21
Alexandrium angustitabulatum Taylor in Balech	PSP		wSE	NZ?	Native?	17
Alexandrium catenella (Whedon & Kofoid) Balech	PSP	Cysts	seES	Atl?/ Pac?	Bal?	25
Alexandrium leei Balech	?	Cysts	NL	Korea	Bal?	1
Alexandrium minutum Halim	PSP	Cysts	NO, wSE, n&w DE, NL, Ire, FR	Atl?/ Med?	Curr?/ Bal?	2, 17
Alexandrium monilatum (Howell) Taylor	Fish kill	Cysts	Black Sea	Atl	Bal?	4, 23
Alexandrium pseudogonyaulax (Biecheler) Horiguchi	PSP	Cysts	neES, IT	Atl?	?	22, 25
Alexandrium ostenfeldii (Paulsen) Balech & Taylor	PSP	Cysts	seSE	Ska-Kat	Curr?/Bal?	18
Alexandrium tamarense (Lebour) Balech	PSP	Cysts	sFR, neES, sw&eIT	Atl?	?	21, 22, 25
Dicroerisma psilonereiella Taylor & Chattell			wSE	ne Pac	Native?	17
Gonyaulax grindley Reincke	Yesso-toxin	Cysts	neES, sw&neIT	Atl?	?	8, 21, 25
Gymnodinium aureolum (Hulbert) G. Hansen " "	Fish kill		s&wNO, wSE, DK, wDE, NL, BE, GB, s&wIre, nwFR, PT, nES, swIT	Pac?/ e US?	Bal?	1, 2, 17, 19, 21
Gymnodinium breve Davis	NSP		ES, GR	e Atl	Bal?	23, 25
Gymnodinium catenatum Graham	PSP	Cysts	PT, n, sw, ne & seES, IT|Si|	Japan	Bal	1, 3, 7, 9, 10, 13, 21, 25
Gymnodinium uberrimum (Allman) Kofoid & Swezy		?	Black Sea	w Eur. (freshw)	?	4
Gyrodinium corallinum Kofoid & Swezy			wSE	Calif	Native?	17
Ostreopsis ovata Fukuyo	(toxic?), benthic		neES?, wIT	trop-subtrop	Native?	24, 25
Prorocentrum minimum (Pav.) Schiller	(toxic?)		Baltic	Atl?	Bal?	14
Prorocentrum redfieldii Bursa			wDE, NL	e N.Amer	Bal?	1, 2
Pyrodinium bahamense Plate	PSP	Cysts	PT	w N.Atl?	Bal?	5
RHAPHIDOPHYCEAE						
Chattonella antiqua (Hado) Ono	Fish kill	Cysts	NL	Japan	Bal?	1
Chattonella marina (Subrahmanyan) Hara & Chihara	Fish kill	Cysts	NL	IndO	Bal?	1
Chattonella verruculosa Hara & Chihara	Fish kill	?	w&sNO, wSE, DK, wDE	Japan	Bal?	1, 17
Fibrocapsa japonica Toriumi & Takano	Fish kill	Cysts	wDE, NL, nwFR	Japan	Bal?	1, 2, 6

CLASS, Genus, species & Author	Notes	Resting stages	Established in:	Origin	Vector	References
PRYMNESIOPHYCEAE						
Mantoniella squamata Manton & Parke		?	Black Sea	Crypto-genic?	?	4
Phaeocystis pouchetii (Hariot) Lagerheim	Mucus, toxic	?	Black Sea	N.Atl	?	4
BACILLARIOPHYCEAE						
Asterionella japonica Cleve & Möller			Black Sea	Atl?/Pac?	?	4
Corethron criophilum Castracane			wSE, wDE, N.Sea, EngCha,	Atl	Curr	2, 17
Coscinodiscus wailesii Gran & Angst " "	Mucus		wNO, wSE, wDK, wDE, NL, BE, seGB, nwFR	Pac/ IndO	Mol	2, 3, 11, 12, 17
Odontella sinensis (Grev.) Grunow			N.Atl, Baltic	IndO-Pac	Bal	2, 3, 12, 17
Pleurosira laevis f. *polymorpha* (Kütz.) Compère	benthic		eSE	?	Eel-Stock	16
Pleurosigma simonsenii Hasle			GB	IndO?	Bal?	3, 12
Pseudosolenia calcar-avis (Schultze) Sundström			Black Sea	?	?	4
Rhizosolenia indica Peragallo			wDE, NL	Atl	Curr	2
Stephanopyxis palmeriana Grunow		Spores	wSE, N.Sea, EngCha	warm seas	Curr	2, 17
Thalassiosira hendeyi Hasle & Fryxell			wDE, N.Sea, EngCha	Atl	Curr	2
Thalassiosira kuschirensis Takano			nGB	Pac?/ e US?	Native?	15
Thalassiosira nordenskioeldii Cleve		Spores	Black Sea	N.Atl	?	4
Thalassiosira punctigera (Castr.) Hasle			Ska-Kat, N.Sea, EngCha	IndO-Pac	Mol	2, 3, 12, 15, 17
Thalassiosira tealata Takano			NO, N.Sea, EngCha	Japan?	Mol	3, 12, 15

References: 1) Elbrächter 1999; 2) Nehring 1998; 3) Wallentinus 1999a; 4) Alexandrov this volume; 5) Amorim & Dale 1998; 6) Billard 1992; 7) Blackburn et al 2001; 8) Boni et al. 2000; 9) Bravo & Ramilo 1999; 10) Carrada et al. 1991; 11) Edwards et al. 2001, 12) Eno et al. 1997; 13) Giacobbe et al. 1995; 14) Hajdu et al. 2000; 15) Harris et al. 1995; 16) Jansson 1994; 17) Karlson & Kuylenstierna 2001; 18) Larsson et al. 1998, 19) Macdonald 1999; 20) Montresor et al 1998; 21) Montresor et al. 2000; 22) Occhipinti Ambrogi 2001; 23) Taylor et al. 1995; 24) Tognetto et al. 1995; 25) Vila et al. 2001

3.3.3 Harmful species and competition

As stated above the main concern with microalgae is to prevent harmful algal blooms (HAB), which can be witnessed by the huge amount of literature on that topic. Nearly half of the listed species are potentially toxic or bloom-forming (Table 4). However, this high percentage could be due to the many programmes focusing on HAB species, since in reality those species are a small part only of all phytoplankton species. The HAB introductions do not only include toxic species, but some species also secrete mucus (e.g. *Phaeocystis pouchetii, Coscinodiscus wailesii*). Furthermore, it is emphasized that the majority of the harmful bloom events in Europe still is caused by native species. It is even more difficult to control established phytoplankton populations than for macroalgae, especially for species having benthic resting stages which may act as a "seed bank" in the sediment. Many studies are devoted to looking for organisms breaking down the HAB species (e.g. Lovejoy et al. 1998; Skerratt et al. 2000). Perhaps in the future it could be possible to use such organisms to decrease the effect of HAB algae locally as in e.g. isolated aquaculture sites or ballast tanks, but probably not on a regional scale. There are numerous studies on factors regulating competition between phytoplankton species, involving both introduced and native species, but no clear pattern of the role of introduced species versus native harmful algae have evolved.

4 Conclusions

A large number (more than 680 spp.) of introduced plants and algae occur in European aquatic environments, the figure being highest for vascular plants. More of those vascular plants occur in or close to freshwater than in the sea or on the seashores. For algae such comparisons are not possible since no limnic species were included in this review. For macro- and microalgae it is often difficult to state if a species is introduced or not, due to incomplete knowledge of taxonomy and that many geographical areas and/or habitats are less well studied. The high proportion of toxic microalgae may be an overestimation due to the many programmes monitoring such algae.

The dominant vector for vascular plants has been gardening including crop plants and for macroalgae as non-target organisms in aquaculture movements of shellfish. The contribution by the aquaria trade is substantial but not as high as the other vectors. Lessepsian immigrants are not uncommon among macroalgae, but only one vascular plant has been brought from the Red Sea to Europe. Different vectors have also been responsible for introducing the same species in different areas, both for vascular plants and algae. For phytoplankton ballast may be the main dispersal mechanism, but difficult to prove. However, the high numbers of microalgal cells and cysts recorded in ballast tank samples support this assumption. Shellfish movements may also be important to spread microalgae. Secondary dispersal by currents and winds, as well as by migrating seabirds, is important for algae, but can also occur for vascular plants.

If established weed species may be quite hopeless to eradicate, especially in the sea and in large inland waters and acclimatizing processes may help species survive in colder climate than in their original ranges. Hybridization with native species or polyploidization may increase the risks further, as may climate changes, allowing more warm-dependent species to survive in more northern countries. Thus precaution should always be taken to minimize the risk for new introductions, intercontinental as well as those within Europe, not to loose the biodiversity in our aquatic environments. However, it is more difficult to minimize the risks in gardening, when seeds and fruits may be dispersed further by wind and animals and more and more new species are on sale. Probably few of us would like to stop having foreign species in our gardens, but this has been emphasized in Canada and the US. Also we wish plants are hardy enough to survive in our climate and search for such strains, which increases the risk of their establishment in the wild. For the non-target organisms, great care should be emphasized both in aquaculture and wetland cultivation to reduce the number of accidental introductions. Thus the ICES Code of Practice (Anon. 1995) and similar regulations should always be implemented. However, Verlaque (2001) pointed out that illegal imports of shellfish still bring in new species, which emphasizes the need not only of legal restrictions but also better information. Information is also needed to stop the disposal in nature of live organisms from aquaria.

I also would like to stress the importance of including information on the non-native origin of a species in all publications, especially for floristic lists and in databases, where such information often is omitted.

Appendix 2. Marine introduced macroalgae in the different European areas, incuding also occasional and drift species, field experiments and some disjunct, likely introduced species. Names in bold denote invasive species, bold and underlined highly invasive species. [a] after a species denotes used in aquaria trade, [c] cultivated and [ex] used in field other species in the genus used in the aquaria trade. Underlined and bold superscripts as above. For areas the international ISO codes are used (capital letters) except for (Balt) Estonia, Latvia, Lithuania and Kaliningrad, (Ire) Ireland and Northern Ireland, (Yug) former Yugoslavia. (|Az|) the Azores, (|Bal|) the Baleares, (|Co|) Corsica, (|Sa|) Sardinia, (|Si|) Sicily, (c) central, (n) northern, (e) eastern, (s) south, (w) western, (Atl) the Atlantic, (Pac) the Pacific, (IndO) the Indian Ocean, (Carib) the Caribbean, (Medit) the Mediterranean, (Eur) Europe, (Afr) Africa, (trop) tropical, (subtr) subtropical. Vectors: (Aqu) aquaria trade, (Bal) solid ballast, (Cul) cultivated, (Exp) experiments, (Fi) fishing activities (e.g. baits or nets), (Les) Lessepsian immigrant, (Mol) mollusc transports, (Oy) oyster transports, (Shi) shipping activites (i.e. hull fouling or ballast). References are explained at the end. For authors and taxonomic affinities see reference 1.

CLASS and *Genus*	*species*	Established in: ()=occasional or drift	Origin	Vector	For reference see
BANGIOPHYCEAE					
Porphyra	*yezoensis*[c]	sFR	Japan	Oy	2, 3, 4
Porphyra	cf. *insolita*	(wDE)	?	?	1, 11
FLORIDEOPHYCEAE					
Goniotrichiopsis	*sublittoralis*	nFR, ES|Bal|	nw Pac?	?	1, 3, 29
Acrochaetium	cf. *codicolum*	sFR	N.Atl?	Host?	4
Acrochaetium	*balticum*	NL	Baltic S?	Shi?	56
Rhodophysema	*georgii*	sFR	N.Atl?/ N.Pac?	Oy?	1, 4
Devaleraea	*ramentacea*	(wDE)	N.Atl	Exp	2
Ganonema	*farinosum*	sES, IT	IndO	Les	1, 3
Gracilaria	*multipartita*	(Ire)	N.Atl	?	1, 2
<u>Asparagopsis</u>	***<u>armata</u>***[c]	(NL), GB, Ire, n-sw&sFR, PT	AU	Shi?	1, 2, 3, 5, 10, 16,
"	"	n&sES&|Can|, nw-s&eIT, MT, s&nGR			21, 22, 31, 36, 39
Asparagopsis	*taxifolia*	It|Elba|	trop & subtr	Les	1, 3
Bonnemaisonia	***hamifera***	nwIS, FO, NO, wSE, DK, nwDE, (NL),	Japan	Shi?	1, 2, 3, 5, 9, 31
"	"	GB, Ire, n-swFR, PT|Az|, n&sES, IT|Si|			
Pikea	*californica*	GB	California	Shi?	1, 2 but see 17, 28
Grateloupia	***doryphora***	NL, sGB, nFR, n&sES, sw&eIT	nw Pac?	Mol/Shi	1, 2, 3, 4, 5, 23, 31, 50
Grateloupia	*filicina*				
	var. *luxurians*	GB, n&sFR, nwES	AU?/Japan?	Oy	1, 2, 4, 10, 12, 23
Grateloupia	*lanceolata*	sFR	Japan	Oy	4
Grateloupia	cf. *turuturu*	sFR	nw Pac	Oy	4
Grateloupia	sp.	sFR	nw Pac?	Oy?	4
Prionitis	*patens*	sFR	Japan?	Oy?	4
Lithophyllum	*yessoense*	sFR	Japan?	Oy?	4
Caulacanthus	*ustulatus*	nwFR	nw Pac	Oy?	1, 2, 42
Chondrus	*giganteus*	sFR	Japan?	Oy?	1, 4
Hypnea	*esperi*	sGR	RedS/IndO	Les	1, 3, 6
Hypnea	*spinella*	sES, sIT, neGR	Carib?	Shi?	1, 3, 6, 31
Hypnea	*valentiae*	sFR, sGR	Pac?, RedS	Oy?/Les	1, 3, 4, 6
Predaea	***huismanii***	ES|Canl|	AU	Shi	1, 2, 44

CLASS and Genus	species	Established in: ()=occasional or drift	Origin	Vector	For reference see
Mastocarpus	*stellatus*[ex]	wDE	N.Atl	Exp	2, 11
Ahnfeltiopsis	*flabelliformis*	sFR	Japan	Oy??	4
Agardhiella	*subulata*	NL, GB, sFR, nES, sIT	N.Atl?/IndO?	Shi?/Mol?	1, 4, 5, 19, 31, 33, 56
Sarconema	*filiforme*	sFR	IndO	Les?	3
Sarconema	*scinaioides*	GR	IndO	Les?	3
Solieria	*chordalis*	GB, nFR	Atl	?	1, 2
Solieria	***filiformis***	sIT	trop & subtr Atl?	?	3, 31, 33
Plocamium	*secundatum*	IT|Si|	sub-Antarctic?	?	3, 31
Lomentaria	*hakodatensis*	s&nwFR, nES	Japan	Oy	1, 2, 3, 4, 10
Botryocladia	*madagascarien-sis*	seIT&|Si|, MT	s IndO	?	1, 3, 16
Chrysymenia	*wrightii*	sFR	nw Pac	Oy	2, 3, 4
Acrothamnion	***preissii***	sFR, ES|Bal|, nw&swIT	IndO-Pac	Shi	1, 2, 3, ,31 35
Aglaothamnion	*feldmanniae*	sFR, nw&neIT	N.Atl	Shi?	1, 3, 31, 32
Anotrichium	*furcellatum*	(NL), sGB, nwFR, nES&|Can|, PT	n Pac?/ Medit?	Oy?	1, 5, 10, 21, 27
Antithamnion	*amphigeneum*	n&seES&|Bal|, wIT	e AU?	Shi?	1, 3, 31, 38, 39, 40, 47
Antithamnion	*decipiens*	ES|Bal, Alb|, FR|Co|, w,sw&seIT, nGR	Native?	?	1, 3, 8
Antithamnion	*densum*	GB, Ire, nFR, nES&|CanI|,	IndO-Pac?	?	1, 2, 7, 10
Antithamnion	*diminuatum*	PT|Az|, ES|Can|	s AU?/S.Atl?	Shi?	1, 9, 43
Antithamnion	***pectinatum***	sFR, PT|Az|, se&neIT	Japan	Oy	1,2,3,4, 9,18,31
Antithamnionella	***spirographidis***	NL, BE, GB, Ire, nw&sFR, PT|Az|,	n Pac	Shi?/Oy?	1, 2, 5, 10, 31
"	"	n&sES, s&eIT			
Antithamnionella	***ternifolia***	(nwDE), (NL), BE, GB, Ire, nFR,	South Hemisph?	Shi?	1, 2, 3, 5, 10
"	"	PT, nES			
Grallatoria	*reptans*	ES|Can|	Carib?	Native?	1, 43
Griffithsia	*corallinoides*	sFR, sIT	Japan?	Oy	1, 3, 4, 31
Gymnophycus	*hapsiphorus*	ES|CanI|	e Austr	Shi	1, 44
Pleonosporium	***caribaeum***	n&sFR, PT, sES&|CanI|	Antilles?	Shi?	1, 2, 3, 21
Scageliopsis	*patens*	PT|Az|	s AU	Shi?	1, 9
Dasya	*baillouviana*	NO, wSE, DK, NL	Medit/ne Atl	Shi?/Oy?	1, 2, 5
Dasya	sp.	sFR	nw Pac	Oy?	4
Dasysiphonia	(?) **sp.**	NO, NL, nw&sFR, nES	Japan?/Eur	Oy/Shi	4, 5, 25
Platysiphonia	*caribaea*	ES|CanI|	S.Atl?	?	2,
Radicilingua	*thysanorhizans*	s&eIT	N.Atl	?	1, 31
Acanthophora	***nayadiformis***	FR|Co|, s&seIT, MT, e&nGR	RedS/IndO	Les	1, 3, 6, 16, 22, 31, 37
Chondria	*coerulescens*	sFR	Atl?/ Medit?	Oy?	4
Chondria	*curvilineata*	sFR&|Co|	Carib?	?	1, 3
Chondria	*polyrhiza*	sIT, neGR	?	?	1, 3, 6, 13
Chondria	*pygmaea*	s,w&eIT, MT, AL	RedS/IndO	Les	1, 3, 16, 31, 41
Dipterosiphonia	*dendritica*	FR|Co|, sES	pantrop	?	1, 2, 52
Herposiphonia	*parca*	sFR	Pac/Japan?	Oy	4
Laurencia	*brogniartii*	nwFR	Carib?/ Japan?	Oy?	1, 2, 12
Laurencia	*coronopus*	sFR, GR	BlackS	Oy?	1, 3
Laurencia	*okamurae*	sFR	e Asia	Oy?	4
Lophocladia	***lallemandii***	sES, nw-seIT, MT, sGR	Pac?	Les?	1, 3, 6, 16,31,36
Polysiphonia	*atlantica*	sFR,sES, seIT	N.Atl?/ N.Pac?	Shi?/Oy?	1, 4

CLASS and Genus	species	Established in: ()=occasional or drift	Origin	Vector	For reference see
Polysiphonia	**fucoides**	sFR, sES, IT, SI, GR?	N.Atl	Fi?	1, 2, 3, 31
Polysiphonia	**harveyi**	sNO, DK, nwDE, NL, BE, GB, Ire,	n Pac/nwAtl	Host?	1, 2, 5, 10, 44
"	"	n&sFR, nES&|CanI|, nwIT			
Polysiphonia	*morrowii/senti-ouloaa*	sFR	n Pac	Oy?	4
Polysiphonia	*morrowii*	neIT	nw Pac		32
Polysiphonia	*paniculata*	sFR	e Pac		1, 4
Polysiphonia	**senticulosa**	NL	n Pac	Oy?	5, 56
Pterosiphonia	*pinnulata*	GB, nFr	?	Shi?/Oy?	1, 27
Pterosiphonia	sp.	sFR	Pacific?		4
Symphyocladia	**marchantioi-des**	PT|Az|	AU?	Shi?	1, 2, 9
Womersleyella	**setacea**	sFR, ES|Bal|, nw-sw&seIT, MT, ne&sGR	trop	Shi?	1, 2, 3, 6, 16, 22, 31, 32, 35,39,45
PHAEOPHYCEAE					
Ectocarpus	*siliculosus*				
	var. *hiemalis*	IT	Atl?	?	31
Pilayella	*littoralis*	sFR	N.Atl?/ N.Pac?	Oy	3, 4
Padina	*boergesenii*	IT	Red S?	Les	3, 31
Stypopodium	*schimperii*	sGR	Red S	Les	3, 45
Acrothrix	*gracilis*	sFR	Japan	Oy?	4
Cladosiphon	*zosterae*	seFR, sIT	Japan?/?	Oy?	1, 4
Sphaerotrichia	*divaricata*	sFR	Japan	Oy	1, 2, 3, 4, 34
Halothrix	*lumbricalis*	sFR	Japan?	Oy?	3, 4
Corynophlaea	*umbellata*	GB	Japan?/ Medit?	Host?	20
Leathesia	*difformis*	sFR	N.Atl?	Oy	1, 2, 3
Leathesia	*verruculiformis*	NL	nw Pac	?	1, 56
Colpomenia	**peregrina**	s&wNO, wSE, DK, NL, GB,	Pac	Oy	1, 2, 3, 4, 10, 31
"	"	Ire, n&sFR, PT, n&sES, w&sIT			
Endarachne	*binghamiae*	PT|Az|	S.Atl?	Native?	1, 9
Scytosiphon	*dotyi*	sFR, nES, neIT	Pac	Oy/Mol	3, 4, 10
Asperococcus	*scaber*	NL, (sGB)	Adriat	?	1, 56
Punctaria	*tenuissima*	neIT	N.Atl?	?	31
Desmarestia	**viridis**	sFR, sES, neIT, CR, BlackS	N.At?/ nw Pac?	?/Oy?	1, 2, 3, 32, 57
Alaria	*esculenta*	(nwDE)	N.Atl	Exp	2
Undaria	**pinnatifida**[c]	NL, BE, sGB, s&nwFR	Japan	Oy, Cul	1, 2, 3, 14, 17,
"	"	nES, s&neIT		& Shi	20, 24, 31,32,54
Chorda	*filum*	sFR, GR	N.Atl?/ nw Pac?	Oy?	1, 3
Laminaria	*japonica*[c]	sFR	Japan	Oy	1, 2, 3, 4
Laminaria	*ochotensis*[ex]	(DE)	Japan	Exp	2
Macrocystis	*pyrifera*[ex]	(nwFR)	Chile	Exp	2
Fucus	*evanescens*	sNO, wSE, nDE, DK	n N.Atl	Fi?/Shi?	1, 2, 46
Fucus	*spiralis*	(sFR), seES	N.Atl?	Fi?	1, 2, 3
Sargassum[a]	**muticum**	s&wNO, wSE, nwDE, nDK, NL, BE,	n Pac	Oy	1, 2, 3, 18, 24,
"	"	s&swGB, neIre, n&sFR, PT, nES, neIT			31, 32, 51, 55
CHLOROPHYCEAE					
Monostroma	*obscurum*	sFR	Atl?/Pac?	Oy?	3, 4
Ulva	*fasciata*	sFR	Medit?/ nwPac?	Shi/Oy	4

CLASS and *Genus*	*species*	Established in: ()=occasional or drift	Origin	Vector	For reference see
Ulva	*pertusa*	sFR	Japan?	Oy?	4
Codium	*fragile* ssp. *atlanticum*	NO, NL, GB, Ire		Shi?	1, 2, 48
Codium	*fragile* ssp. *scandinavicum*	NO, wSE, DK		Shi?	1, 2, 49
Codium	***fragile*** ssp. ***tomento-soides***	IS, NO, wSE, DK, wDE, NL,BE, GB, Ire, n&sFR, PT|Az|, n&sES&|Canl|, s&neIT, GR	Japan	Shi/Oy	1, 2, 3, 22, 24, 31, 44
Derbesia	*rhizophora*	sFR	Japan	Oy?	4
Caulerpa[aa]	***racemosa***[a]	sFR, ES|Bal|, nw-sw&IT&|Sa & Si|, GR	IndOc?	Les	1, 3, 22, 30, 31, 32, 36, 38, 53,
Caulerpa	***scalpelliformis***	IT	IndOc?	Les?	31
Caulerpa	***taxifolia***[a]	FR, ES|Bal|, nw-swIT, CR	trop Aquar	Aqu, Shi & Fi	1, 2, 3, 31, 32,
CHAROPHYCEAE					
Chara	*connivens*	eSE, Balt, PL, nDE	w Eur, n Afr	Bal	2, 26

References: 1) Guiry & Nic Dhonncha 2001, 2) Wallentinus 1999a, 3) Verlaque 1994, 4) Verlaque 2001, 5) Maggs & Stegenga 1999, 6) Athanasiadis 1987, 7) Athanasiadis 1990, 8) Athanasiadis 1997, 9) Athanasiadis & Tittley 1994, 10) Bárbara & Cremades 1996, 11) Bartsch & Kuhlenkamp 2000, 12) Cabioch et al. 1997, 13) Cecere et al. 1996, 14) Cecere et al. 2000, 15) Chapman 1999, 16) Cormaci et al 1997, 17) Cremades 1993, 18) Curiel et al. 1998, 19) Farnham 1997, 20) Fletcher & Farell 1999, 21) Guadelupe et al. 1995, 22) Haritonides & Lazaridou 2001, 23) Irvine 1983, 24) Kerckhof 2001, 25) Lein 1999, 26) Luther 1979, 27) Maggs & Hommersand 1993, 28) Maggs & Ward 1996, 29) Magne 1992, 30) Modena 2000, 31) Occhipinti Ambrogi 2000, 32) Occhipinti Ambrogi 2001, 33) Perrone & Cecere 1994, 34) Peters & Breeman 1992, 35) Piazzi & Cinelli 2000, 36) Piazzi et al. 2001, 37) Por 1978, 38) Ribera & Pagès 2001, 39) Rindi & Cinelli 1995, 40) Rindi et al. 1996, 41) Rindi et al. 1999, 42) Rueness & Rueness 2000, 43) Sansón 1994, 44) Sansón & Reyes 1995, 45) Sartoni & De Biasi 1999, 46) Schaffelke et al. 1995, 47) Secilla et al. 1997, 48) Silva 1955, 49) Silva 1957, 50) Simon et al. 2001, 51) Staehr et al. 2000, 52) Verlaque 1981, 53) Verlaque et al. 2000, 54) Wallentinus 1999b, 55) Wallentinus 1999c, 56) Wetsteyn 2000, 57) Zaitsev & Mamaev 1997.

COSCINODISCUS WAILESII - A NUISANCE DIATOM IN EUROPEAN WATERS

IAN LAING[1*] & STEPHAN GOLLASCH[2]
[1]*CEFAS Weymouth Laboratory, Dorset, United Kingdom*
[2]*GoConsult, Hamburg, Germany*
[*]Corresponding author *i.laing@cefas.co.uk*

Abstract
The diatom *Coscinodiscus wailesii* is native to the Indian and Pacific Oceans. It was introduced to European waters and was first detected in the English Channel, near Plymouth, Devon, in 1977 (as *Coscinodiscus nobilis* Grunow). Its tolerance to abiotic factors enabled it to become established and it spread rapidly. It had reached the Atlantic coast of France by 1978 and Norway by 1979. Today it can be found along the Atlantic coast of France, through the English Channel and north to the Friesian Islands, Helgoland and Norway.

1 Introduction

As with many other introduced phytoplankton species, *Coscinodiscus wailesii* was probably introduced with discharge of ships' ballast water (Hasle 1990; Fig. 1). It can form resting cells and these have been found in sediment samples from ballast tanks (Macdonald 1995). These cells can survive long dark periods (Nagai et al. 1995a). It has also been suggested that importation of shellfish (oysters) could be a vector, with the cells being carried within the gut/pseudofaeces of the animals (O'Mahony 1993; Essink 1994).

It is a nuisance species as it forms dense blooms which produce copious amounts of mucilage and is inedible to most grazing zooplankton.

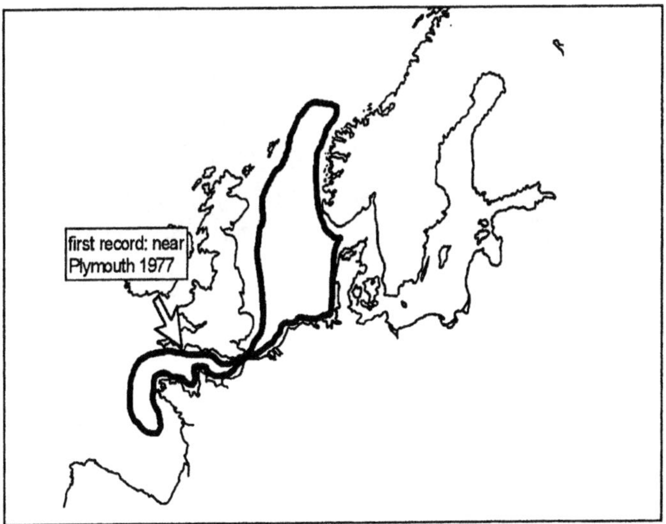

Figure 1. Map of northwestern Europe with location of known blooms of *Coscinodiscus wailesii*, not the overall distribution.

E. Leppäkoski et al. (eds.), Invasive Aquatic Species of Europe, 53–55.

2 Biology

Coscinodiscus wailesii is a large (320-350 µm diameter) centric diatom. The cell is shortly cylindrical, giving it a markedly rectangular outline in girdle view (Boalch & Harbour 1977; Semina & Zernova 1985) (Fig. 2). The valves are circular, striated, with bi-labiate processes. The valve is flat with a mantle size of 50 µm, meeting the valve face at right angles.

Resting cells are known and may be found in sediments. They can survive under dark conditions for long periods (at least 15 months). The resting cells are distinguished from vegetative cells as their cytoplasm is partially separated from their frustules and is concentrated in the centre of the cell. These resting cells can rejuvenate rapidly given adequate light, temperature and nutrients (Nagai et al. 1995a). When isolated resting cells are incubated in the laboratory at 20°C under continuous illumination of 65 µmol $m^{-2}s^{-1}$, they change rapidly into vegetative cells and most finish their first cell division within 48 h. They then continue to multiply by binary division. In the sea a doubling of biomass in 70 h has been estimated, based on data from four successive field surveys in the German Bight, in 1991. Results from laboratory cultures under comparable conditions confirm this rate (Rick & Durselen 1995). The field data also showed that this species has a high tolerance to heavy metals, with 8-10 times lower copper and 10-20 times lower cadmium and zinc accumulation in *C. wailesii* cells compared to concentrations found in native phytoplankton species. As blooms develop cells produce copious mucilage and this species can eventually reach a biomass of 1,400 µg carbon l^{-1}.

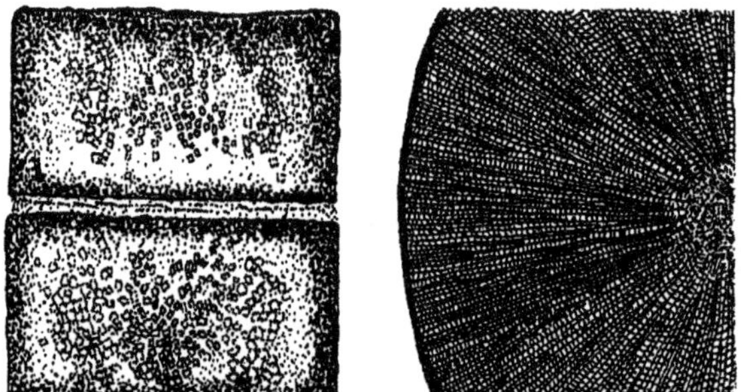

Figure 2. Coscinodiscus wailesii (a) Girdle, (b) valve face with scattered labiate processes (after http://www.marbot.gu.se/SSS/Coscinodiscus_wailesii.htm).

3 Impact

Blooms may occur in coastal waters, or they may be advected from frontal areas further offshore. Blooms can form up to 90% of the total algal biomass (Rick & Durselen 1995). Benthic habitats are threatened, including those in fishery and aquaculture areas, as this species produces copious mucilage, which can aggregate, sink and blanket the seabed (Boalch & Harbour 1977). More direct effects on fisheries and aquaculture, including seaweed culture, are also noticed, as this mucilage has been found to cause extensive clogging of fishing nets and other equipment, such as cages (Boalch & Harbour 1977). The cells are so large that they are not easily grazed by zooplankton (Roy et al. 1989; Rick & Durselen 1995), and the eventual decay of a bloom is likely to cause anoxia (Manabe & Ishio 1991; SCOPE Newsletter 1998) giving further problems especially for benthic organisms. *Coscinodiscus* probably also displaces other algae species, including useful food species for cultivated shellfish, both physically and by stripping nutrients from the water (Nagai et al. 1995b).

With blooms of this species widely established in European waters it is possible that this species may be carried by transport in ships' ballast water to other parts of the globe. Further areas considered being at risk of colonisation by this species are the coastal habitats of Australasia and the eastern seaboard of America, both of which have important fishery and aquaculture industries.

THE COMB JELLY *MNEMIOPSIS LEIDYI* IN THE BLACK SEA

AHMET E. KIDEYS
Institute of Marine Sciences, Middle East Technical University, Erdemli, Turkey
kideys@ims.metu.edu.tr

Abstract
Following its transportation in ballast waters, the western Atlantic ctenophore *Mnemiopsis leidyi* caused a striking damage to the ecosystem of the Black Sea in the early 1990s. The concentration of food zooplankton and simultaneously catches of anchovy and other pelagic fishes sharply decreased during the peak levels of the ctenophore. There have been many studies on the distribution and physiology of this ctenophore in its new environment due to its substantial impact on the marine ecosystem. Later in mid 1990s its biomass was at moderate levels till the appearance of its predator, another ctenophore *Beroe* sp. in 1997. The predator decreased *Mnemiopsis* biomass to very low levels which was confirmed by both laboratory and field observations. Now *Mnemiopsis* is causing very similar problems in the Caspian Sea following its transport from the Black Sea.

1 History and impact of *Mnemiopsis* invasion in the Black Sea

The North American comb jelly *Mnemiopsis leidyi* (also identified as *M. mcradyii*, Zaika & Sergeeva 1990; Fig. 1) is a member of phylum ctenophora (order Lobata, family Bolinopsidae). The invasion case of this species in the Black Sea is one of the most drastic among marine bioinvasions. It caused a great damage to the entire pelagic ecosystem of the Black Sea at the end of 1980s and at the early 1990s devastating valuable pelagic fishery. As a matter of fact similar events are now taking place in the Caspian Sea after *Mnemiopsis* inadvertently transported here from the Black Sea.

It is believed that *Mnemiopsis* reached the Black Sea in the ballast waters of ships. In the spring of 1986, this ctenophore was first recorded off the Bulgarian coast (Bogdanova & Konsoulov 1993), then in offshore waters (Vinogradov et al. 1989). Increasing substantially following years, the total biomass of *Mnemiopsis* in the Black Sea was calculated as 800 million tons (live weight) in 1989 summer (Vinogradov 1990). A great alteration of the structure in the planktonic community has resulted from this mass development (Kideys 1994; Niermann et al. 1994; Kovalev et al. 1998a,b; Kideys et al. 1999, 2000). The sharp decrease in the concentration of anchovy eggs and larvae observed since 1989 has been suggested to be due not only to overfishing, eutrophication or pollution but also to *Mnemiopsis* predation on fish eggs and larvae or competition for food in the northern Black Sea (Niermann et al. 1994). The catch of anchovy (*Engraulis encrasicolus*), being the most abundant fish species of the Black Sea, decreased sharply too in this period (Fig. 2), as well as other pelagic fish species. The economic loss for fishermen of the Black Sea was hundreds million dollars during this period.

2 Biology and ecology of *Mnemiopsis*

Mnemiopsis is a transparent animal with gelatinous soft body. The locomotory apparatus is composed of eight symmetrical comb rows, distinguished by fused ciliary plates (ctenes) on the body surface (Mianzan 1999). Bioluminescence is a common feature in

E. Leppäkoski et al. (eds.), Invasive Aquatic Species of Europe, 56–61.

ctenophores occurring along the comb rows. Its body is laterally compressed, with large lobes arising near stomodeum. It may reach to a total (lobe) length of 14 cm in the Black Sea. Early life stages of this ctenophore are shown in Fig. 3.

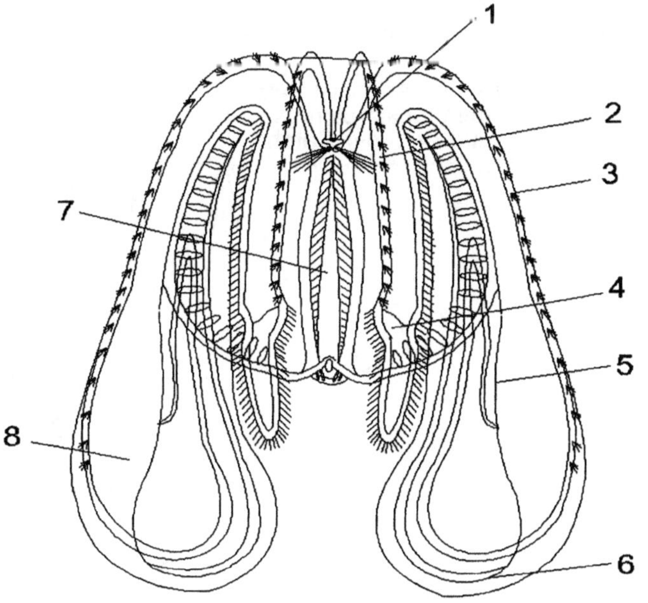

Figure 1. Mnemiopsis leidyi. (1) aboral organ, (2) subtentacular row of comb flappers, (3) subsagital row of comb flappers, (4) auriculus, (5) subsagital tube, (6) translobal tube, (7) tentacular tube, (8) lobe (from Shiganova 2000).

2.1 FEEDING

Mnemiopsis is a voracious predator on zooplankton. Unlike medusae, *Mnemiopsis* is a macrophage capable of consuming fairly large prey (up to 1 cm or more in length). *Mnemiopsis* has also been shown to feed on eggs and larvae of fishes in the Black Sea (Tsikhon-Lukashina et al. 1992; Mutlu 1999). Mutlu (1999) observed the following groups in the stomach of *Mnemiopsis* from 1991 to 1995 in the southern Black Sea: copepods (50%), mollusks (40%), cladocerans (1%), fish eggs and larvae (1%), and others (8%). *Calanus euxinus* was the most frequently consumed copepod.

2.2 GROWTH AND REPRODUCTION

Mnemiopsis, like all other ctenophores, are hermaphroditic with a very high reproductive capacity. *Mnemiopsis mccradyi* is able to produce 8,000 eggs within 23 days, after 13 days of its own birth (Baker & Reeve 1974). The growth rate of this species is comparable to that of phytoplankton (daily doublings, Reeve et al. 1978). Naturally such high growth rates can only be achieved with a high feeding rate.

In the Black Sea, mature specimens of *Mnemiopsis* spawn at night (Zaika & Revkov 1994) in summer temperatures of 19 to 23 °C. Embryonic development takes about 20 h (Fig. 3). The maturing of gonads and the subsequent spawning can take place only at a reasonably high food concentration of medium size copepods (up to 100 copepods l^{-1}).

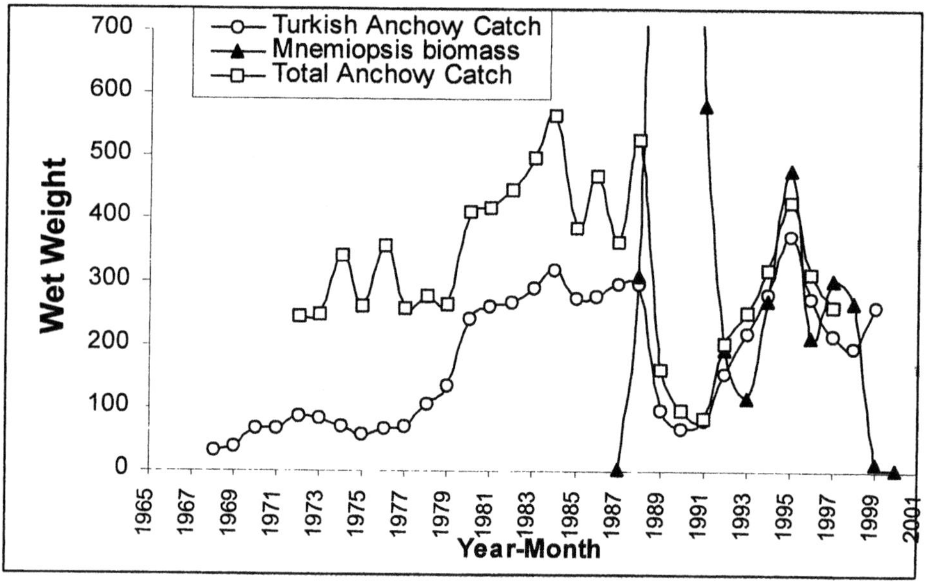

Figure 2. Relationship between *Mnemiopsis* biomass (g m^{-2}) and anchovy catch (as thousand tons) values from the Black Sea (modified from Kideys et al. 1999).

2.3 RESPIRATION

Minkina & Pavlova (1995) determined an equation of R (μl O_2 h^{-1}) = 3.20 dry wt $(g)^{0.89}$ for 23 °C for the Black Sea *Mnemiopsis*. Kideys's (1996) equation obtained for 20 °C (R (μl O_2 h^{-1}) = 2.88 dry wt $(g)^{0.86}$) was close to this one, but lower rates for 8 °C (R (μl O_2 h^{-1}) = 1.48 dry wt $(g)^{0.87}$). Anninsky et al. (1998) described the respiration rate of *M. leidyi* with the equation R (μl O_2 h^{-1}) = 6.73 wet wt $(g)^{0.83}$) for freshly captured ctenophores at 22 °C. Rate of respiration is important in explaining nearly absence of *Mnemiopsis* from the suboxic water layers in the Black Sea. For example, the intercept value obtained for *Mnemiopsis* is much lower than that obtained for the other ctenophore occurring in suboxic layers, (i.e. *Pleurobrachia pileus* R=1.2 to 2.2 $W^{0.625}$ measured at 17-20 °C; Lazareva 1961). This indicates a higher metabolism in *Mnemiopsis* compared with *P. pileus*. Keister et al. (2000) also reported that when the dissolved oxygen levels were below 1 mg l^{-1} (approx. 30 μM), *Mnemiopsis* were nearly absent in the bottom layers of Patuxent River, Chesapeake Bay.

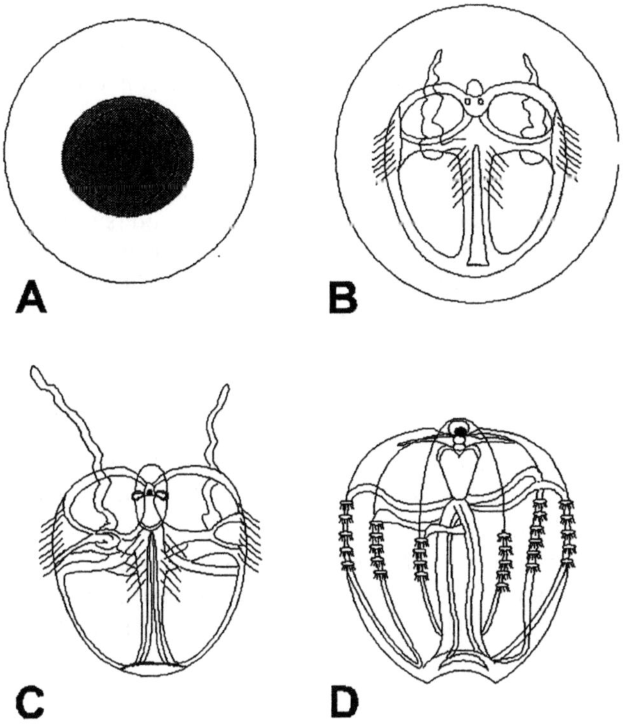

Figure 3. Early life stages of *Mnemiopsis*. (A) newly laid egg, (B) embryo with tentacles, about 30 hours old, still within the egg envelope, (C) newly hatched embryo in cydippid stage, (D) embryo about 32 hours old showing muscular strands (from Shiganova 2000).

3 Spatial and temporal distributions

In the shallow well mixed northeastern American waters (e.g. Narragansett Bay) *Mnemiopsis* was usually fairly evenly distributed vertically (Kremer & Nixon 1976). In the Black Sea, where the oxic layer may extend below 200 m, *M. leidyi* was usually in the upper mixed layer, or in and above the seasonal thermocline, with only a few individuals found in deeper layers (Mutlu 1999; Kideys & Romanova 2001). The major reason for the very low occurrence of this ctenophore in deeper layers could also be the low levels of oxygen concentration (< 20-30 μM) as pointed out above.

Horizontal distribution of *Mnemiopsis* in the Black Sea is much more patchy compared to the native ctenophore *P. pileus*. There are hot spots for the abundance of *M. leidyi*, mainly occurring in the coastal regions. Overall density of *Mnemiopsis* seems higher in the northern Black Sea compared to the south.

Abundance of *Mnemiopsis* has been observed to fluctuate throughout the year in USA waters with higher values in summer (Kremer & Nixon 1976). In the Black Sea too, high values are generally observed in summer-autumn though a considerable stock was present during the winter months until the appearance (i.e. 1997) of its predator *Beroe* in the Black Sea. Since then, winter population of *Mnemiopsis* is very low both in northern and southern Black Sea.

In the open sea, the maximum biomass values for *Mnemiopsis* (1.5-2 kg m^{-2}) were obtained in autumn 1989 (Vinogradov et al. 1989; Fig. 2). After 1990, the biomass decreased sharply. Comparing data only from the warm period (summer and early autumn when these ctenophores are known to increase their abundance and biomass) to eliminate seasonal variability, one can see that after the high values at the end of the 1980s, the biomass of *Mnemiopsis* was very low with a mean value of 131 g m^{-2} in June 1991. There was again an increasing trend during the summers of 1992, 1993 and 1995 when the biomass increased steadily up to 465 g m^{-2}. However, another decrease trend has occurred since 1995, that is prior to the appearance of the predator *Beroe*. One of the most important factors for the decrease in the biomass of *Mnemiopsis* in this period was the food limitation as was demonstrated by Anninsky et al. (1998). Under conditions of starvation, the level of glycogen, which is accumulated energy in ctenophores (Shulman 1972), decreases and this characteristic can be used to understand the condition of ctenophores. Glycogen content of *Mnemiopsis* in 1996 varied from 21 to 44 μg/g fresh weight and accounted for 52.5 ± 14.2% of total polysaccharide content. Thus the condition of *M. leidyi* observed during the conducted survey corresponded to that expected after a two-day fasting period (determined experimentally) implying that the ctenophores were starving. Earlier measurements (in 1991-1992) displayed high glycogen levels for *Mnemiopsis* indicating that the Black Sea had provided the ctenophores with a good food supply (Anninsky 1995).

4 Control of *Mnemiopsis*

The unprecedented impact caused by *Mnemiopsis* in the Black Sea attracted great attention from the scientific community so that UNEP intervened to develop a strategy to control this ctenophore in the Black Sea (GESAMP 1997). *Beroe* arose as one of the best agents for the aim. Interestingly, by October 1997, this ctenophore (*Beroe ovata*) had already appeared in shallow waters of the Black Sea (Konsoulov & Kamburska 1998), in September 1998 in the Sea of Marmara (A. E. Kideys, unpubl.) and in August-September 1999 in Sevastopol Bay and adjacent water regions as well as in the north-eastern Black Sea (Finenko et al. 2000). Since genus *Beroe* occurs in the Mediterranean, there is a possibility that this invasion was a spontaneous one from the Mediterranean.

Several studies performed since then have shown that *Beroe* feed entirely on *Mnemiopsis* (Shiganova et al. 2000; Finenko et al. 2001). The latter investigators calculated the mean daily ration as 45% of *B. ovata* wet weight.

Since the appearance of *Beroe* in the Black Sea, population of *Mnemiopsis* decreased gradually to almost zero values as found during the most recent (May-June 2001) cruise to the Black Sea (Fig. 3). This shows the effectiveness of *Beroe* in controlling *Mnemiopsis* in the Black Sea ecosystem. This is a very important event displaying that the impact of a holoplanktonic alien species could be reverted via biological control.

Mnemiopsis appeared in the Caspian as was expected (Dumont 1995; GESAMP 1997) since 1997 (pers. comm. with Caspian fishermen). Ivanov et al. (2000) suggested that this ctenophore was transported with ballast water taken aboard in the Black Sea or the Sea of Azov (where *Mnemiopsis* occurs in warm months) and released after ballast-loaded ships passed through the Volga-Don Canal and the shallow freshwater North Caspian Sea, into the saltier Central or South Caspian (Ivanov et al. 2000). In 2001 summer abundance of *Mnemiopsis* was as high as 1,393 ind m^{-2} in the southern Caspian Sea (Kideys et al. 2001), and about 50% reduction in kilka fishery compared to two-three years earlier. Kideys et al. (2001) suggested introducing *Beroe* to control *Mnemiopsis* in Caspian Sea after necessary laboratory experiments performed and conditions set by the FAO guidelines on the precautionary approach on inland fisheries and species introduction as well as the ICES (1995) code of practice on the introductions and transfers of marine organisms.

Acknowledgements

The present investigation was carried out with the support of a NATO Linkage Grant (ENVIR.LG.974491) and by the Turkish Scientific and Technical Research Council (TUBITAK YDABÇAG 100Y017).

THE PREDATORY WATER FLEA *CERCOPAGIS PENGOI* IN THE BALTIC SEA: INVASION HISTORY, DISTRIBUTION AND IMPLICATIONS TO ECOSYSTEM DYNAMICS

IRENA V. TELESH[1*] & HENN OJAVEER[2]
[1]*Zoological Institute of the Russian Academy of Sciences, St. Petersburg, Russia*
[2]*Estonian Marine Institute, University of Tartu, Tallinn, Estonia*
[*]Corresponding author *itelesh@zin.ru*

Abstract
Cercopagis pengoi, a predatory cladoceran from the Ponto-Caspian basin, invaded the north-eastern Baltic Sea in 1992 due to ballast water discharges, and by the end of the 1990s formed several permanent populations in different gulfs and lagoons of the sea. The present-day distribution of *C. pengoi* witnesses for its high ecological plasticity. Evaluation of predation pressure of *C. pengoi* on its potential preys (younger larval stages of copepods, *Daphnia* spp., *Bosmina* spp. and rotifers) demonstrated significant impact of this alien species on native zooplankton communities and the pelagic food webs in general. A considerable dietary overlap of *C. pengoi* and planktivorous fish may contribute to a decline of food resource for important commercial fish - herring and sprat. On the other hand, during the population maximum *C. pengoi* forms an important portion of the diet of these fish. This invasion has increased species diversity of the pelagic fauna of the Baltic Sea and incorporated an additional element to the trophic link of carnivores in the pelagic food web. Due to high population densities in some localities and considerable predation pressure on plankton communities, invasion of *C. pengoi* is likely to cause important structural and functional ecosystem alterations in the Baltic Sea.

1 Introduction

In 1992, a carnivorous species of the Ponto-Caspian origin - the cladoceran *Cercopagis* (*Cercopagis*) *pengoi* (Ostroumov, 1891) - invaded the north-eastern Baltic Sea (Ojaveer & Lumberg 1995). During the second half of the 1990s, the species expanded its range by forming several permanent populations in different sub-basins of the Baltic Sea. In the last decade, *C. pengoi* was studied for morphology, body carbon content, population structure, distribution and spatio-temporal dynamics, impact on native zooplankton community, role in fish diet and the invasion risk assessment. In this chapter, we will concentrate on some aspects of invasion history and ecological significance of *C. pengoi* in the Baltic Sea.

2 Invasion history

C. pengoi was first observed in two large gulfs of the Baltic Sea: Gulf of Finland and Gulf of Riga, in 1992, most likely as a result of discharge of ballast waters of ships (Ojaveer & Lumberg 1995). Unlike in the Gulf of Riga, the cladoceran was not encountered in the Gulf of Finland in 1993 and 1994 but appeared again in 1995 in several sites at the southern coast of the gulf (Ojaveer et al. 2000). Occurrence of *C. pengoi* in the Gulf of Finland in 1995 was documented also in the eastern basin (e.g. Avinski 1997) followed by a range expansion to open and western Baltic (Gotland basin, west coast of the Baltic proper, 1997) and to north (Gulf of Bothnia, 1999) and south (Gulf of

E. Leppäkoski et al. (eds.), *Invasive Aquatic Species of Europe,* 62–65.

Gdansk, Vistula Lagoon, 1999; Gorokhova et al. 2000; Naumenko & Polunina 2000; Zmudzinski 1999; K.E. Persson, pers. comm.). Six years after the first record in the Baltic Sea, the species was detected in the North American Great Lakes (MacIsaac et al. 1999), the Baltic Sea being likely the donor region (Cristescu et al. 2001).

3 Distribution and density dynamics

Actual distribution of *C. pengoi* in the Baltic Sea allows to conclude about high ecological plasticity and the continuing range expansion of this Ponto-Caspian species. This is possible due to high reproduction rates (average egg ratio 16.6, maximum 24 eggs per female, Litvinchuk et al. 2001) and significant adaptability of these crustaceans to variable water temperature and salinity (Avinski & Telesh 1999).

The wide range of optimal salinity (2-10 PSU) allows spreading of *C. pengoi* throughout major part of the Baltic Sea. However, water temperature and food availability can be considered the basic factors limiting the distribution of the species. Additionally, stability of the water column may be important: higher abundance of these crustaceans has been recorded in sheltered locations, less affected by wind-induced water movements, and in the areas with low currents whereas the species was encountered at minimum abundance (1 ind m^{-3}) at one station in the region with strong currents (Väinameri Archipelago) in 1994 (Avinski 1997; Ojaveer et al. 2000). Vertical distribution of the species in the Baltic Sea is mostly confined in the upper, warmer water layers (Krylov et al. 1999; Avinski & Telesh 1999).

Since the invasion to the Gulf of Riga in 1992, the annual mean abundance of *C. pengoi* increased linearly reaching the peak of 173 ind m^{-3} in 1999. No significant correlation was found between its population density and water temperature during the warm season. The cladocerans are present in the pelagic community in the Gulf of Riga for 7-20 weeks annually having 1 to 3 abundance peaks per year. As a long-term mean, *C. pengoi* reaches peak densities in the first half of August (mean 231 ind m^{-3}) by exhibiting significantly higher abundance in the first half of the month than most of the remaining vegetation period (Ojaveer et al. 2000). In the Gulf of Riga, as well as in the Gulf of Finland, the species usually appears in the pelagic zooplankton community at water temperatures over 15 °C and disappears when temperature falls below 8 °C. In the Gulf of Finland, the crustaceans are present in plankton from late June through mid-October. Population density of *C. pengoi* peaks twice during the vegetation season: in July-August at water temperatures 18.0-19.5 °C and in September at 12.5-14.0 °C. The highest population density of 2,325 ind m^{-3} was recorded in the Gulf of Finland (Telesh et al. 2000), usually abundance of *C. pengoi* remains below 500 ind m^{-3}.

4 Impact on pelagic communities: structural and functional aspects

In general, zooplankton in the eastern Gulf of Finland (EGF) was taxonomically diverse and abundant, typical of mesotrophic water bodies (Telesh et al. 1999). The copepods *Eurytemora affinis*, *Eudiaptomus gracilis*, *Mesocyclops leuckarti* and *Thermocyclops oithonoides* form the basis of the community contributing ca. 50% to the total mesozooplankton biomass. Besides, high abundance of cladocerans *Bosmina longirostris*, *B. crassicornis*, *Daphnia cristata* and rotifers *Keratella cochlearis* , *K. c. baltica*, *K. quadrata*, *Synchaeta* spp. were registered during the peak population densities of *C. pengoi*.

Average biomass of *C. pengoi* in the EGF during the study period in 1996-1998 formed 1-24% of the total zooplankton.

Along with other water fleas, *Leptodora kindtii* and *Bythotrephes* sp., *C. pengoi* is known to be a large-bodied planktonic predator. During the population maximum of *C. pengoi,* it eliminated a considerable part of the assemblage of small planktonic crustaceans. In the Gulf of Finland, this resulted in strong positive correlation (R^2=0.816) between the calculated consumption rate of *C. pengoi* and mean individual body mass of herbivorous cladocerans thus witnessing for the community structuring effect of the *C. pengoi* population. Calculations showed that the size structure of zooplankton community and trophic webs in the ecosystem have been modified as a result of the elimination of small planktonic filtrators and sedimentators by the population of *C. pengoi* (Telesh et al. 2000).

Long-term dynamics of population density of the small-sized cladoceran - *Bosmina coregoni maritima*- in the Gulf of Riga evidences that significant differences occur prior and after the invasion of the predatory *C. pengoi.* Although a declining tendency in the abundance of *B. c. maritima* was established before the invasion of *C. pengoi* there were no such long-lasting low population density values recorded since 1973. Low abundance of the small-sized cladoceran - *Bosmina*, accompanied by the invasion of *C. pengoi* since 1992 could be the direct effect of predation by this exotic species. In contrast, there were no evidences of similar drastic changes in the numbers of other small-sized zooplankters (cladocerans *Evadne nordmanni* and *Podon polyphemoides*) neither nauplii of copepods (Ojaveer et al. 2000). According to Gorokhova (1998), copepods (nauplii and copepodites of *Acartia*, *Eurytemora* and *Temora*), rotifers (*Synchaeta*) and cladocerans (*Evadne*) constituted 60, 20 and 20% in the diet of *C. pengoi* in the northern Baltic Proper, respectively. Uitto et al. (1999) argued that if copepod predation by *C. pengoi* is the case in the eastern Gulf of Finland, it may result in the indirect positive effect on ciliates which are regulated by mesozooplankton predation.

Recently, a model for evaluation of predation pressure and overall impact of this predatory cladoceran on its potential preys (younger larval stages of copepods, *Daphnia* spp., *Bosmina* spp. and small rotifers) was proposed (Telesh et al. 2001). The algorithm for this evaluation allows global assessment of the invader's impact on the aquatic communities of the Baltic Sea. It is based on knowledge of dynamics of the population range size, or total area occupied by *C. pengoi* in the water body, population density of the species, and predation pressure of *C. pengoi* during its peaks, i.e. relation of its consumption rate to production rate of potential prey organisms.

Calculations showed that even though average population density of *C. pengoi* in the plankton community of EGF was relatively low, its consumption rate in 1997-1998 peaked at 77.3% (±SYMBOL2.8, n=4) of the total production of herbivorous zooplankton. The model demonstrated that calculated values of impact of *C. pengoi* on the zooplankton were relatively low, however, there was a significant increase of the overall influence of this alien species on the native community during the second and third year after the invasion (Telesh et al. 2001). Therefore, possible increase of population density of *C. pengoi* and, consequently, the growth of its predation pressure on planktonic filtrators in the estuary in future may result in alteration of pelagic food webs in the

EGF. The latter event is most likely to be followed by changes of fluxes of matter and energy through the zooplankton community and the ecosystem in general, thus enhancing eutrophication processes in the water body.

5 Role in fish diet

Study of preferable food of *C. pengoi* revealed high potential possibility of considerable dietary overlap of *C. pengoi* and abundant planktivorous fish in the Baltic proper (Gorokhova 1998). This may contribute, in case of high population density of *C. pengoi*, to a decline of food resource for important commercial fish - herring and sprat.

Long-term fish stomach investigations in the Gulf of Riga in 1994 through 1998 demonstrated that except bleak, *Alburnus alburnus* the mean share of *C. pengoi* in fish stomachs was low (< 10% of wet mass). However, strong seasonality was evident: the average contribution of this cladoceran to the stomach content of herring, *Clupea harengus membras* in this period varied from 0-0.1% in June and July to 11.3-16.6% in September and August. However, the situation was different in the Gulf of Riga in July 1999 when 59% of wet mass of the herring' diet consisted of *C. pengoi*, and stomachs of 66% of fishes contained this exotic species. In general, *C. pengoi* was not a favourite prey for planktivorous fish in the Gulf of Riga, evidenced by negative values of mean Ivlev selectivity index for herring, smelt and sticklebacks (Ojaveer et al. 2000). Antsulevich and Välipakka (2000) have, however, found that in the Gulf of Finland, herring shows a neutral to very high positive prey selectivity for *C. pengoi* and this cladoceran made over 40% of the herring stomach content in the warm period (August-September).

6 Conclusions

Structural transformations in aquatic communities cause changes in biological diversity and induce modifications in functioning of aquatic ecosystems. Invasion of the predatory cladoceran *C. pengoi* has increased species diversity of the pelagic fauna of the Baltic Sea and, consequently, the overall stability of the system by incorporating an additional element to the trophic link of carnivores in the pelagic food web. This invasion has also resulted in the enhancement of relative importance of the warm-water planktonic invertebrates in the energy flow to cold-water bentho-pelagic fish communities of the Baltic Sea through direct predation. Although the impact of *C. pengoi* in some parts of the Baltic Sea (e.g., the open Baltic proper) remains generally insignificant, in large gulfs (Gulf of Finland and Gulf of Riga) and lagoons *C. pengoi* is likely to cause important structural and functional ecosystem changes, due to higher population densities and considerable predation pressure on pelagic communities. In this respect, progress in methodology for evaluation of the impact of alien species on the aquatic ecosystems is essential for the purpose of predicting possible global changes and alterations in the ecological status of the recipient water bodies.

Acknowledgements

The work was supported by the Russian Foundation for Basic Research (grants 00-15-97825, 01-04-49560), Finnish Ministry of Environment, Maj and Tor Nessling Foundation, and the Estonian Science Foundation (grant 3604).

HISTORY AND SUCCESS OF AN INVASION INTO THE BALTIC SEA: THE POLYCHAETE *MARENZELLERIA* CF. *VIRIDIS*, DEVELOPMENT AND STRATEGIES

MICHAEL L. ZETTLER[1*], DARIUS DAUNYS[2], JONNE KOTTA[3] & ANDREAS BICK[4]
[1]*Baltic Sea Research Institute, Rostock, Germany*
[2]*Coastal Research and Planning Institute, Klaipeda University, Klaipeda, Lithuania*
[3]*Estonian Marine Institute, Tallinn, Estonia*
[4]*University of Rostock, Department of Zoology, Rostock, Germany*
[*]Corresponding author *michael.zettler@io-warnemuende.de*

Abstract
In the mid 1980s, a new polychaete appeared in coastal waters of the Baltic Sea and rapidly became an important faunistic element. The identification of this spionid as *Marenzelleria* cf. *viridis* indicated North America as the most likely area of origin. The history of invasion into the Baltic Sea is outlined. Sound taxonomic experience and competence was and is necessary to elucidate the background when dealing with bioinvasions. A review of taxonomic problems of the genus *Marenzelleria* is given. Based on long time series we could draw a picture on the species strategy in stressed habitats. Three coastal areas were selected to demonstrate the role of environmental factors facilitating *Marenzelleria* invasion in the Baltic Sea in the last decade. This review deals with the expansion and niching of an allochthonous species and its influence on the indigenous macrozoobenthos. Both the life history and physiology of *Marenzelleria* make it well adapted to life in brackish water conditions, occupying a niche in an environment with low species diversity.

1 Introduction and invasion history

In the late 1970s and early 1980s, a new polychaete appeared in brackish waters of the North and Baltic Seas and rapidly became an important faunistic element. Identification of this spionid was beset with problems from the start. First it was identified as *Marenzelleria wireni,* later as *M. viridis*, until genetic studies revealed differences between populations from the North Sea and the Baltic Sea and the species were named as *M.* cf. *wireni* (North Sea) and *M.* cf. *viridis* (Baltic Sea and Elbe Estuary) (Bick & Zettler 1997). The identification of the Baltic species as *Marenzelleria* cf. *viridis* led justifiable to assume that it had very probably been introduced from North America. Genetic studies showed the correspondence with animals from the Arctic (Canada) and several coastal waters along U.S. coast (New Hampshire to Georgia) (Bastrop et al. 1997).

M. cf. *viridis* prefers oligo- to mesohaline conditions. The species colonizes soft bottom habitats and inhabits vertical mucus-lined burrows down to sediment depths of 35 cm. The species reaches a size up to 120 mm in length (250 setigers) and has a life span of approximately 3 years (Zettler 1996b, 1997a). Fertilized eggs occur in the water column and the larval development is entirely pelagic for 4-12 weeks (Bochert 1997).

The way of introduction into the Baltic is not fully clear yet. The transport as larvae in the ballast water is likely. The enormous production of larvae is probably the reason for the rapid speed of dispersal. The species was probably first introduced to the harbors of

E. Leppäkoski et al. (eds.), Invasive Aquatic Species of Europe, 66–75.

Swinoujscie or Stralsund (Gruszka 1999). In 1985 *M. viridis* was first reported from the Baltic Sea in a German coastal water (Bick & Burckhardt 1989). Since then the species spread effectively and established practically along all coasts of the Baltic to the Åland Islands, the Gulf of Finland, the Gulf of Riga and the Bothnian Bay (Fig. 1).

In 1986 *Marenzelleria* was observed in Polish waters (Gruszka 1991; Maslowski 1992) and in 1988/89 it reached Lithuanian, Latvian and Estonian waters (Olenin & Chubarova 1992; Lagzdins & Pallo 1994; Zmudzinski et al. 1997; Kotta & Kotta 1998). In the beginning of the 1990s the species occupied Swedish and Finnish coastal areas (Persson 1990; Stigzelius et al. 1997). By the early 1990s *M. viridis* managed to colonize almost the whole Baltic Sea. At the end of the century the polychaete found its way into the northernmost part of the Baltic, the Bothnian Bay (Leonardsson 2001). Recently the highest densities (> 5,000 ind m^{-2}) of *M. viridis* were observed in the southwestern part of the Baltic (Darss-Zingst Bodden, Greifswalder Bodden, Oderhaff) (own observations) and in the Gulf of Riga (Cederwall et al. 1999) (Fig. 1). Medium abundances (500-5,000 ind m^{-2}) occur in the Bothnian Bay (Umeå area) and Vistula Lagoon (Zmudzinski et al. 1997; Leonardsson 2001). All other Baltic waters are colonized in low abundances from some few specimen to 500 ind m^{-2} (e.g. Kube et al. 1996; Boström & Bonsdorff 2000; Daunys et al. 2000).

M. viridis belongs to the most successful recent invasion species of the Baltic Sea. In the last decade this polychaete has become a dominant element (in respect to abundance and biomass) in the local zoobenthic communities of several coastal waters. This study deals with the expansion and niching of an allochthonous species and its influence on the indigenous macrozoobenthos.

2 The systematic state of *Marenzelleria*

During an investigation on the conditions of the sea fisheries of the south coast of New England (northeast coast of North America) two species of *Scolecolepis* were recorded by Verrill in 1873: *Scolecolepis viridis* and *S. tenuis*. Both species occurred on sandy shores of the bays and sounds in large abundance (Verrill 1873). In the beginning of the 20th century two further species, important in this context, were described from the Arctic region by Augener (1913) and Chamberlin (1920): *Marenzelleria wireni* and *Scolecolepides arctius*.

Maciolek (1984) revised the genera *Marenzelleria* and *Scolecolepides*. As we know now, she was wrong in some details. These mistakes were one cause of following uncertainty about the taxonomic state of specimens recorded in European coastal waters at the end of the 1970s. She transferred *S. viridis* to *Marenzelleria* and synonymised *S. tenuis* with *S. viridis* (=*M. viridis*). Maciolek described a third *Marenzelleria* species, *M. johnesi* (according to Rodi & Dauer (1996a) junior synonym of *M. viridis* and for that reason not further regarded here). As a result of this revision only two species are recognized as belonging to *Marenzelleria*: *M. wireni*, occurring in waters around the Arctic Circle, and *M. viridis*, occurring in the coastal water of the east coast of North America. According to Maciolek, *M. wireni* differs only lightly from *M. viridis*.

The occurrence of *Marenzelleria* in European coastal waters resulted in numerous ecological and taxonomical investigations. Finally the immigrant was identified as *M. viri-*

dis (Fig. 2) on the basis of Maciolek's revision and the assumed area of distribution (Atkins et al. 1987; Bick & Burckhardt 1989).

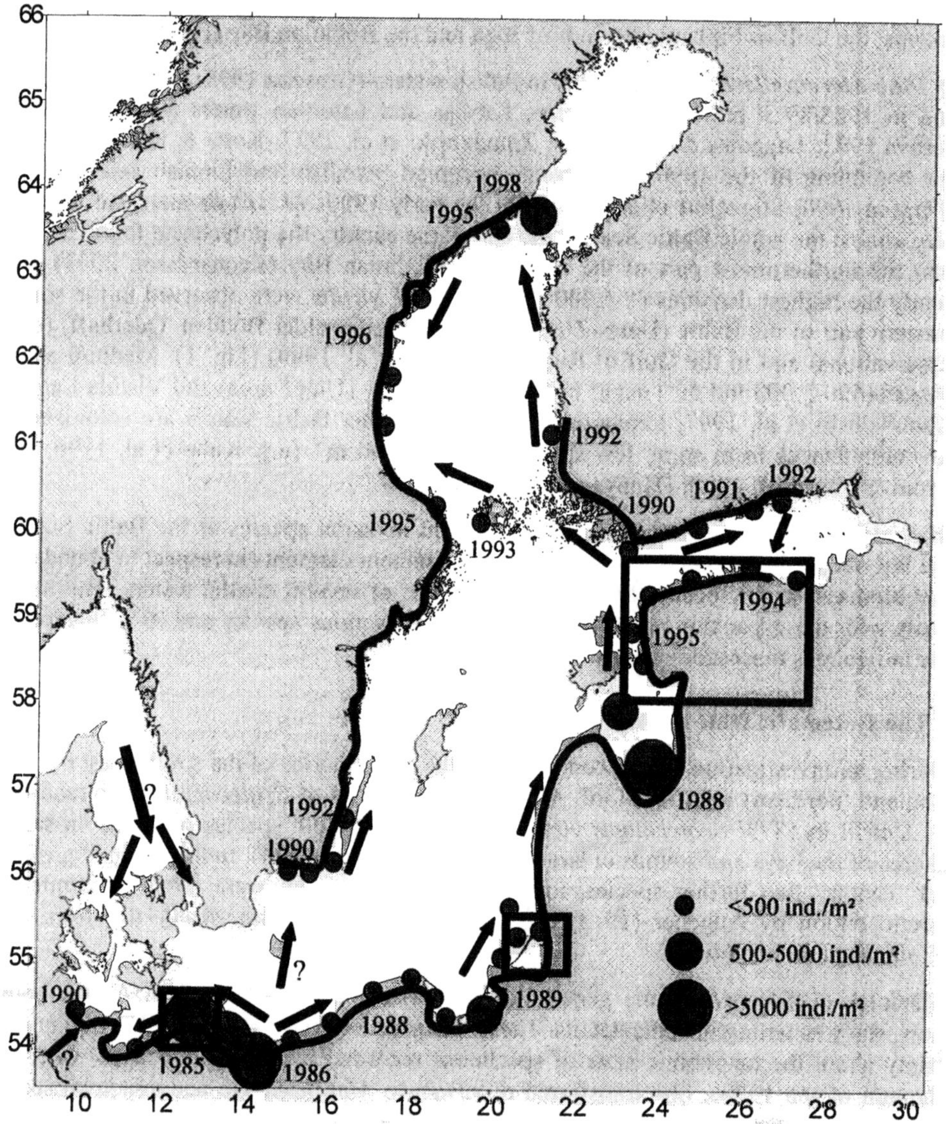

Figure 1. Invasion history of *Marenzelleria* cf. *viridis* in the Baltic Sea. The bold line indicates the recent distribution. The size of the circles correlates with the mean abundance recorded in the sub-areas. For data see Zettler (1997b) and references. The frames show the locations of the reference areas (from left to right: Darss-Zingst Bodden, Curonian Lagoon, Estonian waters).

Population genetic studies by Bastrop et al. (1995) showed the presence of genetically distinct forms in the North and Baltic Sea as well as in different regions of the northeastern coast of America. The morphological studies undertaken against this background allowed good discrimination between these species. Most important diagnostic characters are: length of nuchal organ, number of branchiate setigers and the first appearance of hooded hooks (Bick & Zettler 1997). These characters vary with growth and make it more difficult to identify juvenile worms (Bick 1995). Owing to the lack or poor condition of type material the specimens were named as *M.* cf. *wireni* and *M.* cf. *viridis*, according to Maciolek's description (Bick & Zettler 1997).

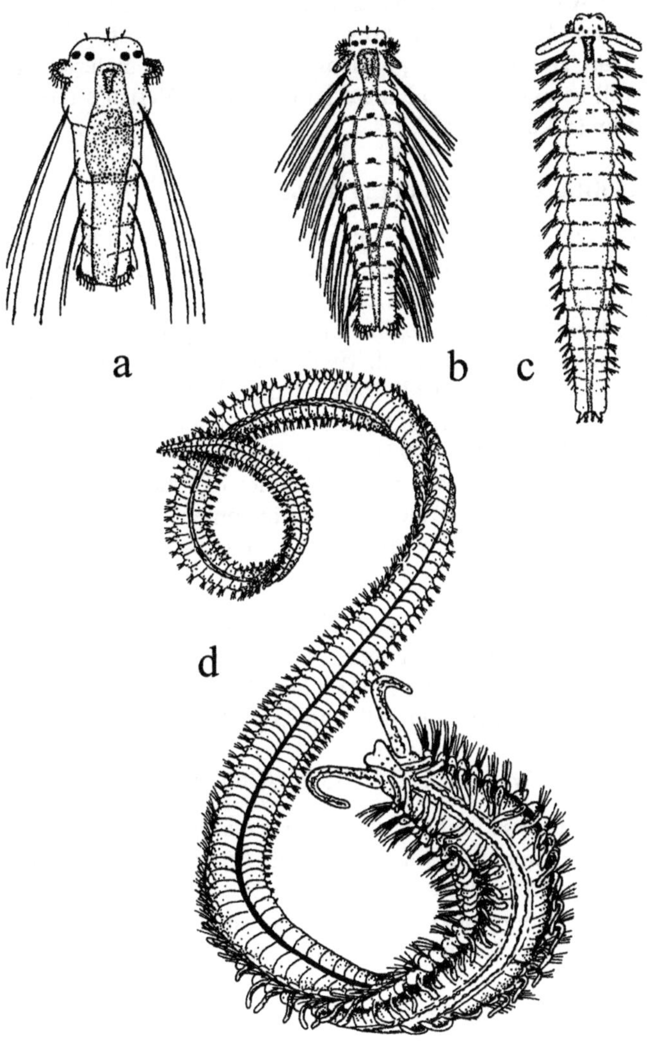

Figure 2. Marenzelleria cf. *viridis*, (a) 4-setiger larvae, (b) 15-setiger larvae, (c) juvenile benthic stage, (d) habitus of an adult worm [a-c] from Bochert & Bick (1995), (d) from Bick (1995).

In 1998, Russian polychaetologists examined types of *Scolecolepides arctius* Chamberlin and *Laonice annenkovae* Zachs, 1925 (Sikorski & Bushinskaya 1998). They synonymised both species and suggested *S. arctius* (valid name: *M. arctia*) to be separate from *M. viridis* and *M. wireni*. According to these authors four species of *Marenzelleria* exist of which two occur in the Arctic (*M. wireni* and *M. arctia*) whereas the other could be found in coastal waters of North America and Europe (*M. johnesi* and *M. viridis*).

According to Bastrop et al. (1997) and Bick & Zettler (1997) at least three species have been found in North America, in the North Sea and in the Baltic. This means that we know 5 valid species of *Marenzelleria,* however, usage of corresponding species names is not absolutely clear. To resolve this problem, a revision with material from Arctic and boreal zones, as well as available type material (including Verrill's rediscovered material), is in preparation (Sikorski et al.).

3 Life strategies in different waters of the Baltic

M. viridis has been studied intensively since it was discovered that it has become an important component of the coastal macrozoobenthos of the Baltic Sea. Field studies and laboratory investigations have been carried out to analyse the life cycle (e.g. Bochert & Bick 1995; Bochert 1997), the distribution and physiological capacities according to ecological factors (e.g. Kube et al. 1996; Bochert et al. 1997; Burckhardt et al. 1997; Schiedek 1997; Stigzelius et al. 1997; Zettler 1997a; Zmudzinski et al. 1997; Cederwall et al. 1999; Daunys et al. 2000) and morphological and genetic features (e.g. Bastrop et al. 1997; Bick & Zettler 1997). *M.* cf. *viridis* was discovered in a wide range of brackish habitats, where it showed different modes of establishment. Environmental tolerance, high genetic variability, short generation time, early sexual maturity, high reproductive capacity and a broad diet are among attributes of *M. viridis*.

Three coastal areas were selected in the Baltic to demonstrate the role of environmental factors facilitating species invasion. Darss-Zingst Bodden might be characterized by favourable habitat features which benefited species establishment in the past. In contrast, both Curonian Lagoon and the Gulf of Finland/Gulf of Riga can serve as examples of stressed areas, where the species meets the limits of its environmental tolerance.

3.1 GERMAN COASTAL WATERS (DARSS-ZINGST BODDEN)

Since 1985 *M. viridis* dispersed rapidly throughout the boddens south of Darss-Zingst and became one of the dominant faunistic elements (Zettler 1996a). It has colonized a range of habitats varying from almost limnic (0.5 PSU) to brackish water regions up to 10 PSU. The brackish eutrophicated environment of the Darss-Zingst Bodden has provided excellent opportunities for the species to quickly develop a dominant population within few years (Fig. 3).

Since the first appearance of *M. viridis* in the mid 1980s, its abundance has steadily increased rising from a few hundred ind m^{-2} to about 10,000 ind m^{-2} in 1994. Maximum densities were revealed in the middle of the bodden (about 28,000 ind m^{-2}, 30-40 g AFDW m^{-2}). Afterwards the densities of the spionid did strongly decrease to approx. 100 ind m^{-2} in the innermost area (salinity < 3 PSU in the last 5 years) and to medium abundance of 3,000 ind m^{-2} in the outer region (never < 5 PSU). A very similar development was observed in the Dollart (The Netherlands) of the sibling species *M. wireni* by Essink & Dekker (2000). They discussed three phases of population development: an initial increase, a stabilization period and a period of decline. The developmental stages of Darss-Zingst populations are kinds of stable equilibrium density (high salinity area) and an example of an extremely stressed one (low salinity area). Due to the high variability of recruitment success, caused by high or low salinity extremes and variations of nutrition supply (sensu Bochert 1997; Burckhardt et al. 1997) in this transition area, the abundance of *M. viridis* fluctuated or decreased.

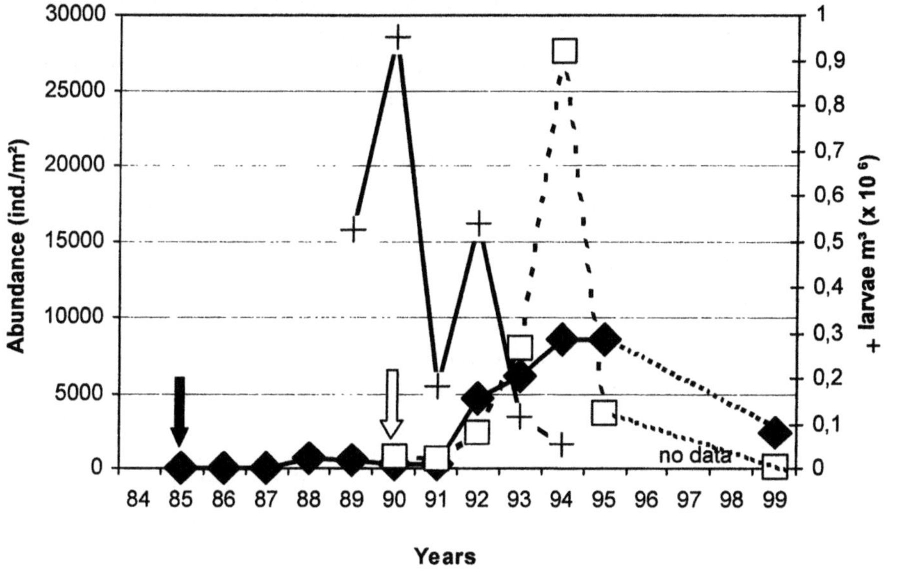

Figure 3. Mean annual abundance of *Marenzelleria* cf. *viridis* in two different (salinity) sub-areas of the Darss-Zingst Bodden from 1984 to 1999. The abundance of larvae (+) are averaged (Bochert 1997). The arrows indicate the beginning of colonization.

In the Darss-Zingst Bodden *M. viridis* occurred at greater sediment depths than other species in this area. In respect to that and in view of its rapid increase in terms of both abundance and biomass, a negative impact on the indigenous fauna was expected. In contrast the sibling species (*M. wireni*), studied in North Sea estuaries (Atkins et al. 1987; Essink & Dekker 2000), where partly drastic changes of benthic colonization were observed, comparison with pre-invasion data for Darss-Zingst Bodden revealed no negative consequences. On the contrary, colonization of the substrate by the spionid appears to encourage the proliferation of other organisms. With increasing spionid abundance an increase of most indigenous species (e.g. the polychaete *Hediste diversicolor*, chironomid larvae and hydrobid gastropods) was observed. Only the amphipod *Corophium volutator* showed a negative correlation with *M. viridis*. Several experiments (predation and competition) substantiated these results.

The lower limit of 5 PSU for successful development (metamorphosis) of *M. viridis* larvae (Bochert 1997) and the strong dependency on both food concentration and quality (Kube et al. 1996; Burckhardt et al. 1997) clarify the adaptation to eutrophic conditions prevailing in brackish waters.

3.2 LITHUANIAN COASTAL WATERS (CURONIAN LAGOON)

The benthic population of *M. viridis* also successfully established in the shallow eutrophicated Curonian Lagoon since it was first observed there in 1990 (Olenin & Chubarova 1992). The species colonized the oligohaline lagoon areas (down to mean annual salinity of 0.2 PSU) being affected by seawater inflows and does not penetrate further into permanently freshwater habitats. As in many other estuaries or lagoons of the Baltic, the species established in sandy/muddy habitats previously occupied by poor communities of oligochaetes and chironomids. Even with low densities, usually < 500 ind m^{-2}, the species increased to total benthic biomass of a few 100 g m^{-2} (wet weight).

Curonian Lagoon differs from many enclosed water bodies in the Baltic by wind driven irregular salinity fluctuations during autumn and winter months. Salinity is frequently varying from freshwater up to 7 PSU at the time scale of hours to few days. The species spawning and development of pelagic larvae, which take place during cold period in the Baltic, are critical stages in *M. viridis* life history. Oosorption of ripe gametes in *M. viridis*, a typical phenomenon in the lagoon benthic population, is most likely caused by accidental salinity drops. Nevertheless, the density of pelagic stages during reproduction season may occasionally reach up to 50,000 ind m^{-3} in lagoon waters. The outbursts of *M. viridis* in the Curonian Lagoon plankton community are strongly related to the seawater inflows, whereas the pelagic larvae became completely extinct under freshwater runoff conditions (Fig. 4). Lecitotrophic stages (eggs to 3-setigers) usually are most abundant among larvae contributing at least to 60% of their total density found in the plankton. In opposite, late planktotrophic stages, which are well towards metamorphosis to benthic life mode, are exceptionally rare. Development boundary at the 10-segment stage was found to be characteristic for the pelagic larvae in the lagoon waters during three years of observations (Fig. 4).

The population outside the entrance of the lagoon obviously acts as a pool of recruits: a part of its offspring is directed to the maintenance of populations in the adjacent oligohaline habitats. According to the structure of the pelagic stages, only 0.1-1.0% of those larvae that potentially "visited" the lagoon underwent metamorphosis to benthic juveniles. This recruitment is negligible compared to the average density of some few hundreds of ind m^{-2} being found in the lagoon benthic population. It is more likely that migration of juveniles during post-reproductive periods contributes more significantly to population maintenance. Migration capability of *M.* cf. *viridis* juveniles, reported for the Atlantic coast of North America was confirmed in the Baltic, too (see Zettler 1997b for references). Probably this mode of dispersal ensures the permanent existence of local populations for a decade since its invasion. This also coincides with steady deficiency of early benthic stages among bottom macrofauna in post-reproduction seasons.

The continuous turnover of age classes is of exceptional importance for a long-lasting establishment of species with relatively short life span. It seems that the ecological roulette principle (Carlton & Geller 1993) fits perfectly to *M. viridis* survival and establishment in extremely stressed habitats as well.

3.3 ESTONIAN COASTAL WATERS (GULF OF FINLAND/GULF OF RIGA)

The first observation of *M. viridis* at the northern coast of the Gulf of Finland was made in 1990 (Stigzelius et al. 1997). During 1990-1993 *M. viridis* expanded its distribution into the eastern parts of the Gulf (Stigzelius et al. 1997). However, anti-clockwise circulation of the currents would not permit *M. viridis* to spread from the northern side of the Gulf towards its southern side. In addition, the larvae of the polychaete are unable to complete their development at salinities < 5 PSU (Bochert 1997), which may frequently occur in the easternmost part of the Gulf of Finland. Only one specimen was recorded in the south-eastern coast of the Gulf of Finland in 1994. Until 1997 this species was not observed along the southern coast of the Gulf.. Some occasional findings of *M. viridis* in the westernmost bays suggest the Väinameri (NW Estonia) as a donor region. *M. viridis* is expanding its distribution range towards the eastern parts of the Gulf of Finland, being established as far as in Narva Bay (NE Estonia) (Kotta 2000).

Depth did not correlate with the abundance and biomass of *M.* cf. *viridis* while sediment type was a significant factor for both. In shallower areas (< 10 m) the species preferred sand or gravel bottoms. Deeper down (> 10 m) it was confined to silty clay bottoms (Kotta & Kotta 1998). In the more eutrophicated regions of the Gulf of Riga, i.e. off the mouths of Daugava and Pärnu rivers, the densities of *M.* cf. *viridis* were fivefold higher than in adjacent sea areas. Besides nutrient concentrations, higher salinity values increased the population densities of *M. viridis* in these areas.

M. viridis is a deposit-feeder sharing the same food resources with the majority of the native benthic invertebrates. Hence, the addition of *M. viridis* into the benthic system of the northern Baltic is likely to offset a stress on the native fauna due to possible competitive interactions for food and/or for space. There exists circumstantial evidence that after the invasion of *M. viridis* the densities of the polychaete *Hediste diversicolor* (Kotta & Kotta 1998) and the deep-water amphipod *Monoporeia affinis* dropped considerably in the northern Baltic Sea (Cerderwall et al. 1999). Manipulative experimental studies have demonstrated that *M. viridis* has a significant effect on the growth of *M. affinis* (Kotta 2000). The reduction in the growth of amphipods due to the competitive interaction between these species is expected to change the duration of amphipod life cycle: its recruitment is prolonged and fecundity is reduced. In the shallower areas *M. viridis* reduced the survival of *H. diversicolor*. On the other hand, the bivalve *Macoma balthica* caused a significant mortality of *M. viridis*. We suggest that the availability of food limits the population growth of *M. viridis* in the deeper parts whereas competitive interactions between *M. viridis* and *M. balthica* are a possible key factor determining the distribution pattern and abundance of *M. viridis* in the shallower parts of the northern Baltic Sea. *M. viridis* should prevail only in these biotopes where the polychaete can escape from the competitive interactions with *M. balthica*. These are for example the sea-areas adjacent to river mouths and municipal wastewater outflows where the food for deposit feeders is in excess.

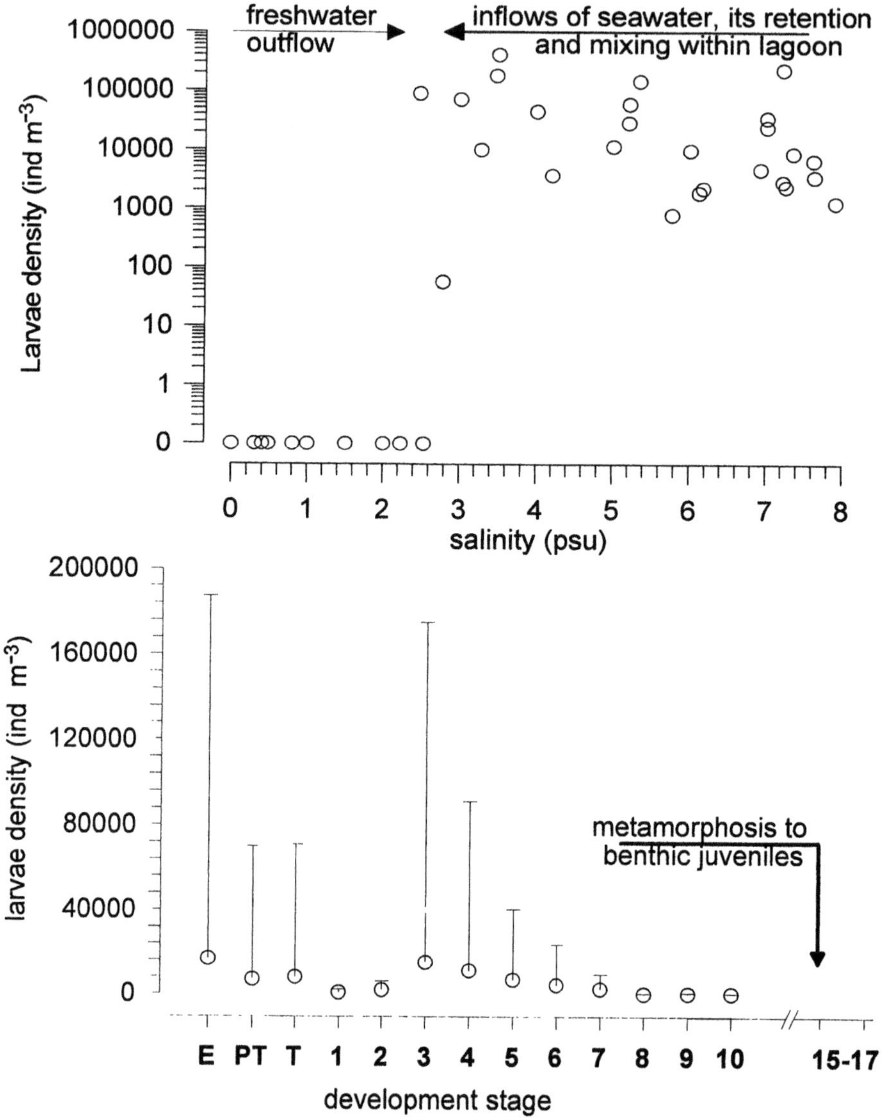

Figure 4. Density of *Marenzelleria* cf. *viridis* pelagic larvae *versus* salinity (top: temperature range 3-15 °C) and contribution of different development stages (bottom: averages and max. reached) to the total species abundance found in the Curonian lagoon in 1996-1999. ((E) eggs, (PT) pre-trochophores, (T) trochophores; numbers indicate stage of larvae according to the number of setigers).

4 Conclusions

Due to eutrophication, primary production is extraordinary high in most estuaries, inhabited by *Marenzelleria*. Thus, food shortage will not limit the phytoplankton feeding spionid because of the nearly absence of filter-feeding bivalves in these horohaline waters. In these areas with low species diversity no or weak indications on interspecific competition were observed. Moreover, positive correlations were found between densities of the new invader and local (native) species. Competition for food with the indigenous fauna is likely to occur in oligohaline waters with freshwater mussel beds (e.g. Oder Lagoon) or in more oligotrophic offshore waters with high densities of filter-feeding bivalves, like *Mytilus edulis*, *Macoma balthica* and *Mya arenaria* (e.g. Pomeranian Bay, off-shore waters) where the species exploits the same and limited food resource.

The results obtained in studies referred to in this review indicate an extremely high tolerance of the *M. viridis* adults to many environmental factors such as depth, salinity, temperature, eutrophication and sediment quality. *M. viridis* shows a Baltic wide distribution within favourable habitats in near coastal areas. It propagated in semi-enclosed shallow lagoons as well as in coastal zones and off-shore bays. Most success was observed in enclosed coastal areas, like boddens and lagoons. The success of its establishment is probably facilitated by higher summer temperature, homogeneity of habitat and eutrophication in coastal areas of the Baltic Sea. However, spawning event and development of pelagic stages were found to be critical phases in the species life cycle in stressed environments. Most likely the species is limited by salinity fluctuations rather than its average value.

Long lived though less tolerant pelagic larvae, being the species advantage to disperse over extended distances, do not ensure expansion into stressed (marginal) habitats. The key force driving the species expansion into adverse environments might be dispersal of relatively more tolerant benthic stages. However, this way of population maintenance leads to the accidental nature of the species dynamics (ecological roulette) in stressed habitats.

ALIEN CRAYFISH IN EUROPE: NEGATIVE AND POSITIVE IMPACTS AND INTERACTIONS WITH NATIVE CRAYFISH

KAI WESTMAN
Finnish Game and Fisheries Research Institute, Aquaculture, Helsinki, Finland
kai.westman@rktl.fi

Abstract
Worldwide, about 500 species of freshwater crayfish are known. Europe has only five, all of which belong to a single family, Astacidae, with two genera: *Astacus astacus*, *A. leptodactylus* and *A. pachypus*, and *Austropotamobius pallipes* and *A. torrentium*. The widespread elimination of native populations by crayfish plague, *Aphanomyces astaci*, and human intervention has led to the introduction of plague-resistant North American crayfish in an effort to restore Europe's once highly profitable crayfish industry. Crayfish have also been imported to occupy a vacant niche, for aquaculture purposes, and for home aquariums and ornamental pools. The main alien species established in Europe are two cambarids, *Orconectes limosus* (introduced in the 1890s, currently in 15 countries) and *Procambarus clarkii* (1970s, 11 countries), and one astacid, *Pacifastacus leniusculus* (1960s, 25 countries). Recently, *Orconectes immunis* and *Procambarus zonangulus* have been released into natural waters in Germany. The main benefits of introductions have been new productive fishery in plague-stricken waters, new opportunities for crayfish culture and an additional source of food. The socio-economic aspects and recreational values are also important. The most disastrous impact has been on native crayfish populations. Alien species can be vectors of the plague fungus and thus present a persistent risk to native species. The fact that these species are highly competitive alien species - even without the plague - gives rise to detrimental competition and eventually replace native species when sympatric. Despite the increasing interest in conserving Europe's unique native crayfish fauna and its biodiversity, the threat of crayfish plague, habitat changes, continuous expansion of the ranges of alien species by human activity and the removal of trade barriers hamper efforts to ensure the future of indigenous species.

1 Introduction

Crayfish are the largest mobile invertebrates inhabiting freshwaters. There are about 500 known species of freshwater crayfish worldwide. The majority occur in North and Central America, about 100 in the southern hemisphere, four in Asia and five in Europe (Hobbs 1988; Fitzpatrick 1995). In addition, three North American species have become established in European natural waters: two cambarids, the spiny-cheek crayfish, *Orconectes limosus*, and the red swamp crayfish, *Procambarus clarkii,* and one astacid, the signal crayfish, *Pacifastacus leniusculus*. The first of these exotic species (*O. limosus*) was introduced into Europe in the late 1800s, the other two not until the second half of the 1900s.

Crayfish have attracted considerable gastronomic attention in Europe for centuries and are harvested in the wild and farmed (astaciculture). According to Ackefors (1999), the annual output of freshwater crayfish from natural waters in Europe totals 4,000-4,500 tons in addition to which culture production accounts for some 150 tons. The value of the total catch of all species in Europe amounts to several tens of million Euros. Crayfish trapping also embodies recreational values that are impossible to assess in purely economic terms. Mainly due to losses caused by crayfish plague, *Aphanomyces astaci*,

E. Leppäkoski et al. (eds.), Invasive Aquatic Species of Europe, 76–95.

production in Europe is unable to satisfy the continuously growing demand for this delicacy and some 3,300 tons are imported every year. The total global annual harvest from the wild and farms may exceed 100,000 tons. Production is currently greatest in North America but it is increasing rapidly in China.

Native crayfish in Europe have suffered and are under continuous threat from many factors, diseases (e.g. Cerenius & Söderhäll 1992; Vogt 1999), pollution, habitat modification, overfishing (Westman 1985; Westman et al. 1990; Gherardi & Holdich 1999) and alien species. By far the most devastating event was the arrival of crayfish plague (*A. astaci*) in Italy in the early 1860s and its subsequent spread throughout the continent. The fungus most likely reached Europe with the transportation of affected North American freshwater crayfish in ballast water (discussed e.g. by Unestam 1969a, 1969b; Alderman 1996; Vogt 1999). Since then the disease has probably been re-introduced several times with the importation of various North American crayfish species. European crayfish are highly susceptible to *A. astaci*, whereas North American species are resistant but may serve as vectors (Unestam 1969a, 1972; Smith & Söderhäll 1986; Alderman and Polglase 1988; Svärdson 1992). As observed in *P. leniusculus*, not all populations harbour crayfish plague and mixed populations including native European crayfish can develop and live in sympatry (e.g. Fürst & Boström 1978; Lowery & Holdich 1988; Nylund & Westman 2000; section 4.3).

The economic losses caused by the plague far outweigh those due to the effects of all fish diseases combined. Numerous fresh outbreaks are recorded every year both in previously plague-free waters and in waters where annihilated populations have been replaced by restocking.

The disappearance of crayfish can have adverse environmental consequences, but equally their introduction into waters where they were previously unknown can affect the aquatic ecosystem in many ways (Holdich 1988, 1999; Nyström 1999a,b). This latter fact appears to be widely ignored: whereas the importation and stocking of alien species are heavily restricted, many countries do nothing to curb the spread of native species into waters formerly without crayfish and in whose ecosystem they are "alien" species.

Here, the term alien (nonindigenous, exotic) is used for any introduced crayfish that has not occurred naturally in historical times and the term native (indigenous, endemic) refers to a species that is native to, and restricted to, a particular geographic area (see Holdich & Gherardi 1999). Thus, for instance, the east European narrow-clawed crayfish, *Astacus leptodactylus*, is frequently considered to be an alien species in northern and western Europe; indeed in many countries it is regarded as more damaging to native species than American species. As the aim of the present study is to look only at crayfish species introduced into Europe from other continents and their impacts, the term alien (nonindigenous, exotic) is restricted to species alien to Europe unless otherwise mentioned (as, e.g., in section 3.6).

Introductions of marine decapods (such as the Chinese mitten crab *Eriocheir sinensis*, red king crab *Paralithodes camtschatica*, and the American lobster *Homarus americanus*) are discussed in other chapters of this volume.

2 Native crayfish species in Europe

With less than 1% of the world's crayfish species, Europe has far fewer native freshwater species than either America or Australia. The few that do exist occur nowhere else and hence anything that threatens their survival is viewed with mounting concern. There is only one family, Astacidae. This has two genera, *Astacus* and *Austropotamobius*, which comprise only five recognised species (nomenclature according to Holdich & Gherardi 1999): the noble crayfish, *Astacus astacus*, in northwestern Europe and Scandinavia; the narrow-clawed crayfish, *A. leptodactylus*, in eastern and southeastern Europe; the thick-clawed crayfish, *A. pachypus*, in southeastern Europe; the white-clawed crayfish, *Austropotamobius pallipes*, in western and southwestern Europe; and the stone crayfish, *A. torrentium*, in central Europe. The natural range of many native species has been extended by translocating them to areas where they did not occur before (Holdich & Lowery 1988; Westman et al. 1990; Holdich et al. 1999a). In Finland, for example, the distribution of *A. astacus* has been expanded by stocking to about three times the natural range of the species (Westman 1991).

Despite these efforts, mainly due to the crayfish plague Europe's once uniform native populations have been fragmented and the production of native crayfish is only a fraction of what it used to be. The annual catch of *A. astacus*, once the most important species, is estimated to amount to no more than 300 tons today (Westman et al. 1990; Ackefors 1999), which is only a few percent of production before the arrival of the plague. For example, in the Nordic and Baltic countries, the pre-plague harvest of *A. astacus* exceeded 2,110 tons annually, whereas now it is only about 170-180 tons (Skurdal et al. 1999). In Finland, the catch of *A. astacus* was at its height in 1900, when some 20 million individuals were caught. Equivalent to 850 tons, this is probably the largest harvest of *A. astacus* ever recorded in a European country (Westman & Westman 1992). The total catch of *A. astacus* in Finland today ranges from 2.5 to 3.7 million individuals a year (about 80-130 tons).

The current catch of *A. leptodactylus* totals 1,200 tons (Ackefors 1999). The catastrophic impact of crayfish plague is illustrated by the decline in Turkey's annual harvest from 8,000 tons in the peak year 1984 (Köksal 1988), to less than 500 tons since the plague entered the country. Catches of other native crayfish species are low, being only about 3 tons for *A. pallipes* and even less for *A. pachypus* and *A. torrentium* (Ackefors 1999). Although the plague fungus has now been present in Europe for over 140 years and selection pressure has been intense, no resistant strains of any of the five indigenous species have been identified (e.g. Unestam 1969a,b, 1972; Alderman & Polglase 1988; Svärdson 1992; Vogt 1999).

In water systems and large lakes the plague may retain a chronic presence. As a result, despite the large-scale stockings carried out in many European countries, new catchable populations of native species have seldom developed in such waters as shown, for instance, in Finland (Nylund & Westman 1992) and in Sweden (Fürst 1995).

The plague fungus does not require its host in order to spread, as the spores may be transmitted e.g. via damp surfaces. Thus, in Europe, crayfish plague has, and is continuing to, spread via the hundreds of thousands of crayfish trappers and their gear but also by the trade and stockings of crayfish. In many areas, the release of plague-resistant alien crayfish carrying the disease has exacerbated the situation by both spreading the plague and acting as a reservoir of the fungus.

Re-introduction of native crayfish may lead to the re-establishment of healthy populations, but exposure to crayfish plague followed by collapse of the populations is a persistent risk. Thus the plague remains the most severe threat native crayfish populations and fisheries are facing.

3 Non-native crayfish species in Europe

With the exception of some small-scale experiments, alien species introduced into Europe for stocking in natural waters originate from North America, where crayfish show natural plague resistance, a trait lacking in European crayfish and those of other continents (Unestam 1969a,b, 1972). The plague fungus being ubiquitous in North America, crayfish there have presumably adapted to life with the fungus parasite (Svärdson 1992).

Efforts to combat the plague by importing plague-resistant species for stocking experiments were started in the 1960s, first in Sweden and later in other countries. The main purpose was to replace lost native populations and to restore the once highly profitable European fisheries to pre-plague levels. Crayfish have also been imported to occupy a vacant niche, for aquaculture purposes, for home aquariums and ornamental pools, and for snail and weed control (references in Gherardi & Holdich 1999). Crayfish have also been moved outside their natural range accidentally, e.g. in ballast waters, via canals, as bait and through escapes from holding facilities. Crayfish can be transported for relatively long periods, even days, in moist containers without standing water what permitted their distribution and rapid (both legal and illegal) expansion.

Species imported into and established in Europe are the spiny-cheek crayfish, *Orconectes limosus*, the signal crayfish, *Pacifastacus leniusculus* (which belongs to the same small family, Astacidae, as European species) and the red swamp crayfish, *Procambarus clarkii*. When first imported, it was not known that all three species were potential vectors of the crayfish plague. Imports of some other species have failed (see 3.4). Recently, Dehus et al. (1999) reported that the "paper shell crayfish", *Orconectes immunis,* and the white river crayfish, *Procambarus zonangulus*, have been released into natural waters in Germany from aquariums and ornamental ponds.

3.1 THE SPINY-CHEEK CRAYFISH, *ORCONECTES LIMOSUS*

Orconectes limosus is a hardy, aggressive, temperate-water species well adapted to life in permanent lentic and lotic waters. It occurs in northeastern USA and southeastern Canada. *O. limosus* grows rapidly and may attain a weight of 40 g; however, stunted species weighing less than 15 g are commonly encountered. Unlike native species it seems to tolerate deoxygenated, eutrophic or polluted waters (reviewed by Momot 1988). Gherardi et al. (1999) point out that gourmets do not rate *O. limosus* as highly as other species owing to its small size and narrow chelipeds. Its meat, too, is considered less of a delicacy than that of other alien species.

The first introduction of *O. limosus* into Europe was into a fish pond in Germany in 1890 followed by several introductions into natural waters in France and Poland in the early 1900s. These continued later in the century (Momot 1988; Westman et al. 1990). Under favourable conditions, *O. limosus* may reach very high densities. According to Henttonen & Huner (1999), *O. limosus* is currently the dominant freshwater crayfish in most permanent water bodies in France, Germany, the Czech Republic, Slovakia and Poland. It is also encountered in The Netherlands, Belgium, Switzerland, Spain, Italy, Lithuania and Belarus, and probably also in Great Britain, Russia and Ukraine (references in Gherardi & Holdich 1999). It is still extending its range, particularly in central Europe.

3.2 THE SIGNAL CRAYFISH, *PACIFASTACUS LENIUSCULUS*

Pacifastacus leniusculus is indigenous to cool temperate regions of northwestern USA and southwestern Canada (Lowery & Holdich 1988) and comprises three subspecies: *P.l. klamathensis* (Stimpson), *P.l. leniusculus* (Dana) and *P.l. trowbridgii* (Stimpson). Of these, two at least, *P.l. leniusculus* and *P.l. trowbridgii* and their hybrids, have been introduced into Europe (Agerberg 1993).

P. leniusculus is the largest of the species introduced into Europe from North America, commonly attaining a length of 17-18 cm (references in Westman 2000). According to Lowery (1988), it is the fastest growing northern temperate zone crayfish species.

Due to its origin in cool temperate regions, *P. leniusculus* is well adapted to life in permanent lentic and lotic waters throughout Europe. As an astacid, *P. leniusculus* appears to have developed convergently and, in many respects, in an ecologically homologous manner to its European relatives. For example, the morphology and life history of *P. leniusculus* are similar to those of *A. astacus* and, further, these species share omnivorous food preferences, have a nocturnal activity pattern and favour the same types of habitat in lakes and streams, primarily in the littoral zone on stony bottoms or other substrates offering some form of refuge (references in Cukerzis 1988; Hogger 1988; Lowery & Holdich 1988; Söderbäck 1993; Westman 2000). It is thus no wonder that *P. leniusculus* quickly acclimated to areas into which it was introduced, first in Sweden (Abrahamsson 1973; Fürst 1977; Svärdson 1995) and then in Finland (Westman 1973).

Importation into Sweden was prompted by the seemingly insuperable problem posed by crayfish plague due to which, in the 1980s, only 5% of the original population of the single native species, *A. astacus,* remained (Fjälling & Fürst 1988). To compensate for this loss and to meet market demands, crayfish were imported into Sweden in increasingly large quantities and in the 1970s imports exceeded 2,000 tons annually.

With the launch of domestic production in view, a study had been conducted in Sweden in the late 1950s on the feasibility of restoring the numerous plague waters with North American crayfish species. A survey of 16 American species finally resulted in *P. leniusculus* (and *Orconectes virilis,* for details see 3.4) being imported into Sweden in 1960. *P. leniusculus* established a local population in the small test lake (Svärdson 1965). Encouraged by this promising start, the government allowed stocking experiments to continue in a few other lakes (reviewed by Svärdson 1995). *P. leniusculus* is now best established in Sweden, occurring in some 1,050 lakes, 320 streams and rivers, and 1,340 ponds (Skurdal et al. 1999).

Reports from Sweden regarding plague resistance of *P. leniusculus,* its high growth rate, large size and broad claws, good flavour and similar appearance to the native *A. astacus* (Svärdson 1965) aroused hopes in Finland (and subsequently in many other European countries) that this species might enable the crayfish fishery to be revived (Westman 1973, 1995). No adverse factors had emerged in Sweden, neither had anything negative come to light in studies conducted in its original range (Abrahamsson 1973). In fact, quite the contrary: in every respect the species seemed to be almost ideal for replacing the native crayfish species in plague-affected waters.

In 1969 Austria, in 1971 Poland, in 1972 Germany and France followed Sweden and Finland in importing *P. leniusculus*. Later in the 1970s a number of other countries, too, followed suite (Westman et al. 1990; Ackefors 1999; Henttonen & Huner 1999). Thus far *P. leniusculus* has been introduced into water bodies in around 25 European countries and stockings are continuing to expand.

3.3 THE RED SWAMP CRAYFISH, *PROCAMBARUS CLARKII*

Procambarus clarkii is a "warm water", hardy, aggressive species native to northeastern Mexico and the south central USA. It grows rapidly, may mature at weights as low as 10 g and attain a weight of over 50 g in 3-5 months. It is cultivated widely in Louisiana, USA, and its production, processing, home consumption and export have developed into a flourishing multi-million dollar industry (reviewed by Huner 1988; Huner & Barr 1991). Unlike *O. limosus*, *P. leniusculus* and native European crayfish species, *P. clarkii* is a burrowing species adapted to life in seasonally-flooded wetlands that may lack standing water for some months. It can tolerate sharp changes in environmental factors, e.g. temperature, and is well able to survive in oxygen-depleted waters. Consequently, it may invade habitats where native crayfish do not usually occur. Thus it comes as no surprise that *P. clarkii* thrives in southern Europe, where there are extensive warm, shallow natural and agricultural wetlands (references in Gherardi & Holdich 1999). Nevertheless, *P. clarkii* also thrives in colder, high latitude locations the USA and also in Japan, where it was introduced in the late 1900s (Huner & Barr 1991). As pointed out by Lindqvist & Huner (1999), *P. clarkii* has unique life history characteristics that further enhance its ability to invade new habitats: populations include individuals that are incubating eggs or carrying juveniles at any time during a year and therefore, whenever environmental conditions are favourable, there are always young crayfish available to perpetuate the population. Due to this ecological plasticity, *P. clarkii* is the most widely distributed crayfish species in the world, occurring on every continent with the exception of Australia and Antarctica.

The first *P. clarkii* in Europe were imported into Spain from Louisiana in 1973 (Habsburgo-Lorena 1979). By 1979, the species was also established in southern Portugal (Gutiérrez-Yurrita et al. 1999). Commercial success encouraged illegal introductions, and the practice of selling live *P. clarkii* as food and as garden pond pets contributed to its expansion into natural waters in many other European countries. It is now also known from Austria, Italy, France, Cyprus, Belgium, The Netherlands, Switzerland,

Germany, Majorca and England (Huner 1988; Hobbs III et al. 1989; Gherardi & Holdich 1999). In these cooler areas of Europe, *P. clarkii* is most abundant in small, isolated permanent ponds (Henttonen & Huner 1999).

P. clarkii is the dominant freshwater crayfish species in the world, over 85,000 tons being commercially harvested in the USA, the People's Republic of China and Spain annually (Huner 1994; Henttonen & Huner 1999).

3.4 OTHER ALIEN CRAYFISH SPECIES

The French authorities made an unsuccessful attempt to introduce the cambarid, the northern crayfish, *Orconectes virilis*, into France in 1897 (Arrignon et al. 1999). In 1960 the same species was introduced into Sweden at the same time as *P. leniusculus*. However, it did not thrive and any further attempts to import it were abandoned (Svärdson 1965). As far as is known, *O. virilis* does not exist in natural waters in Europe.

The North American white river crayfish, *Procambarus zonanqulus* (= *P. acutus acutus*), was introduced into Spain in 1973 (Habsburgo-Lorena 1979) but it disappeared (Henttonen & Huner 1999).

Three parastacid species of the genus *Cherax*, the marron, *Cherax tenuimanus,* the red claw, *C. quadricarinatus*, and the yabby, *C. destructor*, were introduced into Europe from Australia for aquaculture purposes during the 1980s and 1990s. *C. tenuimanus* requires fairly warm permanent waters and *C. quadricarinatus* tropical conditions. Only *C. destructor*, which is a temperate species adapted to seasonally flooded wetlands, would appear likely to thrive in natural waters, and then only in southern Europe (Henttonen & Huner 1999). The species is indeed encountered in natural waters in Spain (Gutiérrez-Yurrita et al. 1999). All parastacids are considered to be highly susceptible to *A. astaci*, suggesting that no *Cherax* spp. could succeed in natural waters as aquaculture escapees.

3.5 OTHER POSSIBLE INTRODUCTIONS INTO EUROPE

There are more than 300 crayfish species in North and Central America, and most of them have several features that can have an adverse effect in a new environment: they are mobile, invasive, aggressive and able to tolerate a wide range of habitats and thus can spread rapidly; they are omnivorous and so have a wide trophic spectrum; and they are frequently competitive and cannibalistic (Hobbs 1988; Holdich 1988, 1999). These species occupy habitats of all types, from wet meadows to wetlands and from small to large permanent water bodies. Moreover, some are primary or secondary burrowers (Huner 1988) and occupy niches thus far totally unoccupied in Europe. Many North American species have very small, localised distributions but, according to Lindqvist & Huner (1999), there are about 20 species with extensive distributions that might be described as generalists. Some of these grow large enough to be considered as food. With the management of crayfish stocks and, especially, the repopulation of plague-devastated waters in mind, interest in introducing these North American species into new areas is growing in Europe. Some species are unique in appearance or otherwise and so are attractive as ornamental pets for aquariums or artificial pools. Imports of crayfish for these purposes, and thus the risk of escapes into natural waters, can be expected to increase. Consequently, the likelihood of further colonisation by exotic crayfish in Europe cannot be discounted.

3.6 CRAYFISH TRANSPORTATION WITHIN EUROPEAN COUNTRIES

Many native species restricted to a particular geographic area in Europe have been moved outside their natural ranges either naturally (migrations, floods etc), accidentally (via canals etc) or deliberately by humans (stocking etc) and many have become established in new surroundings (Westman et al. 1990). For example, in Britain, three of the introduced nonindigenous crayfish species are now established in the wild and officially classified as pests. Two of these, *A. astacus* and *A. leptodactylus*, are native European species; only one, *P. leniusculus*, originates outside Europe. Their classification as pests is based on the fact that they have the potential to damage stocks of the single native crayfish species, *A. pallipes*, through competition (Holdich et al. 1999b). *A. leptodactylus* in particular is now widely established outside its natural range in Europe and is considered by many to be an alien species in northern and western parts of the continent (Westman et al. 1990; Holdich 1999). According to Dehus et al. (1999) in Germany, for example, imports of alien species fall far short of those of *A. leptodactylus*. As the crayfish are imported and sold live, they are often released into natural waters either intentionally or accidentally.

4 Impacts of non-native species established in Europe

The environmental impact of an established introduction can be either positive, neutral or negative (Fig. 1). However, as Holdich (1999) has pointed out, what might be a positive feature from one point of view (e.g. that of the crayfish farmer) might be negative from another (e.g. that of the conservationist). This is certainly true of the three North American species (*P. leniusculus, P. clarkii, O. limosus*) introduced into and established in Europe. Many countries have profited from their introduction into waters that have often for long been devoid of crayfish due to the plague and in some countries alien crayfish are cultivated by extensive and semi-intensive methods (Skurdal et al. 1989; Huner 1994). At the same time, however, alien species have caused environmental problems, particularly in lakes where they displace co-occurring native species.

In the case of the three North American species mentioned, this is due to the fact that in addition to plague resistance, they appear to have at least some of the characteristics of successful invaders.

Characteristics of successful invaders include, e.g. a larger body and chela size, faster growth, more aggressive behaviour, greater fecundity and overall better thermal tolerance than the native species (discussed e.g. by Butler & Stein 1985; Mather 1990; Lodge & Hill 1994; Lindqvist & Huner 1999; Westman 2000). Many of these characteristics encourage the spread of these species into less favourable habitats. Further, when these introduced species colonise, their rapid population increase and their feeding habits may have important consequences for their environment (Nyström 1999a,b). Plague-resistant alien species can also be vectors of the plague and thus contribute to the decline in native species.

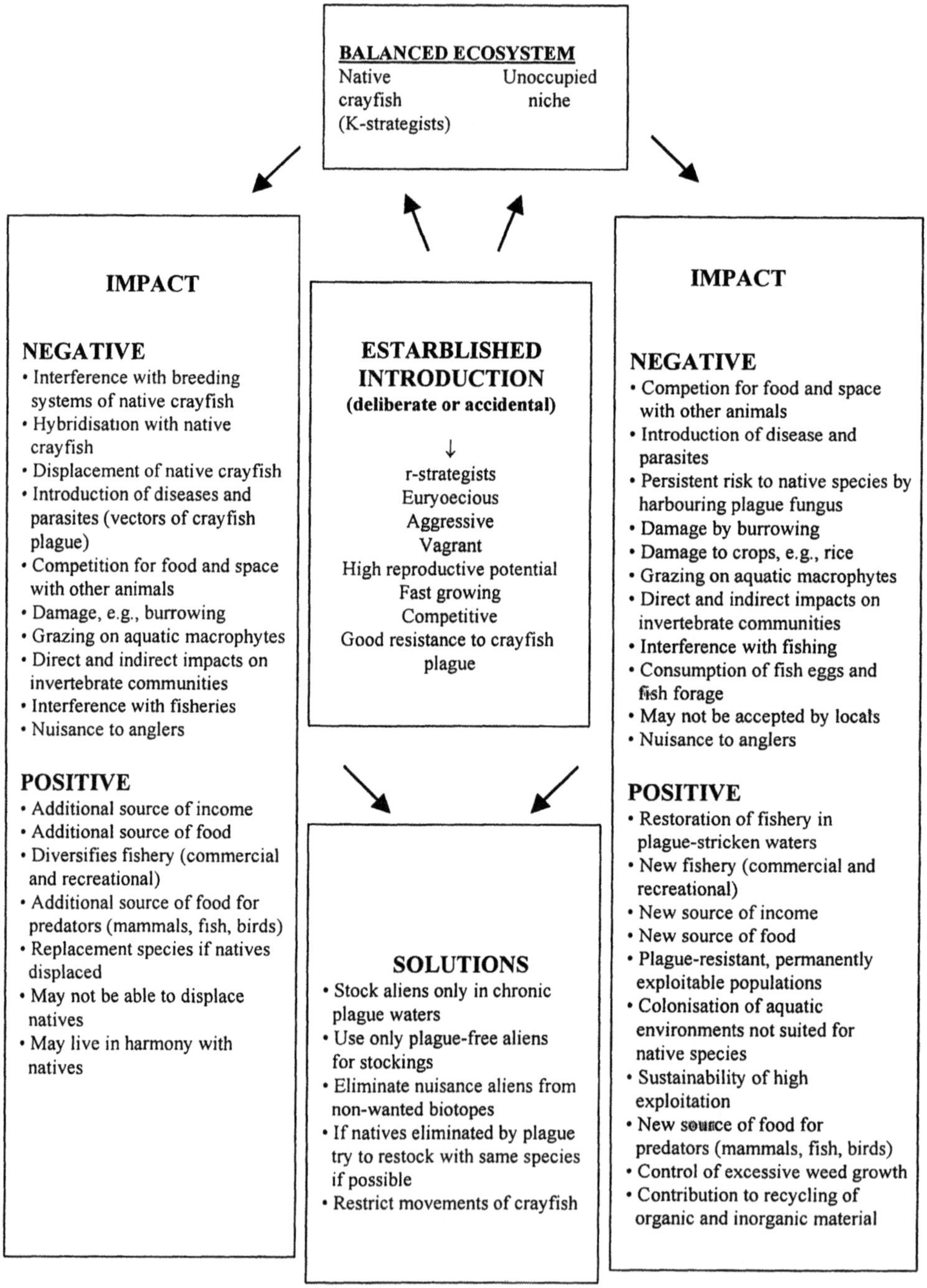

Figure 1. Summary diagram to illustrate the negative and positive impacts of the established introduction of alien crayfish on two ecosystems, one inhabited by native crayfish and one without native crayfish (modified from Holdich 1999).

Although the volume of research into crayfish has increased markedly since the 1970s, the effects of both introduced and native species on the structure and diversity of freshwater communities has, as pointed out by Nyström (1999b), received surprisingly little attention. Questions not yet answered concern, for example, whether the need for increased crayfish production and the maintenance of biodiversity in freshwater ecosystems will eventually come into conflict, whether introduced crayfish species have the same impact on the ecosystem as do native crayfish species and whether we should encourage the stocking of native species into as many new habitats as possible.

Crayfish are omnivorous and may act as predators, herbivores and detritivores; they are thus important consumers and key organisms in many aquatic food webs and may dominate the invertebrate biomass in lakes and rivers (e.g. Momot 1995; Nyström 1999a,b). The loss of a crayfish population or the introduction of crayfish into a new habitat may therefore have a major impact on the aquatic environment.

4.1 NEGATIVE IMPACTS

Only a small proportion of the 500 freshwater crayfish species are known to cause environmental problems, and then usually only when they are moved outside their natural geographic range (Holdich 1988, 1999). The situation in Europe varies from one country to another. In some, the economic benefits of an alien species have far outweighed any damage done as, for example, shown by *P. leniusculus* in Sweden and Finland (Svärdson 1995; Westman 2000) and *P. clarkii* in Spain (Habsburgo-Lorena 1983; Huner 1988). In many other countries, though, the effects have been negative.

Holdich (1988, 1999) and Nyström (1999a,b) have examined the negative implications of an introduced species becoming established in a freshwater environment both with and without an existing native crayfish fauna (Fig. 1). As the biomass of crayfish species may vary considerably between habitats (Lodge & Hill 1994), so may the impacts, depending on the size of the population. The good invasive capabilities of the North American species introduced into Europe result in both physical and biological environmental degradation. Possible physical effects include damage due to burrowing. This has been noticed in particular with *P. clarkii*, which, as a burrowing species, has caused problems in agricultural and recreational areas (e.g. rice plantations, lawns, dams, dykes and river banks; Huner 1988; Arrignon et al. 1999; Gutiérrez-Yurrita et al. 1999).

Biological problems are caused by the fact that the three alien species mentioned often reach much higher population densities than native crayfish. Their feeding on macrophytes, algae, insects, molluscs, amphibians and fish may then lead to changes in food webs and even to the elimination of some species.

Aquatic macrophytes play an important role in freshwater ecosystems. They influence water chemistry, are the primary source of littoral detritus and are also an important food source for many invertebrates, crayfish included; they may also influence the distribution of vertebrates and invertebrates (reviewed by Nyström 1999b). Like native species, introduced crayfish graze on aquatic macrophytes and may significantly reduce their biomass and species richness. The effects may be both direct, e.g. on animals' food, and indirect, e.g. on water chemistry. Studies suggest that even though the preference of introduced crayfish species for macrophytes may be similar to that of natives, the impact on the macrophyte biomass may differ. For example, Nyström & Strand (1996) showed that *P. leniusculus* consumed significantly more *Chara* spp. than did *A. astacus*. Due to the greater consumption and more rapid population development of alien species, they can therefore be expected to have a greater impact on the macrophyte biomass than native species. The reduction in macrophyte biomass accompanied by the consumption of detritus may affect the ontogeny of shallow lakes and ponds.

P. clarkii, which is also encountered in habitats where no native crayfish have ever been able to survive, may, as a new species, have a very pronounced effect indeed on macrophytes and invertebrates in these environments, as observed in Spain (Gutiérrez-Yurrita et al. 1999). Likewise in Italy, *P. clarkii* has caused damage to submerged vegetation, destroying, for example, the only European cultivation of lotus lilies. The monetary loss has been considerable (Gherardi et al. 1999).

Introduced crayfish species are also reported to have had both direct negative impacts (by grazing and predation) and indirect ones (by competition and through habitat changes) on macroalgae, on a number of invertebrate species (molluscs, leeches and insect larvae), on amphibians and on some fish species (reviewed by Nyström 1999b). In Spain, the exploitation of *P. clarkii* has led to the death of birds and aquatic vertebrates in crayfish traps. According to Gutiérrez-Yurrita et al. (1999), more than 42,000 birds and 750,000 aquatic vertebrates of various species died in the traps in two natural parks in the 1989 season.

The greatest negative impacts are, however, caused by *P. leniusculus, P. clarkii* and *O. limosus* displacing co-occurring native species through competitive exclusion. Alien crayfish may also act as vectors of crayfish plague fungus, *A. astaci*, as already mentioned and also of other diseases to which the native crayfish are more susceptible (discussed by Holdich 1999; Vogt 1999). They may also colonise aquatic environments not occupied by native crayfish. In many cases, however, this may turn out to be beneficial in so far as it leads to new fishery and new sources of income.

P. leniusculus at least is known to be very tolerant of saline conditions (Holdich et al. 1997) and has been recorded in the Baltic Sea far from the nearest river outlet (Skurdal et al. 1999). It may, therefore, invade estuarine environments in Europe as it has done in North America (Miller 1960). This may have many unpredictable effects on the estuarine fauna and flora. Research into the subject appears to be lacking, however. Other possible damage includes nuisance to anglers, interference with fishing (e.g. eating fish from nets and other gear) and consumption of fish eggs. Note, however, that native crayfish may do similar damage when introduced into new areas. The release of any crayfish species into waters previously without crayfish will always have far-reaching biological consequences, e.g. on food webs.

Regarding the impact of individual species, *O. limosus* has displaced native European crayfish in many areas, particularly in eutrophic waters in its geographic range, which extends over vast areas of eastern, central and western Europe; competing with native species for living space and food, it poses a continuous threat to their survival (Huner 1988; Laurent 1988; Momot 1988; Arrignon et al. 1999). The recent discovery of populations of *O. limosus* in England could further jeopardise the survival of the already threatened only native species, *A. pallipes* (Holdich et al. 1999b).

P. leniusculus, too, has displaced European crayfish species (Lowery & Holdich 1988; Söderbäck 1995) although this may take a long time, even in small water bodies (Westman et al. 2002). *P. leniusculus* is known to dig long passages in suitable substrata (Guan 1994; Kirjavainen & Westman 1999) and so to cause damage to lake shores and river banks. It can attain higher densities than native crayfish (Svärdson et al. 1991; Guan & Wiles 1996; Westman 2000) - in terms of fishery a positive feature - and thus exerts a greater effect on the other biota (Nyström 1999a,b; Nyström et al. 1999).

Through its burrowing activities and rapid colonisation capability, *P. clarkii* has caused environmental damage, e.g. to earthen dykes and water control structures in irrigation systems in agricultural and recreational areas and, in Spain and Portugal, to rice cultivations (Huner 1988; Gutiérrez-Yurrita et al. 1999).

4.2 POSITIVE IMPACTS

The main benefits of introduced North American crayfish species have been the source of income, food and recreation for human beings (Fig. 1). The alien species have contributed to the increasing commercial value of capture fishery and culture in Europe. According to the conservative estimate of Ackefors (1999), fishermen and aquaculturists earn about USD 13 million annually. However, the value for the whole fishing industry is many times higher. Further, the socio-economic value of crayfish production is difficult to assess in monetary terms but, for rural dwellers, it is considerable (e.g. Habsburgo-Lorena 1983).

Even though the small size of *O. limosus* had prevented it from achieving any real commercial value despite its abundance, it is still an acceptable food for humans, especially in areas suffering from food shortages or economic problems (Henttonen & Huner 1999). Estimated to be in the region of only 15 tons a year, production of *O. limosus* is nonetheless rather small (Ackefors 1999).

P. leniusculus, in contrast, has turned out to be very productive (Svärdson et al. 1991; Westman 2000) and has allowed countries whose stocks of native crayfish have been devastated by crayfish plague to develop an alternative species for human consumption. According to data assembled from various sources, the total annual yield is about 500 tons, of which 50 tons are produced by cultivation (Ackefors 1999). The catch from natural waters (about 450 tons) is equivalent to 15-20 million 10 cm specimens. Catches of *P. leniusculus* are greatest in the first country to import the species, Sweden, where they amount to about 250-300 tons a year (Skurdal et al. 1999) (equivalent to 10-12 million 10 cm specimens). The catch exceeds that of the native *A. astacus*, 30-50 tons, many times over (Skurdal et al. 1999). Sweden is followed by France, with annual production of around 50 tons (about 2 million specimens). Finland's production amounts to no more than a few hundred thousand individuals a year but it is growing rapidly with the harvesting of the populations that have developed from crayfish stocked in large lakes in the 1980s (Mannonen & Westman 1998).

P. clarkii, the dominant species in Europe today, yields an annual harvest of at least 3,000-4,000 tons. Introduction of the species has been particularly beneficial in Spain, where annual production has ranged from 2,000 to 3,500 tons, in 1986 reaching 5,000 tons. The productive populations have given rise to an economically important export industry, too (Habsburgo-Lorena 1983; Ackefors 1999; Gutiérrez-Yurrita et al. 1999). Portugal produces a smaller (700 tons) but steadily growing volume (Gutiérrez-Yurrita et al. 1999).

The importance of introduced species for crayfish production in Europe is demonstrated by the fact that 2,500-3,000 tons of the 4,000-4,500 tons harvest from natural waters, that is, about two-thirds, already comes from alien species, mainly *P. clarkii* and *P. leniusculus*, and only about 1,500 tons from native species, primarily *A. leptodactylus* and *A. astacus*. These two alien species have thus rapidly restored crayfish fishing in many parts of Europe.

Alien crayfish species introduced into natural waters are assuming an ever more meaningful role in Europe, permitting a new productive fishery, new sources of income and an additional food source. The income from the fishery of alien species is especially important for rural people, as it helps to keep the countryside alive. In Spain, for example, the *P. clarkii* industry has revitalised the local economy in certain districts, enabling hundreds of fishermen to make a living from this fishery year round (reviewed by Gutiérrez-Yurrita et al. 1999).

Of particular importance, as shown in Sweden, is the capacity of alien species to make Europe's innumerable plague-stricken waters productive once again. Thus they also increase the biological diversity, which in itself may be regarded as desirable. As the North American species are highly resistant to crayfish plague, they permit the development of more permanent, continuously exploitable populations. Production of the rapidly growing and strongly reproducing *P. clarkii* and *P. leniusculus* is greater than that of native species, which, in addition to their plague-resistance, is yet another reason why their culture is so attractive. Socio-economic aspects and recreational values, including the associated tourism, are also important (Habsburgo-Lorena 1983; Gutiérrez-Yurrita et al. 1999).

Alien species introduced into Europe are not without positive impacts on their environment, too. Abrahamsson (1966) noted that when the native *A. astacus* was eradicated by the plague from small Swedish lakes, the bottoms were soon covered by *Chara* spp. However, when *P. leniusculus* was introduced into the same ponds, it eliminated submerged vegetation (Nyström 1999b). During eutrophication, many shallow lakes develop emergent vegetation, which eventually results in swamp and marsh conditions. Consumption of macrophytes and detritus by crayfish may prolong this succession. As pointed out by Ackefors (1999), crayfish also contribute to the recycling of energy, and organic and inorganic material and thus promote the production of other organisms. Crayfish are also important sources of food for mammals, fish and birds (reviewed by Foster & Slater 1995). For example, the invasion of *P. clarkii* is found to have been beneficial for wading birds that use crayfish as food (Gherardi et al. 1999; Nyström 1999b).

4.3 INTERACTIONS BETWEEN NON-NATIVE AND NATIVE SPECIES, *PACIFASTACUS LENIUSCULUS* AND *ASTACUS ASTACUS* AS AN EXAMPLE

For 30 years, the Finnish Game and Fisheries Research Institute has been monitoring the development of *P. leniusculus* and *A. astacus* in a small, enclosed Finnish lake,

Slickolampi (Westman 2000; Westman et al. 2002). In 1971, a small batch (900) of cultured 2nd stage (10 mm) *P. leniusculus* juveniles were released into the lake, in which the *A. astacus* population was increasing after having been decimated by crayfish plague (Fig. 2).

That both species have lived side-by-side right up until the present year (2001) without any signs of the plague indicates that the *P. leniusculus* population must be plague-free.

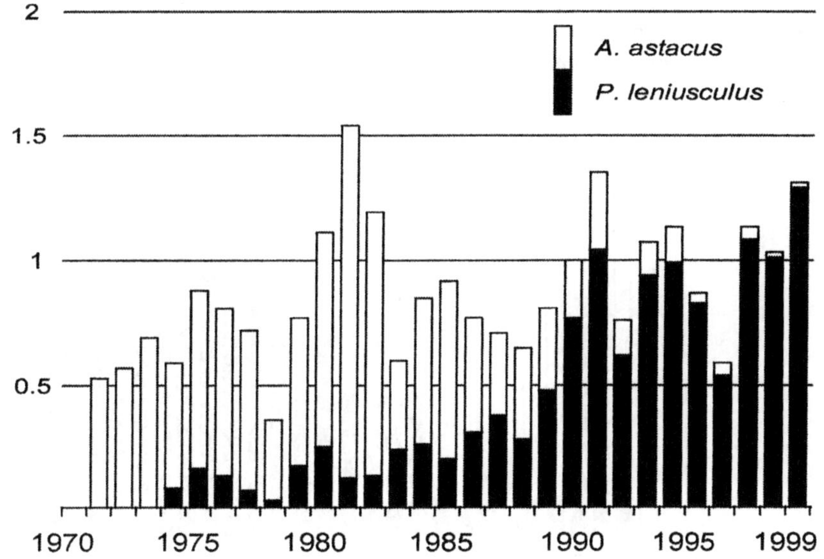

Figure 2. Mean trap catches (crayfish/trap/night) of *Astacus astacus* and *Pacifastacus leniusculus* in Slickolampi in late summer in 1970-1999. In 1970-1975 only some of the shores were trapped. Harvesting (≥ 100 mm individuals) began in 1979. No *A. astacus* were removed after 1994, the year the catch size limit of *P. Leniusculus* were lowered to 90 mm.

According to annual trap catches and population size estimates *A. astacus* was clearly dominant in the 1970s and most of the 1980s (Fig. 2). At the end of the 1980s, however, there was a shift in the relative abundances of the two species, and in the 1990s *P. leniusculus* became dominant. In the late 1990s it accounted for > 98% of total catches (Fig. 3). Consequently, *P. leniusculus* has not augmented the existing fauna in this lake but instead has almost completely replaced *A. astacus*. Both species seem largely to favour the same types of biotope but *P. leniusculus* is distinctly more demanding and has been encountered less often than *A. astacus* on gently-sloping soft shores. The proportion of *A. astacus* with chelae injuries (16 years, mean 17.3%) was nearly twice that of *P. leniusculus* (9.3%), suggesting that agonistic interspecific encounters do occur and that *P. leniusculus* is much more competitive (Westman et al. 2002). As a result, *A.*

astacus is presumably displaced, at least to some extent, from preferred food resources and shelter-providing habitats to inferior biotopes, which would in turn increase predation and other risks (Lodge & Hill 1994). Competition for food and habitat may also have contributed to the slowdown in growth of *A. astacus* in Lake Slickolampi.

P. leniusculus is more successful than *A. astacus* in competitive encounters because it is innately more aggressive and also more active than similar-sized *A. astacus* (Cukerzis 1988; Svärdson et al. 1991; Söderbäck 1991). Further, it has both a higher individual growth rate (Söderbäck 1991; Westman 2000) and usually also larger chelae than *A. astacus* (personal observations); consequently, *P. leniusculus* individuals in any one cohort tend to be larger than *A. astacus* of the same age, a factor that enhances their competitive dominance (Momot & Leering 1986; Söderbäck 1991, 1993; Edsman & Jonsson 1996).

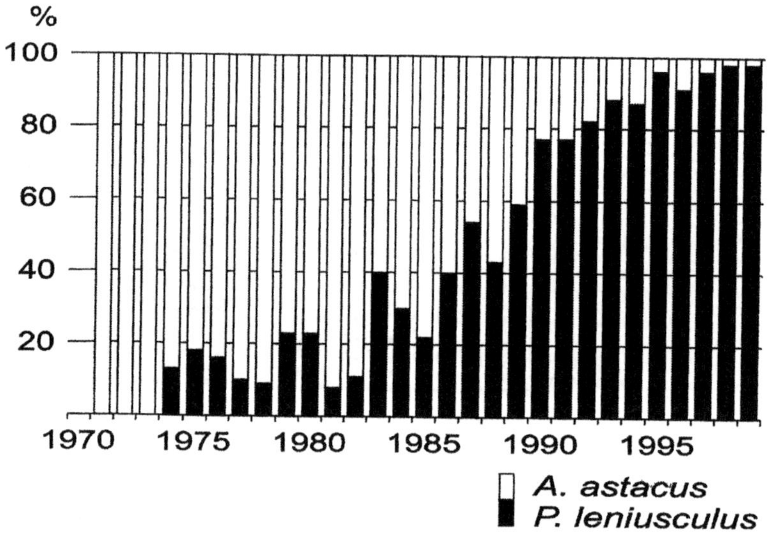

Figure 3. Changes in relative abundance of *Astacus astacus* and *Pacifastacus leniusculus* in Slickolampi in 1970-1999.

In other characteristics, too, *P. leniusculus* seems to be more competitive than *A. astacus* because a stocking as small as that in Lake Slickolampi, which corresponded to the production of only 5-10 females (Savolainen et al. 1997), was sufficient for a *P. leniusculus* population to expand even when all the niches in the stocking area were already occupied by *A. astacus*. The earlier hatching and greater growth rate of young-of-the-year *P. leniusculus* than of *A. astacus* (Jonsson 1995; Söderbäck 1995) suggests that *P. leniusculus* might be less vulnerable to winter conditions than *A. astacus*. *P. leniusculus* is also apparently well able to tolerate high temperatures (Firkins & Holdich 1993). Overall, *P. leniusculus* appears to have many of the characteristics, e.g. larger body and chela size, faster growth, aggressive behaviour, greater fecundity and overall thermal tolerance, found in the successful crayfish invader (discussed, among others, by Butler & Stein 1985; Mather 1990; Lodge & Hill 1994; Lindqvist & Huner 1999).

However, the consistent weakening of *A. astacus*, even at sites with only a low density of *P. leniusculus*, indicates that its elimination is not adequately explained by competitive exclusion alone. Westman et al. (2002) suggest that its disappearance is in fact governed by a combination of several interacting mechanisms, of which harvest (> 100 mm specimens of both species) and competition with *P. leniusculus* were initially the main reasons for the population decline. Ultimately, however, the collapse of *A. astacus* seems to have been due to the almost complete cessation of successful reproduction, presumably as a result of reproductive interference between the two species. Interspecific mating results in females laying sterile eggs. Although both species suffer from the ensuing loss of recruitment, the consequences are less serious for *P. leniusculus*, which has a higher capacity for population increase than *A. astacus* (Savolainen et al. 1997).

When the two populations were equally abundant, intermating presumably did not play a very important role. As the *A. astacus* population continued to decline, however, reproductive interference assumed greater importance as a replacement mechanism. In the present situation, in which the catch of *A. astacus* accounts for only a few percent of the total, the likelihood of finding conspecific mates is so low that practically all matings are with *P. leniusculus*. It is indeed likely that *A. astacus* will be driven into extinction unless the management actions introduced succeed in reversing the trend.

In conclusion, the long-term data presented by Westman et al. (2002) indicate that introduced *P. leniusculus* - even when plague-free - cannot co-exist with native *A. astacus*, as the ecological niches of these species overlap too much. Similarly, Söderbäck (1993, 1995) has attributed the near-extinction of *A. astacus* co-occurring with *P. leniusculus* in a Swedish lake to the combined effect of interspecific competition, predation and reproductive interference. The study of Westman et al. (2002) confirms earlier claims (e.g. Lodge & Hill 1994) that crayfish replacement is difficult to study, has been inadequately understood and is a complex interaction of multiple factors. A prime task for crayfish research today is therefore to compare the characteristics of alien and native species and study their interactions in order to establish the major factors governing replacement. Once these are known, it will be easier for us to predict the future spread and impact of invading species and so devise ways of preserving native species.

5 Conclusions and management measures

The elimination of many European crayfish populations by crayfish plague and human activities has led to measures to conserve and revive the remaining native crayfish stocks but also to restore the once so flourishing and economically important crayfish industry. One such measure has been to introduce plague-resistant alien crayfish into Europe. Two species in particular, *P. clarkii* and *P. leniusculus*, have proved to be economically successful, and in many waters crayfish production has returned to, or even far exceeded, the levels achieved in the late 19th century. On the whole, though, production in Europe is still below pre-plague levels. As well as economic benefits, the introductions have brought socio-economic, recreational and biological benefits.

Unfortunately, the introduction of alien species has also had many adverse effects. Abundant fast-growing alien crayfish populations affect the biodiversity of freshwater communities by reducing the biomass and species richness of macrophytes and invertebrates. As earlier mentioned, in the long term, crayfish grazing on macrophytes may have serious consequences for the diversity of and interactions within lake communities and for their ontogeny. These consequences must be taken into account when the stocking of alien species is considered. The in-depth review of Nyström (1999b) on the ecological impact of introduced and native crayfish on freshwater communities reveals the serious shortcomings in our knowledge of the effects of both alien and native species. Consequently, Nyström stresses the urgent need to determine the effects of *P. clarkii* on invertebrates, macrophytes and amphibians. Information is also lacking on the ecological impact of *O. limosus*.

The most injurious effects of the introduction of alien crayfish species have been, however, on native crayfish species. In efforts to conserve these species it is important to bear in mind that crayfish spread slowly by themselves and cannot move between water bodies without human intervention. Even dispersal upstream may be prevented by barriers such as dams, waterfalls and unfavourable stretches of water. However, water owners, the holders of crayfish harvesting rights, crayfish trappers and tourist enterprises are all keen to promote the continuing expansion of plague-resistant alien species, as they are also more productive than natives. Human activity is thus responsible for the spread, often at an alarming rate, of alien crayfish in many European countries.

Imprudently stocked, all alien crayfish species pose a threat to the increasingly few remaining native populations, as not only may the plague spread with these resistant species but by harbouring the plague fungus they may also present a persistent risk to native species in the waters in which they occur. Moreover, being highly competitive even without the plague, alien species established in Europe give rise to detrimental competition and even replacement of native species when sympatric. According to Taugbøl & Skurdal (1999), without effective measures to conserve native crayfish populations, and controlled stocking of alien species, many European countries may before long be faced by a situation in which the majority of waters are dominated by alien crayfish species, and the native species, considered critically endangered, survive only in protected areas.

It has been suggested that plague-resistant "strains" of native species should be developed by genetic techniques. Predicting when such strains would be created is pure guesswork but even these modified native species would be vectors of the plague fungus and thus also pose a threat to the original native species. Furthermore, the release of genetically modified organisms (GMOs) into the natural environment is attracting increasing concern from international organisations.

Realistically speaking, due to the extensive spread of crayfish plague and the difficulty of combating the disease, it is no longer possible to revitalise the crayfish industry in plague-stricken waters or to meet the high demand for crayfish for human consumption without plague-resistant and productive *P. leniusculus* and *P. clarkii*. Thus, since stockings of introduced species will be promoted in any case, we should endeavour to direct them solely to habitats in which the production of native species is known to be impossible, e.g. in chronic crayfish plague waters. Stockings should also be carefully planned, rigorously controlled and carried out only with populations known to be plague-free. To reduce the amount of new releases into the habitats of native species, we need strict regulations meticulously enforced and better public education.

If we want to control or eradicate nuisance alien species in some way, we have legislative, mechanical, biological, physical and chemical means at our disposal (reviewed by Holdich et al. 1999c). The relevant authorities of most countries, however, are unlikely to permit other than environmentally friendly methods, e.g. intensive trapping, and we do not know whether these methods regulate the excessive abundance of populations of alien species sufficiently effectively.

As pointed out by Gherardi et al. (1999), the Earth Summit in Rio de Janeiro (1992), requires its signatories, among other things, "to prevent the introduction of, control and eradicate those alien species which threaten ecosystems, habitats or species" (Article 8h), offers good opportunities to regulate the introduction of alien crayfish species, too. Moreover, Council Directive 92/43/ECC requires Member States to ensure "that the deliberate introduction into the wild of any species which is not native to their territory is regulated so as not to prejudice natural habitats within their natural range or the wild native fauna and flora and, if they consider it necessary, prohibit such introduction".

Unfortunately, as shown by the recent country-by-country review (see Gherardi & Holdich 1999), international and European regulations and agreements referring to the introduction and stocking of exotic species have been taken little notice of in Europe. Moreover, the removal of trade barriers encourages the translocation of crayfish, both native and alien. The fact that live crayfish for human consumption can be sold between member states of the European Union represents a potential for the spread of undesirable non-native species as it is impossible to ensure that crayfish are not released into natural waters. The legislation should be tightened and harmonised throughout Europe.

It is particularly important to stress that crayfish plague, which is overwhelmingly the greatest threat facing native species, has been – and still is – spread not primarily by alien species but by Europe's hundreds of thousands of crayfish catchers and their traps and other gear, and also by fishing tackle, translocations of native crayfish etc. Due to the damage it does, crayfish plague should be a reportable disease – irrespective of how it is spread. A central disease register should be established, as suggested by Nylund & Westman (1995). Dehus et al. (1999) point out that Council Directive 91/67/EEC concerning fish diseases provides the legal possibility for areas or regions that have no history of diseases to become acknowledged disease-free zones with defined import-export rules. Crayfish plague is also covered by this Directive, and plague-free regions occupied by native crayfish could be protected by allowing the import of live crayfish into them only from areas certified plague-free. To date, however, no zones in the EU have been declared "disease free". Vogt (1999) has stressed that there are still great gaps in our knowledge of crayfish diseases (in particular viral and rickettsian diseases and the geographical distribution of crayfish pathogens) and that much more research needs to be done in this area.

There are many reasons, both environmental and economic, for sustaining Europe's unique native crayfish fauna and its biodiversity. Due to the degradation of populations *A. astacus*, *A. torrentium* and *A. pallipes* are now all identified as threatened species in the Red List of the IUCN (The World Conservation Union) (Groombridge 1993). They are protected to some extent by the Bern Convention on the Conservation of European Wildlife and Natural Habitats and the European Habitats Directive (Council Directive 92/43/EEC) (e.g. Holdich et al. 1999b). Before uniform and effective conservation regulations applying to all Europe have been achieved and protected areas set up, the conservation of native species should be a national aim in all countries, as is indeed implied in the ratification of the Rio Convention. For example, *A. pallipes* is one of 116 organisms identified as of key importance in the UK Biodiversity Action Plan (Holdich et al. 1999b). Taugbøl & Skurdal (1999) propose that individual countries should establish "Native Crayfish Areas" to protect indigenous species.

In a questionnaire sent to correspondents of the FAO European Inland Fisheries Advisory Commission (EIFAC), Westman & Westman (1992) sought to establish what regulations and measures had been introduced in various countries to protect and manage native crayfish. Although by far the greatest damage to native crayfish populations has been done by crayfish plague, measures taken to prevent the spread of the fungus have still been astonishingly limited. Several international organisations, including ICES, EIFAC, IUCN and the International Association of Astacology (IAA), have issued recommendations concerning the introduction of new species, and advised that stocking be subjected to tighter control. Even so, many countries do not appear to pay much attention to these recommendations.

Native populations should not only be protected from crayfish plague and competition with alien species but they should also be revitalised through various management measures. Encouraged by the rapidly growing interest in - and the importance of – the management of native crayfish stocks, EIFAC organised a workshop on crayfish management and stocking in 1991. The workshop report (Finnish Game and Fisheries Research Institute 1992) and other reports (Lindqvist 1988; Westman 1992; Holdich et al. 1999a; Taugbøl & Skurdal 1999) proposed a variety of management measures covering, among others, the control of crayfish diseases, habitat improvement, crayfish releases, and guiding the catching of harmful species and reducing their numbers to promote the maintenance and restoration of native crayfish stocks.

It is clear from the above reports that, with the exception of stocking, very little research has been done into the contribution of management measures to the restoration of crayfish populations. To ensure that research is targeted appropriately, there is urgent need for a review of management measures already carried out and of their benefits for the revival of native crayfish stocks. Before large conservation and restoration programmes are initiated we must also know the current distribution of native and alien species, regionally, nationally and Europe-wide.

It is particularly important to increase public awareness of the dangers of crayfish plague and the risks posed to natives by alien species. With this in mind, it might be worth setting up national crayfish information centres that would be in contact with a pan-European crayfish centre and database and that would regularly monitor the state of crayfish stocks, catches, outbreaks of disease – crayfish plague in particular, give instructions about the management of crayfish stocks and assist the authorities. Institutions and individuals doing crayfish research in Europe (reviewed by Westman & Manninen 1999) should be in closer contact with one another and should coordinate their research programmes to avoid overlaps and close information gaps.

As reviews of native and alien crayfish in various European countries (Westman et al. 1990; Westman & Westman 1992; Gherardi & Holdich 1999) show, despite the increasing interest in conserving native crayfish populations, the threat of crayfish plague, habitat changes, continuous expansion of the ranges of alien species and the removal of trade barriers continue to hamper efforts to ensure the future of indigenous species. Alien species already widely established in Europe have to be accepted as part of the European fauna. For this reason, we should make the best of the situation by exploiting alien crayfish as intensively as the carrying capacity of the stocks permits.

Even though *P. clarkii* and *P. leniusculus* have already been an economic success in many water areas and they have still greater economic potential (Ackefors 1999), numerous studies show clearly that the stocking of alien North American species may pose serious risks to the existence of native species. Not only can the resistant alien species spread the plague but plague-free stocks are able to outcompete co-existing native species. It is therefore vital that water owners and authorities take the protection of native species into account and carefully weigh the cost and benefits of introducing alien species into a new environment. Under a number of regulations (those of the EU) and conventions (e.g. Rio, Bern) every European nation bears responsibility for conserving aquatic diversity and thus also for preserving native crayfish species. Their preservation is also of great importance for fishery as only then can their continued participation in crayfish catches and trade be secured. As a first step each country should draw up an action plan for the conservation of its native species similar to that made for *Austropotamobius pallipes* in the United Kingdom (Palmer 1994).

INVASION HISTORY, BIOLOGY AND IMPACTS OF THE BAIKALIAN AMPHIPOD *GMELINOIDES FASCIATUS*

VADIM E. PANOV* & NADYA A. BEREZINA
Zoological Institute of the Russian Academy of Sciences, St. Petersburg, Russia
*Corresponding author *gaas@zin.ru*

Abstract
During the 1960s and 1970s the Baikalian amphipod *Gmelinoides fasciatus* (Stebbing) was intentionally introduced into more than 20 lakes and reservoirs outside its native range in Siberia and European Russia, in order to enhance fish production. Abilities of *Gmelinoides* to spread within the basins and to compete with native amphipods were neglected. In the European Russia this species successfully established in the Volga River basin, in such large lakes as Lake Ladoga, Lake Onega, Lake Peipsi, Lake Ilmen and their basins, and in the Neva Estuary (Baltic Sea). In most cases the native amphipods were completely replaced by *Gmelinoides*, and negative impact on other aquatic invertebrate species is also likely because direct predation by *Gmelinoides*. Studies of *Gmelinoides* biology, including experimental estimation of its salinity and temperature resistance, showed that this invasive amphipod tolerates wide range of environmental conditions and potentially is able to invade other parts of the Baltic Sea and inland waters within its basin. Considering intensive shipping activity in the Neva Estuary, and high densities of *Gmelinoides* in the St.Petersburg harbour area, introduction of this species into the North American Great Lakes and estuarine ecosystems with ballast waters of ships via existing invasion corridor is likely.

1 Introduction

During 1960-1985 intentional introductions were one of the main vectors of aquatic species invasions into inland waters of Siberia and European Russia. Facilitation of fish production was the principal reason for these large-scale introduction efforts (Zadoenko et al. 1985). High productivity and growth potential of the selected species were considered as the most important justifications of species introductions into aquatic ecosystems with "empty" niches. Possible adverse impacts of introduced species, like competition with native species and transfer of parasites, were generally neglected. Studies of biology of the organisms, considered for intentional introductions ("acclimatization"), were specifically focused on the species productivity and, in less extent, on their role in fish diet. Studies of the functional role of these species in the communities, and even studies of their foraging behavior and interactions with other species were lacking. Such examples of large-scale headlong efforts to fill an "empty" niche are well known for North America and Europe (Nesler & Bergersen 1991).

Large-scale introductions of the Baikalian amphipod, *Gmelinoides fasciatus*, in the lakes and reservoirs in European Russia can serve as one more example of headlong intentional introduction with initially unexpected, long-term and generally unwanted consequences. The present paper provides a first attempt to review existing published and, in some extent, author's unpublished data on the *Gmelinoides* invasion history, biology and impacts.

Taxonomy of *Gmelinoides* is described elsewhere (Bazikalova 1945; Panov & Berezina 2001). This species belongs to the gammaridean crustaceans and was first described for

E. Leppäkoski et al. (eds.), Invasive Aquatic Species of Europe, 96–103.

Lake Baikal in 1874 by B. N. Dybowsky as *Gammarus zebra* for its specific striped colour pattern (Panov & Berezina 2001).

2 Invasion history

Gmelinoides fasciatus, a gammaridean amphipod of Baikalian origin, is a common species in freshwater ecosystems in Lake Baikal basin. In Siberia this species is found in Lake Baikal, and also in lakes and rivers in basins of the Angara, Lena, Yenisey, Irtysh, Pyasina, Tunguska, Selenga and Barguzin (Bekman 1962). *Gmelinoides* was considered as one of the most suitable species for intentional introductions aimed to enhance fish production in lakes and water reservoirs, mainly because of its high environmental plasticity and generally high abundances within its native range (Bekman & Bazikalova 1951; Greze 1951). Abilities of *Gmelinoides* to spread within the basins and to compete with native amphipods were neglected.

During the 1960s and 1970s, hundreds of millions of *Gmelinoides* specimens were introduced into 22 lakes and reservoirs outside its native range in Siberia and European Russia (Zadoenko et al. 1985). In the European Russia, *Gmelinoides* was introduced intentionally into Gorkovskoe Reservoir in the Volga River basin (Volkov & Potina 1977; Mordukhai-Boltovskoi & Chirkova 1979), several Karelian Isthmus lakes (Arkhiptseva et al. 1977) and Lake Ilmen (Savateeva 1985) in the Lake Ladoga basin (Fig. 1). Also, this alien amphipod was introduced accidentally in Lake Peipsi during headlong attempts to acclimatize *Gammarus lacustris* from the Siberian population in 1970-1975 (Timm & Timm 1993). *Gmelinoides* was found first in Lake Peipsi already in 1972, and by 1990 it established itself successfully in the whole littoral zone of this lake (Timm & Timm 1993; Panov et al. 2000).

First intentional introductions of *Gmelinoides* in European inland waters were conducted in 1962-1965 in Gorkovskoe Reservoir (Fig. 1), and by 1975 it successfully colonized the whole 434 km long reservoir. In 1977 *Gmelinoides* was first found in Kuibyshevskoe Reservoir (Fig. 1), around 400 km downstream from the site of intentional introduction in the Gorkovskoe Reservoir (Borodich 1979). Upstream movement of *Gmelinoides* took longer, and it was first found in Rybinskoe Reservoir 300 km upstream only in 1986 (Skalskaya 1994). By 1990 *Gmelinoides* colonized the whole Rybinskoe Reservoir and in 1994 was first found in the south part of the Beloe Lake, located in the northernmost part of the Caspian Sea drainage basin (Shcherbina et al. 1997) (Fig. 1).

In the Baltic Sea drainage basin, *Gmelinoides* was first introduced intentionally in 1971 in Lake Otradnoe (Nilova 1976), and between 1971-1975 more than 10 million specimens were introduced into five other Karelian Isthmus lakes, located close to the Lake Ladoga western shore (Fig. 1) (Arkhiptseva et al. 1977). Via rivers, connecting some of these lakes with Lake Ladoga, *Gmelinoides* invaded the Lake Ladoga. In late 1980s it was found established along western and northern shores of the lake (Panov 1996), and by 1996 it successfully colonized the whole littoral zone of this largest European lake (Panov et al. 1999). In the 1990s, via the Neva River, *Gmelinoides* invaded the Neva Bay, freshwater part of the Neva estuary (eastern Gulf of Finland), the largest estuary in the Baltic Sea (Fig. 1). In 1999 *Gmelinoides* was first registered in the inner Neva estuary, the very first record of the Baikalian amphipod in the Gulf of Finland. Results of our field survey in 2000 indicate that *Gmelinoides* established permanent population in the littoral zone of the brackish Neva estuary, almost 30 years after the first intentional introduction of this species into Karelian Isthmus lakes, located around 250 km upstream in the Neva estuary drainage basin.

Also, our field survey of Lake Onega littoral zone in August 2001 showed, that *Gmelinoides* established dense populations along the western shore of this second largest European lake. However, the route of *Gmelinoides* invasion to Lake Onega is not clear, it might be via Svir River, connecting Lake Onega with Lake Ladoga, or the Volga-Baltic waterway, connecting Lake Beloe with Lake Onega (Fig. 1). Thus, at present *Gmelinoides* is widely distributed in the inland waters of the European part of Russia, and also in the Neva estuary (Fig. 1). Present range of *Gmelinoides* is extensive and limited by latitudes 48°-74°N and longitudes 25°-110°E (Panov & Berezina 2001).

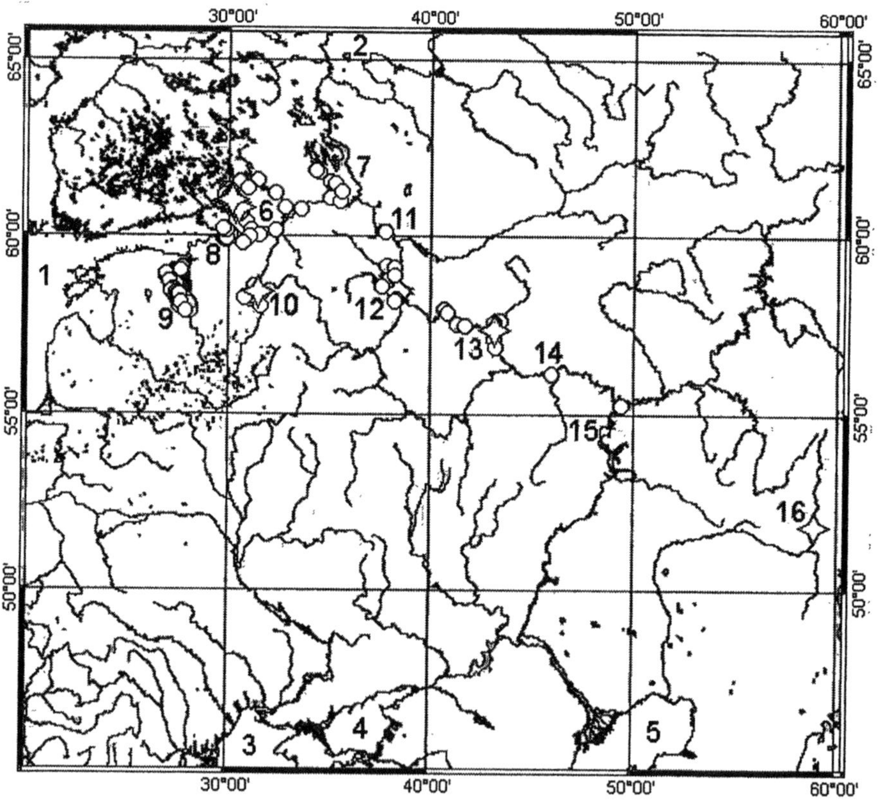

Figure 1. Distribution of *Gmelinoides fasciatus* in Europe. Asterisks indicate sites of intentional introductions, open cycles records of range expansions. (1) Baltic Sea, (2) White Sea, (3) Black Sea, (4) Azov Sea, (5) Caspian Sea, (6) Lake Ladoga, (7) Lake Onega, (8) Neva estuary, (9) Lake Peipsi, (10) Lake Ilmen, (11) Lake Beloe, (12) Rybinskoe Reservoir, (13) Gorkovskoe Reservoir, (14) Cheboksarskoe Reservoir, (15) Kuibyshevskoe Reservoir, (16) Iriklinskoe Reservoir.

3 Biology

3.1 REPRODUCTION

Reproduction in *Gmelinoides* starts early spring in shallow littoral zone at water temperatures 4-5 °C and terminates in late autumn at <10 °C. Fecundity of *Gmelinoides* averages 3-45 eggs per female, depending on female body size. Fecundity is highly variable in different ecosystems, and this likely reflects differences in local feeding conditions. Highest fecundity was found in the eutrophic Neva Bay (Fig. 2).

Our experimental studies showed that egg developmental time (hatching time τ, days) in *Gmelinoides* (period between release of fertilized eggs in brood pouch and release of new born juveniles from the pouch) is strongly temperature-dependent. This relationship at water temperatures (T, °C) range 12-24 °C can be described by equation $\tau = -26.5\ln(T)+87.8$ ($n=45$, $R^2=0.98$). At high temperatures, egg developmental time in *Gmelinoides* is rather short and averages 3-4 days at water temperature 24 °C, while at temperature 12 °C fertilized eggs develop into juveniles as long as in 21-23 days.

Laboratory observations revealed an important feature of *Gmelinoides* reproduction: after copulation females are able to lay fertile eggs in brood pouch immediately after release of developed juveniles, without copulation with males (Nilova 1976; Panov unpubl.). Along with high fecundity and high rates of eggs development at summer, it makes this invasive species enable to fast population growth during favourable conditions.

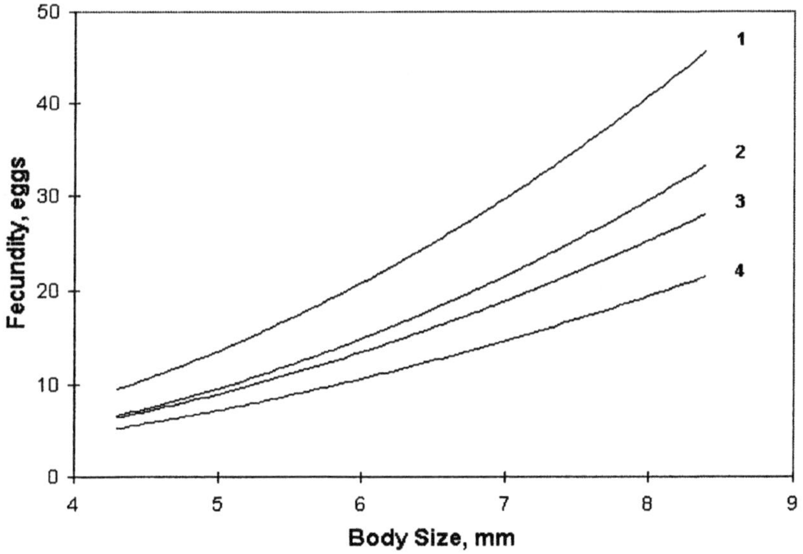

Figure 2. Mean fecundity in females of *Gmelinoides fasciatus*, related to body length: (1) – in the Neva Bay, Gulf of Finland; (2) in Baikal and Ladoga lakes; (3) in Rybinskoe Reservoir and Lake Otradnoe; (4) in Gorkovskoe Reservoir and Lake Ilmen (calculated from Bekman 1962; Mitskevich 1980; Savateeva 1985; Skalskaya 1996; Volkov & Potina 1977; Panov & Berezina unpubl).

3.2 LIFE CYCLE AND ABUNDANCE

Sizes of newborn specimens of *Gmelinoides* ranged from 1.2 to 1.4 mm. Sizes of mature females ranged from 4 to 10 mm in different locations, with smallest fecund females found at higher temperatures. Sizes of mature males ranged from 4-16 mm. As in females, at high temperatures males develop in mature specimens faster and reach maturity at smaller sizes. For instance, in spring, at mean water temperature 11 °C, 6-7 mm is minimum size of mature *Gmelinoides*, while in summer, at mean water temperature 25 °C, the minimum size of mature individuals decreases to 4-5 mm (Vershinin 1967).

For development of newborn amphipods into adults, around 1,000-1,250 day-degrees is needed (around 55-65 days at water temperature 18.5 °C). Because strong temperature effect on the development, number of generations per season is temperature-dependent and ranges from 1-3 (Fig. 3), with 1 generation in aquatic habitats with summer day-degrees < 1,200 (Lake Baikal littoral zone), 2 generations in locations with 1,500-2,000 day-degrees (Lake Ladoga, Lake Peipsi and Lake Otradnoe), and 3 generations in locations with > 2,200 day-degrees during season (Upper Volga reservoirs).

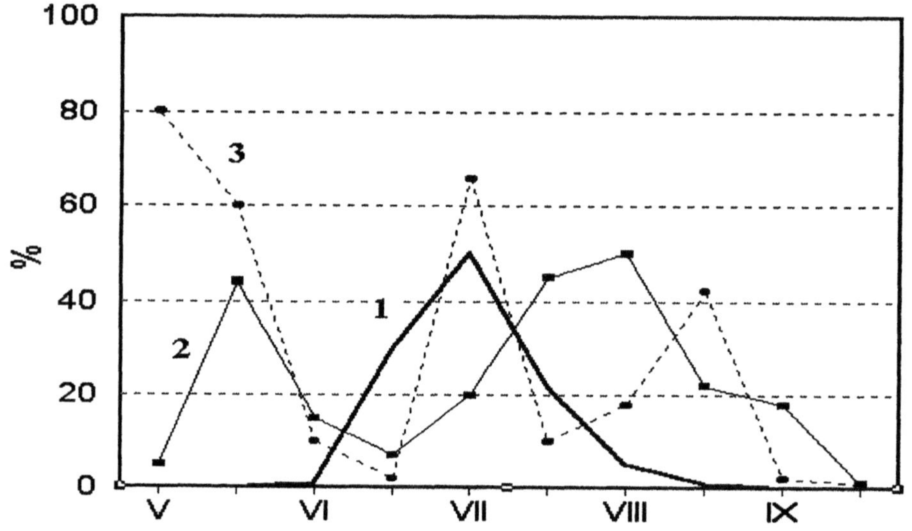

Figure 3. Ratio (%) of immature specimens (body length 1.5-3.0 mm) in *Gmelinoides* populations during season in Lake Baikal (1), Lake Otradnoe (2), and Rybinskoe Reservoir (3) (calculated from Bekman 1962; Nilova 1976; Skalskaya 1994, 1996).

Life cycles in *Gmelinoides* can be of 4 different types, varying in the amphipod life duration and number of generations per season. First, a three-year life cycle with 1 generation per season and newborn amphipods developing into mature amphipods in 17-18 months (Lake Baikal) (Bazikalova 1945; Bekman 1962). Second, a one-year life cycle with 2 generations per season, 1 summer peak in abundance and 2-3 sequential breeds in females (lakes and reservoirs in Lake Baikal basin) (Bekman & Bazikalova 1951; Bekman 1962; Mekhanikova 1982). Third, a one-year life cycle with 2 summer peaks in abundance and 6-8 sequential breeds in females (Lake Ladoga and Lake Otradnoe,

Neva Bay) (Nilova 1976; Zadoenko et al. 1986; Berezina & Panov unpubl). Fourth, a one-year life cycle with 3 summer peaks in abundance and 8-10 sequential breeds in females (Upper Volga reservoirs) (Volkov & Potina 1977; Skalskaya 1996).

During reproduction period, most part of *Gmelinoides* population concentrates in shallow near shore habitats. Maximal abundances of this amphipod during summer were found at depths less than 2 m (Table 1). During winter period, abundance of *Gmelinoides* decreases 4-5 fold (Mekhanikova 1982).

Table 1. Maximal abundances and biomasses of *Gmelinoides* in different locations.

Location	Abundance, ind m^{-2}	Biomass, g m^{-2}	Reference
Lake Baikal	10,000-20,000	63-100	Bekman & Bazikalova 1951; Bekman 1962
Lake Ladoga	37,000-54,000	121-160	Panov 1996
Lake Otradnoe	1,600	4,3-4,7	Mitskevich 1980
Lake Peipsi and Narva River	14,000-29,500	57-102	Timm & Timm 1993; Panov et al. 2000
Rybinskoe Reservoir	6,800	19,8	Skalskaya 1994, 1998
Gorkovskoe Reservoir	15,000	66	Volkov & Potina 1977

3.3 FEEDING

Gmelinoides is a nektobenthic omnivorous amphipod, and able to prey effectively upon small benthic and zooplankton organisms. Some authors, who suggested *Gmelinoides* as object for intentional introductions, considered this amphipod as mainly detritivorous animal, feeding on detritus, filamentous and diatom algae, and consuming animal food only accidentally (Nilova 1976). However, other authors found *Gmelinoides* to be an active predator, effectively foraging on small zooplankton crustaceans (Bekman 1962; Panov 1996).

Our study of gut content in *Gmelinoides* revealed significant difference in food composition for small (5-7 mm) and large (> 9 mm) amphipods. Filamentous, diatom and unicellular green algae along with remnants of small oligochaetes comprised the main part of gut content in small amphipods. In large amphipods, guts were empty or filled with mixture of algae and remnants of oligochaetes, chironomids and caddis flies, with dominance of animal food (up to 80% of gut content biomass) (Berezina unpubl.).

Quantitative estimates of food selectivity, feeding rates, functional response and food assimilation rates for *Gmelinoides* are not present in the literature and need further study.

3.4 EFFECTS OF ENVIRONMENTAL FACTORS

Gmelinoides is able to survive at variable range of temperatures, oxygen content in water and water mineralization (total content of salts in water). However, soft water with calcium content less than 5-7 mg l^{-1} and low pH (< 6.0) terminate normal moulting in *Gmelinoides* and can be limiting factors (Bekman 1962; Berezina unpubl). *Gmelinoides* colonizes a wide range of bottom substrates and macrophytes, reaching highest abundances in shallow waters (< 2 m). *Gmelinoides* tolerate moderate pollution, and were found among the first invertebrates re-colonizing previously lifeless location of pulp-mill discharges in Lake Ladoga (Panov 1996). However, this amphipod avoids localities with heavy toxic pollution and sites impacted by warm water from power plants in Volga River reservoirs (Skalskaya 1998).

In experiments on salinity resistance in *Gmelinoides*, amphipods tolerated increase in salinity up to 5-7 PSU (at constant temperature 18 °C). Successful reproduction in *Gmelinoides* occurred at salinities from 0.06 to 2 PSU, but development of eggs in fecund females terminated at salinities > 2 PSU. At experimental water salinity 2 PSU, mortality of eggs was significant (60-70%). Most likely expansion of *Gmelinoides* in brackish waters will be limited to salinities < 2 PSU (Berezina et al. 2001).

Also, experimental studies show that survival in *Gmelinoides* at salinities 2-5 PSU decreased significantly with increase in water temperature. At temperatures 22-26 °C, adult amphipods were not able to acclimate to salinities > 2 PSU, and their survival rates ranged from 0 to 30%. Maximum resistance (50%) of *Gmelinoides* to salinity 5 PSU occurred at the lowest experimental temperatures of 12-14 °C. Most likely high water temperatures will limit distribution of *Gmelinoides* in brackish waters (Verbitsky & Berezina 2002).

4 Impacts

In both Lake Ladoga and Neva Bay *Gmelinoides* almost completely replaced the native amphipod, *Gammarus lacustris*. From Gorkovskoe Reservoir, the site of its intentional introduction in the River Volga basin, *Gmelinoides* spread both to upstream and downstream reservoirs, and to some rivers and lakes in the basin. Drastic decline in Caspian amphipod populations was observed in Gorkovskoe Reservoir after introduction of *Gmelinoides* (Mordukhai-Boltovskoi & Chirkova 1971). In Lake Peipsi *Gmelinoides* was first found in 1972, and established itself successfully in the littoral zone of the lake by 1990, replacing completely the native population of *Gammarus lacustris* along with possible successors of Siberian specimens (Panov et al. 2000). Significant negative impact of *Gmelinoides* on local populations of other amphipod species was also found in Lake Baikal basin (Saphronov 1993). For instance, after construction of reservoirs on Angara River, rich fauna of other Baikalian amphipods, formerly characteristic for the river, was almost completely replaced by *Gmelinoides* (Saphronov & Erbaeva 1998). At present *Gmelinoides* is actively replacing local populations of *Gammarus lacustris* in lakes in Lake Baikal basin (Erbaeva pers. comm.), likely as a result of increasing anthropogenic disturbance of their ecosystems.

Mechanisms of replacement of other amphipod species by *Gmelinoides* are not clear and require further study. Most likely this is a result of predation by *Gmelinoides* on juveniles of other species. Predation could be also a main reason of negative impact of *Gmelinoides* on isopod *Asellus aquaticus* in Lake Ladoga littoral zone (Panov unpubl.).

5 Conclusions

Large-scale intentional introductions of *Gmelinoides* into inland waters of European Russia, conducted between 1960-1975, resulted in significant range expansion in this species. By 2000, *Gmelinoides* successfully established in the Upper Volga basin, in

Lake Ladoga, Lake Onega, Lake Peipsi, Lake Ilmen and their basins, and invaded the largest Baltic estuary. In most cases the native amphipods were completely replaced by their aggressive Baikalian congener, and negative impact on other aquatic invertebrate species is likely because predation by *Gmelinoides*. However, mechanisms of effective replacement of relative species by *Gmelinoides* are not clear and require further study.

Recent studies of *Gmelinoides* biology, including experiments on its salinity and temperature resistance, showed that this invasive Baikalian amphipod is potentially able to invade other parts of the Baltic Sea and inland waters within its basin. Considering intensive shipping activity in the Neva Estuary, and high densities of *Gmelinoides* in the St. Petersburg harbour area, invasion of this species to the North American Great Lakes and estuarine ecosystems via existing invasion corridor is likely.

Acknowledgements

Present study was supported by the Russian State Program "Biodiversity", Maj and Tor Nessling Foundation and INTAS grant 99-00674.

PONTO-CASPIAN AMPHIPODS AND MYSIDS IN THE INLAND WATERS OF LITHUANIA: HISTORY OF INTRODUCTION, CURRENT DISTRIBUTION AND RELATIONS WITH NATIVE MALACOSTRACANS

KESTUTIS ARBACIAUSKAS
Institute of Ecology, Vilnius, Lithuania
arbas@ekoi.lt

Abstract

Ponto-Caspian amphipods *Pontogammarus robustoides*, *Obesogammarus crassus* and *Chaetogammarus warpachowskyi*, and mysids *Paramysis lacustris*, *Limnomysis benedeni* and *Hemimysis anomala* were introduced into the Kaunas Water Reservoir located on River Nemunas, during 1960-1961. The crustaceans established sustainable populations there. Later transplanting of non-native species to over 100 lakes and two other water reservoirs was tried. Recently, sustainable populations of Ponto-Caspian amphipods and mysids inhabit these two water reservoirs. Non-native amphipods and mysids were also found in 6 and 8 lakes, correspondingly. Current distribution suggests that Ponto-Caspian amphipods can establish sustainable populations only in large mesotrophic lakes, while mysids can also survive in large and open eutrophic lakes. Both groups of these crustaceans can also live in well-drained lakes. In the habitats suitable for Ponto-Caspian amphipod *P. robustoides*, this amphipod excludes the native species *Gammarus lacustris* and *G. pulex*. There are also indicators suggesting that Ponto-Caspian mysid *P. lacustris* could have a negative effect on the native glacial relict mysid *Mysis relicta*.

1 Introduction

Ponto-Caspian crustaceans are among the most successful invaders of aquatic environments in parts of Europe and North America. The predatory onychopod *Cercopagis pengoi* recently invaded Lake Ontario and the Baltic (MacIsaac et al. 1998). The mysid *Limnomysis benedeni* reached River Rhine basin (Reinhold & Tittizer 1998). Another mysid species *Hemimysis anomala* was found in the Baltic coastal waters of SW Finland and in the inland waters of the Netherlands (Salemaa & Hietalahti 1993; Ketelaars et al. 1999). At present Ponto-Caspian amphipods are common in Polish waters.

Dispersal of Ponto-Caspian mysids and amphipods in the drainage basin of the Baltic Sea is at least in part associated with introduction of these animals into adjacent freshwater bodies. In the 1960s, these crustaceans were widely used in the former Soviet Union for introduction into various water bodies with intention to improve the feeding conditions for fish. Lithuania was the first place in the Baltic Sea basin, where Ponto-Caspian species were introduced 40 years ago.

The purpose of this work was to review the history of introduction and current distribution of these crustaceans in the inland waters of Lithuania. Another goal was to test the hypothesis about environmental factors limiting the establishment of sustainable populations of introduced Ponto-Caspian amphipods and mysids in lake ecosystems. The inland waters of Lithuania are inhabited also by native species of amphipods. *Gammarus lacustris* is frequent in lakes, and *G. pulex* occurs in flowing waters. For mysids, *Mysis relicta* still lives in large and deep lakes in Lithuania. Thus, the last goal was to analyse the relations of introduced Ponto-Caspian amphipods and mysids in

E. Leppäkoski et al. (eds.), Invasive Aquatic Species of Europe, 104–115.

analyse the relations of introduced Ponto-Caspian amphipods and mysids in relation to native species.

2 History of introduction of Ponto-Caspian amphipods and mysids

The Ponto-Caspian amphipod *Corophium curvispinum* invaded River Nemunas via the channel, which connected Dniepr and Nemunas rivers in 1768, and was first recorded in Lithuania in 1921 (Gasiunas 1963). Another amphipod species *Chaetogammarus ischnus* was found in the lower Nemunas in 1960 (Gasiunas 1963). Other Ponto-Caspian higher crustaceans recently inhabiting Lithuanian waters were introduced intentionally. In 1959, River Nemunas was dammed up, and Kaunas Water Reservoir (WR) formed (WR of a hydroelectric power plant). In May and October 1960, two batches of Ponto-Caspian amphipods and mysids were transferred from Dniepr WR to Kaunas WR. The third batch came from Simferopol WR (the Crimea) in May 1961. Six species were introduced: the mysids *Paramysis lacustris* (some 4,400 individuals), *Limnomysis benedeni* (1,600) and *Hemimysis anomala* (came only from Dniepr WR, 3,500); and the amphipods *Pontogammarus robustoides, Obesogammarus (Pontogammarus) crassus* and *Chaetogammarus warpachowskyi* (1,600, all together) (Gasiunas 1972a). Gasiunas (1972a) also refers to *P. robustoides aestuarius*. According to other authors (e.g. Dedyu 1980) this subspecies should be considered as a variation, rather than a subspecies.

The establishment of populations of mysids *P. lacustris* and *L. benedeni*, and all amphipod species was documented in 1961. During a few years these Ponto-Caspian crustaceans spread over the whole Kaunas WR. *P. lacustris* exhibited the highest dispersion abilities. To the end of 1963, this mysid colonized all Kaunas WR and was encountered in different places of River Nemunas below Kaunas WR (Gasiunas 1972a). The mysid *H. anomala* was first recorded in 1967, in the lower part of Kaunas WR at 18-20 m depth. Later this species was also found in Kaunas WR, in the bed of River Nemunas, up to 50 km above the dam (Gasiunas 1968, 1972a).

After the establishment of sustainable populations of Ponto-Caspian mysids and amphipods in Kaunas WR, animals were introduced from here into other Lithuanian inland waters (e.g. the Curonian Lagoon in 1963). In the end of 1963, it was observed that introduced crustaceans reached this lagoon naturally via River Nemunas, and introduction was terminated. *P. lacustris, L. benedeni, P. robustoides, O. crassus* and *C. warpachowskyi* had successfully acclimatized in the Curonian Lagoon. The mysid *H. anomala* has also been encountered in this water body (Razinkovas 1990), but permanent population of this species was not documented.

During 1963-1989 Ponto-Caspian mysids and amphipods, and during 1995-1998 only amphipods, were transferred into two water reservoirs and 103 lakes in Lithuania (Lazauskiene 1997). In later years, animals for introduction were taken not only from Kaunas WR, but also from the Curonian Lagoon. The establishment of populations of Ponto-Caspian mysids and amphipods was documented by Gasiunas (1972a,b) for both water reservoirs and 10 lakes (Table 1). Later, successful introduction of these animals was reported for another 22 lakes (Lazauskiene 1997).

Table 1. Occurrence of Ponto-Caspian mysids and amphipods in the inland waters of Lithuania. Included those water bodies, which were referred by Gasiunas, and in which non-native crustaceans were found recently. Trophic status (Kavaliauskiene 1996): m-o (mesotrophic with traits of oligotrophy); m-e (mesotrophic with different degrees of eutrophication); eu (eutrophic). Occurrence: (1) 1964-1970 (Gasiunas 1972); (2) around 1989-1992 (Lazauskiene 1997); (3) 1998-2000 (current studies).

Water body	Area, ha	H_{max}, m	Trophic status	Species	Introduction years	Occurrence 1	2	3
Antaliepte WR	1,911	46	m-o	*P. lacustris*	1963-1967	+	+	+
				P. robustoides		+		+
				O. crassus		+	+	
				C. warpachowskyi		+	+	
Elektrenai WR	1,250	32	m-e	*P. lacustris*	1968	+	+	+
				L. benedeni		+	+	+
				P. robustoides		+	+	+
				O. crassus		+	+	+
				C. warpachowskyi		+	+	+
Dusia	1,332	32	m-e	*P. lacustris*	1963-1964	+	+	+
				P. robustoides		+	+	+
				O. crassus		+	+	+
				C. warpachowskyi		+	+	+
Metelys	1,286	15	m-e	*P. lacustris*	1963-1964	+	+	+
				P. robustoides		+	+	+
				O. crassus			+	+
				C. warpachowskyi			+	+
Obelija[a]	584	11	eu	*P. lacustris*	1963-1964	+	+	
				P. robustoides		+	+	
Simnas	243	4	eu	*P. lacustris*	1963-1964	+	+	
				L. benedeni		+	+	+
				P. robustoides		+	+	+
				O. crassus		+	+	
				C. warpachowskyi		+	+	
Dysnai	2,439	6	eu	*P. lacustris*	1964-1967	+	+	+
				P. robustoides		+	+	
				O. crassus			+	
				C. warpachowskyi			+	
Sartai	1,332	21	eu	*P. lacustris*	1968	+	+	+
				P. robustoides		+	+	
Daugai	954	44	m-e	*P. lacustris*	1965 and	+	+	
				L. benedeni	1970-1972			+
				P. robustoides		+	+	+
				O. crassus		+		+
				C. warpachowskyi				+
Arinas[b]	375	18	eu	*P. lacustris*	1966-1968	+	+	
				P. robustoides		+	+	
Plateliai	1,204	46	m-e	*P. robustoides*	1964-1968	+	+	+
				O. crassus			+	
				C. warpachowskyi		+		
Kretuonas[a]	880	11	eu	*P. lacustris*	1966	+		
Zeimenys	455	24	m-e	*P. lacustris*	1965		+	+
				C. warpachowskyi			+	+
Dysnykstis	538	5	eu	*P. lacustris*	1985		+	+
				amphipods			+	

[a] during 1998-2000 not studied
[b] during 1998-2000 samples for mysids not taken

During the bout of introductions of aquatic invertebrates in the former Soviet Union, in 1964-1969, Ponto-Caspian mysids and amphipods were taken from Kaunas WR also for introduction in regions located to the north from Lithuania, in particular Latvia, Estonia and the St.Petersburg region. Successful introductions were documented only for Latvia (Gasiunas 1972a).

3 Factors for survival of Ponto-Caspian amphipods and mysids in lake ecosystems

In their native region, the mysids *P. lacustris, L. benedeni,* and the amphipods *P. robustoides, O. crassus* and *C. warpachowskyi* inhabit the lower reaches of the rivers, deltas, estuaries and lagoons (Dedyu 1967). At least low current velocities can always be observed in these waters. Hence, these mysids and amphipods are ascribed as potamophilic organisms (Dedyu 1980; Komarova 1991). All these crustaceans are also considered to be oxyphilic animals. However, the mysids *P. lacustris* and *L. benedeni* can tolerate a significant decrease of oxygen concentration and, hence, could be characterized as euryoxic organisms (Komarova 1991). The demands for water flow and high oxygen are most likely interrelated. Consequently, the principal factors for the survival of Ponto-Caspian amphipods in Lithuanian lakes are water motion interrelated to sufficient oxygen concentration in the littoral. Mysids also require water motion and high oxygen, but at a lower level than amphipods. Hence, non-native Ponto-Caspian crustaceans should be able to establish sustainable populations only in large and open lakes with strong wind-born water motion in the littoral, or in well-drained lakes.

4 Material and methods

During 1998-2000, usually in August - September, three water reservoirs (Kaunas, Elektrenai and Antaliepte) and 29 lakes ranging in area from 12 to 2,400 ha were studied for the presence of Ponto-Caspian mysids and amphipods. Sampling was performed with a sled net and dredge. These equipments were usually dragged a fixed distance, and numbers of caught animals were used to assess their relative density. Estimates of relative density (i.e. means for the two richest catches in the littoral of a certain water body) were utilized to compare study sites for the abundance of mysids.

The abundance of littoral amphipods (*P. robustoides* and *O. crassus*) was assessed using the relative frequency index (*RFI*). *RFI* indicates the number of individuals counted over 10 min, when searching in substrates where amphipods should live. Substrates (detritus, macrophytes, roots, etc.) were collected in approximately equal volume portions and analysed. Animals were picked one after another and counted. *RFI* (ind·10 min^{-1}) was calculated as follows:

$$RFI = \frac{N}{t-(t_h \times N)} \times 10,$$

where N is the number of counted individuals; t is the total time of sample collection and analysis in min; t_h is the mean handling time of one individual in min. Handling time changed dependent on amphipod abundance. Based on experimental measurements, the relationship between t_h and frequency of amphipods, which was assessed as

$f = N \times t^{-1}$, was determined $t_h = (4.79 + \exp(2.05 - 0.29 \times f)) \times 60^{-1}$. Presented values of *RFI* are means of all estimates for a certain water body.

5 Results

5.1 DISTRIBUTION AND SPECIES COMPOSITION

The location and main characteristics of water bodies inhabited by non-native mysids and amphipods are given in Fig. 1 and Table 1. All species of Ponto-Caspian mysids (*P. lacustris, L. benedeni* and *H. anomala*) and amphipods (*P. robustoides, O. crassus* and *C. warpachowskyi*) were found in Kaunas WR. For other water reservoirs and 8 of 10 lakes, in which acclimatization of non-native mysids and amphipods was documented by Gasiunas (1972a,b) these crustaceans were found in all water bodies, excluding Lake Arinas. When comparing with earlier records, the same species composition was detected only for Elektrenai WR and Lake Dusia (Table 1). Changes for mysids were noted in two lakes. Recently, only a few specimens of *L. benedeni* were caught in comparatively small but well-drained Lake Simnas. For Lake Daugai, a population of *L. benedeni* was detected in this lake, but mysid *P. lacustris* was not found.

Much larger changes were established for amphipods (Table 1). As compared with the 1960s, more species of amphipods were found in lakes Metelys and Daugai. In Antaliepte WR and lakes Simnas and Plateliai only *P. robustoides* was detected. Ponto-Caspian amphipods have recently been absent in lakes Dysnai, Sartai and Arinas.

Another 14 lakes studied for the presence of non-native species were selected from 22 lakes, in which acclimatization of these animals was documented by Lazauskiene (1997). All lakes were investigated for littoral amphipods and three of them for mysids. Ponto-Caspian crustaceans were found in two lakes. Lake Dysnykstis was inhabited by mysid *P. lacustris*. In Lake Zeimenys, the same mysid species and amphipod *C. warpachowskyi* was found (Table 1).

Finally, 7 lakes were studied for which information on introduction results is absent, or introduction of amphipods was performed during 1995-1998. Ponto-Caspian crustaceans were detected only in Lake Asveja (area 1015 ha, maximum depth 50 m).

Large batches of non-native amphipods were released into this lake during 1995-1997. In August 1998, Ponto-Caspian amphipods were absent. The last batch of amphipods was transferred from Kaunas WR into this lake in September, 1998. In the next year non-native amphipods were found in the place of their release (*RFI*=68 ind$\cdot$10 min^{-1}). The most of collected individuals were *P. robustoides*, and only 3% were identified as *O. crassus*. In 2000, the density of Ponto-Caspian amphipods significantly decreased (*RFI*=2.3 ind$\cdot$10 min^{-1}, in August; and 0.3 ind$\cdot$10 min^{-1}, in September). All non-native amphipods found were *P. robustoides*.

5.2 ABUNDANCE

The most numerous population of *L. benedeni* was found in Elektrenai WR (Fig. 2). In Kaunas WR and Lake Daugai densities of this species were lower, and the lowest in Lake Simnas. To collate the abundance between populations, estimates of density of mysids in Elektrenai WR and Lake Daugai excluding neonates are plotted in Fig. 2. The most of *L. benedeni* at the study period (the end of August) in these waters were new-born individuals. The total relative density of this mysid in the littoral of Elektrenai WR was estimated at 44 ind·m^{-2}. Among macrophytes abundance was up to 80 ind·m^{-2}. In Lake Daugai total density was 12 ind·m^{-2}.

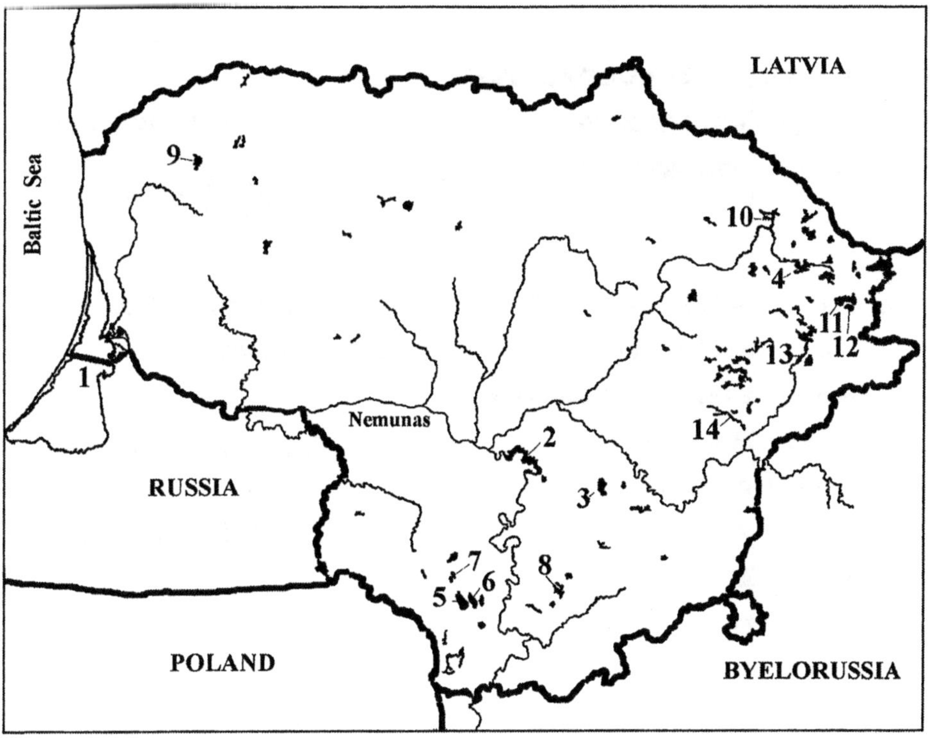

Figure 1. Lithuanian waters inhabited by Ponto-Caspian mysids and amphipods: (1) Curonian Lagoon; (2) Kaunas WR; (3) Elektrenai WR; (4) Antaliepte WR; and lakes (5) Dusia, (6) Metelys, (7) Simnas, (8) Daugai, (9) Plateliai, (10) Sartai, (11) Dysnai, (12) Dysnykstis, (13) Zeimenys, (14) Asveja.

For the mysid *P. lacustris*, the highest densities were determined in Kaunas WR and Lake Dusia. A population twice as less abundant was found in Lake Zeimenys, and more than 4 times lower relative densities were observed in other waters inhabited by this mysid species (Fig. 2). The relative abundance of the third non-native mysid species *H. anomala,* in the lower part of Kaunas WR at 10-12 m depth, comprised around 1 ind·m^{-2}.

Relative frequencies of littoral amphipods (*P. robustoides* and *O. crassus*) are illustrated in Fig. 3. In all studied places *P. robustoides* was obviously the dominant species. In waters inhabited by *O. crassus*, this species usually comprised 1-4% of collected specimens among littoral amphipods. The part of *O. crassus* was higher (22%) only in Lake Metelys. The highest frequencies for *P. robustoides* were found in Kaunas WR and

Elektrenai WR. Close estimates were obtained for lakes Metelys and Dusia. In lakes Daugai and Plateliai, and Antaliepte WR abundance of *P. robustoides* was 2-3 times smaller, and the lowest was in Lake Simnas. The third non-native amphipod species *C. warpachowskyi* in the studied water reservoirs and lakes did not occur among substrates collected for the assessment of *RFI*. Relative densities of this amphipod estimated with dredge were around 1-4 ind·m^{-2}.

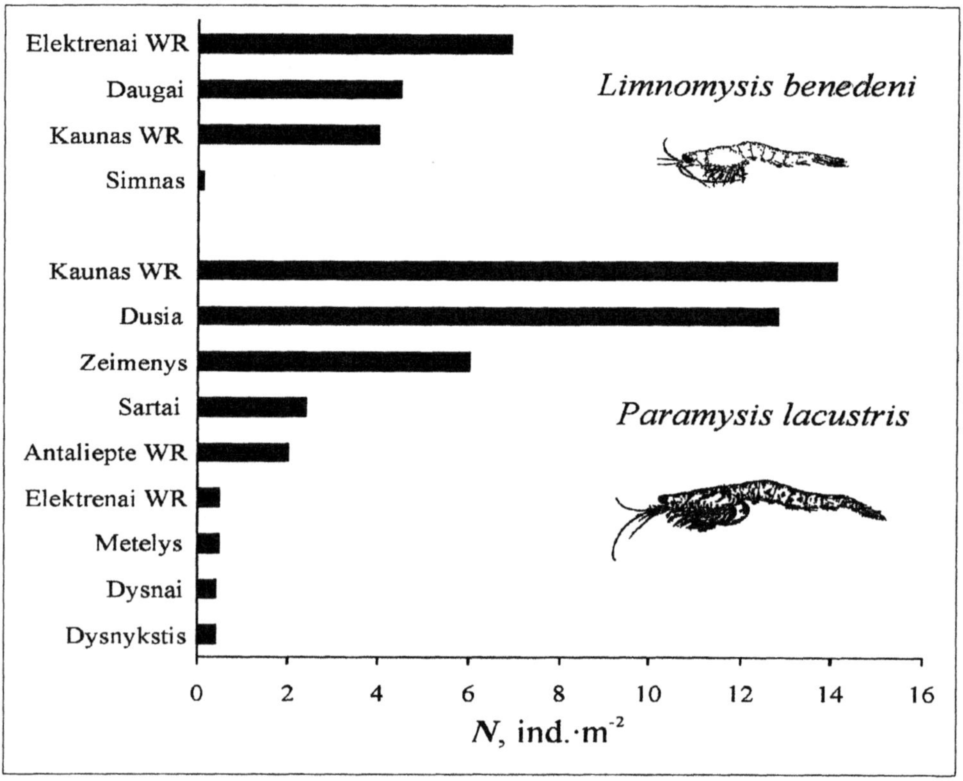

Figure 2. Relative density (*N*) of Ponto-Caspian mysids *Limnomysis benedeni* and *Paramysis lacustris* in the inland waters of Lithuania.

5.3 NATIVE SPECIES

For lakes free from *P. robustoides*, native amphipod *G. lacustris* was found in all, excluding two, studied lakes. Otherwise, native amphipods were usually absent in waters inhabited by *P. robustoides*. Native species were not found in the samples from Kaunas WR, Elektrenai WR and lakes Dusia, Metelys, Daugai and Simnas. Native amphipod *G. lacustris* was also undetected in the open parts of Lake Plateliai and Antaliepte WR, which are occupied by *P. robustoides*. In these water bodies *G. lacustris* was found only in places free from non-native amphipods, in the closed bay located in the southern part of Lake Plateliai (*RFI*=12 ind·10 min^{-1}), and bays of the middle part of Antaliepte WR (5 ind·10 min^{-1}).

Interesting case of relations between non-native and native species was observed in Lake Asveja, in which *G. lacustris* is frequent. In 1999, a year after the last introduction of Ponto-Caspian amphipods, non-native and native amphipods were found actually in the same place, but in different substrates. *P. robustoides* occupied flooded roots of land grasses, substrates that a year before were inhabited by native species, while *G. lacustris* occurred among submerged macrophytes and detritus.

Estimates of relative frequency even for the most numerous studied populations of *G. lacustris* indicate that abundance of native amphipod usually was significantly lower in comparison with that for Ponto-Caspian amphipod *P. robustoides* (Fig. 3). Only two open lakes with a wide littoral, Lake Nedingis and Lake Dysnykstis, were inhabited with populations of native amphipod, which densities are comparable with that for *P. robustoides*.

Among studied lakes Lake Dusia is unique in the aspect that glacial relict mysid *M. relicta* and Ponto-Caspian mysid *P. lacustris* inhabit this lake. In the end of August 1999, during the sampling for Ponto-Caspian crustaceans only one individual of *M. relicta* was caught at 13 m depth. At the same time, at 12 m depth, 2 vertical net hauls were performed at night searching for information on densities of the relict mysid in Lake Dusia. One haul was without mysids, but another one provided estimate of mysid density comprising 60 ind·m^{-2}. However, all caught individuals were non-native mysids. In September 2000, qualitative samples were collected in Lake Dusia from the 10-19 m depth. The most of the caught mysids were Ponto-Caspian mysid *P. lacustris*. Among 70 analysed individuals only two specimens were *M. relicta*.

6 Discussion

Non-native Ponto-Caspian mysids (*P. lacustris, L. benedeni* and *H. anomala*) and amphipods (*P. robustoides, O. crassus* and *C. warpachowskyi*) appeared in the inland waters of Lithuania due to introduction of these crustaceans into Kaunas WR forty years ago, and their later transference into other waters. At present, all species of introduced Ponto-Caspian mysids and amphipods live only in Kaunas WR. Five species of non-native crustaceans, excluding *H. anomala*, were detected in Elektrenai WR. It is noteworthy, that *H. anomala* was also transferred to this water reservoir, but information on results of introduction are absent. The number of Ponto-Caspian species found in one water body for studied lakes and Antaliepte WR was smaller (Table 1).

The mysid *L. benedeni* was encountered in Kaunas WR, Elektrenai WR and two lakes, Simnas and Daugai. Small well-drained Lake Simnas was frequently quoted as an example of a small lake, in which five introduced Ponto-Caspian species successfully acclimatized (Gasiunas 1972a). During 1999-2000, it was heavily polluted with untreated wastewaters of town Simnas, and received a hypertrophic water body status. Most likely, this was the main reason that only a few individuals of Ponto-Caspian mysids and amphipods were caught in this lake. It is quite possible that in the result of pollution Ponto-Caspian species will disappear from this lake. The occurrence of *L. benedeni* in Lake Daugai is rather unexpected result of this study. This mysid species had been transferred to various lakes, however, sustainable populations, excluding Lake Simnas, were not always established. Hence, it was supposed that *L. benedeni* is less tolerant to lake environment in comparison with the other non-native mysid *P. lacustris*. Earlier, *P. lacustris* was also reported for Lake Daugai. At present this mysid species was not found in this lake, however the final conclusion about *P. lacustris* in Lake Daugai warrants additional studies.

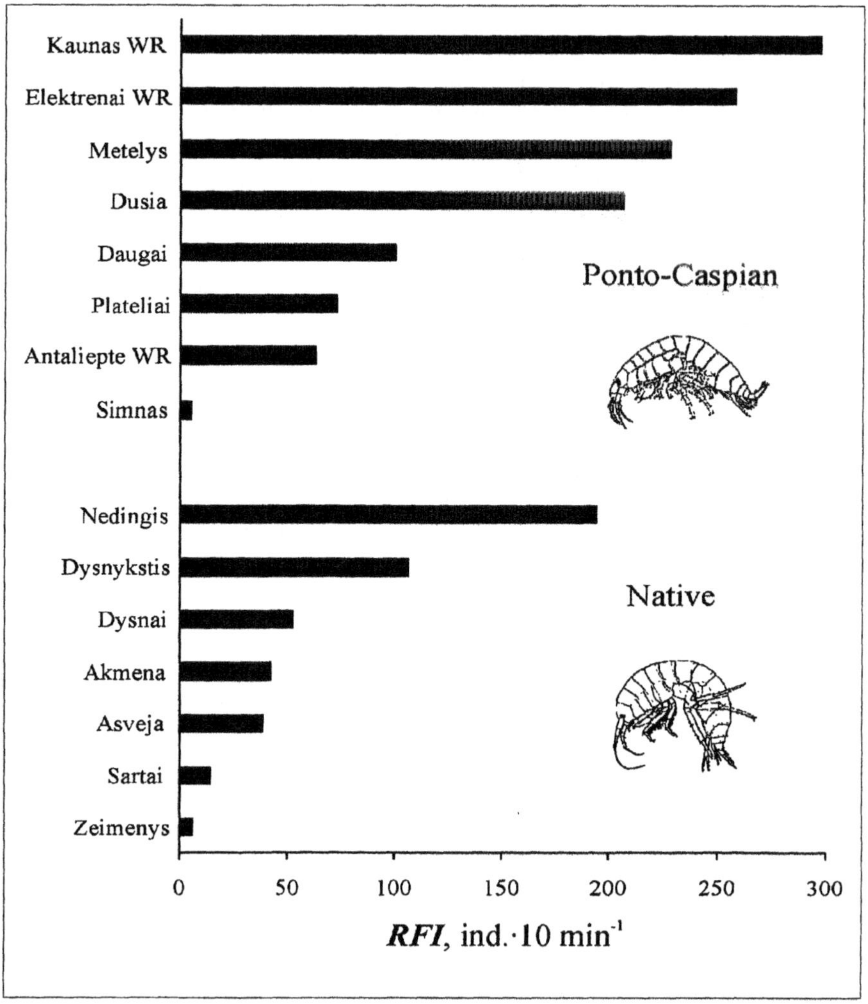

Figure 3. Relative frequency (*RFI*) of Ponto-Caspian littoral amphipods (*Pontogammarus robustoides* and *Obesogammarus crassus*) and native amphipod *Gammarus lacustris* in the inland waters of Lithuania.

In comparison with *L. benedeni*, the mysid *P. lacustris* has wider distribution in the inland waters of Lithuania. This non-native mysid was found in all studied water reservoirs and six lakes, Dusia, Metelys, Dysnai, Dysnykstis, Sartai and Zeimenys. Most likely this species occurs also in a few other lakes of Lithuania.

Presently, the Ponto-Caspian amphipod *P. robustoides* inhabits all studied water reservoirs and 5 lakes, Dusia, Metelys, Daugai, Plateliai and Simnas. Another non-native amphipods *O. crassus* and *C. warpachowskyi* were found together with *P. robustoides* in 2 water reservoirs and 3 lakes (Table 1). *C. warpachowskyi* was also detected in well-drained Lake Zeimenys. In this lake it is the only introduced species of Ponto-Caspian amphipods. Ponto-Caspian amphipods were not found in lakes Dysnai, Sartai and Arinas, in which the establishment of populations of non-native amphipods was reported by Gasiunas (1972a). For Lake Asveja, into which Ponto-Caspian amphipods were transferred during the last few years, the question about the establishment of population of *P. robustoides* is open. According to collected data, abundance of this amphipod in the place of its release had dramatically dropped by the end of 2000.

Amphipod *P. robustoides* showed the widest distribution among non-native amphipods in the inland waters of Lithuania. This species has higher tolerance to oxygen deficiency in comparison with other species (Dedyu 1980), and hence should be better adapted to lake environments than other non-native amphipods.

The most numerous populations of Ponto-Caspian mysids inhabit Kaunas WR, Elektrenai WR, and lakes Dusia, Daugai and Zeimenys (Fig. 2). Real densities of non-native mysids should be a few or even more times higher than estimated. For Kaunas WR reported densities of mysids *P. lacustris* and *L. benedeni* in various biotopes and depths fluctuated between 16-123 and 10-35 ind·m^{-2}, correspondingly. The same characteristic for *P. lacustris* in Lake Dusia comprised 30-112 ind·m^{-2} (Vaitonis 1994). The largest frequencies of introduced littoral amphipods dominated by *P. robustoides* were observed in the same water reservoirs, and lakes Dusia and Metelys. The Ponto-Caspian amphipod *C. warpachowskyi* in the inland waters of Lithuania use to live in the 1-8 m depth. Its documented densities in Kaunas WR and Lake Dusia comprise around 10-50 ind·m^{-2} (Vaitonis 1994). Estimates of relative frequencies for non-native and native amphipods also indicate that abundance of populations of *G. lacustris* is usually lower than that for Ponto-Caspian amphipod *P. robustoides* (Fig. 3).

Distribution data of Ponto-Caspian mysids and amphipods provide the possibility to test the hypothesis about factors limiting the establishment of sustainable populations of non-native crustaceans in lake ecosystems. The hypothesis predicts that non-native species could survive only in large and open lakes with strong wind-born water motion in littoral, or in well-drained lakes. All lakes in which Ponto-Caspian amphipods and mysids were found during this study or were reported earlier, are large and open lakes (Table 1), or well-drained lakes (water retention for lakes Simnas and Zeimenys are around 0.8 per year). This hypothesis explains well data on current distribution of mysid *P. lacustris* in the lakes of Lithuania. However, the large size and openness of a lake alone are insufficient for survival of the non-native amphipod *P. robustoides*.

All lakes, excluding well-drained Lake Simnas, in which Ponto-Caspian amphipods were detected during this study are ascribed to mesotrophic lakes. Lakes in which *P. robustoides* had retreated are eutrophic lakes (Table 1). Hence, the establishment of sustainable populations of Ponto-Caspian amphipods in eutrophic lakes is limited. It is very probable that the limiting factor for long-term survival of non-native amphipod *P. robustoides* in eutrophic lakes can be low oxygen concentration in the littoral during winter, i.e. under the ice cover. Data for Lake Plateliai and Antaliepte WR support this hypothesis. In these water bodies the native amphipod *G. lacustris*, which could tolerate lower oxygen concentrations than that for *P. robustoides*, occurs in bays, while non-native amphipod inhabits open parts of these waters. Oxygen concentrations in bays during winter usually are lower than that in the open part of a lake. On the other hand, the retreat of Ponto-Caspian amphipods from large eutrophic lakes could also be associated with global climate change. During the last decade, water retention in lakes in Lithuania decreased due to higher temperatures and lower precipitation in summer. Consequently, the decrease in water retention led to the deterioration of oxygen conditions in lakes, especially in eutrophic ones (Timinskas pers. comm.).

Another goal of this study was to analyse the relations of non-native to native species. The native amphipod *G. lacustris* was found in 22 of 24 studied lakes, in which Ponto-Caspian amphipod *P. robustoides* is absent. Two lakes do not even contain the isopod *Asellus aquaticus* (one was poisoned to eliminate aboriginal fish fauna in the 1970s, another one was the smallest among studied lakes). Highest frequencies of native amphipods were always observed in the open parts of these lakes. On the contrary, in waters inhabited by non-native amphipod *P. robustoides*, the amphipod *G. lacustris* was not detected in lakes Dusia, Metelys, Daugai and Simnas and the open parts of Lake Plateliai and Antaliepte WR (this water reservoir was formed after the flooding of a few lakes). Another native amphipod, *G. pulex*, inhabiting flowing waters occurred previously in different parts of Kaunas WR (Gasiunas 1972a, Vaitonis 1994), but has recently not been found among amphipods collected in this water body. In Lake Asveja non-native and native amphipods were found in the same place, but occurred in different substrates. All these observations definitely suggest that Ponto-Caspian amphipod *P. robustoides* and native species *G. lacustris* and *G. pulex* do not coexist. In the habitats suitable for non-native amphipod *P. robustoides*, it excludes the native species. The ability of *Pontogammarus* sp. to out-compete *Gammarus* sp. has also been reported earlier (Dedyu 1980).

The relations between the Ponto-Caspian mysid *P. lacustris* and the native glacial relict mysid *M. relicta* in Lake Dusia are not so clear as for non-native and native amphipods. It was assumed earlier that during the vegetation period these mysids occupy different habitats. The non-native mysid inhabits littoral until 10-12 m depth, while the relict mysid lives in deeper cold waters and dominate from 10 m depth. A decade ago, densities of *M. relicta* in deep waters of Lake Dusia had comprised to around 35-47 ind·m^{-2} (Vaitonis 1994). Recently, in the 12-13 m depth only one individual of *M. relicta* was caught, while density of *P. lacustris* in these depths was estimated up to 60 ind·m^{-2}. In qualitative samples from 10-19 m depth the part of the relict mysid among all mysids was less than 3%. Hence, two facts are obvious. Firstly, the density of population of *M. relicta* in Lake Dusia significantly decreased. Secondly, the habitats of the relict and Ponto-Caspian mysid have recently overlapped greatly, as *P. lacustris* extended its habitat into deep waters of Lake Dusia. These facts suggest that the decrease in density of glacial relict mysid could be the result of competition with more productive non-native mysid *P. lacustris*, which produces 3 generations in a season (Vaitonis 1994). The negative effect of non-native mysid on *M. relicta* is most likely fastened by adverse conditions in the profundal waters during summer. The lack of oxygen in the deepest places restricts the habitat of the relict mysid to shallower waters, and habitats of mysids of different origin fully overlap. During certain summers oxygen deficiency occurs in the deepest places, however obvious indications of a rapid eutrophication of Lake Dusia are absent (Joint Research Center, Ministry of Environment unpublished). Unfavorable conditions for crustaceans in the deepest waters of this lake also indicate the absence of crustaceans in qualitative samples taken deeper than 25 m.

During 1999, at the same time as for Lake Dusia, 4 other Lithuanian lakes inhabited by *M. relicta* were sampled. In contrast to Lake Dusia, in the case of presence of relict mysids their populations were always abundant.

Even in lakes with higher trophy than Lake Dusia, in which conditions in the depth over 15 m were most likely hostile for glacial relict crustaceans (*M. relicta* and *Pallasea quadrispinosa* were absent), relative densities for the relict mysid in suitable habitats (e.g. in 2 lakes *M. relicta* was concentrated at 10-12 and even 8-9 m depths, correspondingly) were always estimated at above 100 ind·m^{-2} (Arbaciauskas unpublished data). These data also support the supposition of the negative impact of the Ponto-Caspian mysid on the glacial relict *M. relicta* in Lake Dusia.

Among studied lakes there is one more lake, in which the glacial relict mysid could clash with non-native one. Lake Zeimenys inhabited by mysid *P. lacustris* and amphipod *C. warpachowskyi* is located downstream to a large and deep mesotrophic lake with abundant populations of glacial relict crustaceans (Fig. 1). Earlier, glacial relict crustaceans have been recorded in Lake Zeimenys (Grigelis 1980), but in this study they were not found.

Non-native Ponto-Caspian mysids and amphipods have recently inhabited three water reservoirs and around ten lakes in Lithuania. These crustaceans have always received much attention, and a few conclusions on their ecology and impact on native ecosystems are possible. Now we understand which environmental factors could limit the establishment of their populations in lake ecosystems. We know that non-native crustaceans are actively consumed by different fish species (Bubinas 1979), and negatively effect native amphipods, and likely the glacial relict mysid *M. relicta*. However, the mechanisms of interaction between non-native and native crustaceans, and the impact of Ponto-Caspian species on local communities remain unexplored. Furthermore, the natural dispersal of introduced species, and possibly other crustacean invaders from Ponto-Caspian region, also warrants attention for better understanding of the invasion ecology and prediction of possible changes.

Acknowledgements

This study was partly supported by grants from the Department of Fish Resources, Ministry of Environment of Lithuania.

TEREDO NAVALIS - THE CRYPTOGENIC SHIPWORM

KAI N. HOPPE
CRM Coastal Research & Management, Kiel, Germany
Kai.Hoppe@crm-online.de

Abstract
The shipworm *Teredo navalis* is one of the most effective and harmful marine invaders worldwide. It is not clear, whether it invaded Europe from South East Asia or whether it originated in Europe and invaded the rest of the world from there. Recently it reappeared in the Western Baltic causing damages estimated at 25 - 50 Mio Euro along the Baltic coast of Germany alone. An immense reproduction rate and a high resistance to unfavourable environmental circumstances are the keys to its success. Even though mankind has tried to develop counter measures for thousands of years, still there is no easy solution to the shipworm problem in sight.

1 Introduction

The shipworms, wood boring bivalves of the family Teredinidae, belong to the oldest invaders. They naturally spread with wooden flotsam, but have also settled in all wooden boats and ships from the earliest dugouts to modern sailing yachts. With the trading vessels they travelled literally around the world. In fact, the most prominent shipworm, *Teredo navalis*, spread so early that we do not know where it originates. Even though seafaring nations have tried to protect themselves from shipworms for thousands of years, these animals still inflict immense damages on harbour constructions.

2 Biology

Teredo navalis is a bivalve with a wormlike, elongated body and a short, helmet-like shell at the anterior. It lives completely embedded in wood, with a tiny opening for two retractable siphons.

Teredo eats its way into wood digesting it with the help of endosymbiotic bacteria. *T. navalis* is unique in this regard, because it can survive on a wooden diet only, unlike other shipworms. However, it also filters and digests plankton from the seawater. Its symbionts are able to fix dissolved nitrogen from the water and incorporate it into essential amino acids. The shell is a boring organ with tiny teeth on the valves; it is used like a rasp. Up to 2,300 rasps are performed per hour, with a rotation of 360° during 20-40 rasps (Mann & Gallagher 1985). Undigested sawdust is expelled through the exhalant siphon, often accumulating around the siphons. The tunnel is always slightly greater in diameter than the lateral extension of the valves, so that the shipworm can retract to a certain extent. It usually tunnels along the fibres of the wood and avoids other shipworms. It stops or turns around at the end of the piece of wood it inhabits, or when it approaches another shipworm's tunnel.

The tunnel is lined with a calcareous excretion, and the opening can be closed with two paddle-like calcareous plates called "pallets", which are important in identifying shipworm species. Sealed off the shipworm can survive anaerobically for about three weeks (Lane 1959), with the lining buffering the acid metabolites. Therefore it can withstand

E. Leppäkoski et al. (eds.), Invasive Aquatic Species of Europe, 116–119.

unfavourable conditions like exposure to air or freshwater. It even survives very cold winters, when the tunnelled wood is completely covered in ice. The larvae of *T. navalis* can live and settle in salinities as low as 9 PSU, but the adults withstand even lower salinities. They grow to an average length of 20-30 cm, the largest specimen recorded in the Baltic was 59 cm (Kristensen 1969; 1979).

Teredo navalis attacks wooden structures as a pediveliger larva with a diameter of about 1 mm. It detects wood chemically from a distance and actively swims the last centimetres, before it attaches itself to the surface with a byssus thread. The soft-shelled larva penetrates the wood in an unknown manner, with maternal enzymes playing a role in softening the surface. The tiny borehole will remain the only connection to the outside for the shipworm. *Teredo navalis* is a protandric hermaphrodite, it starts to develop male gametes six weeks after metamorphosis (Nair & Sraswathy 1971). The transition from male to female is gradual (Sordyl et al. 1998); it can even self-fertilise. After reproduction in the female phase it changes back to the male phase and starts another cycle. In warmer climates the shipworm is assumed to develop several cycles per year, however, in the Baltic it is unlikely to have more than one cycle. Fertilisation is internal, the developing larvae are kept for about 8 days until the straight-hinge veliger stage (Calloway & Turner 1988). They are implanted in the tissues of the gill, followed by a "placental reaction" (Lane 1959), whereby the larvae get nourishment from the maternal tissues.

The reproductive rate is immense, up to 2 million larvae are released per cycle under favourable conditions. They are dispersed by currents and are pelagic for two weeks, like other bivalve larvae. The high number of larvae and the resistance to various environmental factors are necessary to find and utilise the few pieces of tree that are naturally transported into the sea. After all, the shipworm genus is about 20 million years old (Turner 1966), and it holds an important niche in the marine environment, where the degradation of wood by bacteria and fungi is very slow. When humankind appeared at the shoreline with dugouts, boats and piers, the adaptations made the shipworm a mightier enemy to seafaring than all pirates and warships.

3 *Teredo navalis* in Europe

Teredo navalis is a cryptogenic animal, it is nor entirely clear where it originated. The species was first recorded in 1731 (Sellius 1733), when it destroyed wooden dyke gates in the Netherlands, causing a terrible flood ("Holland in peril"). According to the Waddensea web page (www.waddensea.org) the Dutch believed it was introduced from Asia and sent as a punishment from God.

However, it is very likely that it was this shipworm that attacked the Spanish Armada, while it was waiting in French and Portuguese harbours for the invasion of England in 1588. It has been proposed to originate in the North Atlantic area (Schütz 1961), its resistance to low temperature supports this theory. Nowadays *T. navalis* is one of the most successful invaders reported from Indonesia, Japan, Australia, Brazil, the Atlantic and Pacific US and Canada. In Europe it is found along the Atlantic and North Sea coast. There were sporadic mass invasions into the Baltic Sea, which lasted for two or three years in the 1930s and 1950s (Schütz 1961). The easternmost border of shipworm settlement in Germany was the Island of Rügen, where a salinity of 9 PSU stopped the establishment of *Teredo* larvae. After three years the animal died out, presumably because no reproduction was possible in low salinity.

However, the latest invasion in the Baltic is taking place since 1993, and the population appears to have established itself for good. Wooden pilings used in harbours, sea bridges and groynes are mainly attacked, with even large pieces being destroyed in two years. Up to 40,000 larvae m^{-2} were recorded in one month of exposure (pers. obs.). It is now proven, that *T. navalis* can reproduce in the Baltic at least near Rostock (Sordyl et al. 1998). It was suggested that shipworm larvae conditioned to live in lower salinities, for example in estuaries, were transported in ballast water tanks from the North Sea area to Eastern Germany and established a foothold there (H. Rosenthal, pers. comm.). From Eastern Germany they spread to the West German Baltic, where they were recorded 3 years later. However, there is a conflicting theory of a small population that was always present and triggered by an unknown combination of environmental factors to mass reproduction and subsequent damage. This is supported by the fact that a survey in 1989 demonstrated the presence of low numbers of shipworm at several places along the Schleswig-Holstein Baltic coast (Schweimanns 1993).

4 Fighting the shipworm

Any wooden ship or structure in the sea has to be protected from marine borers. The earliest strategies included pulling the dugout up on the beach, taking the ship upstream into fresh water for extended periods or charring the bottom of the ship (Nair & Sraswathy 1971). Chemical defences like paints or copper or lead plating were used by early Chinese, Egyptian and Roman seafarers. Unfortunately, there are no reports whether Viking ships were attacked by *Teredo navalis*, as this could indicate the origin of the species.

European medieval seafaring nations used tar, which was quite effective as long as it could be reapplied at least every year. When the merchantmen travelled for a long time in tropical waters, where shipworm attack usually is more severe, more ships were sunk by shipworms than by pirates. Another product called creosote, used in a pressure impregnation process, is probably the most effective shipworm deterrent. However, it is also highly toxic and carcinogenic, so its use is banned in many countries. The modern alternatives are heavy metal salts, containing copper, chromium and arsenic (CCA) or borax (CKB). These salts are often used in high retentions (up to 40 kg m^{-3}) and even though they are supposed to bond firmly with the wood fibres, they are still controversially discussed by environmental biologists (Weis & Weis 1996).

Another modern alternative is to wrap wooden pilings with plastics. Polyethylene or polyvinyl wrappers are very widespread in US harbours, but not in Europe. Canadian logging companies used to detonate dynamite charges close to floating timbers in estuaries to kill the shipworms with the shock waves (Quayle 1992). This method proved to be quite effective, but is not very popular among German harbour masters (pers. obs.). Protection tests with electric currents or the release of certain poisonous chemical into harbour waters have not been successful (Nair & Sraswathy 1971).

Some tropical timbers offer a good natural protection, but with the over-exploitation of rain forests by logging companies, the use of these timbers has become restricted in Germany and other countries. Local oak trees also offer some protection, especially when they are used with the bark still intact (that is where most of the deterring tannic acids are), but there are obviously different qualities to oak piling. Some last for 30 years, others only for two (pers. obs.).

The detection of shipworms is very difficult. Only the 1-2 mm large hole is visible, with the siphons only sometimes extracted. The best detection method is by scuba diving and probing for tunnels with a knife. Also the rasping can be detected with a stethoscope. X-rays are possible, but very complicated especially under water. This method is very helpful to record the advance of the shipworm tunnelling in smaller pieces of wood. In conclusion, it is quite amazing that after thousands of years of fighting the shipworm, man has still to come up with a solution for detection, prevention and protection for wooden installations along our coasts.

INTRODUCTION AND ACCLIMATISATION OF THE PACIFIC CARPET CLAM, *TAPES PHILIPPINARUM*, TO ITALIAN WATERS

PAOLO BREBER
Institute for the Study of Coastal Ecosystems, National Research Council of Italy, Lesina (Foggia), Italy
isecpb09@area.ba.cnr.it

Abstract

In 1983 the exotic clam *Tapes philippinarum* was introduced to the lagoons of the north Adriatic as an aquaculture project. The purpose was to surrogate the insufficient production of the nearly identical native *T. decussatus*. Seed was obtained from European and North American hatcheries. Loose sediment in shallow, non-turbulent but well-flushed waters constitutes the habitat of the species. By 1986 there was evidence that the species was breeding freely and colonising all suitable areas. By 1990 the commercial yield from the culture parcs and natural beds had reached 20,000 tons year^{-1} furnishing the livelihood to a labour force of about 4,000 units. In the most productive areas densities of > 1,000 ind m^{-2} are common. Acclaimed as a great success by the fisheries sector, this initiative has been condemned by scientists and environmentalists as infringing the general principle of keeping out alien species and more specifically because the exotic clam is thought to have caused the disappearance of the native sympatric counterpart and generally upsetting the indigenous benthos. Field surveys carried out before and after the arrival of the alien clam show that the accusations are unfounded or cannot be demonstrated.

1 Introduction

Whereas the circumstances of the arrival of an alien species are usually not well known, in the case of *Tapes philippinarum* in Italian waters the first introduction was made precisely in 1983 at the southern extremity of Venice Lagoon (Breber 1985a). In that year 200,000 seed clams (3 mm) were imported from an English hatchery, and again the year after 1,500,000 from the same source were distributed between Venice and Caleri lagoons (Fig. 1). This Pacific species was not new to European waters having been acclimatised a few years before by the British and the French. Its natural geographic distribution is in the Pacific in the seas between Japan and the Philippines. Early in the 20th century it arrived accidentally on the Pacific coast of North America where it became a commercial species and is now bred in hatcheries. It is from this source that the European stock is derived.

The acclimatisation in Venice Lagoon was an experiment, funded by the Province of Venice, to surrogate the local fishery of the indigenous carpet clam *Tapes decussatus* with that of the nearly identical Pacific species. *T. decussatus* is one of the most sought-for sea foods in Italy. The supply comes from wild stock but is naturally very limited. Before the changes brought by the advent of the exotic, the annual consumption of the indigenous carpet clam amounted to just 1,000 tons, 60 to 90% imported from other Mediterranean countries (Breber 1996). Such scarcity generated a high price and an off-and-on presence on the market.

E. Leppäkoski et al. (eds.), Invasive Aquatic Species of Europe, 120–126.

The encouraging results of the first on-growing experiments with the Pacific clam immediately stimulated private initiative. In 1984, just one year after the first experiment, an entrepreneur imported one million young clams and set up a commercial parc in Caleri lagoon just south of Venice Lagoon (Fig. 1). In 1985 three million were sown in five different locations in the Po Delta (Caleri, Vallona, Scardovari lagoons) and two million in two Sardinian lagoons (S. Antioco and Mistras). In 1986 the activity spread to Marano lagoon (two million seed), at the northernmost end of the Adriatic, and to Goro (10 million seed) another lagoon on the Po Delta. Until 1985 all the seed was coming from the U.K. but in 1986 U.S. and Spain also supplied stock, part of which went to Varano lagoon in S. Italy. In 1987 30 million came from the U.S. and 20 million from France. In this same year *T. philippinarum* was brought to Paola, a lagoon on the Tyrrhenian coast just south of Rome. The year after importations increased to 100 million clam seed sent from U.S., French and Spanish hatcheries.

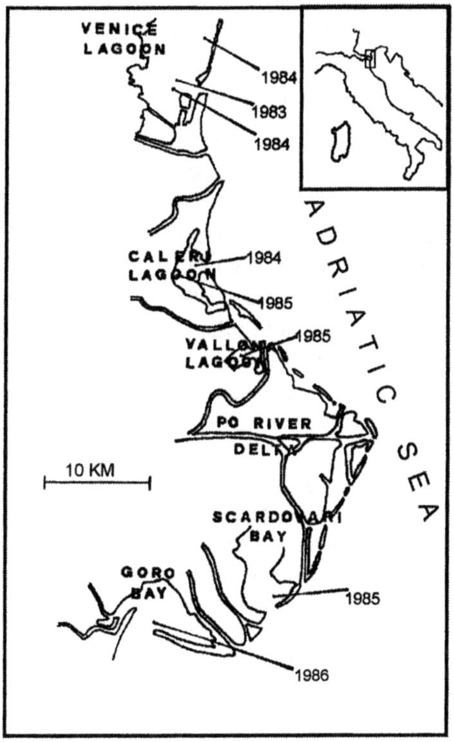

Figure 1. Areas within the Venice Lagoon - Po Delta complex with the place and date where *Tapes philippinarum* was intentionally introduced for the first time.

The first signs that *T. philippinarum* was breeding in nature came from Goro lagoon where clams were found outside and far away from the parcs, probably spawned by those sown just north in Scardovari lagoon the year before (Fig. 1). By 1987 the wild catch at Goro had surpassed the farmed produce. By 1989 wild seed was being gathered in quantities enough to supply the parcs so that importation from abroad stopped. In

1990, just seven years after the arrival of the first seed, between 70 and 100 tons of Pacific carpet clam coming from the various lagoons of the north Adriatic were being marketed daily and this high level of production maintains to this day (Breber 1992).

2 Possible explanations for the ecological success of *Tapes philippinarum*

Although it has had occasion to settle in many other places in Italy, either by intentional introduction or as spawn from the holding tanks of seafood merchants along the coast, the Pacific carpet clam has proved to be truly invasive only in the lagoon ecosystems of the north Adriatic, naturally extending its range southwards from where it was initially introduced at Chioggia and in the Po Delta all the way to Ancona and beyond at the rate of 30 km a year; its presence, however, petering out with the lack of lagoons (Fig. 2). Naturally formed beds in the most favoured areas show densities of > 1,000 ind m^{-2} with a biomass > 10 kg m^{-2}. The commercial size of 40 mm is reached within 24 months from birth. The species seeks the shallow, non-turbulent but well-flushed waters of lagoons and estuaries, and, where these are missing, settles in harbours and in the lee of breakwaters. Within the lagoon ecosystem it prefers the area next to the seamouth occupying the same niche of the native *T. decussatus*.

The lagoons of the north Adriatic are part of river estuaries where salinity values fluctuate strongly. *T. philippinarum* is an euryhaline bivalve and can tolerate 15-50 PSU (Breber 1996). The body fluids balance easily with diluted seawater down to 9.9 PSU, but not beyond. At 21°C the larvae live and grow in salinities from 12-32 PSU, but optimal growth and survival are recorded within the range of 20-28 PSU. During an exceptional spring flood in Goro lagoon no deaths were observed when the salinity fell to < 5 PSU for a few days. But in this as in other similar occasions the clams can escape extreme conditions by lying deep in the sediment where environmental stresses are buffered. Thermal conditions in the lagoons closely follow those of the atmosphere, with minima around 6°C in winter and maxima around 30°C in summer. *T. philippinarum* is considered to be adapted to a colder climate than the native *T. decussatus* and this allows for a spring and an autumn spawning, while the latter spawns only once during summer (Breber 1980). When the temperature over the parcs exposed at low tide dropped to –15°C in January 1985 no losses were noticed. There is no growth below 5°C and fastest growth is observed when the temperature rises above 15°C.

The nutrient levels in the lagoons of the north Adriatic are high, generating intense primary production. The diet of *T. decussatus* and *T. philippinarum* consists in benthic pennate diatoms. In the larval and juvenile stages phytoplankton is taken, but in the later stages the presence of benthic diatoms is essential to the extent that the spatial distribution of the clams coincides with the flats where these microalgae flourish. Benthic pennate diatoms develop as a film on sedimentary substrates where light is strong and wave action is weak.

The presence of a relatively pronounced tidal range (~ 1 m) in the north Adriatic, otherwise absent in most of the Mediterranean, is probably one of the discriminating factors which accounts for the high production of *T. philippinarum* here with respect to other areas colonised by the clam. The flushing caused by the tide creates conditions highly favourable to the productivity of benthic pennate diatoms, which is a prerequisite for the clam. The growth rate cited above of 40 mm in less than 24 months is associated with counts of 100-1,000 pennate diatoms mm^{-2} of substratum.

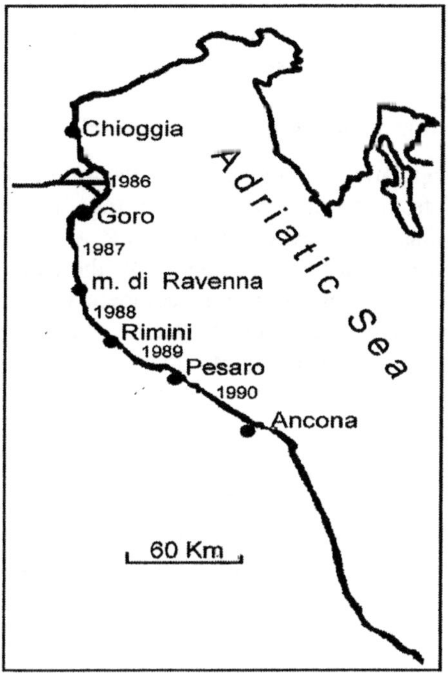

Figure 2. Coastline of Italy south of Venice Lagoon-Po Delta showing the rate of natural diffusion of *Tapes philippinarum* southwards.

The abundant recruitment year after year of the Pacific carpet clam, notwithstanding the intense exploitation by man, is explained by the presence of on-growing culture parcs which in fact constitute highly effective breeding populations. Fertilisation is external, males and females releasing gametes in the water. In order to ensure that ova and spermatozoa have a sufficient chance of meeting, the spawners should be close enough to one other, otherwise the dispersal of the gametes will be such that fertilisation will fail. In wild populations such as those of *T. decussatus* reproductive effectiveness is diminished when dredging subtracts a large portion of sexually mature clams and scatters far apart those which remain. The presence of hectares of culture parcs where the clams are grown at densities of 500-1,000 ind m^{-2} to the size of 4 cm is a sure guarantee that all the open surrounding areas will be systematically repopulated no matter how much overfished.

Another reason which goes to explain the unabated exuberance of the Pacific carpet clam since its introduction is the absence of epizootics. Certainly one of the causes of population loss in the indigenous *T. decussatus* is a long list of parasites and diseases. Just to mention two, the castrating trematode *Bacciger bacciger* was found to have infected up to 80% of samples of *T. decussatus* gathered in Venice lagoon (Breber

1985b) where the fungus *Perkinsus marinum* also causes heavy inroads. *T. philippinarum* on the other hand appears to be resistant to this fungus. These and other pathogens present in Italian waters have not as yet been able to adapt to the exotic clam, and its own parasites which it may have brought with it from its native waters have their life cycles blocked by the lack of the necessary intermediate hosts left behind in the Pacific.

3 Is the exotic a competitor of the indigenous sympatric *Tapes decussatus*?

Contemporarily to the population explosion of the Pacific carpet clam, the indigenous carpet clam disappeared from the market. This fact, together with the observation that the exotic had occupied the ecological niche of the native species and shown more fecundity by spawning twice a year instead of once like the latter, truly gave the impression of a further case of an ecologically more competitive exotic species causing the extinction of its local counterpart. In order to verify whether this widespread opinion corresponded to truth I conducted a survey of two areas that used to furnish most of the national production of *T. decussatus* before the introduction of *T. philippinarum* and which now are densely populated by the latter (Rossi & Paesanti 1992).

The areas surveyed were Venice Lagoon and Goro Lagoon where *T. philippinarum* was introduced in 1983 and in 1986 respectively. Samples were dredged with commercial gear from a given surface and sorted for the presence of *T. decussatus*. The dredge collected only clams larger than 3.5 cm so that the presence of the smaller size classes was not registered. A practiced eye can usually distinguish the two very similar species: the reliefs of the shell of *T. decussatus* are less pronounced and, given the same height, *T. philippinarum* is shorter. There are cases, however, where in order to distinguish the two it is necessary to place the clams in water and observe the extended siphons: the native species shows two separate siphons whereas in the exotic the siphons are joined for two thirds of their length. The abundance of the native clam was then compared to its abundance in that same area before the introduction of the Pacific carpet clam in 1983 (Anon. 1984).

In Venice Lagoon 2% of the clams in the samples belonged to the native species while the rest were Pacific clams. In Goro lagoon 0.1% of the clams were *T. decussatus*. In Venice Lagoon I calculated that the mean density in the sampled area of *T. decussatus* (> 3.5 cm) was 2.5 ind m^{-2}. Unfortunately a comparable datum prior to the advent of *T. philippinarum* is lacking. Venice Lagoon used to yield on average 200 tons $year^{-1}$ of *T. decussatus* (> 3.5 cm) but we do not know the extent of the zones from which it was gathered. The two species do not indiscriminately colonise lagoon sediments but are usually concentrated near the seamouth, so that density cannot be referred to the entire surface of the lagoon but only to these particular zones. In the case of Goro lagoon, on the other hand, data on the density of *T. decussatus* before and after the arrival of the exotic are extant (Anon. 1984). The records show pronounced fluctuations in abundance of the native species. Before *T. philippinarum*, the highest density of the native ever recorded was 13 ind (> 3.5 cm) m^{-2} which corresponds to a yield of ca 200 tons $year^{-1}$ from 120 ha of lagoon bottom. Between 1969 and 1983 such abundance occurred only in 1970, 1971, 1975 and 1981. After these peaks the consistency of the stock (> 3.5 cm) dropped to such minima as 1.3 ind m^{-2} in 1973 and 1977. In a survey of 1984 < 1 ind m^{-2} was found. In recent (2001) samples from Goro I found a mean of 0.4 ind m^{-2} of *T.*

decussatus against 384 ind m^{-2} of *T. philippinarum*. Thus the native has not disappeared but its presence is "hidden" by the prodigious abundance of the exotic.

The widespread impression that *T. decussatus* had "disappeared" is otherwise explained. Initially, the exotic species fetched the same high price as the native so that there was no reason to pick out and sell separately the far less numerous natives from the mass of the exotics. Another reason is that, as the Pacific clam began to supply the market, the seafood merchants stopped importing *T. decussatus* from other countries. Of the ca 1,000 tons of *T. decussatus* consumed in Italy before 1983, only 10-40% actually came from Italian waters (Breber 1996), a fact which the consumer was scarcely aware of. This cut in the supply gave the impression that the stock had dwindled in nature. In recent years, however, the glut of the Pacific clam with the accompanying drop in price and the rallying of gastronomic snobbery have separated the destinies of the two species. *T. decussatus* gathered from Italian waters has now reappeared by itself on the market for exclusive restaurants at four times the price of the exotic.

The great similarity between the two species gave the impression at first that hybridisation was possible but the electrophoretic examination has revealed quite different genetic maps so that this possibility may be excluded (Fava & Meggiato 1995).

In other areas of Italy such as Sardinia and Sicily *T. philippinarum* has not been invasive and has not, therefore, given the impression of having smothered the local *T. decussatus* populations.

4 What has been the impact of *T. philippinarum* on the benthic community?

Has the dense settlement of the Pacific clam changed the characteristics of the benthic community through competition for food and space? The problem of answering this question is that we do not have a clear vision of the situation before the arrival of the alien. The natural habitat of the alien is on the mud flats just on the inside of the channels which link lagoons with the sea. The hydrodynamics of these ecotones are, however, forever changing and the benthic community changes with it. Such circumscribed zones are furthermore dredged assiduously for their rich harvest of various molluscs and this activity certainly modifies the benthic community structure.

Let us take for example the already mentioned area (120 ha) of Goro lagoon and compare the situation before and after the arrival of *T. philippinarum* in 1986. In six surveys of bivalve molluscs carried out between 1982 and 1985 19 species were found, but only eight every time: *Cerastoderma glaucum, Chamelea gallina, Crassostrea gigas, Mactra corallina, Mytilus galloprovincialis, Paphia aurea, Scapharca inaequivalvis, Tapes decussatus* (Rigatti Luchini et al. 1989). This combination of species is not homogeneous. Only *C. glaucum, P. aurea,* and *T. decussatus* are recognisable as belonging to the Mediterranean paralic biocoenosis. *C. gigas* and *S. inaequivalvis* are alien species that have arrived accidentally, the former appearing in the 1960s and the latter in the 1980s. The presence of *M. galloprovincialis* in Goro lagoon is artificial because bottom culture is practiced here. The area has been commercially dredged for *C. glaucum, M. galloprovincialis* and *T. decussatus* during the last 15 years preceding the introduction of *T. philippinarum.* Mediterranean lagoons are also subject to dystrophic crises which wipe out the benthos and Goro lagoon is no exception. During the summer of 1984 the lack of oxygen in the water caused a drastic reduction of bivalve species from 15 to eight, the mean abundance diminished from 208 to 69 ind m^{-2} and the relative biomass from 5,281 to 1,154 g (wet weight) m^{-2}.

It is on this dynamic scenario that *T. philippinarum* was introduced in 1986. Surveys conducted where the Pacific clam settles thickest have revealed that the typically paralic species *C. glaucum, P. aurea* and *T. decussatus* are still present. But a comparison of abundances before and after does not, from the given data, unambiguously tell us whether there has been a permanent decline of the pre-existing species and whether this decline is attributable to the impact of *T. philippinarum.* Trophic competition may be excluded. Competition for space would seem more likely. But the substrate that has since become entirely taken up by the Pacific clam (> 1,000 ind m^{-2}; > 10 kg m^{-2}), at least in the most favourable areas, was far less occupied by the pre-existing benthic macrofauna even in the most abundant sample: 208 bivalves of various species m^{-2} with a biomass of 5.3 kg m^{-2}.

DREISSENA (D.) POLYMORPHA: EVOLUTIONARY ORIGIN AND BIOLOGICAL PECULIARITIES AS PREREQUISITES OF INVASION SUCCESS

MARINA I. ORLOVA
Zoological Institute of the Russian Academy of Sciences, St. Petersburg, Russia
ecology@zin.ru

Abstract
The evolution of *Dreissena polymorpha* took place in changeable conditions in the successive series of Paratethys Great Lakes and Seas and their estuaries. Complicated intraspecific structure of species and sufficient divergence in environmental requirements among its subspecies allows *Dreissena polymorpha* to inhabit a variety of environmental conditions in its native area, the Ponto-Aralo-Caspian region. The only subspecies, *Dreissena p. polymorpha,* is known as an amphiatlantic invader in fresh and oligohaline waters of Europe and North America. Wide and fast dispersal and the successful establishment of this subspecies are facilitated both by biological peculiarities of the zebra mussel (e.g. unusual for freshwater bivalve mode of life, wide limits of environmental requirements, genetic polymorphism) and availability of man-mediated vectors.

1 Introduction

Dreissena polymorpha (Pallas, 1771): Type: Mollusca; Class: Bivalvia; Order: Cardiida Ferussac, 1822; Family: Dreissenidae (Andrusov, 1897); Genus: *Dreissena* Van Beneden; Subgenus: *Dreissena s. str.* (Synonyms: *Mytilus polymorpha* Pallas, 1771; *Dreisenssia polymorpha* (Pall.) Andrusov, 1897)

Available natural and man-mediated vectors of invasion and similarity in environmental conditions of donor and recipient water bodies have promoted rapid dispersal of the zebra mussel. The species is a classic example of "neozoans" (Kinzelbach 1992). It has a remote marine ancestor and hence the success of its invasion to freshwater has been greatly facilitated by unusual biological peculiarities remaining from the species' "marine past". Long-term adaptation to oligohaline and freshwater conditions through the course of evolution in ancient series of the Paratethys Great Lakes and Seas (Zenkevich 1963; Dumont 1998), and their estuaries now succeeded by basins of the Ponto-Caspian region, is also a prerequisite of the success of invasion and establishment in target regions.

2 Origin of two invasive freshwater dreissenids

Dreissena polymorpha is a representative of the Caspian autochthonous faunal group (Mordukhai-Boltovskoi 1960, 1978; Zenkevich 1963). Palaeontological evidence leads to the conclusion that genus *Dreissena* has a polyphyletic origin. Modern representatives of the genus *Dreissena* originated from at least two branches of the genus *Congeria* (Baback 1983). Nevesskaya (1976) studied evolution of different species in genera *Dreissena* and *Congeria* and concluded that *Dreissena* originated from *Congeria* by "fetalization", i.e., by retaining of juvenile features (intact septum and absence of apophysa) of ancestral forms in adults of descendant species. First transitional forms

E. Leppäkoski et al. (eds.), Invasive Aquatic Species of Europe, 127–134.

between *Congeria* and *Dreissena* were found in Late Miocene - Early Pliocene deposits in the Pannonian basin, in Later Maeotice deposits in the Dacian basin and Euxinian basins (Baback 1983), as well as in estuaries of the eastern Mediterranean part of Paratethys (Starobogatov 1994). The appearance of *D. polymorpha* is referred to Late Miocene (approx. 14 mya) or the transition from Miocene to Pliocene. It is suggested to occur along the northern margin of the range of the genus *Dreissena*, in estuarine portions of rivers draining into Paratethys Lakes-Seas. Geological evolution of the region was characterised by a series of transgressive and regressive stages and changes of salinity from fresh water up to 20-25 PSU (Starobogatov 1994; Dumont 1998). The species invaded successively all accessible areas of the lacustrine parts of the Paratethys, where it diverged into a range of endemic species and subspecies, and colonised various habitats along main environmental gradients (e.g. salinity, depth, type of substrate; Starobogatov 1994). Regressive and transgressive phases in the geological history of Paratethys created conditions of temporary isolation and connection between the basins, which facilitated the adaptive radiation of *Dreissena* within basins, and exchange of species and subspecies between basins during latter phases. One more invasive species *Dreissena (Pontodreissena) bugensis* (Andr.) is thought to have evolved in the Euxinian part of the eastern Paratethys. It has evolved rather recently (less then 1 mya in the freshened Chauda lake; salinity < 4 PSU) and later in estuaries of the succeeded Neo-Euxinian Basin (Starobogatov 1994). Until the 1960s, *D. bugensis* distribution was restricted to a few rivers and estuaries of the Black Sea basin (Zhuravel 1967). Since the 1990s *D. bugensis* is known from the Volga River and the northern part of the Caspian Sea (Antonov 1996; Orlova et al.1998). It was reported from the North American Great Lakes some years earlier, in 1989 (Domm et al. 1993).

3 Native range and intraspecific structure

The range of *D. polymorpha* was in early Holocene limited to the Black, Azov, Caspian and Aral Seas, estuaries and lower and middle reaches of the Ponto-Aralo-Caspian rivers. In that time the northern margin of its range was restricted by a line from Mogilev, Kursk, Voronezh and Nizhniy Novgorod to Izhevsk (Fig. 1) (Starobogatov & Andreeva 1994). Meanwhile, during the Mindel-Riss interglaciation, this margin was situated far northward. *D. polymorpha* is considered to be a species that is extremely variable both in morphology and habitat requirements (Karpevich 1955). Currently, the species comprises three living subspecies (the fourth one, *D. p. obtusicarinata* became extinct due to the salinity increase in the Aral Sea in the 1980s) and several varieties, all of them characterised by slightly different environmental requirements. Therefore, the species' range covers both fresh, oligohaline and particularly mesohaline conditions along the native area. For example, in the Caspian Sea basin *D. p. polymorpha*, the most widespread subspecies, occurs in the Volga River and its delta. In the Northern Caspian Sea it inhabits a freshwater zone mainly in the north-western part of the sea. The subspecies is also known from the oligohaline zone of the central and western parts of the Northern Caspian Sea.

Within this subspecies, four main varieties can be distinguished: *D. polymorpha* var. *typica* Karpevitsch, inhabiting now fresh waters lakes, rivers and reservoirs of Europe and North America; *D. p.* var. *fluviatilis* (Locard), occurring in branches of the upper part of the Volga delta, in conditions of fast stream and high turbidity of water; *D. p.* var. *tschapurica* Karpevitsch occurs in freshwater or slightly oligohaline conditions of the northern Caspian Sea in the transition area, where water masses of Volga River are mixed with water masses of the Caspian Sea. Individuals from brackish water conditions (*D. p.* var. *marina* Pallas, a transitional form between *D. p. polymorpha* and the next Caspian subspecies *D. p. andrusovi* (Andr)) are found in offshore areas of the northern Caspian Sea, at salinity as high as 3-8 PSU. This subspecies occurs mostly in the eastern part of the northern Caspian Sea up to 7-8 PSU. The third subspecies *D. p. aralensis* (Andr.) is peculiar for lower portions of rivers, draining into the Aral Sea and inhabits small lakes along the sea shore (Mordukhai-Boltovskoi 1974).

4 Shell morphology

The shell of *Dreissena polymorpha* has a typical heteromyarian form with great reduction of the anterior territory and great elongation of the opistodentic ligament. Ventrally one or both valves bear the byssal notch. The shell is triangle-shaped from anterior view with flattened ventral surface. There is a sharp shoulder ridge ("carina") located ventrolaterally on each valve. Internally there is a well-marked pallial line with large insertions for the posterior adductor and posterior byssal rectractor and, correspondingly, small insertions for anterior muscles. These are situated on a conspicuous vertical shelf, also called septum, which occupies the umbonal region. The shell is ordinary coloured. Patterns are diverse and can change in the same mollusc from one growing season to the next (Biochino 1994; Marsden et al. 1996).

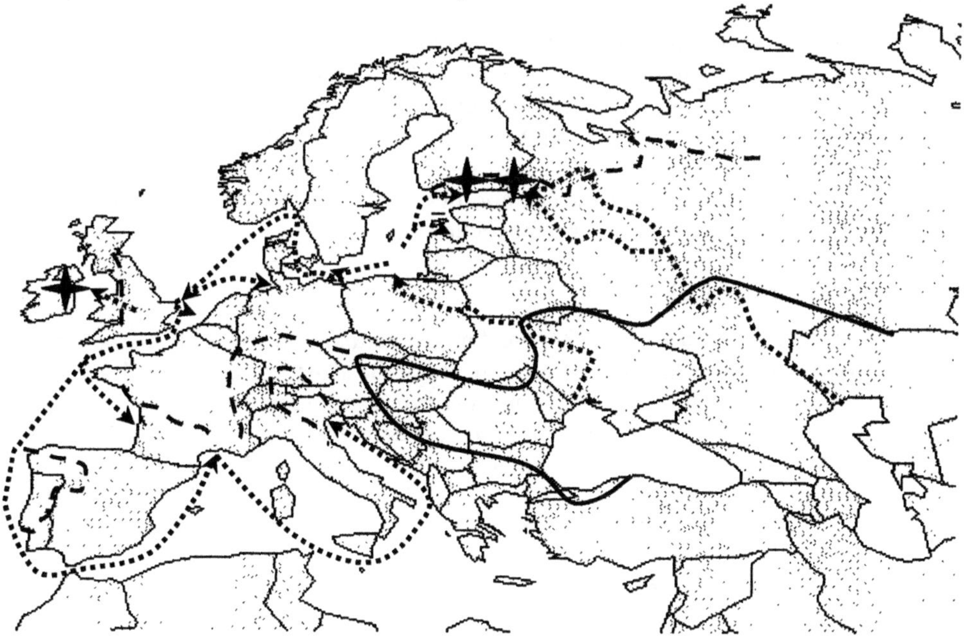

Figure 1. Holocene (solid line) and recent (dashed line) range of *Dreissena p. polymorpha* dispersal in Europe; main routes of invasions (dotted lines); relatively recent (1980s-1990s) records (asterisks).

5 Biological peculiarities of *Dreissena p. polymorpha* as a prerequisite of invasion success

5.1 LIFESTYLE AND REPRODUCTION

Dreissena is a sessile byssate bivalve able to inhabit fresh waters, where the niche of sessile filter feeders was vacant before invasion. Occupying this niche *Dreissena* forms dense settlement in all accessible zones and on all kinds of substrates in fresh and oligohaline parts of invaded water bodies. Due to this attached mode of life it can be transferred by shipping as a member of fouling communities on ships' hulls.

Having a remote marine ancestor *Dreissena* has a life cycle with a pelagic larval stage similar to that of marine bivalves. These stages are termed preveliger, D-stage veliger, post-D-stage veliger, and plantigrade. The presence of free-living stages promotes both remote dispersal in ballast water communities between regions as well as rapid colonisation of accessible habitats within and between localities in target water bodies. *D. polymorpha* is a gamogenetic species with equal occurrence of males and females in populations. The age and body length for maturation (at an age of 0+ or 1+) depend on temperature conditions and the time of the year of the juveniles' settlement in each locality. The zebra mussel in Europe is characterised by annual reproduction, with the exception of the most northern populations. *Dreissena* is an r-strategist; mature molluscs are known to produce over 1 million eggs or 10 billion sperm (Sprung 1991). Early maturation and high generative production are two more prerequisites for its invasion success.

Temperature and salinity are the most important environmental abiotic factors limiting the recruitment and development of *D. polymorpha*. The spawning period is extended from late spring to early autumn at temperatures above 12°C. Duration and the time of the year for the breeding season are highly variable and dependant on the geographical position of the locality (Lvova & Makarova 1994; Lvova et al. 1994). Hence further dispersal and establishment of *Dreissena* northward is limited by reproduction, which is successful only above mean summer temperatures of 12°C.

5.2. HABITAT PREFERENCES

The zebra mussel occurs usually on hard and mixed bottoms. At soft and silty bottoms it can also overgrow empty shells and live molluscs of the same and other species, as well as pieces and debris of other hard substrata. Vertical distribution of the species depends first of all on the availability of hard substrata and accessible food resources. The mussel can live from shoreline down to tens of meters. Its maximal depth of occurrence registered for European waters is 60 m (Grim 1971). Maximum abundance is at 1-5 m depth (Liakhnovich et al. 1994). In water bodies with winter ice-cover the distribution of *Dreissena* toward the shoreline is limited by ice abrasion, in lakes and reservoirs it is often limited by water level fluctuations (Liakhnovich et al. 1994). As a rule, *Dreissena* lakes are mesotrophic with relatively high pH, moderate alkalinity, and moderate amounts of dissolved mineral salts in the water (Liakhnovich et al. 1994).

5.3 OTHER ENVIRONMENTAL REQUIREMENTS

5.3.1 Salinity, Ca^{++}, and pH

D. p. polymorpha occurs in freshwater and oligohaline conditions, its expansion to more saline waters (> 2-4 PSU) is restricted by salinity tolerance of larval stages (Kilgour et al. 1994). Similar to other brackish water euryhaline bivalves, the European zebra mussel is capable of hyperosmotic regulation in range from freshwater to 2 PSU (Karpevich 1952). In experiments with step-by-step increase of salinity, the adults (e.g., the Caspian and the Baltic Sea populations) tolerated salinity up to 12-15 PSU (Orlova et al. 1998). In freshwater Belorussian lakes (524 totally studied) *Dreissena* occurs in a range of 164-388 mg l^{-1} total dissolved solid concentrations. It is absetn in lakes with values below 64 mg l^{-1} (Liakhnovich et al. 1994). Among inorganic ions calcium is important for *Dreissena* dispersal and early development. *Dreissena* is less tolerant to decrease of calcium concentration in water and low pH than other freshwater molluscs (Unionacea, Sphaeriidae). Minimum Ca^{++} concentration in lakes inhabited by *Dreissena* is < 28.3 mg l^{-1} Ca++ at pH 7.3 in Europe, while in North America it occurs at > 15 mg $l^{-1}Ca^{++}$ and forms dense populations at > 25 mg l^{-1} (McMahon 1996). Rearing requires 12 mg l^{-1}, while minimum Ca^{++} concentration for successful development of veliger is 34 mg l^{-1} (Sprung 1987). So far limited dispersal of *Dreissena* in low mineralised lakes and reservoirs of Europe, as well as in dystrophic lakes, may be explained by calcium limitation. Together with ionic composition and dissolved solids concentration, pH is one more important limiting hydrochemical factor, restricting *Dreissena* dispersal in European lakes. The study of 524 Belorussian lakes (pH 3.3-9.6) revealed *Dreissena* occurrence in lakes with pH 6.7-8.9 only (Liakhnovich et al. 1994). In North America the mussel has colonized areas with pH 7.0-9.0; successful development of larvae occurs at pH 7.4-9.7 (Bowman & Bailey 1998).

5.3.2 Temperature and oxygen

The species is considered as low boreal by its origin (Starobogatov 1994) and can thus inhabit accessible waters in temperate climatic zones. Temperature of 0°C is the lower and 30-32°C the higher limit for adult survivorship. Upper limit for cells is 40-42°C (Shkorbatov et al.1994). As mentioned above 12°C is the lower limit for recruitment.

Capacity to tolerate low oxygen conditions varies strongly between different populations (Shkorbatov et al. 1994). In general *D. polymorpha* can be considered as a species relatively resistant to short-term hypoxia, having the critical saturation level as low as 25-26% at 20°C. Individual mussels are able to tolerate experimental anoxic conditions at 20°C for 3-6 days, while the increase of temperature up to 25°C rapidly accelerates the mortality of mussels (Shkorbatov et al. 1994). In nature, however, the distribution of the zebra mussel is usually limited to littoral and sublittoral zones, and slopes of riverbeds, where oxygen conditions are appropriate.

5.3.3 Exposure to air

Dreissena from different European populations can survive in air, in conditions of high humidity (80-100%) for 3-5 days (Shkorbatov et al. 1994). Consequently, it cannot tolerate recession of the shoreline longer than for this period. Therefore the distribution of *Dreissena* toward the shoreline in lakes and reservoirs is limited by water level fluctuations, that do not exceed 0.5 m in the majority of lakes; in reservoirs they can reach 2 m or more. Newly settled individuals only contribute to temporary *Dreissena* settlements in the zone of water level fluctuations (Liakhnovich et al. 1994) in lakes and reservoirs, while populations with normal age-size structure occur deeper.

5.4 FEEDING AND GROWTH

Dreissena polymorpha is a polyphagous seston-feeder. It feeds on particulate organic matter with size 1-500 μm. Content of food consumed is close to the content of seston (Yablonskaya 1967, 1969). Food selectivity is provided by the sorting role of gills and labial palps. Besides of useless inorganic particles *Dreissena* usually avoids eating sticky-looking diatoms, big-sized colonies of Chlorophyta and Cyanophyta (Vanderploeg et al. 1995), emulsions of organic liquids etc. At high concentrations of seston only 10% of the suspension filtered by *Dreissena* was found in its stomach (Mikheev 1964).

Diurnal diet of *Dreissena* depends on the size of molluscs, nutritional value and concentration of seston, and abiotic characteristics of the environment (Mikheev 1964). Diurnal diet can be calculated using the balance equation:

$$C=1/U*(P+R),$$

where C = diurnal diet; P = production; R = rate of metabolism ($R=(0{,}1396W^{0{,}63})*24$ (mgO_2 ind h^{-1}), where W is wet weight (g) (Alimov 1981).

When the concentration of seston does not differ sufficiently from natural values (3-15 mg l^{-1}) and the size of particles is 4-50 μm (Mikheev 1967), then the daily diet can be estimated as the rate of filtration multiplied by seston concentration:

$$C=(F*c)*12,$$

where c = is seston concentration (mg l^{-1}); 12 = number of hours per days, when *Dreissena* is usually actively filtering.

Efficiency of assimilation varies from 0.5 to 0.6, being most often close to 0.6 (Alimov 1981). Filtration normally occurs at temperature range 5-30°C (Kondratiev 1962), and pH 8-9 (Morton 1971). Filtration rate is strongly dependent on body weight and can be described by the equation:

$$F=85.5W^{0.6}$$ (Alimov 1981),

where F= rate of filtration (water clearance) expressed in ml ind h^{-1}; W = wet weigh of body (g).

Linear growth of *Dreissena polymorpha* can be described by Betralanffi equation:

$$L(t) = L\infty e^{-kt}$$

where L(t) = length to time t, $L\infty$ = theoretically maximum length (mm), k = constant of growth (1/year), e-basic of natural logarithm.

Parameters of linear growth for different European populations are calculated by Alimov (1981). By his results $L\infty$ varies from 29 to 51 mm; Measured L_{max} from 23.2 to 42 mm, k varies from 0.122 to 0.492 1/year; calculated maximum age varies from 4.2 to 11 years.

5.6 GENETIC VARIABILITY

The review of the data collected has shown high level of genetic variability in the zebra mussel populations both in Europe and North America. The results of the analyse, estimated by the allozyme electrophoresis method (Marsden et al. 1996), confirm that the zebra mussel is able to adapt to a wide variety of environmental conditions with no loss of genetic diversity. Obviously no bottleneck effect is present trough the process of *Dreissena* ntroductions. This peculiarity of invasion is characteristic for molluscs extending their range via free-living planktonic larvae (in ballast water) and/or by repetitive transfers when founder population consists of large number of individuals.

5.7 SUCCESS AND LIMITATION OF *DREISSENA POLYMORPHA* INVASION

Available natural and man-mediated vectors and similarity of environments in donor, transitional and target areas have promoted rapid expansion of the Ponto-Caspian *Dreissena polymorpha* northward at least twice during the known history of the species. According to Kinzelbach (1992) the first event occurred naturally during the Mindel-Riss interglaciation. The history of that natural expansion was the following: Following the rising water level in the Ponto-Caspian rivers *D. polymorpha* colonized the middle and upper portions of them. When the connection appeared between the ancient Pripiat' and Bug Rivers it invaded the Vistula River. From there the zebra mussel continued expansion as far as to the Netherlands and England via interglacial primary stream valleys. In the beginning of the Wurm glaciation the species recessed back to the Ponto-Caspian region.

The second unremitting natural and man-mediated expansion of *D. polymorpha* took place in Europe beginning in the early 19th century. By the end of the century the species colonized most of the inland waters of West and Central Europe and estuaries of the Baltic Sea and the North Sea. For the second half of the 20th century it was reported as the dominant species in bottom assemblages of reservoirs constructed on large rivers in Eastern Europe and was also found in alpine lakes. By 1986 it was recorded in the eastern Gulf of Finland and by 1994 in Ireland and then, in 1995, along the Finnish coast (Nowak 1971; Kinzelbach 1992; Starobogatov Andreeva 1994; Kharchenko 1995; Valovirta & Porkka 1996; Olenin et al. 1999).

Biological peculiarities of the species are the most important prerequisites of either success of transfer or success of the final step of the invasion process the establishment of a population in the target locality. The attached mode of life and free-living stage (remote marine origin) broadens the ways of introduction of *Dreissena*. It can be transferred both in larval (ballast water) and adult (attached to ship bottoms) stages. Invading fresh water lakes, rivers, artificial water-bodies and oligohaline marine environment (e.g. the Baltic Sea) the zebra mussel becomes the dominant species in fouling communities. In some cases it forms communities of Ponto-Caspian type together with other

Ponto-Caspian invaders (Riccardi & MacIsaac 2000). Being an r-strategist and producing annually numerous larvae, which are able to fast dispersal with currents, the zebra mussel colonises fast all accessible habitats within target localities. As *Dreissena* originates from habitats that had wide temporal and spatial variations of environmental characteristics throughout their geological history (first of all salinity, water level and temperature), it can establish populations in a wide range of freshwater and oligohaline habitats in temperate climates.

In both range expansion events mentioned above, whether they are natural or human-mediated, the biological peculiarities of the species are also defining the limits of potential dispersal. Having relatively vulnerable free-living stages in the life cycle and weaker osmoregulatory capacity (Dietz et al. 1994), than many of native freshwater bivalves, the distribution of Dreissena is limited by low concentration of dissolved mineral salts in water, especially by low calcium concentration, as well as by low pH values. Temperature requirements are also limiting expansion northward by summer isotherm above 12°C. Being active sessile filter-feeder *Dreissena* requires mesotrophic conditions in the habitat with propitious concentrations of seston.

6 Conclusions

The zebra mussel is a classic case of "neozoans", with a remote marine ancestor. So far the success of its invasion to freshwater has been facilitated, first of all, by unusual biological peculiarities of the species received from "marine past", and secondly by long-term adaptation to oligohaline and freshwater conditions, through the course of evolution in ancient series of the Paratethys Great Lakes – Seas and their estuaries succeeded now by basins of Ponto-Caspian region.

Acknowledgements

The paper preparation was supported by Maj and Tor Nessling Foundation, and RFBR Foundation (Project 01-04-48542).

ZEBRA MUSSEL: IMPACTS AND SPREAD

DAN MINCHIN[1*], FRANCES LUCY[2] & MONICA SULLIVAN[2]
[1]*Marine Organism Investigations, Ballina, Killaloe, Co Clare, Ireland*
[2]*Environmental Services Ireland, Aughamore, Lough Allen, Carrick-on-Shannon, Co Leitrim, Ireland*
*Corresponding author *minchin@indigo.ie*

Abstract
Zebra mussels colonise new regions at different rates. Their rate of spread from isolated water bodies is slow but if introduced to upstream navigable regions their colonisation can be rapid and encompass a wide area. Zebra mussels continue to colonise European regions. Its recent appearance in Ireland since 1993/4 has incurred rapid colonisation through the inland navigation routes and it is poised to expand to other isolated regions. The management of controlling further spread has been successful to-date but its subsequent expansion is expected in time with human activities. It is likely that the species may also spread to other regions in Europe and North America with the potential to colonise large rivers and lakes in other temperate regions of the world.

1 Introduction

The zebra mussel, *Dreissena polymorpha* (Dreissenidae) is a bivalve native to the lakes, slow-moving rivers and low salinity areas of the Caspian and Black Sea regions. It was formerly present in the Aral Sea, but due to increasing salinities and reduced lake area it no longer exists there (Karatayev et al. 1998).

D. polymorpha is one of a number of species in Eastern Europe separated by Quaternary events to provide a range of forms. *D. polymorpha* and *D. bugensis*, although distinct on account of morphological and ecophysiological differences (Biochino 1990), may be the same species. Ludyanskiy (1991) considers that *D. bugensis* from Lake Ontario in North America closely corresponds with the more salinity tolerant *D. polymorpha andrusovi* or *D. rostriformis bugensis* from the northern Caspian Sea. Within the several subspecies, *D. polymorpha polymorpha* is especially abundant in alkaline freshwater, may occur in brackish water in salinities to about 5 PSU and is sufficiently plastic to colonise a wide range of habitats.

The zebra mussel produces veliger larvae which remain within the plankton for some weeks and become concentrated by wind and water currents in embayments along rivers or in lakes producing high settling numbers. Mortalities during their free-living stage and metamorphosis are high and they may perish when carried downstream in rivers or estuaries.

Rapid colonisation has been facilitated by various biological characteristics; high fecundity (up to 1 million eggs per female); a planktonic veliger larval stage of some weeks enabling downstream transport and the ability to firmly attach using byssal threads to hold its flattened side close to firm surfaces. This attachment can cause biofouling, similar to that found in marine mussels, and this has created a wide range of impacts for the environment and to industry. The zebra mussel gets its name from the distinctive, striped dark and pale shell colouration, although this is not seen on all specimens, especially older individuals.

E. Leppäkoski et al. (eds.), Invasive Aquatic Species of Europe, 135–146.

2 Expansion in Europe

Without the vagaries of human interference, the zebra mussel would most likely have remained for a long time within its native range. From 1775 connections between waterways facilitated trade, initially in eastern Europe. A canal linked the Dnieper and Zapadnyi Bug rivers in 1775 thus permitting its expansion (Dexbach 1935) from the Black Sea region. In 1804 the Oginskij Canal was completed; this linked the Dnieper and Neman rivers and the Dneiper and Zapadnaya Drina rivers one year later. This provided zebra mussels access to the Curonian Lagoon in the Baltic region. The later development of canals from the Volga River to the Don River provided opportunities of transmission of zebra mussels from the Caspian region. Consequently the early colonisation to western Europe was from the Black Sea (Fig. 1).

The zebra mussel reached The Netherlands by 1826, being found in the Rhine at Rotterdam, most probably carried with timber imports from the Baltic (Kearney & Morton 1970). New canal links as well as overland transport provided opportunities for further expansion of the zebra mussel carried on the hulls of watercraft. Once introduced to a previously uninvaded river system the expansion could rapidly evolve downstream. The species has expanded its range following the closure of Lake Ijsselmeer to the North Sea since 1932. This resulted in a declining salinity from 6 PSU to < 0.5 PSU in less than four years and subsequently the species colonised the whole lake by 1938 (Smit et al. 1993). Since 1994 an increase in their abundance was noticed in Lakes Slotermeer, Heergermeer, Fluessen and Sneekermeer, all shallow Frisian lakes (Maasdam & Claassen 1998). They have also expanded into the shallow lakes in the Eastern Scheldt Estuary, The Netherlands since 1987 (Smit et al. 1993). The zebra mussel appeared in the largest lake in Central Europe, Lake Balaton, Hungary in 1932 although it had been present in the country since 1794 (Clarke 1952). In 1838 it was recorded in Germany in the Elbe (Nowak 1971), in Copenhagen, Denmark by 1840 (Morton 1979), and Sweden in the 1920s (Jansson 1994). On mainland Europe it is found as far west as France (Kinzelbach 1992) and may soon be expected in Spain.

Their spread to the alpine lakes, Geneva, Zurich and Constance in Switzerland, probably occurred in the 1960s (Binder 1965). *D. polymorpha* spread to northern Italy (Giusti & Oppi 1972) and reached Lake Garda in 1969/70 (Annoni et al. 1978). Elsewhere it reached Pärnu Bay, Estonia by the 1840s (Schrenk 1948), and was found in Riga Bay (Latvia) in 1996. More recently the species spread to the Gulf of Finland (Kotta et al. 1998), perhaps from nearby Lake Ladoga in Russia where it had been present for 150 years. It has abruptly expanded it range over a period of two-hundred years in accordance with human commercial and leisure activities, leaving some water bodies within its range uncolonised for several decades (Olenin et al. 1999).

Zebra mussel populations in European lakes, reservoirs, ponds and lagoons tolerate a wide range of environmental conditions. Expansion of this species is notable in lakes where there is an abundance of phytoplankton and alkaline water with calcium and magnesium salts used for shell production. Once established zebra mussels normally remain except following salinity increases or where there has been a serious deterioration of water quality. Once established population abundance may fluctuate due to feedback effects of depletion of phytoplankton and consumption of their own veligers.

Ramcharan et al. (1992) found large lakes with comparatively low calcium levels and high phosphate levels had stable populations. Numbers of zebra mussels under some conditions may exceed 100,000 m^{-2} with a biomass of > 10 kg m^{-2} on firm surfaces. At these densities any movements of small fouled objects to another water body could result in a new colony. Dinghies, with zebra mussels, carried overland, are likely to result in new colonies. There has been a steady spread to isolated European lakes and lagoons especially during the last fifty years, enhanced by the expansion of aquatic leisure activities (Kinzelbach 1992). In western Europe there are currently three regions of expansion, the Gulf of Finland, Northern Italy and Ireland, in addition there is a re-colonisation in some eastern German rivers following improvements of water quality.

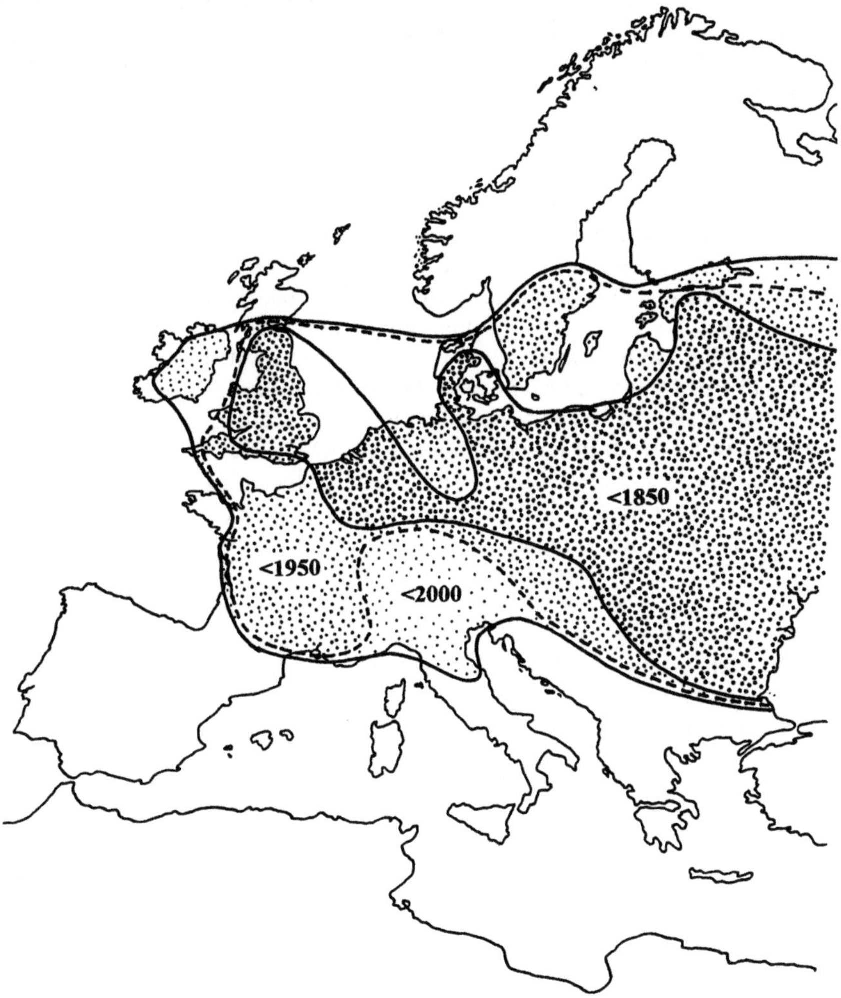

Figure 1. Ranges of the zebra mussel in western Europe. Early expansion < 1850 (heavy stipple), spread to < 1950 and recent expansions (light stipple) in Gulf of Finland, Italy and Ireland.

The establishment of *D. polymorpha* in Britain probably arose following imports of damp timber from the Baltic region. In the early 1800s timber was rafted downriver and exported from Baltic ports including Klaipeda, in the Curonian Lagoon (Lithuania), but also Riga (Latvia), and Gdansk (Poland) (Kerney & Morton 1970). The timber was exported to several other European ports and on arrival may have been further transported by river and canal. Zebra mussels could also have been transmitted on stones used as ships' ballast (Kerney & Morton 1970). By 1924 the zebra mussel was already well established on the Thames. It reached Yorkshire in 1831-33, a year later in the lowlands of Scotland and Exmouth, on the southwest coast of Britain in 1842, all probably with further timber imports. It was present on many midland navigable waters by 1850 (Kearney & Morton 1970). The species continues to spread to isolated water bodies, such as flooded quarries although in several regions it has not become established, either because of marginal or unsuitable conditions for their establishment (Boycott 1936) or due to the absence of an inoculation vector.

3 Expansion in North America

A remarkable expansion of the zebra mussel continues in North America. It was first identified in 1988 from Lake St Clair, near Detroit (Hebert et al. 1989) although probably established since 1986 having arrived either attached to anchor chains or more probably from ships' ballast water originating in Europe. By 1989 it had spread downstream to western Lake Erie and satellite populations had evolved in Lakes Michigan and Huron in 1990. Within seven years of its introduction it spread to eighteen American states and two Canadian provinces (Johnson & Padilla 1996) (Fig. 2). Commercial vessels, leisure craft and larval dispersal downstream facilitated its spread. By 1991 it had expanded to the Hudson River (Strayer et al. 1996), New York's Finger Lakes region and was reported from the Mississippi River (Allen et al. 1998). It is unfortunate the Great Lakes were high in the watershed because this inevitably resulted in veliger dispersal to the great plain region via the Mississippi River and eastwards to the St. Lawrence and Hudson rivers and estuaries. By the end of 1993 dispersal had extended to the Gulf of Mexico. Spread also occurred upstream to uninfested tributaries of the Mississippi probably via river traffic. Colonisation of ponds, reservoirs and other isolated water bodies on the Michigan State peninsula between Lakes Michigan and Huron, has been rapid since 1993 almost certainly due to the overland transmission of boats with aquatic plants snagged on trailers and engines.

By 2000 the zebra mussel had colonised the Canadian provinces of Quebec and Ontario and within 21 of the United States where it occurs in seven major rivers and reservoirs, estuaries and many ponds principally between the Rocky and Appalachian Mountains. Spread to water bodies in British Columbia south to California is likely to take place with overland transport. Prediction models indicate this region is suitable for colonisation (Padilla et al. 1996). Some boats in transit have been found with living zebra mussels and so expansion to this region, is not only possible, but likely at some time in the future. Their establishment in the Mississippi River provides a further opportunity of spread to other parts of the world with LASH barges (**L**eighter **A**board **SH**ip). These are navigable units, which trade upriver from main ports (in this case New Orleans) and are loaded onto a mother ship for sea/ocean crossings. Since these are again deployed following ocean/sea crossings within a further river navigation, transmission of the species back to Europe is possible to the Rhine and Elbe rivers. Under damp conditions zebra mussels can survive at least 18 days and so may easily survive such journeys. The expansion of the zebra mussel within North America has resulted in new physiological and ecological challenges for a short-lived plastic species which may have further consequences once returned to Europe.

4 Ireland, a zebra mussel frontier, patterns and consequences

The zebra mussel was first noticed in Ireland in the spring of 1997 when a blockage occurred to inflow pipes of a fish hatchery on the lower Shannon River caused by detached druses (McCarthy et al. 1998). At this time zebra mussels were undergoing a rapid expansion and had already been broadcast via boating activity to lakes upriver. It was apparently present in the Limerick Docks at the top of the Shannon Estuary in the spring of 1995; many mussels were seen in the flooded dry dock area, which are presumed to have settled in 1994 or earlier.

The most likely means for their introduction was from used craft coming from Britain or the Netherlands. These were imported for private use on the Shannon Navigation and almost certainly were established in the lower part of Lough Derg following the immersion of a zebra mussel fouled hull (Minchin & Moriarty 1998). Larvae from here could be carried 20 km downstream to become entrained within the Limerick Docks.

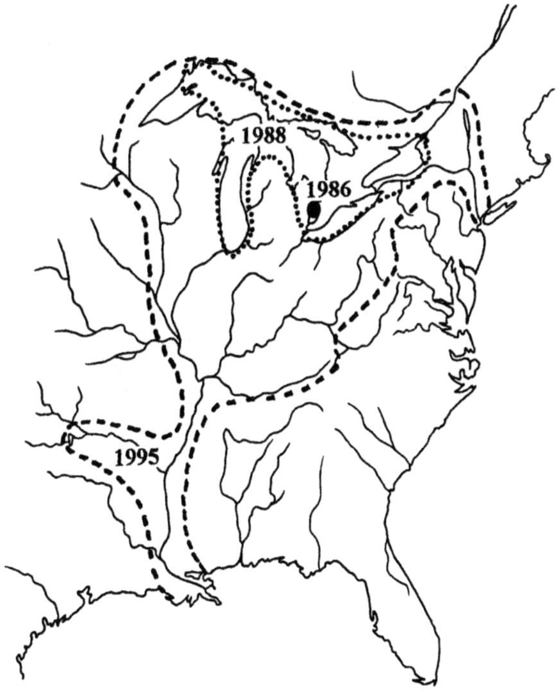

Figure 2. Rapid expansion of the range of the zebra mussel over a ten year period in North America. Seminal inoculation at Lake St. Clair in 1986 (spot), spread by 1988 (dots) and after ten years of its arrival (dashes).

5 Expansion in Ireland

Once established, significant zebra mussel settlements took place on leisure craft and these were carried upstream via locks and swing bridges to the entire navigation on Loughs Derg (11,600 ha), Ree (10,500 ha), Key (900 ha) and several smaller lakes (< 500 ha) by the end of 1996. Following the opening of a recently restored canal in 1994, the Shannon-Erne Waterway, vessels had access to the Erne Navigation. By 1996 zebra mussels had become established in Lower L. Erne (11,000 ha) (Rosell et al. 1999) and in the following year were present in Upper Lough Erne (3,500 ha) (Fig. 3). The 1998 to 2000 summer surveys show general increases in abundance in Shannon lakes (Minchin et al 2002; Lucy & Sullivan 2001). Back-extrapolations of abundance to 1995 (Fig. 4) were based on year classes present in collections made in late 1997 and 1998.

Although the zebra mussel was well dispersed by 1996 it was not apparently established in northern Lough Derg, in most of the west and north of Lough Ree or some smaller Shannon lakes until the following year. It was well established throughout Lough Derg by 1998, but the main expansion in the upper Navigation took place in 1999. Lough Allen, the highest lake on the Shannon Navigation remains uncolonised, although some individuals briefly occurred at its most southern end in 1998 (Lucy 2000). Colonisation may have been compromised by acidic water arising from a catchment with extensive blanket bog.

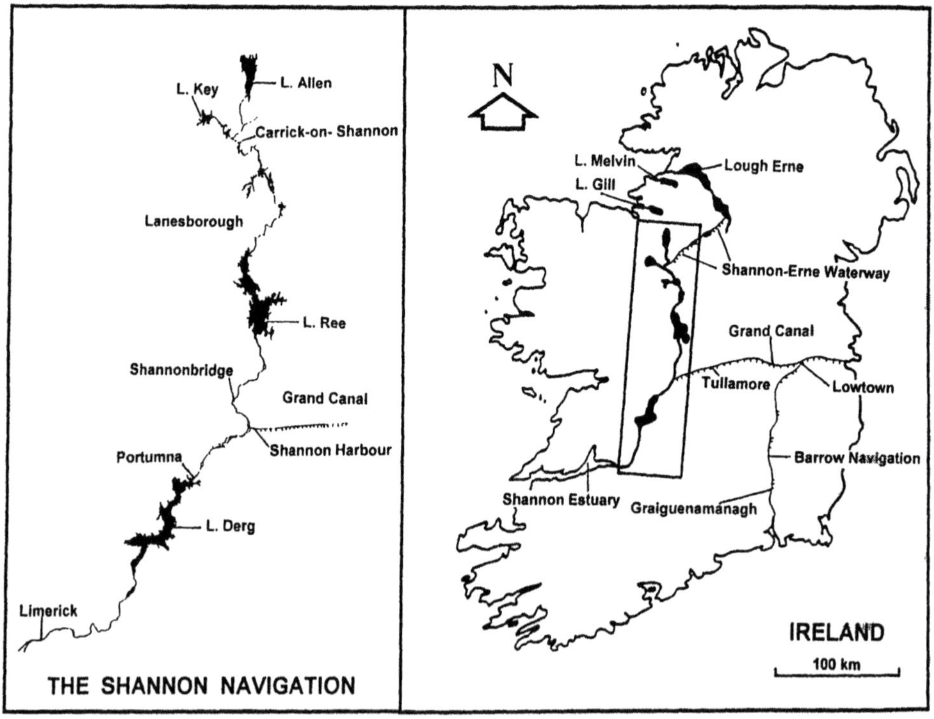

Figure 3. Zebra mussels are abundant on the lakes of the Shannon and Erne navigations and small numbers are found on canals and navigable rivers. Locations appearing in text are shown. Data to end of 2000.

It was inevitable zebra mussels would spread through navigable areas and create new populations either by spawning or from founder clusters dropping from hulls. Boat traffic peaks during the summer months and this coincides with their spawning period (Fig. 5) once temperatures exceed 15 °C. Most boats either congregate at marinas with berthage capacities for ~10-150 vessels and numbers varying according to the time of year. During festival events hundreds of boats can assemble at one location. Should spawning occur on these occasions a large production of larvae is possible. In the Grand Canal and Barrow Navigation canal locks, zebra mussel larvae and settlements are unknown except at Tullamore Harbour. Adults occasionally were found attached to marginal reeds, these probably dropped from the hulls of passing boats. In the late autumn > 50 boats leave the Shannon Navigation to overwinter on the Grand Canal. These carry considerable numbers of zebra mussels as a result of their summer sojourn(s), estimated as ranging from 14 to 40 million on barges 20 m to 30 m length with fewer on cruisers and yachts. In Shannon Harbour on the Grand Canal the overwintering population in 1998 exceeded 120 million. Smaller numbers were found at Lowtown/Robertstown (< 20 million) and Graignamanagh (< 0.1 million) (Fig. 3), yet reproduction in these areas is not presently known. Several classes of craft were fouled with zebra mussels. One 1998 study indicated that 24% of lake-boats used by anglers were infested. Zebra mussels may be spread to different lakes, since anglers regularly move boats on trailers in accordance with the emergence of various aquatic flies (Fig. 6). With the expansion of zebra mussel populations in the Shannon and Lough Erne systems there is an increased possibility of transmission to other important brown trout angling lakes, for example loughs Gill and Melvin, both close to Lough Erne. A public information campaign developed in 1997 used leaflets, posters, seminars and media discussions to make anglers, and other water users aware of the potential of transmitting zebra mussels with plant snagged trailers and by hull fouling (Box 1). This approach appears to have been successful, up to the year 2000 with no further lakes known to have become colonised.

Most boat-owners have developed a higher maintenance routine with more frequent hull cleaning and application of copper based antifoulants. Although more boats were present on the waterways in 2000 the biomass of zebra mussels in transit has been considerably reduced.

Fishing for the European eel *Anguilla anguilla* is licensed in Ireland. Unlicensed captures using fyke nets take place at night. Transmission of zebra mussels to other water bodies is possible should they attach to fouled waterlogged branches which are regularly snagged in nets. Other modes of transmission, such as anglers keep nets, bilge water in small boats have a low perceived risk (Table 1).

Box 1. Management advice provided to boat users and anglers.

(i)	Scour the hull to scrape off mussels
(ii)	Wash down the hull, with hot water if possible
(iii)	Drain the bilge and any other water
(iv)	Remove all weed from boat and trailer
(v)	Dip keep-nets in hot water or keep overnight in a deep freeze
(vi)	Ensure that zebra mussels are not returned to the water
(vii)	Do not leave boats idle over long periods in the water

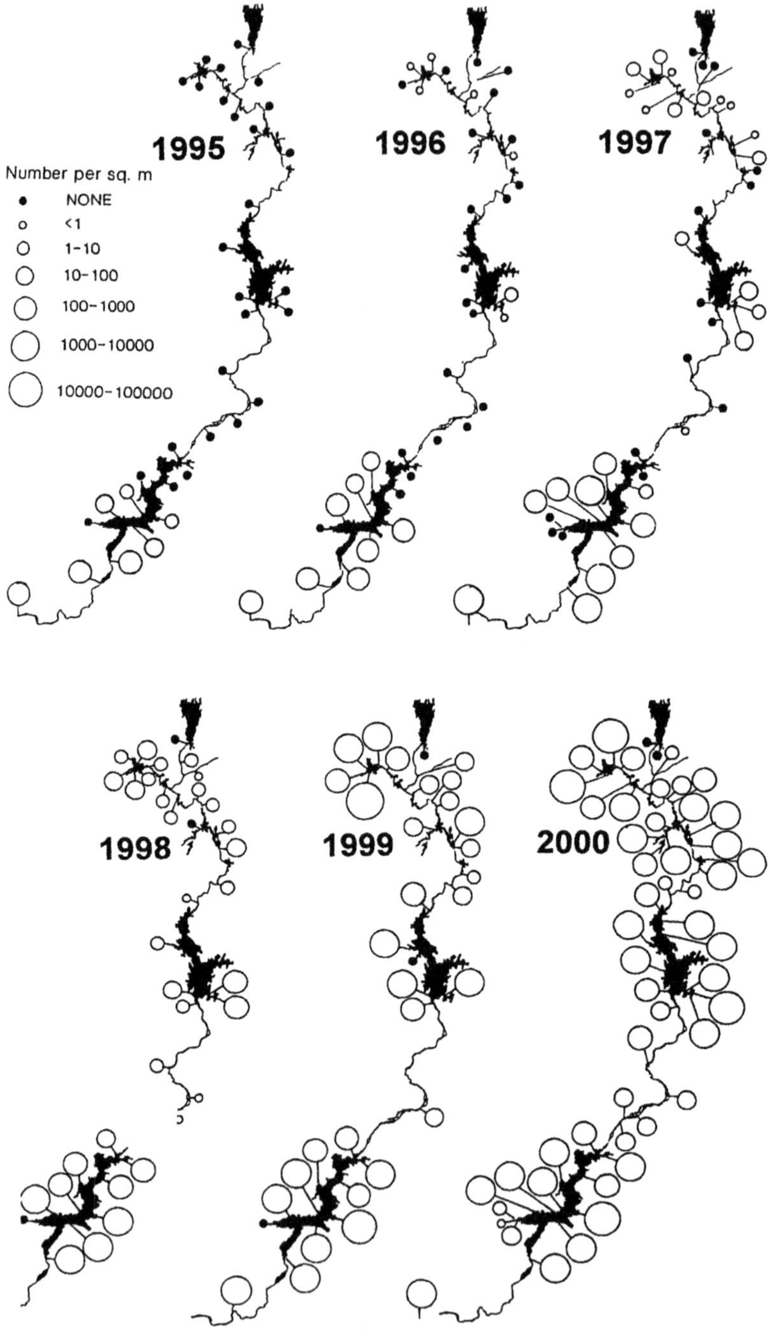

Figure 4. Estimated zebra mussel abundance for July 1995-2000 in the Shannon navigation. Zebra mussel abundance in years 1995 to 1997 were based on back extrapolations of data collected during 1997-2000 and anecdotal information.

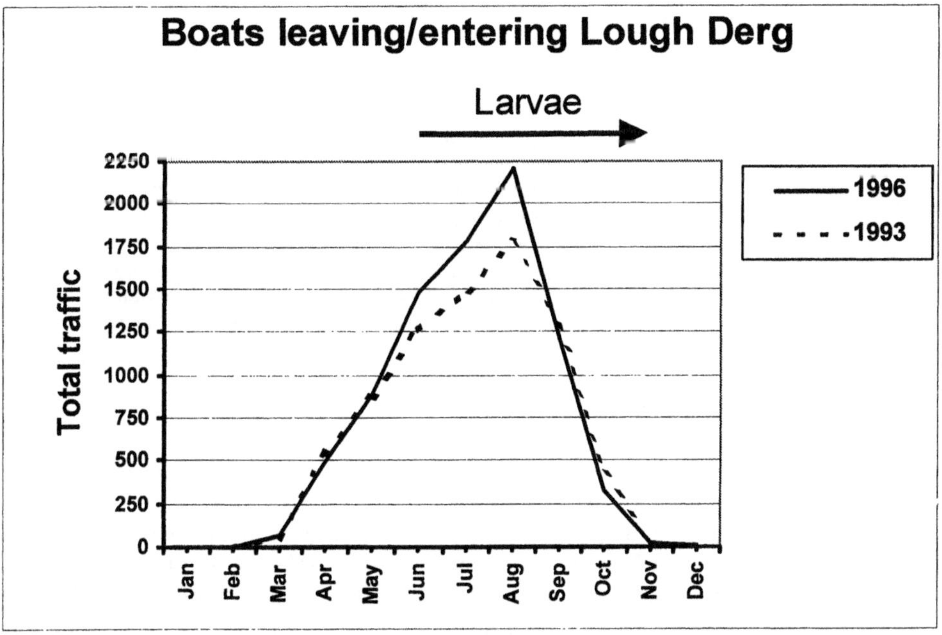

Figure 5. Main periods of boat activity and when zebra mussel larvae are found in the plankton.

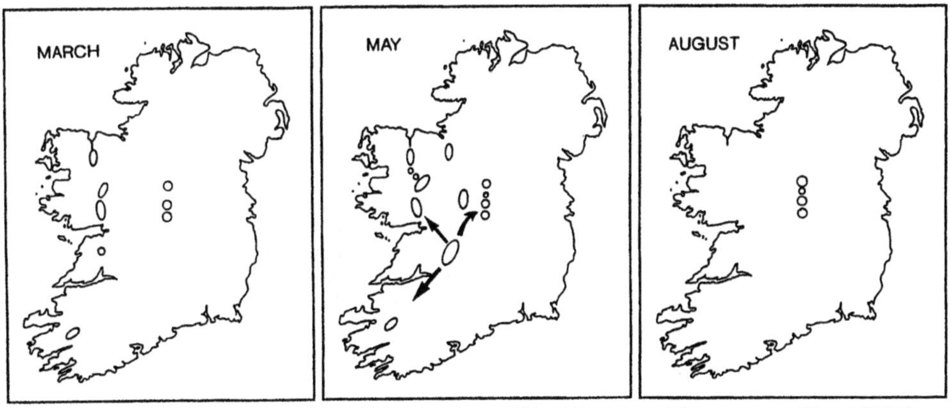

Figure 6. Main fly emergence periods and likely times of angling boat movements.

6 Environmental impacts and natural redistribution

Zebra mussels attained a greater biomass within lakes than in rivers and canals. In some strong current regions of rivers bryozoans and sponges overgrow mussels preventing them from becoming well established; but in lakes zebra mussels form extensive carpets covering most stones, rock surfaces and quay walls attaining densities of 120,000 m^{-2} and biomass of ~5 kg m^{-2}. Their numbers result in large volumes of water being filtered and since their arrival there has been a decline in the levels of chlorophyll and a consequent increase in water transparency.

Table 1. Relative risk of vectors associated with the transmission of zebra mussels.

	Vector	
High Risk	**Some risk**	**Low risk**
Hull fouling of freshwater craft	Aquarium releases	Birds
Ballast water in ships'	Fish stocking	Canoes
Deliberate introduction	Keep nets for coarse fishing	Diving equipment
Overland transport of boats	Research into fish stocks	Bait box water
Jet ski and boat trailers	Anchor and chain	Water in bilges
Eel fyke nets, other nets		Jet skis

Zebra mussels have consequences for lake-populations of bivalves *Anodonta anatina* and *A. cygnea* (Table 2). These unionids bury in sediment, with about < 5 to 15% of the anterior part of their shell exposed. Zebra mussels growing on the exposed shell form dense clusters whose biomass will often exceed that of the unionid. *Anodonta* were abundant in lakes in 1998 and common in 1999; but only three juveniles were recovered from Shannon lakes in 2000 despite a greater sampling effort. Vacant paired juvenile valves were abundant on shallow lake sediments. In the lakes of the Shannon Navigation in 2000 only three living *Anodonta* juveniles were recovered, where in 1998 they were abundant, and in 1999 many living individuals were found. Vacant paired *Anodonta* shells were recovered from shallow lake sediments in areas where formerly living specimens were common in 2000 and paired juvenile valves were commonly cast ashore overwinter. However, it is likely residual unionid populations survive in rivers, riverine areas entering lakes, and in some shallow sandy bays.

Table 2. Impacts in Ireland following the introduction of zebra mussels.

Changes to environment	**Changes to industry & leisure**
Fouling of firm substrata	Pipe blockages
Increased water clarity	Fouling of craft
Increase of submerged macrophytes	Maintenance of navigation marks
Changes to benthos	Lacerations to bathers
Reduction of unionid populations	Municipal water supply
Changes to food-web dynamics	Disturbance from feeding birds
Fouling of marginal aquatic plants	Fouling of lock gate systems
Changes to phosphorus cycling	Water abstraction

Zebra mussels extensively fouled rooted, submerged macrophytes. Those settled on the rhizoids of *Nuphar lutea* in the previous year may ascend the buoyant petioles to the leaves. Two-year classes may be present once settlement has taken place. This burden can cause the leaves to sink to the bottom. Zebra mussels also ascend the stalks of reeds and rushes, the greatest burdens were found on *Phragmites australis* with up to 900 g on a reed at the reedbed margin. Most aquatic plant species could be colonised including exposed roots and submerged branches of trees and on shed leaves and grasses borne into lakes during autumn flooding (Fig. 7). In some shallow muddy lakes, reeds and unionids provide the main substrata. Zebra mussels on plant fragments are redistributed by wind and water currents to different lake regions and also sometimes up-river. Storms, cattle tramping through reedbeds to drink and bird activities result in many reed breakages. The filamentous alga *Cladophora* can become buoyant with trapped gas during the late summer. Its filaments tangle to form floating mats and these can bear away zebra mussels to which it has attached. Some of these may survive but most are washed ashore.

7 Impacts on human activity

The water of the Shannon Navigation is principally used for leisure activities, drinking water and electricity generation. Water abstractors have had pipe blockages and tainted water. Fortunately the main abstraction either takes place on rivers or near river inflow regions of lakes where zebra mussel densities are low. Nevertheless there have been interruptions of drinking water for livestock, private supply and small industry.

The main activities on the Shannon Navigation are boating and angling. Most craft are 5-10 m in length and few exceed 20 m and are based at fourteen marinas, some with haul-out facilities. Blockages of the narrow water intake slots on inboard-outboard engines may take place in the autumn of the first year of settlement but for conventional diesel engines with wider coolant pipes this may occur in subsequent years leading to engine damage from overheating. Instruments on hulls, unless treated, normally become non-functional during the first summer. Fouling increases fuel consumption because of greater drag. Fouling burdens can sink mooring buoys. However, the impacts have been considerably reduced with the greater public awareness, boats are more regularly maintained and many hulls are antifouled with copper-based paints.

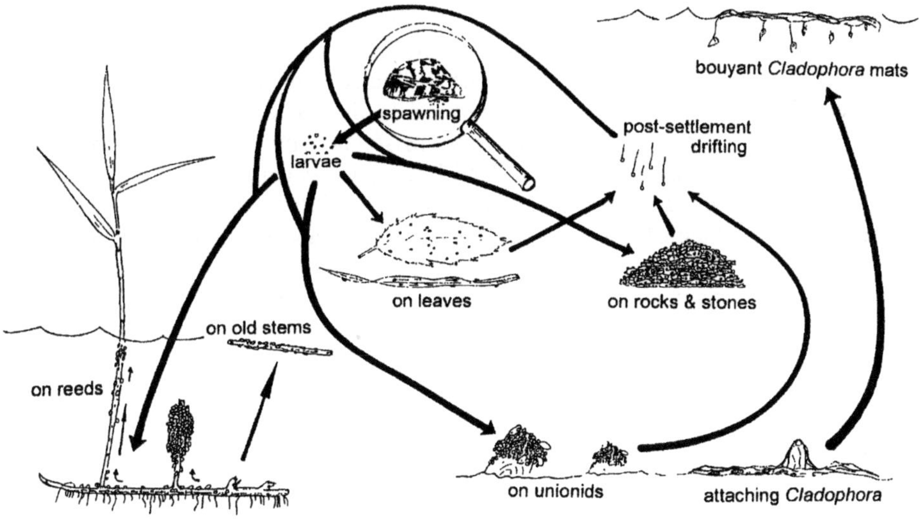

Figure 7. Natural dispersal mechanisms.

Within Europe some notable brown trout *Salmo trutta* angling lakes are found in Ireland. The main fishing period occurs when the burrowing mayfly *Ephemera danica* hatch, other mayfly nymphs have different requirements. Their abundance causes trout to develop a feeding fixation and so are more easily captured. Impacts of zebra mussels on this recreational fishery are unknown. Young trout develop in small streams and rivers and so are unlikely to be affected until they enter lakes where their behaviour could be modified. Zebra mussel druses and 'carpets' provide interstitial refuges and opportunities for detritivores such as amphipods that may become an important food source. Whether the ephemeropteran biomass is likely to change is not presently known. Increases in water transparency, following zebra mussel filtration, leads to an alteration of the abundance and diversity of invertebrates. However, a greater deposition of enriched wastes is likely to lead to complex interactive processes that could modify certain fish populations. Visual predators such as the pike *Esox lucius* may have advantages in targeting prey, yet increases in water clarity will promote macrophyte cover in shallow water areas to provide refuges for small fishes. Pike usually assemble in quiet waters to spawn, where the eggs stick to aquatic plants. The presence of zebra mussels on vegetation, where eggs are normally laid, may effect successful egg attachment and development. Opportunities may arise for the roach, *Rutilus rutilus*; an avid benthophagic predator with pharyngeal teeth that are well suited to feed on zebra mussels (Specziár et al. 1997). Bathing frequency increases in summer when water temperatures may exceed 20 °C. This often takes place close to berthing areas where there may be raw sewage discharges. Zebra mussels in paddling and wading depths result in cuts and lacerations. During July 1999 and 2000 approximately half of the members of each bathing group surveyed had cuts. These wounds may be prone to a wide range of infections that could include Wiel's disease, caused by the bacterium *Leptospira interogans.*

8 Possible future scenarios

It is inevitable the zebra mussel will spread to other water bodies in Ireland. To-date its expansion has been controlled following information campaigns promoting a shared responsibility with known user groups. The difficulty is to maintain this interest over longer periods of time without having to rely on its discovery in a new lake or river and thereby generate a further public awareness. In Ireland, the species has potential for further spread because of extensive midland limestone deposits, providing suitable water quality. Future spread in Ireland will almost certainly result from leisure related activities. The presence of uncolonised lakes near to infested areas carries a high risk and angling associations are well informed about this. In Lough Melvin there are three genetic stocks of brown trout the ferox (*Salmo trutta ferox*), the gillaroo (*Salmo trutta stomachius*) and the sonaghan (*Salmo trutta nigripannis*). These spawn in different parts of the lake, not in rivers (Cabot 1999), with possible reproductive impacts due to zebra mussel colonisation. The whitefish *Coregonus autumnalis pollan,* in Loughs Derg and Erne, may also have reduced reproductive success as a result of fouling of substrata where eggs are laid.

RED KING CRAB (*PARALITHODES CAMTSHATICUS*) AND PINK SALMON (*ONCORHYNCHUS GORBUSCHA*) IN THE BARENTS SEA

VICTOR V. PETRYASHOV[1*], NATALIA V. CHERNOVA[1], STANISLAV G. DENISENKO[1] & JAN H. SUNDET[2]
[1]*Zoological Institute, Russian Academy of Sciences, St. Petersburg, Russia*
[2]*Norwegian Institute of Fisheries and Aquaculture, Tromsø, Norway*
[*]Corresponding author *pvv@aster.zin.ras.spb.ru*

Abstract

In 1961-1969 the red king crab (*Paralithodes camtschaticus)*, brought from the Far East, was introduced into some inlets of the Kola Bay and in the open areas of the Barents Sea. The first red king crab was caught in the Barents Sea in 1974. The crab is now common in all fjords along the northern coast of Norway and coastal areas reaching about 25°E. Total catch for Russia and Norway amounted 200,000 crabs in 2001. Even if 1,500 tons of the north Pacific pink salmon *Oncorhynchus gorbusha* were caught in the Barents and White Sea basins in 1999, it is considered that a stable self-reproducing population has not been established yet in the Barents Sea; it is still necessary to transport eggs of the pink salmon area periodically.

1 Introduction

Barents Sea is the open epicontinental basin, which washes upon the northernmost shores of the European continent. It is the largest sea (1,424,000 km^2) among the Arctic seas (Atlas of Arctic Region 1985). The maximum depth (600 m) is found in the western part of the sea near the border of the Norwegian Sea. Approximately 48% of the sea is shallower than 200 m. The Barents Sea is connected with adjacent seas by rather wide channels and has a good water exchange via them. The warm waters of the North Atlantic penetrate into the southwestern part of the sea, circulate mainly within the southern area and reach finally the northern end of the Novaya Zemlya archipelago. The cold Arctic waters, both formed in the Barents Sea itself and penetrated from the Kara Sea and the Arctic Basin, go from the northeast towards the Bear Island area.

The temperature and ice conditions in the Barents Sea correspond to the allocation of water masses of different origin. In the southwestern part of the sea the water temperature is > 0 °C (0-4 °C) even during the winter period. Off the coast of the Kola Peninsula the sea, as a rule, is free from ice at this time of year. Water temperatures < 0 °C (down to –1.8 °C) are observed in the eastern and northern areas during winter. During summer season the surface water in great part of the sea has a temperature > 0 °C (up to 8-9 °C in the open sea and 12-15 °C in coastal areas). The surface cold water (< 0 °C) remains only in the northern areas, which can be covered by drifting ice.

The salinity of the Barents Sea water is close to normal oceanic values. In summer it varies from 31 PSU (in the southeastern and northeastern parts of the sea) up to 35 PSU (in the southwestern part) and in winter, accordingly, from 33 up to 35 PSU. The river discharge into the Barents Sea is rather small. Therefore marked and extensive freshwater influence takes place only during summer periods in the areas close to the Pechora

E. Leppäkoski et al. (eds.), Invasive Aquatic Species of Europe, 147–152.

River mouth (Atlas of Arctic Region 1985; Oceanography Atlas for the Winter Period 1997; Oceanography Atlas for the Summer Period 1998).

2 Origin and biogeography of the fauna

The recent fauna of the Barents Sea was formed in Holocene when hydrographical conditions close to those prevailing today were established. Continued changes were due to climate fluctuations caused by changes of both volume and temperature of Atlantic waters penetrating the Sea (Galkin 1998).

The Barents Sea fauna includes about 2,700 species (Bryazgin et al. 1981; Sirenko & Piepenburg 1994). The majority of present-day zoologists draw the conventional border between Atlantic Boreal and Arctic faunistic areas in the Barents Sea. It goes from the South Cape (Western Spitsbergen) to the Bear Island and further southeast to the Eastern Murman coast of the Kola Peninsula (Antipova et al. 1989).

3 Red king crab acclimatization

The idea to introduce the Pacific boreal crab *Paralithodes camtschaticus* in the Barents Sea arose in the end of the 1920s. The first experiments on transportation of the red king crab and its larvae from the Pacific coast to the Barents Sea were undertaken in the 1930s and 1950s. Task-oriented efforts to introduce *P. camtschaticus* were carried out in 1961-1969. The transplant material was transported by aircrafts (28 trips) and ships (6 trips), taken mainly in the Bay of Peter the Great Bay (Japanese Sea), a smaller part at the western coast of Kamchatka (Ohotsk Sea). In total 1.5 million larvae, 10,000 young crabs and more than 3,000 adults were released (Orlov 1962, 1977).

In 1961-1969 the crabs, brought from the Far East, were released in some inlets of the Kola Bay and in the open areas of the Barents Sea. At the end of the 20th century the geographic range of *P. camtschaticus* expanded from the coastal zone of the Kola Peninsula and covered an area from western coast of the White Sea Funnel up to Kanin-Kolguev shallow and Goose Bank in the east. In the Barents Sea, the distribution of the crab in many respects reflects the distribution of warm Atlantic waters. The species lives down to 320 m depth on gravel-pebble substrates, shifting with increasing depth to muddy and sandy substrates, at the temperature of 0.4-7.0 °C and salinity at 34.4-34.7 PSU (Kuzmin & Lobodenko 1997). These conditions coincide well with the tolerance range of abiotic factors for Pacific populations of *P. camtschaticus* (Rodin 1985; Klitin 1996). Further spread of the crab in the eastern and northeastern direction has probably reached its limit. It is impossible to say the same about southwestern and northwestern limits of the potential geographic range. After successful colonizing of the Norwegian Sea shelf, the crabs probably have penetrated areas to the south from Spitsbergen.

The first red king crab was caught in the Barents Sea in 1974. During the next years the geographic range and abundance of *P.camtschaticus* increased in the region. Further increase of its abundance is obviously going on, but long-term fluctuations are characteristic. The years with fast increase of abundance are followed by years with a significant reduction. In Norwegian waters the crab was mainly caught during gillnet fishery and until 1992 the catches there were of occasional nature and only small numbers were taken. According to the latest reports the crab is now common in all fjords and coastal areas reaching about 25°E. The abundance of crabs in the Barents and Norwegian seas has increased from 212,000 individuals in 1993 up to 7,263,000 in 1998 (Kuzmin 1999). There are almost three times as many legal size crabs in the Russian part of the Barents Sea compared to the Norwegian part.

Table 1. Stock estimates of red king crab in Russian and Norwegian waters in the period 1993– 2000 and estimates of total stock and legal size males of carapace length larger than 117 mm or 132 mm.

Estimated stock (x 1,000)				
Year	Russia	Norway	Total legal males	
			117 mm	132 mm
1993	117	95	75	
1994	310	62	149	
1995	660	140	374	304
1996	272	165	-	242
1997	510	206	-	426
1998	6 768	495	-	951
1999	4 948	na	-	1 508+na
2000	12 546	na		1513 + 676

The spawning of *P.camtschaticus* in the Barents Sea takes place from the end of March to April, and settling in the second half of June. The females carrying eggs have carapace breadth > 100 mm, half of them even > 130 mm (Oganesyan et al. 1998). Physiological maturity of males is reached at breadth about 80 mm, but they start mass participation in copulation at carapace breadth about 130 mm (Gerasimova et al. 1996). In the Barents Sea males have maximum carapace breadth at 270 mm, females at 220 mm.

4 Ecological and economic consequences

On the basis of stomach contents, important food items for P. *camtschaticus* in the Barents Sea are bivalves (up to 39% of stomach contents), gastropods (35%), and echinoderms (66%). Less common are polychaetes and, in the spring season, fish roe (up to 5%) (Kuzmin et al. 1996). Along the Norwegian coast the king crab feeds mainly upon polychaetes, bivalves and algae (Sundet et al. 2000). In the moulting and mating period the crab inhabits shallow hard bottoms with food objects that differ from those of the deeper soft-bottom areas. This is assumed to be the reason for the observed differences between sexes and season in stomach content.

It is possible that *P. camtschaticus* is a food competitor of the small-sized crabs *Hyas coarctatus* and *H. araneus*, the Atlantic craboid *Lithodes majus*, and hermit crabs. Fragments of red king crab juveniles were repeatedly found in stomachs of cod, catfish (*Anarhichas lupus*), halibut, dogfish and spike scat (*Raja radiata*) (Gerasimova et al. 1996; Rafter et al. 1996). Other details of the crab's impact are still unknown.

Red king crab is becoming an important object for commercial fishery in the Barents Sea. However, until recently it was caught only together with fish and there was no specialized crab fishery. The total quota (TAC) for both Russian and Norwegian part of the Barents Sea increased from 22,000 specimens in 1994 to 75,000 in 2000 (Sundet 2000). According to recent evaluations (Kuzmin 1999), the allowed quota of specialized fishery of the red king crab in Russian waters only could amount 377,000 individuals (about 1,500 tons) at 25% level of fishery mortality (only males > 150 mm) in 2000. The crab seems to have adapted quite well to its new environment and feed on most benthic species available. It is the only species of commercial value in the Barents Sea that is exclusively exploiting the benthic production. Being a high price seafood product the crab is wanted as a fishing resource, but heavy bycatches of it in the gillnet fishery have raised demands in Norway for eradication of the species in this area.

5 Pink salmon

Acclimatization of salmonids of the genus *Oncorhynchus* into the Barents and White Sea basins started in the 1940s. Taibolsky and Onezhsky fishery farms began to incubate eggs of the chum salmon (*O. keta*) from Amur River. About 9 million fries were released in rivers from 1933 to 1939 (Alekseev & Kulachkova 2000). However, only a few adults returned from the sea back into rivers, and acclimatization efforts were stopped. In 1956 acclimatization of *O. keta* and the pink salmon *O. gorbusha* into the White and Barents Seas started again. Large amount of eggs of *O. keta* was transported and incubated, but return of adults was insignificant again, and the main attention was turned to work with the pink salmon. About 189 millions of pink salmon fries were incubated from eggs brought from the Russian Far East during 1958-1979. In the 1970s, fishery of the pink salmon was permitted officially and average annual catches were about 115 tons or 77,000 specimens (Alekseev & Jakovenko 1999).

In order to check if a self-reproducing population of *O. gorbusha* exists in the European North, transportation of eggs from the Far East was stopped in 1979, 1981 and 1983. It became apparent that a stable self-reproducing population of *O. gorbusha* was not generated. At present the pink salmon acclimatization continues successfully, but nevertheless, it is considered that a stable self-reproducing population of *O. gorbusha* has not been established yet (Zubchenko et al. 1998). In order to support the stock at a sustainable level in any climate and hydrographical conditions, it is still necessary to transport eggs from the Magadan area periodically (Alekseev & Kulachkova 2000).

O. gorbusha is an anadromous, mainly boreal species, naturally inhabiting seas of the northern part of the Pacific Ocean and adjacent areas of the Arctic. It occurs also in low numbers in the Lena and Mackenzie rivers. In years of its maximum quantity, pink salmon, acclimatized in the European North, entered rivers from Scotland, Iceland, Svalbard and Finmarken in the west to Enissey River in the east. Spawning migrants have been recorded in rivers of the Novaya Zemlya archipelago (Kalyuzhniy et al. 1990) and of southeastern part of the Barents Sea basin: Volonga, Pechora and its tributary Zil'ma, and Korotaikha. Pink salmon arrived in the Kara Sea in 1963, and more than 1,000 spawning migrants were caught in its southwestern part, from Pyasina River to Enissey River, in 1977. The natural boundary between native Northern Pacific and acclimatized European populations of the pink salmon is the Wilkitsky Strait, covered by ice almost all over the year.

The main spawning grounds of the acclimated pink salmon are located in rivers of the Kola Peninsula and the White Sea basin. The spawning migration takes place from May to October, with maximum in July-August. It spawns in middle and upper streams from August to October. For example, in Keret' River (White Sea basin) the pink salmon spawns from the second half of September to the beginning of October (water temperature 9.3-2.5 °C); in Volonga River the spawning was observed from July 22 to September 30 (maximum in August 15-30); nests were recorded along a stream from Semuzhya River to Beluginskaya Zolotukha River (ca. 20 km). Water temperature at breeding season varies from 7 to 19 °C in different areas. Average fecundity is 1,500-1,800 eggs per female. After spawning individuals die. Larvae leave the nests in May-June, at the end of high-water period, and start to drift downstream. Young fishes enter the sea in June-July; their food consists of water insects and their larvae. In estuaries juvenile pink salmon feed upon freshwater and marine plankton; at the end of autumn they go into the open sea (Surkov 2001). Total average length of the pink salmon in the White Sea is 43-52 cm and weight ca. 1.5 kg.

From 1956 to 1999 the number of spawning migrants varied much in rivers of the Kola Peninsula and the White Sea. During the first period of acclimatization the maximum catch of the pink salmon was ca. 250 tons (in 1973; Gorelov 2001). When the egg transportation from the Far East was stopped in the 1980s, the stock greatly decreased. Since 1985 pink salmon entered rivers only in the odd years, because the generation of the even years was lost during unfavorable hydrological conditions. Since 1991 the stock increased: total catch in the Barents and White Sea basins was 960 tons in 1997, and 1,500 tons in 1999 (Gorelov 2001). For Kola Peninsula the catch was as follows (specimens caught):

Year	Numbers	Year	Numbers
1960	74,000	1977	108,000
1965	48,000	1989	32,000
1971	28,000	1991	60,000
1973	144,000	1993	177,000
1975	104,000	1995	350,000

Pink salmon is numerous also in coastal waters of the west Finmark, where it arrived in 1960. It arrives there at the second half of July and enters rivers for breeding. At Finmark coast, catches increased to a maximum in 1973, but decreased later. In 1977 total catch in this area was 12,800 individuals (> 18 tons), including 5,600 salmons caught in the coastal zone.

Recently spawning migrants of *Parasalmo mykiss* (Walbaum, 1792) (syn. *Salmo gairdneri*, *Oncorhynchus mykiss*) have been recorded in small amounts in a few small rivers of the Murman coast). Obviously specimens escape from fish farms in this region. It is unclear if the breeding and successful development of fry is possible in the wild.

6 Ecological and economic consequences

Sea period of the life cycle of acclimatized pink salmon has not been studied (Karpevich et al. 1991). Presumably it spends about 13 months in the southeastern Barents and Norwegian Seas, feeding probably in the upper 100 m layer. Food comprises of fish and crustaceans (Surkov 2001). When fries are set free into the sea from fish farms, significant amount of juveniles may be eaten by herring. In rivers the pike feeds upon pink salmon juveniles (Pervozvansky & Bugaev 1992).

A question discussed is whether the pink salmon acclimatization is harmful for the stock of the native Atlantic salmon *Salmo salar* or not. Majority of researchers thinks that there is no negative influence. During the spawning period there is no competition for nesting territory with native salmonids (*Salmo salar* and *S. trutta*), and there is no food concurrency with salmon during larval and fry period during feeding of young fishes in estuarine areas and adults offshore (Surkov 2001).

Acclimatization of the pink salmon in the northern parts of Europe increased the fish productivity of the area. This is positive trend on the contrary to a background of general tendency of decreased stocks of main commercial fish species. The most long-term ways of further development of the situation has to be the prolongation of acclimatization works, effective fish farming of fry, control of spawning migrants number for native reproduction and rational fishery.

ALIEN FRESHWATER FISHES OF EUROPE

HANNU LEHTONEN
Department of Limnology and Environmental Protection, University of Helsinki, Helsinki, Finland
hannu.lehtonen@helsinki.fi

Abstract
A total of at least 76 alien freshwater fish species belonging to 21 families have been introduced into European fresh waters. The majority of them originate from North America (34) and Asia (26). Most introductions have been successful at least in one country. Nowadays at least 50 originally non-European freshwater fish species inhabit European fresh waters in self-sustaining populations. Economically most important invasive species are *Oncorhynchus mykiss*, *Salvelinus fontinalis*, *Ctenopharyngodon idellus*, *Aristichthys nobilis* and *Hypophthalmichthys molitrix*. These species may form 20-30% of the catch in several areas and *Oncorhynchus mykiss* is very important for aquaculture. In most cases the ecological impacts of invasive species have been small. However East Asian carps, especially *Ctenopharyngodon idellus* have caused considerable changes in native ecosystems and fishes by eliminating aquatic macrophytes and by resuspending sediment increasing water turbidity. Also predation and competition by invasive species have caused reductions and collapses of native fish populations.

1 Introduction

The history of the introduction of alien fish species into European waters began already over 2000 years ago when Romans moved the common carp (*Cyprinus carpio*) from the Caspian and Black Sea basins for fish culture (Balon 1995). However, large-scale movement of fish from other continents into Europe is a comparatively recent phenomenon and since the 1850s the pattern of introductions has shown a steady increase. Alien fishes have been introduced for a variety of motives including aquaculture, sport, the improvement of wild stocks, ornamental trade, and biological control (Lever 1998). Fishes are also introduced accidentally together with other species (e.g. shortfin eel, *Anguilla australis*, *Opsariichthys uncirostris*, brook stickleback, *Culaea inconstans*). In natural waters the invasive species have in many cases established breeding populations.

Introductions have been continuous since the early part of the 19th century when North American centrarchids (*Micropterus* and *Lepomis*) and salmonids (*Oncorhynchus mykiss* and *Salvelinus* spp.) were imported. During the 1920s and 1930s a large number of alien fish species were moved into European fresh waters for general fishery management purposes. The third period of intensive introduction was during the 1960s and 1970s (Welcomme 1991).

This paper summarizes the introductions of non-European fish species and their effects on ecology of the waters. Europe is considered by its generally accepted eastern boundaries west from Ural mountains and north from Caucasus mountains. The list of introduced freshwater fish species (Table 1) is based on the following sources: Welcomme (1988, 1991), Holcík (1991) FAO (1997), Gerstmeier & Romig (1998), Keith & Allardi (1998), Bianco (1998), Maitland (2000), DIAS (2001) and Froese & Pauly (2001). However, this list does not include the several hundred of species which have

E. Leppäkoski et al. (eds.), Invasive Aquatic Species of Europe, 153–161.

been moved by the aquarium fish trade (Conroy 1975) being candidates of potential introduction into lakes or rivers.

2 Introduction of non-European freshwater fishes

Welcomme (1991) referred to 113 and Holcik (1991) 134 non-native fish species or subspecies which have been introduced into European fresh waters. However, their lists contain also many native European fishes which have been translocated into new water bodies within Europe. When native European species are excluded, at least 76 true exotic species belonging to 21 families are recognized (Table 1). The most numerous introduced species involve cyprinids and salmonids. Altogether at least 28 introduced species belong to these two families and some of them are widely distributed in Europe (e.g. rainbow trout, *Oncorhynchus mykiss*, brook charr, *Salvelinus fontinalis*, grass carp, *Ctenopharyngodon idellus*, silver carp, *Hypophthalmichthys molitrix*). Other numerically important families are cichlids, centrarchids, poecilids, ictalurids and catostomids.

The majority of introduced species originate from North America (34 species) and Asia (26). According to Holcik (1991) the timing of the first attempts to introduce fish species from other continents into Europe are not exactly known. However, it included North American salmonids, catfishes and centrarchids. Brook trouts (*Salvelinus fontinalis*) were transferred to the United Kingdom already in 1869. This species was followed by many other salmonids; chinook salmon (*Oncorhynchus tshawytscha*), rainbow trout (*O. mykiss*), lake trout (*Salvelinus namaycush*) which were first introduced to France, Germany and Switzerland. Catfishes (black bullhead (*Ameiurus melas*) and brown bullhead (*A. nebulosus*)) and several centrarchids (redbreast sunfish (*Lepomis auritus*), pumpkinseed (*L. gibbosus*), smallmouth bass (*Micropterus dolomieui*), largemouth bass (*M. salmoides*)) were also introduced before the end of the 19th century.

The Common carp, native to Asia, was introduced already more than 2000 years ago but some other nowadays important cyprinids, such as bighead carp (*Aristichthys nobilis*), grass carp (*Ctenopharyngodon idellus*) and silver carp (*Hypophthalmichthys molitrix*) were introduced after the Second World War to several countries in Central and Southern Europe. Today they have an important role in the ecosystems of several lakes and rivers in Europe.

3 Establishment of breeding populations

Welcomme (1988) remarked on six possible patterns of behaviour to follow an introduction of a new fish species. The introduced species

(i) never becomes established, disappears without trace or its fate is listed as unknown,

(ii) does not breed under natural conditions and is maintained artificially or by continuous import,

(iii) becomes locally established in an unusual habitat,

(vi) establishes and maintains itself either as widespread populations at very low densities, or as isolated populations in individual water bodies or in aquaculture facilities,

(v) establishes itself and may even increase rapidly in abundance to become a dominant element of the population. Later the species disappears or declines in abundance. The decline may occur naturally or the species may be deliberately eradicated, or

(vi) becomes a significant or dominant element of the host fauna.

All the above mentioned cases can be seen in European fresh waters. Introductions of exotic fish species are usually not based on ecological principles and Holcik (1991) called most introductions a gamble because their potential ecological impacts were not scientifically foreseen. Potential negative impacts include alteration of host environment, disruption of host community, competition, predation, stunting, genetic degradation of host stock, and introduction of disease agents and parasites (Welcomme 1988). Positive impacts of introduced species include increased fish production if the new species is able to fill a vacant ecological niche or gives additional value for aquaculture. The impacts of introductions are, however, not limited solely to ecological effects but may also have both positive and negative socio-economic implications.

The majority of introduced fish species form breeding populations at least in one European country. Of the 76 listed non-European species in Table 1 the introduction of 51 species have resulted in a self sustaining population and 25 have not been successful in any part of Europe.

Only few introductions have resulted in economically important fisheries. Introductions of East Asian species, bighead, grass and silver carps have been successful in some areas of Central and Southern Europe. They now contribute to some 30% of the commercial catches in the Danube delta (Cowx et al. 1996). The culture of the tropical African Nile tilapia (*Oreochromis niloticus*) has produced high yields in cooling waters of Belgian power plants (Welcomme 1991). Also the salmonids rainbow trout and brook charr live throughout Europe. The most remarkable gain from introduced species is the high production of rainbow trout by European aquaculture. Introduction of Siberian peled whitefish (*Coregonus peled*) has also created economically important fisheries in at least Russia, Poland, Germany and Finland.

However, the economically valuable introductions are exceptions and the majority of attempts to establish a breeding population has failed or resulted in economically meaningless populations. Also the few successful introductions have in most cases succeeded only in one or only a few countries probably due to climatological reasons. Introduced species have in some cases become abundant, but they are economically insignificant. For example, an East Asian cyprinid, stone moroko (*Pseudorasbora parva*), was accidentally introduced in Romania in 1961. By the end of 1980s this species was found in all Central European countries (Gerstmeier & Romig 1998) but its presence is still insignificant for fisheries. The same concerns also the introductions of North American ictalurids and centrarchids which today form abundant but from the fisheries point of view insignificant populations in numerous European lakes.

The introductions of Pacific salmons have resulted in breeding populations only in Kola peninsula (Russia) (pink salmon (*Oncorhynchus gorbuscha*) and chum salmon (*O. keta*) and in France (coho salmon (*O. kisutch*) (Keith & Allardi 1998; Froese & Pauly 2001). The three introduced non-European sturgeon species, Siberian sturgeon (*Acipenser baeri*), lake sturgeon (*A. fulvescens*) and white sturgeon (*A. transmontanus*), have not established self reproductive stocks. Rapidly disseminated and favourable species such as East Asian carps and rainbow trout have established reproductive populations only in Southern and Central Europe, whereas in Northern Europe their stocks have to be maintained artificially. These species are maintained today predominantly in aquaculture.

4 Ecological impacts

Introduced fishes have in many cases disappeared without causing any ecological impacts. However, most species have established reproductive populations at least in some parts of Europe and some species have caused complex ecological implications e.g. by changing the structure of food webs (Kolar & Lodge 2000). When established, the alien fish species may have various biological effects (see below). Fortunately, in most cases only minor ecological impacts have been documented so far.

4.1 HABITAT ALTERATION

Alterations in habitats have occurred in many lakes after the introduction of East Asian carps. They usually increase water turbidity, and the increased mud suffocate both macrophytes and benthic invertebrates. More rapid recycling of nutrients contributes to accelerated eutrophication (Welcomme 1988). As a result the composition and abundance of the native fish fauna has been altered. Introductions have also had considerable effects on macrophytes, phytoplankton and invertebrates. For example, in some waters stocking with grass carp resulted in dramatic changes of the species composition and total fish catch, as well as in the physical and chemical properties of the water (Holcik 1991).

The grass carp has been introduced with the intention to eliminate underwater vegetation. However, through selective feeding on more tender species grass carps may promote the development of tougher vegetation, such as e.g. *Phragmites australis*, which is even a greater problem. Moreover, the removal of submerged vegetation reduces the spawning substrates of several phytophilous fishes and may inhibit their reproduction. Also increased turbidity has disrupted the reproduction of native fishes (Taylor et al 1984; Holcik 1991; Mikschi et al. 1996). For example, in Lake Balaton (Hungary) extinctions of native fish and invasion of new alien species have alternated. Alien species, such as east Asian carps have transformed the food webs, causing long-lasting ecological changes (Biro 1997)

4.2 COMPETITION AND PREDATION

Sometimes introduced species have been able to outcompete native fishes resulting in a considerable reduction in their populations or even in their complete disappearance. For example, a number of fishes which were accidentally introduced together with grass carp, resulted in declines in local species through superior growth and fecundity (Rosenthal 1976). In some Italian oligotrophic lakes pumpkinseed (*Lepomis auritus*) supplanted Italian bleak (*Alburnus alborella*). Another predator that has been blamed for the disappearance of local species is largemouth bass (*Micropterus salmoides*). It has been demonstrated to have caused the decline of Italian bleak, pike (*Esox lucius*), perch (*Perca fluviatilis*) and introduced *Lepomis* species in some Italian lakes (Welcomme 1988).

Aggressive behaviour of the introduced species may alter distribution patterns of native fishes as shown by studies on interactions between European brown trout (*Salmo trutta*) and Arctic charr (*Salvelinus alpinus*) (Nilsson 1985). However, in the lakes of the Hautes-Pyrenees region (France) the introduction of rainbow trout, brook charr, lake trout and the hybrid between brook charr and lake trout (*S. fontinalis x S. namaycush*) did not cause any ecological effects on the native brown trout (Delacoste et al. 1996) However, in Lake Ohrid (Macedonia) the introduced rainbow trout have been involved in the reduction of native salmonid populations (Nijssen & de Groot 1974). Rainbow trout may also be a competitor to native salmonids, as they to a large extent have overlapping habitats and dietary requirements (Weidema 2000).

One consequence of the introductions of alien species is the elimination of local species by predation or competition. For example, seven alien fish species, e.g. grass carp and pumpkinseed were introduced into Neusiedler See (Austria) resulting in the disappearance of four indigenous fishes (*Umbra krameri, Leucaspius delineatus, Misgurnus fossilis, Proterorhinus marmoratus*). Concurrently the abundance of several other species (e.g. tench, *Tinca tinca*) declined considerably, while some planktivorous species (e.g. ziege, *Pelecus cultratus* and bleak, *Alburnus alburnus*) increased in population density (Mikschi et al. 1996). Stone moroko (*Pseudorasbora parva*) forms dense populations in some waters and it is blamed for heavy competition with other fish species and for predation on their eggs and larvae (Holcik 1991).

4.3 STUNTING

The quality of the fish stock of any water area can deteriorate through stunting by the introduced competing species. Stunting is a process whereby the population of a species expands rapidly, producing large numbers of individuals which mature and breed at a much reduced size. This behaviour considerably diminishes the usefulness of the population for sport or commercial fishing purposes. Fish species which have been reported as producing stunted populations include: *Alburnus alburnus, Carassius auratus, Lepomis cyanellus, L. gibbosus, Oreochromis mossambicus, O. niloticus, Perca fluviatilis, Salvelinus alpinus, Coregonus lavaretus* and *C. peled* (Welcomme 1988; Salonen & Mutenia 1992).

4.4 DISEASES

Pathogens and parasites associated with the introduced fishes may spread to the receiving area. This is particularly conceivable in areas where the fishes are cultured. They are thus concentrated which increases their susceptibility to diseases. Several examples of outbreaks of new diseases or parasite infestations following fish transfers are documented in the literature (e.g. Welcomme 1988). Accordingly, the parasite fauna of Europe has increased with several alien species. These include, for example, the tapeworms *Botriocephalus acheilognathi* and *Khavia sinensis* which have been widely distributed in fish farms of Central Europe (Holcik 1991).

The appearance of nematode parasites of the genus *Anguillicola* in Europe is also an example of parasite infection caused by an introduced species. The parasite, which is endemic to Australian and Asian eels, was introduced into Europe together with imported oriental eels intended for human consumption and not for stocking into natural waters. Nevertheless the nematode escaped and has since spread rapidly through the waters of Northern Europe (Welcomme 1988). Another introduced disease, probably originating from ungutted Pacific salmon carcasses imported for consumption, is IHN (infectious haemato-poietic necrosis), which has been recognized in Southern and Central Europe (Roberts & Shepherd 1997).

These examples of introductions indicate the potential dangers of transporting any biotic material over frontiers. The ASS disease or *Furunculosis* appeared in Europe following the introduction of rainbow trout from Western North America (Snieszko1973). Similarly, infectious dropsy of cyprinids spread rapidly around continental Europe in the 1930's after carp were transferred from Yugoslavia without adequate sanitary controls (Welcomme 1988).

5 Economic and social consequences

The introductions of alien fish species have accomplished very variable results depending on the species and geographic area. Only few fish species are generally accepted as beneficial at having some positive or socio-economic effects, usually by improving fishing or aquaculture opportunities. Among the 51 exotic species some have proved to be valuable for the fishing industry. For example, European aquaculture has gained benefits from rainbow trout, common carp and East Asian carps. In natural waters the introductions have resulted in economically profitable fisheries in many cases (e.g. common carp, East Asian carps, some centrarchids, peled whitefish) although most introductions have failed or led to unwanted consequences in the form of reduced or collapsed native fish stocks. Most introductions have, however, been harmless from the fishery point of view. Most freshwater fish introductions have not been based on risk evaluation and ecological awareness. Management should therefore address both beneficial and negative aspects of exotic species introductions. However, there is often an inadequate knowledge base on which to base policy and management decisions.

Impacts on the socio-economic level are even more difficult to determine than economical effects. According to Weidema (2000), the most obvious effects are the ones that make people change their way of utilising or relating to their local environment. If the introduced fish species has caused serious changes in native fish assemblages or ecosystems, the lost recreational fishing has a social rather than an economic effect in the local society. In cases where the preconditions of professional fishery have changed, the economic impacts may be considerable.

The dangers and risks of introductions have led to measures to control these. There are two alternatives, either actions at international level to reduce the risks of introductions, or ban all future movements of fish species between countries. The latter alternative has been considered impractical and undesirable because of an existing established trade in live fish that would be difficult to disrupt, and the changing needs of the European aquaculture to import new species or races from time to time (Welcomme 1991). European Inland Fisheries Advisory Commission (EIFAC) has adopted a variant of the existing Code of Practice on the Introductions and Transfers of Marine Organisms 1994 by the International Council for the Exploration of the Sea (ICES 1995).

Table 1. List of alien fish species introduced into European fresh waters. Status: (S) successful introduction at least in one European country, (D) disappeared after introduction. Countries: Albania (AL), Austria (AT), Belgium (BE), Bulgaria (BG), Belarus (BY), Switzerland (CH), Czechoslovakia (CS), Cyprus (CY), Czech Republic (CZ), Germany (DE), Denmark (DK), Estonia (EE), Spain (ES), Finland (FI), France (FR), Georgia (GE), Greece (GR), Hungary (HU), Ireland (IE), Iceland (IS), Italy (IT), Liechtenstein (LI), Lithuania (LT), Luxembourg (LU), Latvia (LV), Moldavia (MD), Malta (MT), Netherlands (NL), Norway (NO), Poland (PL), Portugal (PT), Romania (RO), Russia (RU), Sweden (SE), Slovakia (ST), Soviet Union (SU), Ukraine (UA), United Kingdom (UK), Yugoslavia (YU).

Species	Origin of species	First introd.	Countries where introduced	Status
ACIPENSERIDAE				
Acipenser baeri	Northeast Asia	1956	FR, HU, SU, ES	D
Acipenser fulvescens	North America	1969	SU	D
Acipenser transmontanus	North America	1989	SU	D
POLYODONTIDAE				
Polyodon spathula	North America	1974	HU, SU	D
ANGUILLIDAE				
Anguilla australis	Pacific	1975	IT	D
Anguilla japonica	East Asia	1975	IT	D
SALMONIDAE				
Coregonus clupeaformis	North America	1881	DE, FR, UK, CH, NL	D
Coregonus peled	North Asia	1960	PL, FI, CS	S
Oncorhynchus clarki	North America	1960s	SE, DK, CY	D
Oncorhynchus gorbuscha	North America	1939	FI, IE, NO, SE, PL, UK, RU	S
Oncorhynchus keta	North America	1932	FI, SU	S
Oncorhynchus kisutch	North America	1971	FR, DE, GR, IT, NL, CY, ES, CH	S
Oncorhynchus mykiss	North America	1879	FR, DE, BE, CH, UK, AT, HU, RO, SU, PL, ES, SE, IT, YU, ST, CZ, DK, FI, EE, BG, , IE, LI, NO, CY, PT, IS, NL, LU, GR, AL	S
Oncorhynchus nerka	North America	1933	FI, SE, DK	D
Oncorhynchus rhodurus	Eastern Asia	1976	DE	D
Oncorhynchus tshawytscha	North America	1877	DK, FI, DE, FR, IT, IE	D
Salvelinus fontinalis	North America	1869	UK, NO, DK, CH, CS, SE, DE, BE, PL, YU, FI, EE, PT, SU, HU, BG, ES, AT, CY, RO, NL, IT, GR. LT	S
Salvelinus leucomaenius	Japan	1977	DE	S
Salvelinus namaycush	North America	1886	FR, DE, FI, CH, CS, DK, ES, SE, IT, UK, NO, AT	S
Thymallus baikalensis	North Asia	1959	CS, PL	D
UMBRIDAE				
Umbra pygmea	North America	1913	DE, FR, BE, NL	S
CYPRINIDAE				
Aristichthys nobilis	Southern Asia	1954	SU, RO, HU, YU, DE, PL, CH, FR, IT, NL, BG, CS, AT, AL, GR	S
Barbodes schwanenfeldii	Southeast Asia	?	IT	D
Catla catla	Southern Asia	1966	SU	S
Cirrhinus mrigala	Southern Asia	1966	SU	S
Ctenopharyngodon idellus	Southern Asia	1949	SU, FR, RO, CS, YU, HU, PL, DE, DK, UK, NL, BE, SE, FI, CH, IT, CY, GR, EE, BY, AT, BG, AL, MD	S
Hypophthalmichthys molitrix	Southern Asia	1953	SU, PL, RO, FR, HU, YU, UK, DE, NL, GR, CY, CH, BE, EE, UA, RU, IT, MD, BG, SE, AL, AT, DK, CS	S
Labeo rohita	Southern Asia	1966	RU	S
Megalobrama terminalis	Eastern Asia	1960	HU, AL, RO, YU	D
Mylopharyngodon piceus	Eastern Asia	1960	DE, HU, SU, CS, AL, UA, MD	S
Opsariichthys uncirostris	Eastern Asia	1961	SU	S
Parabramis pekinenesis	Eastern Asia	1960	RO, GR, HU, AL	S
Pimephales promelas	North America	1980	DE, FR, BE	S

Species	Origin of species	First introd.	Countries where introduced	Status
Pseudorasbora parva	Northeast Asia	1961	SU, CS, UK, HU, RO, IT, YU, FR, PL, CH, DE, AT, BG, GR, MD, AL	S
Varicorhinus capoeta sevangi	Southern Asia	1950s	SU	
CATOSTOMIDAE				
Catostomus catostomus	Northeast Asia	1981	SU	?
Ictiobus bubalus	North America	1971	RO, RU, HU	S
Ictiobus cyprinellus	North America	1971	HU, CS, BG, RO, RU	S
Ictiobus niger	North America	1971	RO, RU, PL, CS, BG, HU	S
ICTALURIDAE				
Ameiurus melas	North America	1871	BE, FR, HU, IT, NL, NO, CS, UK, SU, AL, IE, YU, ES, PL	S
Ameiurus natalis	North America	1906	IT	S
Ameiurus nebulosus	North America	1885	AT, CS, UK, FI, FR, DE, HU, IT, CH, NL, PL, ES, RO, RU, YU	S
Ictalurus punctatus	North America	1968	BE, CY, SU, CS, FR, UK, HU, IT, YU	S
CLARIIDAE				
Clarias batrachus	Southeast Asia	?	UK	D
Clarias gariepinus	Africa	1974	NL, CY, FR, HU, PL, CS	D
GASTEROSTEIDAE				
Culaea inconstans	North America	1960s	FI	S
POECILIIDAE				
Gambusia affinis	North America	1921	FR, GR, HU, IT, PL, RO, SU, ES, RU, CY, YU, PT, AL	S
Gambusia holbrooki	North America	1922	IT, ES, FR, PT	S
Poecilia reticulata	South America	1924	CS, HU, IT, UK, NL, AL	S
Poecilia sphenops	South America	?	CS, HU	S
Xiphophorus helleri	Central America	1932	HU, CS	S
MUGILIDAE				
Mugil soiuy	Eastern Asia	1972	SU, UA, BG, RO, GE	S
ATHERINIDAE				
Odontesthes bonariensis	South America	1974	IT	S
FUNDULIDAE				
Fundulus heteroclitus	North America	1970s	ES, PT	S
PERCIDAE				
Stizostedion vitreum	North America	1925	UK	D
PERCICHTHYIDAE				
Morone saxatilis	North America	1965	RU	S
CENTRACHIDAE				
Ambloplites rupestris	North America	1904	FR, UK	S
Lepomis auritus	North America	1895	DE, IT	S
Lepomis cyanellus	North America	?	DE	S
Lepomis gibbosus	North America	1885	BE, DE, FR, IT, HU, NL, ES, RO, PL, CS, PT, YU, RU, GR, BG, AT, BY, UA, LV, MD, EE, UK, CH	S
Micropterus dolomieui	North America	1873	BE, UK, DE, NO, SE, FI, NL, FR, AT, CS	S
Micropterus salmoides	North America	1877	DE, PL, FR, BE, UK, FI, IT, PL, DK, HU, AT, YU, ES, PT, CY, UA, NL, LT, LV, EE, SU, RU, CH, BY, CS	S
CICHLIDAE				
Astronotus ocellatus	South America	?	IT	D
Cichlasoma facetum	South America	1943	PT, ES	S
Oreochromis aureus	Africa	1976	CY, CS, SU, UK	S
Oreochromis mossambicus	Southern Africa	1962	CS, MT, UK, SU	D
Oreochromis niloticus	Africa	1957	BE, CY, PL, UK, DE, FR,	D
Oreochromis urolepis	Central Africa	?	CS, SU	D
Sarotherodon melanotheron	West Africa	?	SU	D
Tilapia guineensis	West Africa	1978	SU	D
Tilapia mariae	West Africa	?	SU	S

Species	Origin of species	First introd.	Countries where introduced	Status
Tilapia zillii	Africa	1963	UK, SU	S
ELEOTRIDAE				
Hypseleotris swinhonis	East Asia	1961	SU	S
Perccottus glehni	East Asia	1950s	RU	S
CHANNIDAE				
Channa argus warpa-chowskii	Asia	1949	CS, RU	S
OSPHRONEMIDAE				
Osphronemus goramy	Southeast Asia	?	IT, FR	D

INTRODUCED SEMIAQUATIC BIRDS AND MAMMALS IN EUROPE

PETRI NUMMI
Department of Applied Biology, University of Helsinki, Helsinki, Finland
petri.nummi@helsinki.fi

Abstract

At least 4 alien semiaquatic mammals and 9 birds have been introduced in Europe. Of these, the muskrat, the coypu, the American mink, and the Canada goose have wide novel ranges. Ecological effects of alien birds and mammals in Europe include coypu and muskrat herbivory on aquatic plants, mink predation on archipelago birds, competition between the American and European mink and the Canadian and European beaver, and hybridisation between the alien ruddy duck and the endangered white-headed duck.

1 The cast

Among some organisms, e.g. insects (Simberloff 1989), it has been noted that Palearctic species have been quite invasible in many parts of the world whereas the Palearctic itself has been relatively immune to invasions of alien species. Semiaquatic birds and mammals introduced to Europe seem to have been successful colonizers (Table 1). Propagule pressure easily explains why there is, for example, 4 alien semiaquatic mammals in Europe but only one (the coypu) in the Nearctic. However, it seems that the North American ecological equivalents for Eurasian species indeed can be stronger competitors, e.g. the American mink *Mustela vison* (Sidorovich et al. 1999) and Canadian beaver *Castor canadensis* (Nummi 2001).

Table 1. Established introduced birds and mammals of Europe.

Species	When introduced	Where established[1]	Origin
American beaver, *Castor canadensis*	1937	Finland, Russia	North America
Muskrat, *Ondatra zibethicus*	1905	Eurasia	North America
Coypu, *Myocastor coypus*	1882	France, Germany Italy, Netherlands	South America
American mink, *Mustela vison*	1920s	Parts of Eurasia	North America
Chilean flamingo, *Phoenicopterus chilensis*	1980s	Germany	South America
Black swan, *Cygnus atratus*	1963	Germany	Australia
Swan goose, *Anser cygnoides*	?	Germany	Asia
Canada goose, *Branta canadensis*	1665	North Europe	North America
Egyptian goose, *Alopochen aegyptianus*	1676	NW Europe	Africa
Ruddy shelduck, *Tadorna ferruginea*	1960	Netherlands, Germany	SE Europe
Wood duck, *Aix sponsa*	1870s	England, Germany	North America
Mandarin duck, *Aix galericulata*	1740s	England, Netherlands	East Asia
Ruddy duck, *Oxyura jamaicensis*	1952	England, West-Europe	North and South America
Cryptogenic species			
Mute swan, *Cygnus olor*			
Barnacle goose, *Branta leucopsis*			

Lever 1985, 1987; Gebhardt 1996; Hagemeijer & Blair 1997; Mitchell-Jones et al. 1999
[1]Small populations may exist in additional countries

E. Leppäkoski et al. (eds.), Invasive Aquatic Species of Europe, 162–172.

Introduced vertebrates of Europe represent some real success stories. The novel distribution of muskrat *Ondatra zibethicus* is the widest for any introduced vertebrate - apart from the black rat *Rattus rattus* and Norway rat *R. norwegicus* and house mouse *Mus musculus* which all live commensally with man. In Eurasia, the muskrat now ranges from Atlantic to Pacific coast (Lever 1985) and its distribution is even wider here than in North America. American mink, which has a muskrat like wide original distribution in North America, does not lag much behind the muskrat, and may eventually spread to quite similar areas also in Eurasia.

Canada goose *Branta canadensis* has been very successful in Nordic countries, like the Canadian beaver in Finland and Russia (Weidema 2000; Nummi 2001). The coypu *Myocastor coypus*, again, has been spreading effectively in more southern parts of Europe (Mitchell-Jones et al. 1999). All these species are widening their ranges.

On the other hand, there are quite a few ducks for which the distribution is confined to restricted areas only; mostly in England and Germany (Table 1). One of these, the ruddy duck *Oxyura jamaicensis,* however, started to increase considerably in the end of the 1970s. This also led to the expansion of ruddy duck range from England to many other western Palearctic countries. The ruddy duck expansion has caused considerable concern since the ruddy duck poses a threat (see 4.4) to the congeneric white-headed duck *Oxyura leucocephala,* which is in Europe only found in Spain (Hughes et al. 1999).

Birds and mammals are not the only non-fish vertebrates introduced to Europe. The bullfrog *Rana cathesbeana* has been brought to Netherlands (Sneep 1999) and the marsh frog *R. ridibunda* and a tortoise *Pseudomys scripta* to Switzerland (Neet 1999).

2 Patterns of spread and population growth

Aquatic birds and mammals have been brought to Europe for various reasons. Alien birds have been released in many places as pond ornaments, and coypu, muskrat and mink were first farmed for their fur. Muskrats were also deliberately introduced in Finland and Russia as fur game animals; for example in Finland, the yearly muskrat catches were on average 150,000 individuals for many decades but in late 1990s the yearly catch dropped to around 20,000 animals. Canadian beaver was brought to Finland as a part of beaver reintroduction program (see 4.3).

The spread of muskrat is known in detail in several areas in Europe (e.g., Ulbrich 1930; Artimo 1960; Danell 1977a), and it has become a classical case of alien invasions (Fig. 1). From 5 individuals introduced near Prague in 1905, the population increased to 2 million individuals in 10 years. Danell (1996) and Williamson (1996) reviewed the invasion studies and concluded that the average spread of the muskrat in different localities was about 10 km a^{-1}. In one case muskrats were found to disperse even at 120 km a^{-1} (Artimo 1960). Although reproducing much slower, the Canadian beaver has also spread at a rate of around 10 km a^{-1} in some parts of its range in Finland (Ermala et al. 1989; Ermala 1996), like the reintroduced European beaver in Sweden (Hartman 1995).

Trapping and habitat suitability have an effect on the rate of spread. In the early phase (1905-1930) of the spread of the muskrat in Central Europe, the rodent widened its range at a rate of 11 km a^{-1}. In 1930-1960, with trapping added, the rate of spread diminished to 5 km a^{-1} (van den Bosch et al. 1992). In order to reach Sweden from Finland, the muskrat had to go around the Bothnian Bay, and pass a relatively unproductive and cold region. In the early phase (1963-1974) the rate of the spread of the muskrat in Sweden, was at 5 km a^{-1}. When the population reached more productive areas with shorter winters, the rate of spread increased to over 10 km a^{-1} (Danell 1996).

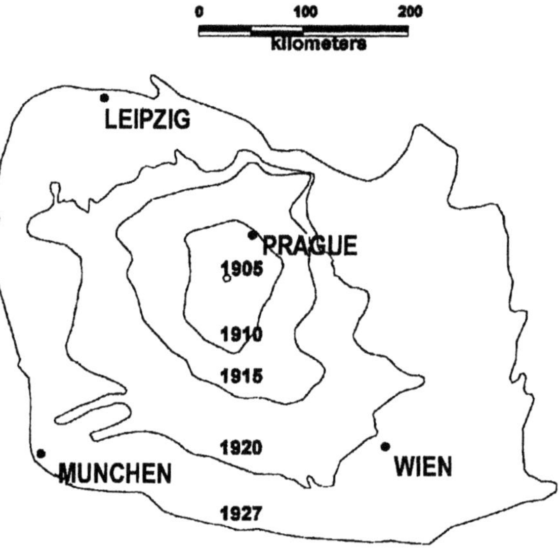

Figure 1. The beginning of an invasion: the spread of the muskrat from the point of introduction of 5 individuals near Prague in 1905 (modified from Ulbrich 1930).

Also the spread of the American mink is well documented, especially in the Nordic countries (Gerell 1967; Bevanger & Henriksen 1995; Kauhala 1996a). The rates of spread are not so clearly determined in this case, because the mink populations originated from feral animals escaping from fur farms from 1928 onwards. The colonization was effective, however, and for example, most of Sweden was colonized by 1960, most of Norway by 1970, and most of Finland by 1980.

In the early phase of an introduction there may be a time lag before a population starts to grow (Williamson 1996).The most pronounced case of the species group presented here is the Canada goose in Britain. From the time of its introduction in 1665 it took about 300 years until the goose numbers reached 10,000 (Lever 1987). In Sweden this took only 40 years - from 1930 to 1970 (Fabricius 1983). Shorter time lags were found, e.g. before the eruption of coypu and ruddy duck in England (Gosling 1989; Hughes et al. 1999)

3 Population dynamics

It is typical for successful aliens to reproduce quite effectively. This is again exemplified by the muskrat. In the north, muskrats have 1-3 litters per year with an average litter size of 4.8 (Artimo 1960; Danell 1978). In Germany muskrats have 2-4 litters per year and the mean size of the litter is 6.5 (Hoffmann 1958; Akkermann 1975).

Muskrats have reached densities of 3-6 animals ha^{-1} in northern Europe (Danell 1979) and densities of 28-55 animals ha^{-1} in central Europe (Pelikan et al. 1970). In France, also the density of the coypu population reached almost 3 individuals ha^{-1} (Micol et al. 1996). In biomass, this resembles the higher muskrat figures because coypus weigh 6-7 kg. High water levels reduced coypu densities because the animals dispersed to wider areas, cold weather affected the density by killing 60% (in England even 90%) of the animals (Gosling 1989; Micol et al. 1996).

The community processes of the muskrat's novel ecosystem are reflected in the species population fluctuations. In northern Sweden, muskrat populations usually fluctuated in 4 year periods - a time length identical to the vole cycle length of the area. The muskrat populations, however, reached their peaks one year after the voles. It seems that North European muskrat fluctuations are linked to the 4-year vole cycles whereas the native populations in northern North America show a periodicity of 10 years connected to the cycle of the snowshoe hare *Lepus americanus* (Danell 1978, 1985).

Muskrat population may be connected to vole or hare populations via predation. In years of high vole densities, foxes (*Vulpes vulpes*) destroyed 0-11% of the muskrat houses, whereas during low vole density years over 20% of the houses were destroyed. Thus, muskrat seems to be an alternate prey for foxes (Danell 1985).

Introduced herbivores may find a predator-free situation in their novel areas, and therefore reach high densities. The effect of density on breeding success is shown in a population of introduced mute swan *Cygnus olor* in Finland. The breeding success of the swan dramatically decreased while the population increased from 2 to 54 pairs (Nummi & Saari 2002, manuscript).

4 Ecological and economic impacts

Invasive semiaquatic species affect ecosystems in many ways. At least six factors can be considered (Ebenhard 1988; Nummi 2001).

4.1 HERBIVORY

The muskrat is known to affect the abundance of certain plants (Danell 1977b), and it has changed species dominance relations in small lakes in Finland: for example, *Phragmites* and *Typha* have been shown to increase at the expense of *Equisetum* and *Schoenoplectus* (Toivonen & Meriläinen 1980). Danell (1996) noted that when considering the impact of muskrat on the vegetation, two aspects should be taken into account: productivity and invasion phase. In productive habitats the impact is moderate during the invasive phase and low in post-invasive phase. In low production habitats the impact is high during invasion and moderate during the post-invasive phase. In eutrophic habitats in Sweden and Czech Republic, muskrats reduced the annual production of *Equisetum* and *Typha* by 4-10% (Pelikan et al. 1970; Danell 1979).

Likewise, in late 1950s when the numbers of coypu in England reached its maximum (estimated to 200,000) some wetland plants, such as the flowering rush *Butomus* and cowbane *Cicuta*, became extremely rare. Coypus also devastated large areas of reed swamps by the rivers (Gosling 1989).

Coypus have caused considerable economic losses to crops. In France, coypus most often damaged corn (41% of the cases); on average 10% of the area of an attacked field was destroyed. The other commonly attacked cultivated plants were cereals and fodder graminoids (Verheyden & Abbas 1996). In England coypus preferred sugar beet, cereals and brassicas (Gosling & Baker 1989), moreover, mostly cereals are also damaged in Italy (P. Genovesi, pers. comm.).

Both coypu and muskrat cause problems with their burrowing activities (Gosling & Baker 1989). The problem is most pronounced with muskrats in the low-lying Netherlands where about 10 million euros per year are used for muskrat control and professional muskrat trappers are employed (Doude van Troostwijk 1976; Gosling & Baker 1989). In France, the coypu's extensive burrowing systems in the banks of ditches and rivers cause considerably more damage than the species' use of the crops (Verheyden & Abbas 1996).

Little is known about the effects of other alien herbivores. Mute swans, which in America are known to affect the amount of submerged vegetation (Cobb & Harlin 1980), live in high densities in many areas. The density-dependent decline in breeding success found in the growing population of introduced mute swans in Finland points to the possibility of overpopulation (Nummi & Saari 2002, manuscript). In some areas in Sweden, moulting mute swan populations have totally eliminated their food *Ulva* in certain areas (Mathiasson 1973).

4.2 PREDATION

The effect of predation is not easily shown, if it is not as dramatic as it has been on oceanic islands (Ebenhard 1988). The effect of mink has in some cases been shown on a local population level, especially on Baltic islands and in Iceland. This is due to the fact that it has colonized areas where no similar predator has existed earlier. In the eastern Baltic, the indigenous European mink did apparently not reach outer islands (Maran & Henttonen. 1995). In other marine areas of Europe, bird colonies nesting in outer islands have faced little mammalian predation (Dunstone 1993; Kauhala 1996b; Hersteinsson 1999).

In Iceland a high probability of mink colonization has been found on islands up to 500 m from the shore (Hersteinsson 1999), and in Finland radio-marked individuals swam a between-island distance of 250 m even twice a day (Niemimaa 1995). Hario et al. (1986) found considerable mink predation on seabirds in a group of scerries of the Finnish archipelago, 3 km from the nearest large islands.

Seabirds appear to differ in their ability to adapt to mink predation. In some areas in Sweden common eiders, *Somateria mollissima,* have gradually returned to islands near the mainland where they disappeared during the initial colonization by mink (Gerell 1985). In other Baltic areas eider populations have increased in spite of the mink (Niemimaa & Pokki 1990). The black guillemot *Cepphus grylle* and the razorbill *Alca torda*, which feed their young in crevice nests for several weeks, are more vulnerable than eiders. In the Finnish archipelago, black guillemots have locally declined or even disappeared due to mink predation (Hilden & Hario 1993).

Mainly herbivorous animals can also create predation problems in their new area. In Germany there is evidence that local destruction of populations of the mussel *Unio crassus,* an endangered species in Germany, was triggered by muskrat predation (Gebhardt 1996).

4.3 COMPETITION

There are at least two species pairs in which the American counterpart seems to outcompete the Eurasian species: European and American mink and European and Canadian beaver.

The American mink seriously threatens the European species. There are several possible factors, including habitat loss and overhunting (see also 4.4 and 4.5) that could have caused the earlier decline of the European mink (Maran & Henttonen 1995). However, the ongoing disappearance of the European mink from its last European resorts in Belarus and Estonia seems to be caused by the American mink. Direct interspecific aggression has been detected. In the first phase of the European mink decline, the remaining European individuals are larger than in the beginning: the European mink is responding to the aggression of the American mink, resulting in smaller European mink individuals disappearing first (Fig. 2, Sidorovich et al. 1999). The two species might also compete for food (Maran et al. 1998).

At least some degree of food competition seems to exist between the European otter *Lutra lutra* and the American mink. In Scottish islands the diets of the two species differed more when the species occurred together than when they were parapatric. The otter seemed to dominate the relationship since the mink diet shifted towards the use of more terrestrial prey when the two predators occurred sympatrically (Clode & Macdonald 1995).

During the beaver reintroduction to Finland in the 1930s, both Canadian and European beavers were brought to the country (Nummi 2001). It was not known at that time that these two were different species. By 1955, especially the population of Canadian beavers of eastern Finland had increased (Linnamies 1956). In the late 1940s and early 1950s, Canadian beavers crossed the eastern border into Russian Karelia (Danilov 1995). Today, there are about 12,000 Canadian beavers and 1,500 European beavers in Finland (Ermala et al. 1999).

In northern, central and southern Finland where both species were once released, only the Canadian beaver exists today. With small populations, chance plays a role in whether they perish or start to increase (Roughgarden 1986). Therefore - and because there are only three replicates - we cannot be absolutely sure that it is the Canadian beaver that has caused the local extinction of the European species.

However, there are some features in the biology of the two species which point to this possibility. Apparently the Canadian beaver uses resources more effectively, and has also larger litters. In the Russian northwest, the mean litter size of Canadian beavers is 3.3, whereas that of the European species is only 1.9. About 66% of the Canadian beaver settlements have a built dam and 75% a lodge; respective figures for the European beaver are only 45% and 34% (Danilov 1995). In Finland, the differences in the building activity seemed to be even larger (Ruusila 1997). The Canadian beaver is causing forestry problems especially in southeastern Finland (Härkönen 1999).

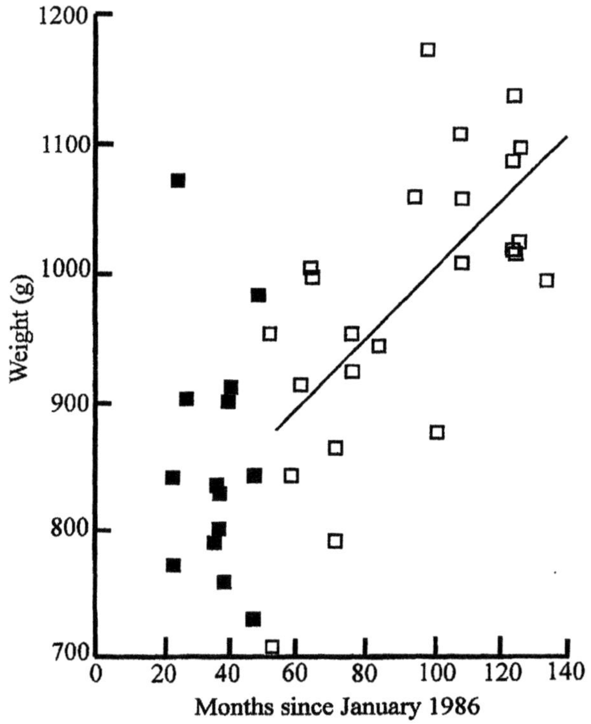

Figure 2. Only the largest survive for a while. The significant body mass change of European mink males after the arrival of the American mink. Solid squares, before arrival of American mink; open squares after its arrival (modified from Sidorovich et al. 1999).

In principle, the populations of Canadian and European beavers should have been prevented from coming into contact. However, this has happened. In western Finland, the two species live in close proximity (Ermala 1996). Moreover, during the last 10-15 years the Canadian beaver has spread from eastern Finland to the northwest at such a rate (Ermala et al. 1989, 1999) that if it continues the beaver will reach Sweden within the next 10 years.

The greylag goose *Anser anser* and the Canada goose seem to be able to coexist since they both have increased in the same areas during the last decades. This appears to be due to the fact that the Canada geese are more aggressive to conspecifics which leaves room for the greylag nests between the Canada goose territories (Fabricius 1983). The situation might, however, become more complicated as the barnacle goose *Branta leucopsis* is also entering the Baltic Sea (Forslund & Larson 1991).

4.4 HYBRIDISATION

Hybridisation is a very difficult part of the alien species problem because it is often hard to even detect (Simberloff 1996).

In Europe, the ruddy duck from America poses a serious threat to the endangered, indigenous white-headed duck. The ruddy duck was brought to England in 1940. Its numbers remained low for 30 years, but in the 1970s the population started to increase steadily. Along with the species' increase in England it started to expand its range to continental Europe. In 1991, first ruddy duck x white-headed duck hybrids were observed in Spain which is the only country in Europe having a small white-headed duck population. The hybrids were also shown to be fertile. In addition to hybridisation, the ruddy duck may threaten the white-headed duck competitively, both via aggression and by being ecologically more flexible (Hughes et al. 1999). Also within Anseriformes species that do not belong to the same genus hybridise easily: in Germany the introduced Canada goose has hybridised with the native greylag goose (Gebhardt 1996). Local gene pools should also be protected (Simberloff 1996). Therefore the practice of introducing alien forms of mallards *Anas platyrhynchos* to certain areas for hunting purposes is not a good policy, and should no longer be practised (Siekkinen & Nummi 1992).

4.5 DISEASES

Introduced parasites and diseases can affect the host-parasite dynamics of their novel environment in various ways. They can simply impair populations of new vertebrate host species (Dobson & May 1986) or they may act indirectly by affecting food resources. A possible example of the latter is the decline of the European mink in Finland in 1920-1940, before the invasion of the American mink. A reason could have been the crash of the noble crayfish *Astacus astacus* due to the introduced fungal disease, the crayfish plague *Aphanomyces astaci* (Maran & Henttonen 1995).

Introduced species can also act as new vectors for indigenous parasites. The coypu, for example, was the main carrier of the sheep liver fluke *Fasciola hepatica* in a rodent-lagomorph assembly in western France. It was concluded that the coypu may play a role in the maintenance and dissemination of the liver fluke (Menard et al. 2000).

It has been noted that introduced hosts may quite easily leave behind parasites with indirect life cycles, however, those with direct life cycles or unspecific intermediate hosts follow their hosts more easily. (Dobson & May 1986). In Volga Delta area it was found that the American minks and muskrats were losing some of their native trematode parasites (which have indirect life cycles) but had acquired new local ones instead. An especially interesting case was the mink parasite *Rossicotrema donicum*; the parasite used *Lithoglyphus* snails - introduced from the Black Sea - as intermediate hosts (Ivanov & Semenova 2000).

4.6 CHANGE IN HABITAT STRUCTURE

Apart from affecting plant abundance *per se*, novel kind of herbivory may affect vegetation structure. In a Swedish lake muskrat grazing created a new kind of short-term plant succession dynamics. Especially the use of emergent hydrophytes to house building by muskrats opened up closed stands of *Equisetum* and *Schoenoplectus*. These open patches can be colonized by submerged plants, e.g. *Myriophyllum* and *Potamogeton*, which are not normally competitive enough to enter closed stands of emergents. During low muskrat population numbers, the emergent stands became more or less closed again (Danell 1977b, 1996). Although the open water area was only 1-4%, the irregular shape of the patches resulted in a high edge effect: there was 400-1,500 m of combined perimeter lengths per hectare. Vegetation patchiness as well as presence of submerged vegetation, harbouring abundant invertebrates is considered beneficial to waterfowl (Danell 1979 and references). The muskrat acts as a keystone species by creating a diverse set of patches in different stages of succession (Danell 1996). At moderate population levels, that is at intermediate disturbance levels (see Rosenzweig 1995), the muskrat can increase plant and invertebrate diversity. When becoming too numerous, the muskrat has the potential to destroy vegetation which can lead to diminishing diversity (Danell 1996).

5 Reasons for success

Williamson (1996) concluded that the properties of different species did not very well predict their success in introductions. The climatic matching, for example, could in some cases but not in others, predict the success of invaders. In the species set presented here, climatic matching explains their success quite well: 4 out of the 5 species of North American origin have widely established themselves or are about to do so (ruddy duck, see 4.4 and 6). The one without larger success yet, namely the wood duck *Aix sponsa*, would be a potential next aquatic bird invader would it only spread from England to suitable beaver habitats of e.g. Fennoscandia. From other areas there is only one successful invader so far, the coypu. In addition to the list presented here there are, of course, also species which have not established at all, e.g. the king penguin *Aptenodyctes patagonius* (Myrberget 1987). However, harsh climate has apparently also prevented the establishment of the coypu in northern Europe (Nummi 1996; climate effects, see Gosling 1989).

As predicted, all successful aquatic bird and mammal invaders of Europe have large, almost or complete continent wide original ranges. However, also 3 out of 7 unsuccessful species are originally widely distributed, therefore, the idea of a wide original range as a predictor of success does not work too well in the present species set. The successful species entering a vacant "functional niche" in Europe are probably the muskrat and the coypu, and maybe at least partly the Canada goose. The functional niche means here that the species does not affect population levels of other species at the same trophic level (see Williamson 1996). The fact that even two species could have entered the aquatic rodent quild, between the water vole *Arvicola terrestris* and the beaver, is interesting (for additional reasons for the muskrat's success, see Danell 1996). The muskrat and the coypu are probably occupying partly the same niche as the "lake beaver" *Trogontherium* did during the Pleistocene (see Mayhew 1977). Introduced populations could suffer from the loss of genetic variability if very few individuals were translocated. The Swedish Canada goose population, founded probably by only 5 individuals, has very low DNA fingerprint variability but has not suffered much from apparent deleterious inbreeding effects (Tegelstöm & Sjöberg 1995).

6 Control and eradication

Both limited control and total eradication projects on alien aquatic birds and mammals have taken place in Europe. In the Archipelago National Park in southwestern Finland a mink eradication project covering a 12 x 6 km area has been carried out to protect birds.

In the Park, minks have been hunted with the aid of a portable air-blasting device (normally used for leaf collection) and a dog. The dog locates the mink's hiding place, and high pressure air is blown into crevices to scare the mink out (Nordström et al. 2002). Effective local mink control is also practised in Iceland around eider colonies, which are harvested for their feathers. There is also an assessment of the feasibility of total eradication of the mink going on (Hersteinsson 1999).

Box 1. Eradication is possible.

Both the coypu and muskrat have been eradicated from England. The muskrat campaign took place already in 1932-1937. In the campaign, muskrat areas were divided into squares of 10 square miles. A trapper was deployed to each of these squares to survey and trap. Most of the population was eradicated by the end of 1934, but some trapping was maintained although only two muskrats were killed in the last 28 months of the campaign. In total 4,400 muskrat were killed (Gosling & Baker 1989).

The eradication campaign of coypu in England in 1981-1989 is the most successful and best documented in Europe, so whenever vertebrate control is planned one should check the original coypu papers (e.g. Gosling et al. 1988; Gosling 1989; Baker 1999). The successful campaign in 1981-1989 was preceded by an unsuccessful one in 1962-1965. The earlier campaign managed to reduce coypu numbers dramatically but failed to eradicate the last individuals. This was partly due to hard winter 1962/63 which removed the opportunity of understanding how the population responded to trapping. From the remaining individuals, the population increased rapidly when the winters became mild again in 1970s (Gosling 1989).

The 1980s campaign was preceded not only by proper monitoring, but also a trial eradication on a 30 km section of a river and trapping simulations at different levels of trapping intensity. Based on these, 24 trappers were employed for a 10 year campaign with expected costs of about 4 million euros (Gosling et al. 1988; Gosling 1989). Trapper motivation was one important part of the project. The question is: why should trappers try to eradicate a species when in doing so they lose their jobs? The scheme was to restrict the funding to a maximum of 10 years, and promise the trappers a bonus of up to 3 times their annual salary if they eradicated the coypu within 6 years. After 6 years the bonus would gradually decline. Most trappers worked hard to achieve the maximum bonus, and the coypu numbers declined from over 5,000 in 1981 to fewer than 40 in 1986. Altogether 35,000 coypus were caught during the campaign. In the final phase it was a question of capturing the last individuals. The last breeding group was found in 1987, and two old males in 1988. The systematic field effort with 3 trappers was ceased in 1992 (Gosling 1989; Gosling & Baker 1989; Baker 1999).

In the muskrat campaign during the 1930s leg-hold traps were used. These caused a large toll of non-target species including moorhens. A more ethical method was needed for the coypu, and in the coypu campaign live trapping with cages on baited rafts was used. With cage traps, very little mortality was caused to non-target species (Baker & Clarke 1988).

In the Archipelago National Park in southwestern Finland, 63 minks were taken in the first year. Since then only on average 4 minks need to have been taken yearly to control the population. The numbers of many birds species have increased after the control started. Among them are black guillemot *Cepphus grylle*, velvet scoter *Melanitta fusca*, tufted duck *Aythya fuligula* and black headed gull *Larus ridibundus*. On the other hand, common eider *Somateria mollissima*, the greylag goose *Anser anser*, common merganser *Mergus merganser* and large gulls did not respond to mink eradication (Nummelin & Högmander 1998; Nordström et al. 2002). The case of the mobile ruddy duck exemplifies the need for a common international policy towards invasive aliens. There is an action plan for 1999-2002 for eradication of the ruddy duck from the western Palearctic. Concerning feral ruddy ducks, the plan contains elements such as adequate bird monitoring, legal provision for control, and commitment to eradication. For aviary management the action plan stresses the importance of ruddy duck registers, release and trade bans, and eventually keeping bans in order to prevent the re-establishment of feral ruddy duck populations. In Spain the most prompt actions has been taken so far: most ruddy ducks and all ruddy duck x white-headed duck hybrids seen until 1999 have been shot (Hughes et al. 1999).

7 Public awareness

Public awareness and training are a necessary part of all alien species management. Among aliens, birds and mammals form a group type of their own because by being soft and furry they can be very appealing to the general public. And, the public may react very emotionally towards the control of appealing alien species (Genovesi 2000; Mack et al. 2000). The ruddy duck control, for example, has been opposed by animal welfare groups in England (Hughes et al. 1999). And in Europe in general, the elimination of animal species for the sake of nature is new for the conservation community. Therefore, a lot of education is needed, both to the general public and to nature managers at different levels. In Finland, for example, an exhibition coupled with a studia generalia symposium was held on alien species in 1999, the exhibition has been touring around the country ever since (Nummi 2001).

Acknowledgements

I am much obliged for Salla Hannunen and Veli-Matti Väänänen for their comments, and for Jani Pellikka and Esa Pienmunne for the preparation of the figures.

GENETICS ON INVASIVE SPECIES

JAKOB C. MÜLLER* & EVA M. GRIEBELER
Department of Ecology, Institute of Zoology, University of Mainz, Mainz, Germany
*Corresponding author *jakob@oekologie.biologie.uni-mainz.de*

Abstract
Genetic markers are especially appropriate to reveal historical processes, as for example regarding biological invasions. Phylogenetic and population genetic methods enable to identify the taxa involved in an invasion (native and non-native species) and often lead to the perception of multiple or cryptic invasions. Phylogeographic and multi-locus approaches have been developed to assign individual invaders to potential source populations. The results may have implications for management measures. However, little is known about the genetic architecture of newly colonised populations. We introduce here genetic models for colonisations in linear habitats like rivers or coastlines. The expectations about the neutral genetic population structure in relation to different invasion modes are compared to European field data of *Dikerogammarus villosus* (Amphipoda) and *Dreissena polymorpha* (Bivalvia). Finally, a potential correlation between genetic diversity and invasion success is discussed.

1 Introduction

Many different types of characters have been applied to identify and describe invasions (Williamson 1996). Different aspects of invasion biology like direct record, differential success of invaders or differential threat and impact to communities have been addressed by the study of morphological, ecological, behavioural, life history and genetic traits. Genetic markers - in particular neutral or nearly neutral markers - are especially appropriate to reveal historical processes, because it is assumed that these marker states can easily be hierarchically structured and change regularly across time (Page & Holmes 1998; Avise 2000). The analysis of the genetic architecture in invasive species plays therefore a central role in understanding the invasion history and the invasion process (Holland 2000). Further insights of the knowledge on the historical process may help to predict future invasions. In this way, genetic studies are able to identify genealogical relationships among the participating taxa, invaders as well as native species. Cryptic invasions of morphologically similar species can be detected. Moreover, genetic studies within a single species are able to infer the history of single individuals or populations during an invasion, which is mostly unknown. There are only rare examples, where an invasion event is continually observed. In most cases, the invasion attracts attention only when the invader becomes a pest.

This chapter outlines current genetic methods in the broad field of invasion biology. Depending on the question under concern, different marker systems are needed. For taxonomic identifications - i.e. older history - strongly differentiating markers like mitochondrial gene loci are best suited. Even within short evolutionary time periods - for example among closely related species - these markers often form monophyletic groups due to their small effective population size, which leads to rapid lineage sorting. There are however no universally valid rules about the level of differentiation upon which species status is indicated. According to the relatively comprehensive cohesion species

E. Leppäkoski et al. (eds.), Invasive Aquatic Species of Europe, 173–182.

concept (Templeton 2001), additional evidence for species status are reproductive and/or ecological boundaries, which are best analysed in syntopic incidences. Nuclear recombining markers of all kinds can test reproductive isolation. The analysis of intraspecific population structure and source population assignment - i.e. young history - requires extraordinarily variable genetic markers. Founder events that are generally involved in invasion processes reduce the genetic diversity in the newly colonised populations (Nei et al. 1975; Hartl & Clark 1989). Depending on the effective population size and the mode of inheritance, mitochondrial gene loci are more affected than nuclear loci (Wade et al. 1994; Davies et al. 1999a). Highly variable nuclear markers like microsatellites, introns, RAPD's and AFLP's are certainly best suited, but others like allozymes may be applied if original polymorphism is high.

2 Taxonomic relations and invasion origin

Invasions are often not perceived or are misinterpreted when invaders are morphologically similar to resident species or previous invaders (May & Marsden 1992; Geller et al. 1997; Jürss et al. 1999; Müller & Schramm 2001). These taxonomic shortcomings are frequent in aquatic communities, because species complexes below the water line are relatively poorly studied (Knowlton 1993). Most cryptic invasions arise when two or more members of a sibling species complex colonise a new area in rapid succession. Genetic studies often supply first evidence for multiple species invasions. Three years after the first record of the zebra mussel (*Dreissena polymorpha*) in North America, an unexpected second non-native dreissenid species, the quagga mussel (*D. bugensis*), has been genetically identified (May & Marsden 1992). During these years of unconsciousness, when zebra and quagga mussel were inadvertently identified as one "species", failures or confusion in the interpretation of results concerning the distribution, population structure and ecological characteristics were certainly produced (Boileau & Herbert 1993; Haag & Garton 1995; Marsden et al. 1996). An European example of a taxonomically uncertain invasion is the *Marenzelleria spp.* invasion into the North and Baltic Seas. About 15 years after the first record of these North American polychaetes in the North Sea, genetic investigations corroborated that two sibling species (*M.* cf. *wireni* and *M.* cf. *viridis*) are involved (Bastrop et al. 1995, 1998). Meanwhile ecophysiological differences have been stated between the two *Marenzelleria* species (Jürss et al. 1999). Another example is the genetic confirmation that two amphipod species (*Dikerogammarus haemobaphes* and *D. villosus*) invaded the river Rhine from the Danube in successive years, the latter displacing the former (Tittizer et al. 2000; Müller & Schramm, unpublished). Recently, a third *Dikerogammarus* species from the lower Danube invaded the upper Danube and could be identified as *D. bispinosus* by molecular markers (Müller & Schramm 2001). Before, it was incorrectly described as a subspecies of *D. villosus*. The three distinct clusters of nuclear genotypes in a natural syntopic population indicate complete reproductive isolation among the three invasive *Dikerogammarus* species (Fig. 1).

Taxonomic assignment of doubtful invaders limits the geographic area of origin to the former distribution range. Of course, the knowledge of the original distribution is here of paramount importance. Further identification of the source population is desirable if the original distribution area is large. It has to be noted that specimens from the whole distribution range have to be sampled representatively, which is rarely done. With non-recombining genes, like mitochondrial genes or small nuclear fragments, standard phylogeographic methods can be applied (Avise 2000). The evolution and the dispersal of the marker gene are directly visible, if the intraspecific haplotype tree is superimposed upon the geography. For recombining nuclear genes likelihood-based approaches have been developed to assign individuals to most likely source populations (Paetkau et al. 1995; Siegismund 1997). The population characterised by its allele frequencies with the highest likelihood of containing the test multilocus genotype is the presumed source population. Issues concerning statistical significance and potential sources of error have been addressed recently (Rannala & Mountain 1997; Marshall et al. 1998; Davies et al. 1999b). A similar method is the mixed stock analysis, which estimates the most likely fraction of potential source populations in a mixed invaded population (Utter & Ryman 1993). Such analyses have been applied extensively in the Pacific Northwest and Alaska where local salmon populations may spawn in separate freshwater streams, but are harvested together in salt- or freshwater fisheries (Scribner et al. 1998; Seeb & Crane 1999). Some refined genetic models even include ancestors of the test population as potential immigrants or do not need an a priori population definition (Rannala & Mountain 1997; Pritchard et al. 2000).

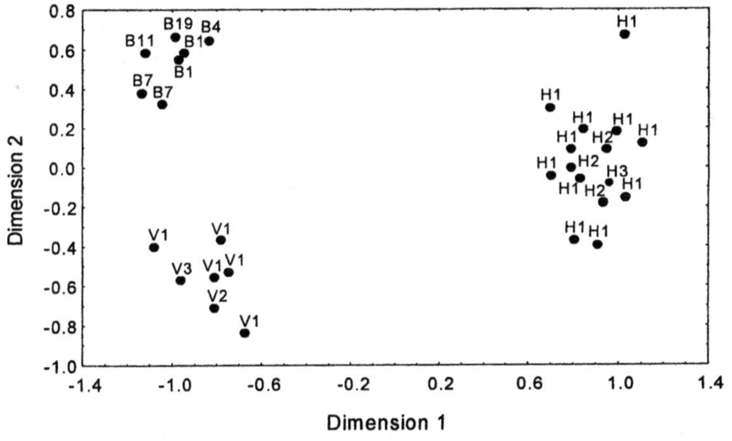

Figure 1. Multidimensional scaling of genetic distances among *Dikerogammarus* individuals in a syntopic population of the Hungarian Danube. Genetic distances are based on multilocus genotypes of 9 allozyme loci. (H) *D. haemobaphes*, (V) *D. villosus* and (B) *D. bispinosus*; the associated numbers indicate the number of typed individuals.

The determination of invasion sources may have significant implications for potential management measures regarding introducing vectors. Further, biological control of a nuisance invader, i.e. the introduction of specific predators or lost parasites from the area of origin, can more clearly be identified knowing the precise area of origin of the invader, especially if its native range extends over a wide region. Moreover, it may be interesting to analyse the ecological situation in the original community (competition, food web position, etc.) to predict the final equilibrium state and impact of the new environment.

3 Invasion modes: Models and the real world

Mathematical theory of biological invasions focuses on population demographics and is a relatively well-developed area. This theory has been extensively tested on multiple invading species (Shigesada & Kawasaki 1997). In contrast, the impact of individual movements on the spatial genetic structure of the invading population is still not well understood. Especially, spatial patterns of genetic variation that are generated during the invasion process are not clearly identified. Detecting general effects on genetics of invading populations can help to identify and to rate range expansion. It has already been suggested that the mode of dispersal has a direct effect on spatial genetic structure (Waser & Elliot 1991; Larson 1984, Hewitt 1993; Templeton et al. 1995; Ibrahim et al. 1996). Therefore, we first describe extreme models for two different invasion modes. Model expectations about the genetic structure within and between the newly colonising populations are given followed by case studies of amphipods and the Zebra mussel.

Analogous to a range expansion model given in Ibrahim et al. (1996), we developed a computer simulation model that organises populations in discrete demes with migration between them. We considered a linear array of 250 demes. Therefore, the results presented herein may be applied to species inhabiting riverine systems, coastal waters or other serial aquatic and even serial non-aquatic ecosystems. The simulation was implemented as an individual based model (de Angelis & Gross 1992). Models of this type represent a population as a set of individuals which have properties of different values (e.g. sex, age, genotype). In our model an individual was characterised by one single property. Two neutral diallelic loci represented the genetic system in a diploid organism. Each deme was assigned a carrying capacity (*K*) of 80 individuals. For each deme, the number of individuals and the frequency of each of the nine multilocus genotypes was stored. At the beginning of the invasion process, eight demes located at one edge of the array were inhabited and the population size of each of these demes was set to carrying capacity. For each deme, allele frequencies were set initially to 0.5 at each locus with gametic phase equilibrium. Maximum recombination between the two neutral loci was allowed. The sequence of three simulation modules (dispersal, growth with random mating, sampling) was repeated in each deme for 20 non-overlapping generations.

Dispersal. The number of migrants from each deme was drawn from a binomial distribution. We chose a low emigration probability $m = 0.05$ (Slatkin 1993; Ibrahim et al. 1996). The displacement of each migrant was possible in both directions and was determined by either a stepping-stone-type dispersal or a distance-dependent dispersal (exponential distribution function $F(x)$ with parameter $\alpha = 0.447$ (Ibrahim et al. 1996), $F(x) = 1-\exp(-\alpha * x)$, where x is the migration distance). In the case of stepping-stone dispersal and colonisation, both neighbours of a deme received migrants with a probability of $0.5*m$ from this deme. This dispersal mode implements only short-distance dispersal of species. In contrast, in the case of exponential displacement many individuals dispersed short distances and a few more long distances.

Growth with random mating. The total number of adults in a deme at generation t (N_t) after dispersal is the sum of non-migrants and immigrants. The total number of offspring at generation t+1 (N_{t+1}) that constitute the next generation was drawn from a Poisson distribution with a mean $N_t + r^*N_t\ (K-N_t)/K$ where r was the intrinsic rate of growth and was kept constant (0.8) for all simulations.

Sampling. In each generation, samples of 50 individuals were randomly drawn from each deme for representation of results. This sample size was also used in field studies.

The contour map (Fig. 2) shows characteristic examples of the spatial distribution of allele frequencies generated by stepping-stone and exponential dispersal after 20 generations. Stepping-stone dispersal (Fig. 2a) results in a slower range expansion than the exponential displacement model (Fig. 2b) since the mean migration distance was smaller in our stepping-stone model than in the exponential model. When a small group of emigrants from a deme founds a new deme, genetic drift alters allele frequencies in the founder population in comparison to the source population. For the stepping-stone dispersal mode, the sampling-subsampling sequence that results from successive foundation of neighbouring demes rapidly ends up in the loss of one of the alleles in the invasion front (Fig. 2a). This observation is supported by Slatkin (1993). He found for stepwise colonisation that genetic similarity between demes increases with an increased distance from the start location of the invasion due to the increased limitation of genetic variability.

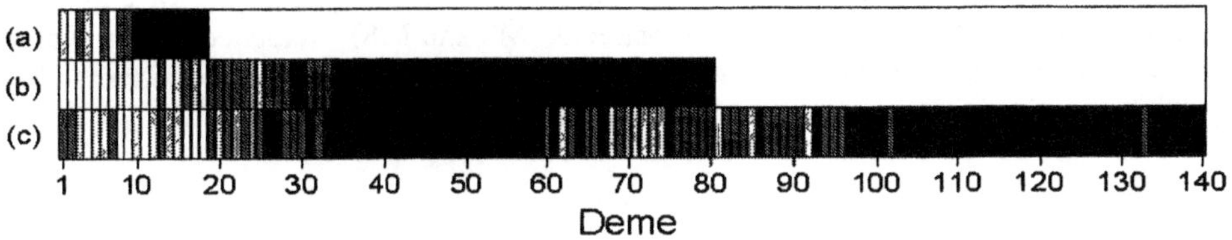

Figure 2 Contour map showing the spatial distribution of allele frequencies after 20 generations of stepping-stone dispersal (a) and exponential dispersal (b and c) at one of two diallelic loci. In these simulations allele frequencies at this locus did not fall short of 0.5. Four levels at intervals of 0.1 are mapped. The four framed colours - white, light grey, dark grey and black - represent frequency intervals of < 0.7 to ≥ 0.9, respectively. Unframed white indicates that a deme is not inhabited.

For the exponential mode a similar loss of diversity occurs when no long-distance migration event randomly occurs (Fig. 2b). If such a rare long-distance dispersal occurs and if there is enough time for the founder population to establish and to expand, a satellite population will arise (Fig. 2c, around deme 80). Such satellite populations are likely to have different genotype frequencies from the migration front due to the effect of random genetic drift. Therefore, if such satellites come in contact with the main front (Fig. 2c around deme 40-55) or with another satellite population two distinct gene pools will meet. Thus, a cline in allele frequencies can built up in the contact zone (not visible in Fig. 2c, because allele frequencies at the single shown locus happened to be similar at the contact populations due to random genetic drift). In addition significant linkage disequilibria are found in the area of contact (Fig. 3), indicating that alleles derived from the same population tend to cluster together in the same individuals (Hartl & Clark 1989, Barton & Gale 1993). The expansion of the main front and of the satellite front before they came into contact is still documented in the genetic patterns (Fig. 4). The genetic similarities to neighbouring populations increase with increasing approach to the area of contact (Slatkin 1993).

Clines in allele frequencies, genetic similarity patterns and linkage disequilibria are general patterns that are moulded during the approach and the contact event of two distinct populations. These indicators of ongoing range expansion of invading species persist over multiple generations and are smoothed over time by subsequent migration between the mixing populations.

Are there any natural examples of the model expectations about the genetic population patterns and may we infer from these the underlying mode of invasion? We found some strikingly congruent patterns in the analysis of the *Dikerogammarus* invasion. *D. villosus* invaded recently (first record 1992) the upper Danube and the Rhine system across the new Main-Danube canal (Tittizer 1996). It is suggested that it disperses both by active swimming and by hitchhiking on vessels being transported in ballast water or in the fouling community of the ship hull (Grabow et al. 1998; Reinhold 1999). About 50 individuals sampled 1997 in each of 16 populations along the invasion route from the middle Danube to the Rhine have been genotyped at 9 allozyme loci. No consistent allele frequency cline, but rather significant linkage disequilibria were found in the river Main (a tributary of the Rhine) (Fig. 5). This indicates recent admixture or contact of genetically different populations at two locations (M3 and M8). Evidence for the mutual approach of invasion fronts came from the among-population similarity patterns (Fig. 6). High genetic similarities clustered around the locality M8.

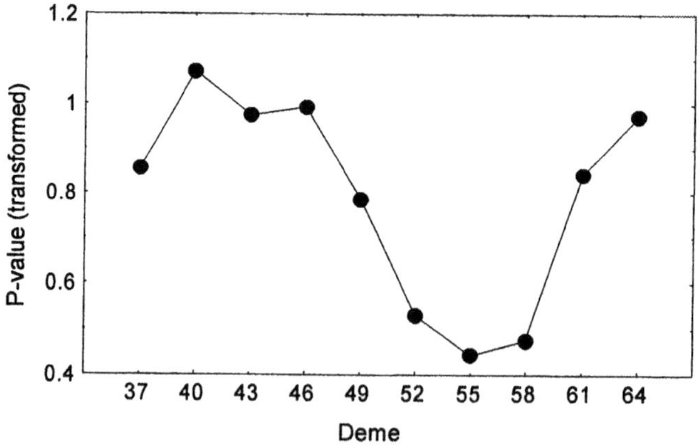

Figure 3. Genetic linkage disequilibria observed in the contact zone of simulation (c) in Fig. 2. The amount of linkage disequilibrium is determined by the p-value of an exact test for genotypic disequilibrium (genetic software GENEPOP; Raymond & Rousset 1995). Low p-values indicate a strong linkage disequilibrium. Given p-values are arc sin √ transformed. P-values of three successive demes are averaged and the number of the mean deme is shown. P-values of deme 48, 56 and 57 are significant at 5%.

According to the model expectations, we hence suggest that the area around M8 represents a contact area of invasion fronts either between established satellite colonies or between the main invasion front and a satellite population. For M3 the genetic similarity pattern is missing. This could be explained by strong long-distance admixture of this locality from outside the study area. Long-distance immigration is probably enhanced near harbours (Reinhold 1999), and indeed M3 lays close to a major harbour. We can therefore conclude that the invasion dynamics of *D. villosus* is similar to the exponential mode of invasion models. Both punctual and contiguous colonization coexist such that some long-distance dispersers break the continual expansion of the active invasion front.

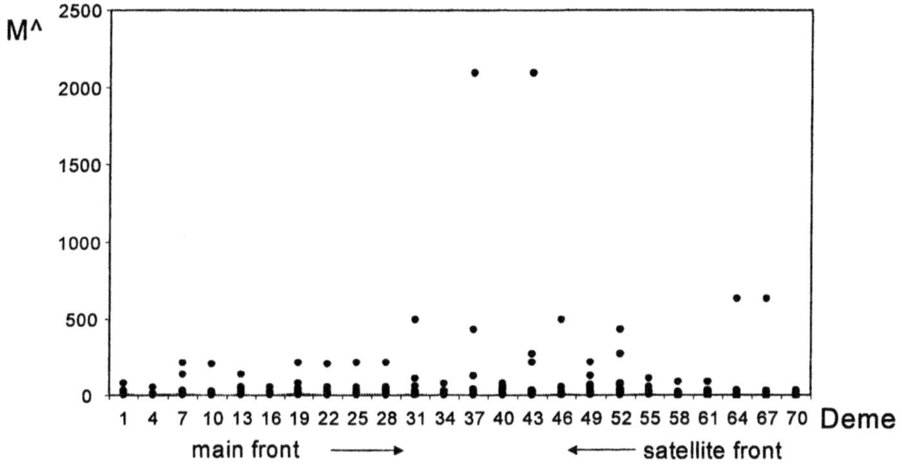

Figure 4. Pairwise genetic similarities in the contact zone of simulation (c) in Fig. 2. Genetic similarity is expressed as the gene flow parameter M^ calculated from G_{ST} (software GENEPOP) among all demes represented for each deme.

The invasion patterns of the zebra mussel *Dreissena polymorpha* show some additional peculiarities. In a geographic population survey we found two major genetic clusters in Western Europe representing the two available corridors of freshwater invasion from the Ponto-Caspian region across the navigable waterways (a northern route up through the Dnjepr river and subsequent canals and a southern route up through the Danube river) (Müller 2001; Müller et al. 2001). The invasion lineages lost significant, although weak, genetic diversity within populations in comparison with the original Ponto-Caspian area, indicating a founder effect. The genetic structure between populations in the invaded area was strongly eroded (Müller 2001), probably due to long-distance transportation along rivers and navigable canals. Boat traffic with adult mussels attached to ship hulls and the downstream swamping of free-swimming veliger larvae is supposed to have a significant effect on the spread (Koutnik & Padilla 1994; Horvath et al. 1996; Schneider et al. 1998). The biology and genetics of the mussel hence suggest neighbouring colonisation (mainly downstream) with substantial long-distance dispersal. Lake populations in contrast are genetically impoverished and/or strongly differentiated (Lewis et al. 2000a; Müller 2001). The long-time separated invasion lineages were given the opportunity for contact since 1992, when the new Main-Danube canal was opened. The characteristic features of admixture were found there: a rapid declining allele frequency cline and linkage disequilibria (Müller et al. 2001). It is predictable that the two invasion lines will be homogenised, in particular when the upper reaches of habitats become interconnected as with the Main-Danube canal.

4 Outlook

Genetic studies on invasive species will certainly reveal more cryptic invasions, in particular of morphologically similar sibling species. It can be argued that the frequency of current invasions is underestimated. This should be seen, however, in the context of cryptic biodiversity in general. Even in the well-catalogued Central Europe, new aquatic species are described frequently after genetic investigations. The native status of the so far undetected taxa is often unknown. Recently reported examples are the bullhead *Cottus gobio* with two taxa restricted to different drainage systems (Hänfling & Brandl 1998) and the amphipod *Gammarus fossarum* with more than eight mostly vicariant taxa in Western Europe (Müller 2000; Müller unpublished).

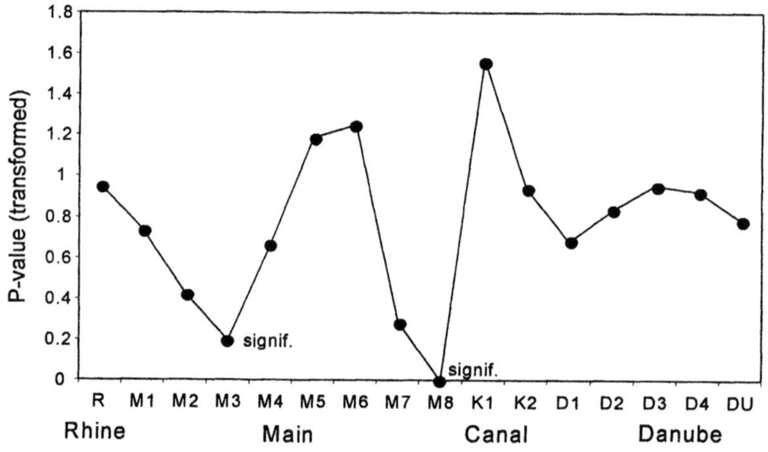

Figure 5. Genetic linkage disequilibria in *Dikerogammarus villosus* populations across the transect between the interconnected rivers Danube and Rhine. Calculation and transformation of p-values see Fig. 3. Significant values (5%) are indicated.

As long as the knowledge about the diversity and distribution of native taxa is incomplete, only invasions of taxonomically uncommon or high-impact species will be detected. Similar species might have been overlooked since a long time. Regional range dynamics and invasions of cryptic taxa often remain unknown. A good example is the Western European *Gammarus fossarum* type B. Independent genetic evidences indicate an eastern range expansion into the river Rhine region, thereby probably displacing the eastern taxon type A (Müller 2001). Evidence for this stepwise invasion came from a pattern of successive genetic loss and increased genetic similarities in the direction of the colonisation (Müller 1998). Moreover a geographic nested clade analysis (Templeton 1998) with a polymorphic mitochondrial marker system revealed a significant recent contiguous range expansion in this area (Müller 2001). This example amply illustrates a general rule of historical analyses: the more independent evidences are found, the more reliable is the description of the past invasion process.

Genetics is not only a tool to uncover past invasion processes in detail, but might also be a significant factor for future invasion ability. As outlined above, an invasion is always accompanied by a more or less strong founder effect that diminishes genetic diversity. On the other hand, range expansion may lead to contact of formerly isolated populations in the original area. Such a mixing of genetically different populations has the potential to produce a genetically divers assortment of offspring. It is argued that genetic diversity within individuals or populations is relevant to the individual or population fitness, respectively (O'Brien & Evermann 1988; Mitton 1997). Inbreeding depression in small and isolated founder populations can mediate the fitness loss. The degree of loss hereby depends on the level of genetic load in the original outcrossing population, which is presumably high in common and ecologically versatile species (Dobzhansky 1964). Potential invaders surely belong to this group. A significant inbreeding effect on relative fitness components like fecundity, viability and growth rate was shown, for example, in the amphipod species *Gammarus duebenii* (Bulnheim 1977). Fitness advantages, on the other hand, may be caused by heterosis effects (Mitton 1997). Heterozygote advantage is reported for benthic life stages of molluscs including survival in post larval oysters (Zouros et al. 1983) and growth rate in adult *Dreissena polymorpha* (Garton & Haag 1991; Lewis et al. 2000b). The situation in *D. polymorpha* may be complicated by a reversal of selection during ontogeny, i.e. heterozygote disadvantage during the larval stage (Haag & Garton 1995).

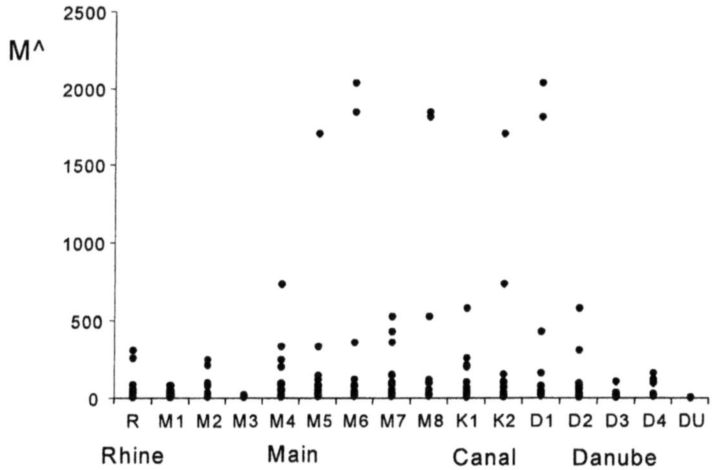

Figure 6. Pairwise genetic similarities among all transect populations of *Dikerogammarus villosus.* Calculation see Fig. 4.

Across taxonomic boundaries, however, no general relationship appears between the genetic variability and the ability to colonise (Duda 1994). The equally successful invading mussels *D. polymorpha* and *Corbicula fluminea* differ drastically in their genetic diversity. The zebra mussel has one of the highest heterozygosities reported, whereas the asiatic clam *C. fluminea* has only low levels of genetic variability (McLeod 1986; Marsden et al. 1996; Renard et al. 2000). The low genetic diversity of the latter species might be due to inbreeding or self-fertilisation of hermaphrodites (Tsoi et al. 1991). The mode of reproduction hence plays an important role for the genetic architecture of populations. Clearly, a better understanding of the correlation between genetic variability and the invasibility is needed.

Acknowledgements

We gratefully like to thank Stephanie Schramm for collecting genetic data.

Vectors

VECTORS - HOW EXOTICS GET AROUND

DAN MINCHIN[1*] & STEPHAN GOLLASCH[2]
[1]*Marine Organism Investigations, Ballina, Killaloe, Co Clare, Ireland*
[2]*GoConsult, Hamburg, Germany*
[*]Corresponding author *minchin@indigo.ie*

Abstract
The opportunities for the spread of exotics increase with the greater movement of goods and people around the world. In early times human exploratory activities were responsible for some selected species being moved as food or for cultural reasons but also inadvertently carried. As trading links and colonisation of distant lands developed a more regular transport evolved. As a result organisms spread to areas beyond their normal range, the ways in which this is done is the subject of this chapter.

1 Introduction

Vectors responsible for exotic species spread arise from either primary or secondary movements. International and national measures need to take account of these pathways so that contingency plans for their management and control and prevention of their spread are possible. However, the great majority of precautions follow at a time when an invasion has been recognised often after it has had some economic impact. As a result proactive measures in the prevention of primary inoculations are likely to be more cost-effective in management plans. In order for this to be successful a good understanding of the vectors involved is needed. On the basis of current information a new species is introduced to a new region worldwide every nine weeks. However, recent indications are that this may be higher with approximately one new exotic species every three weeks over the period 1998-2000 in European waters (ICES/IOC/IMO SGBOSV 2001).

1.2 PRIMARY AND SECONDARY INTRODUCTIONS

A species established in a locality in a different biological province for the first time (usually between continents, from one side of an ocean to another or from one hemisphere to another) result from a primary introduction. More than one primary inoculation may occur at a similar time, however, this seldom occurs and should this happen, it may be difficult to demonstrate, unless associated with deliberate movements. Most primary inoculations have evolved as a result of trading in living organisms or as a

E. Leppäkoski et al. (eds.), Invasive Aquatic Species of Europe, 183–192.

result of inadvertent carriage of species as fouling or in transported water. Shipping, aquaculture activities and aircraft transmissions of living organisms are the main modes of transmission for primary introductions today.

Secondary introductions result from the expansion of the exotic species from its first location of establishment. This secondary spread will normally include a wider range of vectors that may act either separately or together. Some vectors may be cryptic or just not fully understood in the ways they operate. In order to manage the spread of a species the likely modes for spread need to be clearly identified. With the secondary expansion of a species new opportunities are created that enhance the further spread of the species and so expansion can be accelerated. The expansion of both the Japanese brown alga *Sargassum muticum* (Wallentinus 1999) and the Asian shore crab *Hemigrapsus penicillatus* (Noel et al. 1997; Gollasch 1999) are examples of species undergoing secondary spread and have the potential to colonise most north European regions.

1.3 PRINCIPAL VECTORS OF EXOTIC SPECIES TRANSMISSION

There are a wide range of activities that either deliberately or inadvertently result in the transmission of exotic species (Carlton 1994; Carlton et al. 1995; Gollasch 1996; Minchin 2001). The majority of these relate to trade in some way but recreational activities can also be responsible for exotic species spread (Fig. 1, Table 1).

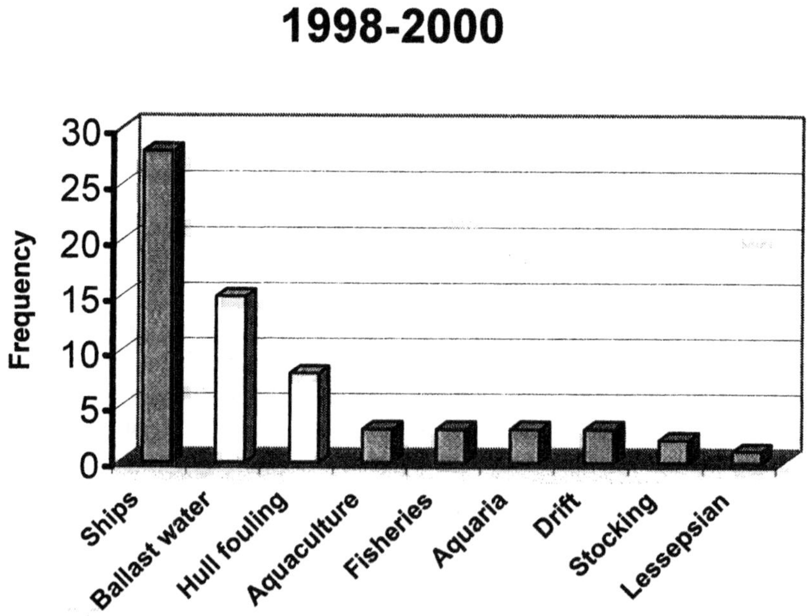

Figure 1. Frequency of first records of nonindigenous species in European waters according to likely vector of introduction in 1998-2000. Ships include two vectors: ballast water and hull fouling (white columns) (after ICES/IOC/IMO SGBOSV 2001).

Table 1. Examples of exotic species spread in Europe other than by natural dispersal mechanisms. Some of the acting vectors will also involve transport by ship, aircraft and land vehicles.

Principal activity	Main vectors or means of transmission/establishment	Species
Shipping and floating structures (drilling platforms, dry docks)	Ballast water and sediment, hull fouling, cargo, ships equipment	*Mnemiopsis leidyi, Elminius modestus, Eriocheir sinensis*
Aquaculture	Imports for culture, transport equipment, untreated shell, host tissues	*Anguillicola crassus, Bonamia ostreae, Orconectes limosus, Crepidula fornicata, Crassostrea gigas*
Fishing activities	Equipment transfer, bait fishes	*Rutilus rutilus, Carassius carassius*
Food processing	Untreated waste disposal of imported produce, exports of tissue	Fish diseases
Live food trade	Escapes, releases, disposal of tissue or contaminated water	*Homarus americanus*
Aquarium trade	Escapes, releases, disposal of tissue or contaminated water	*Caulerpa taxifolia, Elodea canadensis, Carassius auratus*
Stock enhancement	Releases, infested stock	*Oncorhynchus mykiss, Pacifastacus leniusculus*
Recreational activities	Contamination of boats, fishing equipment	*Dreissena polymorpha*
Opening of natural barriers	Opening of new water links, canals	*Cordylophora caspia, D. polymorpha*
Movements of sediment, aggregates	Attached to or living within substrate	*Mya arenaria*
New trade agreements	Erosion of previous legal barriers enabling free movement of goods, new trading routes	*Mytilicola orientalis, Dreissena polymorpha*
Research studies	Releases, disposal	*Mastocarpus stellatus*
Political policy	Production of forage species	*Paramysis lacustris, Limnomysis beneden, Hemimysis anomala*
Cultural preferences	Imports of specific species, festival releases	*E. sinensis, Rapana venosa*
New fisheries development	Releases to the wild and spread	*Paralithodes camtschaticus, Acipenser spp.*, European and Pacific salmonids

2 Shipping

Shipping is implicated in transmission of a great diversity of organisms. This is because ships are capable of carrying a wide range of sessile species, their epibionts and parasites (hull fouling); planktonic species that will include those with even short free-living stages (ballast water); species that bury or are otherwise associated with sediments (ballast sediments). Ships are capable of transmitting large numbers of species and given suitable conditions sufficient numbers may survive to create a viable inoculum in a new region. Successful inoculations may have been relatively rare events in the past but are likely to increase with:

(i) more berths available in ports where there are marine conditions and construction of new ports (providing more suitable targets for primary inoculations to become established; and enhancing opportunities for secondary spread).

(ii) better management of water quality in port regions leading to better conditions for imported organisms in ballast water to become established; and also increase the opportunities of exporting larger numbers of organisms elsewhere (Carlton et al. 1995).
(iii) higher frequency of ship visits, rapid turn-around times in port and changes in trading patterns as new opportunities arise that may not have been present before (with a consequent increased volume of ballast water discharges).

For these reasons it is predicted that further exotic species will become established in European waters.

2.1 SHIPS' HULL FOULING

Ships while in dry-dock are supported on wooden blocks, the hull beneath these blocks does not become painted and so here fouling may freely develop once the ship is reimmersed. Consequently vessels approaching their dry-docking time, despite the use of the toxic anti-fouling ingredient tributyltin (TBT), can have a mature fouling on parts of their hulls. Several invertebrates could spawn, once exposed to temperature fluctuations while entering ports, and release and leave behind a viable inoculum of zygotes that could form a founder population, even following a visit of some hours. Mature molluscs fouling hulls may have the potential to transmit diseases between ports and this topic should be researched. Because of unwanted effects to native biota, aquaculture and fisheries in port regions, it is planned to ban the use of TBT in antifouling coatings by 2008. The new generation of antifouling agents will need to be as effective, or more effective than TBT, if it is to reduce fouling yet the new products should not cause unwanted effects to the environment. Transport of slow moving or otherwise stationary craft do not have the same requirement of effective antifoulants as do fast craft where large savings in fuel consumption can be made. For this reason barges and working platforms may accumulate large fouling communities and if transferred may carry a high risk of spreading exotics.

For some hundreds of years hull fouling was the main means of species transmission but more recently attention has been devoted to ballast water and the organisms carried within it. Nevertheless, recent studies on exotics (North Sea, Australia and USA) indicate that hull fouling continues to be an important vector for invasions although it is often difficult to determine whether these have evolved from hull fouling or from ballast water. Historically the numbers of non-native species likely to have been introduced by hull fouling is greater than for ballast water, and so we may expect a large proportion of established exotics to have arrived in this way.

2.2 SHIPS' BALLAST WATER AND ITS SEDIMENTS

Most of the world trade depends on shipping and to travel safely ships must maintain a correct immersion level by either carrying cargo, ballast water or both.

Early ballasting of ships was undertaken using solid ballast material (sand, gravel and stones), the purpose being to submerge the propeller and rudder thereby providing better control and increased stability. The use of solid ballast was very labour intensive and it was in the late 1870s when ballast water became more regularly used. Ballast water is usually carried in segregated ballast water tanks or in emptied cargo holds and is taken on board in ports, waterways and the open ocean. With the uptake of ballast water organisms, suspended solids (i.e. sediments) and chemicals, including industrial and human wastes are pumped onboard. Vessels almost always carry ballast water when no cargo is carried and even when fully laden some ballast will remain in ballast tanks because the pumps serving the ballast tanks are unable to remove all of the water. Ballast water tanks will have different configurations according to the ship design and will have a complex network of pipes for ballasting and for adjusting water levels between tanks to improve trim. As the tanks will be filled and drained in different sequences either singly or collectively, the ballast water in one tank may be composed of water from several ports (Gollasch 1996).

The amount of ballast water carried can be 30% of the overall cargo carrying capacity of the ship (e.g. Gollasch 1996). Pumping such large volumes of water when undergoing ballast water exchanges at sea, as recommended by the International Maritime Organisation (1996), is time consuming, costly and not always safe. Since the tanks can not be fully drained three 'complete' exchanges are recommended until pump suction is lost. However, under certain swells and other sea-states it may be unsafe to undertake these exchanges because the uneven distributions of ballast water in the tanks can compromise the structural integrity of the ship. In the absence of fully proven sterilisation techniques the mid-ocean exchange (i.e. re-ballasting in mid-ocean) is the only current 'preventive' approach employed by existing ships. Exchanges at sea are probably most effective when freshwater is exchanged for seawater. Up to 10 billion tonnes of ballast water (Rigby & Taylor 1995) and several thousands of species are transported every day (Carlton & Geller 1993; Gollasch 1996). Large numbers of organisms are in transit, estimates of > 50,000 zooplankton and 110 million phytoplankton per m^{-3} have been made by e.g. Lenz et al. (2000). The abundance of some organisms is difficult to quantify, as they may not be evenly distributed within ballast and tank sediments due to the winnowing effects caused by water circulation within tanks. Nevertheless estimates of 150 to 22,500 cysts m^{-3} of sediment were made (Hallegraeff & Bolch 1992). These cysts may remain viable for 10-20+ years. In 14 recent European ballast studies approximately 990 species were recorded from ballast tanks (water and sediment), ranging from bacteria to 15 cm long fishes (Gollasch et al. in review).

To show a cause and effect that exotics transported in ballast water do become established and spread in this way is a difficult task. Nevertheless, the evidence is strong and remains unchallenged. The main sites of introductions are in ports and waterways where shipping is the principle activity and the great numbers of organisms that can be carried in a viable state is strongly indicative that ships are vectors.

3 Aquaculture

Exotic species provide economic opportunities whilst others can impose serious financial loss and unemployment. In the marine environment a small number of exotics are in cultivation or stocked in the wild. These and further exotics are likely to contribute to future production. Species that are tolerant of wide ranges of temperature and salinity, and easily manipulated to produce young and can be maintained at high densities, are most likely to be favoured for production. However, unregulated movements of further exotics intended for culture could provide access for unwanted and harmful pests, parasites and diseases which could subsequently compromise future production. More than 100 species have been transported with living oysters either carried in the packing materials, attached to shells or as parasites and disease agents in the living oyster tissues (Carlton 1992; Sindermann 1992; Minchin 1996). The International Council for the Exploration of the Sea's (1995) Code of Practice on Introductions and Transfers of Marine Organisms provides a procedure whereby the risks of introducing species can be considerably reduced. The introduction of a species by this means is costly because it takes more than one generation of a species before its release from quarantine.

Should an important industry suffer a decline following a serious disease or parasite outbreak a separate exotic strain or species may be imported in large numbers to rapidly replace this decline in production. The politics of such a situation may determine that a precautionary approach is not adopted, because of the time involved using the approved protocols. Direct imports of stock almost inevitably lead to an introduction of unwanted species and some of these may reduce future production or have impacts on the cultivation of other species. For example, the introduction of the Pacific oyster *Crassostrea gigas* to France by aircraft, following the decline in native oyster production, enabled this species to become established but also enabled several associated organisms that included pests to do so despite the use of brine dips as a preventative measure (Gruet et al. 1976). Inevitably species successfully cultivated in one world region will be considered as a suitable species for culture elsewhere. This will include species for commercial and recreational benefit. The stocking of rainbow trout *Oncorhynchus mykiss* has conferred many advantages for recreational angling throughout much of Europe (Lelek 1996; Löffler 1996).

Stocking of the spider crab *Paralithodes camtschaticus* in Russia during the 1960s has resulted in its establishment and expansion to northern Norway (Kuzmin et al. 1996). The species has spread with a combination of planktonic dispersal of its larval stage and by walking. Adults could result in the spread of sessile organisms that may attach to its carapace. Fishes are the most likely species to become widely distributed. Attempts at cultivating the pink salmon *Oncorhynchus gorbuscha* in Russia have resulted in reports of vagrant specimens in Britain and Ireland. Species in marine cage culture may escape and spread in a similar way. Unless there is a sufficiently large inoculum to create a founder population as well as a sufficient knowledge of its biology and interactions with other species, the success of such a programme will have limitations. This is either because of insufficient numbers present to create a reproducing progeny, because of a poor knowledge of their physiological capabilities or because of unexpected interactions with native species.

4 Trade

Agreements involving trade do not normally take into account those associated organisms that may extend their ranges as a result of the trading activity. This is because should there be restrictions on the product the trade may not evolve. In an attempt to overcome this difficulty veterinarians may classify a series of known lower impact diseases that may be moved, and restrict trade for only the most serious diseases. Unfortunately some diseases, listed at a low level of priority, now can become transferred to areas where they did not previously exist. In addition diseases that have yet to be described do not get sufficient attention until they have been spread. Dealing with pests as well as diseases seems to be an additional set of criteria that could compromise the trade activity itself and often appears to receive little or no attention.

The trade in live species, and in particular the trade of half-grown oysters which are relaid for further growth, continues to result in range expansions of molluscan pests and diseases (Minchin 1996). Because oysters survive under cool damp conditions for several days, large consignments are easily transported long distances. Their shells provide habitats for attaching, cryptic and boring species and harmful species can be carried in this way or in molluscan tissues (Bower et al. 1992).

Aquarium species in ornamental trade are a further likely means for the spread of diseases. For example, fish prior to their departure are often held at high densities. Accidental or intentional releases of aquarium fishes that survive in the nature are frequent events in freshwater but rare in the sea. The expansion of epizootic ulcerative syndrome from the Indian Ocean, where it has been responsible for serious declines of fish production, is of concern and movements of aquarium species may aid its spread.

The majority of fishes imported to Europe in ornamental trade are tropical species from freshwater and marine ecosystems. Their management needs to ensure that they are not exposed to prolonged cool periods or sudden temperature changes. Such species are unlikely to thrive in the wild in Northern Europe except possibly in thermal discharges and warm water springs. In tropical and semi-tropical regions exotic fishs have commonly established themselves following release - known from the 1930s (Myers 1940). In temperate waters fish from similar climates are more likely to become established. Some such as the pumpkinseed *Lepomis gibbosus* were deliberately released in central and southern Europe but this species is also an aquarium species and could well become established further to the north.

One of the main concerns is that movements of exotic aquarium species inevitably result in the movement of other species that include diseases associated with the fishes themselves. Robertson & Austin (1994) noted several pathogens associated with introductions of exotic cyprinids, some considered to be harmful to salmon and rainbow trout. Shotts et al. (1976) examined the bacteria associated with imported exotic fishes and the water in which they were carried, coming from Taiwan, Singapore, Hong Kong and Bangkok. Eighteen genera were found associated with the fish and 14 with the associated water demonstrating that transmissions of disease causing bacteria may easily be spread. Tropical aquaria can develop cultures of *Mycobacterium marinum.* Fish normally become moribund and those subsequently cleaning the tanks can develop a sporotrichosis-type condition as a result of infections of this bacterium (Adams et al. 1970).

The ulvophycean alga *Caulerpa taxofolia* may have been accidentally released from an aquarium as it was first discovered in the Mediterranean Sea off the coast of Monaco in near the Oceanographic Museum where it was cultivated for display in the aquaria. The species began to spread rapidly from the northern part of the western Mediterranean to regions further south. This invasive Mediterranean strain of *C. taxifolia* differs from tropical strains of *C. taxifolia* by its great resistance to lower temperatures (Meinesz & Boudouresque 1996).

The aquarium trade needs to consider certification of its products and have regular health inspections. Undue mortalities of stock need to be recorded and explained. Those involved in rearing or collections should be made aware of a protocol of expected standards so as to reduce stresses on collected organisms. They should also be made aware of those species not permitted for trade. This requires a close co-operation between producers and regulators. In many regions of south-east Asia there is a need to treat aquarium fishes as a fishery, conservation of many species will only be achieved with public education (Ng & Tan 1997). According to Davenport (1996) the majority of ornamental fish imported to Europe are exported from Singapore, with less from Israel, USA, Czech Republic, Indonesia, Japan, Brazil, Thailand and Hong Kong. Other countries export smaller amounts. Most of these are imported to Germany, Britain and France.

5 Natural dispersal

Once established within a new locality a species may remain confined to a small region, as do many tunicates (e.g. *Styela clava*) because of their short larval period or because of a lack of a larval stage (e.g. the gastropod *Urosalpinx cinerea*). However the majority of organisms have a pelagic phase that will result in an incremental spread of its range locally. For some species nearest neighbour distances following their planktonic dispersal will be important, barnacles (*Elminius modestus*) practice internal fertilisation and so normally need to be close to each other. Those species likely to be widely distributed include algae with air bladders (e.g. the brown seaweed *Colpomenia peregrina*); thereby allowing water and wind currents to rapidly disperse them. Because marine macrophytes are close to neutral buoyancy these may be easily carried with water currents. Planktonic species may become distributed along coastal fronts and blooms such as those of the naked dinoflagellate *Gyrodinium aureolum*, now known as *Karenia mikomotoi*, have swiftly spread throughout northern Europe since their first appearance in Norway (Tangen 1977).

6 Other vectors

Additional vectors may be associated with trading, for example the use of marine algae used as a packing material with movements of living lobsters or oysters. The algae, epibionts and associated organisms could become established elsewhere, should this material be disposed to the wild. Bait organisms may be exported beyond their normal range and may become discarded alive to the wild. Movements of infested fishing gear may also allow species to colonise new regions (Wallentinus 1999).

The opening up of new waterways and canals, as happened between the Red and Mediterranean seas has resulted in a flow of species (Lessepsian migration) mainly from the Red Sea. Such corridors allow for spread by natural dispersal. Similarly connections between the North Sea and The Baltic (Kiel Canal) have enabled the spread of species (e.g. Chinese mitten crab *Eriocheir sinensis*). In eastern Europe, the building of canals has enabled a transmission and spread of species between the Baltic, Black and Caspian

Seas. There has been a policy to add exotic species to rivers in former USSR under stock enhancement programmes. Mysid and gammarid species were introduced as a food source for commercial fishes.

Scientists and public institutions also have an obligation to act responsibly when disposing wastes. There are many exotic organisms held captive in aquaria in European research institutions. The numbers of species are not known. Unfortunately secondary spread of such organisms to other aquaria is a common feature and some species such as a modified form of the alga *Caulerpa taxifolia* is believed to have spread to the Mediterranean Sea from an aquarium. Likewise, the spread of the golden snail *Pomacea* sp., a serious pest in Southeast Asia and also used in aquaria, are frequently on sale in hobbyist shops. Although unlikely to become established in the wild in Northern Europe they may do so in warmer regions to the south.

The macroalga *Mastocarpus stellatus*, native in Europe, was not occurring in all habitats likely for colonisation. Researchers assessed the reasoning for the limited distribution by planting it into the wild. Today it is well established where it was formerly absent and it is suspected to outcompete some co-existing native species (Wallentinus pers. comm.).

Some species may be imported as live food because they have a cultural value. The recent appearance of the whelk *Rapana venosa* to France may have been a release following imports for the Asian community.

7 Overlapping vectors and risk

Many shipping ports in harbours are situated close to aquaculture activities for reasons of shelter and a nearby market. This proximity of shipping to aquaculture activities poses the unquantifiable threat that some imported organism carried by ships may in some way impair survival, compromise growth, or cause the cultivated product to be unmarketable and this may render the benefits of previous quarantine regulations useless (Rosenthal 1980). Ballasting of water by ships in ports, for example, may result in loading untreated discharges of human sewage and bacteria, such as *Vibrio cholerae*, which on release might enter the food chain in distant ports through cultured filter feeding molluscs (McCarthy & Khambaty 1994).

Small vessels such as yachts and motor boats may develop a fouling compliment on their hulls that may include established exotics acquired in marinas in a shipping port and these may spread to small inlets and lagoons, where ships do not trade.

In ports vectors are likely to overlap because many people normally live in these regions and engage in a wide range of activities. Managing the overlap of vectors in such regions may lead to some hard decisions where some activities may need to be restricted in some way so as to reduce risk. Ports will almost certainly benefit from studies of the exotic species present and when there is a risk of that port acting as a donor to other regions. All relevant activities within the port region should be evaluated where the port may act either as a donor or recipient for unwanted invasives, as was demonstrated for five north-west European port regions by Gollasch & Leppäkoski (1999). Small changes in practice could result in reduced risk, for example, it may be sufficient to reduce the probability of an inoculum becoming established by extending the ballast water discharge trails of ships on entry to a port.

8 Conclusions

This chapter demonstrates how vectors act in the movement of exotic species. Examples are drawn from different taxonomic groups with varying impacts on humans and on ecosystems and will indicate the value of proactive management measures. Some vectors are elusive, and through a better understanding as to how species are spread, together with the knowledge of the critical numbers needed to form new populations and when and where this is most likely to happen, will greatly aid understanding. Today movements throughout the world are continuous; the short transit times by aircraft over large distances are of particular concern because they provide a whole suite of opportunities with reduced challenges. Shipping will continue to be an important vector and those port regions with a large compliment of exotic species may expect to receive more. In the coming century, should predicted changes in climate evolve (global warming), natural ranges of organisms native to northern Europe are likely to change and this will provide new opportunities for exotic species to expand their ranges.

OYSTER IMPORTS AS A VECTOR FOR THE INTRODUCTION OF ALIEN SPECIES INTO NORTHERN AND WESTERN EUROPEAN COASTAL WATERS

WIM J. WOLFF[1*] & KARSTEN REISE[2]
[1]*University of Groningen, Department of Marine Biology, Haren, The Netherlands*
[2]*Alfred Wegener Institute for Polar and Marine Research, Wadden Sea Station Sylt, List, Germany*
[*]Corresponding author *W.J.Wolff@biol.rug.nl*

Abstract

In western and northern Europe there have been deliberate introductions of European flat oyster (*Ostrea edulis*), American oyster (*Crassostrea virginica*), Pacific oyster (*C. gigas*, including the so-called Portuguese oyster *'C. angulata'*), New Zealand oyster (*Tiostrea lurida*), hard clam (*Mercenaria mercenaria*), and Manila clam (*Tapes philippinarum*). Between about 1870 and 1939 tens of millions of *Crassostrea virginica* were introduced from the Atlantic coast of North America. However, *C. virginica* has been unable to establish itself in Europe. For 5 other species it is very likely that they have been introduced with American oysters. Between 1964 and about 1980 *C. gigas* was imported on a large scale from Japan and the Pacific coast of Canada and the USA. It has established itself in Europe permanently. *C. gigas* brought its own parasites and the imports were accompanied by the import of more than 20 species of animals. Most of these observed imports failed, however, and only about 5-6 species seem to have established themselves in European waters. As a vector for the introduction of exotic species into the North Sea area, oyster imports are slightly more important than transport on ship's hulls, and clearly more important than introductions through ballast water. In the Dutch Oosterschelde estuary Japanese oysters interfere with the recreational use of the estuary because of their razor-sharp shells. They also seem to have changed the ecological conditions in the estuary: coinciding with the increase of the oysters, mussels and cockles decrease, as does the oystercatcher (*Haematopus ostralegus)*. It is not yet clear if this is a causal relationship. In the Wadden Sea near the island of Sylt, *C. gigas* established itself as an epibiont on mussel beds, and seems to be at the verge of transforming mussel beds into oyster reefs.

1 Worldwide transport of shellfish

Many species of shellfish have been transported intentionally across the globe to start or to boost aquacultures (Chew 1990). Species included mussels, oysters, clams, scallops, and some less important groups. In western and northern Europe, i.e. the area between Spain and Norway, including the British Isles and the Baltic, deliberate introductions have occurred of European flat oyster (*Ostrea edulis*), American oyster (*Crassostrea virginica*), Pacific oyster (*C. gigas*, including the so-called Portuguese oyster *'C. angulata'*), New Zealand oyster (*Tiostrea lurida*), hard clam (*Mercenaria mercenaria*), and Manila clam (*Tapes philippinarum*). A few other species have been imported for research purposes but never left the research facilities (Utting & Spencer 1992; Flassch & Leborne 1992; Drinkwaard 1999a,b).

Commercial shellfish transfers can be an important vector for the introduction of exotic species. First of all the transferred shellfish species may be an exotic itself. Moreover, live shellfish often contain a variety of parasites and diseases, and uncleaned shellfish

E. Leppäkoski et al. (eds.), Invasive Aquatic Species of Europe, 193–205.

also may host a wide variety of epibenthic algae and animals. Although this is less important for infaunal species of clams, epibenthic mussels, scallops, and oysters may be heavily overgrown with a diverse community of plants and animals (Korringa 1951b). Another factor contributing to the importance of shellfish, as a vector for the introduction of exotic species, are the large quantities that often are shipped. Imports of oysters, for example, often have counted millions of specimens per year, and sometimes even tens of millions (Utting & Spencer 1992; Dijkema 1993).

Before about 1960, the ecological consequences of the large-scale, deliberate introduction of exotic shellfish species were in general disregarded. But the growing awareness that shellfish imports could be accompanied by the import of pests, parasites and devastating diseases as well as the observed effects on native communities, led to a number of measures after 1960. The International Council for the Exploration of the Sea agreed on a Code of Practice (e.g., ICES 1995). Codes of conduct were introduced in several countries and these were translated into license systems and other legal measures. Quarantine measures have been introduced as well. In addition, hatchery production of marine bivalves became technically and economically feasible (Walne 1974), thus diminishing the necessity to import seed shellfish from the wild and often from other parts of the world.

However, within Europe large quantities of shellfish are still being transported from one culture area to another and consequently shellfish, especially oysters, still serve as vector for the introduction of plant and animal species into areas where they did not occur before. The European Common Market even encourages this practice. Moreover, oysters present an extra difficulty. Because of their rugose shells oysters are difficult to clean completely from epiflora and -fauna, provided one wishes to take that precaution.

Certainly in the past shellfish imports have been a very important vector for the introduction of exotic species in Northern Europe. We know numerous examples of species that most likely have been introduced into Europe with imported oysters. In our opinion, before 1975 oyster imports have been of similar importance as ship's hull fouling and ballast water as vectors for introduction. Nowadays, the worldwide scale of oyster imports has become less important for Europe, but on a European scale oyster transfers still constitute a means of transport comparable to the role of ballast water.

2 Transports of European flat oysters between European countries

In Europe, human consumption of shellfish dates back to prehistorical times. Overfishing of shellfish beds is probably a very old custom, but the first information on this is from the 17th and 18th centuries when oyster beds were reported to have been exhausted. This led to attempts to restock the beds with oysters from elsewhere resulting in a long history of oyster transfers in north-western Europe. Already in 1570 a vessel carrying French flat oysters (*Ostrea edulis*) paid dues in Zeeland in The Netherlands (De Jonge 1990), although it is not clear whether these oysters were to be relaid.

As early as 1714 imports of seed oysters from Denmark for relaying in the western Dutch Wadden Sea were reported (Hoek 1911; van der Visch 1975). Baster (1759) reported import of seed oysters from the British Isles into the Oosterschelde estuary in the SW Netherlands. Wehrmann et al. (2000) reported how oyster beds in the East Frisian Wadden Sea, Germany, between 1773 and 1806, were reinforced with oysters imported from elsewhere. There were several attempts to introduce *O. edulis* into the Baltic Sea. Fossil shells are a common feature, and this may have inspired 'reintroductions'. Möbius (1877) mentions imports from the Kattegat to the southern Baltic shore in 1753, 1830, 1843, and later. All these oysters died within 2 years.

Oyster imports for restocking overfished beds but also for fattening of lean oysters on fertile beds continued in the 19th century. Dijkema (1993) reported imports from Denmark, England, and Scotland into the western Dutch Wadden Sea. After the discovery that oysters could be cultivated, cultures were set up in many European countries following the French example in 1860. This, however, did not diminish the demand for oysters for relaying. Thus, in the 19th century flat oysters were exchanged frequently between culture areas in Denmark, Germany, The Netherlands, Belgium, Scotland, England, Ireland, and France (Brittany) (Neudecker 1985; De Jonge 1990). Korringa (1951b) reported import of French *Ostrea edulis* from Morbihan and Arcachon for relaying into the Dutch Oosterschelde estuary in 11 years between 1890 and 1950. From France, some 6.5 million oysters were also introduced into the northern Wadden Sea at the end of the 19th century (Hagmeier & Kändler 1927). Thereafter the Dutch Oosterschelde served as a source for further restocking until 1926.

Korringa (1957) drew attention to the assumed genetic loss due to this exchange of oyster stocks. Natural dispersal might have been rather limited due to the release of rather developed larvae with a short pelagic phase. This may have promoted the selection of local adaptations.

Korringa (1951b), in search of an animal vector for the dispersal of the 'shell disease' of flat oysters, made an exhaustive description of the epifauna living on the shells of *Ostrea edulis* in the Oosterschelde estuary, The Netherlands. Altogether he recorded 134 macrobenthic animal species (Table 1) occurring on the shells of oysters. During this study he discovered a fair number of species that did not occur in Dutch coastal waters. He concluded that these species had been imported with oysters, mainly from Brittany, France. Some of these species (e.g., *Calyptraea chinensis, Anomia ephippium, Polydora hoplura, Sabellaria spinulosa*) were able to maintain themselves for a few years in the Oosterschelde estuary, but in the end all species disappeared again. Some of the species were temporarily even common enough to be recognized by the oyster growers; they called *Anomia*, for example, 'French slippers'.

With imports of Dutch oysters in 1907 to 1926, the brown algae *Dictyota dichotoma* and the green algae *Codium tomentosum* as well as the slipper limpet *Crepidula fornicata* became introduced into the northern Wadden Sea where they did not occur before (Hagmeier 1941). Only the latter established a permanent population. Schodduyn (1931) studied the epiflora and -fauna on flat oysters imported into France by an oyster firm at Ambleteuse, near Calais, from culture areas in Sussex and Cornwall in England, and from Stranraer in Scotland. These oysters were brushed clean before being dispatched, and were 1-4 days underway. Schodduyn, nevertheless, found a long list of species also because he cultivated detritus from the oyster shells as well as the faeces and pseudofaeces (Table 1).

Table 1. Numbers of plant and animal species found on European flat oysters (*Ostrea edulis*). Schodduyn (1931) studied micro- and macroflora and –fauna on oysters transferred from the British Isles to Ambleteuse, France, and cleaned before transport. Korringa (1951b) investigated the macrobenthic fauna on oysters collected from the Oosterschelde estuary, The Netherlands.

Taxonomic group	Number of species found	
	Schodduyn	Korringa
Cyanobacteria	6	-
Chlorophyta	3+	-
Rhodophyta	9	-
Phaeophyta	2	-
'Diatoms'	2 (36?)	-
Foraminifera	0	2
Ciliata	12	1
Porifera	3	5
Hydrozoa	2	3
Scyphozoa (benthic phase)	1	1
Anthozoa	1	2
Turbellaria	0	1
Rhabdocoela	1	0
Nemertini	0	2
Nematoda	2	17
Nematohelminthes (?)	1	0
Polychaeta	10	34
Oligochaeta	0	1
Tardigrada	1	0
Ostracoda	0	7
Copepoda Harpacticoidea	4	27
Cirripedia	1	1
Cumacea	0	1
Tanaidacea	1	0
Amphipoda	0	7
Decapoda	0	2
Acari	1	1
Pycnogonida	0	3
Polyplacophora	0	1
Gastropda	1	3
Bivalvia	3	4
Bryozoa	2	1
Entoprocta	1	1
Echinodermata	0	2
Tunicata	4	4

Schodduyn's (1931) and Korringa's (1951b) studies show both the potential of oysters as a vector for other species and demonstrate that such transports on a European scale did occur. However, we want to point out that their studies do not concern exotic species as discussed elsewhere in this chapter. Exotic species apparently can live in their new environment and are prevented from reaching this new environment by natural barriers. Korringa's species, however, potentially could reach their new environment by natural means (Wolff in prep.), but apparently could not live there because of an unfavourable combination of environmental factors.

In later years also the potential of *Ostrea edulis* as a vector for oyster diseases became clear. The parasite *Bonamia ostreae*, causal agent of the devastating oyster disease bonamiasis, was imported into Spain with infected European flat oysters from California, USA (Cigarría & Elston 1997). Within Europe the pathogens *Marteilia refringens, Minchinia armoricana* and *Bonamia ostreae* were imported into the Dutch culture areas through shipments of infected flat oysters from Brittany, France (Van Banning, 1988).

3 Import of American oysters in the late 19th and early 20th century

The American cupped oyster *Crassostrea virginica* is native to the Atlantic coast of North America. It was imported into Europe for the first time in the early 1870s to restock overfished *Ostrea edulis* beds in England and Wales but also for direct consumption. They were shipped across the Atlantic as deck cargo during winter and early spring. The import came to an end in 1939; it was not resumed, except for a brief episode in 1984, after the Second World War. Quantities involved were large; for example, in 1879 nearly a 100 million American oysters were imported (Utting & Spencer 1992; Eno et al. 1997). Attempts were also made to introduce *C. virginica* from Chesapeake Bay into the Baltic near Haderslev (Schleswiger Nachrichten 1880).

Late in the 19th and early in the 20th century American cupped oysters from English oyster beds were transferred to the Grevelingen estuary in The Netherlands in several years (Korringa 1942; Van Benthem Jutting 1943). In 1913-14 the species was introduced to the German Wadden Sea near Norddeich in Lower Saxony (Neudecker 1985; Wehrmann et al. 2000).

Remarkable enough, *C. virginica* was unable to establish itself anywhere. After the cultures in the UK, The Netherlands and Germany ended, the remaining stock had disappeared after a few years. The reason for this is not clear.

At least five exotic species have reached the British Isles with imports of American oysters: the slipper limpet *Crepidula fornicata,* the American tingle *Urosalpinx cinerea*, the false angel wing *Petricola pholadiformis,* the polychaete *Clymenella torquata*, and the ostracod *Eusarsiella zostericola.* These arrived in the 19th to 20th century. *Urosalpinx, Clymenella* and *Eusarsiella* remained restricted to the estuaries in Essex and Kent (Hedgpeth 1980; Utting & Spencer 1992), but the other two species colonized large parts of the European seas.

The first known occurrence of *Crepidula* in Europe was in 1872 in Liverpool Bay, but the populations in this area have died out since then. It has been subsequently introduced in Essex between 1887 and 1890 (Eno et al. 1997). Live specimens were found at the Belgian coast in 1911 (Nehring & Leuchs 1999). The first live specimens in The Netherlands were collected on a large piece of wreckage washed ashore at Zandvoort in October 1926 (Korringa 1942). In May 1929 two specimens of *Crepidula* were found in the Oosterschelde estuary. Next year hundreds were found and in 1941 4 million kg of *Crepidula* from the Oosterschelde estuary were processed for human consumption (Korringa 1942; Werner 1948). In the Dutch Wadden Sea, the slipper limpet was introduced with oysters for relaying from the Oosterschelde estuary (Van Benthem Jutting 1933; Korringa 1942). German and Danish coastal waters were colonized through oyster imports from The Netherlands in the 1930s (Korringa 1942). France was reached in 1949 and subsequently all major French oyster producing regions were colonized (De Montaudouin et al. 2001). Today, the European distribution ranges from the Mediterranean to southern Norway (Blanchard 1997). Consequently, *Crepidula* has colonized a large part of the European seas through a combination of larval transport, transport by floating objects (seaweeds attached to the shell), and transport with oysters for relaying.

Also *Petricola* is likely to have been introduced with North-American oysters, probably not later than 1890; it was first found in the River Crouch, Essex (Eno et al. 1997). Schlesch (1932) records *P. pholadiformis* from the northern German Wadden Sea near Föhr in 1896, the Belgian-Dutch coast in 1899, and the Danish Wadden Sea in 1905. Van Breemen (1907) and Denker (1907) record this species from the Dutch coast in 1905. This pattern of rapid dispersal suggests transport of planktonic larvae by marine currents although transport with oyster transfers cannot be completely excluded (Nehring & Leuchs 1999).

4 Import of Portuguese oysters into other European countries

The Portuguese oyster ('*Crassostrea angulata*') actually is a strain of *C. gigas* from Taiwan (Boudry et al. 1998; O'Foighil et al. 1998). It was described from Portugal where it occurred for example in the Tagus and Sado estuaries, in 1819. Most likely it arrived into that country as a fouling organism on a ship's hull, somewhere between the early 16th century and the end of the 18th century. In this respect it is significant that the Portuguese maintained trading contacts with Taiwan from the early 16th until the early 17th century.

In the second half of the 19th century '*C. angulata*' was imported into France in years when the oyster farmers lacked sufficient flat oysters. This led to a flourishing culture on the Atlantic coast of France based both on locally grown stock and oysters imported directly from Portugal for relaying (Drinkwaard 1999a). In the 19th century Portuguese oysters were imported into the Netherlands as well. However, after 1885 it was forbidden to introduce these oysters into the Oosterschelde estuary and a few years later the same applied to the Grevelingen estuary. After the termination of the imports the Portuguese oysters disappeared again (Korringa 1965). Since 1936 '*C. angulata*' has been regularly imported accidentally into the Dutch estuaries with French *Ostrea edulis* seed oysters from Brittany. After the Second World War it was again several times deliberately introduced into the Zeeland estuaries from Brittany, as well as from Portugal, , but it did not establish itself (Van Benthem Jutting 1937, 1943; Korringa 1965). In England, Portuguese oysters were first introduced from Portugal into the River Blackwater, Essex, in 1926 (Utting & Spencer 1992). This colony was thought to have died out in 1965 because importations had ceased in 1962, but there was still a substantial population in the River Blackwater in 1970 (Eno et al. 1997). In Germany the species was introduced into the Wadden Sea near Norddeich in 1913-14 and in the Jadebusen in 1964. This did not lead to lasting cultures or establishment of the species (Wehrmann et al. 2000).

Between 1967 and 1973 '*C. angulata*' became the victim of disease in France. Especially due to an iridovirus almost the complete stock disappeared (Grizel & Héral 1991; Drinkwaard 1999a). The species also disappeared from Dutch waters around 1970 (Moerdijk 1986; Drinkwaard 1999b). This disease demonstrated that '*C. angulata*' and *C. gigas* apparently are slightly different. '*C. angulata*' almost died out due to the disease, whereas it was replaced by *C. gigas* in the same areas without any problem (Grizel & Héral 1991).

No examples of introductions of other exotics with Portuguese oysters are known.

5 Import of Japanese oysters in the 20th century

5. 1 IMPORT OF OYSTERS

Starting in 1902 spat and adults of the Japanese or Pacific oyster (*Crassostrea gigas*) were imported into the north-western United States near Seattle and into British Columbia, Canada, from Japan. In 1930 natural spat settlement occurred for the first time and recurred in later years. This, and the later development of hatcheries, formed the basis for a widespread culture of 'Pacific' oysters along the American Pacific coastline (Quayle 1969; Drinkwaard 1999a,b).

The very severe winter in 1962-63 reduced the Dutch flat oyster (*Ostrea edulis*) stock to about 3% of its earlier size. To make up for the losses spat of *C. gigas* was imported from British Columbia into The Netherlands in 1964. In the following years more imports of spat followed, starting in 1966 also from Japan, because the results were promising. In 1971 also adult specimens were introduced into The Netherlands from British Columbia. In 1975 and 1976 *C. gigas* appeared to be able to reproduce in the Oosterschelde estuary. This happened again in 1982 and since that time *C. gigas* has developed explosively in this estuary; 33% of the shellfish biomass now consists of *C. gigas* (Dr. A.C. Smaal pers. comm.). In the 1980s other Dutch estuaries started to be colonized and in 2000 *C. gigas* occurred along the entire Dutch coast (Wolff in prep.).

In 1966 in France, the first spat of the Japanese oyster was probably introduced by an oyster farmer in the Bay of Marennes-Oléron directly from Japan. This was followed by more imports of Japanese spat by oyster farmers in the following years. In 1969 government officials became involved. As a result more than 500 tons of adult *C. gigas* were imported from British Columbia into all major oyster areas on the west coast of France in 1971-75. Imports of *C. gigas* spat from Japan continued; between 1971 and 1977 slightly more than 10,000 tons were imported, mainly to the French Atlantic coast, but also to the Mediterranean (556 tons) and the Channel coast of Brittany. In 1982 all imports from the Pacific were stopped because of a new haplosporidian parasite found in Japan (Grizel & Héral 1991; Boudouresque et al. 1994; Drinkwaard 1999a).

The introduction of the Japanese oyster in Great-Britain started in 1965, when a controlled breeding programme was started with oysters from British Columbia at the Shellfish Culture Unit at Conwy. This has led to *C. gigas* cultures all around the British Isles based on hatchery-reared spat. Remarkable enough, the Britisch *C. gigas* populations hardly show natural reproduction. Local spatfall occurred in the River Blackwater, Essex, and light spatfalls registered in some estuaries of SW England (Dart, Teign, and

Exe) and the Menai Strait following unusual warm summers in 1989 and 1990. Genetic evidence shows that spatfall in the River Teign originated from French stock. However, it is unclear whether this was from adult specimens discarded at English sites or from larvae that crossed from the French side of the Channel (Utting & Spencer 1992; Eno et al. 1997; Drinkwaard 1999a,b). The species was also introduced into Ireland where growth trials were carried out in the 1970s. The cultures in Ireland are based on British (including Northern Ireland) hatchery-reared spat and on half grown oysters imported from the continent. Production in Ireland has increased from 60 tons in 1980 to 4,500 tons in 1996 (Drinkwaard 1999a; Steele & Mulcahy 1999).

In Germany the culture of *C. gigas* started at Sylt in 1971, initially with Scottish hatchery-reared spat and later with oysters from various sources. Activities were also started in the Baltic in the Flensburger Förde. Later experimental cultures were set up in the Wadden Sea at Neuharlingersiel (1974), in the Jade (1976/1982), near Wangerooge (1982) and near Norderney (1987). At this moment, a culture is still in operation only at Sylt. *C. gigas* appears to reproduce naturally in the German Wadden Sea. A wild stock has developed around Sylt, and the East Friesland area seems to have been colonized from the Dutch Wadden Sea (Reise 1998; Drinkwaard 1999a; Wehrmann et al. 2000). On the other hand, Dutch mussel farmers have reported that Japanese oysters have occurred in transports of seed mussels from the North-Frisian Wadden Sea to the Dutch Wadden Sea. In Denmark oysters from the Flensburger Förde, Germany, were planted in the Little Belt in 1979. In 1983 also oysters from England, The Netherlands, and France were imported (Drinkwaard 1999a).

Summarizing, Japanese oysters have been imported directly from Japan as well as from the Pacific coast of America into France, The Netherlands, and Great-Britain from 1964 to about 1980. The material imported into Britain was subject to proper quarantine measures and was used to develop an oyster culture based on hatchery-reared spat. This spat was also later used to set up cultures in Ireland, Germany, and Denmark. The oysters imported into France and The Netherlands were at least partly put directly into the estuarine environment, thus constituting a potential corridor of introduction. At a later stage Japanese oysters grown in France and The Netherlands were transferred for relaying to other European countries, such as Ireland and Germany, thus offering a potential vehicle for intra-European transport for other exotic species.

5.2 IMPORT OF DISEASES AND PARASITES

The Japanese oyster brought at least part of its parasite fauna with it. The parasitic copepods *Mytilicola orientalis* and *Myicola ostreae*, both native to East Asia, were observed both in France and in The Netherlands. In France the species did not occur in every estuary with Japanese oysters, at least not in 1977. The copepods did occur also in a few other species of shellfish (e.g. *Mytilus edulis*) but apparently caused little harm (His 1977a,b; Stock 1993).

It has been speculated that shellfish diseases have been imported with Japanese oysters too. However, there is no strong evidence for this supposition.

5.3 IMPORT OF OTHER ORGANISMS

Gruet et al. (1976) provided firm evidence for the introduction of exotics with *C. gigas* imported from Japan into oyster culture areas at the Atlantic coast of France. They studied the fauna imported on seed oysters transported by plane, which before departure from Japan and after arrival in France, had been immersed in fresh water for one hour to kill predatory turbellarians.

A few species found by Gruet et al. (1976) had not survived transport and immersion in fresh water. This included the ascidians *Ciona intestinalis* and *Ascidia sydneiensis*. Table 2 lists the species that did survive transport and freshwater treatment. Gruet et al. (1976) also recorded "des Algues" but gave no indication of the species involved. Altogether, at least 23 animal species were imported alive from Japan.

Gruet et al. (1976) also investigated which species survived on the oyster beds by inspecting oysters and collectors one year later. At least seven species proved to be still alive (Table 2); this conclusion could be drawn only for those species, which did not occur beforehand in France (e.g. *Lepidonotus squamatus*). Grizel & Héral (1991) listed the five species still occurring 15 years later (Table 2); none of these species seemed to have enlarged its area of occurrence. In addition two Japanese algae had established after import of Japanese oysters into Thau lagoon on the Mediterranean: *Laminaria japonica* and *Undaria pinnatifida* (Grizel & Héral 1991).

For many other species we only have circumstantial evidence for their introduction with *C. gigas* in northern Europe from Japan or the Pacific coast of North America. Such evidence may be based on the following criteria:
(i) the species is known to occur in Japan or on the Pacific coast of North America;
(ii) the species arrived in northern Europe after the introductions of Japanese oysters started in 1964;
(iii) the species can be imagined to be transported with Japanese oysters;
(iv) the first European finds are sufficiently close to the areas in France and The Netherlands where imported Japanese oysters have been relaid, so that natural transport processes can explain their location;
(v) the species is known to be insensitive to dry transport and immersion in fresh water.

From the list by Reise et al. (1999) we can extract the following species occurring in the North Sea and the Channel and fulfilling the first three criteria: *Gymnodinium mikimotoi, Alexandrium leeii, Fibrocapsa japonica, Chattonella* sp., *Thalassiosira punctigera, Coscinodiscus wailesii, Sargassum muticum, Undaria pinnatifida, Grateloupia dorifera, Dasysiphonia* sp., *Anotrichium furcellatum, Polysiphonia senticulosa, Pileolaria berkeleyana, Hydroides ezoensis,* and *Ammothea hilgendorfi.* However, several species do not meet the fourth criterion: for example, the first observation of *Gymnodinium mikimotoi* was in SW Norway in 1966 and the sea spider *Ammothea hilgendorfi*, which has very limited possibilities for dispersal, was found in 1978 in Southampton Water, England. Also the well-known *Sargassum muticum* was originally discovered at the "wrong" place. Therefore, for these species other transport vectors should not be ruled out.

Table 2. Species imported into France with seed oysters (*Crassostrea gigas*) directly from Japan; their survival after one year, and their occurrence after 15 years. After Gruet et al. (1976) en Grizel & Héral (1991).

Species	Group	Found at arrival	after 1 year	After 15 years
Aiptasia pulchella	Anthozoa	x	x	X
Stylochidae	Turbellaria	x	x	
Lepidonotus squamatus	Polychaeta	x	?	?
Syllidae	Polychaeta	x		
Phyllodocidae	Polychaeta	x		
Nereis multignatha	Polychaeta	x		
Platynereis bicanaliculata	Polychaeta		x	
Terebellidae	Polychaeta	x		
Hydroides ezoensis	Polychaeta	x	x	X
Jania pseudocorrugata	Polychaeta	x		
Balanus albicostatus	Cirripedia	x	x	X
B. amphitrite amphitrite	Cirripedia	x	x	X
B. improvisus	Cirripedia	x	?	?
Cymodoce sp.	Isopoda	x		
Corophiidae	Amphipoda	x		
Amphipoda	Amphipoda	x		
Anomia chinensis	Bivalvia	x	x	X
Musculus senhousei	Bivalvia		x	
Diffabala picta	Gastropoda	x		
Bryozoa	Bryozoa		x	
Polyandrocarpa sp.	Tunicata	x	x	
Styela sp.	Tunicata	x	?	?
Molgula sp.	Tunicata	x	?	?

Nevertheless, transports of Japanese oysters for relaying within Europe have contributed to the spread of both European and exotic species. Reise (1998) observed on Sylt, Germany, that concurrently with the oysters the colonial ascidian *Aplidium nordmanni* arrived for the first time in the Wadden Sea. It is common at the British coast. The tube-building polychaete *Pomatoceros triqueter* occurred once on the beds of the native oyster *Ostrea edulis* and then vanished when these became extinct by overexploitation. Apparently, this worm was now re-introduced together with the Japanese oysters from Britain. Also the barnacle *Verruca stroemia* arrived as an attachment to the Japanese oysters but did not spread. Two exotics, the seaweed *Sargassum muticum* and the ascidian *Styela clava*, became established but may as well have arrived by other vectors. In The Netherlands it is probably no chance event that many recent introductions of exotic species that earlier had not been found elsewhere in Europe, occur in the Oosterschelde estuary, the centre of the Dutch shellfish trade (Maggs & Stegenga 1999; H. Stegenga, pers. comm.; Wolff in prep.).

5.4 THE IMPACT OF *CRASSOSTREA GIGAS* ON ITS NEW ENVIRONMENT

The developments in the Oosterschelde estuary, The Netherlands, may serve to demonstrate the impact of *C. gigas* on its environment.

In 1966 the Dutch oyster farmers were informed that the introduction of the Pacific oyster as seed stock was acceptable since these oysters could not reproduce at the latitude of the Dutch coastal waters. However, in 1975 a spatfall occurred during a very warm summer. The real breakthrough came with another spawning in 1976. It resulted in 20-30 million so-called weed oysters in the estuary. In 1982, a new natural larval outburst occurred with up to 50 *C. gigas* larvae per 100 l Oosterschelde water as against 5 *O. edulis* larvae (Drinkwaard 1999b). From that time on researchers from outside the shellfish community start to notice the oysters. Hoeksema (1983) records five live specimens found in the Oosterschelde near Zierikzee. Moerdijk (1986) states that the species "..felt very well at home in the Oosterschelde estuary; in no-time the entire estuary had been colonized." Heinis et al. (1995) record that *C. gigas* in 1989 was established at four out of five investigated locations in the Oosterschelde estuary. Strong expansion occurred in the period 1989-1993, perhaps also due to the warm summers of 1989 and 1992. In 1994 more than 2000 tons of marketable oysters were produced in the Oosterschelde estuary (Drinkwaard 1999b). Consequently, in about 20 years *C. gigas* has become a very important part of the Oosterschelde biota. At the same time as *C. gigas* increased, the stocks of blue mussels *Mytilus edulis* and cockles *Cerastoderma edule* decreased in the Oosterschelde estuary. This decrease was accompanied by a decrease of the population of a shellfish feeding bird: the oystercatcher (*Haematopus ostralegus*), which in spite of its name hardly feeds on Japanese oysters. Within The Netherlands the Oosterschelde estuary is a 'hot spot' for exotics. About 30% of all Dutch introduced marine and estuarine species occur only in this estuary, and about 50% of all species occur only in this estuary and neighbouring waters (Wolff in prep.).

6 Other oysters imported into northern Europe

Tiostrea lutaria has been introduced to Britain from New Zealand. It entered the quarantine procedures at the Fisheries Laboratory at Conwy, Wales, in order to produce seed oysters. These have been introduced into the wild in Menai Strait, Wales, where the species now forms a small stock. It is a slow disperser and, moreover, it is sensitive to cold winters. Hence, *T. lutaria* is not expected to become a dominant species. It has no accompanying other exotic species (Utting & Spencer 1992; Eno et al. 1997).

Crassostrea denticulata has a few times washed ashore in The Netherlands. The specimens were attached to drifting material (Rizzi 1992; Wolff in prep.). Whether this constitutes long-range transport of this West-African species, or denotes a European occurrence, is unclear.

7 Discussion and conclusions

North-western Europe has experienced two periods of significant introductions of exotic oysters and accompanying species. Between about 1870 and 1939 the American oyster *Crassostrea virginica* was introduced from the Atlantic coast of North America. Despite the import of tens of millions of oysters, *C. virginica* has been unable to establish itself in Europe. For five species it is very likely that they have been introduced with American oysters and we may expect that in addition a contribution has been made to the European pool of cryptogenic species. These American imports have not been followed well, however.

Between 1964 and about 1980 the Japanese or Pacific oyster *Crassostrea gigas* was imported on a large scale from Japan, British Columbia in Canada, and the NW Pacific coast of the USA. Also this species has been imported by the millions and in contrast to its American congener it has established itself in Europe permanently. Nowadays important populations occur in France and on the southern shores of the North Sea. It has been established that *C. gigas* brought its own parasites and there is concrete evidence that oyster imports were accompanied by the import of more than 20 species of animals. Most of these observed imports failed, however, and only about 5-6 species seemed to have established themselves in European waters. For a larger number of exotic species we have circumstantial evidence that they may have been introduced or secondarily spread further with Pacific oysters, but other vectors cannot be ruled out. It is unlikely that imports of Japanese oysters have led to the presence of cryptogenic species.

Reise et al. (1999) gave a list of introduced species in the North Sea and the Channel. Of the 80 species listed 32 are believed to have been imported with American or Japanese oysters. Wolff (in prep.) shows that oyster imports form the most important vector for introduced species in the coastal waters of The Netherlands. Oyster imports are slightly more important than transport on ship's hulls, and clearly more important than introductions in ballast water.

Imports of American, Portuguese, and Japanese oysters were undoubtly beneficial to oyster cultures in several European countries because they provided new oyster stocks when earlier stocks had dwindled due to overexploitation or to diseases. However, how many of these diseases were imported with exotic oysters?

Negative consequences occur through the imported oysters themselves and because of other introduced species. In the Dutch Oosterschelde estuary Japanese oysters have expanded enormously and they nowadays interfere with the recreational use of the estuary because of their razor-sharp shells. They also seem to have changed the ecological conditions in the estuary: coinciding with the increase of the oysters, mussels and cockles decrease, as does the shellfish-feeding oystercatcher. It is not yet clear if this is a causal relationship. In the northern Wadden Sea near the island of Sylt, *C. gigas* established itself as an epibiont on densely packed mussel beds, and seems to be now at the verge of transforming mussel beds into oyster reefs.

Accompanying exotic species are not known to have caused positive effects. The predatory oyster tingle *Urosalpinx cinerea* causes damage in one English oyster culture area but does not seem to expand. The slipper limpet *Crepidula fornicata*, however, has colonized a large part of the European coastal waters, where it competes with oysters and other shellfish for food. The shellfish farmers are able to manage the pest by keeping it at a sufficiently low level. The seaweed *Sargassum muticum*, if introduced with oysters, interferes with recreational use, aquaculture, fisheries and intake of cooling water.

Three approaches have been taken in the past with respect to the introduction of exotic species through oyster imports:

(i) total disregard of the possibility that harmful exotics could be introduced;

(ii) measures to prevent introduction of species which might harm the oyster culture;

(iii) quarantine measures to prevent the introduction of any exotic species (except for the oysters, of course).

Disregard of harmful consequences was the rule before about 1960, hence in the period when American and Portuguese oysters were introduced. In incidental cases, however, care was taken to prevent particular introductions such as the oyster drill *Ocenebra erinacea* (Hoek 1882).

The French imports of Japanese oysters in the 1970s were subjected to immersion in fresh water to kill predatory turbellarians. Gruet et al. (1976) clearly demonstrated that this method was inadequate to kill most of the other epifauna. Another practical measure applied was to forbid import of oysters from areas known to be contaminated by oyster diseases or pests. However, within Europe such a measure can slow down introductions but not prevent them because the hydrographical situation poses no absolute boundaries.

The British approach appears to be by far the best. By importing limited amounts of brood stock to produce larvae and subsequently destroying the brood stock, seed oysters without accompanying exotics can be obtained. In combination with commercial hatchery production of oyster spat this method ensures that exotics have no chance to be introduced. Utting & Spencer (1991) reported that indeed no exotics have been introduced into the British Isles with shellfish from outside Europe.

The Rio Convention (1992), which has been subscribed by all countries in northern Europe, explicitly seeks to prevent the introduction of exotic species. Some countries, such as the United Kingdom, comply with this convention because they have imposed strict regulations on the import of foreign species (Eno et al. 1997). Other countries, however, put hardly anything in the way of traders importing oysters. Moreover, the European Common Market stimulates free trade between European countries; the risk of introduction of harmful exotic species from one part of Europe to another has proven to be no reason to prevent such trade.

Finally, the Code of Practice of the International Council for the Exploration of the Sea (ICES) and the European Inland Fisheries Advisory Commission (EIFAC) should be mentioned (ICES 1995). This Code of Practice provides a practical set of rules to prevent harmful introductions through the import of oysters and other non-native organisms. It is mainly based on quarantine measures to be carried out in the exporting and importing countries.

EXOTICS FOR STOCKING AND AQUACULTURE, MAKING CORRECT DECISIONS

DAN MINCHIN[1*] & HARALD ROSENTHAL[2]
[1]Marine Organism Investigations, Ballina, Killaloe, Co Clare, Ireland
[2]Neu Wulmstorf, Germany
*Corresponding author minchin@indigo.ie

Abstract
Exotic species form an important contribution to the economies, fisheries, recreational angling and food production in Europe. In general, culture operations have targeted a small number of very productive species, many of these are exotic or derived from transfers of different native population strains. Stocking has taken place for about 2,000 years with the majority being introduced over the last century. There is an awareness that other species may either form suitable substitutes or may even be more productive and so many further exotic species and hybrids are likely to be tested experimentally in freshwater and brackish and marine areas. Some species, where their biology is not well known and presently not in culture, may be utilised in the future as new technological developments arise. Introduced species will carry other associated species with them and some of these will have unwanted consequences for some native species and natural assemblages.

1 Introduction

Aquaculture is the fastest growing sector within fisheries due to increasing demand for aquatic products. Most of the world's freshwater fish production is based on the traditional and well-tried methods of Chinese carp pond utilisation, whereas the development of modern intensive cultivation in coastal waters was generally restrained by a poor understanding of biological requirements and appropriate husbandry. Presently the culture of organisms has been improved with a better management in hatcheries and culture or restocking areas and aided by rapid transport. However, on account of competitive pricing some operations will take short cuts in the production process and this may have unfortunate consequences.

Stocking aquatic species has enhanced population size in regions with limited natural reproduction by supplying seed for culture, supplementing recruitment lost through habitat alterations, providing a forage species as well as increasing opportunities for recreational fisheries. Such stocking involves movements between specialised facilities and separated regions. Introductions of exotic species will invariably result in establishment of some diseases, pests and parasites associated with shellfish (Bower et al. 1994) and teleosts (Blanc 1997).

Candidate species for culture will require some special criteria such as having: (i) a traditional high value market, (ii) a perceived high quality of flesh, (iii) a tolerance to handling stresses, (iv) a high tolerance to key environmental variables, (v) a controllable reproductive period (vi) a high survival throughout production or following stocking. In principle, species fulfilling these characteristics are considered prime aquaculture species. However, because of the capital costs of equipment to retain stock, such culture species need to be retained at high densities if profits are to be made. Some exotics

E. Leppäkoski et al. (eds.), Invasive Aquatic Species of Europe, 206–216.

cultivated in Europe either commercially or experimentally are discussed in this account. Other reviews have been undertaken by Holcik (1984) and Welcomme (1991).

2 Oyster movements: we should learn from history

Oyster production is important for many coastal communities and the stocking of oysters has evolved following a number of production declines arising from high exploitation, diseases, predators and other factors. With the introduotion of exotic oysters new problems evolved including previously unknown diseases. With the building of railways there was a depletion of British and Irish natural native flat oyster *Ostrea edulis* beds, because oysters could be carried alive over long distances, particularly during the cooler winter months. High exploitation rates and changes in the usage of estuaries (where much of the cultivation took place) arose from encroachment of port developments and increased urbanisation resulting in production declines. Some depleted oyster beds were restocked with oysters from the reliable settlement regions of western France, Brittany and the Netherlands. This followed legislation in the 1870s enabling private management of oyster beds thereby allowing controls on production. With the advent of more regular and rapid crossings of the Atlantic by steam-ships, imports of American half-grown oysters *Crassostrea virginica*, which could now survive these journeys as deck cargo in barrels, could be relaid on the shore for a summers growth before their sale. This evolved as a regular trade to Britain and Ireland from the 1870s to 1939, with an annual relaying of millions of oysters. Some associated species surviving similar emersion periods also became established.

Importations of *Crassostrea angulata* (now considered a form of *C. gigas*, and may have been an early intended importation from the Pacific to Portugal (Menzel 1974) from Portugal to France began about 1865 and were the basis for a thriving industry that gradually grew to a peak in the 1930s and 1950s. There was a steady decline that followed. This was due to a gill disease then followed by viral infections that ultimately led to the collapse of production (Heral 1990). Neither *C. virginica* nor *C. angulata* successfully recruited in Britain or Ireland due to insufficiently high summer temperatures for the spawning of this genus. This was despite attempts to raise them in ponds at a time when shellfish hatchery techniques were unknown. For this reason restocking reverted to transfers of *O. edulis* from the closer and more intensive settlement regions in France during the 1940s-1960s.

In Conwy, Wales, in 1965, the Pacific oyster *Crassostrea gigas* was cultured in quarantine and following this period was released within Britain and to Ireland. Although production took some years the industry became reliant on hatchery produced seed because of the inability of the species to produce significant local settlements. Following this development a series of events took place that resulted in further production declines of native oysters.

A haplosporidian *Martelia refringens* was isolated from moribund *O. edulis* in France (Comps et al. 1975). This disease caused malfunctions to the digestive gland. The disease was not as prevalent after 1979 but soon a further problem evolved following imports of flat oysters originally grown on the Pacific coast of North America. A sporozoan *Bonamia ostreae* disease of the blood of *Ostrea* spp. led to a serious decline of *O.*

edulis production. It was first noticed in France (Pichot et al. 1979). Bonamiosis was most prevalent in older oysters that suffered mortalities of 50-80%. The relaying of disease-free juveniles and their culture in more open and deeper water at reduced densities has since enabled some production. Although *B. ostreae* now occurs from Spain to The Netherlands and also Britain and Ireland, it has localised populations and a further spread to other bays is expected. In the application for approved zone status for Great Britain and Ireland under the European Union Directive 91/67/EEC, all stocks of *Ostrea edulis* are regularly sampled for the diseases *B. ostreae* and *Marteilia refringens*.

The rapid decline in production of *O. edulis* threatened employment in French coastal communities and an urgent solution was sought. Consignments of *C. gigas* were flown directly from Japan to France and on entry were subject to brine dips to kill those organisms attaching to the shell. However, this approach was not effective because many associated species also became established in aquaculture areas. Pests, parasites and diseases are not confined to the shell exterior alone but also occur within the mantle cavity and tissues as well as within the vacant spaces of dead oysters. As a result unwanted organisms gained entry to Europe and spread. Despite this, these imports of *C. gigas* produced dense natural settlements that enabled a rapid rebuilding of the oyster industry (Grizel & Héral 1990).

Since the 1960s it was the policy in both Britain and Ireland to refuse requests of shellfish and fish imports from unapproved areas, which included oyster growing areas in France. However, the EU directive 67/91/EEC (designed to improve trade within Europe from 1993) did not take ecological matters fully into account and so unwanted species were spread with half-grown oysters movements (Minchin et al. 1993). Two copepods *Mytilicola orientalis* found in the oyster gut, and *Myicola ostreae* found in the mantle cavity subsequently spread to Ireland and the Netherlands. Other exotic species are expected to spread with further movements of half-grown oysters. Some of these will have arrived in Europe by different means but have become established in oyster culture areas. Oysters sourced from these colonised areas may now spread the Asian shore crab *Hemigrapsus penicillatus* (which arrived and is spreading from France since the mid 1990s (Gollasch 1999) and the slipper limpet *Crepidula fornicata* (Minchin 1999). After the 1993 EU Directive unexplained summer mortalities of *C gigas* occurring in France started to appear in south-coast Irish bays from the 1994. Similar events are known during warm summers in Japan and the west coast of North America (Cheney et al. 2000).

Unauthorised movements of molluscs pose a risk of introducing unwanted organisms and disease agents that could compromise future shellfish production and because of the ease of transplanting stock using modern transport this may regularly occur. Occasionally such movements are intercepted, for example, in 1997 a seized importation of *C. virginica* at Shannon airport, Ireland, coming from Long Island Sound, USA and flown from New York, contained hundreds of slipper limpets *Crepidula fornicata* and *C. plana*, eggs of the oyster drill *Urosalpinx cinerea* and the commensal pea crab *Pinnotheres ostreum.* Should direct imports of American oysters take place such pests may become established because of the shorter transmission times enabling greater numbers to survive.

3 Transfer of crustaceans for stocking and aquaculture

3.1 FRESHWATER CRAYFISH

Introductions of exotic crayfish to European freshwaters have a long history. Stocking of crayfish in Europe has been prompted by a demand that exceeds local supplies. It was believed that these exotic species, already accepted by European consumers, would thrive once introduced. The crayfish plague *(Aphanomyces astaci)* was probably introduced with stocks of North American crayfish imported to Italy in the 1860s and by 1907 had spread to Scandinavia causing high mortalities to native crayfishes. Restocking of the depleted areas took place from catchments where the disease was not known to occur. However, Hein (1911) warned against such a stocking policy, but this advice was ignored. The lack of awareness of the vectors of the disease was why fishermen themselves contributed to its spread. A fuller knowledge of the fungus biology remained unknown until 1935 and its host range was not known until later (Unestam 1972).

In the UK, the signal crayfish *Pacificastacus leniuculus* may have acted as vector for a rapid spread of this fungus to several river catchments in the UK (Alderman et al. 1991). Here the native *Austropotamobius pallipes* was susceptible to this disease and many populations expired. In Ireland in the mid 1980s *A. astaci* may have been introduced with infested fishing gear but no recent accounts of its presence exist (J. Reynolds, pers. comm.). Because the adult *Pacifastacus leniusculus* is resistant to *A. astaci* infestation (Fürst 1984) except when in culture, it was used to restore a once thriving Swedish fishery in 1960 (Brinck 1988) and by 1982 it was established in about 260 lakes and rivers. Similar introductions elsewhere, such as Finland, have been successful, and yields in some years exceeded those of historical landings of the native *Astacus astacus.* Early Swedish range expansions of the exotic crayfish led to legitimate concerns of the spread of the fungus to Norway. It spread to Norway to the Vrangselven by 1973 (Hastein & Gladhaug 1973).

Movements of exotic crayfish for stocking have resulted in the appearance during 1975 and 1977 of the cryptogenic parasite *Psorospermium haeckeli* (Nylund & Westman 1979). Henttonen et al. (1984) have suggested that the ciliate *Cothurnia* sp. and some ostracods may have been associated with introduced signal crayfish. Managers should note that movements of exotic crayfish could result in the spread of other species; some of these may presently be undescribed. To prevent a further spread of unwanted species will require an open and shared responsibility with regulators and the industry.

3.2 MARINE CRUSTACEANS

Few European marine crustacean stocking programmes involve exotic species, but from 1961 to 1969 young and adults of the king crab *Paralithodes camtschatica* were released in Northern Russia. The species has extended its range into northern Norway (Kuzmin et al. 1996) and this is expected to gradually move southwards.

4 Introductions and transfers of fish

A large number of fish species have regularly been transferred for stocking purposes within regions and between rivers and lakes (Welcomme 1991). An FAO register of international transfers of inland species has been developed in the 1990s to become the Database on Introductions of Aquatic Species (DIAS) now accessible at http//www.fao.org/waicant/faoinfo/fishery/statist/fisoft/dias/mainpage.htm and contains over 3100 records. For fishery enhancement and stocking purposes, many fishes have been cultured and transferred among watersheds within their normal range of distribution, mixing numerous strains and populations. The potential side effects of these activities have rarely been studied and are seldom understood. Lehtonen (this volume) lists 130 exotic species of which 50 have succeeded in establishing themselves. Here we examine some of these fishes used in stocking and aquaculture.

4.1 CARP

The common carp *(Cyprinus carpio)* originates from the Manchurian region of China, having originally been introduced in ancient times to the rivers draining into the Black Sea. In Europe carp keeping was developed by the Romans and was subsequently developed and managed by monks who spread the knowledge of this practice. This husbandry resulted in domesticated forms of pond carp some of which have hybridised, some with backcrosses to the wild. The different strains have varying rates of growth and tolerances as well as resistance to disease (Picl 1995). Carp were probably first introduced to England in 1496, and then to Ireland during the reign of James I (1603-1625). Its cultural importance in England was indicated in records of King Henry VIII: in 1532, various rewards were paid to persons bringing "...carpes to the King..." (Panek 1987). In Holland, carp were often kept in cellars for months and fed by milk and bread to be fattened for the table. Several attempts were also made to introduce the carp to Nordic countries over the last 250 years. In the Middle Ages the monks and other communities in Central Europe, developed carp pond culture not only for food but also as an integral part of the agricultural system. Carp had a remarkable influence on water management supporting agriculture. The Bavarian emperor, recognizing the importance of carp farming in Bohemia, persuaded monks to construct pond systems in Bavaria at the end of the 19th century. Carp continue to provide an important economic role in many parts of southern and central Germany and elsewhere in Central Europe. In addition to conventional pond farming, semi-intensive and large-scale pond culture of carp has been conducted in Hungary, where they have been utilised to manage human sewage in ponds, using well-balanced organic loadings to produce a high quality fish (Szabo 1994). Despite the widespread culture and release activities, the common carp is not a well-established species in many parts of Europe and mainly relies on stocking from hatchery-produced fingerlings.

4.2 SALMONIDS

Although there are many salmonids native to Europe used in stocking and culture, their separate populations have distinct characteristics on account of varying degrees of isolation, behaviour and physiological capabilities according to their habitat and geography. A large number of these originally distinct populations are now genetically mixed on account of stocking and thereby alters the original variability in receiving populations. Some characteristics, such as those of the landlocked Atlantic salmon *Salmo salar* Gullspångsälven strain have been exploited with successes in culture, unless their identity is maintained some of these attributes may become more difficult to obtain in the future. Exotic salmonids have also been used for stocking and culture, *Salvelinus fontinalis* and *S. namaycush* have been stocked in mountain lakes.

The brown trout is a 'plastic' species capable of colonising creeks, isolated lakes, rivers and estuarine regions and can have a migratory phase to the sea (sea-trout). It is clearly a complex species (Baglinière & Maisse 1999). When restocking, cognisance of the stock origin should be taken before release for fisheries enhancement and recreational and commercial fishing. Initially, for almost a century, their fertilised eggs were widely distributed because these are easily transported. In Norway it is assumed that the stocking in the isolated higher forest and mountain regions date back to the Stone Age but rare waterspout events may also have transferred them. In Norway about 3 million fry and fingerlings are transplanted annually and after several years have recapture rates normally between 30% and 50%. In France, commercial and sport fisheries of brown trout over several decades have been reliant on stocking.

Stocking has been particularly successful in supporting populations in rivers where spawning grounds have been damaged arising from waste releases and sedimentation (Sipponen & Hakkari 1984). Rivers devoid of populations from pollutant discharges in Germany were restocked following improvements in water quality and the non-migratory trout form showed the most promising results (Conrad 1991). In the Netherlands brown trout were released in the Zuidersee and the River Ijssel. This took place during the 1920s to provide alternative commercial and recreational fishing after a reduction in the number of freshwater lakes following land reclamation from the sea (Anonymous 1923).

Movements over comparatively small distances can subsequently create impacts for populations. Transfers of the Atlantic salmon from Swedish hatcheries are believed to be responsible for the establishment of the monogean *Gyrodactylus salaris.* This was first noted in Norway in 1974. The parr and smolts of Norwegian salmon populations are susceptible to this skin parasite, unlike those in the nearby watersheds that drain into the Baltic (Johnsen & Jensen 1991).

The North American Pacific coast rainbow trout *Oncorhynchus mykiss* was introduced to Europe to France in 1879. The species is widely distributed on all continents except for Antarctica. They have been introduced to mountain regions in the tropics, islands with irrigation canals and reservoirs (i.e. Madeira) and to cool temperate regions (MacCrimmon 1971) with successful opportunities for fishing and aquaculture. In Europe it accounts for significant landings from culture. This production is dependant on hatchery-produced stock because there are few wild self-sustaining populations known in Europe.

In 1971 fertilised eggs of the Pacific Coho salmon *Oncorhynchus kisutch* were imported to France for culture. Seven tons of 0.3 to 1.0 kg fish were produced in 1974 and by 1980 this had reached 90 tons (Harache 1980). Production was discontinued following concerns about its impact on the growth and behaviour of other salmonids (Heland & Beall 1997). Other Pacific salmon *O. gorbuscha* and *O. keta* have been introduced to the White Sea region where they ascend far up rivers in Russia and vagrants from here occur elsewhere in northern Europe from time to time (Picl 1995).

4.3 EELS

The eel *Anguilla anguilla* occurs along the European Atlantic coast, the Mediterranean and Black seas and the Sea of Azov. It is a highly prized food in Northern Europe and commands high prices. Consequently small landings of eels can be profitable. Eels are captured by longlines or by using nets to capture eels moving downstream to spawn in the ocean. When the subsequent generation of elvers ascend estuaries in great numbers they are collected and redistributed further upstream to lakes and rivers. This process ensures a more even distribution of eels within catchments and promotes survival. The collection and dissemination of elvers has been practiced for at least a century, elvers travel well over long distances with low mortalities. In the early 1900s millions of elvers were exported to different regions of Europe. Such is the demand for eel that sources from as far away as North America and New Zealand were considered for cultivation. Following experimental culture of the imported oriental eel *A. japonica* to Germany in the mid 1970s, it's parasitic nematode *Anguillicola crassus* became established with a high abundance and prevalence in native eel populations causing extensive damage to the swimbladder, whereas it does not cause undue harm to its normal host. The nematode has now reached most North European countries due to transport of live infected eels. Glass eels are not infected because the parasite does not propagate in saline water, consequently farms that avoid transplanting post-elver stages into culture from the wild should avoid contamination. A further parasite appeared in a closed water circulation eel-farm containing native adult eels following a transfer of elvers from the Severn Estuary in Britain. The cultured adult eels developed red spots on their internal organs and after three weeks many died (Stewart 1983). The cause was due to the bacterium *Pseudomonas anguilliseptica* previously described from the Japanese eel *A. japonica*. Eels have been stocked in thermal water discharges on the Baltic coast of Sweden. Here eels tended to remain in the warm plume. The recaptured eels showed greater growth than nearby cooler regions. Andersson et al. (1991) recommended allocating transplants of elvers to such regions.

4.4 OTHER FISHES

Recirculating culture systems or culture in areas with thermal discharges provide opportunities for cultivating species with higher water temperature requirements, for example the channel catfish *Ictalurus punctatus* (Hilge 1980) and various tilapias. Grass carp *Ctenophyrongodon idella* originally from the Amur catchment and eastern Asia was introduced in 1996 to the Netherlands from Hungary and Taiwan. This species has been in cultivation in China for at least 2000 years. In some world regions it has been stocked to eliminate dense stands of submerged aquatic plants especially in canals and reservoirs. The fingerlings are raised in warm water hatcheries and then released to ponds and lakes (Huisman 1981). In southern France releases of the mosquito fish *Gambusia affinis* in Mediterranean lagoons have been used to control the abundance of emerging insects, particularly mosquitos (Kieth & Allardi 1997).

5 The apparent spread of toxic algal events and consequences for culture

The occurrence of toxic algal events world-wide appear to have increased in frequency and range, and although some of the causative species are considered to be native, toxic strains of some of these species may have been introduced. For example the *Alexandrium tamarense* blooms in Cork Harbour, Ireland, are of a toxic strain not normally attributed to European populations. This toxic form may have been introduced to Cork Harbour with the large consignments of half-grown American oysters imported over a Century earlier from Long Island Sound. Cysts of this species can survive passage through the molluscan gut (Laabir & Gentian 1999) and so could have been carried in a viable state with these oyster transmissions. A large reservoir of cysts occurs in the North Channel of Cork Harbour in a region where the American oysters were once laid (J. Silke, pers. comm.).

Harvesting of molluscs may be prohibited from time to time, most usually in the summer and autumn, following toxic events. These toxins occur in certain phytoplankton species (most usually dinoflagellates) and are filtered and concentrated within molluscan tissues, some organs, such as the digestive gland, normally store most of the toxin (Shumway 1990). Some toxins may also accumulate within the tissues of molluscan feeding crustaceans. Should contaminated shellfish be consumed by humans, symptoms such as diarrhetic shellfish poisoning, paralytic shellfish poisoning, neurological shellfish poisoning and amnesic shellfish poisoning may occur, caused by different toxins according to the presence of specific phytoplankton species. New toxins continue to be identified and recently azaspiracid described from Ireland (Ofuji et al. 1999) may be more widespread than presently known. The toxins have different breakdown rates in the shellfish tissues resulting in different durations when the products are prohibited from sale. Monitoring programmes for evaluating the occurrence of these events in Europe are now standard practice and this process protects consumers from illness. When toxin levels are sufficiently high to suspend harvesting, consignments from abroad may be required to fulfil production orders. Such an event in 1998 led to the mussel *Mytilus galloprovincialis* being imported from the Venice Lagoon, Italy to Ireland, for processing. These mussels spawned *in transit* and so were not suitable for processing; the refused consignment was discarded on a shore close to an oyster production site. Although no organisms associated with this consignment are known to have survived, there is a potential risk that such events may spread toxic algal cysts and other unwanted species.

6 The future of aquaculture and stocking in Europe

It is likely further species will be developed and marketed in Europe. The current experimental culture of the savory clam *Nuttalia obscurata* in British Columbia (D. Kieser, pers. comm.) may soon lead to trials in Europe. There is a high dependence on a small number of non-native species in molluscan culture in Europe. Should a serious decline of molluscan production take place, as has occurred in the past, this could lead to decisions to import large consignments of stock with few controls. It is therefore important to consider the development of new culture species in advance of such events. Laboratory studies to search for new candidate species have taken place in France and Britain, and some of these are now in cultivation. 'New' species developments should initially be processed through quarantine and the subsequent grow-out facilities need to be in relatively 'uncontaminated' regions from diseases and parasites. Perhaps appropriate for European development would be the islands of the western European seaboard

(Faroes, Ireland, The Azores) according to the climate where further development is expected.

Finfish aquaculture will also become more important in the decades ahead, yet there is uncertainty about changes of climate as well as other impacts that may affect production. The way in which management approaches current and future developments is important for the sustainability of these resources. It is likely that water resources will need to be managed more carefully and this requirement will extend to the management of water in conventional farming systems (ponds, raceways, extensive coastal lagoon culture, net cage culture in protected coastal areas). Opportunities for spatial expansion are likely to become more limited particularly near centres of population. New technologies are likely to evolve to aid in the management of water circulation, organic recycling and cost-efficient energy methods. There will be a need to practice up to date husbandry procedures to effectively control parasite and disease burdens. Disease-free stocks using modern genetic methods are also likely to evolve so that higher density cultivation is possible. The use of waste heat discharges will almost certainly involve stocking of exotic species.

Stocking and aquaculture in public waters may interact with, or be imposed upon, by other industries in the coastal zone (Rosenthal et al. 2001). For example, there are risks for human health already demonstrated with virulent strains of cholera, released in ship's ballast water, which could be filtered by molluscs, which may then be consumed without cooking. Where several activities overlap there are opportunities for pests, parasites and diseases to be spread by a greater number of vectors to other localities. This could include stock transfers and other aquaculture practices.

Future species introductions should use the advice and procedures indicated in the International Council for the Exploration of the Sea's code of practice on introductions and transfers of marine organisms (ICES 1995 and Box 1). The precautionary measures outlined, make it unlikely that unwanted species would enter a 'new' environment. This code is reviewed from time to time taking account of recent scientific findings and management measures. Should the code not be followed and unwanted impacts arise, those directly involved parties could be considered to have acted inappropriately.

Box 1. Summary of the main principles of the ICES Code of Practice (ICES 1995).

(i) Conduct comprehensive disease and ecological studies in the native habitat in advance of the introduction.
(ii) Transfer the introduced species, as far as can be ascertained as a pest, parasite and disease free population, to a secure quarantine system within the recipient area.
(iii) Maintain and regularly sample the contained population, and the water quality therein.
(iv) Develop a broodstock in quarantine.
(v) Grow isolated F1 individuals in quarantine.
(vi) Introduce small numbers of F1 individuals in a pilot project after clearance following a rigorous histological investigation.
(vii) Determine any potential impacts with native biota during pilot phase.
(viii) Have a contingency plan to withdraw the species should this be necessary.

7 Conclusion

Movements of species intended for stocking and aquaculture are expected to continue and developments in technology will enable new opportunities for further species to become cultured. Development of parasite and disease-free strains may become more important as pathogenic organisms and pests, normally associated with species in cultivation, are spread. Health and safety audits of farm sites are likely to include farm practices and the health of cultivated stock. It is expected that recirculating systems, particularly where there are special advantages provided by heated water, will lead to further exotic species being utilised. Future introductions should take account of the problems that have evolved following inappropriate introductions in the past and use a quarantining process as proposed by ICES.

Table 1. Examples of some exotic species cultivated or used for re-stocking in Europe. Those species used in trials not currently used in industry appear in parentheses. (Q) species in quarantine.

Species	Source	to	Q	Year	Status	Author
Algae						
Macrocystis pyrifera	Chile	(N. France)		early 1970s	experimental	Lüning 1985
Undaria pinnatifida	S. France	(N. France)		1983	in culture	Fletcher & Farrell 1999
Laminaria longicruris hybids	Canada	(Heligoland)	no	1970s	experimental	Lüning 1985
Laminaria ochotensis hybrids	Japan	(Heligoland)	no	1970s	experimental	Lüning 1985
Laminaria abyssalis hybrids	S. Atlantic	(Heligoland)	tank	1970s	experimental	Lüning 1985
Laminaria pallida hybrids	S. Atlantic	(Heligoland)	tank	1970s	experimental	Lüning 1985
Laminaria schinzii hybrids	S. Atlantic	(Heligoland)	tank	1970s	experimental	Lüning 1985
Laminaria setchelli hybrids	N. Pacific	(Heligoland)	tank	1970s	experimental	Lüning 1985
Laminaria bongardiana hybrids	N. Pacific	(Heligoland)	tank	1970s	experimental	Lüning 1985
Asparagopsis armata	Ireland	Ireland	no	1990s	in culture	S.Kraan pers. comm.
Flowering plants						
Spartina anglica	NE Atlantic	Europe	no	1970	plantings	Eno et al. 1997
Gastropoda						
Haliotus rufescens	USA	(UK)	yes	1982	experimental	Anonymous 1990
Haliotus rufescens		Iceland	yes	1988	in culture	Jónasson et al. 1999
Haliotus discus hannai	Japan	Ireland	yes	1985	in culture	Minchin 1996
Bivalvia						
Patinopecten yessoensis	Japan	(France)	yes	1987	stock remains	J.C. Dao pers. comm.
Patinopecten yessoensis	Japan	(Denmark)	no	1985	experimental	Anonymous
Patinopecten yessoensis	Japan	(Ireland)	yes	1988	experimental	Minchin 1996
Chloromytilus chorus	Chile	(Wales)	yes	1965	experimental	Utting & Spencer 1992
Crassostrea angulata	Portugal	(Germany)	no	1913	once cultured	
Crassostrea angulata	Portugal	(Ireland)	no	1910-1920s	once cultured	C.B. Duggan pers. comm.
Crassostrea angulata	Portugal	(France)	no	1865	once cultured	Heral 1990
Crassostrea gigas	Canada	Wales	yes	1965/1972	in culture	Utting & Spencer 1992
Crassostrea gigas	Japan	France	no	1970s	in culture	Gruet et al. 1976
Crassostrea virginica	USA	(UK, Ireland)	yes	1870s-1939	once cultured	Minchin et al. 1995
Crassostrea rhizophorae	Brazil	(Wales)	yes	1980	experimental	Utting & Spencer 1992
Ostrea densallamellosa	Korea	(France)	yes	1982	experimetal	Minchin 1996
Ostrea puelchana	Argentina	(France)	yes	1980s	experimental	Minchin 1996

Species	Source	to	Q	Year	Status	Author
Ostrea angasai	New Zealand	(France)	yes	<1985	experimental	Minchin 1996
Tiostrea chilensis	Chile	(Wales)	yes	1962	experimental	Utting & Spencer 1992
Tiostrea lutaria	New Zealand	Wales	yes	1963, 1966	experimental	Utting & Spencer 1992
Mercenaria mercenaria	W. Atlantic	Netherlands		1965	experimental	Minchin 1996
Tapes philippinarum	W. Canada	France		1973-75	culture	Lucas 1977
Mya arenaria	N. America	Baltic	no	>1200	unknown	Strasser 1999
Arthropoda						
Orconectes limosus	N. America	Europe	no	1890	stocking	Vigneaux 1997
Pacificastacus leniuculus	N. America	UK, France	no	1970s	stocking	Vigneaux 1997
Paneus japonicus	Japan	Spain		1984	culture	Anonymous 1997
Procambarus clarkii	N. America	Europe		1995	experimental	Rosecchi et al. 1997
Hemimysis anomala	Ponto-Caspian	Lithuania	no	1992	stocking	Jansson 1994
Limnomysis benedeni	E. Europe	Lithuania	no	1960s	stocking	Jansson 1994
Paramysis lacustris	E. Europe	Lithuania	no	1960s	stocking	Jansson 1994
Paralithodes camtschatica	W. Pacific	N. Russia	no	1985	stocking	Kuzmin et al. 1996
Pisces						
Abramis brama	Central Europe	Poland	no	1970s	stocking	Zawisza & Cieoielewski 1973
Acipenser baeri	Central Asia	(S. Baltic)		1962-69	stocking	Baltz, 1991
Acipenser gueldenstaedti	Ponto-caspian	(S. Baltic)		1962-69	stocking	Baltz, 1991
Anguilla australis	Australia	(Italy)		1975	experimental	Cowx 1997
Anguilla japonica	Japan	(Germany)		1970s	experimental	Cowx 1997
Aristhichthys nobilis	Asia	Poland/ Germany	no	1960-70s	culture/stocking	Opuszinski 1978
Carassius auratus	Asia	Europe	no	1700s	stocking	Keith & Allardi 1997
Clarias gariepinus	Africa	Europe		1980s	warm culture	Filipak et al. 1993
Coregonus peled	Siberia	(S. Baltic)		1965	stocking	Nyman 1993
Ctenophyrangodon idella	E Asia	Europe		1949	experimental	Cowx 1997
Cyrpinus carpio	E Asia	Europe	no	<1700	culture/stocking	Leppakoski, 1994
Dicentrarchus labrax	France	Germany		1990s	culture	Rosenthal pers. ob.
Gambusia affinis	N America	S Europe	no	1920s	stocking	Keith & Allardi 1997
Hypophthalmichthys molitrix	Asia	Central Europe	no	1960s	culture/stocking	Müller 1984
Ictalurus melas	N. America	Europe		1871	stocking	Keith & Allardi 1997
Ictalurus punctatus	N. America	Germany, Italy	yes/no	1970-80s	experimental	Hilge 1980
Micropterus salmoides	N. America	Europe	no	1950s	stocking	Maitland 1969
Oncorhynchus clarkii	N. America	(Denmark)		1960s	experimental	Cowx 1997
Oncorhynchus gorbusha	N. America	(Europe)		mid 1980s	culture/stocking	Picl 1995
Oncohrynchus keta	NW America	E. Baltic	no	1982-84	stocking	Heikinheimo-Schmid et al 1984
Oncorhynchus kisutch	NW America	(France)		1970-80	experimental	Keith & Allardi 1997
Oncorhynchus tshawytscha	NW America	(France)	no	1877	experimental	Keith & Allardi 1997
Oncorhynchus mykiss	N. America	Europe	no	ca 1890s	culture/stocking	Leppakoski 1994
Salvelinus fontinalis	N. America	France	no	1904	stocking	Keith & Allardi 1997
Salvelinus namaycush	N America	France	no	1886	stocking	Keith & Allardi 1997
Salvelinus namaycush	N America	Finland	no	1982-84	stocking	Heikinheimo-Schmid et al 1984
Stizostedion lucioperca	Central Europe	UK	no	1878	stocking	Maitland 1969

LIFE IN BALLAST TANKS

STEPHAN GOLLASCH[1*], ELSPETH MACDONALD[2], SARA BELSON[3], HELGE BOTNEN[4], JENS T. CHRISTENSEN[5], JOHN P. HAMER[6], GUY HOUVENAGHEL[7], ANDERS JELMERT[8], IAN LUCAS[6], DANIEL MASSON[9], TRACY MCCOLLIN[10], SERGEJ OLENIN[11], AGNETA PERSSON[12], INGER WALLENTINUS[12] LAMBERTUS P. M. J. WETSTEYN[13] & THOMAS WITTLING[14]
[1]*GoConsult, Hamburg, Germany*
[2]*Food Standards Agency, Aberdeen, Scotland*
[3]*Maritime Research Centre, Southampton Institute, Southampton, UK*
[4]*UNIFOB, Section of Applied Environmental Research, High Technology Centre, Norway*
[5]*Dept. of Marine Ecology, Institute of Biological Sciences, University of Aarhus, Aarhus C, Denmark*
[6]*University of Wales, Bangor, School of Ocean Sciences, Menai Bridge, Anglesey, UK*
[7]*Université Libre de Bruxelles, Brussels, Belgium*
[8]*Institute of Marine Research, Austevoll Aquaculture Research Station, Storebø, Norway*
[9]*IFREMER, Station de La Trembalde, France*
[10]*FRS Marine Laboratory, Aberdeen, United Kingdom*
[11]*Coastal Research and Planning Institute, Klaipeda University, Lithuania*
[12]*Department of Marine Botany, Göteborg University, Göteborg, Sweden*
[13]*National Institute for Coastal and Marine Management, Middelburg, The Netherlands*
[14]*Institut für Hydrobiologie und Fischereiwissenschaft, Universität Hamburg, Hamburg, Germany*
*Corresponding author *SGollasch@aol.com*

Abstract
The abundance and diversity of species in ballast water, a recognised vector for the accidental introduction of nonindigenous organisms, has been examined through many studies around the world over the last 25 years. The results of European research activities are summarised in this contribution by outlining the objectives of some of these studies, and by focusing on the diversity of taxa determined from ballast water and tank sediment samples. In total 1508 samples (1219 ballast water, 289 tank sediment) were collected on 550 ships. A total of 990 taxa were identified during the 14 European shipping studies. The diversity of species found included bacteria, fungi, protozoans, algae, invertebrates of different life stages including resting stages, and fishes with a body length up to 15 cm. Crustacean, molluscan and polychaete invertebrates and algae form the majority of species found.

1 Introduction

The introduction of nonindigenous organisms to new areas has resulted in populations of many species being established outside their native ranges, with, in some cases, po-

E. Leppäkoski et al. (eds.), Invasive Aquatic Species of Europe, 217–231.

tentially deleterious effects on native species, the receiving habitats and their economies (Hedgpeth 1993). Whilst some species are introduced deliberately to new areas for aquaculture purposes, shipping is often regarded as one of the main vectors for the accidental transfer of aquatic organisms. Organisms may be transferred by ships as hull fouling and also inside ballast tanks. Since the introduction of steel hulled vessels in the late 19th century, water is the preferred ballast material used by ships to stabilise and trim the vessel and to submerge the propeller when ships are not fully loaded. The global nature of the shipping industry makes it inevitable that many ships must load ballast water in one area and discharge it in another. It is most likely that the modern shipping industry, with faster ships transporting large quantities of ballast between geographically separate areas, has an increasing potential to transport nonindigenous species to new areas. It has been estimated that major cargo vessels annually transport nearly 10 billion tonnes of ballast water world-wide, indicating the global dimension of the problem. On average, 3,000 – 4,000 species (Carlton and Geller 1993; Gollasch 1996) have been estimated to be transported between continents by ships each day. Organisms discharged with ballast water in coastal waters and ports of call beyond their native range may have the potential to impact native species and ecosystem functions, fishing and aquaculture industries, as well as public health. However, the introduced organism must first survive the ballasting and subsequent deballasting processes in addition to the period in transit inside the ballast tanks (Carlton 1985). Conditions upon discharge must also be suitable for survival, and the organism must then find sufficient food and resources, suitable habitat, and in some cases, other individuals with which it can reproduce before a population can become established (Gollasch 1996). The inoculation stage is simply one step in the complex process of invasion biology (Carlton 1985).

The first account attributing the introduction of an aquatic organism to ballast water transport was published by Ostenfeld (1908) after a mass occurrence of the Asian phytoplankton algae *Odontella* (*Biddulphia*) *sinensis* in the North Sea in 1903. The first documented studies to sample ships' ballast water were carried out 70 years later by Medcof (1975) followed by e.g. Hallegraeff and Bolch (1991), Locke et al. 1991, Lucas et al. (1999). This contribution summarizes the methods used and the diversity of taxa transported in ballast water and tank sediments sampled in completed and ongoing European shipping studies.

2 Material and methods

The European research considered here consists of practical studies in which ballast water and, in some cases, sediments, were sampled from ships' ballast tanks (Table 1). Ballast water samples were taken using a variety of nets, hoses and pumps operated via tank openings (manholes), pumps operated via sounding pipes or air vents connecting the ballast tanks to the ships′ surface and by extracting water at the ships′ ballast pump (in line sampling). The net haul depth varied according to tank design from 1.5 m to > 15 m. Sterile samples were collected by immersing sterile 1L bottles from opened manholes and by collecting sediment in sterile cut-off syringes (Table 2). Ballast water sampling methods vary greatly world-wide, and the particular method used will often depend upon the main taxa that particular researchers are interested in, and the practical and logistical requirements of the ships to be sampled. This variety in methods has been addressed by a European research initiative entitled "Concerted Action Introductions with Ships" and is reported elsewhere (Rosenthal et al. 2000).

Table 1. Number of ships sampled, number of samples and duration of shipping studies carried out in European countries.

Shipping study	Number of ships sampled	Number of samples taken: Ballast water	Number of samples taken: Tank sediment	Duration of study	Type of sampling
Belgium	5	26	6	1995-1998	in ports en-route
Denmark	1	8		ongoing, start 2000	en-route
England	20	49		1998 – 2000	in ports
England & Wales	112	114	102	1996 – 1999	in ports
France	0	0	0	ongoing, start 2000	in ports
Germany 1	189	131	71	1992 – 1996	in ports
Germany 2	9	1	12	1998 – 1999	in ports
Lithuania	11	22		1999 – 2000	in ports
Netherlands	17	21	2	ongoing, start 1999	in ports
Norway 1	45			1996 – 1997	in ports
Norway 2	6	8	4	1998 – 1999	in ports
Scotland	127	134	92	1994-1997	in ports
Sweden	3			1996	in ports
EUCA	5	705		1998 – 1999	en-route
Total	550	1219	289		

Vessels sampled ranged from small cargo vessels of < 1,000 deadweight tonnes (dwt) to very large crude carriers (VLCCs) of > 300,000 dwt. The ballast water sampled was from more than 200 different origins world-wide, but predominately from the northern hemisphere. The principal objective of all studies was to identify the variety of species transported with ships ballast water.

In most case studies, ballast water samples were preserved by the addition of formaldehyde or glutaraldehyde or other fixative (e.g., Lugol's iodine). In some studies unpreserved samples were taken for culturing purposes and flow-cytometrical analysis. The organisms were identified to the lowest taxonomic level possible using stereomicroscopes, microscopes, selective culture media, and in one case study, DNA-based tools (Polymerase Chain Reaction followed by gene sequencing). Wherever possible, the individuals were determined to species level. Due to unfavourable conditions (see below) some individuals were not determined to the species level, but are included as "cf.". In other cases where a species identification was not made, only the genus was determined and this is noted by "sp.". Specimens that could not be ascribed to a genus level were assigned to the lowest appropriate taxonomic unit (Family, Order etc.) with the addition "indet.". Reasons why some organisms could not be identified to the species level included the following. Some resting stages (phyto- and zooplankton) and invertebrate larvae can only be determined to species level by rearing the specimens to motile, juvenile or adult stages. As a majority of the samples were fixed, such cultivation was not possible. Second, cultivation of unknown specimens is difficult, as the optimal culture conditions are generally unknown, and in many cases can only be attempted by trial and error. This is time consuming and costly, with no guarantee of success. Third, it can be very difficult to culture all species in a sample considering that many are present at low densities. Fourth, in some samples only one individual was found. For some species (e.g. copepods) it is essential to investigate both sexes or adults to confirm the identification. Fifth, organisms were frequently damaged or in poor condition thus rendering difficult culturing the species. Sixth, it is very difficult to identify specimens which may have originated from a wide variety of locations as taxonomic keys might not adequately represent the range in variation of the species. Finally, for the microbial community, in some cases the percentage of bacteria that could be cultured was low compared to total numbers, and in some cases when molecular methods were used, identification was hampered by the incomplete library of genomic sequences for marine bacteria.

Bray-Curtis similarity analysis. Several shipping studies considered here sampled the ballast water for zooplankton species. Results from these studies were selected for a Bray-Curtis similarity analysis as e.g., surveys where the main focus was to collect bacteria should not be compared with a survey focussing sediment dwelling species, fish or zooplankton. Comparing studies that targeted different types of organisms would likely result a very low similarity. Species found in sediment samples were omitted unless they were found both in sediment and water samples.

A short description of each practical study follows below. While this list attempts to summarise all the European studies, it should not be assumed to be fully comprehensive (Table 1).

Belgium, Universite Libre de Bruxelles (total number of taxa identified: 28). The investigation entitled "Study of the Potential Role of Transportation of Ships Ballast Water on the Geographical Extension of Blooms of Toxic Algae" was carried out at the Université Libre de Bruxelles. The main results of the study were that risks do exist concerning the introduction of nonindigenous toxin-producing phytoplankton species into European waters with ballast water or sediment discharges. They recommended the implementation of ballast water management guidelines at an international level (Vanden Boeck 1995).

Denmark, University of Aarhus, Institute of Biological Sciences, Dept. of Marine Ecology (total number of taxa identified: 4). A joint project between Lithuania and Denmark has recently started to regularly sample the Ro-Ro Ferry "Urd" (Scandlines), running between Aarhus/Aabenraa (Denmark) and Klaipeda (Lithuania). The main objective of the study is to assess the survival of planktonic organisms en-route. It is planned to regularly sample the ship before departure in the port of Aarhus and after arrival in the port of Klaipeda (and vice versa). Additionally, it is planned to carry out port profiles of the Danish harbours Aarhus and Fredericia, and to sample ships that arrive in these ports from outside the Baltic Sea (J. Christensen, University of Aarhus, Denmark, pers. comm.).

Table 2. Ballast tank sampling methods used by each European shipping study.

Shipping study	Methods: Ballast water samples	Methods: Tank sediment samples
Belgium	In line, sampling at ships' ballast pump Pump, used to collect water samples using a hose via air vents	Trowel, scoop and bottle to collect sediment from the bottom of emptied and ventilated tanks
Denmark	Water sampler, lowered down through tank opening. On deck the sample volume was concentrated by using a 60 µm plankton net (zooplankton). Phytoplankton samples were not concentrated.	
England	Bottle, repeatedly lowered down through tank opening to below the water surface In line, sampling at ships' ballast pump	
England & Wales	Hose, 2.5 cm internal diameter, weight on the end, lowered through manholes to collect integrated water samples Hand-hauled plankton net, 53 µm mesh 30 cm diameter, 75 cm length lowered through manholes Pump, used to collect water samples using a hose via manholes or sounding pipes. In line, sampling at ships' ballast pump	Pump, used to pump sediment slurry using a hose via manholes or sounding pipes Trowel, scoop and bottle to collect sediment from the bottom of emptied and ventilated tanks
France	Bottle, repeatedly lowered down through tank opening to below the water surface	
Germany 1	Hand-hauled plankton net, cone shaped opening, cone diameter 9,7, net diameter 25 cm, length 95 cm, mesh size 55µm (zooplankton) and 10µm (phytoplankton), Pump, small hand pump operated via sounding pipes, hose diameter 14 mm, flow rate 1l/min. In line sampling at the ship´s ballast pump	Small trawl, width 30 cm, mesh size 250 µm, Trowel, scoop and bottle to collect sediment from the bottom of emptied and ventilated tanks
Germany 2	Hand-hauled plankton net (net diameter 25 cm, length 95 cm, mesh size 55µm)	Trowel, scoop and bottle to collect sediment from the bottom of emptied and ventilated tanks
Lithuania	Hand-hauled plankton net, cone shaped opening cone diameter 9,7, net diameter 25 cm, length 95 cm, mesh size 55µm (zooplankton) and 10µm (phytoplankton)	
Netherlands	Hand-hauled plankton net, net diameter 25 cm, length 50 cm, mesh size 20 µm (phytoplankton) and 55 µm (zooplankton), In line sampling at the ship´s ballast pump, Pump, hand pump (Whale bilge pump) operated via sounding pipes, inner hose diameter 13 mm (outer 19 mm), flow rate ca. 30 l/min.	Trowel, scoop, to collect sediment from the bottom of emptied and ventilated tanks
Norway 1	Hand-hauled plankton net, opening diameter 20 cm, length 100 cm, mesh size 20-30 µm (zooplankton) Ruttner water sampler, 1.5 l (phytoplankton)	
Norway 2	Sterile bottle lowered down through manholes	Sterile "decapitated" syringe collecting sediment plug in sediment at bottom of emptied and ventilated tanks
Scotland	Vertical hose (2.5 cm internal diameter through manhole or 1.25 cm hose through sounding pipe) for phytoplankton. Hand-hauled plankton net (0.5 m diameter ring net, 68 µm mesh size) or by filtering a known volume of water through a 68 µm mesh. In line sampling at the ship´s ballast pump	Pumping sediment slurry from bottom of tank by hand-operated mono-pump or battery operated bilge pump.
Sweden	Pump, small hand pump operated via manholes Bucket lowered down below the water surface through manhole	Trowel, scoop and bottle to collect sediment from the bottom of emptied and ventilated tanks
EUCA	Hand-hauled plankton net, cone shaped opening, cone diameter 9.7, net diameter 25 cm, length 95 cm, mesh size 55µm	

England, Maritime Research Centre, Southampton Institute, United Kingdom (total number of taxa identified: 68). A research studentship assessed the transport of phytoplankton in the ballast water of vessels using the Port of Southampton. The project aimed to establish whether ballast water was a potential source of exotic phytoplankton species to the Southampton Water estuary, and also whether the number of cells and taxa could be correlated to factors including the area of origin, age and selected physical and chemical properties of the water. All of the samples collected originated within the northern hemisphere, and two-thirds were from European waters (S. Belson, Southampton Institute, United Kingdom, pers. comm.).

England & Wales, School of Ocean Science, University of Wales, Bangor (total number of taxa identified: 252). Between 1996 and 1999, samples were taken from ships arriving at English and Welsh ports. The aim of the study was to investigate the range of organisms present in ships` ballast tanks and assess variation in ballast water organisms in relation to geographic origin and season (McCollin et al. 1999). A further objective was to assess the potential risk of the introduction of organisms in ballast water and sediment to English and Welsh waters. Ships were sampled at 20 ports throughout England and Wales. Some culturing was carried out to investigate the viability of organisms (Hamer et al. 1999). The majority of ballast water sampled originated from ports in the northern hemisphere (McCollin et al. 1999).

France, IFREMER, Station de La Tremblade (no taxonomic results available yet). A study started recently in La Tremblade, involving sampling of ballast tanks to identify the microbial component of species found (D. Masson, IFREMER, France, pers. comm.).

Germany 1, Institute for Marine Research, Kiel, and University of Hamburg (total number of taxa identified: 502). The first European shipping study was undertaken in Germany (Gollasch 1996), the purpose of which was to conduct a thorough taxonomic assessment of planktonic and benthic organisms found in ballast water tanks with additional samples taken from ship hulls (Lenz et al. 2000). The vessels investigated were selected according to type of vessel (e.g., container ships and bulk carrier) and sea areas covered by their voyages. The majority of samples originated from tropical and warm-temperate regions (Gollasch 1996; Lenz et al. 2000).

Germany 2, Institute for Marine Research, Kiel, and University of Hamburg (total number of taxa identified: 19). A follow up study was carried out from 1998-1999. The key objective was to sample ballast tank sediments (T. Wittling, Universität Hamburg, Germany, pers. comm.).

Lithuania, University of Klaipeda (total number of taxa identified: 90). A study started recently in Klaipeda, involving sampling ballast water and examining hull fouling on ships in dry docks (S. Olenin, University of Klaipeda, Lithuania, pers. comm.).

The Netherlands, National Institute for Coastal and Marine Management, Middelburg and AquaSense/Tripos, Amsterdam (total number of taxa identified: 88). During a pilot project in the years 1995 and 1996, in cooperation with the Smithsonian Environmental Research Center (USA), phytoplankton and zooplankton samples were taken in cargo holds of bulk-carriers travelling between the ports of Maryland and Rotterdam. One key objective was to assess the viability of organisms after an oceanic voyage (Tripos 1997). In 1998 questionnaires were sent to incoming ships and literature reviews were carried out to estimate the amount and origin of ballast water carried to and from The Netherlands and to assess the risks of the introduction of nonindigenous species into Dutch coastal waters (AquaSense 1998a, b). Since 1999 samples have been collected on board ships arriving in the ports of Rotterdam, Vlissingen-Oost and Amsterdam. The objectives include documentation of the diversity of plankton organisms and their survival potential in Dutch waters. Culture experiments were carried out at 10°C and 20°C and at salinities 5 PSU, 15 PSU and 32 PSU and also in port water and in GF/F filtered port water. Finally, number, size and pigment distributions of particles in live ballast water samples are analyzed by flow-cytometry; this method also provides information on autotrophic viability (chl-a fluorescence) of plankton upon arrival as function of voyage duration.

Norway 1, UNIFOB, Section of Applied Environmental Research, Bergen (total number of taxa identified: 181). In 1996, a project entitled the Sture Project was launched. The aim of the study was to investigate the potential of unintentional introductions of nonindigenous aquatic species to Sture via ships' ballast. All ships sampled arrived from ports outside Norway (Botnen et al. 2000).

Norway 2, IMR, Bergen and University of Bergen (total number of taxa identified: 3). Between July 1998 and April 1999, ballast water and sediment samples were collected from 6 ships entering the harbour of Mongstad (one ship redirected to Sture) in Norway to identify the microbial component of species in ballast tanks. The faecal bacterium *E. coli* was found in one vessel out of six, while the cholera bacterium *Vibrio cholerae* was not detected in the two ships analysed for its presence (A. Jelmert, Institute of Marine Research, Norway, pers. comm.).

Scotland, FRS Marine Laboratory, Aberdeen (total number of taxa identified: 327). Between May 1994 and December 1996, ballast water was sampled from vessels arriving in Scottish ports. The aim of the study was to investigate planktonic organisms in ballast water and sediment, with special emphasis on harmful and potentially toxic phytoplankton (Macdonald 1998; Macdonald and Davidson 1998).

Sweden, Department of Marine Botany, Göteborg University (total number of taxa identified: 41). A pilot project entitled "Risks associated with introduction of nonindigenous organisms to Swedish waters by water/sediment in the ballast tanks of ships" was carried out in 1996. The project also included a regional survey of dinoflagellates (Persson 2000) along the province of Bohuslän on the Swedish west coast as a baseline study. The study focused on phytoplankton and on culturing sampled raw material. Concentrated (by filtration, 10µm) ballast water samples and untreated sediment samples were incubated in a culturing chamber (16°C, 80 $\mu Em^{-2} \cdot s^{-1}$, 12:12 h light:dark) in enriched (with f/10 nutrients) deep sea water from the Swedish west coast. This treatment was performed to compare initial findings in the ship samples with cultures established from resting stages undetected in the untreated samples (Persson 2000; Persson et al. 2000).

European Union Concerted Action (total number of taxa identified: 67). This recently completed European wide Concerted Action (EUCA) included five Ocean Going Workshops (OGW) that were undertaken to assess the survival of zooplankton organisms in ballast water en-route. The OGW were undertaken in European waters and during inter-oceanic voyages. In total, 705 samples were collected during more than 100 days at sea (Gollasch et al. 2000b, Olenin et al. 2000, Rosenthal et al. 2000).

3 Results

During the 14 European ship sampling studies considered here, a total of 1508 samples (1219 ballast water, 289 tank sediment) were collected on 550 ships (Table 1). The total number of taxa identified during all completed shipping studies varied between 3 and 502 per study (Fig. 1) and the number of taxa identified overall was 990 (Table A1, available at http://www.ku.lt/nemo/EuroAquaInvaders.htm). The diversity of species found included e.g. bacteria, fungi, protozoans (summarized as "other taxa"), algae, invertebrates of different life stages including resting stages, and fishes with a body length up to 15 cm (Table A1). More species were identified in those studies which sampled more vessels (Fig. 1).

The most frequently collected taxa identified by eight shipping studies were diatoms, harpacticoid copepods, rotifers (but all were not identified to species level) and the diatom *Skeletonema costatum* (Table A1). The second most frequent identified taxa in seven studies were *Ditylum brightwellii, Chaetoceros* sp., *Navicula* sp., *Thalassionema nitzschioides, Thalassiosira* sp. (Bacillariophyceae (Diatoms)), and *Temora longicornis* (Copepoda, Calanoida), calanoid copepods, larvae of Gastropoda, Bivalvia and Polychaeta (Table A1).

3.1 FLORA

Most of the 497 taxa found in the sampled ballast tanks were determined to species level (66%) (Fig. 2, Table A1)). In total, taxa of 13 higher taxonomic groups (i.e., phyla and orders) were identified. Diatoms (250 taxa) and dinoflagellates (126 taxa) were the predominant groups found. Taxa of all other 11 higher taxonomic groups were found in frequencies less than 70 (Table 3). Rarely found, with five or fewer records were the taxonomic groups Cryptophyceae, Prymnesiophyceae, Xanthophyceae, Ulotrichales, Rhodophyta and filamentous green alga. Resting stages of Xanthophycaea (1 taxa) and dinoflagellates (64 taxa) were found rarely (Table A1).

3.2 FAUNA

Fifty-five percent of the 425 taxa were determined to species level (Fig. 2, Table A1). In total, taxa of 18 higher taxonomic groups were determined. Crustacea (240 taxa) and Mollusca (73 taxa) were dominant faunal members of the ballast tank community. Taxa of all other 16 higher taxonomic groups were found in frequencies less than 20. Rarely found, with five or fewer records were the taxonomic groups Ctenophora, Turbellaria, Nemertea, Oligochaeta, Pantopoda, Insecta, Arachnida, Chaetognatha and Echinodermata (Table 4).

In the most commonly encountered group (Crustacea), Calanoida (54 taxa), Harpacticoida (33 taxa) and Cladocera (28 taxa) were dominant (Table 4). In the second frequently identified group (Mollusca), the Bivalvia clearly prevailed with 42 taxa, followed by Gastropoda with 30 taxa. Resting stages of seven cladoceran taxa were found (Table A1).

The similarity between studies is low (10% -33%), which indicates that there are only few commonly found species within the surveys. The similarity between e.g., the Scottish and English survey is about 25%. The samples from the Lithuanian and German study 1 have the highest similarity with 33% indicating the high number of species commonly collected during both studies. The English, Scottish, Dutch and German 2 surveys form one group which has low similarity (20% -25%), but show even a lower similarity to the other studies (10%) (Fig. 3).

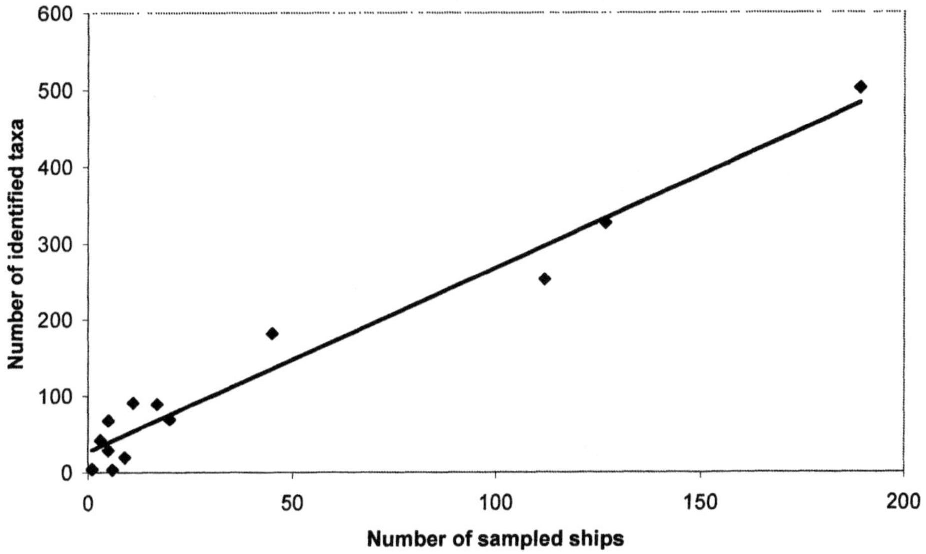

Figure 1. Number of identified taxa according to number of sampled ships (all shipping studies).

3.3 OTHER TAXA (BACTERIA, FUNGI AND PROTOZOANS)

Most of the 68 taxa found were not determined to species level (22.1%) (Fig. 2, Table A1). In total taxa of ten higher taxonomic groups were determined. Foraminifera (19 taxa) and Ciliata (16 taxa) were dominant in this category. Taxa of all other eight higher taxonomic groups were found in frequencies less than ten. Rarely found or less frequently studied, with five or fewer records, were the taxonomic groups Bacteria, Radioloaria, Heliozoa and Fungi (Table 5). Resting stages of one protozoan taxon (Ciliata) were found (Table A1).

4 Discussion

The purpose of this paper is to document the diversity of taxa transported in ballast tanks sampled in European shipping studies. The enormous amount of data generated by these studies is too vast to be considered in great detail here. It was not attempted to identify native and nonindigenous species, but rather to look at the diversity of species transported in ships ballast tanks. Within Europe, it is often not clear for many species whether they are indeed native or not (i.e., they are cryptogenic (Carlton 1996)), particularly for the smaller organisms (e.g., bacteria, algae and protozoa) as they are generally not so well studied or described.

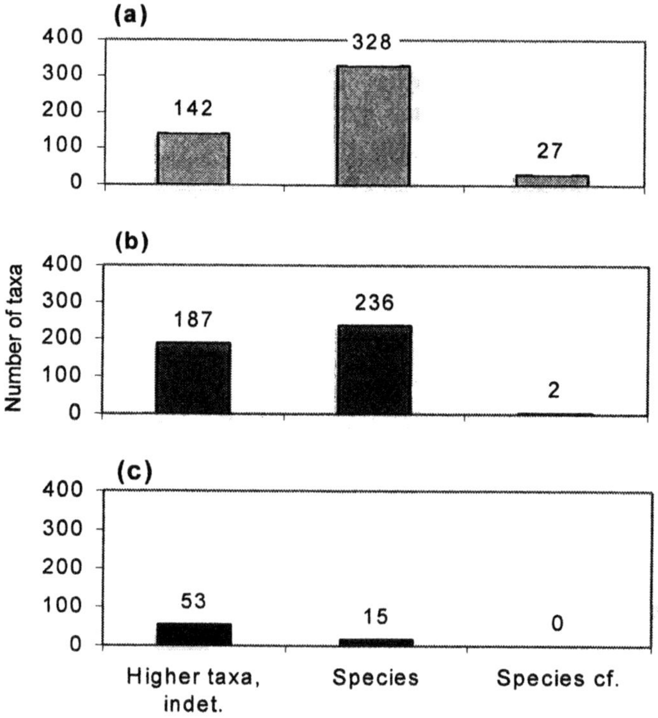

Figure 2. Species, including species with unconfirmed identification (species cf.) of the flora (a) (total number of taxa 497), fauna (b) (total number of taxa 425), and bacteria, protozoa and fungi, summarised as "other taxa" (c) identified during all European shipping studies. Specimens that could not be ascribed to a genus level, were assigned to the lowest appropriate taxonomic unit (Family, Order etc.) summarised here as "higher taxa" with the addition "indet.".

In total, the 990 species found during the shipping studies considered here belong to a highly diverse range of taxonomic groups including bacteria, protozoa, phytoplankton, zooplankton and zoobenthos. The largest specimens found inside a ballast tank were *Petromyzon marina* (Cyclostomata, Pisces) of 15 cm length. However, the total number of taxa identified may be assumed to be an underestimate. One reason is that some studies targeted specific taxonomic groups. Furthermore, individuals listed as "sp. " might not be the same in different studies, due to the range of taxonomic expertise available in different institutes. It is likely that more species are hidden in the category "indet." although some taxa may also be the same as others have identified to genus or species level. Even more species are probably not identified because many taxonomists have their specialities within certain taxonomic groups. It is also most likely that many more organisms occurred in the sediments which were often not possible to sample and it also is more difficult to find microscopic organisms in the sediments than in the water.

Table 3. Flora: Number of taxa identified during European shipping studies according to higher taxonomic unit.

Taxa	Number	%
Cyanophyceae	24	4.8
Chrysophyceae	6	1.2
Dictyochophycaea	6	1.2
Prymnesiophyceae	5	1.0
Bacillariophyceae (Diatoms)	250	50.3
Xanthophyceae	2	0.4
Dinophyceae (Dinoflagellates)	126	25.4
Cryptophyceae	5	1.0
Prasinophyceae	6	1.2
Chlorophyceae	63	12.7
Ulotrichales	1	0.2
Rhodophyta	1	0.2
Filamentous Clorophyta	2	0.4
Total	497	100.0

Sampling ships is different from sampling the environment, especially considering that sampling access is not always straightforward, and is highly variable between ships. One needs to deal with several obstacles, such as cargo overlaying the manhole preventing direct access to the ballast water, and support frames installed to stabilise the ballast tanks that make it difficult to access to the very bottom of the ballast tank. Sampling equipment can also become stuck inside the tanks (Sutton et al. 1998; Rosenthal et al. 2000).

Due to the problems encountered when sampling ballast water (Sutton et al. 1998; Rosenthal et al. 2000), results can be regarded as a minimum qualitative and quantitative assessment of the biological content of ballast water. This is apparent when considering the increase in identified species when ballast water already sampled with nets is additionally sampled with traps or the samples cultured in the laboratory. Further, some shipping studies focus ballast water and do not sample the ballast tank sediments or vice versa. By selecting the type of samples (i.e., ballast water or tank sediments) some species present in the ballast tanks were not collected. The application of a wider range of sampling methods would possibly increase the range of species found in ballast water, but most shipping studies are limited by practical constraints and time available for sampling. The choice of the methods will further depend upon a number of operational procedures, on ship design, location of and access to the ballast tanks and also the amount of water in these tanks. The objectives of sampling (e.g. qualitative or quantitative samples, target organisms or all taxa) are other criteria for method selection. The preferred access to ballast tanks for sampling is via opened manholes. This would normally result in the use of short nets which are more easily manipulated and can be operated in ballast tanks which often restrict the depth of sampling tows.

Cone-shaped nets may be regarded as a suitable way of easily and efficiently sampling a ballast tank. The main reason for the high sampling efficiency is the particular net configuration that increases the filtration rate by limiting the overflow of water from the net caused by mesh resistance. It is recommended that the cod end of a net should be made of a cup with filtration panels on its side and a tap at the base of the cup. If the cod end is metallic no additional weighting is required to sink the net and this will reduce the risk of entanglement in structures in the ships ballast tanks. For phytoplankton sampling nets, it is recommended that relatively small mesh-sizes (e.g., 10 µm) be used. Larger mesh sizes will exclude smaller species and may result in lower species richness estimates however, fine mesh nets may clog quickly if organisms are very abundant, so a degree of compromise may be required. In zooplankton studies, nets with mesh size of 55 µm will capture the youngest stages of Mollusca and Crustacea as well as many of the other taxonomic groups commonly found in ballast water.

Table 4. Fauna: Number of taxa identified during European shipping studies according to higher taxonomic unit. Crustacea broken down in taxonomic groups.

Taxa	Number	%
Cnidaria	11	2.6
Ctenophora	2	0.5
Tentaculata	7	1.6
Turbellaria	2	0.5
Rotatoria	16	3.8
Nemertea	1	0.2
Nematoda	16	3.8
Mollusca	73	17.2
Polychaeta	16	3.8
Oligochaeta	3	0.7
Pantopoda	1	0.2
Crustacea	240	56.5
Cladocera	28	
Ostracoda	12	
Copepoda	1	
Calanoida	54	
Cyclopoida	21	
Harpacticoida	33	
Poecilostomatoida	16	
Siphonostomatoida	1	
Cirripedia	1	
Balanomorpha	21	
Lepadomorpha	1	
Mysidacea	11	
Isopoda	6	
Amphipoda	14	
Cumacea	1	
Euphausiacea	3	
Decapoda	15	
Insecta	4	0,9
Arachnida	1	0.2
Chaetognatha	5	1.2
Echinodermata	2	0.5
Tunicata	6	1.4
Pisces	19	4.5
Total	425	100.0

Cone-shaped nets were the most effective in terms of number of specimens and number of taxa collected and therefore such nets are recommended in future ballast water surveys. However, this net will not capture all taxa and its exclusive use would lead to an underestimate of species diversity. The use of pumps, operated via sounding pipes, exhibited similar effectiveness to the cone-shaped net. However, some pumps are unable to lift water from more than 8 meters depth, consequently ballast tanks with low water levels or in deep location within the ship are unlikely to be sampled at all (Rosenthal et al. 2000).

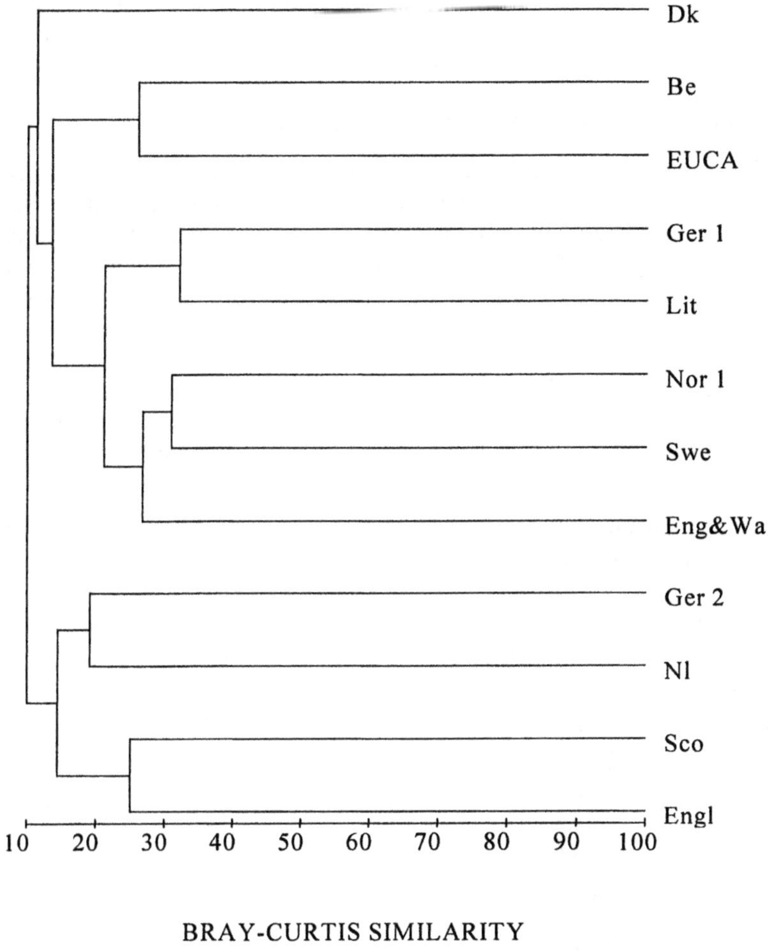

Figure 3. Bray-Curtis similarity analysis based on studies that collected fauna in ballast water. Species found in sediment samples are omitted unless they were found both in sediment and water samples (Abbreviations see Table A1).

Very few taxa were found in more than four studies and the high number of taxa found during one or two studies alone may indicate the unique character of different shipping studies (i.e. according to the inclusion of both sediment and ballast water samples). However, it should be taken into account that some studies have sampled more than 100 ships and some less than ten, and this is likely to have had an impact on the range of taxa found. Eventually this may result in more species being identified in those studies which sampled more vessels, as more samples will have been examined. Furthermore, the diversity of ballast water origins and the seasonal spread of the times when the ballast was loaded contributes to the specific character of each shipping study. Additionally, some shipping studies focused on a range of target groups from both ballast water and sediment, whilst others were more restricted in their scope. The apparent differences in taxa encountered when employing different sampling methods is the major reason for not including quantitative data on the various taxa in this review, and the comparison of the studies concentrated on qualitative results.

4.1 CONCLUSIONS AND RECOMMENDATIONS

While a wide range of taxa have been identified from preserved ballast tank samples, this gives no firm conclusions whether specimens were alive, healthy, in poor condition or dead at the time of the investigation. However, vital stains can be used to identify organisms that were alive at the time of collection, even if samples cannot be immediately examined in the laboratory. However, it is recommended that while fixed samples give valuable results, it would also add to our knowledge to examine living material and to undertake culturing experiments in order to document taxa that grow in cultures, as was done in some of the studies considered here. This approach also has limitations, because if taxa do not develop in cultures, it is not always known whether this was due to the poor condition of the organisms at the time of culturing, or due to inappropriate culture conditions.

It is recommended that several sampling methods should be employed to sample the full range of organisms in ballast tanks.

Table 5. Bacteria, protozoans and fungi, summarised as "other taxa" identified during all European shipping studies according to higher taxonomic unit.

Taxa	Number	%
Bacteria	5	7.4
Euglenophyceae	6	8.8
Zoomastigophora	8	11.8
Rhizopoda	9	13.2
Radiolaria	1	1.5
Foraminifera	19	27.9
Heliozoa	1	1.5
Ciliata	16	23.5
Fungi	1	1.5
incertae sedis taxa	2	2.9
Total	68	100.0

The diverse range of nonindigenous species introduced to many regions around the world - by a range of vectors - and the diversity of organisms transported with ships′ ballast, illustrates the need to develop treatment options in order to minimise the risks of species introductions in ballast. While retention of ballast water on board for prolonged periods may reduce the risks of ships transporting viable aquatic organisms between different areas, it is unlikely that it will completely eliminate these risks. Effective, safe, economically viable and environmentally sound treatment methods and/or management options for ballast water are necessary to minimise the risks associated with ballast water releases. The presence of organisms in ballast tanks that may potentially affect public health is also of particular concern, and is further justification for research efforts into effective treatment of ballast water to minimise the unwanted transport of these organisms on a global scale.

This study has demonstrated that virtually all forms of life are found in ballast water, and concerted efforts must be made to reduce further invasion.

Acknowledgements

The authors gratefully acknowledge the cooperation of the shipping industry in all these studies. Funding was provided by: Maritime Research Centre of Southampton Institute, UK; Ministry of Agriculture, Fisheries and Food, UK; German Environment Protection Agency; North Sea Directorate of the Ministry of Transport, Public Works and Water Management, The Netherlands; Norsk Hydro, Norway; Research Council of Norway (132453/230); Russian State Hydrometeorological University (St. Petersburg); Scottish Office Agriculture, Environment and Fisheries Department, UK Marine Safety Agency and Scottish Natural Heritage; Swedish Environmental Protection Agency and WWF Sweden; European Union, MAST-Programme and UNESCO Intergovernmental Oceanographic Commission (IOC). We are grateful to all student researchers and colleagues that assisted within these studies, who are too numerous to mention individually. We express our thanks to more than 200 taxonomists world-wide for the support in species identification.

BALLAST TANK SEDIMENTS

JOHN P. HAMER
School of Ocean Sciences, University of Wales Bangor, Anglesey, UK
j.p.hamer@bangor.ac.uk

1 Introduction

When a vessel takes on ballast water in shallow areas, resuspended sediment and the associated benthic organisms and resting stages are also taken onboard. Once inside a ballast tank, the sediments settle out of suspension and begins to accumulate. The volume of accumulated sediment present in a tank is a result of the ships ballast management practices, the type of ballast tanks involved and the time since the tanks were last cleaned in dry dock. During a ballast water sampling programme carried out in England and Wales, numerous dedicated ballast tanks of ships in dry dock (ferry, container, cruise ships) were entered and the sediment sampled (Lucas et al. 1999). The volumes of accumulated sediment were found to vary considerably within ballast tanks, between ballast tanks and between ships. Sediment accumulations varied from a few cm to more then 30 cm depth which translate to 10's and even 100's of tons of sediment in the ballast tanks of larger vessels. Although the sediment type varied between ships, most of the sediment found was fine mud with a mean particle size of < 20 μm.

2 Biota in Sediments

In order to effectively manage ballast water, with the aim of minimizing the transport of organisms in ships ballast tanks, it is important to have an understanding of the biology of the sediment that accumulates in ballast tanks. The biota of ballast tanks sediments is not well known but studies have demonstrated the presence of adult and juvenile stages of numerous invertebrates and also the occurrence of various resting stages. Copepod eggs, tintinnid cysts, diatom spores and dinoflagellate resting cysts have been reported in ballast tank sediments (Hallegraeff & Bolch 1992; Hamer et al. 2001). Because of potential toxic effects, the occurrence of diatom spores and dinoflagellate cysts in tank sediments has been studied in most detail.

Dinoflagellate resting cysts have been recorded in ballast tank sediments of ships arriving at ports in Australia (Hallegraeff & Bolch 1991, 1992), America (Kelly 1993), Canada (Harvey et al. 1999), Scotland (Macdonald & Davidson 1998), England and Wales (Hamer et al. 1998, 2000, 2001) and Germany (Lenz et al. 2000) (Table 1).

All of these studies found dinoflagellate cysts to be a common component of ballast tank sediments. It is not clear at this stage what proportion of cysts are due to taking on board dinoflagellate cysts present in the water column compared to taking onboard live cells which then encyst. Hallegraeff & Bolch (1992) suggest that dinoflagellate blooms in the water column rather than resuspended sediments are the most common origin of ballast water cysts. However, both scenarios are possible.

Much of the published work describing dinoflagellate cysts in the sediments of ships' ballast tanks has focused on samples taken from bulk cargo carriers (Hallegraeff &

E. Leppäkoski et al. (eds.), Invasive Aquatic Species of Europe, 232–234.

Bolch 1991, 1992; Kelly 1993). Such vessels fill the cargo holds with ballast water when unladen and empty and clean the tanks out prior to loading the next cargo. Any sediment in these tanks therefore originates solely from the port where the vessel last ballasted. However, the majority of European shipping traffic uses dedicated ballast tanks (wing (side), double bottom, fore peak and aft peak tanks); the sediment present in these tanks has typically accumulated over several years of vessel operations and consequently, tanks frequently contain a mixture of sediment from several different ports/regions. Such sediments may therefore be expected to contain a more diverse assemblage of dinoflagellate cysts, the origin of which is considerably more difficult to determine.

Macdonald & Davidson (1998) first reported the presence of dinoflagellate cysts in ships' ballast tank sediments in Europe. Cysts were found in 62% of sediment samples with a number of potentially toxic or non-native species being identified (Table 1). A similar study in England and Wales found dinoflagellate cysts in 69% of sediment samples and again identified cysts of potentially toxic or non-native species (Hamer et al. 2001). However, considering the difficulty in obtaining representative samples from ships, the potential for between tank variability (Hamer et al. 2000) and the appearance of un-recorded species during germination experiments, it was considered that the reported percentage occurrence and diversity of cysts was probably a considerable underestimate.

Although we have established that in European waters ships' ballast tank sediments frequently contain the resting cysts of potentially harmful dinoflagellates, the fate of the accumulated sediments is less clear. We still have very little appreciation of the volumes of sediment that may become re-suspended and discharged with the ballast water or the volumes of ballast tank sediments that are released into the marine environment whilst ships' are cleaned in dry dock.

Dinoflagellate cysts may be long-lived with many species capable of surviving for two years and some species capable of germination after nine years (Lewis et al. 1999) and are remarkably tolerant to adverse environmental conditions, making them a useful model organism for testing the effectiveness of ballast tank sediment management practices. Hallegraeff (1998) outlines the options for possible management strategies for reducing the risk of introducing potentially toxic dinoflagellates via ships' ballast water and considers that an international warning network for algal blooms in ports may be an effective method. Various chemical and physical treatment options have been trialled on toxic dinoflagellate cysts (reviewed by Hallegraeff 1998) and there is certainly scope for developing some of these methods further. Generally, chemical treatment options are likely to be impractical because of their excessive cost and due to concerns over the environmental effects of the treatment products themselves. Heat treatment utilising waste heat from the ships may be the most practicable option in combination with ballast water management such as not ballasting in areas where a phytoplankton bloom is occurring or in shallow water areas of high turbidity.

Current International Maritime Organisation guidelines recommend the exchange of ballast water in oceanic waters during certain voyages. However, Hallegraeff & Bolch (1992) note that ballast exchange does not necessarily remove the cyst-containing sediments present in ballast tanks and within Europe there is limited scope for effective ballast exchange during short intra-European voyages.

3 Conclusions

Although we have a basic understanding of the occurrence and behaviour of sediments and dinoflagellate cysts in ballast tank sediments, a number of important questions remain largely unanswered, including:

(i) How much sediment is actually discharged during de-ballasting?

(ii) How does the volume of suspended sediment vary with factors such as journey time and sea-state during the voyage?

(iii) Will the ballast water exchange be effective in reducing the accumulation of sediments and cysts?

(iv) Do other ballast management strategies significantly decrease the uptake of sediment and cysts?

Table 1. Summary of studies reporting the occurrence of dinoflagellate resting cysts in ships' ballast tank sediments.

Country of study	No. of samples (and type)	No. of dinoflagellate cyst species	% occurrence	Viable cysts ?	Harmful species ?
America [a]	6	n.r.	n.r.	Y	n.r.
Australia [b]	83	n.r.	37	Y	Y
Canada [c]	9	23	~50%	N	Y
England & Wales [d]	113	48	69	Y	Y
Scotland [e]	92	29	62	Y	Y
Germany [f]	40	16	20%	Y	Y

n.r. = not reported; [a]Kelly (1993); [b]Hallegraeff & Bolch (1991); [c]Harvey et al. (1999); [d]Hamer et al. 2001; [e]Macdonald & Davidson (1998); [f]Reise et al. (1999).

Regional Overviews

BIOLOGICAL INVASIONS IN THE WHITE SEA

VICTOR JA. BERGER* & ANDREW D. NAUMOV
Zoological Institute Russian Academy of Sciences, St. Petersburg, Russia
*Corresponding author *wsbs@online.ru*

Abstract
The Chinese mitten crab *Eriocheir sinensis* penetrated West European rivers early in the 20th century. It has recently been caught in the White Sea (Northern Dvina River estuary). In 1957 the transportation of eggs of the Pacific pink salmon, *Oncorhynchus gorbuscha*, from the Far East to the White Sea, began. As a result of a successful acclimatisation, the catches of pink salmon in the White Sea now add up to some hundred tons per year. The steelhead *Parasalmo mykiss*, native to the northern Pacific Ocean, has become one of the most important species of aquaculture in the White and Barents Seas. Natural populations of steelhead, registered in the Barents Sea, have not been found in the White Sea area yet.

1 Abiotic characteristics of the White Sea

The White Sea is a semi-enclosed water body, connected with the Arctic Ocean by the Voronka and the shallow (about 40 m) and narrow (about 60 km) Gorlo Strait. Its total surface area is about 89,600 km^2 and volume 5,400 km^3. The average depth is 60 m and maximum depth 340 m (Babkov & Golikov 1984). Mean surface water temperature in summer is 9-12 °C; in winter it is below 0 °C. The temperature in the deep-water layers is about 1.4 °C below freezing all year long. In winter (November-May), the coastal area and the inner parts of the bays are covered by ice. The central part of the sea is free of ice cover, but some floating ice may occur. The salinity of the surface water in summer is 24 PSU in the coastal area, and about 26 PSU in the open sea. In spring, a strong decrease of salinity occurs due to ice melting and fresh water run-off. In winter, it increases up to 28 PSU. The salinity of the deep-water layers is stable at 29-30 PSU. The seasonal fluctuations reach to about 100 m depth. The hydrological processes taking place in the Gorlo Strait affect the structure of the White Sea water-masses: the deep water is formed in winter and the surface water in summer (Timonov 1950).

The bottom sediments of the central part of the White Sea are muddy, whereas most parts of Onega and Mezen' Bays are sandy. The coast in Kandalaksha and Onega Bays are mostly rocky, while in Dvina and Mezen' Bays these are sandy. The coastline of Kandalaksha and Onega Bays is indented with many small inlets and islands. In contrast, the coast of the central part and Dvina and Mezen' Bays is rather even.

E. Leppäkoski et al. (eds.), Invasive Aquatic Species of Europe, 235–239.

2 Origin and biogeographic characteristics of native fauna and flora

The White Sea is geologically young. It was formed about 10,000 years ago in the end of the last glaciation. At that time the fresh-water lakes in the area of the Onega and Dvina bays were connected to the Barents Sea via the Gorlo Strait (Kvasov 1975). Arctic (*Portlandia arctica*) and boreal species of Atlantic origin (*Mytilus edulis*) were the earliest benthic animals found as fossils in the White Sea sediments (Nevessky et al. 1977). Later other species penetrated the White Sea using the same route. The main features of the White Sea biota achieved their present status ca 6,000 years ago.

The fauna of the shallow part of the White Sea belong to the Norwegian Province of the Boreal biogeographic area, whereas the fauna of the deepest part belongs to the Arctic biogeographic area (Fedyakov 1984). In the inner parts of Dvina and Mezen' Bays several representatives of relict brackish-water fauna are found. The current hydrological regime of the Gorlo Strait rather prevents the exchange of fauna between the White and Barents Seas. Nevertheless, there are no endemic species in the White Sea, however, local forms are known for a number of fishes and benthic invertebrates. A few nonindigenous species occur in the White Sea.

3 Chinese mitten crab

The Chinese mitten crab *Eriocheir sinensis* (Milne-Edwards, 1854) has been described from the coastal zone of the Yellow Sea. It is a catadromous species. Its spawning takes place in estuaries. The larvae and young crabs migrate for long distances up the rivers where they spend most part of their life before they reach sexual maturity. When individuals are ready for reproduction, they migrate to the coastal areas of the sea.

The Chinese mitten crab has actively and widely spread outside its original area. It was first found in Europe in the beginning of the 20th century (Gollasch 1999a). Primarily it occurred in some estuaries in Germany, where it apparently had been brought by ship's ballast waters or as component of ship's fouling communities. Through the Kiel Canal it penetrated the Baltic Sea (Gollasch 1999a), where it spread up to and all over the Gulf of Riga, Gulf of Bothnia and Gulf of Finland (Haahtela 1963). It has been found also in Mediterranean Sea (Zibrowius 1991) and in North America (Gruner et al. 1993). In the 1990s, *E. sinensis* was occasionally found in the internal waters of Northwest Russia. In 1992, an adult male was caught in the Onega Lake near the Pukhtinskie Islands. In the mid-1990s, it was found in the Vuoksa River and in the Ladoga Lake. In 1998, a big (72 mm wide) individual was caught in the pre-estuary part of the Northern Dvina River (Berger et al. 1999). Today, crabs are regularly caught in fishing nets in Archangel water area (Bushueva unpubl.). According to the above data, this species is in a phase of biological progress: its area of distribution expands rapidly and the number of separate populations increases steadily. In Northern Europe its ecological niche is not occupied and its rivals are missing. Therefore, it is possible that in a short time it may occupy the White Sea first and then other northern water-bodies.

4 Pink salmon

The pink salmon *Oncorhynchus gorbuscha* (Walbaum, 1792), a representative of the Pacific salmonids, was introduced in the White Sea in the 1950s. Spreading rather extensively in a number of northern seas, it is known from the basins of the Ob and Yenisei Rivers up to Scotland and Iceland.

Transportation of pink salmon eggs from the Far East began in 1957. Eggs in the stage of eye pigmentation were delivered by airplane to a number of fish farms in the White Sea area. After incubation juveniles were set free; however, during several years, only single individuals returned. This has been explained by low survival of the juveniles. Since 1960, the return of pink salmons to the rivers of the Barents and White Sea areas has increased sharply. Mass return of *O. gorbuscha* was registered in 1960, 1965, 1971, 1973, 1975 and 1977. The catch was about 77,000-262,000 fishes or 115-400 tons per year. The actual catch was, apparently, 1.5-2 times higher, taking into account the illegal catch (Yakovenko 1995).

During this period (1960s to 1970s), the contribution of industrial versus natural reproduction of pink salmon was not equal. The natural reproduction was efficient only in 1967, 1975 and partially in 1973.

The sharp fluctuations in the number of the pink salmon populations and low efficiency of natural reproduction were a consequence of specific features in the reproduction of the introduced fishes, which did not meet the conditions in the rivers of the White and Barents Sea basins. In addition, due to early frosts the exposure of the eggs during days with ambient temperature did not reach the critical value for the eggs of the late-spawning individuals, and this caused their death. Only embryos from early sets survived, laid no later than in the first half of September. In 1979 delivery of eggs from the Far East ceased and the number of pink salmon began to fall sharply, which testified the low efficiency of its natural reproduction in these areas. The population was maintained only due to the eggs delivered from outside. Hence, no naturalisation of the species occurred and, thus, the fish-breeding activities can be regarded only as a successful example of sea ranching.

Low survival of pink salmon juveniles and poor efficiency of its natural reproduction, lead the Onega fish farm to bring about $1x10^6$ pink salmon eggs from the Sea of Okhotsk in 1984, rather than from the southern region of its native area as they had done before. Unfortunately, this was not repeated any more. From 1987, the rate of natural reproduction of *O. gorbuscha* began to increase noticeably, marking the completion of its naturalisation, though more time and further investigations are required to prove this.

The catch of pink salmon in the White Sea increased rapidly during the late 1990s up to 525 tons in 1995, and 958 tons in 1997 (Alimov et al. 1998). Thus, if the situation does not change, the pink salmon will gain a stable leading position in the catch of fishes of market value in the White Sea.

The entry of brood stocks into numerous salmon rivers of the White and Barents Sea basins takes place in June-October. The whole population is represented by two-year-old fishes, which spent slightly more than one year in the sea, their gonads being normally developed at the 3rd-4th and, less frequently, at the 4th-5th stages of maturity. Spawning takes place in August-September at water temperature 13.0-9.5 °C. The average fertility is $1.5\text{-}2.0x10^3$ eggs and maximum fertility is more than $3.3x10^3$ eggs per female, which is higher than in the Far East (Nekhludov 1992). Sex ratio is close to equilibrium, though it varies in different periods of spawning migration, with females usually dominating in the beginning and males at the end of the migration.

Fishes entering the rivers of the European North of Russia are bigger (33-61 cm) than those of the Far East. The average body weight varies in different years and different rivers from 1.1 up to 2.2 kg, with some individuals reaching 3.3 kg. The pink salmon from the White and Barents Seas does not differ from that of the Far East in most meristic parameters and is in accordance with taxonomic description (Berg 1948). The fishes from the White Sea can be reliably distinguished from those from the Pacific Ocean only by the number of pyloric appendages (Nekhludov 1992).

Their spawning is similar to that of other salmons. Females make hillocks in the low part of pools of rivers at the depth of 0.2-1.0 m. Eggs (from several dozens to $1\text{-}2\times10^3$) are laid in the hillocks at the depth of 15-35 cm. After spawning the fishes die. Migration of pink salmon juveniles to the White Sea takes place in June-July at any time during the polar day. Compared to the Far East, larvae stay longer in the spawning hillocks and leave them at later stages of development. Active feeding on plankton begins already in the region of spawning grounds, where chironomid larvae constitute the main item of the diet. Further on various insects become dominant.

Our data is not sufficient to discuss the complete naturalization of pink salmon in the White Sea or estimate the impact of it on the ecosystems of the White Sea and on the populations of the Atlantic salmon in different rivers.

5 Steelhead

Primarily the steelhead *Parasalmo mykiss* (Walbaum, 1792) (syn. *Oncorhynchus mykiss*, wrong name *Salmo gairdneri*), commonly known as the farmed rainbow trout, was distributed in the northern Pacific Ocean, rivers of North America and Kamchatka coast only. Its introduction into European waters started in the end of 19th century. Beginning from the 1980s, the steelhead became one of the most important species in fish farming in the White and Barents Seas. In 1983 researchers from the Polar Institute for Fishery and Oceanography, Murmansk, started the growing of the salmon in net cages in the Kandalaksha Bay of the White Sea. The results were promising (Vorob'eva et al. 1985). In 1986 similar experiments were initiated in the Kem' region of Karelia. The fishes of market weight (about 350-400 g) were obtained after 120 days of cultivation during summer-autumn period in the sea cages.

In 1988-1996, steelhead was cultivated in sea cages in the Dvina Bay and in the Onega Bay near Solovetsky Archipelago (Kulida et al. 1998). Steelhead from Onega and Solza fish farms, originally weighing 60-100 g and more, were taken for stocking. Various Russian and Finnish pellets and alimentary additions were used. Farms differed in hydrological and, primarily, temperature conditions. The length of cultivation varied in different years and different places from 75 to 120 days. Relative increase of weight varied from 1.6 to 28 times.

Attempts to cultivate steelhead in the sea cages were also undertaken in the 1990s in different parts of the Chupa Inlet of the Kandalaksha Bay. The best results were obtained in 1995-1997 by Norwegian (Akvaplan-NIVA, Tromsø) and Russian (Petrozavodsk State University) researchers in 1996. Three thousand steelhead smolts, weighing about 200 g, were brought from a fish farm in Karelia. The obtained results surpassed all optimistic expectations. The total product yield was 2,574 kg. The average weight of the steelhead at the end of the cultivation period was 1,350 g (which exceeded the expected value of 700 g almost twice) and weight increase (1,136 g) was twice that of the expected (500 g). The mortality was extremely low.

The cultivated steelhead resisted parasite infections. However, individuals infested with *Diplostomum spathaceum* (Digenea) localized in the eyes of the trout and sometimes resulting in blindness, were encountered.

Thus, it is obvious that the farming of steelhead in sea cages in the White Sea is rather promising and will be developed quite intensively in the future.

As a result of the technology used for the cultivation of steelhead some individuals leave the cages and live in the sea up to the spawning in the rivers. However, natural populations of steelhead have not yet been registered in the White Sea. Some mature individuals of steelhead were caught in recent times in a number of Eastern Murman rivers (Muraveiko et al. 2000). Unconfirmed data about such cases has been received from White Sea fishermen. In spite of these data, we cannot confirm the naturalization process of steelhead in the White Sea. Nothing is known about the impact of this fish on the ecosystems. However, there is a possibility that all these questions will appear very significant in the nearest future.

INTRODUCED MARINE ORGANISMS IN NORWEGIAN WATERS, INCLUDING SVALBARD

CHRISTOPHER C.E. HOPKINS
AquaMarine Advisers, Åstorp, Sweden
aquamarine@telia.com

Abstract
This chapter documents the various alien marine organisms and their associated pathogens in Norwegian waters, including Svalbard. It estimates that about 45 alien marine species have become established in these waters, comprising about 22 plant species (ca 12 macroalgae and 10 phytoplankton), 22 invertebrate species including parasites and pathogens, and a single fish species. Attention is drawn to the ecological and socio-economic damage that has been or potentially may be caused by introductions and transfers of free-living organisms and their associated parasites and pathogens/diseases in Norway.

1 Introduction

There have been no previous reviews in Norway that have addressed the issue of marine alien organisms in depth. As part of a study concerning the OSPAR Commission's Convention area, a list of marine alien species was prepared for the region by Member States using information collected from national contacts, including Norway (OSPAR 1997). The collected data was not subjected to an extensive verification process. A total of 109 'probably established' alien species was reported from the whole OSPAR Convention area (i.e. northeast Atlantic and North Sea), not counting cryptogenic species and incidental species. The data for Norway (36 species in total) indicated (i) 31 probably established alien species (14 plants and 17 animals), (ii) 4 cryptogenic species (i.e. of uncertain origin) and (iii) 1 incidental species (not established).

2 Overview of alien marine organisms in Norwegian waters and their impacts

This paper is based on a report prepared for the Norwegian Directorate of Nature Management (Hopkins 2001). The results are summarized in Table 1. It is emphasized that this overview is not intended to be definitive, as studies generally overlook the alien species that have become established longest, e.g. a substantial component of the cryptogenic species. Of the ca. 4,000 species of macrobenthic flora and fauna found in Norwegian waters, 211 species have an unknown geographical distribution in the world (Brattegard & Holthe 1997), and may qualify as cryptogenic. Thus, there will be obvious uncertainty as to the extent that cryptogenic species should be included with those that clearly are alien. This report aims, however, to illustrate the nature of the continuing and, in many cases, escalating problems concerning alien species in Norway in particular and Europe in general.

2.1 MACROALGAE

About 12 macroalgae species or sibling species may be considered to be alien in Norwegian waters (Table 1). In most cases the effects of these species on the environment, including other species, and commercial interests are not described. The green alga

E. Leppäkoski et al. (eds.), Invasive Aquatic Species of Europe, 240–252.

Codium fragile ssp. *tomentosoides* was first recorded from Norway in 1946 (Silva 1957), and during the following 30 years it had spread all along the coast from the east Skagerrak to North Troms (Rueness 1977). It probably eliminated the native species *C. vermilara* in Norway (Silva 1957). *C. f.* ssp. *scandinavicum* is probably a recent immigrant to Europe, first recorded in Norway in 1929. *C. f.* ssp. *atlanticum*, probably introduced from the Pacific Ocean around Japan into Europe in historical times, has been found in Norway since at least 1895 (Silva 1957). Its ability to form rafts and float helps its secondary dispersal.

The first record of the brown alga *Colpomenia peregrina* was in 1933 outside Bergen (Braarud 1950). In the early 1970s it was found as far north as the area of Ålesund, and north of Trondheim (Wiik & Nerland 1972). In the 1980s it had also spread to the Skagerrak coast (Rueness et al. 1990). Warm winters appear to favour the abundance and dispersal of this species in Norway. The species was first introduced into Europe via France from the Pacific coast of North America with juvenile oysters *Crassostrea virginica.* The species appears to have few negative effects on other species and the environment. However, when growing attached to oysters it may float away with the oyster when the air-filled thalli grow large enough, hence its name of oyster thief.

Fucus evanescens was introduced to the Oslofjord about 100 years ago (Bokn & Lein 1978) and has become a quite common plant in that area, especially in harbours and nutrient enriched waters such as the inner part of the Oslofjord (Munro et al. 1990).

The Japanese large brown alga *Sargassum muticum* was first recorded in Europe in the early 1970s. This alga was first found as drift plants in southern Norway in 1984, and in 1985 the first attached plants were found (Rueness 1989). The lower lethal temperature for the species is ca. -1 °C while growth and reproduction appears to be limited by a lower temperature of ca. 12 °C, which led Rueness (1985, 1990) to predict that the species would not establish itself northwards in Norway beyond the Trøndelag coast. The species has now spread along the Norwegian Skagerrak coast, as well as westwards and northwards along the coast to Hordaland, reaching sizes of over 2 m (Rueness 1989). Expansion rates have been recorded of 100s of km per year, reaching from the littoral down to ca 10 m depth. It has a range of opportunistic features, being particularly able to dominate (i.e. outcompete and exclude) the native algal flora in bays, marinas and similar recreational areas. It can become a nuisance factor with clogging of outboard motors and accumulation of substantial amounts of detached and drifting rotting-fronds. The large canopies hinder light penetration and water circulation, and the plant may cause problems with operating diverse fisheries gear. On the positive side, this species attracts a rich epiphytic fauna.

The red alga *Bonnemaisonia hamifera* was first recorded in Norway in 1902 and now occurs all along the Norwegian coast (Rueness 1977). The gametophytes have been encountered only sporadically from the west and southwest coasts, the females being first seen in the mid 1960s and the males some years later (Breeman et al. 1988). This species may have been introduced unintentionally with shellfish from Japan. The success of *B. hamifera* is probably due to its lack of grazers due to brominated bioactive compounds in its cells, rapid growth rate, and its opportunistic qualities.

The red alga *Dasya baillouviana* was first recorded in south Norway in 1966, and was found during the 1970s at some new localities along the southern coast (Røsjorde 1973). It is now found along the Skagerrak coast from Oslofjord to Vestfold. The red alga *Gracilaria gracilis* has been known in Norway since the mid-1930s. A novel species was reported for Norwegian waters in 1999, a red alga *Dasysiphonia* sp. (Lein 1999). It was probably accidentally transported from the North Pacific Ocean (via ballast water or hull fouling). The first confirmed collection of the red alga *Polysiphonia harveyi* from the British Isles was in 1908. It is regarded as an alien in the northeast North Atlantic where it seems to be spreading rapidly. *P. harveyi* has now been recorded from Norway (ca 1985, J. Rueness pers. comm.) and is now common in the Oslofjord and Skagerrak (Rueness 1998). It possibly came to Europe from Japan with Pacific oysters. It can also probably spread through secondary dispersal through drifting with larger seaweeds as an epiphyte. The species is an opportunist, with a rapid growth rate and high tolerance of temperature changes. It can become very abundant and thus has the ability to displace native species. It is a fouling agent and can become abundant in marinas on artificial structures, although its small size precludes it from being a big problem (c.f. *S. muticum*). The red alga *Sphaerococcus coronopifolius* was noted from Norwegian waters in 1994 by Karlsson (1995). This species was found for the first time in Sweden in the northern part of the Swedish west coast in 1990 (Karlsson et al. 1992), so its further spread to Norway was expected.

Table 1. Alien, cryptogenic and incidental marine species recorded in Norwegian waters. A alien; C cryptogenic; I incidental; E established; ✓ included in the count of probably alien species. For references, see Hopkins (2001).

	Area of origin	First observed	Status	Vector
Macroalgae				
Codium fragile ssp. *atlanticum*	Indo-Pacific, Japan	1895 ?	A✓ E	
C.f. ssp. *scandinavicum*	Indo-Pacific	1929	A✓ E	
C. f. ssp. *tomentosoides*	Indo-Pacific, Japan	1946	A✓ E?	
Colpomenia peregrina	Pacific	1933	A✓ E	Aquaculture
Fucus evanescens	N Atlantic/Pacific	1900	A✓ E	Shipping
Sargassum muticum	Japan	1984/1988	A✓ E	Aquaculture
Bonnemaisonia hamifera	Japan	1902	A✓ E	Fouling
Dasya baillouviana	Mediterranean & NW Atlantic	1966	A✓ E	Shipping
Dasysiphonia sp.	North Pacific	1996	A✓ E	Shipping
Gracilaria gracilis	?	1935	A/C?✓ E	?
Polysiphonia harveyi	Pacific	ca. 1985	A/C? ✓ E	?
Sphaerococcus coronopifolius	?	1994	A/C? ✓ E	?
Phytoplankton				
Coscinodiscus wailesii	Pacific	1979	A✓ E	Aquaculture/BWT& Secondary transfer
Odontella sinensis	Indo-Pacific	1903	A✓ E	BWT & Secondary transfer
Thalassiosira punctigera	?	1979	A✓ E	Aquaculture/BWT & Secondary transfer
T. tealata	?	1968	A/C✓ E	Aquaculture/BWT & Secondary transfer
Chattonella cf. *verruculosa*	Japan	1998	A✓ E	BWT & Secondary transfer
Heterosigma akashiwo	Japan?	1964	C✓ E?	?

	Area of origin	First observed	Status	Vector
Olisthodiscus luteus	?	1999	C✓ E?	?
Alexandrium tamarense	?		C✓ E ?	BWT & Secondary transfer
Gymnodinium aureolum	?	1966	A✓ E	BWT & Secondary transfer
G. mikimotoi				
G. nagasakiense				
Gyrodinium aureolum				
Prorocentrum minimum	?	1979	C✓ E	BWT & Secondary transfer
Animals				
Scolelepsis cf *bonnieri*	?	1995	A✓ E?	BWT
Alkmaria rominji	?		C? E ?	
Marenzelleria cf. *wireni/viridis*	NW Atlantic/Arctic		C? E ?	
Caprella mutica	Japan	1999	A✓ E?	Shipping
Balanus improvisus	America	1900	A✓ E	Fouling?
Lepas anatifera	Atlantictropic/ sub-tropic	<1900	A I?	Fouling/BWT
Corophiumsextonae	New Zealand?/ Mediterranean	1985	A✓ E	?
Eriocheir sinensis	SE Asia	1976	A✓ E	BWT, Sec. dispersal
Homarus americanus	NE America	1999	A✓ E?	Restaurant cuisine
Paralithodes camtschatica	W Pacific	1985	A✓ E	Stocking (Russia)
Cordylophora caspia	Ponto-Caspian	1985	A✓ E	
Gonionemus vertens	W Pacific	1921	A✓ E	Shipping/Aquaculture
Crassostrea gigas	Japan & SE Asia	1979	A I	Aquaculture
Ensis americanus	NE Atlantic	1989	A✓ E	Secondary dispersal
Mya arenaria	N America	ca. 1000	A✓ E	Bait, food or bilge water
Petricolaria pholadiformis	N America	1955	A✓ E	Aquaculture & Secondary dispersal
Tapes philippinarum	SE Asia	1987	A I?	Aquaculture
Teredo navalis	W Pacific	ca. 1700s	A✓ E	Shipping
Crepidula fornicata	NW Atlantic	1958	A✓ E	Aquaculture
Potamopyrgus antipodarum	New Zealand	1952	A✓ E	BWT
Molgula manhattensis	America		A✓ E ?	
Oncorhynchus mykiss	N America	1902	A✓ E	Sport fishing, Aquaculture
Gyrodactylus salaris	Baltic	1975	A✓ E	Aquaculture
Anguillicola crassus	SE Asia	1994	A✓ E?	?
Pseudodactylogyrus anguillae	?	Ca. 1990	A✓ E	Aquaculture
P. bin	?	Ca. 1990	A✓ E	Aquaculture
Bacteria				
Aeromonas s. salmonicida	?	1986	A✓ E	Aquaculture

None of the above-mentioned species is noted by Gulliksen et al. (1999) as being found in Svalbard (including Bear Island) and Jan Mayen waters.

2.2 PHYTOPLANKTON

Phytoplankton species can display different characteristics in different environments, thereby making it very difficult to differentiate between native, introduced and newly discovered species (Weidema 2000). However, it is highly probable that many species have been introduced into the European area via ballast water, and to a lesser extent by aquaculture activities. Once introduced into the European area in general, and the North Sea in particular, there is a further likelihood of secondary distribution of alien phytoplankton cells by transport with water currents. The Norwegian coastal current, starting in the Skagerrak with water exiting from the Baltic Sea, acts as a means of transport of plankton with the current along the coast towards northern Norway. It is also necessary to underline that there have been several misidentifications and some confusion regarding the identity of alien species of phytoplankton in Norwegian waters, and that there has been a tendency for such problems to be perpetuated in the literature (G. R. Hasle, J. Throndsen, pers. comm.).

Several nonindigenous plankton species have been reported in the North Sea since about 1900 (Carlton 1985, 1989; Hallegraeff et al. 1990). About 10 phytoplankton species may be considered as alien to Norwegian waters (Table 1). The diatom *Coscinodiscus wailesii* was first noticed, due to its mucus production, in Europe 'off the coast of south-west England' (i.e. Plymouth) in 1977 and misidentified as *C. nobilis* (Boalch & Harbour 1977). It is probably correct to describe the area of origin of *C. wailesii* as the Pacific Ocean but it is probably incorrect to extend this to the Indian Ocean (G. R. Hasle, pers. comm.). It is likely that *C. wailesii* was introduced to England via ballast water; its introductions elsewhere in the North Sea have likely taken place by secondary dispersal by currents, and the import and transfers of nonindigenous oysters (e.g. Rincé & Paulmier 1986). The first record of *C. wailesii* from Norwegian waters is from 1979 (Hasle 1983, 1990). It forms dense blooms and can account for ca. 90% of the total phytoplankton biomass, and produces copious mucilage that aggregates and may sink to the bottom and coat the seabed. The extensive mucus produced has been known to cause clogging of fishing nets and aquaculture cages.

Odontella (=Biddulphia) sinensis, native to the China Sea and Indo-Pacific area, was considered an immigrant to the North Sea in 1903 (Ostenfeld 1908), and has become widely distributed in the North Atlantic, the North Sea and Skagerrak, and the Baltic Sea (Boalch & Harbour 1977; Leppäkoski 1984; Rincé & Paulmier 1986). It has become abundant in Norwegian coastal waters (e.g., in the Skagerrak; Lange et al. 1992).

The spread of the diatoms *Thalassiosira tealata* and *T. punctigera* into the North Sea has been well documented (e.g. Hasle 1983, 1990; Rincé & Paulmier 1986). *T. tealata* was recorded in Norway in 1968, and *T. punctigera* in the Skagerrak and Oslofjord since 1979 (Hassle 1983, 1990). It is clear that *T. tealata* has been present in the North Sea long before it was first described in the area, as Hasle (pers. comm.) has noted that it was present in samples collected in 1950 from Blakeney, England. Thus it appears that *T. punctigera* can be considered a true alien species in Norway, whereas *T. tealata* is probably best considered as cryptogenic.

In April-May 1998 and at the same time of the year in 2000, an algal bloom of *Chattonella* sp. cf. *C. verruculosa* - a species previously known from Japan - occurred in the Skagerrak and northern Kattegat waters and adjacent parts of the North Sea. Fish kills as a result of this species have been reported from the Swedish west coast, the Norwegian south and south-west coasts and the Danish coasts. This is probably the first record of the species in Europe, and it is highly likely that it was introduced via ballast water discharge. In Norway, it has been registered from the border with Sweden to Stavanger, and has had effects on fish farms in the Farsund and Flekkefjord areas where a total of 350 tonnes of large sized salmon died. Small sized salmon apparently survive better. In an outbreak starting in March 2001, this alga destroyed about 700 tonnes of fish raising questions by the Norwegian marine aquaculture insurance industry as to the consequences for future insurance should the alga be considered as resident in Norwegian waters. Other ecological effects have not been reported. The actual toxicity of the class is poorly known. The mortality of fish is apparently caused by the plankton cells easily clogging the gills of the affected individuals. Raphidophyceans contain slime that is exuded when the algae are on the gills.

The story of *Heterosigma akashiwo* is a 'study in confusion and suppositions' as emphasized by Throndsen (1996). This species was mistakenly identified as *Olisthodiscus luteus* from Oslofjord in 1964 (Braarud & Nygaard 1967; Throndsen 1969), resulting in disinformation (c.f. Smayda 1990) and confusion (c.f. Throndsen 1990) that have frequently been perpetuated in the literature regarding the phytoplankton and alien species in Norwegian waters. The 'true' *Olisthodiscus luteus* (originally described from England in 1937) was only first registered in Norway in 1999 (Grimsrud & Throndsen 2000) in the Oslofjord, where it was one of the most common heterotroph flagellates sampled from the surface layer of marine bottom sediments.

Alexandrium tamarensis is a cryptogenic species in Norwegian waters. It produces paralytic toxins that can be fatal for a number of biota, including humans. It has caused outbreaks of Paralytic Shellfish Poisoning (PSP) on several occasions (e.g. blue mussels) in Norway and caused human health problems as well as temporary bans by the authorities on eating shellfish from the contaminated areas of the coast.

In 1966, a massive dinoflagellate bloom, accompanied by mortality of caged sea trout, occurred along the Skagerrak coast of Norway (Braarud & Heimdal 1970), with the causative agent being identified as *Gyrodinium aureolum*. This species has become one of the most commonly reported blooming dinoflagellates in northern temperate waters (Hansen et al. 2000). Much taxonomic confusion has been connected with this species (e.g. Partensky et al. 1988), although it became generally accepted that the European '*G. aureolum*' is very closely related to or even synonymous with the earlier described *Gymnodinium mikimotoi* (=*G. nagasakiense*) (Partensky et al. 1988). Hansen et al. (2000) - based on analyses involving light microscopy, nuclear-encoded genetic sequencing and pigment isolates of five geographically separate isolates of *G. mikimotoi* - concluded that the European isolates, formerly identified as *Gyrodinium aureolum*, *G.* c.f. *aureolum*, or *Gymnodinium* c.f. *nagasakiense*, are conspecific with the Japanese *Gymnodinium mikimotoi*. As a result of this, and comparing the nuclear sequence from material originating from what is believed to be close to the type locality of *Gyrodinium aureolum* (Hulbert) with *Gymnodinium fuscum* (the type species of *Gymnodinium*), Hansen et al. (2000) have renamed *Gyrodinium aureolum* (Hulbert) *Gymnodinium aureolum* (Hulbert) G. Hansen, comb. nov. Accordingly, the species will be referred to as this in the current paper. It has bloomed on numerous occasions since 1966 in Norway, most frequently from August to September, and has been recorded from the border with Sweden to Sør Trondelag causing the same effects, the most serious occasion being in 1981 when it caused the greatest loss to Norwegian fish farmers as a result of harmful algal blooms.

The dinoflagellate *Prorocentrum minimum* was, first recorded in the North Sea in 1976 (Smayda 1990), and has a wide environmental tolerance being found in brackish as well as fully marine water. It was first registered in northern Europe during a massive bloom in the outer Oslofjord in 1979. It has become annually common in the summer in near-shore areas of Østfold within the influence of the river Glomma, where it discolours the water yellow-brown and reduces light attenuation in the water. A number of reports connect it to accumulation of toxicity in bivalve shellfish and effects on other marine species, but the actual toxicity of *P. minimum* has not been convincingly documented and this alga often appears not to be toxic. The species has extended its distribution into the Kattegat and southwest Baltic Sea in the 1980s (Granéli 1987).

There is currently no evidence to indicate that the phytoplankton species listed in Table 1 are found in Svalbard (including Bear Island) and Jan Mayen waters.

The impact of harmful algal blooms (HABs) caused by alien phytoplankton species in Norwegian waters has not been comprehensively estimated in terms of socio-economic costs. However, there is little doubt that HABs can represent a very substantial threat in terms of toxic effects on living marine resources. A gross starting point for examining the possible socio-economic impacts of HABs caused by alien phytoplankton species is to consider some of the possible tainting and mortality effects on commercially important marine species that are found in the area of greatest likely impact, i.e. the shelf zone. The potentially impacted biota include wild species that form high value harvests (e.g. via 'capture' fisheries), such as blue mussels *Mytilus edulis*, scallops *Chlamys islandica*, crustaceans (pink shrimp *Pandalus borealis*, Norway lobster *Nephrops norvegicus*) and fish (cod *Gadus morhua*, saithe *Pollachius virens*, haddock *Mellanogrammus aeglifinus*, herring *Clupea harengus*, Atlantic salmon *Salmo salar*, turbot *Psetta maxima*, plaice *Pleuronectes platessa*, and halibut *Hippoglossus hippoglossus*), to name but a few.

Norway is one of the foremost fishing and aquaculture nations of the world, thereby making the potential socio-economic impacts of HABs also proportionately great. The export value of all Norwegian fish products in 1999 was NOK 30 billion, representing 8.7% (i.e. similar to the value of natural gas) of all Norwegian exports. Of this farmed Atlantic salmon accounted for about 36% (NOK 10.8 billion), with whitefish (e.g. cod), pelagic fish (e.g. herring) and other species making up the remainder in order of economic importance. Although it is improbable that all this value can be eradicated by HABs, the potential for major socio-economic impacts via alien HABs (e.g. *Alexandrium, Chattonella, Gyrodinium*) can be illustrated by reference to the toxic bloom of the indigenous flagellate *Chrysochromulina polylepis* in May–June 1988 that harmed and killed a large number of marine species in the upper 20 m of the sublittoral along most of the Norwegian Skagerrak coast. This resulted *inter alia* in 800 tonnes of farmed fish being killed in Norway and 100 tonnes in Sweden, resulting in an economic loss of about US$ 11 million or equivalent to more than NOK 70 million at that time (Skjoldal & Dundas 1991). The 2001 outbreak of *Chattonella* in Norway caused over US$ 10 million loss to the fish farming community (Hopkins unpubl.). There is also great optimism in Norway concerning the potential for farming and enhancement of shellfish, and the number of licences granted for such purposes by the authorities has increased substantially over the last several years. The blue mussel is currently the most important species, with 700 tonnes of mussels having been sold for NOK 5.8 million in 1999. Given the full range of potential effects from HABs, the possible socio-economic consequences may amount in a worse case scenario to hundreds of millions of NOK if toxic blooms spread along large stretches of the coastal shelf. This is a substantial enough socio-economic threat when posed by indigenous species without adding to it by introductions of harmful phytoplankton.

2.3 INVERTEBRATES

Table 1 records a count of about 22 invertebrate species, including parasites and diseases, which may be considered as alien to Norwegian waters. The instances of free-living invertebrates, as well as the parasites and pathogens/diseases connected with invertebrate hosts, are further described below.

Annelids. Currently there have been few records of clearly alien polychaete worms in Norway. *Scolelepsis* c.f. *bonnieri* is undergoing taxonomic verification by specialists (H. Botnen, pers. comm.). If its identification is correct, this would clearly emphasize its presence as an alien species for Norway. *Alkmaria rominji* is known from Østfold, but may be considered a cryptogenic species. The effects of these polychaetes on the environment and commercial interests are not described. *Marenzelleria wireni* is considered as cryptogenic in Norway. This is due to the current discussion about the identity and/or taxonomy of *M. wireni* and *M. viridis*.

Cnidaria. *Cordylophora caspia* and *Gonionemus vertens* are the two prime examples of undoubted alien hydrozoans in Norway. The anthozoan *Rhizogetum nudum* has been identified as a possible cryptogenic species in Norway and the UK, with a likely chance of being alien, in an OSPAR questionnaire (Eno 1996; OSPAR 1997). *C. caspia* is unique among hydroids in its ability to tolerate salinities from fresh water up to 30 PSU, but self-sustaining populations of *Cordylophora* spp. are reported only in brackish or freshwater. *C. caspia* is generally considered to be native to the Caspian and Black Seas, but has now been found worldwide. *G. vertens* has been found at several localities in Norway from Oslofjord (Kramp 1922 as *G. murbachi*, but likely to be a misidentification of *G. vertens*) to Trondheimsfjord (Gulliksen 1971). Transport on ships' hulls in the polyp stage (Carlton 1985) from the western Pacific Ocean to Europe in the 19th century is a possible manner of introduction. However, Edwards (1976) suggested that it may have arrived much earlier from Japan with importations of Japanese oysters *Crassostrea gigas* 500 or more years ago. *G. vertens* probably originates from the Pacific (China, Korea and Japan) (Edwards 1976). It was probably introduced to Europe in Portugal and exported from 1867 onwards from Portugal to France, again with oysters in the polyp stage. This allowed the dispersal to other European countries via major French oyster exports. It can also disperse in the hydromedusae stage in water currents and ballast water.

Crustacea. Crustaceans make up one of the largest groups of alien invertebrate species found in Norway. The barnacle *Balanus improvisus* became established in western Europe in the early 1800s, probably having been transported on the hulls of ships from North America (Walford & Wicklund 1973). It causes substantial economic expenses in the need to treat ships to counteract the negative effects of fouling. The goose barnacle *Lepas anatifera* is found quite frequently in Norwegian waters attached to flotsam and jetsam originating from more southerly areas, as well as being introduced as larvae with ballast water. However, this species is unlikely to become fully established by natural breeding and recruitment in this area.

The two alien amphipod species registered in Norway are *Caprella mutica* and *Corophium sextonae*. The former species was only first found recently in Norway by Heilscher (2000). Four individuals of *C. sextonae* were found in 1985 from a single grab station in Aust-Agder (Wikander 1986). The origin of this species has generally been viewed as New Zealand (c.f. Eno et al. 1997; OSPAR 1997); a possible Mediterranean origin could equally well be considered (see Hopkins 2001 for further discussion). The species has spread very strongly throughout European waters and the Skagerrak specimens are probably part of this European dispersal.

The Chinese mitten crab (*Eriocheir sinensis*) was found in 1977 in the estuary of the river Glomma in the Oslofjord (Christiansen 1977). In 1986-1997 further specimens were recorded in the same district (Hardeng & Viker 1997); the species has probably become fully established in the above-mentioned area. In 1999, the American lobster (*Homarus americanus*) was found in Oslofjord, having probably been discarded from the restaurant trade. *H. americanus* is affected by only one internal bacterial disease, gaffkemia, caused by *Aerococcus viridans*, a pathogen to only two hosts, the American lobster, and the European lobster, *H. gammarus*. American lobster have developed a resistance resulting in a small percentage being able to survive and carry the disease with them, while this disease causes 100% mortality for European lobster. American lobster thus can transmit the disease onwards. None of the diseases that damage American lobster are otherwise recorded in European lobster.

A gross estimate of the potential economic value of the indigenous lobster in Norway is of the order of NOK 12-320 million based on landings of 30-800 tonnes and using a kg price of NOK 400. This can thus represent a simple estimate of the socio-economic damage on the native lobster fisheries that could be affected via pathogens transmitted by the American lobster.

The red king crab *Paralithodes camtschatika,* introduced by Russian scientists to the Kola Peninsula of the Barents Sea from the Russian northern Pacific, has spread westwards and southwards towards Lofoten, although its main distribution is east of the Tana River, since its first registration in 1976 in the Varangerfjord of northern Norway. Successful reproduction occurs and many large specimens have been found, many occurring as by-catch in the long-line and net fisheries. The king crab fishery is regulated as a 'research fishery' with a TAC set for equal division between Norway and Russia. At this stage of the population development and encroachment in northern Norway, it is difficult to determine what the environmental effects and additional commercial effects may be in the future on the coastal marine ecosystem. Although being economically valuable for a fishery, the crab may have an ecological impact by feeding on and competing with both benthic invertebrates and demersal fish (e.g. eggs of capelin and lumpsucker) and destroying fish nets and eating the bait off long-lines (Olsvik 1996; Öberg 1997). Further, potentially serious effects might occur from parasites and pathogens associated with the crab; the egg eating nemertean *Carcinomertes* can destroy the eggs of berried females, and there is concern that this nemertean may be passed on to native crabs that have not built up defences against infections. Further, a trypanosome parasite has been identified in the blood of this king crab and there is a concern that the crab may act as an intermediate host in the transfer of the trypanosome to commercial fish with possible harmful effects (J-H. Sundet pers comm.).

Mollusca. Molluscs make up the largest group of alien invertebrates species found in Norway. Seed from *Crassostrea gigas* was imported from the UK to Norway until about the mid 1980s. Subsequently the Norwegian industry became self-sustained for the seed of this species, and started to export surplus seed. One company is still in operation, and mainly produces seed (< a million) according to the market every two or three years. Due to the normally low temperature, *C. gigas* has little chance of establishing self-reproducing populations in Norway. However, elevated ambient temperatures caused by global warming may eventually allow *C. gigas* to form self-sustaining populations. The oyster microcell disease (*Bonamia ostreae*) has not been recorded in Norway (S. Mortensen, pers. comm.), but it can potentially be transmitted with devastating results with imported oysters of the genus *Ostrea*. This was shown with the reintroduction of infected *O. edulis* from the Pacific back to Europe resulting in the production of European *O. edulis* falling to about 10% of the levels that were present prior to infection by *B. ostreae* in 1979 (Mortensen 1993). Thus, it is vitally important that intensive screening of all live imports occurs within the framework of an appropriate quarantine period, e.g. as recommended by ICES (1995).

The North American razor shell *Ensis americanus* was introduced to the German North Sea coast in 1978 and first found in Norway in 1989. The Vikings probably transported the soft-shelled clam *Mya arenaria* to Europe from the Atlantic coast of North America as early as the 1200s (Petersen et al. 1992). In Norway, this species has not caused significant detrimental effects on the environment or on other species, whereas beneficial effects include it being used as bait for recreational and commercial fishing.

The false angelwing or American piddock *Petricola pholadiformis* was introduced to Europe from the USA by the end of the 1800s (ICES 1982), probably with oysters (*Crassostrea virginica*) from the USA. European populations are found from Norway to the Mediterranean and the Black Seas. In some parts of its new range (e.g. Belgium, Netherlands) it has almost completely replaced the native piddock species *Barnea candida* (ICES 1982). Its impact on commercial and socio-economic interests is not known, while its beneficial benefits have not been identified.

In 1987 brood stock of the Manila clam *Tapes philippinarum* were introduced to Norway from the UK for shellfish culture purposes. Large mature specimens of Manila clams have survived at three sites where cultivation trials were carried out in 1987–1991. There is apparently no evidence of successful recruitment as yet.

The alien 'shipworm' *Teredo navalis* established itself in Norway about 300 years ago, probably from having bored into and been transported by wooden hulled sailing vessels as well as floating driftwood. It has not been possible to determine the socio-economic problems caused by this species in Norway, although these must have been substantial.

The slipper shell limpet (*Crepidula fornicata*) was first recorded in Norway in 1958. This species with a high reproductive rate is easily spread attached to bivalves and shipping to new localities. Once established, they usually remain in abundance. Its success is probably due to a lack of predators and the method of reproduction (relying on individuals settling upon each other and reproduction assisted through very close proximity); and a pelagic larval stage aids the spread once introduced. The species is rarely found in abundance below 30 m depth. It can become a pest on commercial oyster beds, competing for space and food, while depositing mud on them so that the substrate is unsuitable for spat settlement. The alien gastropod *Potamopyrgus antipodarum* has been established in Norway since about 1952. In 1889 it was first recognised in Europe in the Thames estuary, England, but it is likely that the species was present in England since about 1850. *P. antipodarum* originates in New Zealand. It was introduced to Britain from southern Australia or Tasmania in drinking water barrels onboard ships (Ponder 1988). The species can reproduce rapidly by parthenogenesis, aiding its colonization. It thrives in fresh and brackish water. In Norway, it has colonized a significant part of the country over 50 years, but it is still primarily confined in coastal areas.

Other invertebrates. *Molgula manhattensis* is apparently the only alien ascidian to become established in Norway, but as yet its distribution is limited. The species is colonial and adults may become abundant fouling organisms on marine structures such as floats and wharf piles. They can also attach themselves to oysters and reduce by their filtration the availability of particulate food for oyster growth. The import of live oysters from Japan has led to the establishment in European waters of the parasitic copepod *Mytilicola orientalis* that infects not only European oysters but also a range of other bivalve molluscs. As yet this parasite has not been recorded in Norway, but it is very probable that it is only a question of time before it is.

2.4 VERTEBRATES

The rainbow trout (*Oncorhyncus mykiss*) is the only alien fish species in Norway (Table 1). It was introduced in 1902 for sport fishing and aquaculture purposes (Hindar et al. 1996). It can form both inland stationary as well as anadromous populations. Until recently the magnitude of *O. mykiss* introductions and transfers were substantial, with a large number of fish escaping from the confines of aquaculture. Although rainbow trout were registered in 55% of Norwegian municipalities only 3-4% could be described as self-sustaining populations (Hindar et al. 1996).

The question arises as to what prevents the establishment of this species in Norway specifically and in Europe generally? The physical and chemical habitat conditions in Norwegian rivers and lakes are well-suited to rainbow trout, a highly flexible and adaptive species (Hindar et al. 1996). Their habitat requirements overlap with those of the endemic Atlantic salmon and brown trout. It is known that rainbow trout can readily coexist with these species of salmonid in other regions of the world, and rainbow trout may actually be competitively superior to *Salmo* species. Hindar et al. (1996) have tentatively put forward the hypothesis that the paucity of establishment of rainbow trout in Europe may be due to the endemic parasite fauna of Europe, particularly myxosporean parasites, that are not native to North America. However, extensive work remains to be carried out to examine this hypothesis.

In the 1970s, Russia transferred substantial numbers of Pacific salmon (*Oncorhyncus keta* and *O. gorbuscha*) ova to the Kola Peninsula and juvenile fish were liberated into the sea. In the late 1970s and 1980s, Norwegian fishermen frequently caught Pacific ('pink') salmon in commercial nets and adults were observed spawning in some rivers in Finmark. After the cessation of the Russian introductions about 1980, the Norwegian catches decreased until it was reported in 1990 that no pink salmon were found in Norwegian coastal waters or rivers. However, this experiment posed a threat to Atlantic salmon in Norway through possible competition for food and habitats.

2.5 PARASITES AND DISEASES

The Atlantic salmon aquaculture industry in Norway has developed to be the largest of its kind anywhere. In the early 1980s this resulted in the need to import large numbers of smolts, but the home production of smolts has grown to account for almost 100% of those used by the industry. Various diseases have affected the cultured salmon industry. Some of these are native, e.g. vibriosis, cold water vibriosis and IPN virus, while some are of unknown origin, e.g. infectious salmon anaemia and pancreas disease. However, others are very probably introduced, e.g. furunculosis caused by *Aeromonas s. salmonica* and the freshwater monogenean skin parasite *Gyrodactylus salaris*. The latter poses a very serious threat to wild salmonids in freshwater and can also survive in brackish water, while the IPN virus can also affect salmon living in seawater and freshwater. Furunculosis is believed to have been introduced to Norway with smolts from Scotland in 1986, and has caused mortality in caged fish as well as some mortality in wild fish.

Fish products accounted for 8.7% or about NOK 30 billion of the total Norwegian exports in 1999. Of these products, farmed salmonids (almost exclusively Atlantic salmon) accounted for NOK 10.77 billion. The Atlantic salmon aquaculture industry in Norway, with its production of about 420,000 tonnes in 1999 has developed to be the largest of its kind in the world, capturing 53% of the total global market. Thus, given the size of the overall production of farmed Atlantic salmon in Norway, it is clear that alien parasites and diseases (e.g. furunculosis) have probably caused socio-economic damage to the industry for several 100s of million NOK.

The wild stocks of Atlantic salmon in Norway have since 1975 been seriously depleted in freshwater by the monogenean skin parasite *G. salaris*,. The ability of this parasite to survive in brackish water (salinities up to 20 PSU for as much as 18 hrs) makes it possible for its dispersal between closely situated river systems along the coast. The catastrophic mortality of most wild parr in over 30 Norwegian rivers has been caused by *G. salaris* that was probably introduced via stocking parr and smolts of Baltic origin, as this parasite is not normally part of the Norwegian fauna (Johnsen et al. 1999). During its maximum distribution and incidence, *G. salaris* reached 40 salmon watercourses in the Counties of Troms, Nordland, Nord-Trøndelag, Møre & Romsdal, Sogn & Fjordane, and Buskerud. By May 2000, this had been reduced to 21 salmon waterways due to rotenone treatment. As rainbow trout can live in both freshwater and brackish water, it can act as a host for the parasite *G. salaris* that in turn can be transmitted to indigenous salmonids. A gross national estimate of the value of all Norwegian wild salmon is likely to be more than NOK 10 billion, so the introduction of alien parasites and pathogens to which the indigenous salmon are not adapted may potentially result in Norwegian wild Atlantic salmon being a severely threatened species, and a substantial value in natural capital being put at risk (Hopkins 2001).

Another area of possible impact via introductions and transfers of alien organisms is that where human pathogens are involved. In 1993, a case of serious human intestinal infection occurred in Norway where the two patients had not travelled outside the country. One of the two persons infected had eaten a substantial amount of crabs where the infection was shown to involve a non-native type of *Vibrio cholerae* that is likely to have been transferred by the discharge of ballast water (Henriksen et al. 1993).

The swimbladder nematode *Anguillicola crassus* arrived in Europe in the 1980s with shipments of eels (*Anguilla japonica*) from Asia. This species is now found in most European countries (except Ireland) including in the Baltic Sea (Kennedy & Fitch 1990) and Iceland. *A. crassus* has caused major problems with both farmed and wild European eels *A. anguilla*. Currently many wild stocks of European eels have been seriously depleted causing major economic losses for the eel fisheries. If infected by this parasite, *A. anguilla* may be more susceptible to bacterial infections. The wall of the swim bladder may thicken and inflammation may occur. Growth may be slowed down, the swimbladder may burst in bad infestations, and swimbladder damage may prevent the spawning migration to the western Atlantic (Køie 1991). A variety of crustacean intermediate hosts and fish parasitic hosts are known for the swimbladder nematode, increasing the chances of its survival. There is high resistance of the sheathed, second stage, larvae to adverse conditions and the species has a well-developed colonizing ability (Kennedy & Fitch 1990).

An absence of native swimbladder nematodes is also a factor in the success of *A. crassus* as there is a lack of competitors and resistance of the host. *A. crassus* has been recorded in the open sea and in brackish coastal localities. The European eel appears to be more susceptible to *A. crassus* than are their original hosts (Køie 1991). The first observation of *A. crassus* in Norwegian waters was published by Mo & Steien (1994). Within three years the parasite was found in an eel farm near Kristiansand, far south in Norway, but no investigations were carried out on wild eels in that area. There is a possibility that the above-mentioned finds of this parasite are connected with transfers of eels from Denmark to Norway (T-A. Mo, pers. comm.).

The alien eel monogenean gill parasite *Pseudodactylogyrus anguillae* was found in two eel farms long before *A. crassus* was observed there (Tor-Atle Mo, pers comm.). In 1998, *P. anguillae* and the related *P. bini* were found in wild eels just south of Oslo (Mo & Sterud 1998). Buchmann et al. (1987) and Køie (1991) provide information on *Pseudodactylogyrus* infections of eels and their effects on the host.

There is currently no evidence to indicate that the alien fish species, or associated pathogens, listed in Table 1 are found in Svalbard, Bear Island and Jan Mayen waters.

Acknowledgements

This study was conducted for the Norwegian Directorate of Nature Management under Contract No. 00040051.

THE BALTIC SEA - A FIELD LABORATORY FOR INVASION BIOLOGY

ERKKI LEPPÄKOSKI[1*], SERGEJ OLENIN[2] &
STEPHAN GOLLASCH[3]
[1]*Department of Biology, Åbo Akademi University, Turku/Åbo, Finland*
[2]*Coastal Research and Planning Institute, Klaipeda University, Klaipeda, Lithuania*
[3]*GoConsult, Hamburg, Germany*
[*]Corresponding author *eleppakoski@abo.fi*

Abstract
Since the early 1800s, about 103 NIS have been recorded in the Baltic Sea. In this brackish sea (salinity range from less than 2 to approximately 20 PSU) horizontal and vertical gradients provide the nonindigenous species of different origin an extended repertoire of hospitable conditions. Several ecological functions of the non-native animals are new (and hence unique) for the species-poor Baltic Sea ecosystem. A review of the research into invasion biology in the Baltic Sea countries reveals a timeline from first records of single new species toward more sophisticated studies in invasion biology.

1 Introduction

The Baltic Sea is a young, and in a historical perspective, environmentally unstable sea. After the latest deglaciation, freshwater periods alternated with conditions, slightly more marine than what is prevailing today. Around 7,000 BP, the Baltic became brackish. Consequently, most animal and plant species living in the Baltic Sea are postglacial immigrants, many of them living close to their salinity tolerance limits.

The Baltic is the world's largest brackish-water sea area (382,000 km^2 or 415,000 km^2 with the Kattegat included), and this semi-enclosed water body is isolated from the North Sea by both geographical and ecological barriers: sill depth is 18 m; the salinity drops from 20-24 PSU in the Kattegat to 6-8 PSU in the Baltic Proper and further to 1-2 PSU in the inner parts of the large gulfs (Gulf of Bothnia and Gulf of Finland). The water is strongly stratified, especially in the Baltic Proper. Temperature conditions vary from boreal Atlantic in the southwestern areas of the sea to sub-Arctic in its northernmost part. The Baltic Sea area (i.e. Kattegat and Belt Sea, Arkona Basin, Baltic Proper, major gulfs, adjacent brackish-to-fresh water lagoons and inlets) represents a chain of sub-regions, well defined by their geomorphological configuration, hydrological regime and biogeographical composition of their biota. These horizontal and vertical gradients influence not only the native biota, but provide also the nonindigenous species (NIS) of different origin an extended repertoire of hospitable conditions. There is scientific proof of inoculation events that have taken place at particular sites, along the whole salinity gradient of the Baltic Sea, from Kattegat in the west to the diluted, innermost parts of the Baltic. Thus, the present Baltic Sea represents an ecological continuum, being a result of significant natural alterations during the past 10,000 years of its postglacial history. During the last two centuries, human-mediated introductions have added a new dimension to this continuum (Leppäkoski & Olenin 2001). In the Baltic Sea countries

E. Leppäkoski et al. (eds.), Invasive Aquatic Species of Europe, 253–259.

NIS were added to the environmental policy agenda in the mid-1990s only, although both intentional and unintentional introductions had existed long before their environmental and/or economic impact was recognized and formulated.

There are several recent reviews available on the NIS, their origin, spread and impacts in the Baltic Sea (Jansson 1994; Gollasch & Mecke 1996; Gollasch & Leppäkoski 1999; Olenin & Leppäkoski 1999; Jansson 2000; Leppäkoski & Olenin 2000a). This paper briefly describes the history of human-mediated biological invasions and their consequences in the Baltic Sea area and outlines the history of invasion biology in the region.

2 Invasion history of the Baltic Sea

As with communities on islands, the structure of species assemblages in semi-enclosed seas is the product of immigration and extinction. No species are yet known to have become extinct in the Baltic Sea in historical times, whereas immigration continues through both natural dispersal and human-mediated introduction of species. The invasion rate for the Baltic Sea region was approximately one NIS every year over the period 1998-2000 (Baltic Sea Alien Species Database 2002).

Several invasion corridors, other than ship traffic, open into the Baltic Sea. The sea and its drainage area are connected to the Ponto-Caspian brackish seas (Black, Azov and Caspian Seas) by rivers and canals. Some 250 rivers discharge fresh water into the Baltic from a drainage area that is 4 times greater than its sea surface itself. Consequently, every single NIS, released into the wild somewhere in the drainage basin, can be transported to the sea or its most diluted coastal areas. Today the Baltic is exposed to other brackish- and fresh-water biota of the world, due to reduced geographical barriers (i.e. ships' traffic and the century-long history of intentional introductions). In brackish-water conditions, the ability of organisms to live and reproduce at the low salinity is a key factor, also in the Baltic. Highly euryhaline species of both marine and fresh-water origin are potential invaders. Consequently, there is a substantial pool of both intra- and intercontinental NIS already established in adjacent water bodies. In the peripheral parts of the drainage area of the Baltic, they consist mainly of the Ponto-Caspian element whereas the northwest European river mouths harbor a number of marine and brackish-water NIS native to other seas. Because of the brackish conditions and low winter temperatures, the Baltic Sea is often thought to be relatively well protected against species invasions. It is however important to keep in mind that this protected status is only hypothetical. Many major harbors, even along fully oceanic coasts, are situated in brackish waters (e.g. river mouths and estuaries). Ballast water may be loaded in the brackish part of a harbor, to be discharged later somewhere in the brackish Baltic Sea; thus, the risk of successful species introductions appears to be relatively high.

The Baltic has a long history of human mediated invasions (see Baltic Sea Alien Species Database 2002 for references). The bivalve *Mya arenaria* probably appeared already in the 12-13th centuries in Danish waters. First benthic species recognized as non-native was the barnacle *Balanus improvisus* found in 1844 in former Königsberg (now Kaliningrad) area of the Vistula lagoon. Introduction for stocking purposes of lobster, crabs, oysters and blue mussels from Denmark to the Finnish and Estonian coasts was planned by Russian authorities as early as in the mid-1700s (Hamel 1852). Delivery of settlings failed, however, and further efforts were questioned by scientists because of the low salinity. Ponto-Caspian species (the bivalve *Dreissena polymorpha*, and probably the gastropod *Lithoglyphus naticoides* and the hydrozoan *Cordylophora caspia*) first appeared in the southeastern lagoons of the Baltic Sea in the early 1800s after opening of the canals (e.g., Neman - Dniepr and Vistula - Bug in the end of the 1700s). In the early 1960s, a number of Ponto-Caspian mysids and gammarids (*Paramysis lacustris*, *Limnomysis benedeni*, *Hemimysis anomala*, *Corophium curvispinum*, *Pontogammarus robustoides*, *Obesogammarus crassus*, *Chaetogammarus ischnus*, *C. warpachowskyi*) were successfully intentionally introduced into the Lithuanian waters, including the Curonian Lagoon (Gasiunas 1963).

Since the early 1800s, about 103 NIS have been recorded in the Baltic Sea (the Kattegat included) most of them being introduced by shipping (ballast water or hull fouling), or spread from their primary sites of introduction in adjacent freshwater bodies (Leppäkoski & Olenin 2000a; see also Baltic Sea Alien Species Database 2002 for full list of species). It is assumed that some 70 species have been able to establish and maintain self-sustaining populations (Table 1). NIS are abundant and even dominant throughout the shallow benthic and fouling communities of the Baltic Sea - at present, no shallow-water habitat is entirely free of human-mediated invaders. Their number is lowest in the northern part of the Gulf of Bothnia and highest in the coastal lagoons of the southern and southeastern Baltic as well as in the Belt Sea and Kattegat area (Leppäkoski & Olenin 2001). The most important source areas for these species have been western European waters, the Atlantic coast of North America and the Ponto-Caspian realm. In addition, at least 5 species are listed as cryptogenic to the area (e.g., the dinoflagellate *Prorocentrum minimum* and the ship worm *Teredo navalis*).

Table 1. Number of nonindigenous species recorded in the Baltic Sea (Kattegat included) 1800-2001. W = in the most parts or in the whole Baltic; R = within one of the Baltic sub-regions (see Gollasch & Mecke 1996; Leppäkoski & Olenin 2000a and Baltic Sea Alien Species Database 2002 for further data).

Ecological or taxonomic group	Number of species recorded	Established species W(R)*	Ecological or taxonomic group	Number of species recorded	Established species W(R)*
Phytoplankton	8	0? (8)	Mollusca	12	9 (5)
Phytobenthos	9	9 (3)	Bryozoa	1	1 (1)
Invertebrates			Tunicata	1	1 (1)
Cnidaria	6	2? (1)	**Vertebrates**		
Platyhelminthes	2	2 (2?)	Pisces	30	10? (4)
Nematoda	1	1 (1)	Aves	1	1 (1)
Annelida	7	7 (2)	Mammalia	2	2 (2)
Crustacea	21	19 (5)			
			TOTAL	103	68? (33?)

There are very few primary introductions known from the Baltic Sea (e.g., the fish hook water flea *Cercopagis pengoi* and some Pacific salmonids *Oncorhynchus* spp.); the Baltic has been and still is subject to secondary introductions from both the North Sea area and adjacent inland waters. Based on published first findings, the minimum rates of secondary spread within the Baltic basin were estimated (Leppäkoski & Olenin 2000a) for the barnacle *Balanus improvisus* from Königsberg (1844) to Turku (1868) 30 km a^{-1}, for the North American polychaete *Marenzelleria viridis* (German Boddens 1985 to Lithuania 1989) 170 km a^{-1}, further to the south coast of Finland (1990) 480 km a^{-1}, and to the Northern Quark (1996) 90 km a^{-1}, and for the mud snail *Potamopyrgus antipodarum*, native to New Zealand (from Wismar Bight, Germany to Gotland, Sweden 1920) 20 km a^{-1}, to the Åland Islands (1926) 50 km a^{-1}, and to Bothnian Bay (1945) 30 km a^{-1}.

Until very recently (up to mid-1980s), most of the abundantly occurring invaders in the Baltic Sea were benthic organisms. Since then, several planktonic (e.g., *Cercopagis pengoi*) and benthic species having pelagic life-stages (*Marenzelleria* cf. *viridis*), appeared in the Baltic Sea. Receiving areas in the Baltic Sea, which are at high risk from NIS introductions, are the Gulf of Finland, the Gulf of Riga, the coastal lagoons and the German boddens. All these areas are known as centres for xenodiversity (Gr. *xenos* = strange; Leppäkoski & Olenin 2000b), i.e. areas that host many well-established NIS (Gruszka 1999; Panov et al. 1999; Olenin et al. 1999; Leppäkoski & Olenin 2000a, 2001). These "hot spots" serve not only as entrance gates for many invasions into the Baltic but they also function as bridgeheads for secondary introductions that are carried towards the inner parts of the Baltic either naturally, by currents, or by regional ship traffic and recreational craft.

Few estimates are available for the proportion of NIS in relation to the total number of species in different parts of the Baltic. In the brackish waters of the German Baltic Sea coast, about 450 bottom-living species have been recorded; of these 15 species or 3% are non-native (Nehring 1999). In a benthos study in the eastern Bothnian Sea, 22 species were recorded, among them four NIS (18%). In the Curonian Lagoon (Lithuania), 16 (17%) of the about 95 benthic animal species recorded are nonindigenous; in the oligohaline part of the lagoon, the ratio would be 16 to 55 (29%) (Olenin & Leppäkoski 1999). An updated list of species found in the Baltic and its sub-regions is available as an interactive database (Baltic Sea Alien Species Database 2002; see Jazdzewski (1980), Kinzelbach (1995), Tittizer (1996) and Jazdzewski & Konopacka (2000) for detailed information of NIS in the rivers and lakes of the drainage area of the Baltic).

3 Impacts and consequences

Functional (ecological) and economic aspects of biological invasions are yet insufficiently studied in the Baltic Sea, therefore it may seem that many of nonindigenous species found in the region have had no obvious effect on the native environment and, consequently, there has been no known impact on human uses of the sea. However, it is clear that at least some functions of the non-native animals are new (and hence unique) for the species-poor Baltic Sea ecosystem. As examples of novelty in functions serve: (i) the mud snail *Potamopyrgus antipodarum* (surface deposit feeding on extremely soft bottoms where native gastropods *Hydrobia* spp. do not occur); (ii) the zebra mussel *Dreissena polymorpha* (filter feeding in oligohaline and freshwater parts of the coastal lagoons where the blue mussel *Mytilus edulis* is absent); (iii) the barnacle *Balanus improvisus* (suspension filter feeding in the uppermost hydrolittoral zone); (iv) the polychaete *Marenzelleria* cf. *viridis* (deep bioturbation of the sediment); (v) the hydroid *Cordylophora caspia* (sessile raptorial suspension feeding); (vi) the crabs *Rhithropanopeus harrisi* and *Eriocheir sinensis* (epibenthic invertebrate predators and scavengers in the diluted parts of the inlets where native marine decapod shrimps do not occur); (vii) the mysidacean *Paramysis lacustris* (nektobenthic crustacean in the inner parts of some lagoons where native marine mysids are absent); (viii) B. *improvisus* and *D. polymorpha* play a role as microhabitat engineers - their empty shells create patches of hard substrate for sessile species on uniform soft bottoms (Olenin & Leppäkoski 1999).

Studies on selected species, especially in recent years, proved that NIS in the Baltic Sea area compete with native species for food and/or space (e.g. case studies on *Cercopagis pengoi*, *C. caspia*, *D. polymorpha*, *Gammarus tigrinus*, *M.* cf. *viridis*, *Pontogammarus robustoides*), they became numerically dominant in native communities (*Acartia tonsa*, *B. improvisus*, *C. pengoi*, *M.* cf. *viridis*, *Mya arenaria*), they change energy/matter flows between pelagic and benthic compartments and modify trophic structure of invaded ecosystems (*A. tonsa*, *C. caspia*, *Coregonus nasus*, *D. polymorpha*, *M. arenaria*, *Neogobius melanostomus*), transfer parasites and diseases to local species (*Coregonus nasus*, *Pacifastacus leniusculus*).

NIS have no direct value of food resources in the Baltic as none of them support commercial fisheries and invertebrates are not harvested for food because of their small size. Some planktonic invaders (e.g., the fishhook water flea *C. pengoi*) and planktonic larvae of several benthic NIS have a high value as food source for commercially harvested fish such as the Baltic herring (e.g. Antsulevich & Välipakka 2000). However, on the basis of existing knowledge approximately 20 NIS (i.e., less than 30% of all introduced species) can be classified as nuisance organisms in the Baltic; only 7 of them have caused significant damage (Leppäkoski 2002). These are three Ponto-Caspian species (*C. caspia*, *C. pengoi*, and *D. polymorpha*), two North-American species (*B. improvisus* and the American mink *Mustela vison*), the Japanese swim-bladder nematode *Anquillicola crassus* and the "ship worm" mollusc *Teredo navalis*, believed to be of Indo-Pacific origin. The economic impacts of NIS have rarely been quantified. The clogging of reels and fouling of nets makes *C. pengoi* a potential nuisance species. This may cause substantial economic loss in fisheries. The estimated loss in one fishery enterprise in the eastern Gulf of Finland, in average during 1996–1998, was at minimum USD 50,000, caused by the drastic decline in fish catches in the coastal zone due to fouling of fishing equipment by *C. pengoi* (Panov et al. 1999). During the exceptionally warm summer in 1999, clogging of fishing equipment by *C. pengoi* became a serious problem in the eastern Gulf of Finland, the inner parts of the Archipelago Sea, Finland, the northern Bothnian Sea and Lithuania (Gasiunaite & Didziulis 2000; Leppäkoski 2002). The cryptogenic ship worm *T. navalis* is now fully established in the southwestern Baltic region. It caused approximately USD 15 to 25 million damage to submerged wooden installations along the German Baltic coast, since 1995 (K. Hoppe pers. comm.). The ship worm also causes damage to marine archaeological objects.

Introduction of alien species interferes with research and monitoring. Establishment of any single NIS in a novel region and ecosystem opens opportunities to ecological research. These "transplantation experiments" can be used for studies of concepts such as adaptive strategies, niche dimensions, interspecific relationships, dispersal mechanisms etc. The Baltic Sea with its low number of indigenous species and relatively simple food webs offers excellent opportunities for detailed studies in invasion biology. However, species introductions may also result in reduced research possibilities in causal biogeography (it is difficult to explain causalities behind present distribution of species) and population genetics. Benthic communities of the Baltic Sea have been monitored by quantitative methods since the 1910s. The results of this effort, based on international sampling programmes, may be invalidated by the establishment of any successful NIS, which becomes dominating, utilizes the space and energy resources in different manner and rate, and re-structures the food webs. For example, the soft-bottom community was totally changed by the polychaete *M. viridis* in the Vistula Lagoon, where it became a dominant species in sandy and muddy habitats in mid-1990s, reaching 216 g m^{-2} and making up to 95% of total benthic community biomass (Zmudzinski et al. 1997). Positive economic impacts include the recreational resources provided for sport fishing by some nonindigenous fish, such as the rainbow trout (escapees from fish farms along the Finnish coast), the round goby (in the Gulf of Gdansk) and hunting (the muskrat, trapped for their fur, and the Canada goose).

4 The Baltic Sea as a donor area of nonindigenous species

The Baltic Sea acts as donor of alien species for e.g., the North American Great Lakes. The predatory cladoceran *Bythotrephes longimanus*, native to Lake Ladoga and the Neva Estuary, invaded the Great Lakes in ballast water of a ship from the port of Leningrad (St. Petersburg) in the early 1980s (Berg et al. 1998). Likewise, *Cercopagis pengoi* was introduced into the Great Lakes in 1998 (MacIsaac et al. 1999), almost certainly carried in ballast tanks the eastern Baltic (Cristescu et al. 2001). *C. pengoi* was among the species of Ponto-Caspian origin, which were predicted to invade the Great Lakes based on ships' traffic from key European to Great Lakes' ports (MacIsaac 1999). The diatom *Thalassiosira baltica* (first found in Lake Ontario in 1988) likely originated from the Baltic Sea, though other sources were also possible (Edlund et al. 2000).

5 The formative years of the biology of invasions in the Baltic Sea

This brief review demonstrates that the biota of the Baltic Sea, in spite of its isolated state, is closely linked with the oceans of the world. The present knowledge of introduction of non-native species into the Baltic Sea has developed through a stepwise process from first records of single new species (e.g. Luther 1927; Schlesch 1937; Gislén 1950) toward more sophisticated studies in invasion biology (Fig. 1). Nikolaev (1951) and Segerstråle (1957) were the first to draw attention to the changes in the Baltic Sea fauna, related to human-mediated introduction of NIS. The first inventories (Leppäkoski 1984; Jansson 1994) revealed 35-40 established NIS. The first PhD worldwide based on ballast water sampling of ships was published in Germany (Gollasch 1996). Further, in 1999 the German group Neobiota was formed dealing with exotic species in Germany, including the Baltic shores.

International cooperation in the field of invasion biology started in 1994, when a working group on estuarine and marine nonindigenous organisms was established by the Baltic Marine Biologists (BMB), a non-governmental scientific organisation. Its systematic work on comprehensive inventories resulted in the first Internet-based inventory, issued in 1997, where 78 nonindigenous species (both established and occasional) were listed.

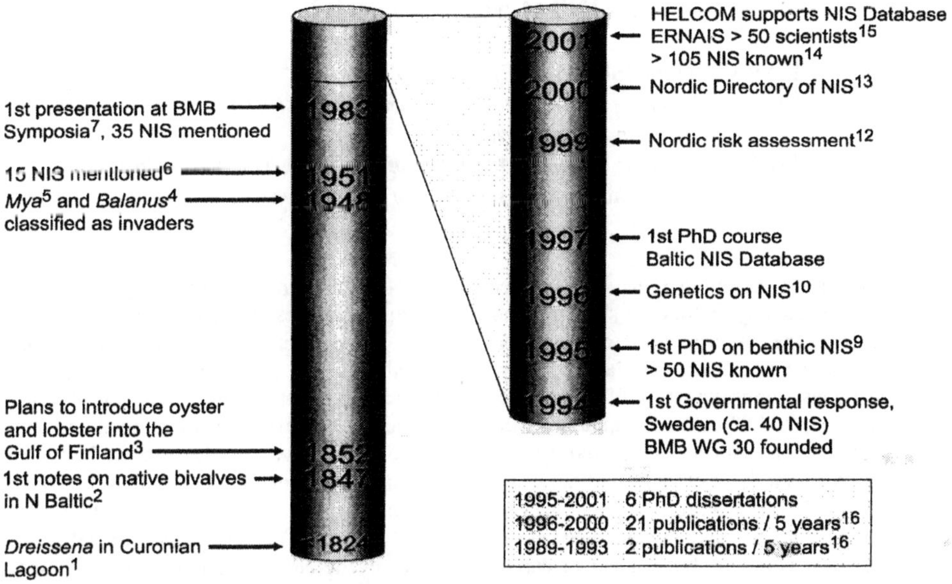

Figure 1. Timeline of research into invasion biology in the Baltic Sea countries. 1) Nikolaev (1963), 2) von Siemaschko (1847), 3) Hamel (1852), 4) Hägg (1948), 5) Hessland (1946), 6) Nikolaev (1951), 7) BMB - The Baltic Marine Biologists; Leppäkoski (1984), 8) Jansson (1994), 9) Fritzsche (1995), 10) Röhner et al. (1996), 11) Gollasch & Leppäkoski (1999), 12) Weidema (2000), 13) Baltic Sea Alien Species Database (2002), 14) ERNAIS (2001; European Research Network on Aquatic Invasive Species), 15) Baltic Marine Environment Bibliography 2002.

Traditionally a major part of marine biological research in this area has been focused on studies of distribution of species, structures of populations and communities, as well as links within and between different sub-systems in relation to natural environmental gradients. A major task for invasion biology is to investigate how these gradients modify the effects of invaders on the indigenous biological systems. The invasion of the North American polychaete *Marenzelleria* cf. *viridis* in the mid-1980s and the Ponto-Caspian predatory fishhook water flea *Cercopagis pengoi* in the early 1990s, in addition to tens of historical and recent invaders already present in the Baltic, have had a major impact on the structure and function of both the benthic and pelagic subsystems. These recent invaders have also increased both the common and scientific awareness of bioinvasions. Increased understanding of the ecological and economic impacts of NIS and effective monitoring of their spread are essential elements in the study of xenodiversity. Interpretation of functional changes related to NIS at a regional or basin-wide scale is a difficult, but essential task for the near future to assess the total impact of NIS. Due to its low number of native species and simplified food webs, the Baltic Sea offers unique opportunities for detailed studies on the autecology of NIS and, in particular, their interspecific relations with indigenous fauna and flora. Even if the modes of impacts are understood for a number of these species, quantitative data on impacts are rare and very little is known about their economic consequences and impact on human activities.

INTRODUCED MARINE SPECIES OF THE NORTH SEA COASTS

KARSTEN REISE[1*], STEPHAN GOLLASCH[2] & WIM J. WOLFF[3]
[1]*Alfred-Wegener-Institut für Polar- und Meeresforschung, List, Germany*
[2] *GoConsult, Hamburg, Germany*
[3]*Rijksuniversiteit Groningen, Mariene Biologie, The Netherlands*
*Corresponding author *kreise@awi-bremerhaven.de*

Abstract
Approximately 80 nonindigenous species occur in the North Sea in self-sustaining populations. The main introducing vectors are shipping and aquaculture activities. Most invertebrate invaders originate from the Atlantic coast of America and were predominantly introduced by shipping, while most algae stem from the Pacific being introduced with live oysters imports. At the open North Sea coasts approximately 6% of the macrobenthic species are exotics, while in estuaries their share is up to 20%. About a quarter of the established non-natives are widespread and some inshore habitats are entirely dominated by exotics. However, the overall effect on the North Sea ecosystem seems to be more additive rather than one of displacement, suggesting that the coastal habitats of the North Sea are capable of accommodating newcomers. It has to be noted that this is no guarantee that the next invader may pose a negative impact. There is a need to reduce the number of new invaders, but current research on exotics in the North Sea is regarded as inadequate.

1 Introduction

Non-native species of the North Sea are here defined as species distributed outside the Atlantic coast of Europe (Gibraltar to North Cape). We further restrict this overview to species likely being introduced via anthropogenically supported vectors (intentional and unintentional introductions). North Sea boundaries are defined following the North Sea Task Force (1993), which includes the Channel region in the south, the Skagerrak and Kattegat in the east, and the Shetland Islands in the north. Species in brackish habitats were included when a salinity of > 5 PSU is regularly encountered at the sites.

This contribution briefly summarises invaders in the North Sea. A more extensive review is given by Reise et al. (1999).

2 Exotics in the North Sea

The number of invaders that established in the North Sea amounts to about 80 species (Table 1). This total number is lower compared to estuarine regions in North America, i.e., Chesapeake Bay (116 species), Great Lakes (137) and San Francisco Bay (212) (reviewed in Ruiz et al. 1997), but some of these studies include organisms from freshwater as well as terrestrial shores and wetlands, while this report focuses on marine and brackish water organisms. The majority of exotics in the North Sea are invertebrates (47), primarily crustaceans, molluscs, polychaetes and hydroids (Table 2). Introduced macroalgae comprise 20 taxa, mostly red and brown ones. *Chattonella* cf. *verrucolosa* (Raphidophyta) was first recorded from the North Sea in 1998 but it is not proven that the species forms a self-sustaining population in the region (Wallentinus pers. comm.).

E. Leppäkoski et al. (eds.), Invasive Aquatic Species of Europe, 260–266.

Only 12 species of protists are among the assumed exotics of the North Sea. Of the 80 species (Table 1), 22 occur preferably in brackish waters; "hotspots" of introduced species are ports and the vicinity of aquafarms. A frequent mode of life is the occurrence as epibionts, mostly on molluscs and algae (27 species). Six introduced species are parasites. Dwelling in or on coastal mud and sand are 17 of the exotic species, and about 10 occur at rocky shores of the open coast. Only 10 holoplanktonic species have been recognized as exotics. However, in this group a distinction between introduction and natural drift by oceanic currents is often impossible.

Almost all invaders were introduced accidentally including the marsh grass *Spartina anglica* and the amphipod *Gammarus tigrinus*. Both species were further distributed deliberately. The bivalves *Mercenaria mercenaria*, races of *Ostrea edulis, Crassostrea gigas, C. angulata* and *C. virginica* were deliberately introduced but the latter two did not establish reproducing populations. This is different from the situation in the adjacent freshwater habitats of central and western Europe where nearly half of the 80 known invaders were intentionally introduced, particularly fish (Kinzelbach 1995).

The share of exotics in the North Sea biota increases from the offshore part towards the coast, and there it increases further from the open coast towards the estuaries. In the latter, the percentage of exotic species is about 20 (Wolff 1998). In the waters around the island of Sylt, which are not estuarine, the macrobenthos of approximately 300 species includes 20 exotics (Reise & Lackschewitz 1998 and unpubl. data). A share of roughly 6% is also estimated for the Dutch coast (Wolff 1998). In offshore areas, only a few exotic species occur in the plankton. An explanation for this pattern is to be found in the availability of vectors. Mariculture and ships transport mainly organisms from coast to coast.

3 Origin and arrival

In terms of homoclimatic conditions, distance and frequency of seafaring, one would expect a clear dominance of exotics emanating from the other side of the Atlantic. This is the case with the invertebrates, particularly molluscs and polychaetes (Table 2). In contrast phytoplanktonic and macroalgal exotics are almost exclusively of Pacific origin. The macroalgae of the cold and warm temperate coasts are much poorer in endemics in the western than the eastern Atlantic (Lüning 1990). Thus, there are not many new species to receive from America. The North Pacific region, on the other hand, is very rich in endemic algal species, and the substantial imports of Pacific oysters (Chew 1990) provided a suitable vector for attached algae and their spores. The algae usually appeared first around French oyster farms, and from there most spread by their own means into the North Sea (Maggs & Stegenga 1998).

Although some invertebrates came by the same route, most travelled by ship (Table 2), either as larvae in ballast tanks or as adults attached to hulls. In these cases, the shorter distance and higher frequency of shipping across the Atlantic to the North Sea compared to the long voyages from the Pacific may be decisive. Furthermore, voyages from the Pacific region to Europe need to pass tropical areas, which may affect the survival rates of species adapted to cold-temperate regions such as northern Japan. In addition, several invertebrates were introduced with American oysters (*Crassostrea virginica*) to the British Isles and the North Sea (Hedgepeth 1980; Eno et al. 1997). In contrast to macroalgae, the macrobenthic fauna along the North American Atlantic shores do not seem to be less diverse than its counterpart on the European Atlantic shore (Briggs 1974). Up to 32 species are assumed to be accidentally introduced with American and Pacific oysters' shipments (Table 1). Almost all others were probably unintentionally introduced by shipping. About 20 species have also invaded other coasts outside the European Atlantic, and thus seem to have an intrinsic aptitude to become introduced and then established. These species are tramping from port to port and estuary to estuary by ship or have followed the worldwide trade with Pacific oysters (*Crassostrea gigas*).

Table 2. Number of introduced species which became established in the North Sea, and their assumed origin and mode of transport. Where two alternatives are assumed to be equally alike, a species is counted twice, if unknown no entry is made (after Reise et al. 1999).

Major group	Number of species	Origin		Imported with	
		Atlantic	Pacific	ship	aquaculture
Phytoplankton	9	1	8	5	3
Macroalgae	20	1	18	5	16
Poaceae	1	1	-	1	-
Protozoa	3	3	-	-	3
Cnidaria	8	2	3	7	2
Mollusca	11	8	3	4	5
Annelida	9	5	3	8	3
Crustacea	15	7	8	10	3
Other invertebrates	6	1	4	5	1

The majority of introduced exotics became first established in the southern North Sea and about half of them have not been recorded in the north. This asymmetry may be caused by three phenomena. (i) Nearly all important ports of the North Sea are in the southern half. (ii) Species associated with introduced oysters tend to originate from regions with higher temperatures than in the northern North Sea, and most of these introductions occurred through the gateway of the French and southern British coasts (Korringa 1976a,b; Chew 1990). (iii) The cold-temperate flora and fauna in the eastern and western provinces of the northern Atlantic have many species in common, while the warm-temperate biota have no close relationship between the opposite coasts because both have been geographically isolated so much longer than the biota further north (Briggs 1974; Watling 1979; Lüning 1990). Thus, there is a considerably larger species pool in the warm-temperate region from which recognizable introductions may be drawn. Consequently, under climatic warming the importance of exotics in the entire North Sea could increase. There is a peak of new records in the 1970s, which may partly be explained with imports of *Crassostrea gigas* and its associated organisms (Fig. 1). These imports occurred to France, Britain and the Netherlands in the 1960s to 1970s (Korringa 1976b). Also the components of antifouling paints have been changed in the 1970s. The new hull coating which contains tributyltin (TBT) might be more effective compared to previously used compounds. This may have prevented a further increase in the rate of introductions since the mid 1970s (Minchin & Sheehan 1995).

4 Impact

Up to now, there is no evidence that exotics have driven natives to extinction. Thus, the species number in the North Sea would have been increased, if there were no extinctions caused by other means, e.g. fishing, hunting, habitat alterations or toxic substances. Locally, the numerical net balance may come close to zero (e.g. Reise 1982), but the type of organisms is not congruent. As has been reviewed elsewhere (Williamson 1996), not all but most of the successful invaders are generalists. Those taken up by ships should be common and abundant at harbour sites. Accordingly, they are often estuarine and well adapted to human-modified environments. To survive a long voyage, high physiological tolerance is needed, at least for some stage of the life cycle.

The main effect of species introductions into the North Sea seems to be a species addition rather than an exchange. Some local displacements have occurred, however. A clear example is the marsh grass *Spartina anglica*, establishing monostands where other saltmarsh vegetation or seagrass had been before, and which greatly modifies the habitat (Gray et al. 1991; Meesenburg 1975). In the Dutch Delta region *S. anglica* even displaced the former stands of *S. maritima* (almost) completely. In England, however, *S. anglica* and *S. maritima* rarely co-occurred, and the latter is assumed to have declined since the 1930s because of land reclamation, coastal erosion and displacement by *Halimione portulacoides* (Gray et al. 1991). *S. maritima* may itself be an early introduction to Europe from Africa (Chevalier 1923). It hybridised with the North American *S. alterniflora* near Southampton around 1870, where the latter was first recorded in 1816. From the sterile hybrid *S.* x *townsendii,* the new fertile species *S. anglica* originated by chromosome doubling. This is at present a wide-spread species in the North Sea region, but remnant populations of *S. maritima, S. alterniflora* and *S.* x *townsendii* still exist.

Near the island of Sylt in the eastern North Sea, the introduced razor clam *Ensis americanus* first occurred in 1979. In a sandy sediment close to spring low tide line, macrofaunal biomass comprised 118 g dry organic weight m^{-2} in March 1993 (Armonies & Reise 1998). Of this biomass, 66% was contributed by the razor clam and another 30% by the ragworm *Nereis virens*, suspected to be an introduced species as well. The site was formerly occupied by a dense eelgrass meadow which completely fell victim to an invasive disease caused by the protist *Labyrinthula zosterae* in the 1930s (Wohlenberg 1935; den Hartog 1987), and never came back. The former benthic fauna was composed mainly of *Lanice conchilega, Psammechinus miliaris* and *Littorina littorea* and its biomass was probably not much higher than the 28 g m^{-2} of an adjacent lugworm flat in March 1993 (Reise unpubl.). This example illustrates the pervasiveness of introductions on coastal North Sea biota, and the history-dependence of present-day assemblages.

About 20 species of those listed in Table 1 are today widespread in the North Sea and are at least locally abundant. Except for some parasites and the shipworm *Teredo navalis*, economic implications of the known introductions into the North Sea are low up to now. Early in the 18th century the wood-boring *T. navalis* caused a revolution in Dutch coastal defence. Before its supposed introduction, seawalls were reinforced with wooden posts, while since then stones and other means had to be used. *T. navalis* continues to cause trouble in wooden structures build into the sea.

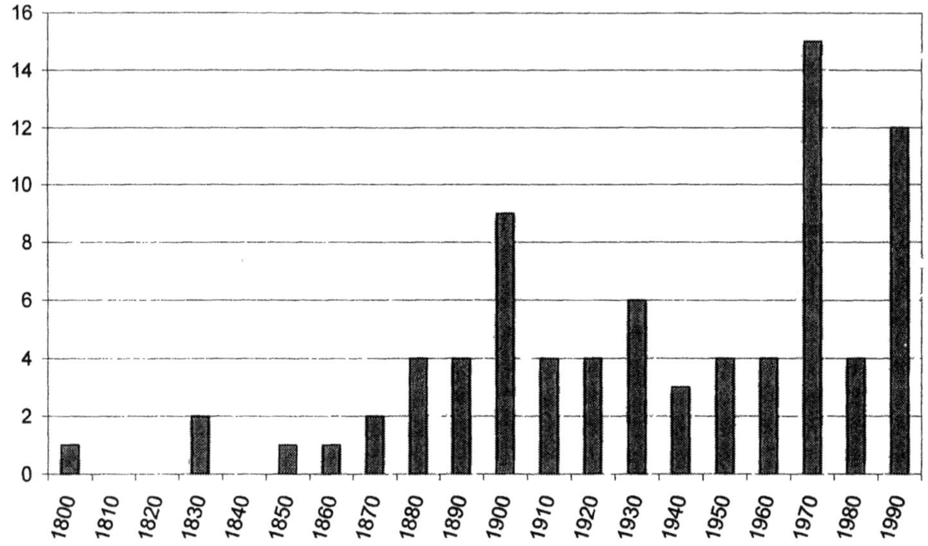

Figure 1. Timeline of recorded North Sea introductions (N = 80).

5 Consequences

Inadvertent species introductions have irreversibly modified the North Sea ecosystem. This is a cumulative process much faster than any natural change could ever accomplish, and enforces a progressive dethronement of the native organisms from the centre of their accustomed ecological communities. Qualitatively and may be also quantitatively, the combined effect of introduced neophytes and neozoans exceeds the more often considered environmental effects of eutrophication and toxic substances, and may rival those of the fishery in the North Sea and of habitat loss along its coastline. In view of the fact that in the North Sea exotic species have been primarily additions rather than the cause of displacements or even extinctions, problems with the exotics may reside less in ecology but more in economy or human health. As introductions elsewhere have shown, there is always an element of risk when new species arrive. To minimize these risks, existing quarantining protocols for maricultural organisms and treatment measures for ballast water should be strictly followed. It is false to say that every species that could have been introduced would be in the North Sea by now. The timeline of introductions does not suggest a saturation level (Fig. 1). The chance of a species to become introduced, established and then to become a serious problem for the environment or economy at the recipient coast is small. But, one single introduced species may be able to cause severe ecological change and economic damage, and this might be the next species about to arrive.

Acknowledgements

We wish to thank the participants of the workshop on *Exotics of the North Sea Shore*, held on Sylt in February 1998, for contributing facts and ideas to this review.

Table 1. List of non-native species established in the North Sea including the Channel and Kattegat. Origin: eastern Atlantic outside North Sea (EA), western Atlantic (WA), southern Atlantic (SA), northern Pacific (NP), southern Pacific (SP), Indo-Pacific (IP); year in brackets: no longer present because habitat has changed; Transport known or assumed to have occurred with aquaculture (A) or ships (S); Status: occurring along most of the coasts (+++), in part of the area (++), at one or a few localities (+), primarily in freshwater but extending into brackish estuaries and lagoons (b); ? species of uncertain taxonomic status or which may have arrived also by natural means; References are limited to 1 or 2 sources where further information may be obtained: (1) Chapman 1999, (2) Elbrächter 1999, (3) Eno et al. 1997, (4) Essink 1999, (5) Fletcher & Farrel 1999, (6) Hayward & Ryland 1990, (7) Lauckner 1983, (8) Maggs & Stegenga 1999, (9) Nehring 1998, (10) Stock 1993, (11) Thiel 1968, (12) Wolff 1999, (13) Zibrowius & Thorp 1989, (14) Gollasch & Riemann-Zürneck 1996, (15) Nijland 2000, (16) Dahl 1946, (17) Dyrynda et al. 2000 (after Reise et al. 1999).

Taxon	Origin	First record	Vector	Status	Reference
Dinophyceae (dinoflagellates)					
? *Gymnodinium mikimotoi* Miyake & Kominami (syn. G. aureolum, G. nagasakiensis)	NP	1966		+++	2,9
Alexandrium leeii Balech	NP	1995		+	2
Raphidophyceae					
Fibrocapsa japonica Toriumi & Takano	NP	1991		++	2,9
Chattonella sp.	NP	1991		++	2,9
Bacillariophyceae (diatoms)					
Odontella (*Biddulphia*) *sinensis* (Greville)	NP	1903	S	+++	3,9
Thalassiosira punctigera Castr.	NP	1978	A/S	+++	3,9
Thalassiosira tealata Takano	NP	1950	A/S	++	3,9
? *Thalassiosira hendeyi* Hasle & Fryxell	SA	1978	S	++	3,9
Coscinodiscus wailesii Gran & Angst	NP	1977	A/S	+++	3,9
Phaeophyceae (brown algae)					
Sargassum muticum Fensholt	NP	1960s	A	+++	3,5
Undaria pinnatifida Suringer	NP	1986	A/S	++	3,5
? *Fucus evanescens* C.Ag.	NP	1902		+	5
Colpomenia peregrina Hamel	IP	1905	A	+++	3,5
Corynophlaea umbellata Kützing	NP	1990	A	+	5
Rhodophyceae (red algae)					
Bonnemaisonia hamifera Hariot	NP	1890	A	+++	3,8
Asparagopsis armata Harvey	SP	1950	A	+	3,8
Grateloupia doryphora Howe	NP	1969	A	+	3,8
Grateloupia luxurians Gepp	IP	1947	A	++	3,8
? *Agardiella subulata* Kraft & Wynne	?	1973	A	+	3,8
Antithamnionella ternifolia Lyle	SP	1926	S	++	3,8
Antithamnionella spirographidis Schiffner	NP	1906	S	++	3,8
Dasya baillouviana Montagne	WA?	1950	A	+++	8
? *Dasysiphonia* sp.	NP	1994	A	+	8
Anotrichium furcellatum Baldock	NP	1976		+	8
Polysiphonia senticulosa Harvey	NP	1993	A	+	8
Polysiphonia harveyi Bailey	NP	1908	A	+++	3,8
Chlorophyceae (green algae)					
Codium fragile ssp. *atlanticum* Silva	NP	1839	A	++	1,3
Codium fragile ssp. *tomentosoides* Silva	NP	1900	A/S	+++	1,3
Codium fragile ssp. *scandinavicum* Silva	NP	1919	A/S	++	1
Poaceae (grasses)					
Spartina anglica Hubbard (hybrid)	WA/EA	1890s	S	+++	3
Ascetospora (parasitic protozoans)					
Bonamia ostreae Pichot et al.	WA	1982	A	+	7
Marteilia refringens Grizel et al.	EA	1970s	A	+	7
Haplosporidium armoricanum (van Banning)	EA	1970s	A	+	7
Hydrozoa (hydroids)					
Cordylophora caspia (Pallas)	Ponto-casp.	1884	S	b	12
Gonionemus vertens Agassiz (syn. *murbachi* Mayer)	NP	1913	A/S	+	3
Garveia (*Bimeria*) *franciscana* Torrey	?	1920	S	+	12
Clavopsella navis (Millard)	?	1973	S	+	3
Nemopsis bachei Agassiz	WA	1905	S	+	11
Anthozoa (sea anemones)					
Nematostella vectensis Stephenson	WA	1935		+	6
Haliplanella lineata (Verrill) (incl. *H. luciae* Verrill)	NP	1896	S	++	3,12

Taxon	Origin	First record	Vector	Status	Reference
Diadumene cincta (Stephenson)	?NP	1925	A/S	++	9,14
Bivalvia (lamellibranchs)					
Crassostrea gigas (Thunberg) (incl. *C. angulata*)	NP	1964	A	++	3
Ensis americanus (Binney) (syn. *directus* Conrad)	WA	1978	S	+++	3
Mytilopsis (*Congeria*) *leucophaeta* (Conrad)	WA	1835	S	b	12
Mya arenaria (L.)	WA	1250?		+++	3
Mercenaria mercenaria (L.)	WA	1864	A	+	3,12
Petricola pholadiformis Lamarck	WA	1890	A	+++	3
Teredo navalis L.	IP	<1800	S	+++	12
Gastropoda (snails)					
Corambe batava Kerbert (*C. obscura*)	WA	[1886]		+	12
Crepidula fornicata (L.)	WA	1887	A	+++	3
Urosalpinx cinerea (Say)	WA	1900	A	+	3
Potamopyrgus antipodarum (Gray) (syn. *jenkinsi*)	SP	1883	S	b	3,12
Polychaeta					
Clymenella torquata (Leidy)	WA	1936	A	+	3
Marenzelleria cf. *viridis* (Verrill)	WA	1996	S	+	4
Marenzelleria cf. *wireni* Augener	WA	1982	S	++	4
Janua brasiliensis (Grube)	WA	1974	S	+	3,13
Pileolaria berkeleyana (Rioja) (syn. *rosepigmentata*)	NP	1974	S	++	3,13
Ficopomatus (*Mercierella*) *enigmaticus* (Fauvel)	SP	1921	S	++	12,13
Hydroides dianthus (Verrill)	WA	1970	A/S	+	3,13
Hydroides ezoensis Okuda	NP	1976	A/S	+	3,13
Hydroides elegans (Haswell)	?	1937	S	+	12,13
Crustacea					
Balanus amphitrite Darwin	IP	1937	S	+	3,12
Balanus improvisus Darwin	WA	1850	S	++	12
Balanus eburneus Gould	WA	1900	S	+	12
Elminius modestus Darwin	SP	1943	S	+++	3,12
Eriocheir sinensis Milne-Edw.	NP	1912	S	++	3,12
Brachynotus sexdentatus (Risso)	EA	?	S	+	6
Rhithropanopeus harrisii (Gould)	WA	1870s	S	b	3, 12
Callinectes sapidus Rathbun	WA	1932	S	+	12
Hemigrapsus penicillatus (de Haan)	NP	2000	S	+	15
Caprella macho Platvoet et al.	?	1995	A/S?	+	12
Corophium sextonae Crawford	SP	1930s	A/S?	+++	3
Platorchestia platensis (Kröyer)	WA?	1931	S?	++	16
Eusarsiella (*Sarsiella*) *zostericola* (Cushman)	WA	1940	A	+	3
Acartia tonsa Dana	WA,IP	1916	S	+++	3
Mytilicola orientalis Mori	NP	1992	A	+	10
Mytilicola ostreae (Hoshina & Sugiura)	NP	1992	A	+	10
Pycnogonida (sea spiders)					
Ammothea hilgendorfi (Böhm)	NP	1978	S	+	3
Bryozoa					
Bugula neritina (L.)	WA	1973	S	+	6
? *Victoriella pavida* Saville Kent	?	1870	S	b	6
Tricellaria inopinata d'Hondt & Occhipinti Ambrogi	SP	1998	S	+	17
Nematoda (here: Swim-bladder nematode)					
Anguillicola crassus Kuwahara et al.	NP	1982	A	+++	3
Ascidiacea (sea squirts)					
Styela clava Herdman	NP	1952	S	+++	3

EXOTICS OF COASTAL AND INLAND WATERS OF IRELAND AND BRITAIN

DAN MINCHIN[1*] & CLARE ENO[2]
[1]*Marine Organism Investigations, Ballina, Killaloe, Co Clare, Ireland*
[2]*Countryside Council for Wales, Ffordd Penrhos, Gwynedd, Wales, UK*
*Corresponding author *minchin@indigo.ie*

Abstract

A greater number of exotics are known from Britain than Ireland, almost certainly arising from the greater levels of trade in Britain and its proximity to the European continent, where a greater number of species are present. This trend is reflected for freshwater, brackish and marine species. Further range expansions from Europe are expected from similar latitudes and many from the Indo-Pacific region.

1 Introduction

Following the last glacial period the British Isles were colonised by a wide range of species. This process still continues by natural dispersal, but the transmission of a wider range of species by man's activities, either intentionally or by accident, has given access to species that otherwise would not have arrived by natural means. Apart from some specific taxonomic groups few studies on exotic species have been undertaken and many species almost certainly await recognition. The general decline of available taxonomic expertise and an incomplete knowledge of marine communities has often made the recognition of exotic species difficult. Often species cannot be clearly ascribed as being native or introduced; therefore, they remain as cryptogenic (hidden-origin) (Carlton 1996). In all probability some taxonomic groups such as the Copepoda have native species yet to be described (Costello et al. 1996). For species that are described for the first time it is tempting to presume that they have recently arrived and are introduced, but this may not be the case. Vagrants, those that occur over long cycles, seasonally or rarely by natural means, may be confused as being exotic, but by definition are excluded in this account.

The British Isles, consisting of Britain and Ireland and including their adjacent associate islands, lie off the North European continent and channels that separate them can act as barriers to the spread of some exotic species. In this account we examine those marine and brackish water species that have become established, those that are likely to spread and species that may eventually arrive at some future time and some notable freshwater invaders.

2 Established exotics and their origin

In Britain and Ireland 79 marine and brackish water exotics are recorded; five present in Ireland are unknown in Britain, whereas in Britain there are 39 unknown in Ireland. Thirty species occur on both islands (Table 1). Some of those already established are expected to expand their ranges further (Table 2).

E. Leppäkoski et al. (eds.), Invasive Aquatic Species of Europe, 267–275.

Table 1. Estuarine and marine species that are either known to be established or recruit in Ireland (underlined) in Britain (italics) or both (bold). Species not known as established have been omitted. Cryptogenic species are in parentheses.

Taxon	Species		
Algae	**(Heterosigma akashiwo)**	**(Alexandrium tamarense)**	*Coscinodiscus wailesii*
	Gyrodinium c.f. aureolum	*Odontella sinensis*	*Thalassiosira punctigera*
	Thalassiosira tealata	*Agardhiella subulata*	*Anotrichium furcellatum*
	Antithamnion densum	**Antithamnionella spirographidis**	*Antithamnionella ternifolia*
	Asparagopsis armata	**Bonnemaisonia hamifera**	**Codium fragile ssp. atlanticum**
	Codium fragile ssp. tomentosoides	**Colpomenia peregrina**	Cryptonemia hibernica
	Grateloupia doryphora	*Grateloupia filicina* var. *luxurians*	*Pikea californica*
	Polysiphonia harveyi	**Sargassum muticum**	*(Scytosiphon dotyi)*
	Solieria chordalis	*Undaria pinnatifida*	
Angiosperma	**Spartina alterniflora** hybrids		
Porifera	*Suberites massa*		
Coelentrata	*Gonionemus vertens*	*Clavopsella navis*	*Haliplanella lineata*
Nematoda	**Anguillicola crassus**		
Annelida	**Ficopomatus enigmaticus**	*Hydroides dianthus*	*Hydroides elegans*
	Hydroides ezoensis	*Janua brasiliensis*	*Marenzellaria cf. wireni*
	Pileolaria rosepigmentata		
Pycnopoda	*Ammothea hilgendorfi*		
Mollusca	**(Calyptraea chinensis)**	*Crepidula fornicata*	**Potamopyrgus antipodarum**
	Urosalpinx cinerea	**Crassostrea gigas**	**Dreissena polymorpha**
	Ensis americanus	*Mercenaria mercenaria*	**Mya arenaria**
	Mytilopsis leucophaeta	*Petricola pholadiformis*	**Teredo navalis**
	Tiostrea lutaria		
Crustacea	**Balanus amphitrite**	**Balanus improvisus**	**Corophium sextonae**
	Elminius modestus	*Eriochier sinensis*	(Herrmannella duggani)
	Limnoria tripunctata	Myicola ostraea	**(Mytilicola intestinalis)**
	Mytilicola orientalis	*Eusarsiella zostericola*	*Pilumnus perlatus*
	(Porcellidium ovatum)	*Rithropanopeus harrisii*	
Bryozoa	*Bowerbankia gracilis*	*Bugula stolonifera*	*Tricellaria inopinata*
Tunicata	*Perophora japonica*	**(Phallusia mammilata)**	**Styela clava**
Teleostei	**Oncorhynchus mykiss**		

For example, the Chinese mitten crab *Eriocheir sinensis* arrived in Britain from Europe after its initial establishment in Germany about 1912. It was recognised in Britain in 1935 and has spread northwards to Northumberland. It has undergone a recent expansion on the River Thames (Clark et al. 1998).

There are more than 60 freshwater species introductions to Britain including freeliving and parasitic crustaceans, fish, molluscs, amphibia and higher plants, some flatworms and an insect (J. Bratton pers. comm.). Fewer freshwater exotics are known in Ireland.

It can be difficult to identify a species' origin, particularly when it is first described from outside of its native range, but as far as it can be reasonably deduced the majority of established marine exotics originate from the Indo-Pacific region (> 60%) followed by those from North America (ca 20%) (Table 3). The component that has come from elsewhere in Europe is difficult to determine because many species will have been carried with the extensive trade that has taken place over the centuries in advance of the studies of taxonomy and ecology.

About 15 of the Indo-Pacific species were first recorded in Britain, whereas in Ireland there is only one likely primary introduction: *Cryptonemia hibernica*, recorded from the south Irish coast (Cullinane & Whelan 1981). The high frequency of exotics from the Indo-Pacific almost certainly relates to the long established trade between Britain and North America and Commonwealth countries. However, the spread from nearby countries, including the secondary spread of Indo-Pacific species, comprises the greater component of the exotics present (ca 60%) (Table 3). Spread between adjacent countries sharing the same coastline, and who have also had extensive trading histories throughout the world, would have been rapid because there would have been few topographical barriers. The proximity of Britain to the north European continental coastline is almost certainly an important feature in their spread to Britain.

3 Patterns of exotic species spread

Generally exotic aquatic species establish themselves in sheltered inlets, docks, marinas, lagoons, large estuaries and lakes. Very often these are regions near to shipping, aquaculture or recreational boating, important vectors in the transmission of species. In freshwater, spread will normally take place with overland transmissions or using interconnected water bodies, including canals. Species with entirely sedentary life-history stages are likely to have locally confined distributions, e.g. the Chinaman's hat limpet *Calyptraea chinensis* in Ireland (Minchin et al. 1987). Sessile species with short planktonic stages may also have locally confined distributions such as the Tunicata (e.g. *Styela clava*; Minchin & Duggan 1988) and those with progressively longer planktonic stages, for example barnacles (O'Riordan 1996) may be expected to be established over a greater area. Microscopic plankton (*Gyrodinium aureolum;* Macdonald 1999) and organisms with air bladders (*Sargassum muticum;* Wallentinus 1999) or other forms of buoyancy, may be expected to spread rapidly according to their exposure to current and wind vectors. Some are able to disperse by walking (*Eriocheir sinensis*) or swimming (most fishes) and spread over greater distances, but unless these pioneers congregate to reproduce or reproduce by parthenogenesis, they will not form established populations.

A large proportion of the known established species appear in port regions. Eno et al. (1997) note concentrations in the Solent region in Britain and Minchin & Sheehan (1998) in Cork Harbour, Ireland. These are sheltered inlets with a wide range of habitats and opportunities for colonisation and water retention. Ports in areas with high tidal ranges may have berths within a basin where sluices retain high water levels. Such conditions may allow for a high retention of released organisms and their progeny, such as has happened in Cardiff Dock where the crab *Rhithropanopeus harrisii* is abundant. For some species preferring warmer conditions thermal discharges may enable their establishment as occurred at an earlier time in this same general area (Naylor 1965).

Table 2. Species that may become established in Irish and British coastal waters, invasives in bold font.

Species	Taxon	Likely vector	Nature of impact
IRELAND & BRITAIN			
Gymnodinium catenatum	dinoflagellate	ships ballast water	paralytic shellfish poisoning
Pfiesteria piscicida	dinoflagellate	ships ballast water	fish mortalities, toxins
'Dasysiphonia sp'	red alga	fouling ships/boats	Fouling cage netting
Pseudostylochus ostreae	flatworm	oysters	molluscan predator
Gryrodactylus salaris	monogenean	salmonids	reduced salmonid stocks
Rapana venosa	gastropod	ballast, oysters	molluscan predator
Ocinebrellus inornatus	gastropod	oysters	eolluscan predator
Corbula flammula	bivalve	aquatic plants?	extensive fouling
Nuttalia obscurata	bivalve	aquaculture	productive species
Caprella mutica	caprellid	ship fouling	fouling fish culture nets
Hemigrapsus penicillatus	crab	ships, oysters	omniverous, habitat changes
Hemimysis anomala	mysid	ships ballast	high biomass in estuaries
Cercopagus pengoi	ostracod	ships ballast	high biomass in estuaries/ lakes
Homarus americanus	decapod	trade, releases	hybridisation with *H. gammarus*
Paralithodes camtschaticus	crab	ships ballast	bivalve mortality, fishery
IRELAND			
Coscinodiscus wailesii	diatom	ships, naturally	covers nets with mucilage
Grateloupia doryphora	red alga	natural spread	may have commercial use
Undaria pinnatifida	brown alga	fouling ships/ boats	competition
Marenzellaria viridis	polychaete	ships or naturally	high biomass
Hydroides ezoensis	tube-worm	ship fouling	extensive fouling
Janua brasiliensis	tube-worm	ship fouling	can foul *Zostera marina*
Crepidula fornicata	gastropod	oysters	competition, habitat changes
Ensis americanus	bivalve	natural spread	high biomass, fishery
Mytilopsis leucophaeta	bivalve	hull fouling	extensive fouling
Rithropanopeus harrisi	crab	ships	high biomass locally
Eriocheir sinensis	crab	ships ballast	predator, habitat changes

4 Pathways and modes of transmission

For several centuries ships and boats have carried exotic species. The soft-shell clam *Mya arenaria* is purported to have been re-introduced to northern Europe by returning Viking explorers from North America (Strasser 1999). Whether deliberately carried for food or unintentionally is unclear, but its subsequent spread from the Baltic region to the British Isles could have been from a gradual spread. Sailing ships before the usage of effective antifouling preparations would have carried species living both in the excavated galleries in the hull and attached to the hull surface. One of these, the 'shipworm', a boring bivalve, *Teredo navalis* became extensively distributed in Europe. This species was present in Cork Harbour but no recent records exist (Minchin & Sheehan 1998), perhaps as a result of reduced habitat. Most boat hulls are now well protected and few wooden structures are present. Serpulid and spirorbid polychaetes have spread extensively as fouling organisms both on ships and oysters (Zibrowius & Thorp 1989).

During the Second World War the Australasian barnacle *Elminius modestus* became established on the south coast of Britain, brought on warships returning from the Pacific (Crisp 1958). It was first found in Chichester Harbour and has spread and become abundant in many sheltered estuaries in Britain, Ireland and northern Europe. The sea squirt *Styela clava* also became established on the south coast of England following the Korean War (Houghton & Millar 1950) and spread. Once established, species such as these are likely to continue to expand their ranges further until their reproductive capacity is impaired by specific physiological conditions. Ships' fouling organisms can occur in great numbers but since the usage of tributyltin (TBT) as a toxic ingredient in hull paint coatings from the early 1970s, no further exotic species that can be attributed to hull fouling have occurred in Cork Harbour (Minchin & Sheehan 1998).

Table 3. Origin of exotic species thought to be established in estuarine and coastal waters of Ireland and Britain. (?) indicates possible origin and bold font indicates a primary introduction. Normal italics indicate a secondary introduction.

ORIGIN	SPECIES	
Northern *Europe*	*?Calyptraea chinensis*	
Southern Europe	*?Mytilicola intestinalis*	*Loxosomella kerfersteinii*
Baltic Sea	none known	
Ponto-Caspian region	*Dreissena polymorpha*	
North America	*Gyrodinium c.f. aureolum*	***Urosalpinx cinerea***
	Spartina alternifolia hybrids	*Goniadella gracilis*
	Hydroides dianthus	***Marenzellaria wireni***
	Crepidula fornicata	***Mercenaria mercenaria***
	Enis americanus	*Mya arenaria*
	Petricola pholadiformis	*Rithropanopeus harrisii*
	Oncorhynchus mykiss	
South Atlantic	*?Cryptonemia hibernica*	*?Balanus improvisus*
	Janua brasiliensis	
Indian and Pacific Oceans	***Sargassum muticum***	***Coscinodiscus wailesii***
	Thalassiosira tealata	*Odontella sinensis*
	Pleurosigma simonsenii	*Asparagopsis armata*
	Bonnemasonia hamifera	***?Grateloupia doryphora***
	Agardhiella subulata	***Pikea californica***
	Antithamnionella spirographidis	*?Antithamnionella ternifolia*
	?Cryptonemia hibernica	*Polysiphonia harveyi*
	Colpomenia peregrina	*Undaria pinnatifida*
	?Scytosiphon dotyi	*Bonamia ostreae*
	Gonionemus vertens	*Haliplanella lineata*
	Anguillicola crassus	*Hyroides ezoensis*
	Ficopomatus enigmaticus	*Pileolaria rosepigmentata*
	Potamopyrgus antipodarum	*Crassostrea gigas*
	Mytilopsis leucophaeta	***Elminius modestus***
	Myicola ostrea	*Mytilicola orientalis*
	Acartia tonsa	***Corophium sextonae***
	Eriochir sinensis	*?Tricellaria inornata*
	Perophora japonica	***Styela clava***

Ships also transmit and release organisms from their ballast-water tanks between ports. In Ireland, ballast water discharges mainly originate from British and other European ports. Species that expand their ranges and colonise greater numbers of ports will subsequently increase the risk of their spread to Britain and Ireland on established trading routes. Although Britain is a net exporter of ballast water (Laing 1998; Macdonald 1998), several non-native species have arrived by this means (Eno 1998a). The Asian shore crab *Hemigrapsus penicillatus* has rapidly spread from France to Spain (Gollasch

1999) and is now found on the southern coast of the North Sea. Its further spread in Britain and Ireland is inevitable and their larvae may be spread with ballast water (or juveniles and adults with oyster movements). A study on ballast discharges in Ireland in 1994/5 showed that the greatest volumes were discharged in Cork Harbour and the Shannon Estuary (Minchin 1996a). Trade with Baltic Sea ports to the Irish and British estuaries could result in introductions of some Ponto-Caspian species that have colonised parts of the Baltic Sea. Opportunities for establishment in a British port are greater because of Britain's proximity to the European continent and the greater volume of shipping traffic.

The possible spread by coastal vessels of the virus causing infectious salmon anaemia (ISA) (Anonymous 2000) in ballast water is of some concern. Salmon farms near to ballast discharge areas may be at risk. The knowledge of micro-organisms transmitted in ballast water is poor and there may be several other viruses and disease agents that could be spread that have implications for aquaculture and human health. Slow-moving craft may have well developed fouling communities that can include mussels and oysters, species used in extensive cultivation. It is suspected that barges spread the oyster blood disease *Bonamia ostreae* on the SW coast of Britain (Howard 1994) and other molluscan diseases could be transported over greater distances; perhaps the arrival of *B. ostreae* in Cork Harbour in Ireland was with oysters fouling ships' hulls. Smaller craft may also spread species, for example the bryozoan *Tricellaria inopinata* is thought to have come from the Venice Lagoon to Britain (Dyrynda et al. 2000) and the Japanese kelp *Undaria pinnatifida* (Fletcher & Manfedi 1995) to the marinas on the south coast of Britain. The tube-worm *Ficopomatus enigmaticus* spread from Cork Harbour to a marina on the Shannon Estuary (Minchin 2000a). An American flatworm *Phagocata woodworthi* may have been carried on an imported mini-submarine in 1977 to Loch Ness (Reynoldson et al. 1981).

Shellfish movements have a long history of being associated with introductions. Several have arrived following different historical events in the trading of oysters. Imports of half-grown American oysters *Crassostrea virginica* began in Ireland and Britain almost 120 years ago and lasted for about 40 years. This followed the depletion of native oyster *Ostrea edulis* beds. The imports were ongrown on the SE coast of Britain and in some Irish bays and harvested following a summer's growth. The associated oyster drill *Urosalpinx cinerea* and the slipper limpet *Crepidula fornicata* were also imported and became established in Britain (Eno 1998b) but are not presently established in Ireland (Minchin 1996b).

During the 1940s and 1950s the native oyster was imported as spat from regions of dense natural settlements in France. However, this practice was discontinued both in Britain and Ireland in the 1960s. In Ireland this followed the finding of the fungal shell disease *Ostracoblabe implexa* (which died out some years later) on imported oysters (C.B. Duggan pers. comm.). In the following years exotic molluscs were used in trials with the intention of increasing aquaculture production. These passed through quarantine so as to produce a disease-free stock. Following these trials the New Zealand oyster *Tiostrea lutaria* became established in the Menai Strait. It remains confined there probably because of its short larval stage (Eno et al. 1997). Oyster production increased in Britain and Ireland following the trials with the Pacific oyster *Crassostrea gigas* which began in the late 1960s (Utting & Spencer 1992). However, in 1993 the European Free-Trade Agreement enabled a wider movement of Pacific oysters cultivated in France. These oysters had originated as direct imports from the wild which had been flown-in from Japan in the 1970s, together with several associated species (Gruet et al. 1976). Following the 1993 agreement two Asian copepods *Mytilicola orientalis* and *Myicola ostreae* became established on the south coast of Ireland with imports of half-grown oysters (Holmes & Minchin 1995). In addition, unexplained summer mortalities of Pacific oysters during warm summers appeared in Ireland for the first time.

Movements of Pacific oysters have been implicated in the spread of the Japanese brown alga *Sargassum muticum*. Boaden (1995) suggested a spread from the Channel Islands to Strangford Lough and its recent appearance in the Menai Strait was probably with oysters from the south coast of Britain. This species will certainly spread further in Ireland and Britain. There have been some serious freshwater introductions through aquaculture and stocking. The most notable one is the crayfish plague, caused by the fungus *Aphanomyces astaci*, which was introduced to Europe with the Signal crayfish *Pacifastacus leniusculus* from America and from there to Britain (Palmer 1994). In the mid-1980s the plague reached Ireland (Environment Agency 1999) and destroyed most of the stock of *Austropotamobius pallipes* in one river system and the disease has not been noted since. This crayfish may have been introduced to Ireland from Britain or the European continent and is susceptible to this disease, and there have been notable population declines of *Austropotamobius pallipes* throughout Europe. Two other crayfish species occur in Britain, *Astacus astacus* and *A. leptodactylus*. A further introduction through aquaculture is the crustacean *Gammarus pulex* - native to Britain but introduced to Ireland for fish food in 1958 (Dick et al. 1990).

Irish native freshwater fish are either anadromous, catadromous or evolved from anadromous species. All others were introduced, some since monastic times in the 1600s (Moriarty & Fitzmaurice 2000). The accidental introduction of the roach *Rutilus rutilus* in 1889, to a river in the south of Ireland, was followed by an escape from an ornamental pond to the Foyle River (Went 1980). They entered the Erne River system in the 1960s and have since spread to most of the interconnected navigable regions, as has the more recently arrived mollusc *Dreissena polymorpha* imported as fouling on second hand water-craft in 1993/4 (Minchin 2000b). The tench *Tinca tinca* also continues to expand its range in Ireland. Some fish such as the exotic pumpkinseed *Lepomis gibbosus* are imported as aquarium species to Ireland and may become established if released to the wild. This is because they are adapted to temperate conditions, whereas tropical species are less likely to survive unless they congregate within thermal water discharges, as is known for the molly *Poecilia reticulata* in Britain (Welcomme 1988).

Temperate aquatic plants, such as the Canadian pondweed *Elodea canadensis* are now widely distributed in the British Isles and this is partly being replaced by *Lagarosiphon major* in Britain and the Australasian *Crassula helmsii* is abundant in some lowland water bodies (Dawson 1991). Other exotic aquatic plants used in aquaria or by garden enthusiasts are also expected to become established and spread (Preston & Croft 1997).

Trade in living organisms carries the risk of their establishment. Airfreight provides new opportunities for them to survive over long distances because of the short period of time in transit, for example imports of marine polychaete worms imported for angling baits from Korea, Africa and USA. The American lobster *Homarus americanus* is imported for human consumption, and either because of their condition or on account of market prices, and although not permitted, may be held in storage in tidal ponds. Such an action in the 1960s led to gaffkaemia, a bacterial disease, spreading to native lobsters *H. gammarus* in a pond on the south coast of Ireland and all lobsters were destroyed (Gibson 1961). Escapes of American lobsters have also led to concerns that these may hybridise with native species. Imported stocks of the oriental eel *Anguilla japonica* without proper quarantine led to its nematode bladder parasite *Anguillicola crassus* becoming established in Northern Europe where it has subsequently spread through movement of infected European stocks of Common eels *Anguilla anguilla*. Its arrival in Ireland (Evans & Mathews 1999) may have been with infested eels in a viviere truck, whose water was exchanged to maintain them. This parasite may have long-term consequences for North European eel populations. Some freshwater introductions have arrived inadvertently through trade, notably with damp timber imports.

5 Future trends and exotic species

Further species are likely to arrive in the British Isles, and many of these will come via northern Europe as has been the past pattern. Introduced species that do not show range retractions elsewhere in Europe but continue to expand, albeit slowly, may extend their ranges to Britain and Ireland (Table 4). Some of these may have ecological and economic consequences (Eno 1996).

The frequency of exotic species introductions is expected to increase because of the expansion of exotic species ranges elsewhere. New trading links present opportunities for colonists from new port regions. Ports are providing more shipping berths further downstream, thereby providing more suitable conditions for the survival of marine species once released. With the general trends of control of industrial discharges and the planned banning of some toxic antifoulants, port conditions will become less inhospitable to species discharged in ballast water or from the recruitment from spawning fouling organisms on ships' hulls.

The planned expansion of aquatic leisure activities through tourism, boat charter, angling and special events will enhance the risks of transmission of some species. The widespread interest in gardening and aquarium species will inevitably lead to the spread of further aquatic garden plants and fishes and their associated biota.

In the development of new aquaculture products, further species will be used in culture for biotechnology or food production. This may include the introduction of species used for forage. Exotic phytoplankton species are already freely used in the cultivation of molluscs and those that are not consumed become discharged into the sea. However, at present, no adverse effects of such practices are known.

The predicted mean annual temperature increases will provide important changes that may enable further species to become established either because reproductive opportunities enable the maintenance of populations or because of reduced winter mortalities.

The current measures to deal with ballast water at best deal with managing a primary introduction route. Many ships find themselves unable to undertake mid-ocean exchanges of water to replace the organisms present in their ballast water and sediments and so 'leakage' of exotics is likely. The main difficulties are with the secondary spread of exotic species where geography and journey times make exchanges impractical and may even compound problems (Macdonald & Davidson 1998). Proposed sterilisation techniques are not yet at the stage of being practicable for use, either because of the expense, the efficiency or because the active ingredient on release may be considered toxic for the environment. For these reasons shipping is still likely to act as a main vector transmitting species.

Table 4. Exotic species already present and likely to spread or recruit in the wild in either or both Britain or Ireland. Cryptogenic species are underlined.

Taxon	Species		
Algae	<u>*Alexandrium tamarense*</u>	*Sargassum muticum*	*Cryptonemia hibernica*
	Undaria pinnatifida		
Protozoa	*Bonamia ostreae*		
Trematoda	*Anguillicola crassus*		
Annelida	*Ficopomatus enigmaticus*	*Gonidella gracilis*	*Marenzelleria wireni*
Mollusca	*Calyptraea chinensis*	*Dreissena polymorpha*	*Crassostrea gigas*
	Ruditapes semidecussata	*Ensis americanus*	*Mytilopsis leucophaeta*
	Crepidula fornicata		
Crustacea	*Mytilicola orientalis*	<u>*Mytilicola intestinalis*</u>	*Myicola ostreae*
	Corophium sextonae	*Gammarus pulex*	*Gammarus tigrinus*
	Crangonyx pseudogracilis	*Eriocheir sinensis*	*Rhithropanopeus harrisii*
	Balanus improvisus	*Elminius modestus*	
Tunicata	*Styela clava*	<u>*Phallusia mammilata*</u>	*Perophora japonica*
Bryozoa	*Tricellaria inornata*		
Teleostei	*Cyprinus carpio*	*Rutilus rutilus*	

6 Conclusions

Exotic species in Britain and Ireland require further investigation and some taxonomic groups remain unstudied. With further investigation more will be found. A small number of the exotic species component will have impacts on economics, environmental and human health. Unfortunately these normally appear unexpectedly. Britain, on account of its extensive trading network and proximity to the European continent, has a greater number of exotics than Ireland. In accordance with previous patterns, many of the exotics established on the northern European continent will expand their range to southern Britain and some time later these may occur in Ireland, and are likely to include species from the temperate regions of the NW Atlantic and the Indo-Pacific. Where several activities overlap, as in a port region, the rate of spread may take place more rapidly.

OPEN ATLANTIC COAST OF EUROPE - A CENTURY OF INTRODUCED SPECIES INTO FRENCH WATERS

PHILIPPE GOULLETQUER[1*], GUY BACHELET[2], PIERRE GUY SAURIAU[3] & PIERRE NOEL[4]
[1]*Shellfish Aquaculture Research of Poitou-Charentes, IFREMER Ronce Les Bains, La Tremblade, France*
[2]*Laboratoire d'Océanographie Biologique, UMR 5805 CNRS, Université Bordeaux 1, Station Marine d'Arcachon, Arcachon, France*
[3]*CREMA (UMR 10 CNRS-IFREMER), Place du Séminaire, L'Houmeau, France*
[4]*Laboratoire de Biologie des Invertébrés marins (ESA CNRS 8044), Muséum National d'Histoire Naturelle, Paris, France*
*Corresponding author *Philippe.Goulletquer@ifremer.fr*

Abstract

This paper aims to establish a comprehensive list of exotic species introduced into Atlantic French waters over the last century as well as their origin, and the vectors involved in the process. At least 104 nonindigenous species have been recorded in the Atlantic and Channel coastal waters. Among those species, only a few (*Bonamia ostreae*, *Crepidula fornicata*) have resulted in highly significant negative impacts and side-effects on the coastal activity. In contrast, the introduction of the Japanese oyster *Crassostrea gigas* can be considered to be successful, sustaining a large industry counting more than 4,000 companies and 150,000 tons of yearly production. Over the last decade, French and European laws have strenghtened the controls over exotic species introductions, limiting their impacts. These new regulations and increased awareness from the scientific community with regard to coastal industry and biodiversity have reduced significantly the cases of deliberate introduction, while the risk analysis requirement has improved the overall management. However, in spite of these improvements, cases remain significant, mostly by accidental and/or increased vectors of introduction such as the ballast waters and sediment. Moreover, global changes should now be considered as a major issue, likely inducing future geographic distribution changes over a large European coastal range. Understanding the invasion population dynamics for exotic species remains a challenging research issue currently addressed by the French INVABIO multidisciplinary program.

1 Introduction

Comprehensive reviews of introduced species along the Atlantic coastline and the English Channel are not yet available. Although preliminary, attempts were made by Gruet & Baudet (1997) and Belsher et al. (1997), focusing on invertebrates and macroalgae of the Atlantic French waters, respectively. Therefore, the present review represents a significant update to this matter, as it includes descriptions of the likely vectors responsible for the present situation and species status. This review is likely to be an underestimate, while multiple introduction vectors are also likely, and only a few cryptogenic species have been determined (Rosenfield & Mann 1992; Carlton 1996; Reise et al. 1999).

Due to its geographic range, biodiversity is particularly high along the Atlantic coasts; e.g., > 3,000 species have been recorded at Roscoff. Most of the introductions (e.g.,

E. Leppäkoski et al. (eds.), Invasive Aquatic Species of Europe, 276–290.

their timing, localization, and expansion pattern) have remained difficult to track, while several cases are well known (e.g., the Japanese brown alga *Sargassum muticum*).

1.1 PHYSICAL AND GEOGRAPHICAL FEATURES OF THE AREA

Of the 5,500 km of French coastline, 3,800 km concern the Atlantic seaside and the English Channel. Most of the French coast is located along the traditional European waterways, and therefore well exposed to the navigation trade effects. Several harbours of international importance are located within these boundaries (Bordeaux, La Rochelle, Nantes-St Nazaire, Brest, Cherbourg, and Le Havre). While 51% of the coasts are urbanized, 800 km are located in highly productive areas, where traditional shellfish farming has been developed (Fig. 1). Moreover, 580 km are island coastlines. The highly diverse French coastline is characterized by 30% rocky shores, 40% sandy beaches, and 30% of salt marshes of particular interest for biodiversity. The effects of the Gulf Stream along the Atlantic coastline result in a biogeographic barrier for several phyla around Brittany, limiting species exchange between the northern and southern areas. In northern Brittany, seawater temperatures range from 6-10°C to 15-17°C in February and summer, respectively. In contrast, summer seawater temperatures can reach 22°C along the Atlantic coastline. The global change has induced a 1.5-2°C seawater temperature increase over a 25 year span in the southwestern coastal areas, likely affecting species distribution range and physiology (Soletchnick et al. 1998). Salinities range from 5 PSU in wetlands (e.g., oyster ponds) to 20 PSU in coastal areas in winter and to 30-35 PSU in summer. This spatial variability of salinity and temperature is one of the main critical factors explaining species distribution, including the exotic ones (Goulletquer et al. 1998). The Atlantic coastline as well as the English Channel are characterized by two tidal cycles a day (i.e., 12 h per cycle). Neap tides alternate with spring tides on a weekly basis. The tidal amplitude varies around 8 m and 4 m on the western northern part of Brittany and Atlantic sides respectively, while it averages 10 m during spring tides. The 15.5 m European high record in tide range is recorded within the Gulf of Saint Malo in northern Brittany (English Channel). Combined with rocky shores, those environmental characteristics result in highly diverse fauna and flora for those areas.

1.2 ORIGIN OF THE NATIVE FAUNA AND FLORA

Along the open Atlantic coasts of continental Europe, from Dunkerque to Gibraltar, the western entrance of the English Channel has long been considered a geographical boundary (Ekman 1953), separating Lusitanian and boreal faunas and limiting the geographical distribution of pelagic crustaceans and fish. However, soft-bottom macrofauna data do not support this assumption as suggested for instance by Glémarec (1978, 1991) for bivalve molluscs and onuphid polychaetes in the Bay of Biscay. Similarly, the occurrence of the boreal *Macoma balthica* community in most estuaries and sheltered bays of France, Spain and Portugal, does not match this assumption. Several migration routes have been recognised between the sub-arctic and Mauritanian areas (see e.g. Castel et al. 1997), but the southwards or northwards spread of boreal, temperate, Lusitanian and Mediterranean species can locally be modified by sedimentological and/or hydrodynamical features of a given route, as the case is in the English Channel (Cabioch et al. 1977), the eastern part of the Bay of Biscay and the Portuguese coast. During the last four decades, northwards spread of some benthic and pelagic species has also been documented (Glémarec 1979; Sauriau 1991; Quéro et al. 1998) and this phenomenon could be linked to the warming of coastal waters (Koutsikopoulos et al. 1998; Soletchnick et al. 1998) and/or modification of offshore current regimes in the vicinity of the Bay of Biscay.

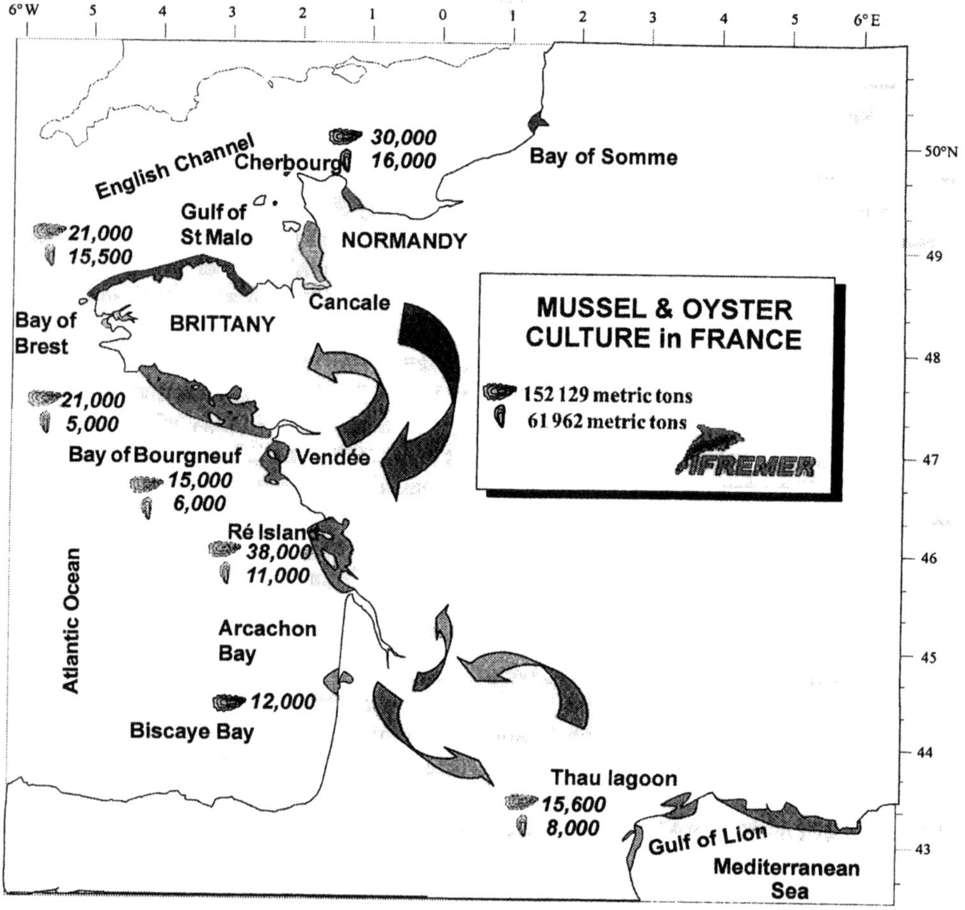

Figure 1. Shellfish production in France and principal stock transfers among French rearing areas (from Goulletquer & Héral 1996).

2 Historical cases

2.1 GENERAL REVIEW OF INTRODUCED SPECIES

Except those species which were maintained in laboratory facilities (quarantine) for evaluation of their culture potential (the oysters *Crassostrea rivularis*, *C. sikamea*, and *Ostrea denselamellosa*) thus being not found in the wild, or in isolated fishponds for aquaculture purposes (the prawn *Marsupenaeus* (=*Penaeus*) *japonicus*), 104 nonindigenous species have been recorded in the Atlantic and Channel coastal waters of continental Europe between the French-Belgian borderline and Gibraltar (Table 1). Those figures are likely underestimated due to multiple introductions (e.g., *Alexandrium* strains), cryptogenic species and also permanent but undetected introductions such as those resulting from ballast water discharges in large harbours. Invertebrates are the major group of introduced species in this area, accounting for 63% of the total (65 species), followed by benthic algae (20%, 21 spp.), phytoplankton (8%, 8 spp.), higher plants (4%, 4 spp.), tunicates (3%, 3 spp.), sporozoans (2 spp.), and fish (1 sp.). Among the invertebrates, arthropods (25 spp.), molluscs (22 spp.), and cnidarians (8 spp.) are the dominant phyla. More than half of all introduced species are native to the Pacific Ocean (49% and 10% from North and South Pacific, respectively). The other native regions are the North West Atlantic (16%), the Mediterranean (7%), the North East Atlantic outside the studied area (5%), the Indo-Pacific region (5%), the South West Atlantic, the Indian Ocean, and the Ponto-Caspian area (2% each). It should be noted that the 'source regions' of the invaders in the studied area are different from the 'native regions', because 45 out of the 104 nonindigenous species invaded the open Atlantic coast of France and Iberian Peninsula secondarily from previously invaded regions of Europe (excluding the Mediterranean).

Table 1. Introduced marine and brackish water species on the Atlantic and Channel coasts of France, Spain and Portugal. Origin: (N,S,W,E)A = Atlantic (NEA = North-East Atlantic outside the present area), C = cryptogenic species, IO = Indian Ocean, IP = Indo-Pacific, MED = Mediterranean, (N,S,W,E)P = Pacific, PC = Ponto-Caspian area. First record: Date = year of the first record, or probable time of introduction, in the area; Location = place of the first record in the area; * = first record in European waters (excl. Mediterranean). Mechanism of introduction: BW = ships' ballast water, COI = accidental release with commercial oyster industry, HP = hatchery production, IR = intentional release, P = parasite introduced with its host, PR = private release, SF = ships' fouling. Present status: established (E) or not established (NE) in the wild.

Taxa/species	Origin	First record (Date / Location)	Mechanism of introduction	Present status
Ascetospora				
Bonamia ostreae Pichot et al. 1979	NEP	*1978-79/Brittany	P (oyster)	E
Haplosporidium nelsoni Couch et al. 1966	NWP, then NEP?	*1970's ? 1993	P (oyster)	E?
Bacillariophyta				
Coscinodiscus wailesii Gran & Angst, 1931	NEP, NWP	1978/N Bay of Biscay and Normandy	BW or COI	E
Odontella sinensis (Grev.) Grunow, 1884	IP, NWP	≤ 1930	BW	E
Pleurosigma planctonicum Simonsen, 1974	IO?	*1966/Ouessant island (W Brittany)	BW	E
Dinoflagellata				
Alexandrium leei Balech, 1985	NWP	*1995/Douarnenez (W Brittany)	BW	E?
A. minutum Halim, 1960	SWP	*1985/NW Brittany	BW	E
Gymnodinium catenatum Graham, 1943	NWP	*1976/Galicia	BW	E
G. cf. *nagasakiense* Adachi & Fukuyo, 1979	NP	≤ 1976/NW Brittany	BW?	E
Raphidophyta				
Fibrocapsa japonica Toriumi & Takano, 1975	NWP	1991/Brittany	BW?	E
Chlorophyta				
Codium fragile (Suringar) Hariot, 1889 ssp. *tomentosoides* (van Goor) Silva, 1955	NWP	1946/Glénan islands & Concarneau (SW Brittany)	?	E

Taxa/species	Origin	First record (Date / Location)	Mechanism of introduction	Present status
Phaeophyta				
Colpomenia peregrina (Sauvageau) Hamel, 1937	NEP	*1905/Vannes (S Brittany) and St-Vaast (Normandy)	COI	E
Macrocystis pyrifera (Linnaeus) C. Agardh, 1820	SEP	*1972/Roscoff (N Brittany)	IR	NE
Sargassum muticum (Yendo) Fensholt, 1955	NWP	1976/Barfleur (Normandy)	COI	E
Undaria pinnatifida (Harvey) Suringar, 1873	NWP, then MED	*1983/Rance estuary and Ouessant island (N & W Brittany)	COI, then IR	E
Rhodophyta				
Anotrichium furcellatum (J. Agardh) Baldock, 1976	NP, then MED	*≤ 1922/Channel	?	E
Antithamnion densum (Suhr) Howe, 1914	NP	*1960s/Brittany	?	E
Antithamnionella spirographidis (Schiffner) Wollaston, 1968	NP	?	COI or SF	E
A. ternifolia (J.D. Hooker & Harvey) Lyle, 1922	SP	1910	SF	E
Asparagopsis armata Harvey, 1855	SWP	*1925/Guéthary (SE Biscay)	COI or SF	E
Bonnemaisonia hamifera Hariot, 1891	NWP	1898/Cherbourg (Normandy)	SF	E
Caulacanthus ustulatus (Turner) Kützing, 1843	NWP (also known from NEA)	*1986/Roscoff (N Brittany)	COI	E
Dasysiphonia sp.	NWP	*1990/Galicia	COI	E
Grateloupia doryphora (Montagne) Howe, 1914	NP	1989/Lorient (S Brittany)	COI	E
G. luxurians A. Gepp & E. Gepp, 1906	IP	1990/Galicia	COI	E
Hypnea muciformis (Wulfen) Lamouroux, 1813	NEA	*1900s/St-Vaast (Normandy)	COI	NE
Laurencia brongniartii J. Agardh, 1841	NWP	*1989/Brest (W Brittany)	COI	E
Lomentaria hakodatensis (Yendo, 1920)	NWP	*1984/Roscoff (N Brittany)	COI	E
Pikea californica Harvey, 1853	NEP, NWP	1991/Galicia	SF (flying boat)	E
Pleonosporium caribaeum (Børgesen) Norris, 1985	IP, NWA	*1967/Brest (W Brittany)	?	E
Polysiphonia harveyi Bailey, 1848	NWA	1980s/Roscoff (N Brittany)	COI?	E
Magnoliophyta				
Spartina alterniflora Loiseleur-Deslongchamps, 1807	NWA	<1960/Brest (Brittany), Hossegor and Hendaye (SE Bay of Biscay)	?	E
Spartina anglica C.E. Hubbard, 1968	NEA	1894	?	E
Spartina x *townsendii* H. & J. Groves, 1881	NEA	1906/Brevends (Normandy)	IR?	E
Spartina versicolor Fabre, 1850	NWA	Arcachon Bay, 1901	?	E
Cnidaria: Anthozoa				
Aiptasia pulchella Carlgren, 1943	NWP	*1973-74/Bourgneuf Bay and Le Croisic (NE Biscay)	COI	NE?
Diadumene cincta Stephenson, 1925	NP	1963/Ré island (E Biscay)	BW or SF	E
Haliplanella lineata (Verrill, 1869)	NWP	End of 19th century + 1970s/Bay of Biscay	COI and/or SF	E
Cnidaria: Hydrozoa				
Blackfordia virginica Mayer, 1910	C	*1971/Loire estuary (NE Biscay)	BW or SF	NE?

Taxa/species	Origin	First record (Date / Location)	Mechanism of introduction	Present status
Cordylophora caspia (Pallas, 1771)	PC	< 1901/Loire estuary (NE Biscay)	SF	E
Gonionemus vertens (Gosse, 1853)	NWP	19th century/Portugal	BW or COI or SF	E
Maeotias marginata (Modeer, 1791)	PC	*1971/Loire estuary (NE Biscay)	BW or SF	NE?
Nemopsis bachei Agassiz, 1849	NWA	1953/Gironde estuary (E Biscay)	BW or SF	E
Platyhelminthes				
Pseudodactylogyrus anguillae (Yin & Sproston, 1948)	NWP	1984/South of France	P. (eel)	E
Pseudostylochus ostreophagus (Hyman, 1955)	NWP	*1970s	P (oyster)	?
Bryozoa				
Victoriella pavida Saville Kent, 1870	C	1960s/Arcachon Bay (SE Biscay)	?	E?
Watersipora aterrima (Ortmann, 1890)	NP	*1983/Arcachon Bay (SE Biscay)	COI	E?
Nematoda				
Anguillicola crassus Kuwahara, Niimi & Itagaki, 1974	NWP	1980s	P (eel)	E
Annelida: Polychaeta				
Boccardia semibranchiata Guérin, 1990	MED	*1999/Arcachon Bay (SE Biscay)	COI	E
Ficopomatus enigmaticus (Fauvel, 1923)	SWP	*1921/Caen (Normandy)	SF	E
Hydroides dianthus (Verrill, 1873)	NWA	*≤ 1927/Ré island (E Biscay)	BW or COI or SF	E
H. ezoensis Okuda, 1934	NWP	*1973-74/Bourgneuf Bay and Le Croisic (NE Biscay)	COI	NE
Streblospio benedicti Webster, 1879	NWA	*1982/Loire estuary (NE Biscay)	SF	E
Mollusca: Gastropoda				
Corambe obscura (Verrill, 1870)	NWA	1973/Fouras (E Biscay)	SF	E?
Crepidula fornicata (Linnaeus, 1758)	NWA	1949/Brest (W Brittany) and Hermanville (Normandy)	SF and COI	E
Cyclope neritea (Linnaeus, 1758)	MED, NEA	*1976/Arcachon Bay (SE Biscay)	COI	E
Gibbula albida (Gmelin, 1791)	MED	*1986/Arcachon Bay (SE Biscay)	COI	E
Ocinebrellus inornatus (Récluz, 1851)	NWP	*1995/Marennes-Oléron (E Biscay)	COI	E
Potamopyrgus antipodarum (Gray, 1843)	SWP	?	Freshwater barrels	E
Rapana venosa (Valenciennes, 1846)	NWP, then MED	*1998/Morbihan Gulf (NE Biscay)	PR	E?
Urosalpinx cinerea (Say, 1822)	NWA	1960s/Arcachon Bay (SE Biscay)	COI	NE
Mollusca: Bivalvia				
Anomia chinensis Philippi, 1849	NWP	*1973-74/Bourgneuf Bay and Le Croisic (NE Biscay)	COI	NE?
Corbicula fluminalis (Müller, 1774) + *C. fluminea* (Müller, 1774)	NWP	*1980 Dordogne river (SW France) and Tagus estuary (SW Portugal)	?	E
Crassostrea angulata (Lamarck, 1819)	NWP, then NEA	*1866/Arcachon Bay (SE Biscay)	IR	E, then extinct
C. gigas (Thunberg, 1793)	NWP, then NEP	*1966/Marennes-Oléron (E Biscay)	IR	E

Taxa/species	Origin	First record (Date / Location)	Mechanism of introduction	Present status
C. rhizophorae (Guilding, 1828)	SWA	*1976/Gironde estuary and La Tremblade (E Biscay)	IR (+ HP)	NE
C. rivularis (Gould, 1861)	NWP, IP, then NWA	*1994/La Tremblade (E Biscay)	HP	NE
C. sikamea (Amemiya, 1928)	NWP	*1994/La Tremblade (E Biscay)	HP	NE
C. virginica (Gmelin, 1791)	NWA	*1861/Arcachon Bay (SE Biscay)	IR	NE
Ensis directus (Conrad, 1843)	NWA	1991/North Sea coast	BW	E
Mercenaria mercenaria Linnaeus, 1758	NWA	1861/Arcachon Bay (SE Biscay), Seudre estuary, St-Vaast Normandy	IR	NE
		1936-39/Morbihan Gulf (S Brittany)	IR	E
Mya arenaria (Linnaeus, 1758)	NWA	13th-17th centuries	IR?	E
Ostrea angasi Sowerby, 1871	SWP	*1985/Crach and Etel (S Brittany)	IR	NE
O. denselamellosa Lischke, 1869	NWP	*1982/Barfleur (Normandy)	HP	NE
O. puelchana d'Orbigny, 1846	SWA	*1989/Arcachon Bay (SE Biscay)	IR	NE
Patinopecten yessoensis (Jay, 1856)	NWP	1986-88/Brest (W Brittany)	IR	NE?
Tapes philippinarum Adams & Reeve, 1850	NWP, then NEP	*1973-74/Brittany-Normandy	IR	E
Tiostrea chilensis (Philippi, 1845)	SEP	1981-82/Morbihan Gulf (S Brittany)	IR	NE
Crustacea: Copepoda				
Acartia tonsa Dana, 1849	IP, NWA	1927/Caen (Normandy)	BW	E
Eurytemora pacifica Sato, 1913	NWP	*1985/Charente estuary (E Biscay)	BW or COI	E?
Myicola ostreae Hoshina & Sugiura, 1953	NWP	*1972/Marennes-Oléron, Gironde estuary and Arcachon Bay (E Biscay)	P (oyster)	E
Mytilicola intestinalis Steuer, 1902	MED	1949/Ailly (Normandy)	P (mussel)	E
M. orientalis Mori, 1935	NWP	*1977/Arcachon Bay (SE Biscay)	P (oyster)	E
Pseudomyicola spinosus (Raffaele & Monticelli, 1885)	MED	*1963/Arcachon Bay (SE Biscay)	P (mussel)	?
Crustacea: Cirripedia				
Balanus albicostatus Pilsbry, 1916	NWP	*1973-74/Bourgneuf Bay and Le Croisic (NE Biscay)	COI	NE?
B. amphitrite amphitrite Darwin, 1854	IO, SWP	*1934/San Sebastian (SE Biscay)	BW or SF	E
B. eburneus Gould, 1841	NWA	< 1940/La Rochelle (E Biscay)	BW or SF	E
B. improvisus Darwin, 1854	NWA	< 1872/between Charente and Capbreton (SE Biscay)	BW or SF	E
Austrominius (Elminius) modestus Darwin, 1854	SWP	≤ 1953/N and S Brittany	BW or SF	E
Megabalanus tintinnabulum (Linnaeus, 1758)	?	?/Loire Atlantique (NE Biscay)	BW or SF	NE?
Solidobalanus fallax (Broch, 1927)	?	?/Houat island (NE Biscay)	BW or SF	NE?
Crustacea: Mysidacea				

Taxa/species	Origin	First record (Date / Location)	Mechanism of introduction	Present status
Diamysis bahirensis (G.O. Sars, 1877)	MED	*1995/Ria de Aveiro (NW Portugal)	BW	E
Crustacea: Isopoda				
Synidotea laevidorsalis (Miers, 1881)	NWP	*≤ 1975/Gironde estuary (SE Biscay)	BW or COI or SF	E
Crustacea: Amphipoda				
Corophium sextonae Crawford, 1937	SWP	1936/Rance estuary (N. Brittany)	SF	E
Crustacea: Decapoda				
Brachynotus sexdentatus (Risso, 1827)	MED	1997/La Rochelle (E Biscay)	?	NE?
Callinectes sapidus Rathbun, 1896	NWA	*1901/Rochefort (E Biscay)	?	NE
Eriocheir sinensis H. Milne-Edwards, 1853	NWP	1930s/Boulogne (E Channel)	BW	E
Hemigrapsus penicillatus (De Haan, 1835)	NWP	*1994/La Rochelle (E Biscay)	BW or COI	E
H. sanguineus (De Haan,1835)	NWP	*1999/Le Havre (Normandy)	BW or SF	E?
Macromedaeus voeltzkowi (Lenz, 1905)	IO	1910/St Vaast La Hougue (Normandy)	SF	NE
Marsupenaeus (Penaeus) japonicus (Bate, 1888)	IP	*1980/Noirmoutier (NE Biscay)	HP	NE
Parapilumnus malardi (De Man, 1913)	IP	1910/St Vaast La Hougue (Normandy)	SF	NE
Rhithropanopeus harrisii (Gould, 1841)	NWA	1955-56/Tancarville, Caen (Normandy) and Gironde estuary (SE Biscay)	BW or SF	E
Chordata: Tunicata				
Molgula manhattensis De Kay, 1841	NWA?	1840's (N. Brittany)	SF?	E
Perophora japonica Oka, 1927	NWP	*1982/Lézardrieux (N Brittany) and St-Vaast (Normandy)	COI	E
Styela clava Herdman, 1881	NWP	1968/Dieppe (Normandy)	BW or COI or SF	E
Chordata : Osteichthyes				
Acipenser baeri Brandt, 1869	NWP	*1999 Gironde estuary (SE Biscay)	Escape from HP	?

Invasion rates appear to have increased dramatically in the past 40 years: 61% of the first records of introduction date from 1960 or later (Fig. 2). This is mainly due to species that entered Europe via the Atlantic and Channel coastal waters of France and Iberian Peninsula (primary introductions). Secondary introductions by dispersal from other European areas, into which the species were previously introduced, also increased during the last four decades of the 20th century, but at a much lower rate. The highest invasion rate occurred in the 1970s (20 spp.); however, the number of introduced species was still consistant in the 1990s (14 spp.), in spite of a more rigorous legislation. The main area of species introductions (88% of the primary introductions, 84% of the secondary introductions) extends from Normandy (English Channel) to Arcachon Bay (SE Biscay), i.e. in the areas with extensive oyster farming. The NE and E coasts of the Bay of Biscay (Morbihan Gulf, Loire estuary, Bourgneuf Bay, Marennes-Oléron Bay, Gironde estuary) and Brittany (northern and southern coasts) were the areas where most nonindigenous species were recorded for the first time (36% and 26% respectively), followed by Normandy (17%) and Arcachon Bay (14%). Thirteen species (13% of the species listed in Table 1) were deliberately introduced for aquaculture research, and/or production, or trade. Of the accidental introductions (87 spp.), 28% of the species are presumed to have been brought in association with oyster shipments (mainly *Crassostrea gigas* in the 1970s): 4 of these species were oyster parasites, while an additional protozoan (*Haplosporidium* sp. *nelsoni)* is currently under evaluation, and likely originating from the 1970s' introduction (Renault et al. 2000). The most recent accidental introduction occurred in December 1999, when the southwest French coastline was hit by a severe hurricane, which resulted in several flooded areas. Several finfish farms were partly or totally destroyed. One of them, located in the Gironde estuary has been specialized in Siberian sturgeon *Acipenser baeri* farming. The Martin hurricane resulted in a 20% loss of the rearing biomass including mature adults, with an unknown number of individuals released into the nearby Gironde estuary. Although professional fishermen were contracted to catch the released animals, the escapees might interbreed with the local and endangered listed European species *Acipenser sturio.* Although likely underestimated, shipping activities accounted for 42% of the inadvertent introductions of nonindigenous species: 21% are presumably spread by ballast waters/sediments and 21% as hull fouling. Other plausible initial vectors of introduction include private release (1 sp.), transport in freshwater barrels (1 sp.), and introduction as parasites of mussels and eels (4 spp.). For 12% of species, the mechanisms of introduction are still unknown. Of the 104 nonindigenous species recorded in the wild in the studied area, 70 have been clearly documented as established. Fourteen species do not appear to be established and one species (*Crassostrea angulata*), following a large expansion by shellfish culture, is now restricted to Southern Portugal. For the remaining 20 species, their present population status is uncertain (few records of isolated individuals or no recent records).

2.2 SPECIFIC CASES

2.2.1 The protist Bonamia ostreae

The acetosporan *Bonamia ostreae*, pathogenic to bivalves was first noticed on Tudy Island (Brittany, France) in June 1979. This protist induces flat oyster gill ulceration and breakdown of connective tissues, massive haemocytic infiltration resulting in animal death (Renault 1996). Following this first sighting, the disease spread quickly to almost all the European rearing areas, occurring now in Spain, the Netherlands, England, Ireland, and inducing severe impacts on natural populations as well as on cultured stocks. The French flat oyster production was drastically reduced by this pathogen (Comps et al. 1980). A microcell disease, similar to bonamiasis was described in California in the 1960s and was known to occur in several flat oyster populations along the western coast of North America (Elston et al. 1986). Monoclonal antibodies and studies on ultrastructural characteristics and host susceptibility later demonstrated that the Californian microcell was actually *B. ostreae.* The vector is well known and directly linked to the shellfish industry by exports of infected oysters from a Californian hatchery to French and Spanish waters (Grizel 1997; Cigarria & Elston 1997).

2.2.2 The serpulid polychaete Ficopomatus (Mercierella) enigmaticus (Fauvel, 1923)

This species was first noticed in France in 1921 (Fauvel 1923). Local population outbreaks showing rapid build-up were recently reported in the Bay of Veys and Honfleur harbour (Normandy), in southern Brittany (Lorient and Vannes harbours) and on the Atlantic southwest coast of France (Poitou-Charentes). Although without significant environmental impacts reported, these outbreaks had several impacts on harbour management and structures (e.g., pipe clogging, blocking tide-gates) as well as on ships. The latter vector facilitates spread by dispersal of mobile adults on ships' hulls. A national monitoring and enquiry was carried out in 1999 to assess the species distribution and impacts. Following its first sighting, this species spread quickly to the western coastline (Rance estuary, Brittany, in 1925) as well as to the eastern part (Tancarville channel in 1937). Meanwhile the species was observed at La Rochelle (NE Bay of Biscay) (1937). In 1949, individuals were observed in several places in northern Spain, while in 1954 the species was largely distributed along the Atlantic coastline from southern Brittany (St Nazaire, Loire estuary), Les Sables d'Olonne, Ré Island, to Le Boucau, St Jean de Luz and the Bidassoa River. Moreover, the species was firstly observed in the Mediterranean lagoons in 1999, thus demonstrating a large adaptation capacity to various environmental conditions. This species is now well distributed along the English Channel, Atlantic coastline and the Mediterranean Sea.

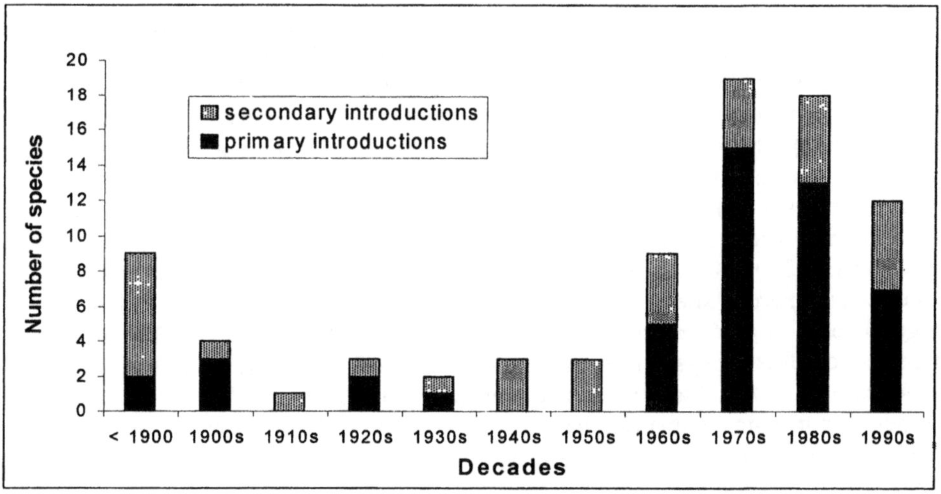

Figure 2. Numbers of recorded nonindigenous species on the Atlantic and Channel coasts of France and Iberian Peninsula by decades. 'Primary introductions' = first European records (excl. Mediterranean); 'secondary introductions' = dispersal from other European areas into which the species were previously introduced.

2.2.3 The slipper limpet Crepidula fornicata (Linnaeus, 1758)

Its first European occurrence was from Liverpool Bay, England, in 1872 and its first record as established species at Brightlingsea, Essex, in 1893 (Minchin et al. 1995). Numerous authors have documented its spread in European coastal waters during the last century. *Crepidula fornicata* currently extends from SW Norway to NW Spain, with no records from Portuguese waters (Blanchard 1997). Due to its detrimental effects on oyster beds (Korringa 1951a) and great-scallop habitats (Grall et al. 1996), *C. fornicata* stocks assessments have been performed in most French bays along the coasts of Brittany and the Bay of Biscay. Several ecosystems such as the Mont-Saint-Michel Bay, the Bay of St Brieuc and the Bay of Brest appear to be largely impacted by this epifaunal suspension-feeder (Grall et al. 1996; Hamon 1996; Ehrhold et al. 1998). On the contrary, the Marennes-Oléron Bay seems to be moderately colonised (Sauriau et al. 1998), due to mechanical cleaning with dredges that local fishermen organisations have performed on a yearly basis since 1980, i.e. 10 years after the record of the first live stocks. Surprisingly, live stocks of *C. fornicata* have been estimated to be extremely modest in Arcachon Bay (de Montaudoüin et al. 2001) although the first specimens were sampled in 1969 in the bay, which could not be considered as a southern limit for the species. Actually, presence of large *Zostera marina* beds in subtidal areas and prohibition of bottom-trawling within the Arcachon lagoon potentially limit the natural or man-made dispersal of the species.

It was often suggested that *Crepidula fornicata* causes deposition of mud in turbid waters of sheltered and estuarine bays, thus rending the beds unsuitable for oyster culture, modifies benthic faunal assemblages and creates a new community with low diversity and competes for food with other filter-feeders (Korringa 1951a; Blanchard 1997). Impact of *C. fornicata* on the seabed of the Mont-Saint-Michel Bay was recently quantified by Ehrhold et al. (1998), who suggested that appearance of large subtidal biogenic mud banks matched the proliferation of *C. fornicata* populations. However, their influence on the diversity of benthic fauna assemblages appeared to be more controversial: a moderate colonisation of *C. fornicata* may increase point diversity (α species richness) as suggested by de Montaudoüin & Sauriau (1999) in the Marennes-Oléron Bay. On the contrary, when colonies spread over large areas of a given ecosystem, decrease of γ species richness and modifications in trophic structures are likely to occur. This process was primarily described in the Bay of St Brieuc, where *C. fornicata* has reached very high biomass (250,000 tons) interacting significantly with the habitats of other species, including commercial species such as the great scallop *Pecten maximus* (Hamon 1996). Large increase in *C. fornicata* biomasses over the last 50 years in the Bay of Brest is suspected to affect ecosystem functioning through feedback in silicic acid recycling and dynamics of phytoplankton blooms (Chauvaud et al. 2000). Finally, recent *in situ* experiments on trophic competition between *C. fornicata* and the oyster *Crassostrea gigas* failed to demonstrate any negative effects of *C. fornicata* on the oyster growth rate (de Montaudoüin et al. 1999).

2.2.4 The new introduced oyster drill Ocinebrellus inornatus (Récluz, 1851)

The Japanese oyster drill, the muricid gastropod *Ocinebrellus inornatus* was reported for the first time in the centre of the Marennes-Oléron Bay in April 1995 by de Montaudoüin & Sauriau (2000). Since then, the species has been regularly collected from oyster parks in the bay, where Pigeot et al. (2000) suggested that its ecological niche seemed to be similar to that of the native oyster drill *Ocenebra erinacea*. Presently, this species has shown a significant northwards spread due to unregulated shellfish transfers between French growing areas while the Marennes-Oléron population has locally exploded. *O. inornatus* is actually recorded in Bourgneuf Bay (2001) and Morbihan Gulf (since 2000), and its spread is likely to increase again in the near future by shellfish transports. The last mild winters have facilitated the species reproduction and juvenile survival, and major impacts on shellfish rearing beds are currently observed in the Marennes-Oléron Bay. Thus, a research program was carried out in the Bay to estimate the species distribution, ecology and habitat features. A complementary approach aims to develop specific molecular markers in order to compare the genetic structure of invasive and native populations, and to determine the invasion history and routes used by *O. inornatus* to colonise French Atlantic coasts from the Pacific coasts (Garcia-Meunier et al. 2002). With a native NW Pacific range, *O. inornatus* is distributed in the NE Pacific as well, after its accidental introduction to Washington State from Japan in the 1920s through inter-continental shellfish trade (Mueller & Hoffmann 1999). However, vectors of introduction in Europe remain speculative and might be related to either ballast waters or uncontrolled overseas shellfish trade.

2.2.5 The predatory gastropod Rapana venosa (Valenciennes, 1846)

This Japanese species was introduced into the Black and Marmara Seas during the 1940s and it has also spread to Aegean and Adriatic Seas. Since its first sighting in 1998 in France, 11 individuals have been observed on shallow (< 5 m) sandy bottom in the same location, a subtidal area of the Bay of Quiberon (Southern Brittany). No geographic expansion has been observed, although local reproduction was recently demonstrated. Except egg cases caught in 2000, all individuals were 5-7 year old adults, and no juveniles have been sighted. The sandy bottom is likely impeding the species spread since the lack of hard substrate limits the spatfall capacity, egg settlement and therefore survival rate. The introduction resulted from international shellfish trade with Italy (Adriatic Sea) for human consumption. Information campaigns at the public and shellfish industry levels were carried out to support an eradication campaign. Although unsuccessful, day and night divings, dredging campaigns and harvest trials by using crab pots have been performed to test an eradication process. Associated risks are linked to the near vicinity of extensive Pacific cupped oyster rearing beds in deep waters (10 m), which are exported to other French rearing areas on a regular basis, being a possible vector for further expansion.

2.2.6 The oysters Crassostrea gigas and C. angulata (Thunberg, 1793)

The Pacific cupped oyster *Crassostrea gigas* was massively introduced between 1971 and 1975 to sustain the oyster industry (Grizel & Héral 1991) as the local *C. angulata*, initially transferred from Portuguese waters, was decimated by a viral disease (Comps 1969). However, unofficial imports from Japan were carried out in 1966 and resulted in seeding in the Bay of Marennes Oléron (Le Borgne et al. 1983). This may have been the vector for disease transfer, which later on affected the *C. angulata* population. Historically, *C. angulata* were imported in 1860 into French waters to maintain landings affected by overfished *Ostrea edulis* natural beds. A shipment had to be jettisoned during a storm in 1868; survivors colonized the Gironde Estuary and spread further north up to the Loire estuary. When *C. gigas* were introduced to reverse the ailing production in several rearing areas along the Atlantic coastline, brood stock was imported from British Columbia (Canada) and seed from Japan. Pathological controls were performed to assess their status in native areas. Controls were carried out at the customs clearance, including histological analysis, presence of predators and commensal species. Depending of the results, the whole sample could be destroyed. Otherwise, before seeding, spat on collectors were immersed in freshwater to destroy fouling organisms and predators. In spite of those measures, several concomitant exotic species were introduced (Table 1) (Gruet et al. 1976). A total of 562 tons of adults and more than 5 billion spat were introduced over a 5 year time span. As soon as 1975, natural spatfall was sufficient to sustain the French oyster production. More recently, Boudry et al. (1998) demonstrated by genetic markers that *C. angulata* was actually a *C. gigas* strain of Taiwanese origin. The massive *C. gigas* imports were therefore a second introduction into European waters. It should be noted that such a massive importation would not be possible again due to international rules such as the enforcement of the ICES Code of Practice (1994).

2.2.7 The decapod crustacean Hemigrapsus penicillatus (de Haan, 1835)

Hemigrapsus was first reported in 1994 around La Rochelle harbour (Atlantic coastline), from where it spread quickly northward to the Loire estuary and southward to Laredo (Spain) (Noël et al. 1997; Noël 1998; Udekem d'Acoz 1998; Gollasch 2000; Breton et al. in press). In Northern Spain, strong populations of *Pachygrapsus marmoratus* tend to limit the *Hemigrapsus* spread (habitat competition). For unknown reasons, the northern distribution limit, Southern Brittany, seems a significant obstacle for further natural colonization. However, a new sighting was reported from wet docks in Le Havre (Normandy, English Channel) and the species has been proliferating there since 1997 (Vincent & Breton 1999). This subtidal population, sampled by scuba diving, is likely to facilitate further extension in nearby waters by shipping activities (English Channel, North Sea). This disjunctive distribution suggests spread by remote dispersal of mobile adults.

2.2.8 The marsh grass Spartina anglica (including S. townsendii and S. alterniflora)

The common cord grass species, initially resulting from the crossing of the North American species *Spartina alterniflora* with the native *S. maritima* (fertile hybrid - occurrence in the UK prior to 1870 - *S. townsendii* being the sterile hybrid) was first observed in France in 1906 (Bay of Veys - Normandy). The first sighting in the Bay of Arcachon (southwest of France) occurred in 1985 to spread quickly since then on the mudflats. Presently, hundreds of hectares have been colonized by this species. Moreover, a second North American species *Spartina versicolor* has been occasionally reported within the Bay since 1901, while *S. alterniflora* was sighted in Bay of Brest, Arcachon Bay and Hendaye (Biscay Bay). This species was intentionnally released in Archachon Bay during the 1970-1980s' and is still present. Those species induce increased sedimentation rates that affect the entire ecosystem. A pilot study was conducted in 1997 to limit the species colonization: quicklime was injected into the mud (15 cm deep) over one hectare to destroy the rhizome. Although not eliminated, it induced a significant decline of the *Spartina* population over time. However, quicklime concentration required for a total destruction (40 t ha^{-1}) is too high for ecological purposes.

3 Administrative and legislative measures to control introductions

The last century has seen many regulations to limit side effects resulting from exotic species introductions at the national, European and worldwide levels. One of the most recent and critical French laws on that matter, called the Barnier Law, was enacted in 1995 to limit those problems. The basic principle in this regulation is '*to avoid any side effect on the environment, wild flora and fauna, it is illegal to introduce any individual of alien species in planned, negligent or risky ways. However, a planned introduction can be authorized by the State following a risk assessment study to evaluate impacts.*' This law modified the Rural Code (Art. L. 211-3). Moreover, a law proposal concerning the prevention and control of the spread of *Caulerpa taxifolia* from the Mediterranean coast is currently under review. If enacted, this law would complement the N°92-3 Law on Water and the N°95-101 Law on natural risk prevention, which already includes regulations to forbid any exotic species introduction into natural ecosystems. However, the main difficulty to enforce such principle is the international trade, which facilitates goods imports-exports for human consumption with logically no release into the environment. Meanwhile, the French law N°C356 was enacted in 1999 to address the 1999/C356/01 EU decision, specifying the management organizations and scientific authorities designated by the State members in agreement with the IX article 1, Convention on the International Trade of Endangered Flora and Fauna Species and concerned by the 12th article of the N°338/97 Council law (December 1996). Similarly, the 1999/C356/02 decision concerns the authorized sites for introduction and exportation specified by the State Members for endangered species described in the 13th article of the previously cited Council law. From a practical point of view and to optimise controls, foreign imports into France have to be declared to the customs in specified locations called PIF (for 'Points d'Introduction France') under the Veterinary Services management. Over the last decade, European laws and Directives have strengthened the controls over exotic species introductions and limit their impacts, learning from severe past events (e.g., bonamiasis) (Berthe & Thébault 2000). As *Bonamia ostreae* is considered a serious pathogen for the European industry, shellfish originating from an infected area should not be moved into a disease-free area. This is the basic principle of the Council Directive 91/67/EC, article 8. This obviously led to a EU monitoring network and rearing area classification system with regard to *Bonamia* occurrence. Moreover a list of exotic diseases of concern for shellfish defined in accordance with OIE organization is annexed to the Council Directive 95/70/EC. A 1998 State Decree (N°98-391) has updated the N°95-100 (01-95) decree regulating the zoosanitary conditions for aquaculture of living molluscs and crustaceans. This decree specifies that every farmer involved in rearing and marketing shellfish must be able to provide a register containing the information on shellfish deliveries (inputs and outputs), total weight, number and origins so as to facilitate tractability of living shellfish products. The registers should be kept at least 4 years and can be controlled by EU experts with State representatives. With regard to shellfish transfers, State decrees were enacted in 1997 to complement previous decrees relative to sanitary conditions for shellfish production and marketing (EU Directive 91/492/CEE). They specify the shellfish transfer modalities with regard to the sanitary zonation, requiring tagging and maximal size for shellfish juveniles allowed in case of transfers among rearing zones and marketing, therefore facilitating tracing the transfers of living shellfish products. As an ICES member, France has participated to the Working Group on Introduction and Transfers of Marine Organisms, and to the specifications of the 'ICES Code of Practice' aimed to specify appropriate management for those exotic species (ICES 1994). With regard to the ballast waters as a vector for exotic species introduction, France has not yet specific regulations, while participating and relying on the IMO organization, which plans to enact a treaty on that matter in 2002 (IMO 1995).

4 Conclusions

The review of species introductions demonstrates that at least 104 species can be listed as exotic - actually a limited number compared to the possible vectors of introduction and to the number of native species (more than 3,000 in Northern Brittany). However, this limited number remains a major issue with regard to impacts on economic activity and environment sustainability: only a few species have demonstrated significant and major side–effects. Appropriate examples are the cases of *Bonamia ostreae*, drastically affecting the flat oyster farming at the European level, and *Crepidula fornicata* in northern Brittany areas. In contrast, one deliberate introduction (*Crassostrea gigas)* can be considered as successful, sustaining a large industry of more than 4,000 companies, 150,000 tons yearly production (yearly exchange value 300 million Euros) in spite of an outdated massive process and several commensal species simultaneously introduced. Eventually, impact on the environment for this case has remained low with a few species permanently established. In contrast, it should be noted that new regulations and increased awareness from the scientific community, with regard to biodiversity and coastal industry, e.g. shellfish and aquaculture, have reduced significantly the cases of deliberate introductions, while the risk-analysis requirement has improved the overall management. However, in spite of those increased efforts, we should emphasize that following the 1970s introduction peak, cases have remained consistent over the last two decades, mostly by accidental or/and increased vectors of introduction.

Presently, the ballast water and sediment can be considered as one of the major issues due to increasing commercial activity and international trade. Besides short-term effects and direct anthropogenic factors, long term impacts resulting from global changes should now be considered as a major issue, as demonstrated by the increased seawater temperature over the last two decades in the Bay of Biscay. This is likely to induce geographic distribution changes for several species, including exotic ones, over a large European coastal range. From a scientific point of view, understanding the invasive species dynamics and processes involved remains a challenging issue to eventually lead to improved management practices and facilitate future decision makings. This is presently addressed by an on-going multidisciplinary research program (INVABIO) funded by the French Ministry of the Environment.

REVIEW OF NON-NATIVE MARINE PLANTS IN THE MEDITERRANEAN SEA

MARIA A. RIBERA SIGUAN
Lab. Botany, Faculty of Pharmacy, University of Barcelona, Barcelona, Spain
ribera@farmacia.far.ub.es

Abstract

In this review 98 marine plant species probably introduced in the Mediterranean Sea have been inventoried (63 Rhodophyta, 20 Phaeophyceae, 11 Chlorophyta, 3 Dinophyta and 1 Spermatophyta). Of these 9 species are invasive and cause ecological or economic impacts. The main introduction vectors are aquaculture activities, the opening of the Suez Canal and maritime transport. Oyster transfer which is probably responsible for 44 introductions is the most important. The time scale evolution of exotic plants in the Mediterranean is exponential, with an increase of 50 species in the two last decades. The western Mediterranean basin is the most receptive zone with 67 non-native species, the majority of them of temperate or warm-temperate affinity with a Japanese or Pacific Ocean origin. In contrast, the 29 non-native species reported from the eastern Mediterranean are mainly warm-water or tropical species coming from the Red Sea or Indian Ocean. The Mediterranean is therefore turning into a reserve of Indo-Pacific species and in the future may become an important exporter of exotic species to other biogeographical regions.

1 Introduction

Biological invasions of plants in marine environments are a phenomenon that is currently of importance due both to the real increase in the number of exotic species in the entire world and to the growing interest, and also concern, that this increase arouses. This current interest contrasts with the lack of attention that has historically been paid to this phenomenon in comparison with terrestrial environment or with invasions of marine fauna. The reasons for this are: (i) a lack of historical knowledge of marine flora on most of the coasts of the world - most floristic studies on marine vegetation are recent - which makes comparison of the evolution of flora composition difficult. (ii) The presence of new marine plants species may easily go undetected. (iii) The taxonomic problems in a large number of systematic groups of algae have led to their confusion with indigenous species. (iv) In most cases, the consequences of marine plant invasions only appear in the long-term. (v) In general, the direct repercussions of marine plant invasions on human activity are much less significant than those of land or even freshwater invasions (Ribera 2001).

On a worldwide scale knowledge about exotic marine flora is not uniform, but varies depending on the geographical area in question, and is almost nonexistent for some areas (Ribera 2001). The Mediterranean Sea, on the other hand, is one of the geographical areas where exotic plant invasions are best documented. This may be due to: (i) a strong historical tradition of algological studies and the high number of Mediterranean algologists (at least in the western and central Mediterranean). (ii) The well documented floristic knowledge of the marine flora in some localities or biogeographical areas of the Mediterranean. (iii) The opening of the Suez Canal in 1869, which brought with it a

E. Leppäkoski et al. (eds.), Invasive Aquatic Species of Europe, 291–310.

large-scale transfer of organisms, a phenomenon which had already aroused the interest of scientists at the close of the 19th century. (iv) The high population concentration and variety of economic activities in the coastal areas which, on the one hand, may be the cause of the high number of marine biological invasions but can, on the other hand, result in a greater interest in learning about, preventing and eradicating them. (v) The recent invasion of *Caulerpa taxifolia* which has, perhaps for the first time, managed to arouse the collective interest of scientists, politicians, administration and the public in general.

2 Current knowledge of non-native species

As we have commented previously, the earliest comprehensive work on exotic marine species, including plants, corresponds to studies on Lessepsian flora and fauna (Por 1978). However, it was during the last decade of the 20th century that some checklists or reviews on non-native marine plants from the Mediterranean were published (Zibrowius 1991; Verlaque 1994; Ribera 1994; Boudouresque 1994; Boudouresque & Ribera 1994; Ribera & Boudouresque 1995; Galil 2000; Boudouresque & Verlaque 2001). Apart from the very numerous works published on the appearances of exotic marine plants in the Mediterranean, we would like to highlight those referring to coastal lagoons, which are high risk areas for biological invasions (Occhipinti Ambrogi 2000; Verlaque 2001). Two reviews on a global scale, including the Mediterranean, have also recently been published on our present knowledge of nonindigenous marine plants (Wallentinus 1999; Ribera 2001).

This work does not aim to be an exhaustive, critical review of introduced plants into the Mediterranean, but rather to give an overall picture of the present level of knowledge. According to Carlton (1985), we consider a species to be introduced only if its presence in a new area is a direct or indirect consequence of human activities. For this reason, we include in this review all the published taxa that are considered nonindigenous to the Mediterranean benthic flora, including both historical introductions and the most recent, regardless of the level of probability of it being an introduced species; when this probability is low it is indicated in the list of species. To make this review we have considered the following types of introductions: (i) introductions in the global Mediterranean biogeographical zone (remote dispersal). These species usually present a natural marginal dispersion into the Mediterranean. In some cases they spread along all the Mediterranean coasts, such as *Codium fragile* ssp. *tomentosoides* or *Asparagopsis armata*; in other cases they develop a population only in the initial zone of arrival, such as *Fucus spiralis* (Ribera & Boudouresque 1995). (ii) Re-introductions of species by different vectors to the initial, but only if this species is new for the country (marginal dispersal by human activities). The presence of *Caulerpa taxifolia* in the Balearic Islands (Spain) might be an example (Meinesz 1992). (iii) Native species that have been introduced in different localities supposedly from exotic populations (remote dispersal). We consider these species should be included since they represent a high risk of genetic contamination for the native species. In some cases, as for example, *Desmarestia viridis* or *Chondria coerulescens* in the Thau Lagoon (France), their presence near shellfish farms lead us to assume they were introduced by oyster transfer from outside the Mediterranean (Verlaque 2001) but only genetic studies will confirm their origin. We consider that these native species are taken into account in this list only when the record is new for the country. For this reason we have eliminated from this catalogue certain species reported as introduced for some specific localities but already present on the coasts of the corresponding country. For example, *Polysiphonia atlantica* is reported as an exotic species in the Thau Lagoon (France) by Verlaque (2001) but this species is widespread throughout the Mediterranean (Gómez Garreta et al. 2001) including the French coasts (Knoepffler et al. 1990). Similarly, *Radicilingua thysanorhizans*, reported as an alien species in the Venice Lagoon (Occhipinti Ambrogi 2000), is reported along the Italian coast and is widespread in the occidental and central Mediterranean basins (Gómez Garreta et al. 2001). Some taxa regarded as introduced by certain authors, but with a wide margin of doubt, have not been included here:

(i) *Cystoseira myrica*, a single report from Egypt considered dubious by Lipkin (1972); the Lessepsian species *Spatoglossum asperum* considered a probable misidentification by Verlaque (1994) because it does not exist in the Red Sea. In the same way the presence in the Mediterranean of the following two possible exotic species, *Chondria collinsiana* and *C. polyrhiza* (from Greece and Tunisia, and from Italy and Greece respectively, Gómez Garreta et al. 2001), requires confirmation (Athanasiadis, pers. comm.),

(ii) Species whose categorisation as introduced is very doubtful. Only subsequent taxonomic, biological, biogeographical and genetic studies, as well as their progression kinetics, will allow us to confirm it. It is probable that some recently reported species in the Mediterranean, which display an evident geographic discontinuity in their area of distribution, may correspond to exotic species. In these cases dispersion in the Mediterranean giving rise to an increase of new reports is to be expected. On the other hand, there are several species described as new to the Mediterranean that could correspond to non-native species that have not yet been described in their area of origin. This occurred in the case of *Antithamnion amphigeneum* described in the Mediterranean for Algeria as *A. algeriense* (Ribera & Boudouresque 1995). *Ceramium graecum*, described for Greece and recorded later in South Italy, might correspond to a similar case. On the other hand, for some species of recent appearance in the Mediterranean, such as *Laurencia microcladia* and *Dipterosiphonia dendritica*, initially considered them - among other explanations - as introduced species (Wallentinus 1999), its later distribution throughout this sea confirms its wide-world tropical affinity,

(iii) Species considered as Lessepsian by Por (1978) but whose world distribution suggests that in fact they must be considered as indigenous species corresponding to relicts of the Tertiary. These are species of tropical affinity distributed in the eastern Mediterranean but also in the western basin and/or in zones close to the Atlantic (Verlaque 1994), *Ganonema farinosum* (=*Liagora farinosa*), *Acetabularia calyculus* and *Polyphysa parvula*. (iv) In addition, the taxa identified only to the genus level have not been included in this work: *Grateloupia* sp. cited by Verlaque (2001) in the Thau Lagoon and *Sorocarpus* sp. cited by Curiel et al. (1999a) in Venice Lagoon. The determination of this species level will be of great interest because *Sorocarpus* is a new genus to the Mediterranean and the algae corresponding to genus *Grateloupia* is probably a new species to the European coasts (Verlaque 2001).

At present our assessment of possible non-native marine plants in the Mediterranean includes 98 species (Table 1). The composition of this exotic flora is as follows: 63 species of Rhodophyta (64%), 20 species of Phaeophyceae (21%), 11 species of Clorophyta (11%), 3 species of Dinophyta (3%) and 1 species of Spermatophyta (1%) (Fig. 1). The red algae are dominant but these macroalgae percentages are proportional to the total number of species in each group, so we cannot relate higher numbers with taxonomic groups at major risk. In contrast, the number of exotic microalgae is very low and it is probably the most underestimated group. Nevertheless, occasional or periodic blooms of some dinoflagellates, toxic or not, in the Mediterranean are known, but there are very few studies available on their relationships with nonindigenous species. These data reflect the lack of knowledge about non-native micro-algae species; the scarcity of floristic studies on the phytoplankton in some geographical Mediterranean regions and the numerous problems in the taxonomy of planktonic microalgae make it difficult to detect non-native species (Ribera 2001). However, in most areas of the world, the proportion of non-native microalgae is always underestimated. It is only in some areas where harmful algal blooms have had disastrous economic consequences, for example in Australia (Hallegraeff 1993), that studies on the introduction of species have been promoted. On a worldwide scale only 15 introduced phytoplanktonic species (basically Dinophyta and Bacillariophyta) have been registered which represents 8% of the total of exotic species (Ribera 2001).

Table 1. List of possible introduced plants in the Mediterranean Sea.
G: Taxonomic group (C) Chlorophyta, (D) Dinophyta, (P) Phaeophyceae, (R) Rhodophyta, (S) Spermatophyta. Date: date of the first record (if the plant is a native of the Mediterranean the date corresponds to the first record of non-native populations). Origin: (A) Atlantic Ocean, (I) Indian Ocean, (P) Pacific Ocean, (IP) Indopacific, (PAN) Pantropical, (RS) Red Sea, (J) Japan, (???) unknown. Species distribution in the Mediterranean (if the plant is a native of the Mediterranean, the distribution corresponds only to the non-native populations): (WMED) presence in western Mediterranean including eastern Sicilian and Taranto Gulf coasts, (A) presence in Adriatic Sea, (EMED) presence in eastern Mediterranean. Pathway: (AQ) aquaria, (BA) ballast, (CUL) algae cultures, (FIS) fisheries, (FO) fouling, (OYS) shellfish farming, (SUEZ) Suez Canal, (???) unknown. Reference: reference of the first record.

SPECIES	G	DATE	ORIGIN	W MED	A	E MED	PATHWAY	REFERENCES
Acanthophora nayadiformis (Delile) Papenfuss (1)	R	1813	RS	X		X	???	Delile (1813)
Acrothamnion preissii (Sond.) E.M. Woll.	R	1969	IP	X			FO	Cinelli & Sartoni (1969)
Acrothrix gracilis Kylin	P	1998	P/A	X			OYS	Verlaque (2001)
Agardhiella subulata (C. Agardh) Kraft & M.J. Wynne (2)	R	1987	A	X			OYS	Cecere (1989)
Aglaothamnion feldmanniae Halos	R	1976	A	X			FO	Sartoni & Sarti (1976)
Ahnfeltiopsis flabelliformis (Harv.) Masuda	P	1994-98	P	X			OYS	Verlaque (2001)
Alexandrium catenella (Whedon & Kofoid) Balech	D	1983	???	X			BA	Vila et al. (2001)
Antithamnion amphigeneum A. Millar	R	1989	???	X			FO	Verlaque & Seridi (1991)
Antithamnion pectinatum (Mont.) Brauner ex Athanas. & Titley	R	1988	P	X	X	X	OYS	Verlaque & Riouall (1989)
Antithamnionella elegans (Berthold) Price & John	R	1882	J	X			FO	Berthold (1882)

SPECIES	G	DATE	ORIGIN	W MED	A	E MED	PATHWAY	REFERENCES
Antithamnionella spirographidis (Schiffn.) E.M. Woll.	R	1914	IP	X	X	X	FO	Schiffner (1916)
Antithamnionella sublittoralis (Setchell & Gardner) Athanas.	R	1988	P	X			FO	Cormaci & Furnari (1988)
Antithamnionella ternifolia (Hook. & Harv.) Lyle	R	1926	P	X			FO	Hooker & Harvey (1845)
Apoglossum gregarium (Dawson) M.J. Wynne (3)	R	1993	???	X			???	Sartoni & Boddi (1993)
Asparagopsis armata Harv.	R	1923	A	X	X	X	FO, OYS	Sauvageau (1925)
Asparagopsis taxiformis (Delile) Trevisan (4)	R	1813	???	X		X	???	Delile (1813)
Audouinella sargassicola (Boergesen) Garbary	R	1950	IP			X	SUEZ	Aleem (1950)
Audouinella spathoglossi (Boergesen) Garbary	R	1950	IP			X	SUEZ	Aleem (1950)
Audouinella subseriata (Boergesen) Garbary	R	1950	IP			X	SUEZ	Aleem (1950)
Bonnemaisonia hamifera Har.	R	1910	A	X			FO	Petersen (1918)
Botryocladia madagascariensis G. Feldmann (5)	R	1991	I	X			???	Cormaci et al. (1992)
Caulerpa mexicana Sond. ex Kütz.	C	1941	RS			X	SUEZ	Rayss (1941)
Caulerpa racemosa (Forssk.) J. Agardh (6)	C	1926	RS	X		X	SUEZ, ???	Hamel (1926)
Caulerpa scalpelliformis (R. Brown ex Turner) C. Agardh	C	1930	RS			X	SUEZ	Hamel (1930)
Caulerpa taxifolia (Vahl) C. Agardh	C	1984	PAN	X	X		AQ	Meinesz & Hesse (1991)
Ceramium strobiliforme G.W. Lawson & D.M. John (7)	R	1991	A	X			???	Cormaci et al. (1992)
Chondria coerulescens (J. Agardh) Falkenb. (8)	R	1995-97	A	X			OYS	Verlaque (2001)
Chondria curvilineata Collins & Hervey (9)	R	1981	A	X			???	Verlaque (1987)
Chondria pygmaea Garbary & Vandermeulen (10)	R	1991	RS	X			SUEZ	Cormaci et al. (1992)
Chondrus giganteus Yendo f. *flabellatus* Mikami	R	1994-98	J	X			OYS	Verlaque (2001)
Chorda filum (L.) Stackh.	R	1981	P/A	X			OYS	Riouall (1985)
Chrysymenia wrightii (Harv.) Yamada	R	1978	J	X			OYS	Ben Maiz et al. (1987)
Cladophora patentiramea (Mont.) Kütz.	C	1992	IP			X	FO, SUEZ	Verlaque (1994)
Cladophoropsis zollingeri (Kütz.) Reinbold	C	1948	RS			X	SUEZ	Aleem (1948)
Cladosiphon zosterae (J. Agardh) Kylin (8)	P	1988	A	X			OYS	Verlaque (2001)
Codium fragile ssp *tomentosoides* (Goor) P.C. Silva	C	1950	A	X	X	X	OYS	Feldmann & Magne (1954)
Colpomenia peregrina (Sauv.) Hamel	P	1956	P/A	X			???	Mendez Domingo (1957)
Dasya sessilis Yamada (11)	R	1984	J	X			OYS	Ben Maiz (1986)
Derbesia boergesenii (Iyengar & Ramanathan) Mayhoub	C	1954	I			X	SUEZ	Iyengar & Ramanathan (1954)
Derbesia rhizophora Yamada	C	1984	J	X			OYS	Ben Maiz (1986)
Desmarestia viridis O.F. Müll. (12)	P	1978	P/A	X			OYS, ???	Verlaque (1981)
Ectocarpus siliculosus var. *hiemalis* (P. Crouan & H. Crouan) Gallardo (13)	P	1999	A		X		???	Bellemo et al. (1999)
Fucus spiralis L.	P	1987	A	X			FIS	Sancholle (1988)
Galaxaura rugosa (Ellis & Solander) Lamouroux	R	1990	RS			X	SUEZ	Mayhoub (1990)
Goniotrichopsis sublittoralis Sm. (14)	R	1992	P	X			???	Magne F (1992)
Gracilaria arcuata Zanardini	R	1931	RS			X	SUEZ	Aleem (1948)
Gracilaria disticha (J. Agardh) J. Agardh	R	1926	RS			X	SUEZ	Lipkin (1972)
Grateloupia doryphora (Mont.) Howe/*G. turuturu* Yamada (15)	R	1982	P/A	X	X		OYS	Riouall et al. (1985)

SPECIES	G	DATE	ORIGIN	W MED	A	E MED	PATHWAY	REFERENCES
Grateloupia filicina var. *luxurians* A. Gepp & E. Gepp	R	1997-98	P	X			OYS	Verlaque (2001)
Grateloupia lanceolata (Okamura) Kawaguchi	R	1970	J	X			OYS	Verlaque (2001)
Griffithsia corallinoides (L.) Trevisan	R	1984	P/A	X		X	OYS	Ben Maiz (1986)
Gymnodinium catenatum Graham (16)	D	1989	???	X			BA	Bravo et al. (1990)
Halophila stipulacea (Forssk.) Ascherson	S	1894	RS	X		X	SUEZ	Fritsch (1895)
Halothrix lumbricalis (Kütz.) Reinke (17)	P	1985	???	X			OYS	Ben Maiz (1986)
Herposiphonia parca Setchell	R	1997	IP	X			OYS	Verlaque (2001)
Heterosiphonia japonica Yendo (18)	R	1998	J	X			OYS	Verlaque (2001)
Hypnea cornuta (Kütz.) J. Agardh (19)	R	1948	RS			X	SUEZ	Aleem (1948)
Hypnea esperi auctorum	R	1972	RS			X	SUEZ	Lipkin (1972)
Hypnea nidifica J. Agardh	R	1928	RS			X	SUEZ	Forti (1928)
Hypnea spicifera (Suhr) Harv. (20)	R	1993	IP			X	SUEZ	Aleem (1993)
Hypnea spinella (C. Agardh) Kütz. (21)	R	1977	PAN	X			???	Gómez Garreta et al. (1979)
Hypnea valentiae (Turner) Mont. (19)	R	1898	RS/P	X		X	SUEZ, OYS, FO	Reinbold (1898)
Karenia brevis (Davis) G. Hansen & Moestrup	D	1972	???			X	BA	Satsmadjis & Friligos (1983)
Laminaria japonica Aresch.	P	1976	J	X			OYS, CUL	Anonymous (1982)
Laurencia okamurae Yamada (22)	R	1984	J	X			OYS	Ben Maiz (1986)
Leathesia difformis (L.) Areschoug (23)	P	1979	A	X			OYS	Verlaque (1981)
Lithophyllum yessoense Foslie	R	1994-97	J	X			OYS	Verlaque (2001)
Lomentaria hakodatensis Yendo	R	1979	J	X			OYS	Verlaque (2001)
Lophocladia lallemandii (Mont.) F. Schmitz	R	1918	RS	X	X	X	SUEZ	Petersen (1918)
Monostroma obscurum (Kütz.) J. Agardh	C	1985	P/A	X			OYS	Ben Maiz (1986)
Padina boergesenii Allender & Kraft	P	1965	RS	X		X	SUEZ	Ramon & Friedmann (1965)
Padina boryana Thivy	P	1981	IP			X	SUEZ	Nizamuddin (1981)
Pilayella littoralis (L.) Kjellm. (24)	P	1985	P/A	X			OYS	Ben Maiz (1986)
Pleonosporium caribaeum (Boergesen) R.E. Norris	R	1974	PAN	X			FO	Ardré et al. (1982)
Plocamium secundatum (Kütz.) Kütz.	R	1991	???	X			???	Cormaci et al (1991)
Polysiphonia fucoides (Huds.) Grev. (25)	R	1988	A	X			FIS	Verlaque & Riouall (1989)
Polysiphonia harveyi Bailey (26)	R	1958	???	X			???	Lauret (1967)
Polysiphonia morrowii Harv.	R	1997-98	P	X			OYS	Verlaque (2001)
Polysiphonia paniculata Mont. (27)	R	1967	P	X	X		OYS	Lauret (1970)
Porphyra yezoensis Ueda	R	1975	J	X			OYS	Anonymous (1982)
Prionitis patens Okamura	R	1994-96	J	X			OYS	Verlaque (2001)
Pterosiphonia tanakae Uwai & Masuda (28)	R	1998	J	X			OYS	Verlaque (2001)
Punctaria tenuissima (C. Agardh) Grev. (29)	P	1999	???		X		OYS	Bellemo et al. (1999)
Rhodophysema georgii Batters	R	1978	P/A	X			OYS	Verlaque (1981)
Rhodothamniella cf. codicola (Boergesen) Bidoux & F. Magne	R	1952	A	X			OYS	Bidoux & Magne (1989)
Rhodymenia erythraea Zanardini	R	1948	RS,IP			X	SUEZ, FO	Aleem (1948)
Sarconema filiforme (Sond.) Kylin (30)	R	1948	RS	X		X	SUEZ	Aleem (1948)
Sarconema scinaioides Boergesen (31)	R	1980	IP			X	SUEZ	Diapoulis et al. (1985)
Sargassum muticum (Yendo) Fensholt	P	1980	J	X	X		OYS	Pérez et al. (1984)
Scytosiphon dotyi M.J. Wynne (32)	P	1978	P	X	X		OYS	Giaccone (1978)
Solieria dura (Zanardini) F. Schmitz	R	1950	RS		X		SUEZ	Aleem (1950)
Solieria filiformis (Kütz.) Gabrielson	R	1988	A	X			???	Cecere (1990)
Spatoglossum variabile (Fig.) De Not.	P	1950	RS			X	SUEZ	Aleem (1948)
Sphaerotrichia divaricata (C. Agardh) Kylin	P	1981	J	X			OYS	Riouall (1985)
Stypopodium schimperi (Kütz.) Verlaque	P	1982	RS			X	SUEZ	Mayhoub (1989)

SPECIES	G	DATE	ORIGIN	W MED	A	E MED	PATHWAY	REFERENCES
& Boudouresque								
Ulva pertusa Kjellm.	C	1984	IP	X			OYS	Ben Maiz (1986)
Undaria pinnatifida (Harv.) Suringar	P	1971	J	X	X		OYS, CUL	Pérez et al. (1981)
Womersleyella setacea (Hollenb.) R.E. Norris	R	1987	PAN	X	X	X	FO, FIS	Verlaque (1989)

Notes of Table 1

(1) This species may well correspond to a pre-Lessepsian species (Aleem, 1948) although some authors point out the possibility that it may be a relict of the Tethys (Cormaci et al. 1982; Verlaque 1984; Giaccone & Geraci 1989; Ribera & Boudouresque 1995).

(2) Perrone & Cecere (1994) pointed out the possibility that this species may be a Tethys relict. According to Verlaque (2001) the record of *Solieria chordalis* from the Thau Lagoon (Ben Maiz 1986) corresponds to *Agardhiella subulata*. Other records of *S. chordalis* (French coast, Knoepffler et al. 1990) should be confirmed.

(3) Sartoni & Boddi (1993) suggest that if it is a non-native species, its introduction was not recent. In relation to the presence of *A. gregarium* in Spain, Clavell & Polo (1998) consider that it does not correspond to an introduced species because it presents a large distribution and a complete reproductive cycle.

(4) This species, described from Egypt and with a pantropical distribution, could correspond to a relict of the Tethys (Rayss 1954; Verlaque 1994). Ballesteros & Rodríguez-Prieto (1996) suggest, for the specimens found in the Balearic Islands (Spain), a natural dispersion from Atlantic populations via Straits of Gibraltar.

(5) Species previously known only from Madagascar, Natal and probably along the coast of southern Africa (Cormaci et al. 1992). Norris (1989) suggests that *B. madagascariensis* is a species of warm-temperate affinity and that its occurrence in Madagascar is probably due to local cool currents.

(6) This species includes *C. racemosa* var. *turbinata-uvifera*, *C. racemosa* var. *lamourouxii* f. *requienii* and *C. racemosa* aff. var. *occidentalis* (Verlaque et al. 2000).

(7) This species, described from the African coasts, is reported for the first time in the Mediterranean in Sicily (Cormaci et al. 1992) without references to possible introduction. The recent record from the Adriatic Italian coasts (Furnari et al. 1999) might confirm an exotic species spreading.

(8) Native species introduced in the Thau Lagoon, probably by oyster transfer from the Atlantic (Verlaque 2001).

(9) Species reported with doubt. *C. curvilineata* was reported for the first time in the Mediterranean from Corsica in 1981 (Verlaque 1987). Verlaque (1994) indicated the possibility of it being an indigenous or introduced species, but its later presence in other Mediterranean localities (Gómez Garreta et al. 2001) may confirm the spreading of an exotic species.

(10) Species possibly introduced as epiphyte of *Halophila stipulacea*, a non-native Mediterranean seagrass (Cormaci et al. 1992).

(11) Species recorded from the Thau Lagoon as *Dasya hutchinsiae* (Ben Maiz 1986), and later as *Dasya* sp. (Verlaque (2001).

(12) *D. viridis* has occasionally been found in the Mediterranean but its appearance in the Thau Lagoon (new record from France) probably corresponds to the arrival of Atlantic or Pacific strains (Verlaque 1994). Verlaque (1981) suggested that this species should be considered as a relict boreal species confined to the cold parts of the Mediterranean.

(13) Species of Atlantic affinity but recorded only from the eastern Mediterranean (Ribera et al. 1992).

(14) This microscopic Rhodophyceae, only reported in the Mediterranean in Majorca by Magne (1992), can be easily confused with other indigenous species i.e *Stylonema cornu-cervi* (Ribera & Boudouresque 1995).

(15) *G. doryphora* was first recorded along European coasts in England (Farnham & Irvine 1973) and it has spread along the Atlantic European and Mediterranean coasts. This species has been considered an alien species and its probable origin by oyster transfer has been indicated by different authors (Farnham 1980; Verlaque 1984; Gargiulo et al. 1992; Ribera & Boudouresque 1995; Wallentinus 1999; Ribera 2001). Verlaque (2001) indicated that this species does not exist along the coasts of Japan and Korea; on the other hand he pointed out that *G. turuturu*, endemic of these two countries, is in good agreement with the introduced alga until now attributed to *G. doryphora*. While awaiting further studies, we include the two species together.

(16) The presence of this species in the Mediterranean needs to be confirmed.

(17) Species with boreal affinity, cited from the coasts of Sicily and Turkey (Ribera et al. 1992) but considered as introduced in the Thau Lagoon (new record from France) (Verlaque 2001)

(18) Species recorded from the Thau Lagoon as *Dasysiphonia* sp. (Verlaque 2001).

(19) *H. cornuta*, *H. hamulosa* and *H. valentiae* have sometimes been cited in the Mediterranean as synonyms (Mayhoub 1976). Further taxonomic studies are required.

(20) It is a synonym of *H. harveyi*.

(21) It is a synonym of *H. cervicornis* (Haroun & Prud'homme van Reine 1993).

(22) *L okamurae* is a synonym of *L. coronopus* J.

(23) Species with a large distribution, cited in the Mediterranean only from the Black Sea and Tunisia (Ribera et al. 1992) but considered as introduced in the Thau Lagoon (new record from France) (Verlaque 2001). Verlaque (1981) suggested this species should be considered as a relict boreal species confined to the cold parts of the Mediterranean.

(24) Species cited in the Mediterranean (Ribera et al. 1992) but considered as introduced in the Thau Lagoon (new record from France) (Verlaque 2001).

(25) Species widely distributed in the Mediterranean (Gómez Garreta et al. 2001) but considered as introduced in Prévost Lagoon (new record from France) (Verlaque & Riouall 1989).

(26) According to Verlaque (2001) this species is synonymous of *P. mottei*, species widespread in the Mediterranean (Gómez Garreta et al. 2001) and probably a long-standing introduction in the Mediterranean.

(27) Species reported with doubt. *P. paniculata* is considered introduced in the Thau Lagoon but it has not been found since 1967 (Verlaque 2001). This species is cited in the Mediterranean from Corsica, Adriatic and Black Sea (Gómez Garreta et al. 2001).

(28) Species recorded from the Thau Lagoon as *Pterosiphonia* sp. (Verlaque 2001).

(29) This boreal species of wide distribution is reported in the Mediterranean only from the French coast and the Black Sea (Ribera et al. 1992). Recently it has been cited from the Lagoon of Venice (Italy) by Bellemo et al. (1999) and Occhipinti Ambrogi (2000) cited it as a possible introduced species.

(30) Rays (1963) proposed that this species might be a relict of the Tethys.

(31) Records of *S. scinaioides* and *S. filiformis* in the Mediterranean may in fact correspond to the same species (Ribera & Boudouresque 1995).

(32) On the other hand, *S. dotyi*, the single report of which from Trieste (Italy) was questioned by Verlaque (1994) and Ribera & Boudouresque (1995), has been reported in the Venice Lagoon (Curiel et al. 1996) and the Thau Lagoon (Verlaque 2001) confirming thus its presence in the Mediterranean.

3 Introduction pathways of non-native species

It is always difficult to know the arrival pathway of each exotic species into a new area. Except for specific examples in which the introduction vector can be determined with certainty, the vector assigned is the most probable one. The proximity of an introduction source, the kinetics of progression of the species, its biological or ecological characteristics or what has happened to the same species in other areas are taken into account. Nonetheless, the same species may have "used" different pathways depending on the geographic area, or even within the same area. In the Mediterranean some species may have been introduced in different zones by different pathways. For example, for *Caulerpa racemosa*, *Cladophora patentiramea*, *Asparagopsis armata* and *Hypnea valentiae,* two or three vectors of exotic inoculum introduction are described. On the other hand, some species may have an initial arrival vector and then use others for their dispersion in this sea such as *Caulerpa taxifolia* and *Womersleyella setacea.*

In spite of this, the introduction vector for some exotic species remains unknown (15 species, 14%). Within this group we could include the *Caulerpa racemosa* populations which are currently spreading along the Mediterranean, with an invasive behaviour. Verlaque et al. (2000) suggested that three distinct taxa of *C. racemosa* co-exist in the Mediterranean: *C. racemosa* var. *turbinata-uvifera*, *C. racemosa* var. *lamourouxii* f. *requienii* and *C. racemosa* aff. var. *occidentalis*. The first two may correspond with doubt to a Lessepsian species (Por 1978; Giaccone & Di Martino 1995) and the third to a new introduction according to Verlaque et al. (2000). These latter authors suggest three possible introduction vectors: deballasting of ships, transport in other parts of ships or aquaria activities. Genetic studies on *C. racemosa* samples from the Canary Islands, western Australia, Panama and the Mediterranean show a high inter- and intra-individual polymorphism in the rDNA ITS1, which is most likely the result of mechanisms as well as genetic drift (Famà et al. 2000). These authors point out that the intercalation of western Australian samples in the Mediterranean clade of the ITS1 tree suggests that the invasive Mediterranean variety of *C. racemosa* may be an introduction from western Australia.

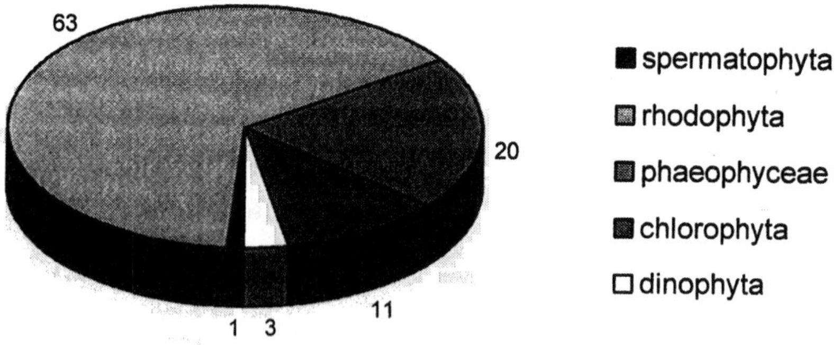

Figure 1. Total number of non-native plants in the Mediterranean Sea, separated into systematic groups.

The main introduction vectors of alien species known in the Mediterranean, taking into account the global dispersal mechanisms of marine organisms proposed by Carlton (1994), are the following: (i) aquaculture activities (44 species, 41%), (ii) the Suez Canal (28 species, 26%), (iii) maritime transport (17 species, 15%), (iv) fishing activities (3 species, 3%) and (v) aquaria trade (1 species 1%) (Fig. 2).

(i) Aquaculture activities provide various different pathways for exotic species introduction: deliberate introduction of species for aquaculture purposes, the escape of individuals from controlled cultures and the accidental introduction of organisms associated with aquaculture species transfer. In the Mediterranean 44 introductions of exotic marine algae were related to this vector (25 species of Rhodophyta (57%), 15 species of Phaeophyceae (34%) and 4 species of Chlorophyta (9%) (Table 1).

Concerning the alga culture it should be added that at the beginning of the 1980s individual of *Chondrus crispus* from the Atlantic coast of France and *Hypnea musciformis* from Senegal were introduced into Corsica for farming in tanks (Mollion 1984). These fish farms no longer exist but we would like to point out the risk involved in introducing geographically isolated strains of species present in the Mediterranean, such as *H. musciformis*, that can lead to a genetic contamination with unpredictable consequences (Ribera & Boudouresque 1995). On the other hand *Laminaria japonica*, besides having been introduced in the Thau Lagoon by oyster transfer, is now cultivated on the open sea along the nearby coasts (Perez et al. 1984; Wallentinus 1999).

The massive importation of the Japanese *Crassostrea gigas* oyster for cultivation has been the principal cause of the transfer of a large species number of Indo-Pacific origin to the Mediterranean. This fact is demonstrated by the direct relation between the location of shellfish farming sites and the number of exotic species. One of the most important centres for the cultivation of this species, the Thau Lagoon in France, is at present one of these sanctuaries of non-native species where 44 non-native taxa have been identified (Verlaque 2001). Other aquaculture sites in the Mediterranean are now developing, such as the Venice Lagoon in the Adriatic Sea, where the number of exotic species is increasing: 8 taxa (Occhipinti Ambrogi 2000), 14 taxa (Verlaque 2001). One risk of this high number of exotics in shellfish culture is that these sites become potential sources of invasion in new areas (Ribera 2001): (i) the nonindigenous naturalised species can increase their distribution in the region by other vectors, such as maritime transport or natural dispersion. In fact some Indo-Pacific species from the Thau Lagoon, *Lomentaria hakodatensis*, *Porphyra yezoensis*, *Agardhiella subulata* (as *Solieria chordalis*), *Leathesia difformis, Sargassum muticum* and *Undaria pinnatifida*, have colonised part of the open sea (Verlaque 1994). (ii) From these sites spats or adult shellfish are transferred to other aquaculture farms. In light of the legislation currently in force, the numerous introduced algae present in the Thau Lagoon have a high probability of spreading throughout Europe and other Mediterranean countries (Verlaque 2001).

(ii) The opening in 1869 of the Suez Canal put the eastern Mediterranean Sea in contact with the Red Sea, and linked two biogeographical marine provinces that had been separated for several million years, the Atlantic-Mediterranean and the Indo-Pacific (Galil 2000). This contact made possible a species migration, called Lessepsian species, that was considered by Por (1989) as the largest bio-geographic phenomenon witnessed in contemporary oceans. The direction of migration has been fundamentally from the Red Sea to the Mediterranean due to the dominant South-North current, but it is not known if the transfer of organisms has been by natural expansion or by maritime transport. The majority of these Lessepsian species have remained localised in the eastern Mediterranean, referred to as the Lessepsian Province (Por 1990) delimited by the water temperatures and the Atlantic current. Only very few species have reach the occidental basin, such as the higher plant *Halophila stipulacea* (Bibliotti & Abdelahad 1990) which is located on the Italian coast, and the red alga *Sarconema filiforme* located on the French coast (Knoepffler et al. 1990). The spread of Lessepsian species throughout the eastern Mediterranean has followed different patterns. In some cases the species spreads in all directions, for example the brown alga *Stypopodium schimperi* has colonised both northern and southern coasts of the eastern Mediterranean (Verlaque & Boudouresque 1991). However, more frequently the expansion is mainly to the north following the dominant current into the eastern basin (Verlaque 1994; Ribera 1994).

A total of 28 marine plants: 17 species of Rhodophyta (61%), 4 species of Phaeophyceae (14%), 6 species of Chlorophyta (21%) and 1 species of Spermatophyta (4%) have been reported with this vector in the Mediterranean (Table 1). We have included in this group the Pre-Lessepsian species, species brought by ship through navigable canals built before the Suez Canal, which linked the Gulf of Suez with the Nile (Por 1971; Aleem 1948; Verlaque 1984).

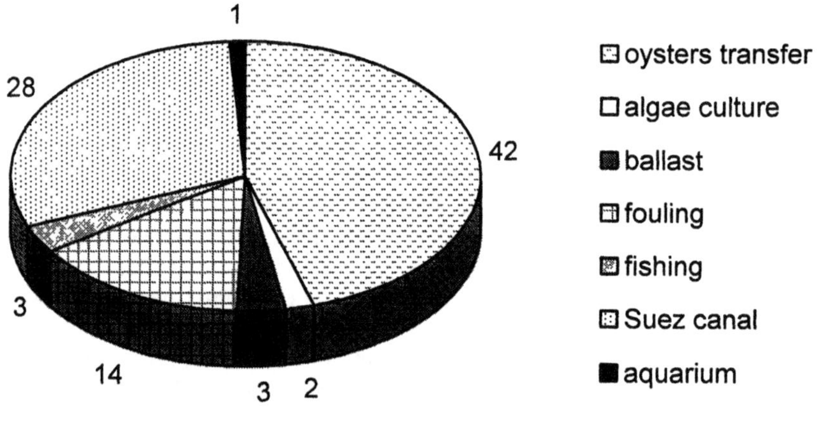

Figure 2. Number of non-native plants introduced by each pathway in the Mediterranean Sea.

(iii) Maritime transport may have been the principal mechanism for global dispersal of marine organisms. This pathway includes different possibilities but with regard to plants in the Mediterranean we can distinguish only the transport on ships' hulls (13%) and the deballasting process (3%) (Table 1). We must also mention the transport on other external parts of the ship that can contribute to the marginal dispersal of marine exotic plants.

The transportation of plants on the hulls of ships is certainly the most ancient vector of non-native species introduction, and it is probable that some cosmopolitan species correspond to ancient introduction by this pathway, such as some Acrochaetiales, Ectocarpaceae or Cladophora species (Ribera & Boudouresque 1995). In the Mediterranean Sea the dominant fouling species belong to Rhodophyta (98%) above all the Ceramiaceae family, such as *Acrothamnion preissii*, *Aglaothamnion feldmanniae*, *Antithamnion amphigeneum* or *Antithamnionella sprirographidis*, and most of them produce blooms in the initial naturalisation phase or periodically, depending on environmental conditions (Ribera 2001).

Concerning other parts of the ship on which plants can be transported, we can mention the anchors and the propellers. Knoepffler-Péguy et al. (1985) suggest that the expansion of *Sargassum muticum* outside Thau Lagoon (France) was caused by small boats. Thus the dissemination of *Caulerpa taxifolia* across the Mediterranean was probably caused by successive anchorage, often over long distances, by pleasure yachts between the French coast and the Balearic Islands or the Italian coast (Meinesz 1992). By following the most frequent pleasure routes in the Mediterranean we can reconstruct the spreading of this species from the French Mediterranean coast (Meinesz et al. 1998). Ecophysiological studies on *C. taxifolia* confirm that this species survives long periods in the prevailing conditions of an anchor locker (Sant et al. 1996). This vector makes it possible to transport vegetative fragments of plants, and if they are tolerant to desiccation, to a wide range of temperatures and to the dark, their survival in a new area is probably ensured.

We have already commented on the scarcity of exotic plants related to ballast transport in the Mediterranean, where maritime traffic has been historically important and is at present very intense. Deballasting may be responsible for the introduction mainly of microorganisms but it is also considered as a possible introduction vector for some macroalgae and higher plants, as indicated for other maritime regions (Novaczek & McLachlan 1989), through the transport of propagules, vegetative fragments or fruits. The absence of species linked to transfer by ballast waters may be attributable to the lack of studies on this subject in the Mediterranean. The presence of *Alexandrium* spp. has been reported from the Adriatic Italian coasts (Cabrini et al. 1996). At the same time, *Alexandrium catenella* seems to present a recent spread along the western Mediterranean coasts: Spain (Vila et al. 2001), France (Masselin et al. 2001) and Sardinia (Lugliè et al. 2001), being always cited in ports or neighbouring zones. This fact induces these authors to consider them an introduced species from discharged ballast waters but at the same time, all of them point out the possibility of a shellfish culture origin. *Gymnodinium catenatum* has been associated with some PSP outbreak in the Mediterranean Sea, in southern Spain, Morocco and western Italy (Bravo et al. 1990). These authors discuss whether the species was possibly native to Atlantic European coasts or if it was introduced; in fact this species presents a disjunct distribution area (Europe, Australia, Japan and Mexico) and for Australian populations an introduction by maritime transport has been proposed (Wallentinus 1999). Finally, another Dinophyta *Karenia brevis* , has been reported from Greek coasts (Satsmadjis & Friligos 1983). This species, known from the Gulf of Mexico and Florida, was found in 1997 as far north as North Carolina causing closure of shellfish harvesting areas (Wallentinus 1999).

(iv) Fishing activities account for three exotic taxa in the Mediterranean (Table 1). Within this pathway we include the use of plants as packing for fish bait, but also involuntary transport by fishing nets. While packing can be responsible for remote transport of organisms, transport by nets is more common as a marginal vector.

The danger of wrapping bait with seaweeds is the risk of their later release into an environment in which they can naturalise. Two examples are known on the French coasts in the Mediterranean. The brown alga *Fucus spiralis* of Atlantic distribution forms a small population in the Gruissan Lagoon (Sancholle 1988) and the red alga *Polysiphonia fucoides* is cited from the Prévost Lagoon (Verlaque & Riouall 1989). The latter species is reported throughout the Mediterranean (Gómez Garreta et al. 2001). On the other hand fishing nets may play an important role in the dispersal transport of invasive species. These sometimes produce a very extensive biomass, either floating or easily detachable, which gets caught in the nets. This pathway involves two species, filamentous red algae belonging to the Ceramiales: *Womersleyella setacea* and *Acrothamnion preissii*. These species are spreading in the western Mediterranean and nets may be the most direct vector of their expansion.

(v) The aquaria industry in Europe, which is regulated by insufficient and inadequate legislation, involves the constant commerce of exotic species. The most sought after species tend to be tropical since they are both exotic and adaptable to aquarium conditions. The aquaria trade can thus act as a pathway for the introduction of exotic species due either to accidental escape through open water renewal circuits or to the deliberate release of aquarium contents. However, although the risk of species escaping from an aquarium is high, only one plant species is identified with this introduction pathway, the tropical green alga *Caulerpa taxifolia*.

The presence of *Caulerpa taxifolia* in the Mediterranean has been recorded since 1984 on the coasts of Monaco, where it probably escaped from an aquarium through a water circuit opening into the sea (Meinesz & Hesse 1991). The individuals of this species used in private or public aquaria in Europe seem to have the same origin: the Wilhelma Zoologish-Botanischer Garten in Stuttgart (Meinesz & Boudouresque 1996). Genetic studies on *C. taxifolia* individuals from several European aquaria and from some Mediterranean populations have demonstrated the ITS rDNA sequence identity of all these specimens (Jousson et al. 1998). These results therefore confirm the hypothesis of the aquarium origin of the Mediterranean population of this tropical species.

The deliberate release of aquarium contents could explain the appearance of *Caulerpa taxifolia* in Des Lecques Harbour on the French coast. The presence of a few blocks of tropical coral reef just beside the population of *C. taxifolia* indicates that the content of a personal aquarium had been emptied into the sea (Laborel 1992).

In conclusion, in the Mediterranean Sea the dominant vector for non-native marine plants is shellfish transfer (42 species), followed by the Suez Canal (28 species) and fouling (14 species) (Fig. 2). The transport of non-native species via Suez Canal, which until the beginning of the nineties was the largest vector (Ribera 1994; Ribera & Boudouresque 1995), has now been relegated to second place, at least for flora, because of the alarming increase of introduced species related to the importation of *Crassostrea gigas*. None of the other pathways are important in terms of the number of species involved.

4 Invasive species

The tropical alga *Caulerpa taxifolia* has given rise to one of the most spectacular marine invasions in the Mediterranean Sea. The Mediterranean specimens have different morphological and physiological characteristics from the tropical populations, which maximise their competitiveness with other species of marine flora, thus favouring their invasive behaviour (Boudouresque et al. 1996). Mediterranean specimens form extremely dense meadows (more than 8,000 leaves m^{-2}) with leaves that may exceed 60 cm in length (Meinesz & Hesse 1991), whereas in warm seas, these meadows are generally patchy and the leaves range from 2 to 15 cm and rarely exceed 25 cm. Today *C. taxifolia* is reported along the coasts of five countries in the northern Mediterranean (Monaco, France, Italy, Spain and Croatia); about 6,000 ha have been affected by this invasion (Meinesz et al. 1998). Recently, a new population of this species has been detected on the Tunisian coast in the southern Mediterranean (Langar et al. 2000). Vegetative reproduction occurs through the ramification of horizontal axes and subsequent fragmentation through natural necrosis or mechanical breakage. Sexual reproduction has been detected in a population of Majorca (Spain) but only male gametes have been observed (Meinesz 1992). *C. taxifolia* contains some potentially toxic terpenoids of which caulerpenyne, specific to the *Caulerpa* genus, is predominant (Lémée et al. 1993). Some of these terpenoids are synthesized in higher concentrations by Mediterranean populations than by tropical ones (Guerriero et al. 1992). The impact of *C. taxifolia* on the Mediterranean communities is pretty well known. The rate of impoverishment of Mediterranean algal communities may reach 75%; most of autochthonous algae tend to disappear dramatically (Verlaque & Fritayre 1994). The number of Polychaeta and especially Amphipoda species decreases in the *C. taxifolia* meadow; on the contrary, the species diversity of Mollusca may increase (Bellan-Santini et al. 1994). Concerning fishes populations, in sites highly colonised by *C. taxifolia*, the mean number of species per census, the mean fish density and the mean biomass are significantly lower (Harmelin-Vivien et al. 1996).

As we indicated before, the spread of *Caulerpa racemosa* along the Mediterranean may be due to an invasive variety of this species. This recent invasion is less known than the *C. taxifolia* one but the first results of its spread are quite alarming. Up to now this variety has been recorded from Spain, France, Italy including Sardinia, Sicily and Lampedusa Island, Greece including Crete, Turkey, Cyprus and Libya (Verlaque et al. 2000). The success of this species can probably be attributed to its ability to colonise a variety of substrates and depths, and to respond to environmental changes through a high degree of morphological plasticity (Modena et al. 2000). This species shows important seasonal dynamics with the greatest stolon growth from June to August -average growth can reach 1.26 cm per day - and an inactive period between December and April (Piazzi & Cinelli 1999). The first studies on its impact on the indigenous benthic communities indicate that macroalgal assemblages present an impoverishment; species number and cover decrease and the community structure changes because the encrusting species are more affected than the erect ones (Piazzi et al. 2001). Concerning the competiviness with the native seagrass *Posidonia oceanica,* Ceccherelli et al. (2000) observed that, although the growth of *C. racemosa* is abundant on the edges of the *P. oceanica* meadows and under *P. oceanica* dead mat, the growth of the alga inside the meadow is greatly influenced by the seagrass density. The results of studies on the interaction between the two invasive species of *Caulerpa* suggest a negative effect of *C. racemosa* on the performance of *C. taxifolia* (Piazzi et al. 1999).

Sargassum muticum, one of the world's most widespread indigenous species (Critchley et al. 1990), occurs in the Venice Lagoon (Italy) (Curiel et al. 1999b) and several localities of the Languedoc-Roussillon coast, including the Thau Lagoon (France) (Belsher & Pommellec 1988). On the Spanish Mediterranean coasts (Catalonian and Balearic), only drift material has been found (Gómez Garreta 2001). The dense *S. muticum* populations inhibit the recruitment and growth of other algae (Critchley 1983). In the Thau Lagoon, the disappearance of much of the macroalgae species under these dense populations has been observed (Gerbal et al. 1985). For example in this lagoon, *S. muticum* has eliminated the indigenous species *Cystoseira barbata* in certain stations (Boudouresque 1994). This invasive species has colonised the aquaculture facilities of the Thau Lagoon forming a very dense canopy that reduces available light and water circulation, thus causing the decrease of oyster growth (Verlaque 2001). At the same time, the proliferation of *S. muticum* hinders the oyster farmers' handling, due to, among other reasons, the weight increase, since the additional load can be as much as 14 kg per rope (Lauret et al. 1985).*Undaria pinnatifida* has not shown the same invasive behaviour in the Mediterranean as in other parts of the world such as New Zealand (Hay 1990). This species occurs in three aquaculture zones on the western basin: the Thau Lagoon (France), the Venice Lagoon and Mar Piccolo (Italy) (Cecere et al. 2000). These authors suggest that the presence of large canopy-forming *U. pinnatifida* might play a positive ecological role in the Mar Piccolo where the marine vegetation is very poor, by providing food and shelter to many organisms. However, on the other hand they point out that *U. pinnatifida* zoospores might compete for the substrate with mussel larvae under the *Mytilus galloprovincialis* cultivation ropes. In aquaculture sites, long individuals of *U. pinnatifida* like *Sargassum muticum* contribute towards reducing light availability and inhibit oyster growth (Verlaque 1994). In the open sea, on the other hand, specimens of *U. pinnatifida* grow less and do not appear to hinder indigenous Mediterranean communities (Boudouresque et al. 1985)

The third large brown alga of Japanese origin, *Laminaria japonica,* occurs only in the Thau Lagoon (Pérez et al. 1984). Initially it had a rapid development and the accumulation and decomposition of large quantities of this species resulted in anoxia in summer (Anonymous 1982). *L. japonica* has not colonised the coasts near the Thau Lagoon as occurred with *Undaria pinnatifida* and *Sargassum muticum,* and has, moreover, not been observed in this lagoon since 1989 (Verlaque 2001)

Some species give rise to invasive processes in very specific zones where they become locally dominant, generally only at certain periods of the year. For example the red alga *Asparagopsis armata,* covers as much as 100% of the substratum in winter in some localities of the north-western Mediterranean, although it virtually disappears for the rest of the year (Ribera & Boudouresque 1995). Associated with *Codium fragile,* it is the dominant species in certain infralittoral communities in the Marseilles area (Ribera & Boudouresque 1995). Similarly *Womersleyella setacea,* an exotic species probably introduced by fouling, forms an almost monospecific layer some 5 cm thick on the coralligenous communities of the Scandola Reserve (Corsica) (Boudouresque 1994). This species produces temporal blooms that clog up fishing nets (Verlaque 1989). The same is true for *Acrothamnion preissii,* which produces invasive processes on Italian coasts and is called "pelo" by Italian fishermen due to its impact on fishing (Cinelli et al. 1984). *A. preissi* and *W. setacea* form dense turfs in some localities of the western Mediterranean, which trap sediments forming a stratum that prevents the development of other algal species (Piazzi & Cinelli 2000). An important proliferation of *Lophocladia lallemandii,* species widely distributed in the Mediterranean, has been observed in the Balearic Islands (Patzner 1998). It is still too soon to evaluate the impact of *L. lallemandii* populations but this author indicated the following repercussions. Several benthic animals such as molluscs and fishes have not been observed any more, benthic algae diversity has decreased and benthic animals have disappeared from the sandy areas covered by dead *L. lallemandii.* During the 60s *Codium fragile* enjoyed a period of expansion on the Marseilles coasts which caused massive accumulations on the beaches where it had to be eliminated mechanically (Boudouresque 1994). In coastal lagoons and closed bays, the accumulation and decomposition of large quantities of *C. fragile* can create seasonal problems in relation to the indigenous communities (Verlaque 1994). The sporadic proliferation of *Cladophora patentiramea* along the coasts of Cyprus caused significant build-up of this plant on the beaches thereby causing negative repercussions on the tourism activities (Boudouresque 1994).

5 Origin and distribution of non-native species

As a consequence of both its complex geomorphological history and variations in environmental conditions, several biogeographical regions are differentiated within the Mediterranean. The principal and most clearly defined are the western basin defined as a warm-temperate region and the eastern basin defined as a sub-tropical region. The two regions are delimited by the Sicily Canal that acts as a biogeographical barrier for species dispersion in both directions. This canal together with the Adriatic Sea, the Messina Detroit, the Alboran Sea and the Gulf of Lions are isolated enclaves of Atlantic-boreal affinities (Cinelli 1985; Verlaque 1994; Bianchi & Morri 2000).

The presence of non-native naturalised species and their subsequent dispersion will depend as much on their focus of introduction as on their ecological requirements. According to this work, 73 species are present in the western basin, 35 in the eastern and 15 in the Adriatic (Table 1). However, taking into account the introduction focus, 67 species have been introduced in the western Mediterranean, 29 in the eastern and 3 in the Adriatic. These figures clearly reveal that the occidental basin is the Mediterranean's main receptive region of exotic marine plants. This could be due to the fact that it is in this basin that we find the previously mentioned high ecodiversity and wide temperature range of sea water, that allows the naturalisation of species belonging to a wide biogeographical affinity range. At the same time we have to consider the greater demographic pressure in this basin that involves an increase in number and strength of the introduction vectors, as we shall see later.

It should be emphasised that only six exotic species occur along all the Mediterranean coasts. *Antithamnion pectinatum*, *Antithamnionella spirographidis*, *Asparagopsis armata*, *Codium fragile* ssp. *tomentosoides* and *Womersleyella setacea* come from the occidental basin and only *Lophocladia lallemandii* comes from the oriental basin. These results will corroborate Verlaque's observation (1994), that each of these two Mediterranean basins contains its contingent of introductions and that exchanges between the two regions are rare, the most frequent being from west to east.

However, eight species occur in both basins and not in the Adriatic, *Acanthophora nayadiformis*, *Asparagopsis taxiformis*, *Caulerpa racemosa*, *Griffithsia corallinoides*, *Halophila stipulacea*, *Hypnea valentiae*, *Padina boergesenii* and *Sarconema filiforme*. All of these, except *G. corallinoides,* are Erythrean species introduced via the Suez Canal. This would suggest that the dispersion from east to west is greater than would be expected, although this may be due in part to the fact that we have included the Gulf of Taranto and the eastern Sicilian coasts in the occidental basin. On the other hand, different introduction vectors are suggested for western populations of two of these species, *C. racemosa* and *H. valentiae.* As is to be expected the majority of these species are warm or temperate-warm species.

In contrast, the species common to the occidental basin and the Adriatic, *Grateloupia doryphora*, *Sargassum muticum*, *Scytosiphon dotyi* and *Undaria pinnatifida,* are cold-temperate species, while only *Caulerpa taxifolia*, a species also present in the western basin and the Adriatic, is a tropical species. However, Mediterranean *C. taxifolia* specimens can resist temperatures below 10 °C (see Invasions chapter).

The identification of the origin of the exotic species is determined as much by its introduction vector as by its geographical distribution. In several cases the exact knowledge of the pathway also permits us to safely assign a geographical area or even a country of origin. A Pacific origin is attributed to the majority of species introduced by oyster transfer from Japan but in some cases they may have arrived from other Atlantic or even Mediterranean aquaculture facilities (Verlaque 2001). In the same way, the majority of Lessepsian species occurs in the Red Sea but for some of them, for example *Audouinella* spp., only an Indian Ocean origin is cited. In the Table 1 we indicate the most probable origin for Mediterranean exotic species but for 10 of them it remains unknown. For the species with various known introduction vectors, we have indicated the origin for each one where different.

According to our data, 26 species come from the Pacific Ocean (17 from Japan), 9 from Pacific or Atlantic Ocean, 15 from Atlantic Ocean, 13 from the Indo-Pacific region, 21 from the Red Sea and 4 have a pantropical distribution. These results are the logical consequence of the main vectors which operate in the Mediterranean: the Lessepsian immigrants correspond to Erythrean or Indian Ocean species, while the species associated to the *Crassostrea gigas* importation correspond to Pacific species. For this reason we can conclude that the Indo-Pacific Ocean is the main reserve of non-native species of the Mediterranean Sea.

6 Increase in non-native species introductions

The increase in number of non-native marine plants is a generalised fact along all the coasts of the world. On the whole this increase is due to human activities involving a movement of goods and/or people which have also increased both qualitatively and quantitatively (Ribera & Boudouresque 1995). However, this increase does not appear to occur at the same rate in all geographic areas. Among the European coasts for example, the temporal evolution of exotic species number presents a linear increase in Britain (Eno et al. 1977) while in the Mediterranean this increase is exponential (Fig. 3). Boudouresque (1994) point out that more than 50% of them have been observed since 1970, but with these new data we can point out that more than 50% of them have been observed since 1980, or in other words, there has been an increase of 50 species in only two decades. Of course these data are to be taken with some caution since the first observation of an introduced species nearly always occurs some time after its real introduction; in addition they may be indicative of an increase in the intensity of observation over the last decades (Boudouresque 1994).

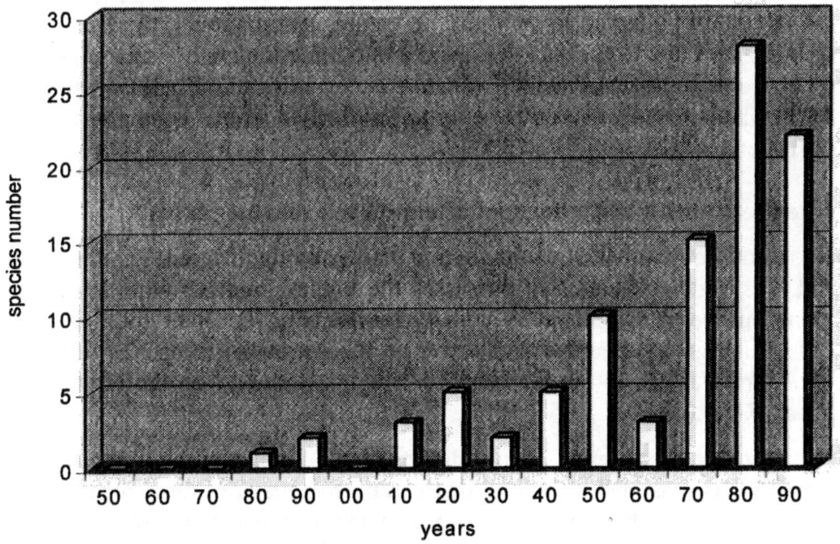

Figure 3. Time-scale evolution of the number of non-native plants in the Mediterranean Sea.

Although it is true that this rise corresponds to a generalised increase in the impact of the introduction vectors, it should be borne in mind that each one presents a different pattern of temporal evolution and that at the same time this pattern can vary according to the geographical zone. In the Mediterranean, as we mentioned above, we are currently witnessing a change in prominence between the introductions caused by the Suez Canal and by aquaculture farming. While the Lessepsian migrations, which up to now had been the predominant force, seem to be levelling off as predicted by Por (1978), the exotic species associated with aquaculture species transfer, which previously occupied second place in importance (Ribera & Boudouresque 1995), are now the dominant vector (Verlaque 20001; Ribera 2001). Fish and shellfish farming are undergoing constant expansion in most countries in the world and in particularly in the western Mediterranean, along the coasts of Spain, France and Italy.

However, the temporal evolution of a pathway will not always necessarily follow the same pattern in the future. For example the Suez Canal vector is decreasing, although any modification in the structure of the Canal could again permit the arrival of new Erythrean immigrant species. At the same time, variations in marine transport routes, such as a change in the location of fishing grounds or a new tourist industry in a small island that had previously had few visitors, can lead to the arrival of new species (Ribera 2001).

Concerning the discharges (intentional or accidental) from aquaria, although this vector has been responsible for the presence of a large number of exotic species in continental waters, it has had little impact (in reference to the species number) in marine waters. However, the behaviour of *Caulerpa taxifolia* in the Mediterranean has brought to light the risk associated with attempting to adapt certain organisms to living in aquarium. In such cases, the strain that best adapts to these conditions may correspond to a unique strain that is not well adapted to surviving in nature. Alternatively, the best strain may originate in aquaria due to the fact that genetic modifications may occur in populations subjected to stress. In this case we will obtain a clonic stock of individuals - since in this environment plants usually reproduce only vegetatively - whose behaviour in nature is unknown (Ribera 2001).

7 Is the Mediterranean Sea a hot spot of non-native marine plants?

A comparison of the number of exotic marine plants among different geographical areas shows that the Mediterranean Sea possesses the highest number with 98 species followed by the European Atlantic coast with 49 species (Fig. 4). Other zones have a lower number of nonindigenous plants: 26 species on the Australian coast, 20 on the North-American Atlantic coast, 20 on the New Zealand coast and 19 on the North-American Pacific coast (Ribera 2001).

Therefore, in comparative terms, the Mediterranean is an exceptional nucleus for non-native species. It is reasonable to think that certain regions are more "receptive" to introductions of species either because they are subject to greater pressure from the vectors of introduction or because they accept potential alien immigrants more easily (Ribera & Boudouresque 1995). According to our results, it seems clear that the Mediterranean can be considered as a zone of high exotic plants' receptivity.

It is obvious that the dense population concentration along the Mediterranean coastlines, especially in the northern part, implies an increase in the strength of the introduction vectors, oyster transfer being the most important. In fact, the Thau Lagoon with 44 exotic species is quickly becoming not only the most important site of macroalgae introduction in the Mediterranean sea, but also in Europe (Verlaque 2001).

What could be the reasons that might favour the naturalisation of new species? The most important would seem to be: (i) high ecodiversity with a great variety of biotopes (coastline structure, types of substrate, light intensity, exposed or sheltered places....).

(ii) Low biodiversity in the eastern basin. This characteristic undoubtedly explains exceptional success of the Lessepsian immigrants (Spanier & Galil 1991). In the occidental basin, low diversity biotopes such as coastal lagoons, ports, estuaries and polluted areas are those which have accepted a greater number of non-native species (Boudouresque 1994). These places are at the same time maximum risk spots since the main species transfer vectors take place there (maritime transport and aquaculture). (iii) Wide range of sea temperature that allows the naturalisation of cold, temperate, warm and tropical species. (iv) Scarcity of both large perennial algae and herbivores. Generally one of the factors that helps the acclimatation of an introduced species is the lack of competitors or predators in the new habitat (Williamson & Brown 1986). For this reason the scarcity of herbivores in the Mediterranean, especially the eastern part, allows the proliferation of nonindigenous plants. (v) Geographical isolation. The relativity high rate of introduced species in New Zealand may be due to the isolation of this region separated from neighbouring regions by natural biogeographical barriers with a high level of endemism (Ribera & Boudouresque 1995).

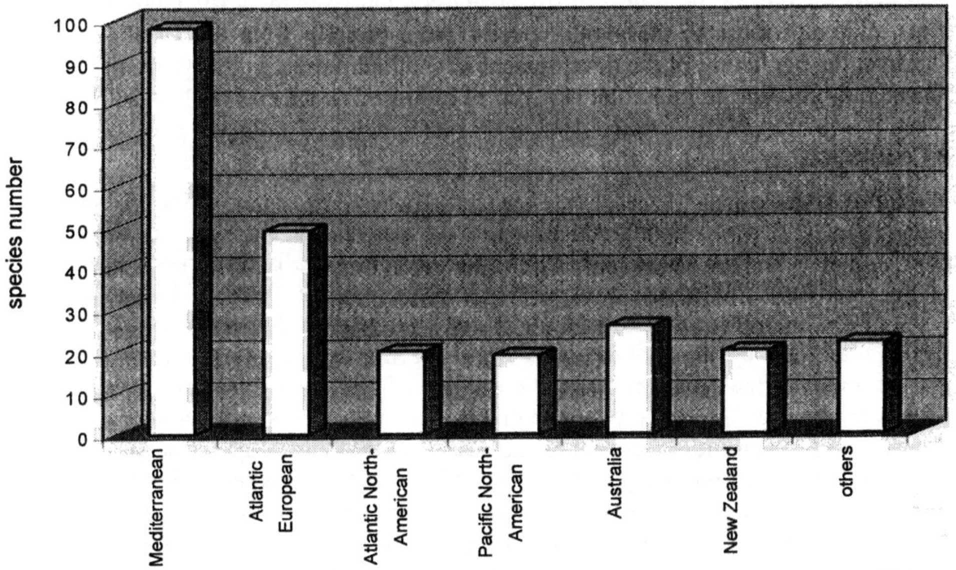

Figure 4. Number of non-native marine plants in each geographical area.

In conclusion the large number of exotic species in the Mediterranean may be the result of a synergy between geological history, environmental factors and anthropogenic activities (Galil 2000). We have already mentioned that the evolution of human activities can modify the intensity of the introduction vectors producing an increase in both the number of arrivals and the size of the inoculum of the species introduced. It should not be forgotten that these human activities can at times give rise to environmental changes that may favour the naturalisation of these exotic species. Among them we should like to emphasize the environmental degradation of some habitats in the Mediterranean that involves a decrease in the biodiversity, the destabilizing of some communities and the proliferation of opportunistic species, all of which facilitate the establishment of new species. At the same time we have to take into account that the waters of the Mediterranean have entered a period of progressive warming related to worldwide climatic changes. This would seem to be the reason not only for the expansion of tropical affinity species in this sea but also for the naturalisation of exotic warm-species.

On the other hand, we have already mentioned the risk of the accumulation of exotic species in the Mediterranean becoming a source of infection for other areas of the world. For the moment there are no explicit references to introduced species originating from Mediterranean exotic populations. However, the presence of *Alexandrium minutum* in Adelaide (Australia) could possibly have arisen from ballast sediment of ships from the Mediterranean Sea (Hallegraeff 1993). RNA sequencing has indicated a remarkable match between Australian and Spanish cultures of this species complex (Scholin & Anderson 1991). However, it has been shown that almost all the species dispersed by shellfish transport on the American coasts are also found on the European coasts introduced by the same vector (Ribera 2001). A clear example is the actual worldwide distribution of non-native populations of *Sargassum muticum*: their expansion kinetics follows the routes of oyster importation (Ribera & Boudouresque 1995). Thus, although most of the exotic species came directly from the Pacific or Indian Ocean at the beginning of the development of shellfish farms, in the future the Mediterranean may become an important exporter of Indo-Pacific species (Ribera 2001).

8 Conclusion

A total of 1,351 marine macrophytes species have been documented for the Mediterranean (Bianchi & Morri 2000). According to our assessment, the total number of possible non-native marine plants (only macrophytes) in this Sea is 95 corresponding to 7% of the total (4.6% of Rhodophyta, 1.5% of Phaeophyceae, 0.8% of Chlorophyta and 0.1% of Spermatophyta). It is difficult to make predictions due to the large number of factors that make up the pool of exotic flora. All that we can say is, that in the case of marine plants in the Mediterranean, the number of introduced species has nearly doubled every 20 years since the beginning of the century (Ribera & Boudouresque 1995). Taking into account more recent data and assuming that this range of increase will continue, the number of nonindigenous species will most likely increase about 100 species over the next 20 years. In fact Boudouresque & Ribera (1994) suggested that in the year 2050 the number of exotic marine macroalgae in the Mediterranean might be between 250 and 1,000; in this latter case the number of nonindigenous species would be of the same order as the number of indigenous species.

Acknowledgements

I would like to thank Dra Amelia Gómez Garreta for her continuous help during the preparation of this manuscript, Dr Marc Verlaque for his valuable comments, Dr. Jordi Rull Lluch for the graphics and Ms Alison Henderson for reviewing the English text.

CURRENT STATUS OF AQUATIC INTRODUCTIONS IN ITALY

ANNA OCCHIPINTI AMBROGI
University of Pavia, Department of Genetics and Microbiology, Ecology Section, Pavia, Italy
occhipin@unipv.it

Abstract
Italy is a crossroads for nonindigenous species introductions into the Mediterranean, both for marine species coming from the Red Sea, favoured by the warm climate of Southern Italy, and for species that find good opportunities in the eutrophic Adriatic Sea. It is also affected by the recent severe invasion in the Mediterranean involving the green algae, *Caulerpa* spp., that are supplanting the native phanerogams constituting the submarine landscape. In the last 40 years, many publications on aquatic species invasions have appeared in the literature; a summary is drawn from scattered material of this overlooked phenomenon. In total 111 nonindigenous aquatic species are know from Italian waters. Also in fresh waters the incidence of introductions is high, as testified by the well-documented cases of fishes, crayfish and zebra mussel. In some cases (the infralittoral beds of *Caulerpa* and the brackish waters of the Lagoon of Venice), the ecological consequences of the invasions have been investigated in more detail, shedding light on the impact of introductions on the Italian aquatic communities. An increasing awareness of the importance of species invasions is slowly emerging from the academic to the legislative and diplomatic arena.

1 Physical and geographical features of the Italian peninsula

The location of Italy in the south of Europe in the western part of the Mediterranean Sea, and the mountainous features of large part of its territory are peculiar and complex. As a consequence of orography and climate the biological communities are highly diverse. Inland waters comprise large pre-alpine lakes of glacial origin and many other smaller lakes, both in the Alps and the Apennines, of which some have volcanic origin. Only one large river, the Po, has a drainage basin comparable to that of the main European rivers, most of the others have stream character. Along the coast some coastal lakes remain as part of the ancient systems of wetlands, the main ones (e.g. Venice) being located in the northern Adriatic Sea. This shallow sea, with a tidal range up to 1 m, receives fresh water from the Po and most of the large rivers coming from the Alps. The Adriatic Sea (Fig. 1) shows peculiar physiographic and hydrodynamic features with a climate more similar to the Atlantic region than other Mediterranean basins. This peculiarity, known as "Northern Adriatic sub-atlantism", is confirmed by the presence of species that have disappeared elsewhere in Mediterranean Sea but are common in the Atlantic (Sacchi et al. 1990).

2 Origin and biogeographical characteristics of the native fauna and flora

Six biogeographical marine sectors surround the Italian peninsula: northern, central, and southern Adriatic to the east, Ionian Sea to the south and Ligurian and Tyrrhenian Seas to the west. All these compartments differ for geological history and environmental conditions (climate and hydrography) and harbour a characteristic fauna and flora giving reason to the high diversity of Italian coastal waters. Over geological time the Mediterranean Sea experienced drastic changes. Since its formation two periods have been

E. Leppäkoski et al. (eds.), Invasive Aquatic Species of Europe, 311–324.

particularly important: (i) the isolation of Mediterranean waterbody from the Indo-Pacific and Atlantic ones since the beginning and at the end of the Miocene (10 to 6 My), which should have nearly desiccated the waterbody probably allowing the establishment of stenoecious and stenothermal species, (ii) and the re-opening of the Straits of Gibraltar at the late Pleistocene (5 My), which repopulated the Mediterranean Sea with species of Atlantic origin. As a result, present marine biota are composed of species belonging to several biogeographical categories which can be summarised as: (i) temperate Atlantic-Mediterranean species (ii) cosmopolitan-panoceanic species, (iii) endemic species (iv) subtropical Atlantic species (remnants of the interglacial period (v) boreal Atlantic species (remnants of the glacial period), (vi) migrants from the Suez Canal (vii) migrants from the Gibraltar Strait (Bianchi & Morri 2000).

3 Role of nonindigenous fauna and flora

3.1 MARINE HABITATS

Introduced marine species are only a small percentage of the total number of species known to the Italian coasts, which are part of the highly diverse Mediterranean ecosystem (Minelli et al. 1995). Introductions have been reported more and more often in the last four decades. These have been algae (32 species), phanerogams (1), invertebrates (61) and fish (17) (Table 1). Their abundance is very high at selected localities, with the exception of *Caulerpa* species that are invasive in large parts of the littoral.

The occurrence of the main invaders is described below, with a particular emphasis on the Adriatic Sea, where the highest densities of nonindigenous species (NIS) are recorded.

3.1.1 Marine fishes

Some fish species have been introduced intentionally through aquaculture, while others are unintentional introductions or part of the so-called Lessepsian migrations. The introduced marine fish species used in aquaculture in Italy are the flounder *Paralicthys olivaceus* (Pleuronectidae), the red drum *Sciaenops ocellatus* (Sciaenidae) and the sea bream *Pagrus major* (Sparidae). *P. major* has been introduced since the end of the 1980s and it is currently reproduced in several commercial hatcheries. Intensive breeding techniques have been adopted in land-based farms, yielding an annual production of 100 tons. The same species is also utilised for hybrid production with other sparids. Brood stock of *S. ocellatus* and *P. olivaceus* have been introduced since 1994. The juveniles, reproduced under controlled conditions, were cultured in intensive and semi-intensive systems. *S. ocellatus*, a rustic species resistant to a wide range of environmental conditions, has recently been observed in semi-natural habitats used for extensive aquaculture, the so called "valli da pesca" in the northern Adriatic. No new pathogens associated with this species have been reported for the time being. Quantitative risk analyses are not available to assess the potential genetic and disease impacts related to introduced species.

Apart from species found only in confined aquaculture plants, 17 NIS have been reported from the Italian coasts (13 from Atlantic region and 4 from Indo-Pacific). At present only one invader, *Seriola fasciata,* is a new target species for fisheries.

The recently recorded *Seriola carpenteri* (Pizzicori et al. 2000) and the saddled snake eel *Pisodonophis semicinctus* (Insacco & Zava 1999) have spontaneously entered the Mediterranean from the Atlantic Ocean. *Stephanolepis diaspros*, a Lessepsian migrant, first recorded in the Ionian Sea (Catalano & Zava 1993), was found again near the harbour of Palermo and at Lampedusa Island in 2000. This species, however, requires further taxonomic studies. The marlin, *Makaira indica*, was first recorded in the 1980s (Orsi Relini & Costa 1986). Two specimens of the toadfish *Chaunax suttkusi* (Ragonese et al. 2001) and one of the tiger shark *Guleocerdo cuvier* (Celona 2000) have recently been found in the Sicily straits. Apparently there is no evidence of human-mediated introduction of these species.

3.1.2 Marine invertebrates

Aquaculture is one of the main ways of introduction, especially for molluscs. The prawn *Marsupenaeus japonicus* was introduced at the end of the 1970s owing to its higher growth rate compared with the native *Penaeus kerathurus*. It is reared in many aquaculture plants, however, since 1989 its production had reduced by 34% until 1997, when the estimated production was 19 tons. The occasionally found *M. japonicus* might form a self-sustaining population, as it is already known from the oriental basin of the Mediterranean (Lessepsian migrant). In fact, it has been recorded in an open habitat only once (Lumare & Casolino 1986).

Penaeus monodon was introduced for aquaculture at the end of the 1980s, but failed to give the expected results, as it did not adapt to the Mediterranean climate. *Penaeus vannamei* is reared in ponds using semi-intensive systems. The introductions were planned under controlled and restrictive conditions for semi-intensive rearing in ponds. This was done especially considering the risk of introducing pathogen agents such as *Baculovirus*, a pathogen which has to be notified to the OIE (Office of International Epizooties) when larvae and postlarvae are utilised. Fertilised eggs are preferred to reduce the risk of disease.

The main species of invertebrates introduced in the open environment for fishery are the bivalves *Crassostrea gigas* and *Tapes philippinarum* (Ghisotti 1971a,b; Cesari & Pellizzato 1985). The Pacific oyster *Crassostrea gigas* was introduced in the 1960s in the northern Adriatic from Atlantic cultures. It is now widespread, colonising all northern Adriatic lagoons, having replaced the native *Ostrea adriatica* (=*edulis*?) whose populations were already rather scanty.

The Manila clam *Tapes philippinarum* is cultivated in ponds, without control of possible release to the open environment, but in the majority of cases it has been planted deliberately into open waters. It was first introduced in the year 1983 as an experiment near one island close to Venice and was afterwards subsequently planted over large areas of the main northern Adriatic lagoons (Pellizzato 1990). The catch of the Manila clams has since then increased, supplanting the catch of the native clams such as *Tapes decussatus* and *Venerupis aurea*, whose populations had been declining already before the introduction. Very dense populations have established in many brackish water areas, especially in the Lagoon of Venice and the Po River Delta. The density of these populations is highly fluctuating as a result of the over-exploitation pressure and the anoxic crises experienced in this area. In the last years, import of adults or seed has ceased, because the natural recruitment in the Italian areas is very strong.

Table 1. List of nonindigenous marine species found in Italian waters.

ALGAE AND HIGHER PLANTS
ALGAE
Acanthophora najadiformis (Delile) Papenfuss
Acrothamnion preissii (Sonder) EM Wollaston
Agardhiella subulata (C.Agardh) Kraft & MJ Wynne
Aglaothamnion feldmanniae Halos
Antithamnion amphigeneum A Millar
Antithamnion pectinatum (Montagne) Brauner in Athanasiadis & Tittley
Antithamnionella spirographidis (Schiffner) E.M. Wollaston
Asparagopsis armata Harvey
Bonnemaisonia hamifera Hariot
Caulerpa taxifolia (Vahl) C Agardh
Caulerpa racemosa (Forsskål) J Agardh
Caulerpa scalpelliformis (R Brown ex Turner) C. Agardh
Chondria pygmaea Garbary & Vandermeulen
Codium fragile (Suringar) Hariot ssp. *tomentosoides* (Goor) PC Silva
Colpomenia peregrina Sauvageau
Desmarestia viridis (OF Muller) Lamouroux
Grateloupia doryphora (Montagne) M Howe
Griffithsia corallinoides (Linnaeus) Trevisan
Ectocarpus siliculosus (Dillwyn) Lyngb. var. *hiemalis* (Crouan frat. ex Kjellm.) Gallardo
Hypnea spinella (C. Agardh) Kützing
Lophocladia lallemandii (Montagne) F Schmitz
Padina boergesenii Allender & Kraft
Plocamium secundatum (Kützing) Kützing
Polysiphonia fucoides (Hudson) Greville
Polysiphonia morrowii Harvey
Punctaria tenuissima (C Agardh) Grev.
Radicilingua thysanorhizans (Holmes) Papenfuss
Sargassum muticum (Yendo) Fensholt
Solieria filiformis (Kützing) Gabrielson
Sorocarpus sp.
Undaria pinnatifida (Harvey) Suringar
Womersleyella setacea (Hollenberg) RE Norris
PHANEROPHYTA
Halophila stipulacea (Forsskål) Ascherson
INVERTEBRATES
CNIDARIA
Garveia franciscana (Torrey)
Diadumene cincta Stephenson
ANNELIDA, POLYCHAETA
Amphicorina eimeri (Lagerhans)
Branchiomma luctuosum (Grube)
Desdemona ornata Banse
Lumbrineris inflata Moore
Lysidice collaris Grube
Mediomastus capensis Day
Metasychis gotoi (Izuka)
Rapana venosa (Valenciennes)
Rissoina spirata (Sowerby)
Sclerodoris cfr. *tuberculata* Eliot
TUNICATA
Botrylloides violaceus Oka
Microcosmus exasperatus Heller
FISHES
Abudefduf vaigiensis (Quoyand Gaimard)
Beryx splendens Lowe
Chaunax suttkusi Caruso
Diodon hystrix (L)
Galeocerdo cuvier (Peron & Le Sueur)
Pristis pectinata Latham
Monticellina dorsobranchialis (Kirkegaard) (=*Tharyx heterochaeta*)
Notomastus aberrans Day
Pileolaria berkeleyana (Rioja)
Rhodine gracilior Tauber (=*Rhodine loveni*)
Spirorbis marioni Caullery & Mesnil
Streblosoma hesslei (Day)
BRYOZOA
Celleporella carolinensis Ryland
Tricellaria inopinata d'Hondt & Occhipinti Ambrogi
Arachnoidea protecta (Harmer)
Electra tenella (Hincks)
PYCNOGONIDA
Ammothea hilgendorfi (Böhm)
Anoplodactylus californicus (Hall)
CRUSTACEA, COPEPODA
Acartia grani Sars
Acartia tonsa Dana
Pteriacartia josephinae (Crisafi)
CRUSTACEA, PERACARIDA
Caprella scaura Templeton
Elasmopus pectenicrus (Bate)
Paracerceis sculpta (Holmes)
CRUSTACEA, DECAPODA
Callinectes danae Smith
Callinectes sapidus Rathbun
Dyspanopeus sayi (Smith)
Marsupenaeus japonicus (Bate)
Percnon gibbesi (H Milne, Edwards)
Portunus pelagicus (Linnaeus)
Rhithropanopeus harrisii (Gould)
Scyllarus caparti Holthuis
Sternodromia spinirostris (Miers)
Thalamita gloriensis Crosnier
MOLLUSCA, BIVALVIA
Anadara inaequivalvis (Bruguière)
Anadara demiri (Piani)
Brachidontes pharaonis (Fisher)
Crassostrea gigas (Thunberg)
Eastonia rugosa (Helbling)
Musculista senhousia (Bensor in Cantor)
Perna picta (Von Born)
Pinctada radiata (Leach)
Saccostrea cucullata (Born)
Tapes philippinarum (Adams & Reeve)
Xenostrobus securis (Lamarck)
MOLLUSCA, GASTROPODA
Bursatella leachii De Blainville
Cerithium scabridum Philippi
Chromodoris quadricolor (Rueppell & Leuckart)
Crepidula fornicata (Linnaeus)
Doris bertheloti (d'Orbigny)
Hipponix conicus (Schumacher)
Odostomia (Megastomia) cfr. *sicula* Philippi
Polycerella emertoni Verrill
Makaira indica Cuvier
Pinguipes brasilianus Cuvier & Valenciennes
Pisodonophis semicinctus Richardson
Pomadasys stridens (Forsskål)
Rhizoprionodon acutus (Rüppel)
Seriola fasciata (Bloch)
Seriola carpenteri Mather
Sphoeroides pachygaster (Muller & Troshel)
Sphyrna mokarran (Rüppel)
Stephanolepis cfr. *diaspros* Fraser-Brunner
Synagrops japonicus (Doderlein)

The data on the catch of the Manila clam are controversial and might be substantially incorrect: yield estimates of 60,000 tons have been made for the years 1995 to 1997. In 1999 yield estimates of 25,000 tons in the lagoon of Venice and 17,000 tons in the area of the Po River Delta were reported. The catch increment between 1987 and 1997 has been about 150 times. No mention of parasites has been reported yet, while many parasites such as the plathelmints *Bacciger* and *Stilochus* and the protozoans *Ancistrocoma* and *Nematopsis* are known for the native clams.

A selection of unintentionally introduced species are commented below in taxonomical order.

The sabellid polychaete *Branchiomma luctuosum*, which origin is probably from the expansion of Red Sea species into the Mediterranean, is common and abundant in harbours and lagoons. Other five species (*Lysidice collaris*, *Metasychis gotoi*, *Monticellina dorsobranchialis*, *Notomastus aberrans*, *Rhodine gracilior)* are likely Lessepsian migrants. Some fouling polychaetes have been introduced, the most common being *Spirorbis marioni* and *Pileolaria berkeleyana* that are found both on ship and boat hulls and on hard substrata in harbours. The mud-inhabiting sabellid, *Desdemona ornata* (native to South Africa and Australia) has been found frequently in brackish environments along the Tyrrhenian coast (Lardicci & Castelli 1986) and later in southern England.

The whelk, *Rapana venosa* (Mollusca, Gastropoda) has been known to occur in the northern Adriatic since 1973. It is widespread along the northern and middle Adriatic coast (including the Lagoon of Venice), both on the Italian side and the Slovenian coast (1997). Its southern limit to date is Civitanova Marche (40 km south of Ancona). Rather surprisingly, no damage to the mussel cultures or exploited clam beds has been reported yet. One specimen has been recorded near the Elba Island (Tyrrhenian Sea).

The soft-bottom clam, *Anadara (=Scapharca) inaequivalvis* (Mollusca, Bivalvia) was first recorded in 1969 (Ghisotti 1973). The way of introduction was probably through ballast water discharge from ships in the harbours of the Emilia Romagna coast. It is very abundant on muddy bottoms of the northern Adriatic, where it can easily survive to the dystrophic crisis due to high resistance of anoxic conditions (presence of haemoglobin as respiratory pigment in the haemolymph). It is widespread in many North Adriatic lagoons, including Venice, and has been caught also in southern Adriatic (Bari). It is a nuisance for the fishery of edible clams, because it entangles the screens of drums of turbo-suction devices. At least two new species of the genus *Anadara* have been recorded in 2000 in addition to *A. inaequivalvis*: *Anadara demiri* has been reported from the middle Adriatic (Morello & Solustri 2001); while another species, taxonomically not yet identified, but different from *A. inaequivalvis, A. demiri* and *A. natalensis* (found in the Mediterranean but not in Italy) has been found in the Lagoon of Venice (Mizzan, pers. comm.).

The mussel *Musculista senhousia* is a Lessepsian species, found in the Mediterranean for the first time in 1973 along the Israelian coasts; starting from the 1980s it was also found in brackish basins near Ravenna (northern Adriatic) (Lazzari & Rinaldi 1994). It is commonly found in association with the native *Mytilus galloprovincialis* cultivated on long lines off the northern Adriatic Sea. In addition, it has been found in the Lagoon of Venice in low numbers and with a scattered distribution. Through the transport of seeds of *Tapes* or *Mytilus* it has also spread into the Tyrrhenian Sea (coasts of Sardinia). Recent unpublished data for the Sacca di Goro (Po river Delta) indicate a patchy distribution of this soft bottom mussel, locally attaining a density of 1,000 ind m^{-2} and consequently competing for space with *T. philippinarum*.

Tricellaria inopinata (Bryozoa) was recorded in the Lagoon of Venice in 1982 (d'Hondt & Occhipinti Ambrogi 1985). Its invasion of the Lagoon has been regularly monitored. It has also been found in the Lagoon of Grado and in southern England (Dyrynda et al. 2000). Another Bryozoa, *Celleporella carolinensis,* has been recorded from the Lagoon of Venice (Occhipinti Ambrogi & d'Hondt 1996), and *Electra tenella* has been recorded twice along the Sicilian coast in 1990 (Rosso 1994), but the area has not been surveyed since then.

Acartia tonsa (Crustacea, Copepoda) was first found in the plankton of the Lagoon of Scardovari (Po river Delta) in 1987 (Farabegoli et al. 1989) and has since then replaced the native congeneric *Acartia margalefi* in other lagoons of the Po River Delta and in the Lagoon of Venice (Belmonte et al. 1994).The way of introduction may be ballast water or, more probably the association with aquaculture products, because the species has been found more often in areas of intense aquaculture than near harbours (Belmonte & Potenza 2001). Other two copepods, *Acartia grani* and *Pteriacartia josephinae* were reported for the Ligurian Sea in 1997 (Sei et al. 1999). The latter was first reported in 1974 from the Tyrrhenian Sea (Crisafi 1974; Belmonte 1998).

The introduced *Dyspanopeus sayi* (Crustacea, Decapoda) from NW Atlantic (Froglia & Speranza 1993) is the most widespread crab in the Lagoon of Venice, it is also known from another northern Adriatic lagoon (Marano). *Callinectes sapidus* from NW Atlantic has been repeatedly recorded in the Adriatic and also in other seas (Cavaliere & Berdar 1975), *Callinectes danae* (native to NW Atlantic) has also been found in the Lagoon of Venice (Mizzan 1993). *Rhithropanopeus harrisi* (native to Atlantic coasts of North America) has been recorded in the Lagoon of Scardovari (northern Adriatic) and is common in the brackish lagoons of the Emilia Romagna coast (Mizzan & Zanella 1996). *Pilumnus inermis* was quoted for the first time in 1987 (Di Geronimo & Fredj 1987) in the Messina Strait; it is probably a cryptogenic species and might have been present along the Mediterranean coasts since a long time. It has entered the Mediterranean through the Gibraltar strait and has now a stable population in Messina and in other southern Tyrrhenian locations (Giacobbe & Spanò 2001). It occurs in association with laminarians and the hydrocoral *Errina aspera*, another species known only in the Messina Strait in the Mediterranean and in Moroccan waters in the Atlantic. Recently *Portunus pelagicus* reached the eastern coast of Sicily where it sustained the local fishery for a short time (Ariani & Serra 1969). Other five species of decapods have been recorded since 1950 with a scattered distribution. The most recent invader *Percnon gibbesi,* found at Linosa Island in summer 1999, has been recorded near another small island off Sicily (Lampedusa) together with the fish *Stephanolepis* cf. *diaspros.* The population is well established, comprising mature adults and young individuals. Few individuals were observed also in the Ustica Island and off the main coast of Sicily (Western sector) (Relini et al. 2000).

3.1.3 Algae and higher plants

There are no indications of deliberate introductions of marine plants or algae; it is likely that most of the species were introduced by ballast water discharge, in the fouling of ships or in connection with the transport of aquaculture products.

Phytoplankton composition of the Adriatic Sea has been studied since the 19th century. The appearance of potentially toxic *Alexandrium* spp. is a recent phenomenon (Honsell et al. 1992; Honsell 1993; Cabrini et al. 1996). The very first appearance of *Alexandrium tamarense (=Gonyaulax tamarensis)* in the Adriatic Sea was in August 1982 (Boni et al. 1986), *Alexandrium pseudogonyaulax* was found in June-July 1987 and *A. lusitanicum* in the same year (Boni et al. 1992). It cannot be excluded, however, that these species are cryptogenic, especially because large numbers of resting cysts occur in the shallow water sediments of this sea. For instance, in samples taken in 1990 *Gonyaulax grindleyi (=Protoceratium reticulatum)*, not known from historical taxonomic lists, was present in plankton samples with small abundances, and in long core samples with peaks of 4,000 cysts $gram^{-1}$ sediment, dating around 12.3 and 9.5 ky BP (Boni, pers. comm.).

The main concern of invading species is probably the benthic green alga, *Caulerpa taxifolia* that has continued its invasion from the French coast to the Ligurian Sea and Italian waters in the years 1991-1992 (Relini & Torchia 1992). Since then it has expanded almost along all Italian coasts and has recently been found in the Messina Strait. It forms very large densities in the infralittoral fringe. Another species of *Caulerpa, C. racemosa*, which is of tropical and subtropical origin, has been recorded in various localities along the Italian coasts since the first finding in Sicily in 1993 (Alongi et al. 1993). Its rapid expansion has been related to the fishery activities as individuals become attached to fishing gear, which is then cleaned in other zones, distributing the species (Modena et al. 2000).

A survey of exotic algae was carried out in the enclosed and brackish sector of the Gulf of Taranto (Ionian Sea), the site of an active shellfish culture. *Caulerpa racemosa* was recorded in small patches in the Mar Grande of Taranto near the Cheradi Islands (1996) and it actively propagated in this zone (Buia et al. 1998). More recently, it was collected in the Second Inlet of the Mar Piccolo (Fig. 1), attached on both hard and soft substrata. In 1998, *Undaria pinnatifida* was collected in the Mar Piccolo of Taranto, where it arrived by means of oyster imports from France. Currently, the species is monitored monthly, to investigate its life cycle and propagation (Cecere et al. 2000). *Agardhiella subulata* and *Solieria filiformis* were reported since the second half of the 1980s in the enclosed basin of Mar Piccolo di Taranto, where they now represent a consistent part of the drifting algal bed of the basin (Perrone & Cecere 1994).

Also the Lagoon of Venice has yielded a large number of immigrants. The brown algae *Undaria pinnatifida* and *Sargassum muticum* have built large populations there. The introduction site has been invariably in the Chioggia sector, near the fish market and the sites where imported seafood (mainly shellfish) is treated. Both container water and part of fish food products are directly discharged in the lagoon canals (Curiel et al. 1999).*Polysiphonia morrowii* is a red alga known from Japan, China and Korea, and was recorded for the first time in the Mediterranean Sea. It was found in spring 1999 in

Chioggia and one year later in several places in the centre of Venice (Curiel et al. in press). *Desmarestia viridis* has been found in the same site of Chioggia as the previous described species (Bellemo et al. 2001). It poses a problem because there are only four records of this species in the Mediterranean. One was in the Adriatic in 1849, another in Croatia in 1948 and the recent ones are from Etang de Thau (France 1981) and from Malaga (Spain 1984). It could also be an introduction from Asia with oysters.

Figure 1. Hydrographic features of Italy: main rivers, lakes and other localities quoted in the text are indicated.

Another algae, *Aglaothamnion feldmanniae* (native to Northwest Atlantic coasts) was previously found only once at Leghorn (Tyrrhenian Sea), but has recently been found in Chioggia (Curiel, pers. comm.).

Grateloupia doriphora, known from the Atlantic and the Pacific Ocean, has been found for the first time in 1969 on the coast of Calabria. It lives in eutrophic waters, especially in calm, shallow sites with boulders (De Masi & Gargiulo 1982). *Lophocladia lallemandii,* a Lessepsian species, native to Red Sea and Indo-Pacific region, was found in Sicily. This red alga has a typical feather-like buoyant growth form, and has been found stranded on the beaches in quantities that have caused nuisance to tourists in summer (Furnari & Scammacca 1971). *Womersleyella setacea*, a pantropical species first found at the island of Lampedusa in 1993, forms thick felts, often on the fronds of *Caulerpa* spp., but also on the surface of sediment, where it prevents the settlement of other species (Airoldi 1998).

The only introduced sea grass is *Halophila stipulacea*, apparently a Lessepsian species, found in shallow subtidal soft-bottoms, especially in harbours (Villari 1988).

3.2. INLAND WATERS

3.2.1 Freshwater fishes

The history of intentional NIS introductions in Italian inland waters can be traced back to 1859, when the first introduction of the European salmon (*Salmo salar*) was planned. The introductions of whitefishes were carried out successfully in successive stages in the lake of Como from 1860 to 1895. Later examples, chosen from a list of both successful and failed attempts to introduce species, in order to restock the fishing grounds of lakes and rivers in both the alpine and peninsular regions, include the introduction of rainbow trout and the beginning of its culture in 1895; the introduction and acclimation of catfish (*Ictalurus* sp.) in waters of Veneto in 1906; the introduction and acclimation in 1922 of *Gambusia holbrooki* in the countryside near Rome in order to combat malarial fevers (Gandolfi et al. 1991). The last officially documented action (1974) concerns the introduction of the Argentinean king fish (*Odonthestes bonariensis*) and acclimation in the Lake of Nemi (near Rome) (Natili et al. 1986). Since then, an increasing number of accidental introductions (mainly from eastern Europe) took place, without any scientific control. The introductions were regulated by a large number of local regulations, issued from the regional and provincial authorities, aiming at preserving fish health through control of parasites and diseases.

Undocumented introductions often take place in private pleasure fishing reservoirs, and whence acclimated individuals often escape to natural habitats. Examples are the wels (*Siluris* sp.) and the brown bullheads (*Ameiurus* = *Ictalurus nebulosus*). Paradoxically, the above mentioned strict regulations have had serious consequences on legal restock activities in many streams threatened by overfishing, because opening of new facilities for fisheries has been complicated by increased requirements on hygienic standards. This in turn increases the risk of genetic problems in native populations, which could be used for restocking. Existing regulations has not prevented, however, attempts of unregulated and unsurveyed introductions of species such as *Padogobius martensi* and *Cobitis taenia* in Appennine lakes (Bianco 1991); *Blicca bjoerkna* (Confortini et al.

1994); the roach, *Rutilus rutilus*, found in Piedmont waters (Delmastro & Balma 1990); and *Misgurnus fossilis* (Groppali 1999). A dedicated effort, backed by profound knowledge of the faunistic and biological basis of this serious matter, has been devoted by the Association of Italian Fresh Water Ichthyologists (AIIAD) since the 1980s. They have produced useful maps of fish distribution in many Italian regions, but have not so far been officially or governmentally supported. Not even their Atlas project has been completed yet, due to lack of economic resources.

3.2.2 Freshwater invertebrates

Among invertebrates, crayfish (Gherardi et al. 1999) and the zebra mussel (Mariani et al. 1992) have been objects of accurate studies in Italy. Two species of crayfish, *Procambarus clarkii* and *Orconectes limosus* have recently appeared as breeding populations in many northern and central Italian freshwater habitats. These newcomers cause numerous risks for the native species *Austropotamobius pallipes* and *Astacus astacus,* and probably also for the river crab *Potamon fluviatile.* At present native species are under threat from alien crayfish, which can locally displace native species by direct or indirect competition and favour the spread of diseases. The crayfish plague, caused by the fungus *Aphanomyces astaci* has been spread by *P. clarkii*, which is resistant to the disease, and has affected *A. pallipes*. *P. clarkii*, originating from North America and having been introduced into all continents, arrived in Italy via Spain and is undergoing a great expansion in some sectors of the Po River drainage basin. In central Italy it is widespread in Tuscany, especially in a small lake that accidentally received the entire cultivated stock of a farm. The deterioration of aquatic vegetation in the lake has been imputed to *P. clarkii*'s voracious feeding behaviour. In addition to crayfish plague, *P. clarkii* contributes to the spread of numerous parasitic helminths of vertebrates. *P. clarkii* has recently been reported from another location in central Italy, the lake Trasimeno (Spilinga et al. 2000). *O. limosus* seems to have been imported accidentally into Italy through the introduction of live fish, and it is not actively farmed. The first Italian record in the wild was made in the Iseo Lake in 1991, and later in the Po River basin. It has recently been reported in a small lake in central Italy. A third species, the East-European *Astacus leptodactylus,* has had a limited expansion in Italian waters, and only a few farms are developing its culture for marketing (Gherardi et al. 1999).

The zebra mussel *Dreissena polymorpha* has been found in different parts of Italy as fossils belonging to the quaternary period, but was probably eliminated by cold temperatures during the Wurmian ice age period (Settepassi & Verdel 1965). In modern times the presence of the zebra mussel in Italy was recorded for the first time in 1971 in lake Garda (Giusti & Oppi 1972). From this lake the species invaded lake Ledro, connected to lake Garda by a hydropower plant and a nearby reservoir, the Valvestino Lake (Bianchi et al. 1976; Annoni et al. 1978). In the following years *D. polymorpha* was found in the outlet of lake Garda, the Mincio river, in the lake Iseo and its outlet, the Oglio river and easternmost in the Adige river (Bianchi et al. 1974; Mariani et al. 1992). The species continued its downstream expansion to the river Po, colonising the lower reach until the delta (Bedulli & Franchini 1978). It has also spread to the Como and Maggiore lakes, the Ticino River and to several artificial canals, constituting a complex hydraulic web in the intensively cultivated Po River plain since the 16th century. In the summer 1999 the species was found in the Lake Trasimeno (central Italy), rapidly colonising all available substrata, comprising an individual of *Procambarus clarkii* and the native bivalve *Anodonta anatina. A. anatina* was once very abundant but has been strongly reduced by oxygen depletion near the bottom, as a consequence of the increased eutrophication of the lake (Spilinga et al. 2000). The presence of *D. polymorpha* in southern Italy is uncertain and should be followed up.

The first appearance of *D. polymorpha* in an Alpine lake, which was not connected to already invaded water bodies, is from lake Caldonazzo in 1992 that is located in the Trentino region. It is likely that it was introduced by transplanting from fishing and pleasure boats. In 1995 the whole lake was colonised by *Dreissena*, which covered all possible hard substrates including the shells of *Unio* and *Anodonta* species. The overgrowth was intense enough to prevent the native Unionidae from opening their valves and eventually caused their death. Predation by the coot *Fulica atra* and the moorhen *Gallinula chloropus* has recently been observed. To date no further expansion of the species has been recorded in Trentino (Maiolini, pers. comm.). No severe impacts on water intakes, due to the clogging by *Dreissena,* have been reported, with the exception of an electric generator of the Como Hospital and a paper mill using the water of Garda Lake (Bronzi, pers. comm.).

Along with *D. polymorpha*, another bivalve, *Corbicula fluminea*, known to cause fouling problems in freshwater, has been recently recorded in lotic environments of northern Italy (Fabbri & Landi 1999).

4 Ecological and economic impact

The consequences of the introduction and spread of non-native species along the coasts of Italy have been object to detailed analysis in at least two cases: the coastal vegetated soft bottoms, occupied by *Caulerpa,* and the large brackish water basin of the Lagoon of Venice.

A recent set of studies on the two species of *Caulerpa* (Relini et al. 1998; Ceccherelli & Cinelli, 1999a,b,c) aimed at establishing the actual impact of the invader on the native communities, mainly through experiments on the relationships with autochthonous plants. Vegetative fragmentation plays a leading role in spreading of the tropical seaweed *Caulerpa taxifolia.* Fragment establishment varies in relation to season and a higher recruitment was found in summer, especially in shallow waters. Competition of *C. taxifolia* with the autochthonous seagrass *Cymodocea nodosa* is enhanced in nutrient-enriched sediments, as the invasive weed has been found to grow faster at high nutrient supply outcompeting the seagrass. As far as the interaction between *C. taxifolia* and the seagrass *Posidonia oceanica* is concerned, dense *P. oceanica* meadows have been shown to be less vulnerable to seaweed invasion, and large meadows with high shoot density are less vulnerable. Competitive success of *Caulerpa racemosa* with *P. oceanica* meadows is also a function of seagrass density and of edge-meadow orientation; unhealthy seagrass beds, characterised by low shoot density, are more susceptible to colonisation and seaweed penetration is higher on seagrass margins oriented towards the coastline (Ceccherelli et al. 2000).

Competition between *C. racemosa* and *Cymodocea nodosa* seems to favour the expansion of another seagrass which usually occurs in the same habitat, *Zostera noltii*; hence, the introduced seaweed modifies the interaction between the two. Areas colonised by *C. racemosa* experiences decrease in cover, number and diversity of native macroalgal assemblages: turf and encrusting weed species are readily affected as soon as colonisation begins, while erect species have been shown to survive until one year after the invasion. Structural changes seem to be irreversible and even during winter, when cover and biomass of *C. racemosa* diminishes, the indigenous macroalgal community does not recover to its initial condition.

The study of the introductions of alien species in the benthic communities of the Lagoon of Venice offers rich material for evaluating such an impact in the context of other anthropogenic changes (Occhipinti Ambrogi 2000a). A large number of nonindigenous species have been described in the last decades in the Lagoon of Venice; 12 out of 34 invaders recently detected represent also the first findings for the Mediterranean. Some of them have formed large populations and have outcompeted native species with similar characteristics. Among seaweeds (Curiel et al. 1999), the Japanese red alga *Antithamnion pectinatum*, an epiphyte on mussels and on other algae, is the most widely spread invader, followed by two other Japanese species, *Undaria pinnatifida* and *Sargassum muticum.* These two large brown algae are more conspicuous. The autochthonous species in the areas covered by *Undaria* have decreased in biomass, but no extinction of species has occurred. The competition for space is won by *Undaria*, due to its large attachment organ. Some of the algal species succeed to overgrow it, but are lost when it detaches from the substrate (usually in July). For an ephemeral few months other native algae find a bare substrate, but are overgrown when *Undaria* returns. The competition for light has damaged photophilous species that were already jeopardised by the increase in turbidity of Lagoon waters. However, *Sargassum muticum* has taken more time to expand. From 1992 to 1999 it was found only in small sectors in the southern part of the Lagoon. A large population was observed in 2000; in 2001 it increased near the central sea entrance of Lido, in direct connection with the city centre of Venice. In that place it has outcompeted almost completely the large native brown alga *Cystoseira barbata.* Until now, the only impact of these algae on human activities in the lagoon is of aesthetic nature, because nobody was used to the sight of such large fronds (up to some meters) along the banks of the canals. The tidal currents are sufficiently strong to clear them from the canals and disperse them into the sea, without causing accumulation or a bad smell.

The most spectacular invasions, however, have been those of the bivalves. *Crassostrea gigas* and *Tapes philippinarum* were introduced for aquaculture and rapidly spread outcompeting the native *Ostrea adriatica*, *Venerupis aurea* and *Tapes decussatus.* It has to be noted, however, that the populations of the latter species were already scanty, probably due to overexploitation. The collection of Japanese clams from shallow areas of the lagoon is now an important economic issue but also poses an environmental problem due to the use of hydraulic dredges, or of other devices disturbing the lagoon's floor, suspending huge quantities of fine sediments. As a result, the turbidity of a large part of the tidal waters has dramatically increased in many canals.

An interesting study of the coexisting populations of the introduced ascidian *Botrylloides violaceus* and the autochthonous *Botrylloides leachi* in the Lagoon of Venice has shown that the competitive advantage of the former resides in the ability of withstanding temperatures as low as 10°C in the active state while the latter undergoes wintering (Brunetti, pers. comm.).

Observed over the years, the history of invasions is not always a long lasting success. For instance *Dyspanopeus sayi*, a crab originating from the North Atlantic, was reported to be the most widespread crab in the Venice lagoon, largely exceeding the numbers of the once ubiquitous and numerous *Carcinus mediterraneus* and *Pilumnus hirtellus* (Mizzan 1995). Recently (Mizzan, pers. comm.) it has been reported to be strongly reduced in many areas and has disappeared from the canals in the centre of the town, where it used to have mass development.

The longest history of an introduction is that of the bryozoan *Tricellaria inopinata* (Occhipinti & Ambrogi 2000b) that was (rather unexpectedly, hence the specific name) discovered in 1982 during regular surveys of the piles marking the navigable canals of the lagoon. The surveys have continued until now enabling the documentation of the different phases of acclimation, spread, expansion and later decline of this species, in comparison with the abundance of other Bryozoa known from the lagoon. However, during the last phase of reduction in abundance, no other bryozoan species has taken advantage of the situation in order to rebuild their original population size. The decline is common for all the species; the causes may be diverse, but one of them is probably the increase in water turbidity.

5 Administrative and legislative measures

Italy has subscribed and ratified the main international treaties concerning alien species, namely the 1973 CITES (Convention on International Trade of Endangered Species), the 1979 Bern Convention, the 1979 Bonn Convention, the 1992 Rio Convention and the 1995 amendment of the Barcelona Convention.

Moreover, the European Union legislation applies to Italy, and Government ratification make them fully applicable. In particular, the Habitat Directive declares that introduction of alien species can be authorised only if an appropriate study has been conducted and only if this study has been evaluated by a competent scientific organisation, assuring that neither natural habitat nor wild flora and fauna will be damaged.

National legislation in Italy delegates the authorisations and control for transfer and introduction of nonindigenous species to the Regions, which have issued a number of Regional Laws. Intentional introductions of marine fish may exclusively occur prior the authorisation of the Regional Authorities and are subjected to sanitary inspection measures and veterinary controls according to: Ministerial Decree (D.M.) N. 454, 1988; Law Decree (D.L.) N. 263, 1997 in accordance with the EEC Directive N. 93/53 (fish diseases). However, the administrative measures to control alien species introductions are still incomplete and an effective policy of prevention is scarcely adopted. Government authorities should promote the increase in scientific knowledge of the biology and impacts of alien species and should use the acquired knowledge for practical measures. For this purpose, the Ministry of Environment has funded a preliminary study (comprising also a survey on invasive species in Italian ports) to the Italian Marine Biological Association in order to get their assistance in preparing legislative guidelines.

6 Regional co-operation

A close link in scientific terms is active through the CIESM (International Commission for the Scientific Exploration of the Mediterranean), the SIL (Societas Internationalis Limnologiae) and other scientific associations, for the exchange of data and information relating to the Mediterranean and the European border states. The Italian Marine Biological Association is actively contributing through its experts to the preparation of the CIESM Atlas on invasive species in the Mediterranean.

The Italian Government promoted the consultation for application of the article 9 of the FAO Code of Conduct for Responsible Fisheries (CCFR) in the Mediterranean region (1997). In general, however, the political regional co-operation is less active, for instance, Italy is not part of ICES, and Italian scientific representatives are only invited observers in the meetings of the ICES WGITMO (Working Group on Introductions and Transfers of Marine Organisms).

Acknowledgements

Most of the colleagues who kindly provided me information are quoted in the text and I warmly thank all of them; I am particularly indebted to Angelo Mojetta for information on inland fishes, Maria Cristina Gambi for polychaetes, Francesco Mastrototaro for tunicates and Giovanna Marino for aquaculture.

A SEA CHANGE – EXOTICS IN THE EASTERN MEDITERRANEAN SEA

BELLA S. GALIL[1*] & ARGYRO ZENETOS[2]
[1*]*National Institute of Oceanography, Israel Oceanographic & Limnological Research, Haifa Israel*
[2]*Institute of Oceanography, National Centre for Marine Research, Agios Kosmas, Hellinikon, Greece*
[*]Corresponding author *bella@ocean.org.il*

Abstract
The eastern Mediterranean is susceptible to biological invasions because of its placement between the Atlantic, Pontic and Erythrean regions, busy maritime traffic, and lagoons and bays that are crowded with fish and shellfish farms. However, the greatest influx of invaders resulted from the opening of the Suez Canal in 1869, which allowed entry of Indo-Pacific and Erythrean biota. Exotic macrophytes, invertebrates and fish are found in most coastal habitats in the eastern Mediterranean. Some invaders have outcompeted or replaced native species locally, some are considered pests or cause nuisance, whereas other invaders are of commercial value. However, at variance with other invaded seas, the invasion into the eastern Mediterranean has increased the region's biodiversity. The rate of marine biotic invasions has increased in recent decades; collectively they have significant ecological and economic impacts in the eastern Mediterranean. Some Erythrean invaders have already spread as far west as Malta and Sicily, and if global warming was to affect the Mediterranean sea-water temperature, then tropical invasive species would gain a distinct advantage over the native fauna.

1 Introduction

1.1 TOPOGRAPHY

The eastern Mediterranean (EMED) comprises the Levantine, Aegean, Ionian and Adriatic Seas. The Ionian Sea connects with the western Mediterranean by the shallow (400 m) Strait of Sicily, with the Adriatic Sea by the Otranto Passage, and with the Levantine Sea by the Cretan Passage. The Aegean Sea is connected to the Black Sea through the Dardanelles, Sea of Marmara and the Bosphorus, and with the Ionian and Levantine Seas at the Cretan Arc Straits. The Levantine Sea is open to the Red Sea and Indian Ocean through the Suez Canal.

The EMED is more active in terms of plate tectonics and is characterized by more complex morphology than the western Mediterranean (Picco et al. 1999). The Ionian, Aegean and Levantine Seas are characterized by deep trenches and raised regions: the Ionian Abyssal Plain is enclosed by the Tunisian Plateau and by the Calabrian and Sirte Rise regions; the Mid-Mediterranean Ridge is bound to the south by the Herodotus Trough, and to the north by the Hellenic Trench. The greatest depth in the Mediterranean, 4,982 m, is located in the Ionian Sea. Recent paleoclimatic data, collected in geologically stable areas, combined with archaeological or historical evidence, indicates that sea level increase might be 30 cm in the 21st century (EEA 1999).

E. Leppäkoski et al. (eds.), Invasive Aquatic Species of Europe, 325–336.

1.2 EARLY HISTORY - IMPLICATIONS FOR THE EVOLUTION OF THE EMED BIOTA

The EMED is a remnant of the Tethys Ocean (Neev et al. 1985). The deep-water communication between the Mediterranean and the Indian Ocean was breached during the early Miocene (Robba 1987), though intermittent marine contacts continued well into the Messinian (Sonnenfeld 1985), and species of Indo-Pacific origin still inhabited the Pliocene Mediterranean (Sorbini 1988). During the Messinian, the Mediterranean was periodically cut off from the Atlantic and the Paratethys, divided into water bodies of varying salinities - freshwater, brackish or hypersaline - or, as some scientists argue, totally desiccated (Hsu et al. 1978). The Messinian Levantine basin may have been a shallow, hypersaline, evaporitic sea, or a brackish, Paratethys-related lago-mare. About 5 million years BP, Atlantic waters supporting tropical and subtropical Atlantic species, refilled the sea. During the Pliocene, the cooling of the sea prompted the substitution of temperate for tropical biota (Ruggieri 1967). The climatic fluctuations of the Pleistocene, accompanied by changes in salinity, temperature, and stratification patterns, have affected the biota as well (Zaccaria 1968). The Suez rift cleaves the once continuous Arabo-African plate. The rift was formed by the Early Miocene, underwent subsidence and was inundated by the Mediterranean, forming the Clysmic Gulf. However, influx of seawater from the Mediterranean must have ceased during the Messinian (Garfunkel & Bartov 1977). In the lower Pliocene the Gulf of Suez and the Red Sea were settled by Indian Ocean biota. The Suez Canal was constructed along the central trough of the Suez rift. Once a lake connected to the Caspian Sea, the Black Sea was joined to the Mediterranean after the opening of the Dardanelles during the interglacials (100,000-150,000 years BP). Following another period of isolation, the Black Sea was reconnected to the Sea of Marmara and the Mediterranean about 6,000 years BP (Zaitsev & Mamaev 1997).

1.3 HYDROGRAPHY, HYDROCHEMISTRY AND PRIMARY PRODUCTIVITY

The building blocks of the eastern Mediterranean upper thermocline circulation are subbasin scale gyres and permanent, or quasi-permanent, cyclonic and anticyclonic structures interconnected by intense jets and meandering currents (Malanotte-Rizzoli et al. 1999). The Levantine Sea is characterized by the cyclonic Rhodes and west Cyprus gyres, the anticyclonic Mersa-Matruh gyre, the Mid-Mediterranean Jet, and the Shikmona eddy south of Cyprus. In the Ionian sea the strong Mid-Ionian Jet crosses the basin in north-south direction before creating the Mid-Mediterranean Jet.

The Nile, the largest river flowing into the EMED, passes 89 $km^3\ a^{-1}$ at Aswan, but only 5 $km^3\ a^{-1}$ flows into the sea. As evaporation exceeds precipitation and river runoff by an estimated 2,500 $km^3\ a^{-1}$ (EEA 1999), the EMED is a 'concentration basin'. The annual salinity average of surface water ranges from 37.5 PSU in the Sicily Straits to 39.0 PSU in the Levantine Sea. Surface water temperatures show high seasonal variations and range from 14 °C during winter to 30 °C in summer in the Levantine Sea.

The EMED is ultra-oligotrophic: data collected in the past decade confirms earlier results concerning the progressive depletion of nutrients and chlorophyll from the Aegean to the Ionian and Levantine Seas (McGill 1961, 1965; Stergiou et al. 1997). Primary production rates are on average three times lower in the eastern basin than in the northwestern basin (Turley 1999); rates integrated over the euphotic zone (maximum depth 120 m) were low (about 40, 78 and 155 mgC m^2 d^{-1}) in the eastern, central and western Mediterranean basins respectively (Gotsis-Skretas unpubl.). In summer, phosphate concentrations in the euphotic zone in the Levantine Sea are mostly below detection limit of 0.008 µmol and nitrate concentrations are less than 0.6 µmol (Krom et al. 1992); chlorophyll-a concentrations are as low as 0.4 µg l^{-1} nearshore, and decrease offshore to 0.05 µg l^{-1} (Berman et al. 1986; Yacobi et al. 1995).

Considerably higher nutrient concentrations (5 µmol nitrate, 0.2 µmol phosphate) are found beneath the nutricline (ca 200 m) and are injected into the upper layers by wintertime vertical mixing. The nutrient distribution and phytoplankton production in the EMED are principally determined by the duration and the intensity of deep water mixing in the quasi-permanent anticyclonic and cyclonic eddies (Yilmaz & Tugrul 1998). Levantine oligotrophy is attributed to the increasingly nutrient-depleted Atlantic inflow, further reduced by the Levantine intermediate water, the arid and semi-arid climate with low riverine discharge of nutrient-rich waters, limited atmospheric input, and the relatively narrow continental shelf that allows little benthic-pelagic coupling for nutrient recycling.

2 Vectors of invasion

2.1 THE ERYTHREAN INTRUSION

The Suez Canal, 163 km long, traverses a series of shallow lakes, the waters of which ranged from brackish to hypersaline: salinity in the Bitter Lakes was as high as 161 PSU right after opening the canal in 1869 (Vadiya & Shenuda 1985). As seawater flowed through the Bitter Lakes the salinity fell to 70 PSU in 1870-1872, and measurements taken a century later were not much higher than in the northern Gulf of Suez (Morcos 1960, 1980). Another obstacle to invasion has been removed with the damming of the Nile at Aswan. The Nile floods coincided with the summer current inversion and the inflow of Mediterranean water into the canal, resulting in markedly decreased salinity, as low as 26 PSU, in the northern portion of the canal and in the adjacent Mediterranean, and 34 PSU as far as Haifa (Oren 1969; Vadiya & Shenuda 1985). Since the completion of the Aswan high dam, the Nile flow has been drastically reduced. When opened in 1869, the canal was barely 8 m deep and 65-98 m wide. Deepened and widened several times the canal is now 345 m wide and its navigational depth is 20.5 m. Each increase in volume decreased temperature fluctuations within the canal, while increasing depth has probably facilitated the passage of infralittoral species.

Though ship passageways had been constructed since pharaonic times, the ancient waterways were routed through eastern branches of the Nile or extinct deltaic lagoons, which constituted a freshwater barrier for marine biota (Sneh et al. 1975). In contrast, the opening of the Suez Canal initiated the invasion of Erythrean and Indo-Pacific biota into the Mediterranean. Despite impediments such as the canal's length, shallowness, current regime, temperature and salinity extremes, hundreds of Erythrean species traversed the Suez Canal and settled in the Mediterranean, forming thriving populations along the Levantine coasts, with some invaders spreading as far west as Tunis, Malta and Sicily (www.ciesm.org/atlas/).

Already Keller (1882) observed that "Der Isthmusdurchstich wurde ja vermuthlich auch als Karawanenstrasse für die thierischen Bewohner beider Meere benutzt", and Steinitz (1919) commented that "The Suez Canal ... connects two oceans having each a very different fauna from the other, which throws up a number of new problems: Does an exchange of fauna take place between the two oceans by means of this canal? Do the immigrants from the other side flourish on this or die? Has such an immigration caused important changes in the fauna of the eastern basin of the Mediterranean?". The 37 scientific publications resulting from the Cambridge Expedition to the Suez Canal in 1924 answered all three questions in the affirmative. Gruvel (1936) added "Mais si l'ensemble de ces mouvements d'espèces animals et végétales constitue, pour la Science, une fait fort intéressant, les passages définitifs de ces espèces à traverse la totalité du Canal présentent un résultat économique également très important".

Calman (1927), Monod (1932), Ben Tuvia (1970, 1983) and others used the term Erythræan to mark the northwards immigrants through the canal. Por (1971) proposed the term Lessepsian immigration in homage to the French engineer Ferdinand de Lesseps, who supervised construction (1859-1869). The term Erythrean invader refers here to marine organisms living in the Red Sea that have crossed the Suez Canal, as larvae or adults, autonomously or by natural phoresy, and have made their way into the Mediterranean. The occurrence of a species in the Red Sea and its distribution along the Levant coast are considered earmarks of an Erythrean invader.

A small mytilid bivalve, *Brachiodontes pharaonis*, widely spread throughout the Red Sea (Oliver 1992), was an early entrant into the Mediterranean - it was already in 1876 recorded from Port Said, where it soon became abundant (Pallary 1912). It has spread along the Levant coast from Egypt to Turkey, Cyprus and Greece and even further (www.ciesm.org/atlas/). Preliminary results of a molecular study have shown that the Mediterranean population possesses both Red Sea genotypes and non-Red Sea genotypes, the latter increasing in frequency with increasing distance from the Canal. This was interpreted as possible indication of ship transport from outside the Red Sea (A. Abelson, pers. comm.). Another early Erythrean intruder, the gastropod *Cerithium scabridum*, was reported from the Suez Canal by 1882, and has established large, stable populations from Egypt to south-eastern Turkey and Cyprus. *C. scabridum* too, has been reported from further afield: Naples, Sicily (Italy) and Tunisia (www.ciesm.org/atlas/). Lavie & Nevo (1986) found high levels of genetic variability in both Red Sea and Mediterranean populations of *C. scabridum*. A more recent, rapidly expanding invader, the gastropod *Strombus persicus*, was first sighted in Mersin Bay, Turkey, in 1978 (Nicolay & Romagna-Manoja 1983), and then in rapid succession off Israel (Mienis 1984), Rhodes (Verhecken 1984), Cyprus (Bazzocchi 1985) and Lebanon (Bogi & Khaizallah 1987). It was postulated that *S. persicus* may have arrived in Iskenderun in discharged ballast water from oil tankers coming from the Persian Gulf (Oliverio 1995). However, the gastropod has been found far from oil terminals, and because tankers at the time lacked segregated ballast tanks, oil-laden vessels would not carry ballast water. Off the Israeli coast the shallow sandy littoral is littered with them: 'one can speak of an invasion...hundreds of dead shells on the beaches and shoals of *Strombus*, of all sizes, colors and patterns, feeding on the sea floor up to 20 m depth' (Curini-Galletti 1988). In recent years it has been sold on the Jaffa fish market (Mienis 1999). The Indo-Pacific rock oyster *Spondylus spinosus* was recorded off Israel (Mienis et al. 1993) and later off Lebanon and south-eastern Turkey. Although it was not recorded from the Suez Canal, it is assumed to be an Erythrean invader; it is unlikely that it spread by shipping as proposed by Lamprell (1998).

Each summer since the mid-1980s huge swarms of the invading jellyfish, *Rhopilema nomadica*, have appeared along the Levantine coast. In 1995 the jellyfish was recorded off the south-eastern coast of Turkey (Kideys & Gücü 1995), and in 1998 a specimen was collected near Izmir (A. Karatas pers. comm.). These massive swarms of voracious planktotrophs, some stretching 100 km long, must play havoc with the limited resources of this oligotrophic sea, and when the shoals draw nearer shore, they adversely affect tourism, fisheries and coastal installations. Local municipalities have reported a decrease in holidaymakers frequenting the beaches because of the public's concern over the painful stings inflicted by the jellyfish. Coastal trawling and purse-seine fishing are disrupted for the duration of the swarming due to net clogging and inability to sort yield. Water intake pipes blocked with jellyfish pose a threat to cooling systems of port-bound vessels and coastal power plants: in summer 2001 Israel Electric removed tons of jellyfish from its seawater intake pipes at its two largest power plants, at estimated costs of 50,000 US$ (M. Cohen, pers. comm.). Yet, the same jellyfish, *R. nomadica*, known to shelter the juveniles of a Red Sea carangid fish, *Alepes djeddaba,* among its nematocyst-laden tentacles (Galil et al. 1990), may have precipitated the sudden population increase of this commercially important species (Grofit 1987).

Erythrean penaeid prawns make up most of the shrimp catches along both Egyptian and Israeli coasts (Galil 1993). The Erythrean prawns are highly prized and a small fleet of Israeli coastal "mini" trawlers has specialized in shrimping since the mid 1980s, bringing in a quarter of the total trawl catch volume and a third of the trawl gross income (Snovsky & Shapiro 1999). *Marsupenaeus japonicus*, *Metapenaeus mono-ceros*, and *M. stebbingi* compose most of the prawn catch off the Mediterranean coast of Egypt and in the Nile delta lagoons (Dowidar & Ramadan 1976; Bishara 1976). An early Erythrean invader, the swimming crab *Portunus pelagicus*, was recorded from Port Said in 1898 (Calman 1927), where it soon became abundant, and by the early 1900s it was offered in the markets of Port Said, Alexandria and Haifa (Fox 1924). Erythrean fish constitute nearly half of the trawl catches along the Israeli coast (Golani & Ben Tuvia 1995). In the 1990s both invading mullids, *Upeneus moluccensis* and *U. pori*, formed 87% of the mullid catch off the coast of Israel at depths of 20 m, and 50% at 55 m, whereas the native mullids were more abundant in deeper waters (Golani & Ben Tuvia 1995). The Red Sea obtuse barracuda, *Sphyraena chrysotaenia*, has outnumbered the native sphyraenids in inshore trawl and purse-seine catches along the Israeli coast (Grofit 1987). In addition, two of the four species of Erythrean clupeids that established populations in the Levant - *Dussummieria acuta* and *Herklotsichthys punctatus* – are of importance in the inshore-pelagic fishery. The growth of these invasive species, to the point where they are harvested commercially, is an excellent index of how prevalent they have become. This abundance is not limited to the south-eastern Levant, as Gücü & Bingel

(1994) have estimated that invaders constitute 62% of the demersal fish biomass in the Gulf of Iskendrun, Turkey.

However, among the Erythrean invaders are several well-documented cases of rapid increase followed by decline. In the late 1940s the invading goldband goatfish, *Upeneus moluccensis*, made up 10-15% of the total mullid catches off the Israeli coast. Following the exceptionally warm winter of 1954-55, its percentages increased to 83% but dropped later to 30% of the catch (Ben Tuvia 1973). Following the same winter, the brushtooth lizardfish, *Saurida undosquamis*, became a commercially important fish: its share in trawl catches rose to 25% in 1979 (Grofit 1987). The population then diminished and stabilized at about 5% of the total trawl catch (Ben Yami & Glazer 1974; Snovsky & Shapiro 1999). *Leiognathus klunzingeri* was "extremely common on trawling grounds at depths of 20-100 m in the early fifties", but its populations were decimated during 1960s (Ben Tuvia 1985). An extreme illustration is the case of the cerithiid gastropod *Rhinoclavis kochi*. It was first recorded off the Israeli coast in 1963, but in the early 1970s was recognized as "one of the abundant offshore species of the Mediterranean coast of Israel " (Barash & Danin 1973), and has spread rapidly to become one of the dominant species at depths between 20-60 m, peaking in the late 1970s (Galil & Lewinsohn 1981; Tom & Galil 1990). Samples collected a decade later contained solely empty shells.

2.2 SHIPPING

The eastern Mediterranean, fringed by major ports (Alexandria, Port Said, Haifa, Beirut, Limassol, Iskenderun, Izmir, Thessaloniki, Peiraias) and innumerable marinas, is susceptible to invasion of ship-borne organisms, whether the nonindigenous species occur in fouling communities or ballast.

The oldest maritime pathway of dispersal and introduction is the transportation of fouling biota, sessile and adherent, on ship hulls. Many cosmopolitan members of the fouling community are quite possibly older introductions into the Mediterranean (Ryland 1967; Boudouresque & Ribera 1994). Scleractinian corals are not ordinarily considered members of ship fouling communities, yet the South American hermatypic coral *Oculina patagonica* was an early transoceanic arrival. Though first noted in 1966 in the Gulf of Genoa (Zibrowius 1974), it has been successively recorded near Egypt (1981), Lebanon (1992) and Israel (1993) (Bitar & Zibrowius 1997), where it is locally abundant. Serpulid polychaete worms of the genus *Hydroides* are frequently found in tropical fouling communities and are among the earliest documented invaders in the Mediterranean: *Hydroides dianthus* was documented in Izmir as early as 1865. Given the state of marine taxonomy in the 19th century, the *Hydroides* species may have arrived many years before they were first detected. They are now well established in ports and lagoons throughout the Mediterranean, where they cause major fouling problems on artificial substrates (Kocak et al. 1999). These worms are however absent from natural marine habitats (Zibrowius 1992).

The American blue crab, *Callinectes sapidus*, native to the Atlantic coast from Nova Scotia to Uruguay, was first recorded in the Mediterranean in the 1940s, in Egypt (Banoub 1963), and successively in Italy (Giordani-Soika 1951, as *Neptunus pelagicus*),

Israel (Holthuis & Gottlieb 1955), the Aegean Sea (Kinzelbach 1965), and the southern coast of Turkey (Kocatas & Katagan 1983). Holthuis & Gottlieb (1955) suggested that *C. sapidus* was transported into the Mediterranean in ballast tanks. The veined rapa whelk, *Rapana venosa*, native to the Sea of Japan, was probably introduced into the Black Sea in the 1940s. It was first recorded in 1947 from the oil-exporting port of Novorossiysk; two specimens were collected near a natural oyster bank in 1986 in Thessaloniki Bay, northern Aegean (Koutsoubas & Voultsiadou-Koukoura 1991). Range extension is possibly mediated by ballast transport of larvae, or of egg masses transported with the products of marine farming. The western Pacific bivalve *Musculista senhousia* was introduced to the northwest coast of America in 1924 and to Australia in 1983, either with hull fouling or in ballast. *M. senhousia* has spread into the French Mediterranean and Adriatic lagoons with imported shellfish mariculture. However, the Egyptian and Israeli records of *M. senhousia* may be either ship-transported or invaders from the Red Sea (www.ciesm.org/atlas). Similarly, the slipper limpet, *Crepidula fornicata,* has been transported into the Mediterranean with shellfish mariculture (see below), though a record from Saronikos Gulf (near Peiraias port) may be attributed to shipping as there is no oyster farming nearby (www.ciesm.org/atlas). A single specimen of the bivalve *Petricola pholadiformis* reported from the south Aegean Sea, is also thought to have been accidentally introduced with shipping (www.ciesm.org/atlas). The bivalve *Anadara demiri*, a native of the China Sea, presumed to have arrived by shipping, was recorded from the central Ionian Sea (Solustri & Morello 2000). *A. demiri* dominated the highly polluted bottoms in Izmir harbour (Demir 1977), and Thessaloniki Gulf (Zenetos 1994), where its population is presently in decline, perhaps due to amelioration of the pollution. Of the 114 invasive molluscs in the EMED, six are presumed to have been ship-borne. Though the true mode of introduction remains unknown, shipping is assumed to be the most probable vector since their occurrence is patchy, they are frequently recorded from ports and some are known as fouling organisms.

Shipping may serve to transport invading Erythrean species further on. A small Erythrean mytilid, *Brachiodontes pharaonis* (see above), common in the Levantine Basin, where it settles in dense clusters on midlittoral and infralittoral rocks, piers and debris (Barash & Danin 1992), has spread as far west as Sicily (Di Geronimo 1971), perhaps through ship fouling. Similarly, the gastropod *Cerithium scabridum* (see above), reported from Naples, Sicily and Tunisia, was possibly transported there by shipping. The pearl oyster, *Pinctada radiata*, was one of the first Erythrean molluscs recorded in the Mediterranean (Monterosato 1878, as *Meleagrina* sp.). Thirty years after the opening of the canal, it was recorded as forming large banks off Tunis, prompting Monterosato (1899) to suggest it might be indigenous to the Mediterranean. *P. radiata* is abundant in the Levantine Basin (Barash & Danin 1992) and has spread as far west as the Tyrrhenian Sea, as well as off Sicily, Malta, Pantellaria Island, and France (Ricordi 1993; Di Natale 1982; Tomlin 1927; Sabelli 1969; Zibrowius 1979). Its rapid dispersal is attributed to ship-borne individuals (Zibrowius 1992), or marine turtles - *P. radiata* was recorded as an epibiont on a loggerhead turtle off Lampedusa Island (Oliverio et al. 1992). The bivalve *Fulvia fragilis* has been recorded from the Suez Canal (Tomlin 1927), Israel (Barash & Danin 1977), south-eastern Turkey and Tunisia (Passamonti 1996). However, its recent finding in Saronikos Gulf (Peiraias port) suggests it may be ship-transported as well (Vardala-Theodorou 1999).

Besides taxa that have established sustainable populations in the Mediterranean, there are ship-transported species that fail to survive: a live specimen of the Brazilian giant barnacle *Megabalanus tintinnabulum* was found on a ship moored in Peiraias port, but not in subsequent surveys of the port (Zenetos 1997).

The Indo-West Pacific portunid crab *Charybdis hellerii* was first sighted in the Mediterranean off the Israeli coast in 1924-1925 (Steinitz 1929), and has since been reported off Egypt, Lebanon, Syria, Turkey and Cyprus (Galil 1993; Zibrowius pers. comm.) - a spread chronology common to Erythrean invaders. Shipping patterns ensure that the Mediterranean exports as well as imports biota. In 1987 *C. hellerii* was collected in Cuba (Gómez & Martinez-Iglesias 1990), and in rapid succession in Venezuela, Colombia, Florida, and Brazil (Campos & Türkay 1989; Lemaitre 1995; Carqueija & Gouvea 1996; Mantelatto & Dias 1999). Transport in ballast tanks is the most probable mode of dispersal because the crab's arrival corresponds to increased shipping contacts with the eastern Mediterranean (Campos & Türkay 1989).

2.3 MARICULTURE

Market-driven demands for exotic fish and shellfish have introduced mariculture farming into lagoons and bays along the shores of the eastern Mediterranean.

The Pacific oyster, *Crassostrea gigas*, native to north-eastern Asia, was introduced to Tunisia, the Ionian Sea (Minelli et al. 1995), to Patraikos, Greece (Dimitrakis 1989), and to Cyprus (Zibrowius 1992). *C. gigas* has established populations outside shellfish farms. A soft-shell clam, the common sand gaper, *Mya arenaria*, native of the North Atlantic, was reported from Saronikos Gulf, probably an accidental introduction as there are no shellfish farms in the vicinity (www.ciesm.org/atlas). The pearl oyster, *Pinctada radiata* (see above) was intentionally introduced to Greece for mariculture purposes, but has since established thriving populations in the wild (Serbetis 1963).

Transport and transplantation of commercially important alien oysters has resulted in numerous unintentional introductions of pathogens, parasites and pest species. Oyster farms have served as gateways into Mediterranean coastal waters for these nonindigenous camp-followers (Ribera & Boudouresque 1995). The slipper limpet, *Crepidula fornicata*, an invasive mollusc on the European Atlantic coast and the North Sea, is assumed to have been introduced with oyster farming. Its occurrence is more local in the Mediterranean. There are presumably separate introductions from the French Atlantic coast to the French Mediterranean lagoons, and from unknown origin to the Italian and Maltese sites (Cachia 1981; Di Natale 1982). The limpet was recorded from Saronikos Gulf (Peiraias port area), probably introduced through shipping as there are no shellfish farms in the vicinity (www.ciesm.org/atlas).

A widely distributed Indo-West Pacific penaeid prawn, *Marsupenaeus japonicus*, that has invaded the Mediterranean through the Suez Canal (see above), and established populations along the Levantine coast from Egypt to Rhodes (www.ciesm.org/atlas), has been successfully cultivated in the Central and Western Mediterranean (Tournier 1972; Lumare & Palmegiano 1980; Rodriguez 1987). Recently, experimental cultivation of the prawn was undertaken in Amvrakikos Gulf, Ionian Sea and in Vistonikos Gulf, N. Aegean Sea (Kevrekidis et al. 996). The widely distributed rock oyster *Saccos-*

trea cucullata was recorded from the Suez Canal, Egypt and Turkey. It had been imported into the Adriatic, but the cultivation efforts were unsuccessful (Cesari & Pelizzato 1985).

The American blue crab, transported to the Mediterranean in ballast tanks (see above) is found mainly near harbours, estuaries or lagoons (Holthuis 1961). The catadromous, euryhaline crab (Williams 1984) is occasionally caught in the Sea of Galilee, a freshwater lake, whence it was accidentally introduced with grey mullet spat collected in the Mediterranean and used to restock the lake (Snovsky & Galil 1990).

3 Ecological impact of bioinvasions in the eastern Mediterranean

Many of the high-abundance invaders constitute neither a nuisance, nor have commercial value, yet their presence is overwhelming: autochthonous species are outcompeted wholly or partially displaced by the invaders. Though there is no documentation of direct competition between Erythrean and indigenous species, there are many instances of sudden changes in abundance; competition is one explanation. Decimation of the indigenous sea star *Asterina gibbosa* populations along the Israeli coast coincided with the rapid increase of its Erythrean cogener *A. burtoni* (Achituv 1973). However, recently, after a long absence, a small population of *A. gibbosa* was detected (Achituv 1999) at a single locality. Between 1992 and 1997 the Erythrean alga *Caulerpa racemosa* has replaced the native sea grass meadows of *Posidonia oceanica* in Moni Bay, Cyprus, and prompted significant change in the benthic macrofauna: the abundance of gastropods and crustaceans decreased, whereas that of polychaetes, bivalves and echinoderms increased (Argyrou et al. 1999). A native penaeid prawn, *Melicertus kerathurus*, was commonly caught by trawlers along the Israeli coastal shelf on sandy or sandy mud bottoms, and supported a commercial fishery throughout the 1950s (Holthuis & Gottlieb 1958). This native prawn has since nearly disappeared, and its habitat has been overrun by the Erythrean penaeid prawns. d'Udekem d'Acoz (1999) reported that *M. japonicus* "has almost evicted the native *P. kerathurus* from the easternmost part of the Mediterranean". The rapid advent of *Metapenaeus monoceros* into the Gulf of Gabes, Tunisia, has raised concerns over the fate of *M. kerathurus* fisheries (Chaouachi et al. 1998). The Erythrean snapping shrimps, *Alpheus lobidens* and *A. edwardsii*are more common now in the south-eastern Levantine rocky littoral than the native *A. dentipes* (Lewinsohn & Galil 1982). The decrease in numbers of the once prevalent indigenous jellyfish, *Rhizostoma pulmo,* along the Levantine coast, has coincided with the massive presence of *Rhopilema nomadica.*

Competitive displacement may also cause bathymetric changes among populations of Erythrean and indigenous species. The local red mullet, *Mullus barbatus* and the native hake, *Merluccius merluccius* were both displaced into deeper, cooler waters by their respective Erythrean competitors, *Upeneus moluccensis* and *Saurida undosquamis* (Oren 1957). The Erythrean mantis shrimp *Erugosquilla massavensis* was recorded from Alexandria, Egypt in 1933 (as *Squilla africana*) (Steuer 1936), and successively from Israel (Steuer 1938), Turkey (Holthuis 1961), Lebanon (Holthuis 1961), Cyprus (Ingle 1963), and Rhodes (1996, Kevrekidis pers. comm.). *E. massavensis* is common in the south-eastern Levant, and considered to be a "most successful Red Sea immigrant" (Por 1971). Though on occasion it is taken together with the indigenous Spottail mantis shrimp *Squilla mantis*, generally the latter occurs in deeper waters than *E. massavensis*. In the north-western Mediterranean, *S. mantis* is fished commercially at 10-25 m, off the Israeli coast at 70-90 m. Lewinsohn & Manning (1980) questioned "whether temperature, bottom type, or pressure from *O. massavensis*, or a combination of these is responsible for its depth distribution".

Increase in the number of Erythrean species has coincided not only with displacement of indigenous species but, in some cases, resulted in apparent competition among the invaders themselves. The prawn *Trachysalambria curvirostris* was first recorded in the Mediterranean in 1924 (Balss 1927), already so abundant that it was sold in the Haifa fish market (Steinitz 1929). It was the most common penaeid on sandy mud bottoms until the mid-1980s (Galil 1986). In 1987 another Erythrean prawn, *Metapenaeopsis aegyptia* was detected in that same habitat (Galil & Golani 1990). By 1993 *M. aegyptia* outnumbered *T. curvirostris* 3 to 1, and by 1996 the native prawn was outnumbered 25 to 1 in samples collected off the Israeli coast. Another Erythrean prawn, *M. mogiensis consobrina*, appeared on the sandy mud bottoms in 1996 (Galil 1997).

It has been suggested that biocontrol, involving the introduction of a predator, parasite or pathogen, may present an option for marine pests management, in particular invasive species (Thresher et al. 2000). Kuris (1997) suggested the use of endoparasitic rhizocephalan castrators, as they may be important regulators of host population density. The Erythrean invasion presented in the last decade an interesting "field experiment". The swimming crab *Charybdis (Goniohellenus) longicollis*, first recorded in the Mediterranean in 1959 off Turkey (Holthuis 1961), has since spread from Egypt to Cyprus (Lewinsohn & Holthuis 1986); it can form as much as 70% of the benthic biomass on sandy-silt bottoms off the Israeli coast (Galil 1986). Of the thousands of specimens collected over three decades, none showed evidence of parasitism until 1992, when a few bore the sacculinid rhizocephalan *Heterosaccus dollfusi* - another Erythrean invader (Galil & Lützen 1995). Within three years *H. dollfusi* had spread as far as the eastern Anatolian coast (Øksnebjerg et al. 1997). The parasite affects the growth, morphology and behaviour of the host, and castrates both male and female crabs. The infection rate in Haifa Bay rose to 77% in 1995 (Galil & Innocenti 1999). Yet, six years later, in spite of high levels of infestation there is no noticeable reduction in the host population.

4 A sea change

Shipping is considered the largest single vector for the movement of nonindigenous marine species across the globe (Ruiz et al. 1997). However, in the eastern Mediterranean, vessel-transported exotics and intentional and unintentional mariculture transfers lag far behind the Erythrean invasion in the number of species introduced – 80%, 87% and 88% of the exotic fish, decapod crustaceans and molluscs respectively are Erythrean invaders (www.ciesm.org/atlas). Por (1978) tallied 30 fish, 31 decapods and 44 mollusc species as Erythrean invaders in the Mediterranean, their numbers at present increased to 56, 40 and 101 respectively (www.ciesm.org/atlas).

With over 300 species, mainly molluscs, fish, decapod crustaceans, polychaetes and algae, the Erythrean intrusion is mainly confined to the Levantine Sea and the south-eastern Aegean Sea. If earlier passage through the canal might have been restricted to euryhaline, eurythermal and generally hardy littoral species, it is now mainly depth-restricted. Lately, investigations of the Suez Canal biota pointed to intensification of the Erythrean invasion (Halim & Messih 1999). The recently announced plans by the Egyptian government to widen and deepen the Suez Canal to permit passage of super tankers are of grave importance (The Times of India, 18.8.2000). The expansion will increase the canal width to 400 m and its depth to 25 m. Following the amelioration of the salinity barriers, depth is the most important physical impediment to invasion. Most Erythrean invaders occupy at present the Mediterranean littoral and infralittoral to a depth of 50 m, and are hardly found in deeper waters. Thus, an effective increase in canal depth of 5 m will allow invasion of species whose upper depth range (as adults or larvae) did not permit passage until now, and cohorts of new invaders will gain admittance to the Levantine Sea.

The Levantine Sea has less than half the number of the benthic species found in the whole of the Mediterranean. This profound faunal impoverishment has been attributed to its comparatively late recolonisation after the Messinian crisis, to pleistocenic climatic fluctuations and to the Basin's extreme oligotrophy (Fredj 1974; Yacobi et al. 1995). Sarà (1985) conjectured that the high temperature and salinity prevailing in the Levant made it unsuitable for many Atlantic-Mediterranean species. The present-day Levantine fauna is not truly representative of the water temperature, since the tropical Atlantic species are mostly barred and their niche is only partially occupied (Türkay 1989), with many of the taxa present presumably at the limit of their ecological tolerance, leaving this part of the Mediterranean vulnerable to invasion. Consequently, when tropical species arrived, there were few ecological obstacles to prevent their successful implantation.

The mariculture introductions are mostly restricted to lagoonar or estuarine habitats, and the vessel-transported exotics to polluted harbours (Zibrowius 1992) - environments known for their low biodiversity. The lagoonar environments differ from the adjacent coastal regions in their hydrography and less diversified biota (Sacchi et al. 1989). The shallow, brackish coastal lagoons, formed by recent changes in sea level and in the pattern of sediment-laden coastal currents, are transitory and unstable environments. This instability may explain their rather limited species diversity. Ionian and Aegean lagoons have been exploited for mariculture of nonindigenous species, and the commercially valuable shellfish have proved successful in establishing reproducing populations. Polluted or physically degraded environments are prone to invasion more than pristine sites. A recent study of macrofouling organisms discovered that many more species were found in a polluted than in a nonpolluted marina, and that the cosmopolitan serpulid worm, *Hydroides elegans,* which comprised 65% of the population in the polluted marina, was only infrequently found in the nonpolluted marina (Kocak et al. 1999).

The eastern Mediterranean, open to the Atlantic, Pontic and Erythrean biotas, is particularly prone to invasions. The littoral and infralittoral biota of the Levantine Sea is undergoing a profound change due to the influx of Erythrean invaders. Already some Erythrean invaders have spread as far west as Malta and Sicily; if global warming were to affect the Mediterranean sea water temperature, then tropical invasive species would gain a distinct advantage over the native fauna.

The rapid increase in human population density and urbanization along its shores has brought about coastal development, increased levels of agricultural run-offs and industrial wastes, unsustainable fishing practices and mariculture. These changes have caused wide spread disruption of the littoral ecosystem and decimation of the biota. Low indigenous biodiversity is certainly another factor in determining the success of these biotic invasions. The large number of exotic species in the eastern Mediterranean may have resulted from synergy between its geological history, environmental factors and anthropogenic activities.

THE MARMARA SEA, A LINK BETWEEN THE MEDITERRANEAN AND THE BLACK SEA

BAYRAM OZTURK
Faculty of Fisheries, Istanbul University, Istanbul, Turkey
ozturkb@istanbul.edu.tr

Abstract
The Marmara Sea represents a unique ecosystem between the Black and Mediterranean Seas due to its hydrographical peculiarities. This sea is one of the busiest water ways in the world. There are 11 exotic species found in the Marmara Sea. These are *Codium fragile*, *Mnemiopsis leidyi*, *Beroë cucumis*, *Rapana venosa*, *Scapharca inaequivalvis*, *Mya arenaria*, *Teredo navalis*, *Penaeus japonicus*, *Callinectes sapidus*, *Mugil so-iuy* and *Gambusia affinis*. Among them, *R. venosa* has been commercially important and *M. leidyi* is the most destructive species for the native species.

1 Introduction

The Marmara Sea, including the Istanbul Strait (Bosphorus), Marmara Sea and Canakkale Strait (Dardanelles), is situated between 40°00' and 41°10'N and 26°15'and 29°55'E. The surface area of the sea is 11,500 km^2 and the volume is 3,378 km^3. The length of the coastline is 927 km. This sea is surrounded by the Anatolia and Trace regions in Turkey. It is one of the busiest water ways in terms of shipping activities between the Mediterranean and Black Sea basins; around 50,000 ships pass every year (Ozturk et al. 2001).

Due to the geographical and hydrographical characteristics of the Marmara Sea, it represents a peculiar ecosystem as it is a transitional zone between the Mediterranean and the Black Sea. As such, it constitutes a barrier, a corridor or an acclimatization zone for living organisms (Ozturk & Ozturk 1996). The Marmara Sea serves as a barrier because it limits the distribution of both warm water marine species of Mediterranean origin and cold water, low saline species from the Black Sea. On the other hand, the Marmara Sea is the most important biological corridor for many species of migratory fish, birds and marine mammals between the Mediterranean and Black Sea. In this acclimatization zone, some Mediterranean species adjust slowly to the new environment of the Black Sea, or the Black Sea species to the Aegean Sea.

2 Hydrographical properties of the Marmara Sea

The Marmara Sea is made up of two layers of water of either Black Sea or Mediterranean origin, separated by a transitional layer of 8-10 m. Therefore, the hydrography of the Marmara Sea is dominated by the conditions of the adjacent basins. The Black Sea water enters the Marmara Sea through the Istanbul Strait as an upper current of 15-20 m depth and exits through the Canakkale Strait. Likewise, the Aegean water enters through the Canakkale Strait in a deeper layer flow, and enters the Black Sea with the Istanbul Strait underflow. The upper layer has a volume of 230 km^3 and an average renewal time of 4-5 months. The deeper layer has a volume of 3,378 km^3 and an average renewal time of 6-7 years (Besiktepe et al. 2000). Life in the upper layer is nourished primarily by brackish water of the Black Sea (Tugrul & Salihoglu 2000). The temperature of the

E. Leppäkoski et al. (eds.), Invasive Aquatic Species of Europe, 337–340.

surface water of the Marmara Sea, which is under the influence of the Black Sea, ranges from 4 to 24 °C. The salinity varies between 10 and 18 PSU. Deeper water shows pronounced changes in salinity and temperature. The salinity at 20 m depth rises to 30 PSU and at 40-50 m depth to 37 PSU. The temperature of the surface water of the Canakkale Strait is 6 to 26 °C and the salinity 24 to 36 PSU. In deeper water, at 70 m depth, the temperature ranges from 14 to 17 °C. The salinity at a depth of 30 m rises to 37.5 PSU and below it to 39 PSU (Kocatas et al. 1993).

3 Bioinvasion by exotic species and their ecological and economical consequences

Exotic fauna and flora of the Marmara Sea have been introduced in two different ways: in ballast water or on ship's hull, e.g. *Mnemiopsis leidyi*, and through man-made introduction, e.g. *Gambusia affinis* and *Mugil so-iuy*. Lessepsian species, which exist in the Aegean and Mediterranean Sea, are not seen yet in the Marmara Sea as it serves as a barrier for these species. On the other hand, non-native species introduced into the Black Sea use the Marmara Sea as a corridor or an acclimatization zone to the Aegean and Mediterranean Sea.

Algae. The origin of ***Codium fragile*** (Suringar, 1867) is the Pacific Ocean; the species was probably introduced accidentally during the Second World War (Mojetta & Ghisotti 1996). This species now inhabits all the Marmara Sea coasts.

Ctenophores. ***Mnemiopsis leidyi*** (Agassiz, 1865) originates from the North Atlantic Ocean. A possible way of introduction is by ballast water. This species was first introduced to the Black Sea and via the surface current to the Marmara, Aegean and Mediterranean Seas. It was first recorded in the Marmara Sea by Artuz (1991). In October 1992, an extremely vigorous outbreak was recorded in the Marmara Sea (GESAMP 1997). The abundance of *M. leidyi* were as high as 4.3 kg m^{-2} near the Istanbul Strait and 9.7 kg m^{-2} near the Canakkale Strait, mostly in 10-30 m deep water (Shiganova et al. 1995). In 1999, at 16 m depth in the Istanbul Strait, a maximum biomass of 2 kg m^{-3} was recorded (Ozturk 1999). This species was also reported from the Turkish coasts of the Aegean and Mediterranean Seas (Kideys & Niermann 1994). *M. leidyi* is a euryhaline organism tolerating a wide range of salinity of 4-75 PSU (Burrel & Van Engel 1976). The fresh water reservoir of the Istanbul City was invaded by this species and it caused a serious economic loss due to the damage of the pipeline (Ozturk et al. 2001). Since *M. leidyi* is a voracious predator, it has caused a decline of zooplankton. Masses of carcasses of this ctenophore caused anoxia in bottom-near waters. They have also been entangled to the fishing nets causing substantial damage. The pelagic fish stocks in the Marmara Sea declined since the pelagic fishes feed mainly on copepods and cladocerans, which are also foraged by *M. leidyi*. Furthermore, *M. leidyi* feeds on fish eggs and larvae, seriously affecting the economically important fishes, such as *Scomber scombrus*, *Sarda sarda, Sprattus sprattus,* and *Engraulis encrasicholus,* which use the Marmara Sea as spawning grounds. The decline of the fish stocks and economic loss of fisheries was estimated at 400,000 USD for Turkey only (Ozturk & Ozturk 2000).

Beroë cucumis (Fabricus, 1780) is a circumpolar species, abundant in the North Sea and the Pacific coast of North America. A possible way of introduction was by ballast water. In the Marmara Sea the density of this ctenophore has been 10-160 ind m^{-3} in summer. The majority have been 5-6 cm long adults on the coasts of the Prince Islands at the depth of 10-15 m (Ozturk 1999). It is known as one of the predators of *M. leidyi* (Burrel & Van Engel 1976).

Molluscs. ***Rapana venosa*** (Crosse, 1861) is native to the Sea of Japan and its possible way of introduction into the Black Sea, is by ballast water and eggs attached to ship hulls. *R. venosa* penetrated the Marmara Sea in the 1960s and is now found in the Aegean Sea as well. *R. venosa* feeds mainly on mussels and oysters on rocky bottoms. In the Marmara Sea, it is quite abundant at 5-25 m depth (maximum density is 15-20 ind m^{-2}). In 1965, the total distribution area of *R. venosa* was only 8 km^2; by 1980 it had increased to 170 km^2 (Ozturk 1999). Due to the high population density of *R. venosa* along the Marmara coasts, oysters and mussels have been exterminated from these areas where fishing of the bivalves used to be commercially important. This gastropod is harvested by diving and by dredging. The dredging method is harmful to benthic ecosystem, as it is a non-selective method unlike diving. For the first time in 1982, a Turkish fisherman discovered the commercial importance of this species and exported *Rapana* meat to Japan. Since then the species has been overexploited and can now be caught only 8 months per year, to protect the stocks in the Black Sea and Marmara Sea according to the Fisheries Law 1380. This species is beneficial to the Turkish fisheries economy; 2 million USD profit was estimated from the export and about 6,000 persons are directly or indirectly involved in this business. Annual production was 4,000 tons in 1997 (DIE 1998).

Probable origin of ***Scapharca inaequivalvis*** (Bruguiere, 1789) is the Pacific Ocean. This species was probably introduced by ballast water. After the first observation in the Marmara Sea in 1989, it now occurs up to the Istanbul Strait and Prince Islands (Ozturk & Ozturk 1996). This species is well adapted to the eutrophic water of the Straits, it inhabits coastal waters to 45 m depth. Its density reached about 60 ind m^{-2} in 1999 (Ozturk 1999).

Mya arenaria (Linnaeus, 1758) originates from the North Sea or the Atlantic coast of North America. A possible way of introduction is by ballast water. *M. arenaria* is found mainly in the sandy and muddy shallow bottoms in the Istanbul Strait and its entrance to the Black Sea. In many parts of the Istanbul Strait, it is a dominant bivalve species (up to 1,300 ind m^{-2}). Its average biomass has been calculated up to 1 kg m^{-2} around the Prince Islands in 1999 (Ozturk 1999). This bivalve is eaten by sturgeon, turbot, goby and mullet. The adults are eaten by *Rapana venosa*. It has no commercial importance.

Probable origin of ***Teredo navalis*** (Linne, 1758) is the Atlantic Ocean. There are two possible ways of introduction, either by transport of pelagic larvae by currents or by floating wood (tree branches, wooden hull, etc.) for adult organisms. This species has been found in the Marmara Sea and the Istanbul Strait shores since the 1950s.

Crustaceans. The Indo–Pacific prawn ***Penaeus japonicus*** (Bate, 1888) was intentionally introduced to the Marmara Sea in the late 1960s from Iskenderun Bay on the Turkish coast of the Mediterranean Sea (M. Demir, pers. comm.). However, its population did not increase as much as expected.

Callinectes sapidus (Rathbun, 1896) is native to the Atlantic coast of North America from Cape Cod to Florida and Gulf of Mexico. This species was found in the Marmara Sea in 1974 (Froglia et al. 1998), probably introduced by ballast water and as a hull fouling organism.

Fish. *Mugil so-iuy* (Basilewsky, 1855), native to the Amu Darya River basin, reached the Turkish Black Sea coast from the Sea of Azov, migrated to the west, reaching the Marmara Sea and later the coasts of the Aegean Sea. This species has potential commercial importance. Annual catch of this species is 15 tons in the Marmara Sea and 10 tons in the northern Aegean Sea (DIE 1998).

Gambusia affinis (Baird & Girard, 1854) originates from the North American wetlands. It was intentionally introduced for the first time by the Ottoman army to control mosquito populations in Amik Lake and its wetlands on the Mediterranean coast. It was later introduced to other wetlands throughout Anatolia by Turkish authorities to combat malaria (Geldiay & Balik 1988). The species is distributed troughout the wetlands of the Marmara Sea, the lagoons of Buyuk and Kucuk Cekmece. *G. affinis* is a euryhaline species which feeds mainly on carp fries.

4 Conclusion

The Marmara Sea is a link between the Mediterranean and Black Seas, which is the reason that exotic species, originally introduced to either of the two seas, are found here. However, for certain species, the Marmara Sea serves as a barrier limiting the distribution, while for others, it serves as a corridor enlarging the distribution.

Compared to the Black Sea, the Marmara Sea has less exotic species. There are 26 species in the Black Sea (Zaitsev & Mamaev 1997) and 11 species in the Marmara Sea (this study). There are two explanations to this. Firstly, due to different biotic and abiotic factors in the Black Sea and in the Marmara Sea, e.g. the higher temperature and salinity of the latter. Secondly, the gap of historical records of exotic species and their scientific studies in the Marmara Sea and in the Black Sea. For example, two more species, a zooplankton species *Acartia tonsa* Dana, 1849 and a phytoplankton species *Rhizosolenia calcar-avis* (M. Schultze, 1858), have recently been added to the list of exotic species in the Marmara Sea (Zaitsev & Ozturk 2001). Consequently, more detailed investigations and monitoring studies are needed.

Interestingly, some species have turned out to be a highly valuable resource, such as *Rapana venosa. Mugil so-iuy* has also a commercial potential without damaging the native ecosystem. On the contrary, some species, such as *Mnemiopsis leidyi,* have turned out to be extremely harmful to the native fauna and flora, creating a considerable economic loss.

THE BLACK SEA - A RECIPIENT, DONOR AND TRANSIT AREA FOR ALIEN SPECIES

MARIAN-TRAIAN GOMOIU[1], BORIS ALEXANDROV[2], NICKOLAI SHADRIN[3] & YUVENALY ZAITSEV[2]
[1]*GeoEcoMar, Constanta, Romania*
[2]*Institute of Biology of the Southern Seas, Odessa Branch, Ukraine*
[3]*Institute of Biology of the Southern Seas, Sevastopol, Ukraine*

Abstract
There is evidence of large-scale biological contamination of the Black Sea ecosystem by nonindigenous species: benthic communities are mostly altered in the western and northwestern parts of the Sea, while the pelagic part are affected in the whole basin. Recent invasion status of the Black Sea is briefly reviewed with special focus on most successful recent invaders.

1 Introduction

The Black Sea ecosystems underwent a number of major changes provoked by invasive aliens. More than 43 new species have become common in the Black and Azov seas (Table 1) and more than 10 occasional species have been recorded (Gomoiu & Skolka 1996; Alexandrov & Zaitsev 2000; Shadrin 2000; Zaitsev & Öztürk 2001). Several reasons may explain their immigration success: (i) the high probability of transfer due to increased ship traffic - from 24,100 vessels (average tonnage 105,500) passing the Bosphorus in 1985 to 49,950 (156,060) in 1996 (Safety precautions against disasters 1996); (ii) the existence of artificial waterways connecting the Black Sea and other seas, (iii) the development of mariculture; (iv) the relatively low ecosystem immunity due to low species diversity and, as a result, reduced ecosystem immunity against bioinvasions (Stachowicz et al. 1999); (v) growing anthropogenic load that leads to environmental degradation and reduction of the immunity to aliens (170 million humans live within the catchment area), and (vi) reduced river inflow due to increased water use in the agriculture and industry.

2 History and recent invasion status of the Black Sea

The shipworm *Teredo navalis* was carried from the Mediterranean by ancient ships in 750-500 BC. (Gomoiu & Skolka 1996). Ancient civilizations were well aware of the shipworm and its ravages. The Phoenicians coated the hulls of their ships with pitch and, later, copper sheathing. An oil mixture of arsenic and sulphur was used by the 5th century BC, and the Greeks used lead sheathing as early as in the 3rd century BC.

The scientifically documented settlers were the barnacles *Balanus improvisus*, introduced around 1844 (Gomoiu & Skolka 1996) and *B. eburneus*, first recorded in 1892 (Ostroumov 1892). The Indo-Pacific polychaete *Ficopomatus* (*Mercierella*) *enigmaticus* was first found in the 1920s, the North American decapod *Rhithropanopeus harrisi* in 1938, and the bivalves *Mya arenaria* in 1966 and *Anadara* (*Scapharca*) *inaequivalvis* in 1982 (Zaitsev & Öztürk 2001).

E. Leppäkoski et al. (eds.), Invasive Aquatic Species of Europe, 341–350.

Table 1. Non-native species in the Black Sea and the Sea of Azov. For references, see Cvetkov & Marinov (1986), Gomoiu & Skolka (1996), Shadrin (2000), and Zaitsev & Öztürk (2001)

Species	Donor region	Year of introduction or finding
Bacteria		
Vibrio cholerae	SE Asia	1970
Phytoplankton		
Mantoniella squamata	?	1980
Alexandrium monilatum	N America	1991
Gymnodinium uberrimum	fresh waters of Europe	1994
Phaeocystis pouchetii	Arctic or N Atlantic	1989
Asterionella japonica	Pacific or Atlantic Ocean	1968
Rhizosolenia calcar-avis	Pacific or Atlantic Ocean	1926
Thalassiosira nordenskioldi	Arctic or N Atlantic	1986
Phaeophyta		
Desmarestia viridis	N Atlantic	1992
Ectocarpus caspicus	Caspian Sea	1980
Higher plant		
Azolla caroliniana	South-Eastern Asia	1989
Azolla filiculoides	South-Eastern Asia	1989
Hydrozoa		
Blackfordia virginica	N America	1925
Perigonimus (Bougainvillia) megas	N America	1933
*Tiaropsis multicirata*** **	N Europe	1990
Eudendrium annulatum	N Atlantic	1990
Eudendrium capillare	N Atlantic	1990
Ctenophora		
Mnemiopsis leidyi	N America	1982
Beroe ovata	N America	1997
Monogenoidea		
Ligophorus kaohsinghsieni	Sea of Japan	1994
Gyrodactylus mugili	Sea of Japan	1995
G. zhukovi	Sea of Japan	1995
Polychaeta		
Ficopomatus enigmatica	Indian Ocean	1929
Hesionides arenarius	Pacific or Atlantic Ocean	1950s
Ancistrosyllis tentaculata	N Atlantic	1950-60
Streblospio shrubsolii	N Atlantic	1950s
Glycera capitata	Pacific Oc. or N Atlantic	1960-70
Nephtys ciliata	N Atlantic or Pacific Oc.	?
Streptosyllis varians	N Atlantic or Pacific Oc.	1960s
Capitellathus dispar	Indian Ocean	?
Magelona mirabilis	English Channel, Mediterranean Sea	1990s
Mollusca		
Mya arenaria	N Atlantic	1966
Anadara inaequivalvis	Indian Ocean	1982
Doridella obscura	N America	1986
Rapana thomasiana	Sea of Japan	1946
Potamopyrgus jenkinsi (antipodarum)	Pacific Ocean	1952
Crassostrea gigas	Sea of Japan	1980***
Teredo navalis	Atlantic, Pacific	750-500 B.C.
Entoprocta (Kamptozoa)		
Urnatella gracilis	N America	1954
Crustacea		
Balanus improvisus	N America	1844
B.eburneus	N America	1890s
Acartia tonsa	N America or Indo-Pacific	1976
Schizopera neglecta	Caspian Sea	1967
Rhithropaneus harrisi	N America	1932
Callinectes sapidus	N America	1967
Eriocheir sinensis	Pacific Ocean	1997
Sirpus zariquieyi	Mediterranean Sea	1980
Euchaeta marina	Mediterranean Sea	1999

Species	Donor region	Year of introduction or finding
Rhinocalanus nasutus	Mediterranean Sea	1999
Pleuromamma gracilis	Mediterranean Sea	1999
Scolecithrix danae	Mediterranean Sea	1999
Philomedes globosa	Mediterranean Sea	1999
Pisces		
Pseudorasbora parva	Sea of Japan	1980
Gambusia affinis	N America (intentional)	1925
Lepomis gibbosus	N America (intentional)	1930s
Mugil soiuy	Sea of Japan (intentional)	1972-82
Oryzias latipes	SE Asia (intentional)	?
Hypophthalmichthys molitrix	E Asia	1950s
Micromesistius poutassou	N Atlantic	1999
Aves		
Vanellochettasia lencura*	Caspian Sea *	1997
*Rubifrenta ruficolis***	Caspian Sea**	1970
Mammalia		
Ondatra zibethicus	North America	1930s
Nyctereutes procyonoides	Easten Asia	1930s
Delphinopterus leucas	Sea of Okhotsk	1992
Callorhinus ursinus	Bering Sea	1990s
Eumetopias jubatus	North Pacific Ocean	1990s

* nesting, ** wintering, *** accidentally introduced by ships on the beginning of the 20th century, and intentionally for cultivation, in 1980 from the Posjet Bay of the Sea of Japan, **** reidentified as *Opercularella nana* (N.Grishicheva, pers.comm.)

The invasion rate estimated for Romania is: one new species every 3-4 years for marine waters and 4-5 years for fresh waters, since the beginning in the second half of the 19th century. In total, 67 alien species, of them 60% of marine and 40% of limnic origin, have penetrated coastal and inland waters of that Black Sea country. Most of these species (up to 93% in marine environment) were introduced accidentally, being Atlantic-Mediterranean immigrants or originating from the North-American and Indo-Pacific regions (M-T Gomoiu unpubl. data).

Several of the most successful invaders were unintentionally introduced in the Black Sea by ships, either in ballast tanks (the soft-shell clam *Mya arenaria,* the rapa whelk *Rapana thomasiana,* the comb jellies *Mnemiopsis leidyi* and *Beroe ovata,* etc.) or attached to ship hulls (the polychaete *Ficopomatus enigmaticus,* barnacles *Balanus* spp., the nudibranch gastropod *Doridella obscura,* etc.).

There are few examples of non-target species associated with intentionally imported species for aquaculture, ornamental trade and even scientific experiments (bacteria and various parasites, e.g. Trematoda, which pose a threat for the native biota, and the fishes *Pseudorasbora parva* and *Lepomis gibbosus*).

For several species, passive or active migration facilitated by natural or man-made waterways has been a key vector. For instance, marine planktonic forms of Copepoda, could enter into the Black Sea either in a natural way, carried by the bottom current from the Marmara and Mediterranean Seas through the Bosphorus Strait, or brought by ships' ballast water. It is not clear, if these species recently immigrated into the Black Sea or if they expanded their natural ranges into a new sea area? The species may have occurred for a long time, however, they were previously overlooked by the scientists before investigation and sampling techniques were improved. Consequently, one cannot quote these as new immigrants but as cryptogenic species (*sensu* Carlton 1996). It is well known that the biodiversity inventory of many areas is far from being complete and the adequate development of research may reveal many new species.

3 The Black Sea as a donor and transit area

Table 2 provides a list of indigenous Black Sea species that have penetrated other seas. Exotic species for which the Black Sea was a transit water body in their further expansion are presented in Table 3. The majority (32 species) of the accidental immigrants in the Caspian Sea originate from the Black Sea; about 60% of the alien free-living animal species found there are not native to the Black and Azov seas. For these species the Black Sea was a bridge on their way to other seas, where step-by-step acclimation to new environmental conditions took place.

Table 2. Indigenous Black Sea species introduced in other water bodies.

Species	Recipient	Year
Nitzchia seriata	Caspian Sea	1990
Acrochaete parasita	Caspian Sea	1955-60
Ectochaete leptochaete	Caspian Sea	1955-60
Enteromorpha tabulosa	Caspian Sea	1955-60
E. salina	Caspian Sea	1955-60
Ectocarpus confervoides	Caspian Sea	1955-60
Entoneum oligosporum	Caspian Sea	1955-60
Acrochaetium daviesii	Caspian Sea	1955-60
Ceramium diaphanum	Caspian Sea	1955-60
Polysiphonia variegatum	Caspian Sea	1955-60
Lecudina sp.	Caspian Sea	1955-60
Chephalodophora communis	Caspian Sea	1955-60
Moeristia maeotica	Caspian Sea	1955-60
Bacciger bacciger	Caspian Sea	1955-60
Cercaria discursa	Caspian Sea	1955-60
Membranipora crustulenta	Caspian Sea	1955-60
Barentsia benedeni	Caspian Sea	1955-60
Mytilaster lineatus	Caspian Sea	1920-28
Dreissena bugensis	Caspian Sea, Great Lakes	1990-1994
*Mytilus galloprovincialis**	Hong-Kong	1979-1983
	Sea of Japan	1988-89
Palaemon adspersus	Caspian Sea	1930-40
Neogobius melanostomus	Great Lakes	1989-1993
Proterorhinus marmoratus	Great Lakes	1989-1993
Nereis diversicolor	Caspian Sea	1939
	Aral Sea	1960-69
Abra ovata (Syndosmya segmentum)	Caspian Sea	1947-1955
	Aral Sea	1960-1969
Calanipeda aquae-dulcis	Aral Sea	1965-70
Liza aurata	Caspian Sea	1930-40
Liza saliens	Caspian Sea	1930-40

*donor sea was the Black or Mediterranean Sea

The Volga-Don Canal, opened in 1952, formed an invasion corridor for Black Sea species to penetrate the Caspian Sea. It would be reasonable to expect a mutually enriching species exchange between the two seas; however this scenario has not come true. Only two Caspian newcomers (*Ectocarpus caspicus* and *Shizopera neglecta*) have reached the Black Sea but they never became mass species there. The influx from the Black Sea into the Caspian Sea has been disproportionately larger; those immigrants rapidly increased their abundance and now they are often dominant in plankton and benthos of the coastal zone of the Caspian Sea (Ardabieva et al. 2000; Orlova 2000; Shadrin 2000; Tinenkova et al. 2000, and others).

Why is an invasion corridor often a one-way street? Two assumptions may serve a working hypothesis: (i) the flux of non-native species will be from water bodies with high ecosystem immunity, i.e. with high biodiversity into water bodies with lower immunity (low biodiversity); (ii) it is more probable that water bodies with higher inertia potential would supply alien species into water bodies with lower inertia potential. The inertia potential is expressed by a morphometric index: m = L/Hmean, where L is the length of the shoreline and Hmean is mean depth of the water body. The relationship between biological characteristics and morphometric index of a water body has been discussed by Shadrin (1985). This scenario may be valid for both the Caspian and Baltic Seas in respect to the Black Sea. It is reasonable to expect that the exotic species, which have established themselves in the Black Sea, may achieve even greater success in the Caspian and then in the Baltic Sea.

4 Spread and impact of some recent invaders

Plants - exceptions among Black Sea invaders. Even if the number of animal invaders is relatively high in the Black Sea, there are only a few nonindigenous plant species recorded in these seas. In phytoplankton, the North American species *Alexandrium monilatum* (Pyrrophyta), the flagellate *Mantoniella squamata* (origin unknown), and two Atlantic species, *Phaeocystis pouchetii* (Prymnesiophyceae) and *Rhizosolenia calcar-avis* (Centrophyceae) are known to cause blooms in the western Black Sea (Zaitsev & Öztürk 2001). Astonishingly, the North American soft sour weed *Desmarestia viridis* is the only alien macrophyte species recorded so far.

The bivalve *Mya arenaria* - a successful molluscan invader. The introduction of the soft clam *M. arenaria* into the Black Sea in 1960s was one of the most drastic and successful bioinvasions into the region before the establishment of the "famous" ctenophore *Mnemiopsis leidyi* in 1980s. First reported in 1966, in the Odessa Gulf and off the estuaries of the Dnieper and the Bug Rivers, the soft clam rapidly spread in other regions: in 1970s, *Mya* became the dominant benthic species along the Romanian coast in zones from mobile sediments of the upper underwater slope to a depth of 30-40 m. After 4-5 years from its colonisation its biomass exceeded that of other molluscs, reaching up to 16,000 g wwt m^{-2} with abundance over 8,000 ind m^{-2} (Gomoiu & Porumb 1969). The huge beach deposits of decaying *Mya* formed after storms attracted and maintained abundant populations of sea gulls, the local people also used them to feed poultry.

The introduction of the soft clam has had multiple consequences on the structure and functioning of the invaded ecosystem. It outcompeted some indigenous species, e.g., by competition for habitat with the small native bivalve *Lentidium mediterraneum*, which avoids sandy bottoms silted by *Mya*. For example, in the Odessa Gulf, the area dominated by *L. mediterraneum* in the 1980s reduced fivefold as a result of the expansion of *M. arenaria* (Zaitsev & Öztürk 2001). Young specimens of *Mya* offer an additional food source for adult bottom-eating fish (gobies, flounder, turbot and sturgeons), gulls and other marine birds. *Rapana*, another exotic mollusc, eats adult specimens of *Mya*. Moreover, *Mya* became an additional biofilter in the coastal zone, which is an important function in eutrophicated coastal waters.

Table 3. Exotic species for which the Black Sea was a transit water body in their voyage to other seas.

Species	Recipient	Year
Blackfordia virginica	Caspian Sea	1955-58
Bougainvillia megas	Caspian Sea	1955-58
Mnemiopsis leidyi	Marmara Sea, Mediterranean Sea Caspian Sea	1989-93 1999
Ficopomatus enigmaticus	Caspian Sea	1956-57
Rapana thomasiana	Marmara Sea, Mediterranean Sea Cheasapeake Bay* Atlantic France	1970-1990 1996-97 1999-2000
*Acartia tonsa***	Caspian Sea	1980-81
Balanus improvisus	Caspian Sea	1955-58
B. eburneus	Caspian Sea	1955-58
Rhithropaneus harrisi	Caspian Sea Aral Sea	1955-65 1970
Mugil so-iuy	Marmara Sea, Mediterranean Sea	1990-1997

* introduced from the Sea of Japan or the Black Sea
** at first identified as *A. clausi*

The effects of eutrophication and pollution caused great changes in the *Mya* populations. Heavy blooms in mid 1970s decimated the soft clam population on the Romanian shelf: in 1982 it vanished from the large areas, especially in shallow water zones, where still in 1974 it occupied all sedimentary bottoms, forming biomasses more than 1 kg m^{-2}. Together with other macrobenthic species, *Mya* was adversely affected by hypoxia in the late 1980s, and its amount decreased from more than 6,000 ind m^{-2} (biomass 3, 000 g wwt m^{-2}) in 1970-1975, to 400 ind m^{-2} (260 g wwt m^{-2}) in 1991 (Leppäkoski & Mihnea 1996). Over the period 1973-1982, the total mortality loss of *Mya* was estimated at 4.1 million tons on sedimentary bottoms down to 30 m depth between the Danube delta and Constanta (Gomoiu & Skolka 1996).

Ups and downs of *Rapana thomasiana*. Native to the Sea of Japan, this gastropod was first observed in the Black Sea (Novorossiysk Bay) in 1946, but it was considered to have settled in the Pontic basin already in the in 1930s-1940s. *Rapana* was found at Yalta and Sevastopol in 1954 and along the Romanian shores in 1963 (Gomoiu & Skolka 1996; Zaitsev & Öztürk 2001). Being very fertile, without serious competitors and endowed with high adaptive capacity to tolerate low salinity, water pollution and hypoxic conditions, the newcomer succeeded in forming rich populations. This largest mollusc in the Black Sea (maximum shell height > 140 mm) became very common and numerous, not only on rocky bottoms down to 30 m depth, where they feed upon oysters, mussels (mainly *Mytilus galloprovincialis*) and other bivalves, but also on sandy bottoms. Mussel beds became largely destroyed by *Rapana*. Having destroyed its preferred food - the large-sized species (*Ostrea edule taurica*, *M. galloprovincialis* and *Modiolus adriaticus*), *Rapana* was forced to attack smaller-sized species such as *Chione* *gallina*, *Pitar rudis*, *Paphia rugata*, and *Spisula subtruncata*. The total biomass of *Rapana* on the Caucasian shelf was estimated at 2,800 tons and on the northeastern shelf in front of the Kerch Strait at 6,000 tons (Grishin & Zolotarev 1988).

As reflected by present composition of beach deposits, *Rapana* populations are in decline at the Romanian coast: trophy shells, caught in important quantities in the beginning of the1990s, have become very rare. Along the Anatolian coast there are no more *Rapana* reserves that can be exported to Japan as in the 1970s (> 800 tons snail meat exported per year). In the northwestern part of the sea the *Rapana* populations are poor and their influence on the marine ecosystems seems to be fairly weak. It is likely that the overfishing of *Rapana* has lead to the population decline (Zaitsev & Öztürk 2001).

Anadara inaequivalvis **- an Indo-Pacific invasive bivalve.** *Anadara* was accidentally introduced into the Mediterranean in the 1960s. It was recorded for the first time in the early 1980s in the Black Sea. Presently it is quite widespread along all coasts and in the southern part of the Sea of Azov (since 1989-1990), inhabiting sandy and muddy bottoms to 40 m depth and reaching densities of 100 ind m^{-2} and biomass > 1 kg m^{-2} (Zaitsev & Öztürk 2001). In areas along the Bulgarian coast, *A. inaequivalvis* became very numerous: within less than five years, this newcomer reached maximum biomass values of up to 4,300 g m^{-2} (Cvetkov & Marinov 1986). This species tolerates both waters with variable salinity and periodical hypoxia and anoxia.

Acartia tonsa **substitutes native planktonic copepods.** This species, previously known as native to the coastal zone of the West Atlantic and Indo-Pacific, was first recorded in the North Sea in 1914, in the Baltic Sea in 1925 and in the Black Sea in 1990, in the area off the Crimea peninsula (Belmonte et al. 1994). *A. tonsa* consumes detritus, phytoplankton, ciliates and small copepods. It rapidly adapts to changing food conditions and propagates even at very low ingestion rates (Kiørboe et al. 1985), being able to survive starvation for 6-10 days (Dagg 1977). Temperature is the factor that controls the geographical distribution of *A. tonsa*: the reproduction rate is low under 10 °C (Jeffries 1962). This species dominates in summer and autumn in warm waters in the upper layers at 0-20 m. For example, in the Baltic Sea mass development of *A. tonsa* occur in June-September at the temperature 16-17 °C and salinity 4-5 PSU (Silina 1989).

There is no information available on any significant impact of *A. tonsa* on the pelagic community, most probably due to its low density in the Black Sea. However the reproduction potential of this species is high: in other regions it is known to develop in mass abundance, e.g., up to 10,000 ind m^{-3} in the Baltic Sea (Silina 1989). Benthic invertebrates can utilize dormant eggs of *Acartia* of high food value; ca. 55% of the egg biomass is protein (Kiørboe et al. 1985).

Presently, the summer mesozooplankton in the Sevastopol Bay (Crimea peninsula) is dominated by two alien species, *A. tonsa* and larvae of the barnacle *Balanus improvisus* (Pavlova & Kemp 1999). *A. tonsa* occupies mainly the ecological niche that was previously occupied by the native species *Acartia latisetosa*. These two copepod species are very similar in size and ecological requirements. A. *latisetosa* had vanished in the bay before the introduction of *M. leidyi* (1986) but several years after the introduction of *A.*

tonsa (1976). The main reason for vanishing was likely increased sedimentation and pollution destroying benthic resting egg stage of *A. latisetosa* (Shadrin et al. 1999).

The mass occurrence of the comb jelly Mnemiopsis leidyi – an unwanted success story. Recent mass occurrence of the newcomer ctenophore *Mnemiopsis leidyi*, first found in the northeastern part of the Black Sea (Southern Crimea coastal waters) in 1982, is one of the most spectacular events in the history of species introductions into marine systems (e.g., Leppäkoski & Mihnea 1996). This species, from the Atlantic coasts of the USA, spread rapidly all over the Black Sea. Introduced further into the Sea of Azov via ships and/or with currents from the Black Sea, it was recorded first in August 1988 after a mass development in the Black Sea.

In 1988 *Mnemiopsis* developed biomasses up to > 1 kg m^{-2} in the open sea, and 5 kg m^{-2} in coastal waters. It remained numerous until 1991-92, when its biomass reached 10-12 kg wwt m^{-2} in some coastal areas. However, it did not exceed 1.5-3 kg m^{-2} in the open Black Sea in 1988-90. Since then, its abundance decreased 4-6 fold (Vinogradov et al. 1993) but increased soon again. In August 1995 there were approximately 12-16 million tons of fresh comb jelly mass in the waters covering the Romanian continental shelf only (Gomoiu & Skolka 1996).

The introduction of *Mnemiopsis* caused severe disturbances in the ecosystems which had already been disturbed by eutrophication and pollution. From the ecosystem point of view *Mnemiopsis* is a dead-end organism in the food chains of the Black Sea. It is a consumer of large amounts of zooplankton and ichthyoplankton and a food competitor of pelagic planktophagous fish, most importantly of anchovy. The zooplankton communities of the Black Sea and the Sea of Azov were severely affected by this invasive predator, which mass development began in 1987. This resulted in drastic decrease of fish production, e.g. for kilka 4-5 times and anchovy > 10 times (Volovik et al. 1993). According to Caddy & Griffits (1990), the damage caused by *Mnemiopsis leidyi* in the late 1980s for Black Sea fisheries reached up to US$ 200,000,000 per year. Much more, according to the same authors, is the damage caused through the inactivity of fishing vessels, fishing ports and factories. In the Sea of Azov, the annual economic losses caused by *Mnemiopsis* (when only the reduction of catches of anchovy *Engraulis encrasicholus maeoticus* and tyulka *Clupeonella cultriventris* were considered) reached up to US$ 40,000,000 (Volovik 2000).

The comb jelly *Beroe ovata* – a cryptogenic invader. *Beroe* (first found in Odessa Bay in 1997) was first identified as *Beroe cucumis* (Zaitsev 1998) in the Black Sea, later on as *B. ovata* (Vinogradov et al. 2000). The possible mechanism of penetration of this circumpolar species into the Black Sea is probably the same as for *Mnemiopsis leidyi,* most likely transferred from the North Atlantic Ocean. Another hypothesis is that *Beroe*, which lives in the Mediterranean and Marmara Seas, was carried by currents and had a chance to acclimatize in the Black Sea during abnormally warm winters 1997/1998 and 1998/1999, which promoted its adaptation.

In May-August 1998, mass development (up to 35-1,500 ind m^{-3}) of *Beroe* was recorded in the northwestern part of the Black Sea along the coastal zone of Romania (M-T. Gomoiu, unpubl.) and Ukraine from the Danube river mouth to Odessa Bay (Nastenko & Polishchuk 1999). In 1999, *Beroe* increased along Crimean coast, in the Sea of Azov and in the coastal water of the northeastern Black Sea (Vinogradov et al. 2000).

Beroe feeds mainly upon other comb jellies during all development stages. In the Black Sea, one individual of *Beroe* consumes 0.3 averaged sized individual of *M. Leidyi* per hour, or ca. 2.5 g of wet weight, equal to 2.5 mg C_{org} (Vinogradov et al. 2000). This might lead to a reduction of the native comb jelly *Pleurobrachia rhodopis* quantity and to an inhibition of *Mnemiopsis*. Obviously these two ctenophores are interacted in a feedback system (an increase of *Beroe* population will result in a decrease of *Mnemiopsis*, followed by a successive increase of zooplankton and pelagic fish). In addition, *Beroe* is a food-web dead-end organism due to the lack of native predators in the Black Sea. Thus, either direct or indirect impact through the entire food web could well be expected and further add to the problem of gelatinous species in the Black Sea.

Non-native fish species. The occurrence of introduced fish species in the Black Sea is restricted mainly to the brackish estuaries and wetlands, among them two North American species, the mosquitofish *Gambusia affinis* (brought from Italy in 1925) and the sunfish *Lepomis gibbosus* (introduced to Europe by aquarists in the 1920 and found in Danube delta since 1930s).

The haarder *Mugil soiuy* is an important commercial species in the Black and Azov Seas; its annual catch in the Black Sea exceeds 10,000 tons. It was transferred from the Amur River estuary in 1972-1982 and released in some coastal brackish lakes (limans). The haarder is highly euryhaline, living from fresh water to 35 PSU in some limans. First breeding of naturalized fish in the Black Sea was noted in 1982. The impact of the haarder on native species is insufficiently known. Its fry feed upon zooplankton and can compete with local plankton-eating fish. Some specific parasites (Trematoda, Monogenea), associated with the haarder, were introduced in the Black Sea and have been found in native grey mullets (Zaitsev & Öztürk 2001).

Mammals in coastal wetlands. The muskrat *Ondatra zibethicus*, native to North America, was introduced in the URSS in the 1930s as a fur animal. It lives in coastal wetlands and is a mass inhabitant of the Danube, Dniestr and Dnieper river deltas. The East Asian raccoon-dog *Nyctereutes procyonoides* was introduced in the Black Sea wetlands in the 1930s as a fur animal. It is omnivorous and feeds mainly upon fish, crayfish and molluscs. In the Danube delta it also eats the soft-shelled clam *Mya arenaria*, which is washed ashore in large numbers on windward sandy beaches (Zaitsev & Öztürk 2001).

5 Conclusions

The consequences of invasions into the Black Sea have been discussed in a number of publications (e.g. Zaitsev 1993; Zolotarev 1996; GESAMP 1997; Zaitsev & Mamaev 1997; Shadrin 1999; Alexandrov & Zaitsev 2000; Leppäkoski & Olenin 2000; Zaitsev & Öztürk 2001).

Invasion is never neutral. Even if an incomer does not replace a native species but occupies the niche, which is or has became vacant owing to some other factors, it excludes the possibility for the native species to recolonise this niche. A species having been accidentally brought into a water body undergoes phenotypic and genetic changes, triggered by the unusual environmental vectors. As a result, the variability and the invasive capacity of the colonizer can increase over time.

In conclusion, the lessons we have learned from the invasion history of Black Sea as a receiving area for exotic species can be summarized: (i) the process of introduction is still going on - we have to pay more attention to the toxic and harmful species; (ii) the impact of alien species is complex and unpredictable; (iii) the problem of invasive non-native species has long gone beyond the bounds of pure science and has gained a socio-economic dimension, (iv) species diversity monitoring is necessary, a special attention must be directed to the proper monitoring of microflora and microfauna; (v) legal measures and regulations to limit the penetration of immigrants are necessary; (vi) training biologists in taxonomy and systematics must be given high priority; the old generation of marine botanists and zoologists is getting reduced and to understand correctly the Black Sea ecosystem and what must be done for its protection, we have to gain deeper knowledge of its biota through young educated marine biologists.

INVADERS IN THE CASPIAN SEA

NIKOLAI V. ALADIN*, IGOR S. PLOTNIKOV &
ANDREI A. FILIPPOV
Zoological Institute, Russian Academy of Sciences, St.Petersburg, Russia
*Corresponding author *aral@zin.ru*

Abstract
The Caspian Sea is the largest lake in the world. All its resident species can be considered as invaders with various time of introduction. The most ancient invaders are regarded as indigenous species. The glacial relicts invaded the Caspian 1-1.5 million years BP. There are also "Atlantic species" and those of fresh-water origin. Human-mediated introductions started in the beginning of 20th century. The construction of the Volga-Don Canal facilitated invasion many-fold. A deliberate introduction of fishes and invertebrates has exerted essential influence on the biodiversity of the Caspian in the 1930s-1970s. The impact of introduced species on indigenous biota can be either positive or negative. The invasion of the comb jelly *Mnemiopsis leidyi* and the bivalve *Mytilaster lineatus* had negative consequences while deliberate acclimatisation as well as a number of accidental introductions of other aquatic organisms had positive effects. The conclusion was drawn that the most ancient representatives of the Caspian flora and fauna should be protected against any negative impact, including the impact of newly introduced species.

1 Introduction

The Caspian Sea, a lake without outlets, washes the shores of five countries; Azerbaijan, Iran, Turkmenistan, Kazakhstan and Russia, and two continents, Europe and Asia (Fig. 1). It is the largest lake in the world. The sea level of the Caspian Sea is 25-29 m lower than the mean sea level; it fluctuates depending on the water balance, rising when the balance is positive and lowering when it is negative. Due to the instable water levels, the surface area is also varying. The Caspian is elongated: its length from the north to the south is 1,204 km (Zenkevich 1963) and the greatest width is 566 km. The surface of the Caspian is 436,000 km^2, its watershed area is 3.1-3.5 million km^2 and total volume about 77,000 km^3. The Caspian Sea is divided into three parts of approximately equal dimension: North, Middle and South. Only a part of the Northern Caspian belongs to Europe, while both Middle and Southern parts are on the Asian side of the sea. The average depth of Northern Caspian is 6 m maximal depths do not exceed 10 m. The mean and greatest depths of Middle and Southern parts are 175-790 and 300-1,025 m, respectively.

A shallow (max depth < 10 m) gulf Kara-Bogaz-Gol lies approximately 3-4 m lower than the sea level and due to this the sea constantly drains into it, making the gulf a large evaporator of the Caspian. Presently, 200-300 m^3 per second of Caspian water is discharged into Kara-Bogaz-Gol carrying some 130-150 million tons of salt from the sea. Average salinity of the Northern Caspian is 5-10 PSU, of the Middle part 12.7, of the Southern part 13 and of the gulf Kara-Bogaz-Gol up to 300-359 PSU. Caspian waters differ from oceanic waters by their salt composition, which is rather poor with sodium and chlorine ions and rich with ions of calcium and sulphates.

Each of the four areas of the Caspian has its own features of the temperature regime. In the Northern Caspian, in the middle of a summer, the average water temperature is

E. Leppäkoski et al. (eds.), Invasive Aquatic Species of Europe, 351–359.

24°C; in the winter this part of the Caspian freezes. Water temperature under ice can drop down to -0.5°C. In the Southern Caspian the temperature never drops below 13°C in wintertime; in summertime it usually increases up to 25 and even 30°C. Constant temperature is maintained at depths of the Caspian Sea both in winter and summer. The water of the Caspian Sea is characterized by high transparency with maximums observed in the open parts of the Southern Caspian. As a result of good turbulent mixing the deep waters have no dead zone and are rich with dissolved oxygen. On the surface the amount of oxygen is close to saturation all over the Sea.

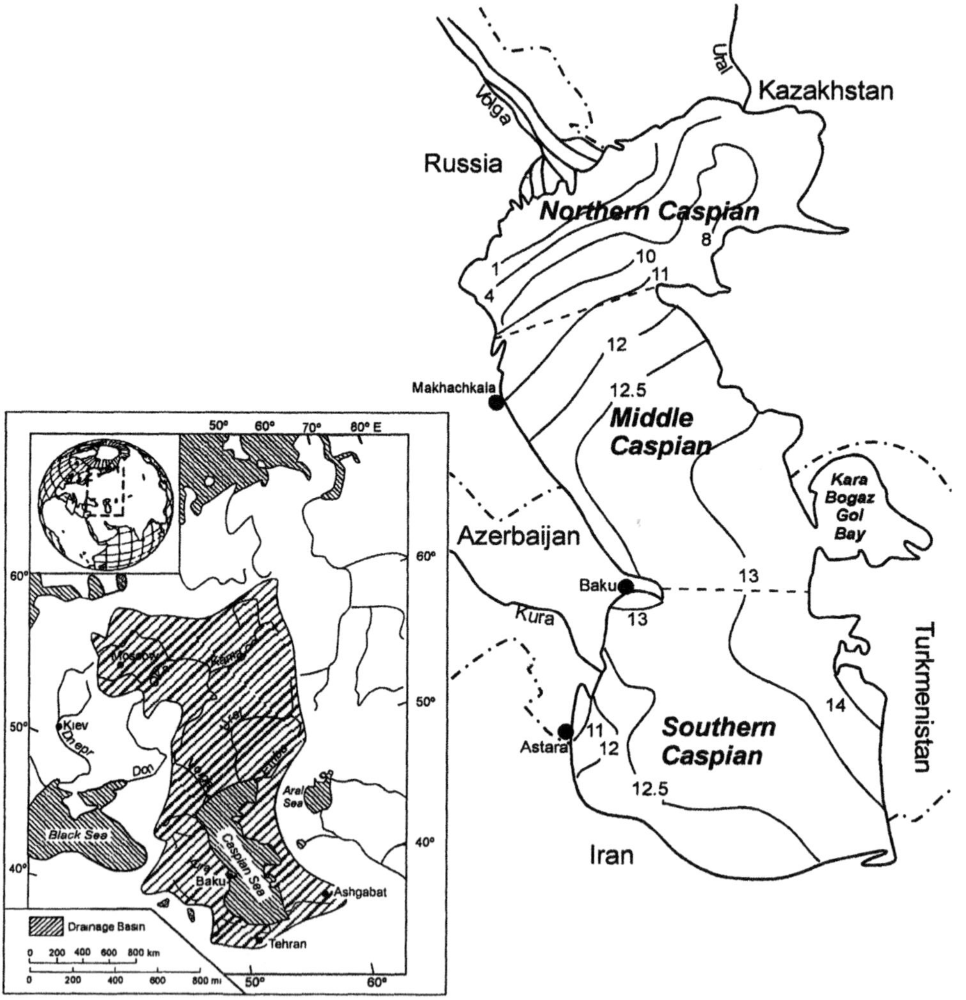

Figure 1. Caspian Sea water area (with isohalines) and drainage basin (by Zenkevich 1963 and Rodionov 1994).

All present resident species in the Caspian Sea can be described as invaders. The only difference is the time of introduction. According to the time of appearance in the Caspian the species can be divided into four groups. (i) The most ancient invaders, which can be regarded as indigenous and are often called "living fossils". (ii) Early invaders from the Arctic seas, the glacial relicts. The ancestors of these species invaded the Caspian 1-1.5 million years BP during the period of melting of the ice sheet that covered almost all Europe, Arctic and coastal areas of the Baltic and White Seas. (iii) Introduced species from the Black and Mediterranean Seas. The most ancient of these "Atlantic invaders" were introduced into the Caspian about 50,000 years ago when the ancient Caspian was connected with the Azov-Black Sea basin through the Manych canal. (iv) Invaders from fresh waters. The most ancient ones are the Caspian gastropods that originated from fresh waters of Pliocene. They probably invaded the ancient Caspian approximately 2-5 million years BP when the sea was most desalinated.

2 Native fauna and flora

2.1 ARCTIC SPECIES

Arctic species are one of the most ancient groups of the Caspian Sea species (Fig. 2), represented by the sabellid polychaete *Manayunkia caspia*, the copepod *Limnocalanus grimaldi*, the isopod *Saduria entomon*, the amphipods *Pseudalibrotus caspius*, *P. platyceras*, *Monoporeia affinis* and *Gammaracanthus loricatus*, the mysids *Mysis caspia*, *M. microphthalma*, *M. macrolepis* and *M. amblyops*, the white salmon *Stenodus leucichthys*, the bulltrout *Salmo trutta*, and the seal *Phoca caspia* (Zenkevich 1963 and references therein). Recently tusks of a walrus were found on the bottom of the sea in the vicinity of the Absheron Peninsula. Probably, this inhabitant of the Arctic region used to live in the Caspian Sea (N. Vereschagin, pers. comm.). Also some plankton crustaceans (Cercopagidae) should probably belong to the northern invaders. According to regulatory and molecular-biological studies of the genus *Cercopagis* they cannot be considered as typical Caspian endemics (Richter et al. 2001).

There are few hypotheses tending to explain how and when these organisms invaded the Caspian Sea. The first of them (Ekman 1916; Sars 1927) stated that there was a direct contact between the Caspian Sea and the Arctic Ocean. At present, this hypothesis is outruled because of lack of evidence. Grimm & Kessler (cit. Zenkevich 1963) deemed that the Arctic immigrants arrived from the north via rivers. Today, this hypothesis is supported by sufficient data. Guryanova & Pirojnikov (cit. Zenkevich 1963) assumed that immigrants originated in the Kara Sea. There are no direct confirmations of this theory. Berg's (1928) hypothesis was based on the existence of an ancient lake, which stretched from the Baltic Sea up to Ladoga and Onega Lakes and further up to the White Lake and to the river Sheksna. This hypothetical lake ensured connection of the Caspian Sea with Arctic and Baltic waters and, consequently, penetration of Arctic immigrants, not directly from the Arctic Ocean, but from the Baltic and White Seas. Today, there is a lot of data supporting this hypothesis. Grosswald (1980) and Dawson (1992) presented the hypothesis that a superflood occurred during the late Valdai period (about 20,000 years ago). The level of the ancient Caspian rose by 2-3 m above the level of the ocean (Lamb 1977), and its waters run through the Azov-Black Sea basin. Waters of a large ice lake that existed in the West Siberian Plain run into the Aral basin and from the Aral into the ancient Caspian.

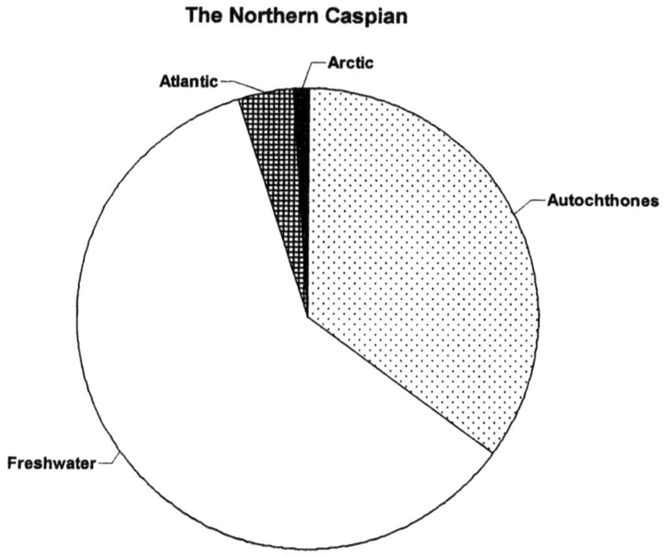

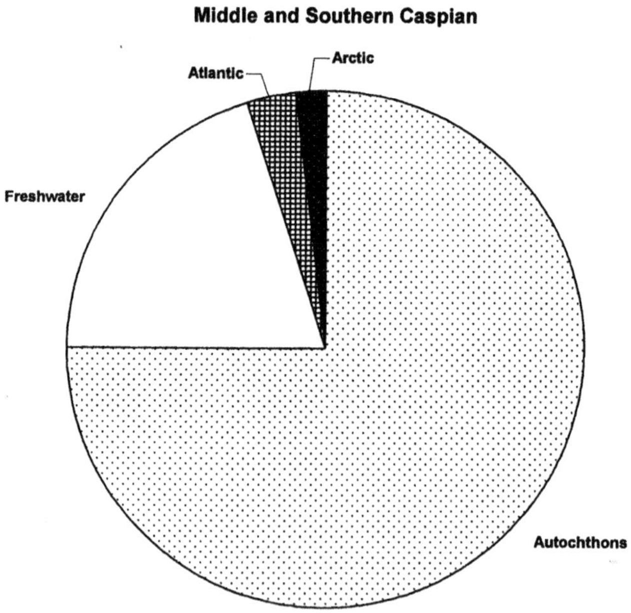

Figure 2. Authochthonous, freshwater, Atlantic and Arctic elements in the northern, middle and southern Caspian Sea fauna.

2.2 FRESHWATER ORGANISMS

The invasion of freshwater organisms (Fig. 2) into the Caspian Sea took place several times during its greatest freshening (see Chuykov 1994 for a comprehensive review). The gastropod molluscs are considered the most ancient freshwater elements. To certain degree this was facilitated by the nature of the salinity of the Caspian, essentially differing from the oceanic water by its composition.

2.3 ATLANTIC-MEDITERRANEAN ELEMENTS

The fauna of Atlantic-Mediterranean origin (Fig. 2) is represented by Turbellaria (1 species) Coelenterata (1), Polychaeta (2), Copepoda (1), Cirripedia (2), Decapoda (3), Mollusca (3), Bryozoa (2), and 6 species of fish (Zenkevich 1963). The first invaders penetrated during the Khvalyn period (approx. 50,000 years BP) in a natural way via the Kumo-Manycheskiy Strait connecting the Black and Caspian Seas. They included 7 species: the seagrass *Zostera nana,* the bryozoan *Bowerbankia imbricata*, the bivalve *Cardium edule,* the polychaete *Fabricia sabella*, and fish species (the silverside *Atherina mochon pontica,* the pipefish *Syngnathus nigrolineatus,* and the goby *Pomatoschistus caucasicus*).

3 Human-mediated unintentional introductions

3.1 EARLY TRANSFERS

In the 20th century the number of introduced species from the Black Sea and the Mediterranean suddenly increased. All the cases of introduction were related to anthropogenic activity. The diatom algae *Rhizosolenia calcar-avis* and the bivalve *Mytilaster lineatus* were brought to the Caspian presumably by merchants, who transported their small wooden boats on carts from the Azov Sea. *M. lineatus* was found for the first time in 1928. It is one of the most numerous benthic species of the Caspian Sea, but it is also known as an important fouling species. Diatom *Rhizosolenia calcar-avis* appeared in 1930. By 1936 it made up 2/3 of the total biomass of phytoplankton.

3.2 THE VOLGA-DON CANAL – AN IMPORTANT INVASION CORRIDOR

The Volga-Don Canal was opened in 1952. Its length is 101 km, the highest water-level is elevated by 88 m above the Volga level and by 44 m above Don level. The canal has 9 shipping locks at the Volga side and 4 locks at the Don side. It has also large reservoirs along its course (Sokolov 1964). After opening the canal, a new group of species invaded the Caspian Sea. Most of them were carried from the Black Sea and the Sea of Azov by ships in ballast water or in biofouling assemblages (Karpevich 1975). At least 10 species of algae penetrated into the Caspian, including the green algae *Enteromorpha tubulosa* and *E. salina,* the brown algae *Ectocarpus confervoides* v. *fluviatilis,* and *Entonema oligosporum,* and the red algae *Acrochaetium deviesii, Ceramium diaphanum, C. tenuissimum*, and *Polysiphonia variegata.* Soon after its introduction *C. diaphanum* became a dominant species in the Northern Caspian.

Two species of barnacles, *Balanus improvisus* and *B. eburneus*, were brought in as vessel biofouling from the Black Sea and the Sea of Azov. *B. improvisus* was first detected in the Northern and Middle Caspian in 1955. Within a year, it spread all over the Caspian Sea. Now it is a dominant fouling organism on vessels and hydro-engineering structures. *B. eburneus* was found in the Caspian Sea in 1956. It is rarely found in fouling of vessels, since it cannot stand strong water currents (Sars 1927). The bryozoan *Conopeum seurati,* the hydroids *Bougainvillia megas* and *Moerisia maeotica,* the kamptozoan *Barentia benedeni,* the polychaete *Ficopomatus enigmatica* and the plankton cladoceran *Pleopis polyphemoides* penetrated in 1950s from the Black Sea and Sea of Azov. Most of these species live in shallow waters and occur in fouling of vessels, buoys, pipes, port structures, offshore oilrigs and piers.

The jellyfish *Blackfordia virginica* was first discovered in the Caspian Sea at the mouth of the river Kura in 1956. This invader originates from the north-eastern coast of North America and came to the Caspian through the Sea of Azov. The North American mud crab *Rhithropanopeus harrisii* was found in the Northern Caspian in 1958. From the Northern Caspian, the crab spread with the aid of vessels and the cyclonic current along the western part of the Middle Caspian and all over the sea. In the 1970s the cladoceran *Penilia avirostris* was found in the Caspian Sea and in 1985 *Podon intermedius* appeared.

The eel *Anguilla anguilla* appeared in the waters of Azerbaijan in 1964. Some years later it was also found in the coastal waters from the Sarah Peninsula down to Iranian waters. Single eels used to be found in the basin of the Caspian Sea. According to Kessler (cit. Zenkevich 1963), they penetrated into the basins of the Black and Caspian Seas from the Baltic. The opening of the Volga-Baltic waterway and transplantation of larval eels (delivered from England and France) into the inland reservoirs of Russia, including basins along the Volga, facilitated its penetration into the Caspian. Eel is rather often caught as by-catch in different parts of the sea, which suggests that not only single fishes have penetrated the Caspian.

Draining of the Kara-Bogaz-Gol Bay facilitates the spread of halophilic cosmopolitan species. Not one of the above-mentioned species can be found in the hyperhaline Kara-Bogaz-Gol Bay. Only halophilous species inhabit this area. They represent a group of cosmopolitan forms originating in arid areas. All these halophilous organisms have a resting stage in their life cycle (delitescence eggs, spores, seeds and cysts) that can survive drying, freezing and other unfavourable conditions (Makrushin 1985). Some cysts could remain viable for tens and hundreds of years. The resting stages are small in size and can be carried with the wind and migrating birds.

In the early 1980s a dam was built in order to isolate the bay from the sea and to "improve" the water balance of the Caspian Sea. In 1900s the water discharge to the Kara-Bogaz-Gol Bay was restored. When the bay dried, resting phases of halophilic organisms remained on its exposed seabed and revived after the bay was refilled with salt water. The dry bay was a source of dust storms, which carried the cysts all over the world. This is the reason why many organisms that inhabit the Kara- Bogaz-Gol Bay are similar to the species from salt lakes of America, Europe, Asia, Africa and Australia.

3.3 MOST RECENT INVADER(S)

More species are currently invading the Caspian Sea, and will be registered in the nearest future (Dumont 1995). In autumn of 1999 *Aurelia aurita* and the comb jelly *Mnemiopsis leidyi* were recorded on video film from the Caspian (Ivanov et al. 2000). No doubt, invaders will continue appearing in the Caspian unless a new balance between native species and invaders will become established. It is very difficult to predict when such a balance will be achieved, due to the current unstable environment of the Caspian Sea, because of climate changes, anthropogenic pollution and other environmental impacts. Until now changes in abiotic and biotic components of the ecosystems of the Caspian Sea has supported invaders.

Table 1. Recent invaders in the Caspian Sea. (B) benthos; (P) plankton; (Bf) biofouling; (P) parasite; (N) nekton; (N-B) nektobenthos; (SH) shipping; (H) with host; (A+) during acclimatization; (A) acclimatization; (Rw) by railway; (Bl-Az) Black Sea – Sea of Azov; (Pac) Pacific; (Okh) Sea of Okhotsk; (NA) Northern America; (Med) Mediterranean; (Ind-Afr) India-Africa.

Species	Ecol. group	Origin	Naturalization	Donor area	Vector	Years
Coelenterata						
Blackfordia virginica	Bf	NA	+	Bl-Az	Sh	1956
Bougainvillia megas	Bf	NA	+	Bl-Az	Sh	1956
Moerisia maeotica	Bf	Bl	+	Bl-Az	Sh	1950s?
*Aurelia aurita**	P	Bl	?	Bl-Az	Sh	1999
Ctenophora						
Mnemiopsis leidyi	P	NA	+	Bl-Az	Sh	1999
Trematoda						
Apophallus muehlingi	Pa	Bl	+	Bl-Az	H	1980-90s?
Plagiorchis entamiatis	Pa	?	+	?	H	1960s
Rossicotrema donicum	Pa	Bl	+	Bl-Az	H	1980-90s?
Nicola scriabini	Pa	Bl	+	Bl-Az	H	1980-90s?
Ancyrocephalus vanbenedeni	Pa	Bl	+	Bl	A+	1930s
Wlassenkotrema longicollum	Pa	Bl	+	Bl	A+	1930s
Saccocoelium obessum	Pa	Bl	+	Bl	A+	1930s
Haplosplanchnus pachymosus	Pa	Bl	+	Bl	A+	1930s
Polychaeta						
Nereis diversicolor	B	Bor?	+	Az	A	1940
Ficopomatus enigmatica	Bf	Pac?	+	Bl-Az	Sh	1950s?
Mollusca						
Mytilaster lineatus	Bf	Med	+	Bl	Rw	1920s
Dreissena bugensis	Bf	Az-Bl	+	Bl-Az	Sh	1994
Abra ovata	B	Med	+	Az	A	1940s
Monodacna colorata	B	Bl	+		Sh	1950s
Lithoglyphus naticoides	B	Az-Bl	+	Bl-Az	Sh	1971
Tenellia adspersa	Bf	Med	+(?)	Bl-Az	Sh	1989
Crustacea (Cirripedia)						
Balanus improvisus	Bf	NA	+	Bl-Az	Sh	1955
B. eburneus	Bf	NA	+	Bl-Az	Sh	1950s
Crustacea (Cladocera)						
Penilia avirostris	P	Bl	+(?)	Bl-Az	Sh?	?
Pleopis polyphemoides	P	Bl	+	Bl-Az	Sh	1957
Podon intermedius	P	Med	+	Bl-Az	Sh	1985
Crustacea (Copepoda)						
Acartia clausi	P	NA	+	Bl-Az	Sh	1981
Crustacea (Decapoda)						
Palaemon elegans	N-B	Med	+	Bl	A+	1930s
P. adspersus	N-B	Med	+	Bl	A+	1930s
Rhithropanopeus harrisii	B	NA	+	Bl-Az	Sh	1950s
Bryozoa						

Species	Ecol. group	Origin	Naturalization	Donor area	Vector	Years
Conopeum seurati	Bf	Med	+	Bl-Az	Sh	1958
Kamptozoa						
Barentsia benedeni	Bf	Med	+	Bl-Az	Sh	1962
Pisces						
Pleuronectes flesus luscus	N		-	Bl	A	1902, 1930s
Scomber scombrus	N		-	Bl	A	1902
Engraulis encrasicholus	N		-	Bl	A	1933-1934
Mullis barbatus	N		-	Bl	A	1933-1934
Rhombus maeoticus	N		-	Bl	A	1930
Mugil auratus	N		+	Bl	A	1930-1934
M. saliens	N		+	Bl	A	1930-1934
Ctenopharyngodon idella	N		+	China	A	1955-1956 1964-1977
Hypophthalmichtys molitrix	N		+	China	A	1964-1977
Aristichthys nobilis	N		+	China	A	1964-1977
Oncorhynchus keta	N		+	Okh	A	1962-1966
O. gorbuscha	N		-	Okh	A	1964
O. mykiss	N		-	?	A	1975-1978
O. kisutch	N		-	?	A	1975-1978
Morone saxatilis	N		?	USA	A	1972
Stenodus leucichthys nelma	N		-	Ob'	A	1965
Mugil so-iuy	N		-	Bl	A	1966, 1979

4 Intentional introductions

In 1902 the first attempt to introduce non-native fish into the Caspian Sea was undertaken. Three species were transferred from the Black Sea: the mullets *Mugil auratus* and *M. saliens* and the flounder *Pleuronectes flesus luscus*. This introduction was unsuccessful. A deliberate introduction of commercial and prey organisms exerted essential influence on the biodiversity of the Caspian. In the 1930s-1970s, at least nine species of fish were introduced into the Caspian Sea (Karpevich 1975). As part of an All-Union acclimatization programme, some invertebrates were also introduced in order to increase the food resources available for commercial fish.

The polychaete *Nereis diversicolor* was brought into the Caspian Sea from the Sea of Azov in 1939-1941. It has successfully naturalised and now occurs all over the Caspian, playing an important role in fish feeding. The bivalve *Abra ovata* was brought from the Azov-Black Sea into the Caspian in 1939-1940 and additionally in 1947-1948. By the end of 1959, *A. ovata* populated the western part of the Middle and Southern Caspian. The shrimps *Palaemon elegans* and *P. adspersus* were introduced from the Black Sea in 1931-1934 together with mullets. Later on these species have occupied the coastal zone of the Middle and Southern Caspian.

Two mullet species, *Mugil saliens* and *M. auratus*, were introduced into the Caspian in 1930-1934 from the Black Sea. Both species are commercially valuable and they are widespread all over the Caspian, especially in the southeastern and southwestern parts. Two species of the mosquito fish, *Gambusia affinis* and *G. holbrooki,* were released into the Caspian basin to combat malaria, as a destroyer of larvae of malaria mosquitoes. Nowadays it is found from Astara in the south up to Khachmas in the northeast and is very abundant in low-lying waters of the Lenkoran natural area. Several other fish species were introduced in 1930s-1970s with various outcomes (Table 1).

4.1 IMPACT OF INTRODUCED SPECIES ON THE CASPIAN SEA ECOSYSTEM

The impact of introduced species on the biological diversity of the Caspian Sea is both chronic (long term) and acute (short-term). Acute impact was usually identified during the first years after the introduction of a new species, after which the effect weakens while impact on the biodiversity becomes chronic. For instance, in 1936 the biomass of the alga *Rhizosolenia calcar-avis* was about 65% of the total plankton biomass. Following the 'biological wave', the abundance of the species reduced. A short-term increase of abundance and its further reduction was recorded for the bivalve *Mytilaster lineatus*, and for two species of shrimps. Both chronic and acute impact can be of a positive or negative character. The introduction of the copepods *Calanipeda aquaedulcis* and *Acartia clausi* can be an example of a positive impact. They form a food base by plankton-feeding fish and thus increase the nutritional value of the Caspian zooplankton. The invasion of the comb jelly *Mnemiopsis leidyi* will probably have a negative impact on the biodiversity of the Caspian Sea (Ivanov et al. 2000). The species preys upon zooplankton and causes starvation for the plankton-feeding fish. According to Aladin & Plotnikov (2000) the ctenophore may cause complete loss of the Caspian population of sprat. A number of accidental introductions have appeared to be commercially useful. Many microscopic crustaceans increased the value of the Caspian plankton as a food base for fishes. A high number of accidental introductions did not have any significant impact on the biodiversity of the water body. However, some of them such as the barnacles and the bivalve *Mytilaster* had a certain negative impact from man's point of view. The above-mentioned organisms cannot be used as food for fish because of their thick shells; they are typical representatives of non-food benthos. In addition, some introduced species can foul bottoms of vessels, port piles and offshore oilrigs. They hamper movement of the vessels and gradually corrode port and oil structures. The role of introduced species in the ecosystem of the Caspian and their impact on its biodiversity has been debated. There is no reason to dramatise the impact of them. They predominantly contribute to the Caspian rich biodiversity. However, there is no doubt that species, such as *Mnemiopsis leidyi,* needs to be managed. It is clear that the most ancient representatives of the Caspian flora and fauna should be protected against any negative impact, including the impact of newly introduced species. The Caspian indigenous flora and fauna are the main value of this continental water body. It is of high biological, ecological, genetic and commercial importance. It has been repeatedly specified (Dumont 1998; Zonn 2000) that biological resources of the Caspian are more valuable than its oil and gas resources.

INVASIONS BY ALIEN SPECIES IN INLAND FRESHWATER BODIES IN WESTERN EUROPE: THE RHINE DELTA

GERARD VAN DER VELDE[1*], IVAN NAGELKERKEN[1], SANJEEVI RAJAGOPAL[1] & ABRAHAM BIJ DE VAATE[2]
[1]*Department of Animal Ecology and Ecophysiology, Section Aquatic Animal Ecology, University of Nijmegen, Nijmegen, The Netherlands*
[2]*Institute for Inland Water Management and Waste Water Treatment, Lelystad, The Netherlands*
[*]Corresponding author *gerardv@sci.kun.nl*

Abstract
Invasions by alien species in freshwater habitats are discussed for the Dutch Rhine delta, as a model of a large Western European river strongly influenced by anthropogenic activities. The barriers to dispersal were lifted in this area by introductions and escapes from aquariums, aquaculture etc. (65% of the alien species), as well as by canals connecting rivers (25%) and invasions from the seaports caused by seagoing ships (10%). Habitat change and deterioration of water quality has led to changes in and weakening of indigenous flora and fauna. Periods of pollution and chemical spills followed by partial restoration of the water quality have led to increasingly successful invasions suppressing the population development of returned indigenous species. This fits in well with the hypothesis that anthropogenic disturbance paves the way for new invasions, as disturbance leads to unused resources from which invaders can profit.

1 Introduction

Rivers are dynamic ecosystems with natural disturbances like unpredictable floods creating circumstances for the co-existence of many species. However, nearly all European rivers have considerably deteriorated as a result of anthropogenic changes (Kinzelbach 1995). This had led to an enormous loss of biodiversity in these rivers, while bioinvasions have increased. The present species diversity in the various rivers includes 8-20% of exotic species versus 8-30% in the connecting canals (Tittizer 1996a). The river Rhine including the Netherlands as its delta area can be regarded as a model of a Western European river in which phases of deterioration and ecological rehabilitation have led to a large number of invasions precluding ecological rehabilitation to its pristine state (Lenders et al. 1998). The share of alien species in the biodiversity in the river channels and floodplain lakes of the Rhine delta ranges from 7-10% among macrophytes, to 5-12% among macroinvertebrates and 17-19% among fish (Van den Brink et al. 1997). The river branches are connected by a network of canals and ditches. Although these watercourses are normally used to drain polders and farming areas river water is used for irrigation during dry periods. Polluted river water has thus led to a deterioration of the biodiversity in the smaller waters as well. The Netherlands is one of the most densely populated countries on earth, highly industrialized and with intensified agriculture which causes various types of stress to the aquatic ecosystems ranging from acidification to eutrophication (Best & Bakker 1993).

In many rivers and lakes, densities of invading species are increasing at an accelerated rate. The total number of species is increasing as a result of continuing invasions showing no signs of saturation. Explanations of this phenomenon include a) incomplete

E. Leppäkoski et al. (eds.), Invasive Aquatic Species of Europe, 360–372.

communities (vacant niches); b) disturbances preventing strong competition (Lozon & MacIsaac 1997; Whittier & Kincaid 1999), and c) the invaders themselves creating disturbance (Vitousek 1990; Ruiz et al. 1999) leading to chaotic fluctuations that allow many species to co-exist (Huisman & Weissing 1999, 2001). In addition, many recently invading species share the same origin and mutual relationships so that interactions are quickly established when these species meet again (Simberloff & Von Holle 1999). The success of invaders is ascribed to a combination of factors such as (i) properties of the species themselves including certain adaptations and biotic interactions (e.g. escape from biotic constraints in the original area) (Kolar & Lodge 2001); (ii) habitat modification; and (iii) creation of new habitats (vacant niches) by human activities promoting invasions (Baskin 1998). Although according to Williamson (1996) all communities are invadable, it is expected that they are not all equally susceptible to invasions. Biodiversity, species composition, disturbance, nutrient supply and climate are considered major controllers of population and ecosystem dynamics and structure and this is exactly how humans impact causes major long-term changes in the functioning of ecosystems worldwide (Tilman 1999).

2 Lifting the barriers

Exotic species can only establish in new areas after human activities, which have lifted the barrier to dispersal. Invaders are entering Western Europe in several ways and via various routes. In Dutch fresh waters 85 species have been identified as exotics including 20 macrophytes, 40 macroinvertebrates and 25 fish and other vertebrates (Table 1). *Ferrissia clessiniana* indicated by Kinzelbach (1995) as an exotic species has been omitted from this list because of its frequent occurrence in holocene deposits in the Netherlands; some other species mentioned by Kinzelbach (1995) as invaders are definitively not immigrants in that area. A number of aquarium, aquaculture, botanical garden and garden pond escapes are known including intentional introductions, some as fish food (*Gammarus tigrinus*), as human food or for angling (*Cyprinus carpio*, *Stizostedion lucioperca*) and some for the control of aquatic macrophytes (*Ctenopharyngodon idella*). All these categories together include 55 species (mostly plants, snails, all fish and other vertebrates). Nine species (especially crabs and bivalves) have been transported by the timber trade or in ballast water, mostly as larvae and managed to invade fresh water bodies via international trade ports and disperse upstream into the rivers. Another 21 invading species (20 macroinvertebrate species of which 12 from the Ponto-Caspian region) reached Western Europe especially from the southern and eastern parts of Europe travelling via canals, either through dispersal by flow or attached to ship's hulls, Of the 85 species referred to above, 65% were introduced or escaped, 25% invaded via canals and 10% by the aid of seagoing ships. 75% of the species originated from North America (34%), Eastern Europe (27%) and East Asia (14%), thus from areas with a climate similar to that of Western Europe. The remaining 25% originated mainly from Southern Europe (9%) and South America (9%). The ten's rule (Williamson 1996), which predicts that on average 10% (5-25%) of the invaders become very successful, also applies in the Netherlands, where 19% of the invading species can be regarded as such (10% plants, 25% macroinvertebrates, 8% fish and other vertebrates). There has been a rapid increase in the number of new invasions (Fig. 1), 63% of the alien species having invaded during the last 50 years (55% plants, 70% macroinvertebrates, 60% fish and other vertebrates).

The migration routes range from simple to complex, as a result of the various origins of the invasions, their directions and means of dispersal. For example the zebra mussel (*Dreissena polymorpha*), formerly endemic to the Ponto-Caspian region, reached Western Europe along several routes (Kinzelbach 1992). It extended its area northwards in the 18th century via the river Volga by means of shipping routes and canals linking the rivers to Lithuania. From there it was dispersed by the Baltic timber trade to the harbours of northern Germany (1838), Amsterdam (1826), and London (1824) from where it dispersed inward. At the same time, westward dispersal also took place via a central corridor connecting the rivers Dnieper, Vistula, Oder, Elbe and Rhine (Bij de Vaate et al. submitted). This route was also successfully used by *Chelicorophium curvispinum*, a Ponto-Caspian amphipod that reached the river Rhine in 1987. Recently, an important southern corridor from the Danube to the Rhine has been created by the opening of the Main-Danube canal in 1992 allowing Ponto-Caspian invaders to reach the river Rhine. This corridor has served as a passageway for at least 9 Ponto-Caspian macroinvertebrate species to the Rhine basin. The exotic macroinvertebrate invasion rate has increased from 0.02 species per year during 1800-1850, to 0.1 during 1851-1900 and 1901-1950, to 0.4 during 1951-1992 and to 1.2 in the period 1993-2001. The last figure is entirely due to Ponto-Caspian species invasions through the Main-Danube canal (Table 1).

3 Paving the way: deterioration and recovery

When people started to settle along rivers and in deltas, their main problem was coping with floods from the sea and the rivers. The first settlers used high natural levees, later on constructing such levees themselves. The rivers and their floodplains were used as a defensive barrier against other people, as a source of food and construction materials (willows, reed, rushes), and as a transport route. In the case of the Rhine basin, the Romans already constructed canals to link rivers and river branches. In the 12th century dike construction started along the rivers branches gradually extending towards the sea to protect people, their farmland and villages against floods. These dikes were vulnerable and often broke creating major floods and leaving deep breakthrough ponds. Therefore, the dikes were regularly strengthened until the river had become constrained to a main channel in summer by means of 'summer dikes' and to a narrow floodplain area (river forelands) in winter by a large 'winter dike'.

As the river Rhine became more and more important as a transport route, it was canalised as much as possible to increase its discharge and maintain its depth. Major ports like Rotterdam and Amsterdam were linked to the river by canals, particularly during the 19th century. During canalisation stone groynes were constructed to strengthen the channel and some parts of the river banks were reinforced by stones (rip-rap). The river forelands were mainly used as meadows in summer and riparian vegetation was removed as much as possible. Open outlets of the Rhine branches to the North Sea were dammed resulting in the disappearance of brackish water habitats. These were replaced by a number of anthropogenic freshwater habitats, with reduced tidal movements and barriers for migratory animals.

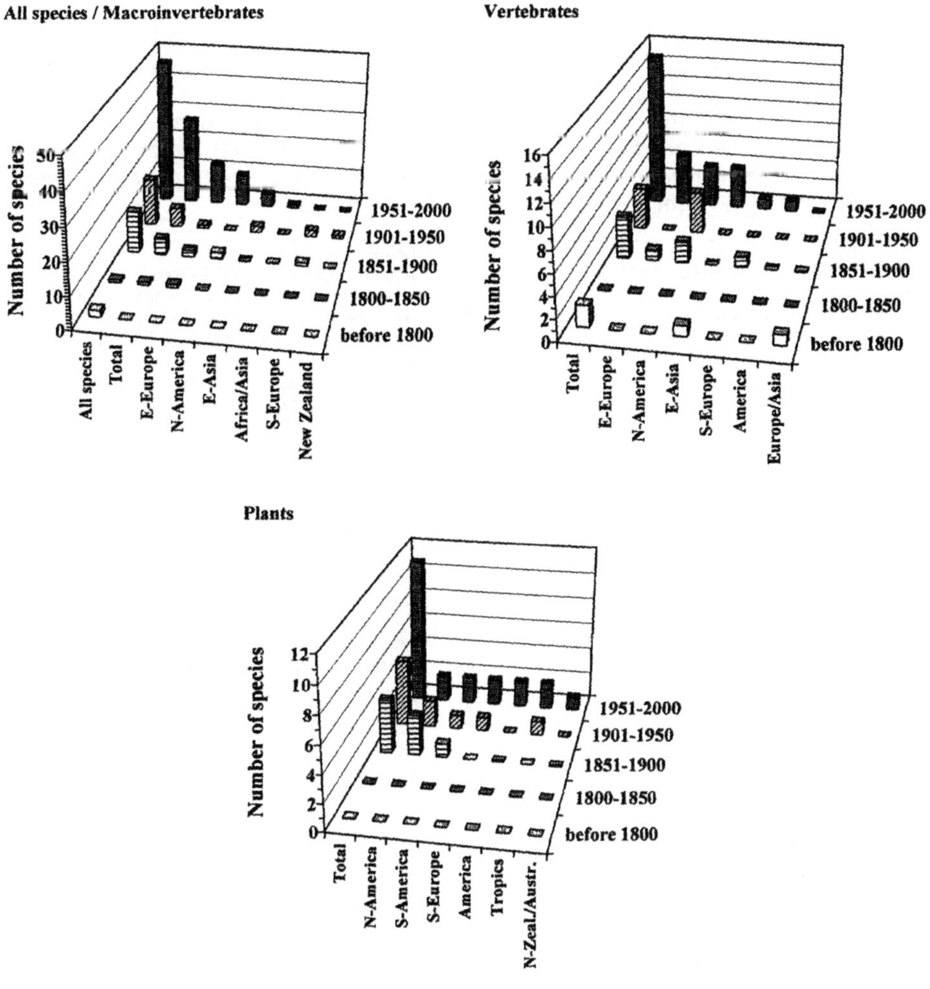

Figure 1. Number and origin of alien species (all species pooled, macroinvertebrates, vertebrates and plants, respectively) newly recorded in the Netherlands over 50 year-periods (based on data presented in Table 1). N-Zeal. = New Zealand, Austr. = Australia.

The river was also important for waste discharge. Since the industrial revolution and the construction of sewage systems, domestic and industrial pollution led to a gradual deterioration of the water quality from the second half of the 19th century to the end of the 1960s. During the period 1950-1970, water quality was very poor with very low oxygen levels, serious eutrophication, high chemical and organic pollution loads, salination caused by the French potassium mines and mining water from brown coal mines in Germany, and thermal pollution. Food webs in the Rhine were very simple and biodiversity was extremely low (Admiraal et al. 1993).

Table 1. List of alien species in inland fresh waters in the Netherlands (excluding parasites and epibionts) based on Gittenberger et al. (1998), Van der Meijden (1996), Van den Brink and Van der Velde (1998), Van der Velde et al. (2000), Mienis (2001). *= in one or a few localities; **= common in part of the area; ***= very successful in the Netherlands, common, widespread. (1) before 1800, (2) 1800-1850, (3) 1851-1900, (4) 1901-1950, (5) 1951-1992, (6) 1993-2001. (a) escapes from aquaculture, aquariums, botanical gardens and garden ponds, (b) by seagoing ships (ballast water, timber trade), (c) dispersal via canals, (h) wood trade ponds, (i) intentionally introduced, (n) natural expansion after creation of the right habitat. PC= Ponto-Caspian invaders.

Taxon		Taxon	
Bryophyta		**Crustacea**	
Octodiceras fontanum S-Europe **	5n	*Astacus leptodactylus* E-Europe**	5a
Pteridophyta		*Atyaephyra desmarestii* S-Europe **	4c
Azolla mexicana N-America **	3a	*Bythotrephes longimanus* N-E-Europe **	5n
Azolla filiculoides N-America ***	3a	*Chelicorophium curvispinum* E-Europe (PC)***	5c
Salvinia natans S-Europe *	4h/a	*Crangonyx pseudogracilis* N-America **	5i?
Salvinia molesta tropics *	4a	*Dikerogammarus villosus* E-Europe (PC)***	6c
Gametophyta		*Eriocheir sinensis* E-Asia ***	4b
Cabomba caroliniana S-America *	5a	*Echinogammarus ischnus* E-Europe (PC)**	5c
Crassula helmsii Australia/New Zealand	6a	*Echinogammarus trichiatus* E-Europe (PC)*	6c
Egeria densa S-America *	4a	*Gammarus tigrinus* N-America ***	5i
Eichhornia crassipes S-America*	5a	*Hemimysis anomala* E-Europe (PC) **	6c
Elodea canadensis N-America **	3a	*Jaera istri* E-Europe (PC)**	6c
Elodea nuttallii N-America ***	4a	*Limnomysis benedeni* E-Europe (PC)**	6c
Hydrocotyle ranunculoides N-S-America**	6a	*Orchestia cavimana* S-Europe **	3b/c
Lemna minuta N-America *	5a	*Orconectes limosus* N-America ***	5a
Ludwigia uruguayensis tropics *	6a	*Proasellus coxalis* E-Europe **	4c
Myriophyllum aquaticum S-America **	3a	*Proasellus meridianus* S-Europe **	4c
Myriophyllum heterophyllum N-America *	6a	*Procambarus clarkii* N-America *	5a
Pistia stratiotes tropics *	5a	*Rhithropanopeus harrissii* N-America ***	3b
Pontederia cordata N-S America*	5a	**Osteichthyes**	
Sagittaria latifolia N-America*	4a	*Acipenser baeri* E-Europe*	6a
Vallisneria spiralis S-Europe *	5a	*Acipenser gueldenstaedti* E-Europe *	6a
Hydrozoa		*Acipenser ruthenus* E-Europe*	6a
Cordylophora caspia E-Europe (PC) ***	3b/c	*Ameiurus melas* N-America **	4a
Craspedacusta sowerbyi E-Asia **	4b	*Ameiurus nebulosus* N-America **	4a
Tricladida		*Aristichthys nobilis* E-Asia *	5i
Dendrocoelum romanodanubiale E-Europe (PC)**	6c	*Aspius aspius* E-Europe (PC) **	5i/c
Dugesia tigrina N-America **	5a	*Carassius auratus gibelio* E-Asia **	1a
Bivalvia		*Ctenopharyngodon idella* E-Asia *	5i
Corbicula fluminalis E-Asia ***	5b	*Cyprinus carpio* Europe/Asia ***	1i/a
Corbicula fluminea E-Asia ***	5b	*Hypophtalmichthys molitrix* E-Asia *	5i
Dreissena polymorpha E-Europe (PC)***	2b	*Lebistes reticulatus* S-N-America**	5a
Musculium transversum N-America *	5b	*Lepomis gibbosus* N-America **	4i/a
Mytilopsis leucophaeata N-America **	5b	*Micropterus salmoides* N-America*	3i
Gastropoda		*Oncorhynchus mykiss* N-America *	3i
Gyraulus chinensis E-S-Asia*	5a	*Pseudorasbora parva* E-Asia **	6a/c
Helisoma nigricans N-America *	5a	*Salvelinus fontinalis* N-America *	5i
Lithoglyphus naticoides E-Europe (PC) **	3c	*Stizostedion lucioperca* E-Europe (PC)***	3i
Melanoides tuberculatus Africa/Asia *	6a	*Umbra pygmaea* N-America **	4a
Menetus dilatatus N-America *	5a	*Vimba vimba* E-Europe (PC)*	5i/c
Physella acuta N-America **	3a	**Amphibia**	
Physella heterostropha N-America *	6a	*Rana catesbeiana* N-America *	5a
Potamopyrgus antipodarum New Zealand ***	4b	**Reptilia**	
Annelida		*Emys orbicularis* S-Europe *	3a
Branchyura sowerbyi E-Asia ***	5a	*Mauremys leprosa* S- Europe*	5a
Caspiobdella fadejewi E-Europe (PC)*	6c	*Trachemys scripta* elegans N-America*	5a
Hypania invalida E-Europe (PC) *	6c	*Chelydra serpentina* N-America*	5a
Acari			
Caspihalacarus hyrcanus E-Europe (PC)**	6c		

During the period 1970 to 1986, waste water treatment plants were constructed along the Rhine. Water quality improved to some extent, with increasing oxygen levels, decreasing organic loads, and the conversion of ammonium to nitrate by oxidation. Concentrations of a number of pollutants, like heavy metals and pesticides have decreased since that time, while oxygen levels have increased to natural levels (Admiraal et al. 1993).

In 1986, a fire at the Sandoz chemical company at Basle was extinguished with large quantities of water, which became heavily polluted and drained into the river Rhine, causing a mass kill of fish and macroinvertebrates over a stretch of hundreds of kilometres. The Rhine states decided to further improve the water quality and to work towards ecological rehabilitation of the river. Water quality improvement schemes have turned Europe's former sewer pipe into one of the more successful examples of environmental rehabilitation. Its water quality is now better than that of the river Meuse, the other large river in the Netherlands, which used to be cleaner than the Rhine. However, the river Rhine has remained eutrophicated, with high salt concentrations, chemical pollution from diffuse sources and thermal pollution (Admiraal et al. 1993; Van den Brink et al. 1997; Van der Velde et al. 2000). Nitrogen pollution in particular is a problem greatly increased by the industrialised agriculture and cattle breeding. Mean quantities of artificial fertiliser used per unit of arable land in the Netherlands are among the highest in Europe (Best & Bakker 1993). Smaller water bodies, like ditches, brooks and other tributaries of the river Rhine also suffer from eutrophication and pesticide pollution, while moorland pools have become eutrophicated or acidified by atmospheric pollution.

All sorts of irreversible changes and human pressure in one of the world's most densely populated areas mean that the Rhine remains a river under anthropogenic stress in spite of improvements over the last decades. Further improvements are also impeded by the river's most important function that of shipping route from the ports of Rotterdam and Amsterdam to the German Ruhrgebiet and further upstream. With the introduction of motor vessels, shipping has increased to such a level that nowadays on average a ship passes the Dutch-German border every two minutes.

4 Invasions in the smaller waters

As a result of polder construction, the shallow lake areas, peat land and salt marshes that used to cover about half of the surface of the Netherlands, disappeared and the newly reclaimed land was used for agriculture (Best & Bakker 1993). Polders and other low-lying areas in the Netherlands are drained by an extensive system of ditches. Drainage, fishing and navigation can all be seriously upset by excessive plant growth and mechanical control has to be carried out once or twice a year. Vegetation removal and site disturbance are the best documented causes of plant invasions (Detenbeck et al. 1999). Since 1977, grass carp (*Ctenopharyngodon idella*) have been released to control macrophyte growth in the ditches. The Dutch fauna lacked such a herbivorous fish and the introduction of grass carp apparently fills an empty niche (De Nie 1996).

Some of the exotic aquatic macrophytes grow to great densities. They profit from eutrophication, being adapted to a quick uptake of nutrients necessary for growth, and avoid turbidity by covering the water surface suppressing all other macrophyte growth. Submersed plants like *Elodea* species are especially successful at moderate levels of eutrophication. Since its introduction by a professor who threw a plant in one of Utrecht canals in 1859, *E. canadensis* has formed such dense stands everywhere that it has become known as water pest in Dutch. Severed sections of the plant can grow out, so fragmentation favours its spread. The plant remains green in winter, and has winter buds that allow it to grow very quickly in spring to fill the available water space to the disadvantage of other macrophytes. With increasing levels of eutrophication it became an indicator of relatively good water quality in the ditches. *E. nuttallii*, present since 1941, can thrive at a higher level of eutrophication and is nowadays much more common than *E. canadensis* (Van der Meijden 1996). Both clonal species have only female flowers, indicating that all plants of each species belong to the same clone. Increasing levels of eutrophication have caused submersed macrophytes to disappear as a result of phytoplankton blooms and increased turbidity; they have been replaced by floating-leaf plants such as duckweeds. This type of vegetation allowed *Azolla* species to invade, particularly *A. mexicana* (formerly called *A. caroliniana*) and *A. filiculoides*. The first species is an inhabitant of fresh, less polluted water, and no fertile specimens have been recorded for the last century. It is nowadays so rare that it is on the red list of endangered plant species in the Netherlands. *A. filiculoides,* by contrast, is an inhabitant of fresh and slightly brackish eutrophic water and colours most ditches red in late summer and autumn, overgrowing duckweed covers. The species is thermophilous and therefore restricted to an Atlantic climate. It builds its population later in the year than the duckweed species. *Azolla* species live in symbiosis with a nitrogen-fixing cyanobacterium *Anabaena azollae* and are therefore efficient phosphorus removers in the absence of nitrate (Costa et al. 1999).

A new successful macrophyte is *Hydrocotyle ranunculoides,* a garden pond species which escaped all over the country and was first observed in 1995. It forms floating rhizomes with roots which reach down half a metre into the water column and can penetrate 15 cm into the sediment. The leaves can lift themselves 30 cm above the water surface. The rhizomes spread from the banks over the water surface, so that it is unaffected by turbid water. It survives cold winters and is expanding its area everywhere, mostly by clonal reproduction, though flowering and seed formation have also been recorded. The sites where *H. ranunculoides* grows are mostly eutrophic, with mostly flowing water strongly influenced by sewage from cities and agriculture. Eradication of the plant seems to be difficult because each part can grow out to a new plant, which also occurs when banks are cleared of vegetation. Plants like *Azolla* and *Hydrocotyle* have an impact on communities by shading the water column, suppressing submersed vegetation, impeding the diffusion of oxygen from the atmosphere and creating anoxic conditions by decomposition of organic matter. Toxic compounds like H_2S, NH_3 and NO_2 accumulate under the vegetation and the water becomes malodorous. The result is an impoverished biodiversity of the communities. The water is not even drinkable for cattle without health risk. These plants also impede the drainage function of the ditches by reducing the flow, while floating biomass causes problems by blocking culverts (Kuenen 1999). Recently *H. ranunculoides* colonized also weakly buffered streams and moorland pools, which are protected as nature reserves, influencing the characteristic Littorellion vegetation in a negative way (De Mars & Bouman 2002). Similar problems are already reported from *Myriophyllum aquaticum* which species became recently widely spread in the Rhine delta.

In 1995 a population of *Crassula helmsii* was discovered. This species establishes on open banks in oligo-mesotrophic moorland pools, which regularly fall dry (Brouwer & Den Hartog 1966). It has become widespread in Britain, where it causes problems by growing to very dense stands (De Waal et al. 1994). It uses runners to penetrate from the banks into other habitats and can thus extend to water depths of up to three metres. Some of the moorland pools have become eutrophicated, while others became acidified. Certain fish species are known to invade these waters; all are escaped aquarium or aquaculture fish originating from North America. *Umbra pygmaea* has profited from acidification; it has adaptated to acid pools down to a pH limit of 3.1. Its food consists of cladocerans for the juveniles and macroinvertebrates (84% chironomids) for the adults (Best & Bakker 1993). By adapting to such an extremely acid environment this slow moving fish escapes competition by other fish. Other alien fish species occurring in soft water moorland pools are *Lepomis gibbosus* and *Ameiurus nebulosus*, but these two species have also been found frequently in other water types (De Nie 1996). Other animal invasions in the smaller water bodies have been rare. The freshwater snails *Potamopyrgus antipodarum* and *Physella* species occur in brooks, while *Gammarus tigrinus* is also found in ditches, especially those that are brackish or ion-rich, where it outcompetes *Gammarus pulex* (Van der Velde et al. 2000 and literature therein).

5 Invasions of rivers and connected waters

Rivers and canals as well as artificial lakes form the most invadable systems. They are subject to environmental stress or are vacant for colonisation or recolonisation. Canals connect the rivers with the ports of Amsterdam and Rotterdam, as well as linking the rivers Meuse and Rhine and forming a network all over the Netherlands transporting river water towards areas outside the Rhine basin. This means that the water quality of all these waters (except the river Meuse) is very similar to that of the river Rhine and that invaders adapted to these conditions have opportunities to spread all over the country. Furthermore, the stone banks and groynes in the Rhine branches provide opportunities for sessile colonisers and other lithophilous species.

The river Rhine has been subjected to continuous invasions from various origins in particular by macroinvertebrates and fish (Van den Brink et al. 1990, 1997). Some of the species entered via the ports and estuaries and then moved upstream, while others moved downstream after entering via canals in the east or south. Several of these species have also penetrated the larger still waters but others seem to be restricted to flowing water and occur only in the major rivers. Many species have profited from the stone banks and groynes constructed along rivers, where hard substrate was originally lacking. *Octodiceras fontanum*, a moss growing on stone and wood is expanding its area in the large ion-rich rivers, canals and lakes. It grows from the water surface to depths of about 60 cm and is usually totally submerged in the same zone as *Dreissena polymorpha* (Touw & Rubers 1989). Fig. 2 provides information on the successive waves of colonisation of stones along the banks of the river IJssel, one of the Rhine branches and on artificial substrates in the Rhine. After a long period of absence of macroinvertebrates from the stones recolonisation started in 1975 with low numbers of species and low densities. Dominant exotic species were the Ponto-Caspian *D. polymorpha*, a filter feeder, and the North-American *Dugesia tigrina*, a predator. These were later joined by the North-American omnivore *Gammarus tigrinus*. This community collapsed during the Sandoz accident in 1986, but later restored itself. In 1987, the Ponto-Caspian species *Chelicorophium curvispinum*, a filter feeder, came in and reached considerable densities from 1990. Since then, it has dominated the communities on the stones with densities of hundreds of thousands per m^2. It has radically altered the communities by covering the stones with a layer of muddy tubes up to 4 cm thick, changing the conditions for colonisation completely (Van den Brink et al. 1993; Van der Velde et al. 2000). It caused a severe decline of *D. polymorpha* by smothering them under the mud layers and depriving their larvae of the bare stone surface they need (Van der Velde et al. 1998). Since 1995 a new Ponto-Caspian gammarid has appeared, viz. *Dikerogammarus villosus*. This species shows various pigmentation patterns, camouflaging the specimens between those of *D. polymorpha*. *D. villosus* turned out to be a predatory species preying on the other macroinvertebrates inhabiting the stones like *G. tigrinus* and *C. curvispinum* (Van der Velde et al. 2000; Dick & Platvoet 2000). Since that time, the densities of *C. curvispinum* have greatly decreased, although it remains the most numerous macroinvertebrate species on stones. The fauna in this habitat has become more diverse because of the availability of more bare stone area for colonisation. The year 1997 saw the arrival of the Ponto-Caspian isopod *Jaera istri*, a grazer on bare hard substrates, which benefited from this development. In 1999, the stones were invaded by the triclad *Dendrocoelum romanodanubiale*, a predator, which completely replaced *Dugesia tigrina*. The Rhine sediments have also experienced invasions. Improving quality led to the return of unionid mussel species, but since 1988 two co-existing Asian clam species (*Corbicula fluminalis* and *C. fluminea*) have dominated the benthic communities (Gittenberger et al. 1998; Rajagopal et al. 2000). The polychaete *Hypania invalida* was the next Ponto-Caspian invader to become very numerous in the sandy sediment. New species have also invaded the nekton of the river Rhine including fish species like *Aspius aspius* and the Ponto-Caspian mysids *Hemimysis anomala* and *Limnomysis benedeni,* which also invaded lakes. Fish species in the Rhine adapted their diet to these rapid turnover of invading species (Kelleher et al. 1998, 2000).

A considerable number of species in the river Rhine is euryhaline and/or thermophilous, and belong to three trophic functional groups, viz. filter feeders, omnivores and predators. All communities are increasingly dominated by Ponto-Caspian invaders for which the invasion by *Dreissena polymorpha* paved the way (Van der Velde et al. 2000). Some exotic species are restricted to cooling water discharge areas or use these areas as refuges during the winter. These species, which originate from the tropics, include macrophytes (*Egeria densa, Eichhornia crassipes, Pistia stratiotes, Salvinia molesta*), snails (*Gyraulus chinensis, Helisoma nigricans, Melanoides tuberculatus*) and fish (*Lebistes reticulatus*). They disappear as soon as the water temperatures become too low for survival mostly when for some reason the cooling water discharges stop in the wrong period. Other thermophilous alien species are able to persist in the larger water bodies like canals, lakes and rivers where they co-exist with cold-water species. The large water volumes provide a temperature buffer which is lacking in shallower waters, where temperatures are too low in winter for the thermophilous species and too high in summer for the cold-water species.

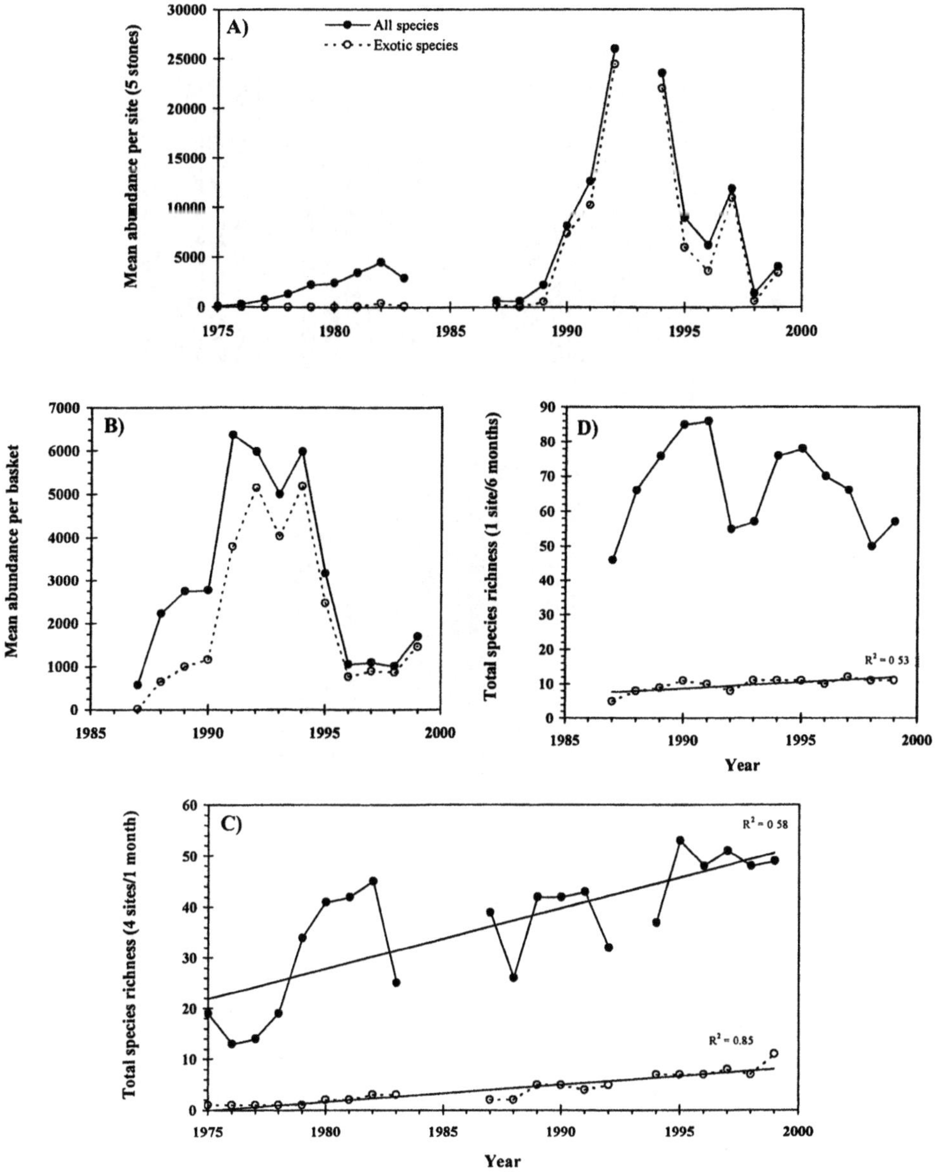

Figure 2 A-D. A and C: all macroinvertebrates and exotic macroinvertebrates on stones in the River IJssel, the northern Rhine branch in the Netherlands during September over the period 1975-1999. B and D: the same for artificial substrate samplers (baskets) in the Rhine near Lobith during the summer months over the period 1987-1999. A,B: mean abundance; C,D: species richness (see next page).

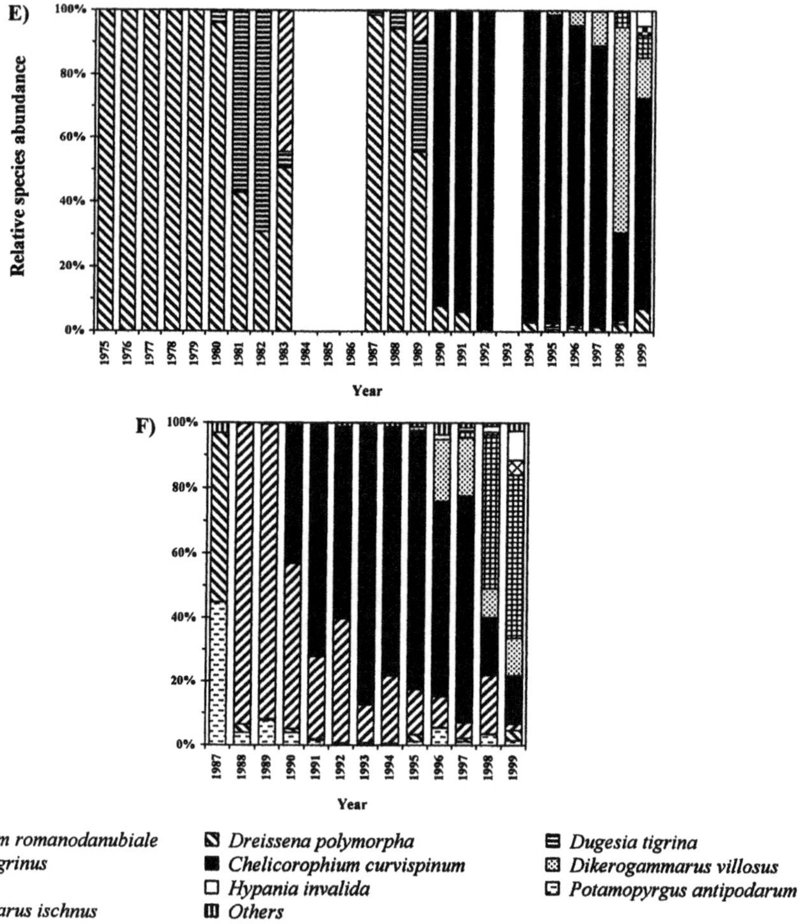

Figure 2 E-F. E: Relative species abundance all macroinvertebrates and exotic macroinvertebrates on stones in the River IJssel, the northern Rhine branch in the Netherlands during September over the period 1975-1999. F: the same for artificial substrate samplers (baskets) in the Rhine near Lobith during the summer months over the period 1987-1999.

Water temperatures in the river Rhine water have increased by some degrees over time as a result of thermal pollution and ice formation no longer occurs in winter (Van den Brink et al. 1996). Population sizes of thermophilous species can fluctuate greatly with stronger or milder winters. This is true for indigenous species, living at the limits of their ecological range as well as for exotics.

The radical environmental changes caused by the damming of the Zuiderzee (De Beaufort 1954) (nowadays IJsselmeer), the Volkerak-Zoommeer and the Haringvliet (Smit et al. 1993, 1997), changing salt-water or brackish estuaries into river-fed freshwater lakes, were very favourable for new invaders originating from the river Rhine. *Dreissena polymorpha* entered all these lakes as larvae making use of seashells on the sediments as substrate. The invader presented a valuable food source for freshwater fish, like eel and for waterfowl, like diving ducks and its presence led to high densities of gammarids like *Gammarus tigrinus* (Noordhuis 2001). Recently, *Dikerogammarus villosus* invaded lake IJsselmeer, exterminating *G. tigrinus* as well as *Gammarus duebeni* (Dick & Platvoet 2000). New gravel, clay and sand extraction pits which have been excavated everywhere in the Netherlands but especially along the rivers, have also created new opportunities for rapid colonisers like *D. polymorpha, Potamopyrgus antipodarum* and *Elodea nuttallii*. The construction of large drinking water reservoirs in the Biesbosch area, storing river water from the Meuse, has also led to invasions of various crustacean macroinvertebrates, which prey on zooplankton such as *Bythotrephes longimanus, Hemimysis anomala* and *Limnomysis benedeni* (van der Velde et al. 2000 and literature therein). *B. longimanus* has profited from the creation of large deep artificial lakes, a habitat not present before in the Netherlands.

6 Invasibility and future developments

Two species actually returned from extinction during the Pleistocene, a period in which European biodiversity impoverished, viz. *Azolla filiculoides* (Van der Meijden 1996) and *Corbicula fluminalis* (Meijer & Preece 2000). Community vulnerability to invasions is ascribed to combinations of several factors like the presence of vacant niches, habitat modification and disturbance before and during invasion. Although the link between the biodiversity of communities and their vulnerability to invasions remains to be proved, invasibility is known to increase if a community lacks certain species, which ought to be present under normal conditions. A new hypothesis linking the various explanations of increased invasibility is that of fluctuating resource availability such as an increased amount of unused resources (Davis et al. 2000).

The Rhine is a good example of all these related factors. Pollution over a long period weakened the original communities and caused the loss of certain species, creating open niches for pollution-tolerant invaders. Water quality improvement led to a partial recovery of the original communities together with the establishment of previously disappeared and new invaders. A major disturbance like the Sandoz accident in 1986 subsequently led to invasions by many new species, which reached unprecedented densities. The fact that filter feeders are particularly abundant can be attributed to intense phytoplankton blooms due to eutrophication. Hardly any macrophytic vegetation is present in the Rhine channel to compete with phytoplankton for nutrients. Recolonisation after partial reduction of pollution in rivers modified by human activities seems to favour invaders more than indigenous species. These invaders then suppress the development of populations of indigenous species, although biodiversity increases (Fig. 2). Smallers water bodies are also subject to invasions mostly related to changes in water quality like eutrophication and organic pollution, or acidification in the case of moorland pools.

Severe pollution can function as a barrier to the dispersal of invaders. An example is the Chicago connection between the Great Lakes and the Mississippi river, where the 1972 Clean Water Act and subsequent improvements in municipal waste treatment improved the water quality to such an extent that the zebra mussel (*Dreissena polymorpha*) and six other non-native 'pest' species were able to spread from the Great Lakes to the Mississippi River (Stoeckel et al. 1996). *Dreissena* returned to the Rhine in the 1970s and 1980s, when cadmium concentrations in the water fell below 1 $\mu g\ l^{-1}$ (accumulation in the mussel at that level was 40 $\mu g\ g^{-1}$ DW) (Van Urk & Marquenie 1989).

Disturbances by human activities are not only favourable for invaders but also for certain categories of indigenous species. Key words in the case of the Netherlands are eutrophication, salination, other pollution (pesticides), cooling water and acidification. The occurrence of thermophilous species is a reflection of propagule pressure because of frequent releases by the garden pond and aquarium lovers all over the country. Plants and animals from various origins can be bought everywhere in garden centres and aquarium shops. Their occurrence in cooling water discharge areas could act as a stepping-stone for successful invasion at the higher temperature expected to result from climate change.

The present day invasions of Ponto-Caspian invaders in the Netherlands and other parts of Western Europe via the Main-Danube canal increases the likelihood that they will reach North America via ballast water transport because of the presence of many major ports in Western Europe (Rotterdam, Amsterdam, Antwerp, Bremen, Hamburg) (Bruijs et al. 2001). These species tolerate high temperatures and brackish water. Several of them have already invaded the Great Lakes (Ricciardi & MacIsaac 2000). The future will bring continued invasions, resulting in unstable communities with an accelerated turnover due to increasing propagule pressure combined with greater anthropogenic disturbance (Nilsson & Grelsson 1995; Stylinski & Allen 1999). This future scenario will cause a shift from battles between invaders and indigenous species towards battles among invaders of various origins.

Acknowledgements

The authors would like to thank Dr. R. van der Meijden for additional information on aquatic macrophytes. CWE publication no. 272.

BIOLOGICAL INVASIONS INTO GERMAN WATERS: AN EVALUATION OF THE IMPORTANCE OF DIFFERENT HUMAN-MEDIATED VECTORS FOR NONINDIGENOUS MACROZOOBENTHIC SPECIES

STEFAN NEHRING
Bundesanstalt für Gewässerkunde, Koblenz, Germany
stefan.nehring@firemail.de

Abstract
To date, 35 nonindigenous macrozoobenthic species have been introduced to the inland waterways of Germany, 26 to the German North Sea and 14 to the Baltic Sea coast. For inland waters and the Baltic Sea, the direct introduction of macroinvertebrates by ocean shipping is insignificant, shipping canals built during the last centuries, being the most important gateway. In contrast, on the German North Sea coast about 60% of the nonindigenous species were introduced by ocean shipping. Prior to the introduction of antifouling paints containing the effective biocide tributyltin (TBT) in 1970, and their subsequent common usage, most of these invasive species came from fouled hulls of overseas trade ships. With the introduction of TBT the transfer of potential invaders via ship fouling was reduced considerably. Today, ballast water and hull fouling are of equivalent importance as vectors. On the German North Sea coast most of the introduced macroinvertebrates established permanent populations in estuaries. The combination of brackish water with its unsaturated ecological niches and intense intercontinental shipping traffic creates the highest infection rate for coastal areas. It is likely that climate change will lead to modifications of species communities in Northern European coastal waters in the near future.

1 Introduction

The species composition of aquatic habitats is subject to continuous change. Species often occur unexpectedly in new regions after an expansion of their natural distribution range as a consequence of natural changes in the area. During the last decades, special attention has been given to nonindigenous species that reached a new area with the active or passive help of man. These species designated as 'aliens' are often regarded as unwanted, in particular, if they cause economic damage, or if the risk for it exists (Carlton 1996). Moreover, the spreading of aliens may result in an increasing uniformity of previously distinct species communities and in a reduction in global biodiversity (CBD 1992).

A comprehensive analysis of the risk potential for various vectors of introduction is a basic requirement for management strategies that could lead to a reduction of anthropogenic dispersal of species. This is focused on the macrozoobenthos of three German aquatic systems: inland waters (especially the Federal water ways), the North Sea, and the Baltic Sea coast and the overall importance of different human-mediated vectors for the introduction of alien species into German waters is evaluated.

2 Alien macroinvertebrates in German waters

Potential areas of origin for species which might become established in new regions, are characterised by comparable environmental conditions. Germany appertains to the bo-

E. Leppäkoski et al. (eds.), Invasive Aquatic Species of Europe, 373–383.

real, the cold temperate climate zone. Similar environmental conditions can be found in the Ponto-Caspian area as well as on the coasts of South and North America, Japan, Tasmania and New Zealand. However, the long distance prevents an intensive natural exchange of species. Crossing the Atlantic Ocean with water currents, which lasts at least 10 months, is generally impossible for larvae of benthic organisms (Thorson 1961). Only attached to drifting substrate such as macroalgae or driftwood, which serves as habitat and food resource for animals, such distances can be overcome by some species. An actual example is the arrival of the North American isopod *Idotea metallica* which was found near the German North Sea island Helgoland in 1994 (Franke et al. 1999). The majority of organisms from remote areas are dependent on other vectors, such as shipping, aquaculture, canals, and intentional release.

2.1 INLAND WATERS

According to Tittizer et al. (2000) 35 macrozoobenthic species are of allochthonous origin in German inland waterways (Table 1). They represent 9% of the total macrozoobenthic species, and their spread is faciliated primarily by the canals built during the last centuries.

The natural range barriers between the catchment areas of the rivers, as they had stabilised themselves in Europe since the end of the Pleistocene, were eliminated by the canals. This enables mobile species as well as organisms that are transported by the drag force of water, by biovectors (birds, fish) or ships, to spread into new water systems. For instance: the early and frequent occurrence of Ponto-Caspian species in northern Europe (like the zebra mussel *Dreissena polymorpha*), the opening of the Dnjepr-Bug-Canal in 1784, which connected the Pripyat system to the rivers Bug and Vistula, was of crucial importance (Fig. 1). Recently, Ponto-Caspian species have been found increasingly in the German rivers Main and Rhine (e.g. the polychaete *Hypania invalida* and the isopod *Jaera istri*, Schöll & Banning 1996). Some species from the Rhine can also be found in the Danube now (e.g. the clam *Corbicula fluminalis*, Tittizer & Taxacher 1997). The spreading of these species can be attributed to the opening of the Main-Danube-Canal in 1992.

The specific role of inland waterway shipping in the faunal exchange between Danube and Main was clarified by examining the macroinvertebrate colonisation of inland vessels (Reinhold & Tittizer 1999). Six alien species were found on hulls and in cooling water filters, but not in ballast water. These investigations show clearly that the passive transport on ship's hulls represents an important vector for limnic macroinvertebrate aliens.

A compensation of the permanent drain deficit of the Main through the canal is necessary. For this reason more than 100 million m^3 of Danube water per annum are transferred through the canal into the Main. As a result of this, mobile species, resting stages of bryozoans and sponges, pelagic larvae of mussels etc. of the Danube ecosystem and the Ponto-Caspian area will arrive without ships' aid directly in the West-European water systems. It is expected that, in addition to the 15 recently established Ponto-Caspian macrozoobenthic species recorded in German inland waters, several aliens from the Danube will arrive in the near future.

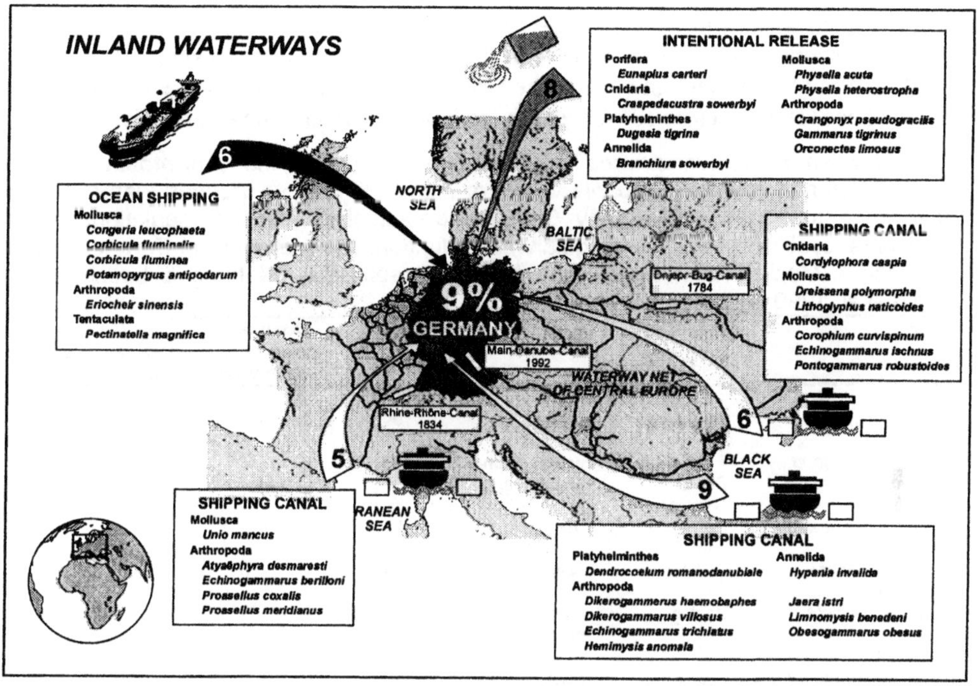

Figure 1. Established alien macroinvertebrates in German inland waterways. Known or probable introduction vectors, number of introduced species and their amount on total species number (in percent), important canals and their opening date. See text and Table 1 for further explanations.

Ocean shipping plays only a subordinate role for the introduction of aliens into German inland waters. The best-known example is the Chinese mitten crab *Eriocheir sinensis*, which was introduced about 1910 from China in ships' ballast water to Germany. In total, only six of the 35 alien species in German inland waters are regarded as introduced by ocean shipping. Four of these species are classified as halophilic or as salt tolerant (in addition to the mitten crab these include the bivalves *Corbicula fluminalis* and *Congeria leucophaeta* and the snail *Potamopyrgus antipodarum*). All these species are also found in brackish water at the German North Sea coast and in parts of the Baltic Sea coast. Up to now, an introduction by ocean shipping into German waters is proved only for two limnic macroinvertebrate species, the Asiatic clam *Corbicula fluminea* and the bryozoan *Pectinatella magnifica*. The minor role that ocean shipping plays for the introduction of limnic alien species, is well explained by the fact, that ballast water is primarily of brackish or marine origin (cf. Gollasch 1996).

2.2 NORTH SEA COAST

Recent summaries of introduced species in the North Sea (including the English Channel and Kattegat region) revealed about 45 alien macrozoobenthic species occurring in self-sustaining populations today (Reise et al. 1999; Nehring 2001). In total, 26 of them have been found recently established on the German North Sea coast (Table 1). The share of alien species in German waters compared to the total number of macrozoobenthic species amounts to 1% in the case of the island of Helgoland (6 species), 3% in the Wadden Sea (10 species) and 7% in the brackish canals and ditches (13 species). Most of the aliens (19 species or 10%) occur in the brackish water zone of estuaries.

Table 1. Alien macroinvertebrates in German waters. Origin - known or probable area of origin; Vector - known or probable vector of introduction into European waters; Date - year of the first record in Germany; Occurrence - actual established populations in German water systems, + rare, ++ regular, +++ abundant (Key references: Eno et al. 1997; Gollasch & Mecke 1996; Leppäkoski 1984; Jansson 1994; Nehring & Leuchs 2000; Reise et al. 1999; Tittizer et al. 2000).

Taxon	Origin	Vector	Date	Occurrence inland water ways	North Sea coast	Baltic Sea coast
Porifera						
Eunapius carteri	Africa Asia	release	1993	+		
Hydrozoa						
Bimeria franciscana	? Indo-Pacific	ocean shipping	1952		+	
Cordylophora caspia	Ponto-Caspian	canal and inland shipping	1858	+++	++	++
Craspedacustra sowerbyi	E Asia	release	1923	+		
Nemopsis bachei	W Atlantic	ocean shipping	1942		+	
Anthozoa						
Diadumene cincta	Pacific	aquaculture product	1928/29		+	
Tricladida						
Dendrocoelum romanodanubiale	Ponto-Caspian	canal and inland shipping	1992	+		
Dugesia tigrina	N America	release	1931	++		
Bivalvia						
Congeria leucophaeta	E Atlantic	ocean shipping	1928	+	+	+
Corbicula fluminalis	E Asia	ocean shipping	1984	++	+	
Corbicula fluminea	Asia	ocean shipping	1987	++		
Crassostrea gigas	Pacific	aquaculture product	1991		++	
Dreissena polymorpha	Ponto-Caspian	canal and inland shipping	1828	+++	+	+
Ensis americanus	W Atlantic	ocean shipping	1979		+++	++
Mya arenaria	W Atlantic	release (?)	1859		+++	+++
Petricola pholadiformis	W Atlantic	aquaculture product	1896		++	+
Unio mancus	S Europe	canal and inland shipping	before 1922	+		
Gastropoda						
Crepidula fornicata	W Atlantic	aquaculture product	1934		+	
Lithoglyphus naticoides	E Europe	canal and inland shipping	1883	+		
Physella acuta	SW Europe	release	1895	++		
Physella heterostropha	N America	release	before 1927	+		
Potamopyrgus antipodarum	S Pacific	ocean shipping	1900	++	++	++
Oligochaeta						
Branchiura sowerbyi	S Asia	release	1959	++		
Polychaeta						
Ficopomatus enigmaticus	S Pacific	ocean shipping	1975		+	
Hypania invalida	Ponto-Caspian	canal and inland shipping	1958	++		
Marenzelleria cf. *viridis*	W Atlantic	ocean shipping	1985		++	+++
Marenzelleria cf. *wireni*	N Atlantic	ocean shipping	1983		+++	

Taxon	Origin	Vector	Date	Occurrence inland water ways	North Sea coast	Baltic Sea coast
Crustacea						
Atyaëphyra desmaresti	Mediterranean	shipping canal	1932	++		
Balanus improvisus	W Atlantic	ocean shipping	1858		+++	+++
Corophium curvispinum	Ponto-Caspian	canal and inland shipping	1912	+++	+	+
Corophium sextonae	S Pacific	ocean shipping	1997		+	
Crangonyx pseudo-gracilis	N America	release	1992	+		
Dikerogammarus hae-mobaphes	Ponto-Caspian	canal and inland shipping	1976	++		
Dikerogammarus villo-sus	Ponto-Caspian	canal and inland shipping	1991	++		
Echinogammarus berilloni	Mediterranean	canal and inland shipping	1924	+		
Echinogammarus ischnus	Ponto-Caspian	canal and inland shipping	1977	+		
Echinogammarus trichia-tus	Ponto-Caspian	canal and inland shipping	1996	+		
Elminius modestus	S Pacific	ocean shipping	1953		+++	
Eriocheir sinensis	N Pacific	ocean shipping	1912	+++	+++	++
Gammarus tigrinus	W Atlantic	release	1957	+++	+	+
Hemimysis anomala	Ponto-Caspian	shipping canal	1997	++		
Jaera istri	Ponto-Caspian	canal and inland shipping	1958	++		
Limnomysis benedeni	Ponto-Caspian	shipping canal	1994	+		
Obesogammarus obe-sus	Ponto-Caspian	canal and inland shipping	1995	+		
Orconectes limosus	N America	release	1880	++		
Pontogammarus robus-toides	Ponto-Caspian	canal and inland shipping	1994	+		+
Proasellus coxalis	Mediterranean	canal and inland shipping	about 1931	++	+	
Proasellus meridianus	W Europe	canal and inland shipping	about 1930	+		
Rhithropanopeus harrisii	W Atlantic	ocean shipping	1936		++	++
Bryozoa						
Pectinatella magnifica	N America	ocean shipping	1883	+		
Victorella pavida	? Indo-Pacific	ocean shipping	1951		+	
Ascidiacea						
Styela clava	N Pacific	ocean shipping	1997		+	
		Total number		**35**	**26**	**14**

The main transport vector was ships hull fouling (e.g. the barnacle *Elminius modestus*). Also, the transport in ballast water tanks (e.g. the American jack knife clam *Ensis americanus*) as well as deliberate commercial introductions (e.g. the Pacific oyster *Crassostrea gigas*) and unintentional introductions associated to oyster imports (e.g. the slimper limpet *Crepidula fornicata*) were important vectors (Fig. 2). Because of the salinity barrier the coastal faunal exchange over freshwater canals plays a subordinate role. Up to now only three Ponto-Caspian species (e.g. the amphipod *Corophium curvispinum*) as well as one Mediterranean species (the isopod *Proasellus coxalis*) arrived in the brackish areas of German North Sea estuaries (Nehring & Leuchs 2000).

The soft clam *Mya arenaria* was the first alien introduced to the North Sea, probably by the Vikings (Petersen et al. 1992). Most of the introductions of macroinvertebrates at the German North Sea coast happened in the 20th century (22 species in total). In the preceding century, only four species (the hydrozoan *Cordylophora caspia*, the brackish water barnacle *Balanus improvisus*, the zebra mussel *Dreissena polymorpha* and the American piddock *Petricola pholadiformis*) were introduced. This discrepancy is certainly a result of recently increased attention to the existence of nonindigenous species, as well as to the global availability of taxonomic literature. Other important factors are the increase in global shipping and the increased use of exotic mariculture products today. It is striking that the majority of the established aliens are epifaunal or vagile species. Representatives of the genuine endofauna, the bivalves *Corbicula fluminalis*, *Ensis americanus* and *Petricola pholadiformis*, and the two polychaetes *Marenzelleria* cf. *viridis* and *M.* cf. *wireni* are strongly under-represented. One reason for this could be the transportation vectors. It is less likely for endofauna species (transported mainly as adults in the sediment of ballast water tanks) to be exposed to a new area, than for vagile species or species attached as epifauna on ship hulls or on mariculture products.

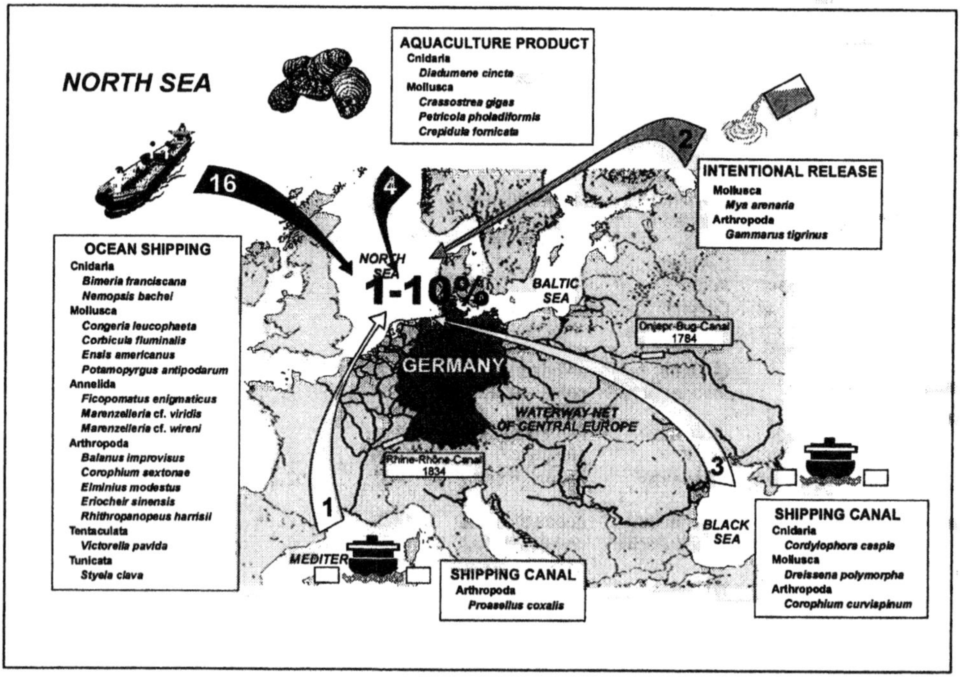

Figure 2. Established alien macroinvertebrates on the German North Sea coast. Known or probable introduction vectors, number of introduced species and their amount on total species number (in percent), important canals and their opening date. See text and Table 1 for further explanations.

In the German sea ports Bremerhaven, Cuxhaven, Emden and Wilhelmshaven are characterised by relatively high salinities, a higher probability for the introduction of species is to be expected because ballast water is mainly brackish or of marine origin as shown by Gollasch (1996). However, it remains remarkable that only a smaller part of the German trans-shipment is made in these ports. In contrast, Hamburg, the most important German port, is characterised by limnic water. In total the introduction of aliens by shipping is regarded to be low because in The Netherlands with Rotterdam as one of the biggest brackish/marine ports of the world, no more alien species were found biggest brackish/marine ports of the world, no more alien species were found than for instance on the German coast (den Hartog & van der Velde 1987).

2.3 BALTIC SEA COAST

Recently, 14 macrozoobenthic species have been identified as aliens on the German coast in the south-western Baltic Sea (Table 1, Fig. 3). They represent 3% of the total macrozoobenthic species. However, a comprehensive study on the status of aliens on the German Baltic Sea coast, is still lacking. Leppäkoski published a first analysis in 1984 about introduced animal and plant species into the central and eastern Baltic Sea. The detailed study of Jansson (1994) about alien species in the Baltic Sea focused entirely on the Swedish west coast. To date, 65 established nonindigenous species, about 40 of them macroinvertebrates, have been reported from the Baltic Sea incl. Kattegat (Leppäkoski & Olenin 2000) but the ecological and economic impacts of these aliens are not yet fully understood. Especially for the German coastal area only for some few cases, e.g. the spreading of the Zuiderzeecrab *Rhithropanopeus harrisii* (Nehring 2000a) or the occurrence of the polychaete *Marenzelleria* cf. *viridis* (Bastrop et al. 1997), closer considerations were made. It seems that only the polychaete *M.* cf. *viridis* was directly introduced by shipping to the German Baltic Sea coast. All other species derive either from the Ponto-Caspian region and are transported through canals with inland waterway vessels, or were at first introduced into the North Sea and from here transported through the Kiel Canal or by natural drift in water currents onto the German Baltic Sea coast (Fig. 3, Table 1). The Zuiderzeecrab *Rhithropanopeus harrisii* represents a special case. The first findings on the German Baltic Sea coast indicate a transfer through the Kiel Canal. Simultaneously with the first finding between 1948 and 1950 in the German Baltic fjord Schlei (Kinne & Rotthauwe 1952), however, the Zuiderzeecrab was also found 560 km further east in Polish waters in 1951 (Pautsch et al. 1969). This occurrence in Polish waters was probably a result of a secondary introduction in the mid/end of the 1940s because prior to this period, the Zuiderzeecrab could not be found further along the Baltic Sea coast. In the following decades, *R. harrisii* spread westward and arrived at the German Baltic Sea coast in the mid 1980s where it was found for the first time in the Wolin-Usedomer Bodden (Nehring 2000a). At present it is not clear if some macroinvertebrate species are actually established aliens, especially on the German Baltic Sea coast. An example is the bryozoan *Victorella pavida*, an introduced species found for the first time in 1880 in the mouth of the river Ryck near the German town Greifswald (Kraeplin 1887). For several decades, no findings of this species on the German Baltic Sea coast were published. Similarly, the recent occurrences of the prawn *Palaemonetes varians* and the amphipod *Corophium multisetosum* in German Baltic waters (Zettler et al. 2000) are possibly results of a natural range extension.

3 Discussion

3.1 MECHANISMS AND EFFECTS

The transportation by ocean shipping is considered as essential for the trans-oceanic introduction of nonindigenous organisms. Recent calculations of Gollasch (1996) about the individual entry through ballast water discharges from overseas areas into the ports on the German North Sea coast revealed that 2.7 million organisms are released daily. Based on increasing shipping intensity, the rate of introductions is estimated to have increased dramatically since 1900, as shown, for instance, for the San Francisco Bay area (Cohen & Carlton 1998). However, the actual data for introductions of alien species in German waters by shipping do not show such a clear trend (Table 1). Also in the entire North Sea, only one species was introduced by shipping on average every 5 years since 1820, with a unique peak of six species in the 1970s (Nehring 2001). However, prior to the introduction and widespread use of antifouling paints containing TBT, about half of the alien fauna introduced into the North Sea came from fouled hulls of overseas trade ships (Nehring 2001). With the introduction of the effective biocide TBT in 1970, there has been a reduction of hull fouling and, consequently, a considerable decrease of introductions via ship fouling. Since then the discussion on the role of ballast water as an important carrier of organisms from overseas has increased (e.g. Carlton 1985, 1999b; Gollasch 1997). However, for the North Sea, ballast water and ship's hulls have been of equivalent importance as vectors since 1970 (Nehring 2001). This is probably based on the fact that fouling still occurs today, especially where the antifouling paint is damaged, difficult to apply, or otherwise ineffective.

Aliens are often regarded as ecological and/or economic threats. Commonly cited spectacular cases come mostly from overseas, as for instance the introduction of the zebra mussel *Dreissena polymorpha* into the Great Lakes (USA) or the occurrence of the European shore crab *Carcinus maenas* on the American Pacific coast (Carlton 1996). However, up to now no real problematic alien species among the macroinvertebrates occur in German waters. An actual unsolved problem is the classification of the borer *Teredo navalis* as an anthropogenic introduced species. Since the 16th century this bivalve is often mentioned in connection with serious damages to wooden structures by aliens in coastal waters of the North and Baltic Sea (Gollasch 1999). But, the origin of *T. navalis* is still unknown (north-eastern Atlantic: Schütz 1961; warm temperate to tropical areas: Gollasch 1996). Of course, it cannot be excluded that this species has an exotic status in Europe and may have been subjected to ship-mediated dispersal. According to Schütz (1961), *T. navalis* has been recorded several times from drift wood, as a result of which the German name of this species is not shipworm but pileworm. Furthermore, in several reports exotic borers are described from wood drifting with the Gulf Stream from tropical waters to Europe (Carlton 1999a). Additionally, several *Teredo* species are described from northern European fossil soils (Moll 1914). Therefore taking into account that its invasion history is still unknown and its human mediated introduction is unconfirmed, at present *T. navalis* can not be regarded as alien in Europe and should therefore not be classified as "an introduced harmful species". In German waters, the predominant majority of macrozoobenthic aliens are rare (Table 1). On the other hand, every fifth introduced species becomes abundant. However, even if displacement of native species by aliens in German waters is sometimes suggested, to date only decreased abundances but no extinctions of native species have been documented. As the autochthonous aquatic species community compositions fluctuate with season, tidal dynamics, drainage regime and other environmental factors, the nature of such communities is distinctly dynamic. Therefore, this leads one to suppose that in view of the stability of the different ecological systems, alien species establishment is only secondary. Nonindigenous species are often simply integrated. It seems that most aquatic systems should be capable of either absorbing or resisting invaders to some extent. However, one should not assume that German waters are immune to harmful bio-invaders at all times, because no region of the world can - *a priori* - be considered safe from significant effects of alien species.

Even if no relevant ecological and economic impacts of introduced species in German waters could be found to date, on an international level the need for action concerning the minimisation of organism introduction has been recognised since the 1980s. Today the spread of alien species is recognised as one of the greatest threats to the global biodiversity (CBD 1992). However, research is needed in order to assess the impact on biodiversity due to introduced species and on what level effects are expressed (populations, species or genetic level). An interesting (positive) example is the extensive establishment of the Pacific oyster *Crassostrea gigas* in Dutch and German coastal waters of the North Sea (Drinkwaard 1999; Nehring 1999). By this alien species a reconstruction of the specific Wadden Sea oyster bank biotope is very likely and could be interpreted as a gain for the biodiversity (Nehring 2000b).

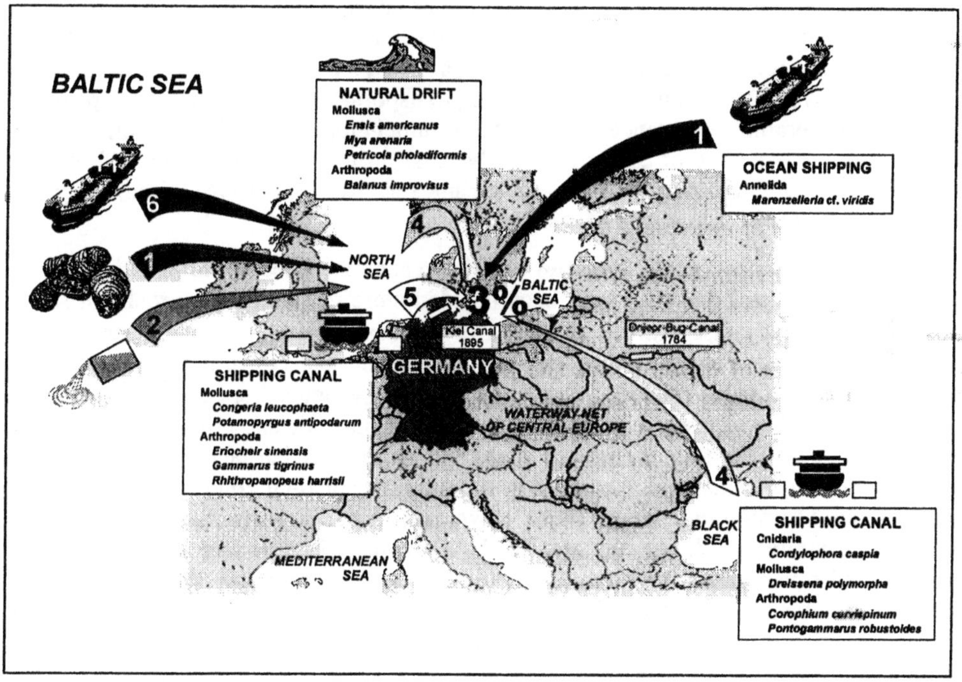

Figure 3. Established alien macroinvertebrates on the German Baltic Sea coast. Known or probable introduction vectors, number of introduced species and their amount on total species number (in percent), important canals and their opening date. See text and Table 1 for further explanations.

3.2 ESTUARIES AS A HABITAT

The majority of alien species has no chance to survive due to the lack of adaptive mechanisms to the different environmental conditions. In German waters, in addition to pollution of water and sediments and engineering impacts (stream straightening, dredging and dumping), another fundamental factor is the changing hydrological regime. Highly variable water runoff in the streams and, especially on the North Sea coast, the tide with all its consequences - above all development of eulittoral areas and of a transition zone between limnological and marine conditions - play an important role for the establishment of permanent populations of alien species. Due to these specific characteristics, generalists should have the greatest chance for settlement. Consequently, among the aliens established in German waters euryoecious and potent competitors predominate. It is remarkable that most of the introduced macroinvertebrates have established permanent populations in the estuaries of the German North Sea (19 species in total). This can be explained by the following reasons:

(i) tolerant limnic species, which were transported through canals with inland vessels, reached the coast first in the estuaries and found fitting salinity conditions somewhere along the estuarine gradient (currently 4 species, see Fig. 1).
(ii) Major harbours are located in estuaries and thus estuaries are characterised by intense intercontinental shipping and have a higher potential infection rate, combined with the fact that ballast water often has estuarine character (Gollasch 1996).
(iii) About half of the introduced macroinvertebrates in the estuaries are genuine brackish water species, characterised by a high tolerance for changing environmental conditions having a better chance of being transported alive than euhaline species.
(iv) Of considerable importance is the natural autochthonous species minimum in the brackish water zone of estuaries, i.e. above all in the mesohaline zone (5 to < 18 PSU) many ecological niches are vacant (cf. Remane 1934), facilitating the establishment of introduced species.

Among aquatic environments, estuaries have been common sites of biological invasions worldwide. It appears that they have been invaded more frequently than rocky or sandy shores of the outer coast (Ruiz et al. 1997). In the German estuaries, the macroinvertebrate communities of mixed native and alien organisms behave in many respects like co-evolved assemblages (Nehring, unpublished). All introduced species appear benign and increase both species and functional diversity. Empty niches, however, seem to be an essential basis for this. In former times, no species of barnacle, crab and filter-feeding mollusc, for instance, occurred in the oligo-mesohaline zone of tidal waters in North-western Europe (e.g. Wolff 1999). Nowadays, the alien barnacles *Balanus improvisus* and *Elminius modestus*, the alien crabs *Eriocheir sinensis* and *Rhithropanopeus harrisii,* and several alien bivalves (e.g. *Congeria leucophaeta* and *Mya arenaria*) are common members of the estuarine benthic communities. It seems that low species richness in aquatic communities is not an inconsiderable factor for invasions, but the frequency and intensity (or size) of inoculation are critical components in colonization success. In conclusion, it can be stated that the combination of brackish water with its unsaturated ecological niches and intense intercontinental ship traffic has the highest potential infection rate for coastal areas, especially in Germany.

4 Outlook

Alhough our present knowledge about the extent, patterns and mechanisms of aquatic bio-invasions is still in ist infancy, it is clear that aliens are a significant force of change in aquatic and especially estuarine communities globally. It is highly probable that in the near future other nonindigenous species will arrive and may establish permanent populations in German waters and other boreal areas. Among them will be several species which, during the last decades, were already introduced to Great Britain and The Netherlands, and which will successively expand their distribution range as did their predecessors, e.g. the barnacle *Elminius modestus* and the American piddock *Petricola pholadiformis*. Potential candidates for future introductions might be the bivalve *Tapes philippinarum* and the polychaete *Hydroides ezoensis*.

The prohibition of the biocide tributyltin (and other organotin compounds) on ships, planned to come into effect world wide in 2003 (IMO 1999), could result in an increase of bio-invasions through shipping, if alternative antifouling paints lack the effectiveness of TBT antifoulants. Especially the biocide-free silicone coatings, from which fouling growth is easily removable, have an extraordinarily high potential for introduction of alien species (Nehring 2001).

In addition to nonindigenous species, introduced directly or indirectly through humans, more and more thermophilic species immigrate into the German Bight and establish themselves here. During the last 20 years 10 exotic phytoplankton species, native to habitats in southern warmer waters, have established permanent populations in the North Sea (Nehring 1998). The trend of expansion northwards was observed also in fish, jellyfish, isopods and cladocerans (e.g. Franke et al. 1999). It is presumed that global warming will play an important role, at least on the northern hemisphere with occurrence of milder winters. Measurements of water temperature yielded an average annual increase of about 1°C during the last 100 years (Houghton et al. 1992). Consequently, the formulated scenarios by experts on global change, according to which European coastal systems will be subject of major ecological changes in the nearest 50-100 years, already can be observed to some extent in the fauna and flora today. Climate change has and will increase the local dynamics of many animal and plant populations, and hence will lead to more extensive alterations in the communities. However, these processes are not linear but complex, variable and highly dynamic. Introduced organisms might be a mere "disturbance signal" of further ecosystem changes to come.

INVASIVE PONTO-CASPIAN SPECIES IN WATERS OF THE VISTULA AND ODER BASINS AND THE SOUTHERN BALTIC SEA

KRZYSZTOF JAZDZEWSKI* & ALICJA KONOPACKA
Department of Invertebrate Zoology & Hydrobiology, University of Lodz, Lodz, Poland
*Corresponding author *kryjaz@biol.uni.lodz.pl*

Abstract
The paper gives a historical review as well as recent information on invertebrate and fish species of Ponto-Caspian origin that have invaded the Vistula and Oder drainage systems and the southern Baltic Sea. Notes on the taxonomy of some species are included.

1 Introduction

Human activities have accelerated the process of range extensions by introductions of alien species to such an extent that their ecological impacts become often harmful, leading to quick global homogenisation of biotas, threatening local communities and biodiversity, and imposing direct costs to national economics (Drake et al. 1989; Lodge 1993; Mills et al. 1993; MacIsaac 1996; Ricciardi et al. 1996; Enserink 1999; Lee & Bell 1999; Olenin & Leppäkoski 1999; Ricciardi & MacIsaac 2000). Range extensions of alien species related to human activity were caused both by intentional introductions as well as by ballast-water transport and constructions of canals connecting different drainage systems (Jazdzewski 1980; Gollasch 1996). Quite recently it appeared that Ponto-Caspian animals constitute an important proportion of hydrobionts invading even very remote ecosystems (Ricciardi & MacIsaac 2000). Already decades ago studies of Ponto-Caspian fauna predicted wide spreading of some species of this faunistic complex, at least in European waters (Mordukhaj-Boltovskoj 1964; Bacescu 1966).

Most of the territory of Poland consists of drainage areas of two major Baltic rivers - Vistula and Oder. Therefore, writing about Polish waters, we mean physiographic entities - Vistula and Oder basins, and the southern Baltic Sea. The whole network of North-European inland waters, including these rivers' basins, together with the whole Baltic Sea, were formed only after the last Würm glaciation, i.e. during last 18,000 years. Polish fresh and brackishwaters were rather early invaded by Ponto-Caspian taxa, owing to the 18th century construction of a network of canals, joining the Dnieper with Vistula and Oder basins.

2 Review of species

2.1 CNIDARIA

2.1.1 Cordylophora caspia (Pallas, 1771)

This athecate hydrozoan is a classical example of brackish water animals originating in the Pliocene Ponto-Caspian basin, and is now one of the widest globally distributed species. *C. caspia* is found in the harbours all over the world, provided that they are

E. Leppäkoski et al. (eds.), Invasive Aquatic Species of Europe, 384–398.

situated at rivers' mouths (Naumov 1960). This is known from Caspian, Azov and northern Black Sea, Baltic Sea, European and North Atlantic coastal waters, harbours of northern Africa, South America, China, Australia, Tasmania and New Zealand. Although *C. caspia* occurs also in freshwaters, it cannot develop properly there (Naumov 1960). This worldwide distribution of *C. caspia* is evidently an effect of ship traffic, however, we do not know if it is only via ballast-water exchange; this hydrozoan might be sufficiently euryhaline to be transported in the ship's hull overgrowth.

C. caspia is probably one of the oldest immigrants to the Vistula and Oder basins. In the southern Baltic first records of this species are from the beginning of 19th century (Leppäkoski 1984). Detailed Polish records of *C. caspia* are scarce: Baltic harbours of Szczecin and Hel, Vistula river mouth and a freshwater locality in Oder near Brzeg (Demel 1933; Sembrat 1953). According to Zmudzinski (1990) (who used a junior synonym *C. lacustris* Allman, 1844) this hydrozoan is a very common inhabitant of southern Baltic lagoons and coastal lakes, overgrowing all submersed objects and disturbing catches with fishing set-nets.

2.2 CRUSTACEA

2.2.1 Cercopagis pengoi(Ostroumov, 1891)

The onychopod cladoceran *C. pengoi* is the only representative of *Cercopagis* Sars, 1897 recorded outside the Ponto-Caspian water basins. In the Caspian Sea, 13 species of this genus are known (Rivier 1998). *C. pengoi* occurred originally in Caspian, Azov and Aral seas and also in some Black Sea lagoons and coastal lakes. After the construction of artificial lakes on Dnieper and Don, *C. pengoi* entered these reservoirs, establishing populations in freshwaters (Mordukhaj-Boltovskoj & Galinskij 1974; Gusinskaja & Zhdanova 1978). Until recently it was not known north of Kiev (Ukraine), but in 1992 it appeared in Baltic Sea. In Polish waters *C. pengoi* was recorded first in Vistula Lagoon by Hornatkiewicz-Zbik (1999, 2000) and by Zmudzinski (1999) in the Gulf of Gdansk near the Gdynia harbour. Further findings were noted near the tip of Hel Peninsula (Duris et al. 2000) (Fig. 1).

2.2.2 Chelicorophium curvispinum (G.O. Sars, 1895)

Recent revision of Corophiidae by Bousfield & Hoover (1997) has placed *C. curvispinum* in the new genus of *Chelicorophium. C. curvispinum* is the oldest Ponto-Caspian amphipod immigrant noted in Polish waters. This species is native to large rivers emptying into the Black and Caspian Seas (Volga, Dniester, Danube and many others). The earliest record of *C. curvispinum* outside the Ponto-Caspian drainage area was that of Wundsch (1912) in the Spree-Havel system near Berlin in Germany. The species was first described as new to science (*C. devium*) and only later synonymized with *C. curvispinum*. Without doubt *C. curvispinum* arrived to the Baltic Sea and North Sea drainage systems through the 18th century canals joining the Dnieper, Vistula, Oder and Elbe basins. Soon numerous localities of *C. curvispinum* were known in Oder in Krosno and near Gryfino, in lower course of Vistula, in Vistula Lagoon, in Notec river, and in Szczecin Lagoon (Jazdzewski 1980 and references herein). It is evident that the species first crossed Polish territory and was present well before its discovery in late 1920s.

Now *C. curvispinum* is common and abundant in many localities along Vistula and Oder as well as in waters connecting these two systems (Jazdzewski 1980; Jazdzewski & Konopacka 1990, 2000) (Fig. 1). *C. curvispinum* seems to be associated with clumps of zebra mussel. At least in the lower Vistula, in our samplings where *Dreissena polymorpha* was present in the dredging, it was always accompanied by *C. curvispinum*. On the other hand the autumn 1999 sampling in the Vistula upstream of the Pilica river mouth, as far as to Sandomierz, revealed an absence of both co-occurring species. The wide-ranging extension of *C. curvispinum* in Western Europe is now permanently observed and its mass occurrence is occasionally noted (Van den Brink et al. 1989, 1993b; Schöll 1990; Pinkster et al. 1992); its far western locality in Great Britain was recorded already in the early 1930s by Crawford (1935).

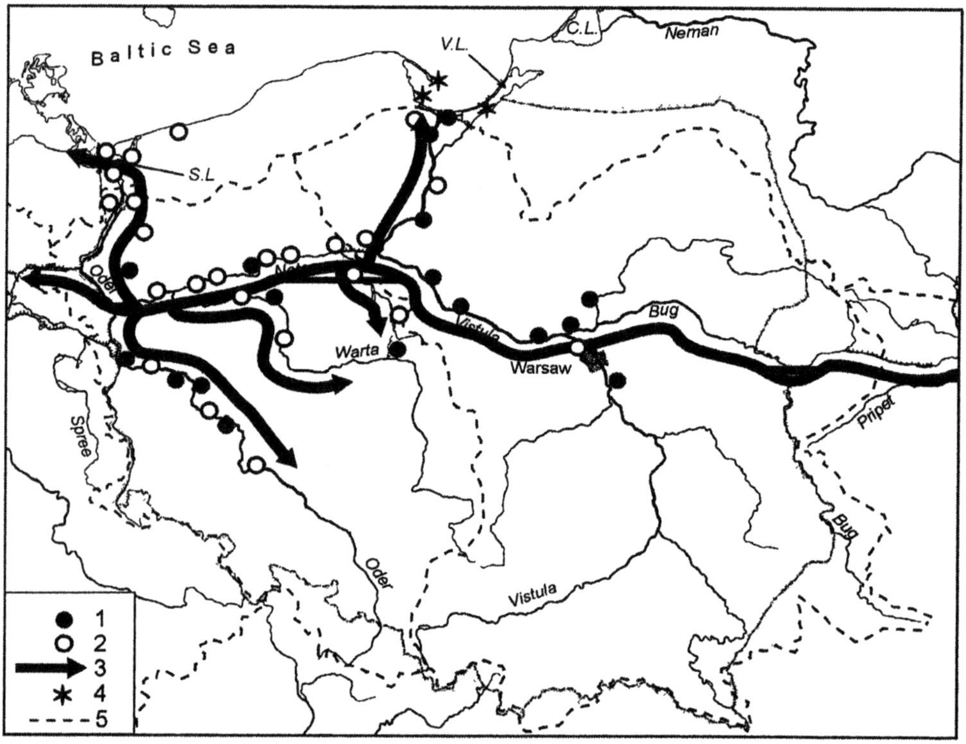

Figure 1. Distribution of *Chelicorophium curvispinum* and records of *Cercopagis pengoi* in Poland. (1) Literature records of *C. curvispinum*, (2) recent authors's records of *C. curvispinum*, (3) probable immigration route of *C. curvispinum* from Dnieper basin, (4) records of *C. pengoi*, (5) drainage basins boundaries.

2.2.3 Chaetogammarus ischnus (Stebbing, 1899)

C. ischnus, another amphipod crustacean of Ponto-Caspian origin, is also a rather old invader of Polish waters. Its name has been spelled in numerous ways (e.g., *Chaetogammarus tenellus*, *C. tenellus behningi*, *C. ischnus sowinskyi*, *Echinogammarus ischnus*). The generic assignment of this species is connected with hitherto not solved nomenclatoric problems with European gammarid taxa of a scale-like endopodite of uropod 3. It was Karaman (1977) who synonymized some formerly used generic names (*Marinogammarus* Schellenberg, 1937, *Chaetogammarus* Martynov, 1924) with the name *Echinogammarus* Stebbing, 1899 uniting all gammarid species with such an uropod 3 (*parviramus* of Schellenberg 1937). This view was then supported by Karaman & Barnard (1979) and Barnard & Barnard (1983), and so the name *Echinogammarus ischnus* was used by Jazdzewski & Konopacka (1988, 1995). However, recently Stock (1995) has returned to the name *Chaetogammarus* for a group of taxa of *Echinogammarus* sensu Karaman, restricting this last name only for the *Echinogammarus berilloni* - group (Pinkster 1973). Stock (1995) used cladistics for this new grouping of the related taxa but his new division is not discussed in detail, evidently left for future, never published paper; his view is not yet commonly recognized, although Karaman's (1977) revision is also not commonly accepted. The subspecific name *Chaetogammarus ischnus sowinskyi* was used by some authors (e.g. Straskraba 1962; Dedju 1967) for the freshwater form of *C. ischnus* in belief that the nominative form inhabits brackish Caspian or Black Sea waters. Birshtejn & Romanova (1968) were even of the opinion that we are dealing with two separate species - a freshwater and a brackish water one. However, the mentioned authors, in order to demonstrate morphological differences between these two forms, used figures of a female from the Caspian Sea (Sars 1896) for one form, and a male from the Danube delta (Carausu et al. 1955) for the second form. However, sexual dimorphism is strongly expressed in these gammarids and these data cannot support the view of specific or subspecific separation of two forms. Without further morphological and taxonomic studies of the rich Ponto-Caspian gammarid complex of cf. *Chaetogammarus* (also other genera), based on original, possibly type material, we will not reach any sound and solid taxonomic base. Therefore, the name *C. ischnus* will be used here. The earliest record of *C. ischnus* outside the drainage systems of the Black and Caspian Seas was from the Vistula (Jarocki & Demianowicz 1931) in 1928. These authors noted *C. ischnus* along the whole middle and lower Vistula flow below Warsaw. The presence of this species in the Vistula mouth was then confirmed by Jazdzewski (1975). Sampling in summer of 1998 in the middle and lower Vistula and in its right side affluent Bug, showed that *C. ischnus* still occurs in these rivers' flows (Fig. 2), but is now clearly outnumbered by the new Ponto-Caspian incomer, *Dikerogammarus haemobaphes*.

Nevertheless *C. ischnus* is evidently an expansive species. In the late 1970s it was recorded already in the North Sea drainage basin, in canals joining the Elbe, Weser and Ems (Herhaus 1978; Herbst 1982). It was later found also in some Mecklenburgian and Pomeranian lakes (Waterstraat & Köhn 1989; Jazdzewski & Konopacka 1990; Köhn & Waterstraat 1990; Zettler 1998), for the first time in stagnant waters outside its natural Ponto-Caspian distribution area, where it occurs in several coastal Black Sea lakes and, artificially introduced, in some Ukrainian dam-reservoirs (Jazdzewski 1980). Finally the species also arrived into the Lower Rhine (Schöll 1990; Van den Brink et al. 1993a). *C. ischnus* is also a recent invader of Great Lakes in North America (Witt et al. 1997).

2.2.4 Pontogammarus robustoides (G.O. Sars, 1894)

P. robustoides originally occurred in lower course of large Ponto-Caspian rivers (Volga, Don, Dnieper, Dniester and Danube) as well as in some freshwater and brackish Black

Sea coastal (inshore) lakes (Jazdzewski 1980). In the 1960s it was introduced to many Ukrainian, Caucasian, and Lithuanian artificial lakes; in the latter case, naturalized in the Neman drainage system, including the Curonian Lagoon, the species has reached a new sea basin - the Baltic Sea (Gasjunas 1968; Jazdzewski 1980).

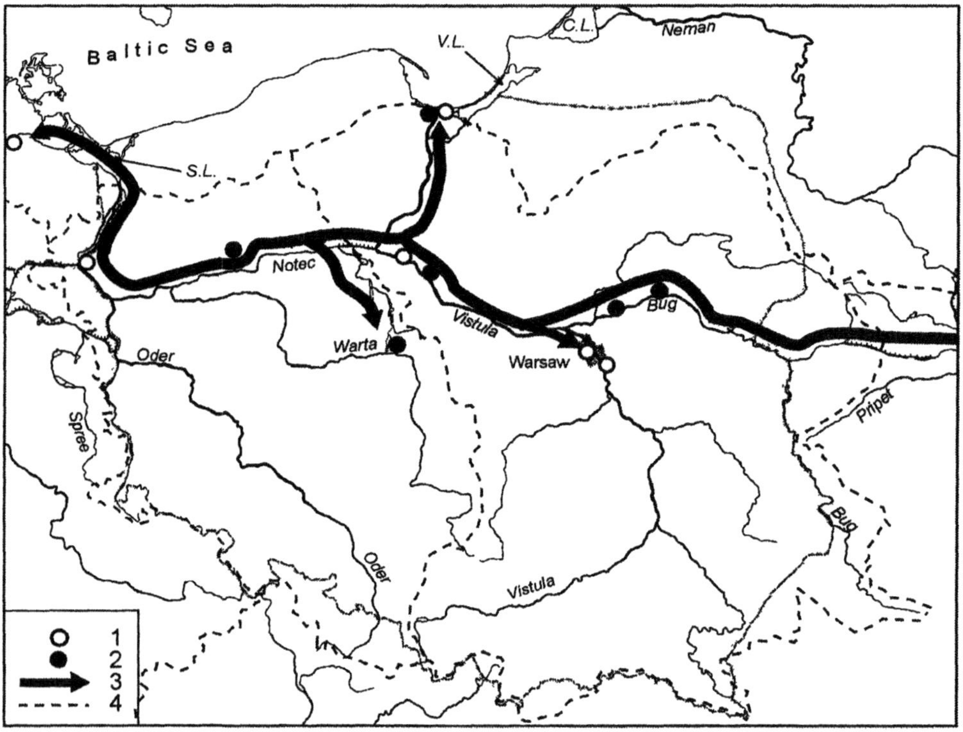

Figure 2. Distribution of *Chaetogammarus ischnus* in Poland. (1) Literature records, (2) recent authors' records, (3) probable immigration route from Dnieper basin, (4) drainage basins boundaries.

Recent records of *P. robustoides* come from Mecklenburgian waters and the Mittelland-canal in Germany (Rudolph 1997; Zettler 1998; Martens et al. 1999), from north-western Poland (Szczecin Lagoon and lower Oder) (Gruszka 1999; Wawrzyniak-Wydrowska & Gruszka 2001), from lower Vistula (Konopacka 1998) and from the Vistula Lagoon (Jazdzewski & Konopacka 2000). It is possible that the species reached the Vistula and Oder delta systems and north-eastern German waters via coastal Baltic Sea waters from the Neman system and Curonian Lagoon, probably by ballast water transport, but recent discoveries of the species in the lower Vistula flow (Konopacka 1998; Jazdzewski & Konopacka 2000 and unpubl. 1998, 1999 and 2000 samplings) cannot exclude that *P. robustoides* could also follow the dispersal route of *C. curvispinum* and *C. ischnus* via the Pripet'-Bug canal from Dnieper to the Vistula system (Fig. 3).

2.2.5 Dikerogammarus haemobaphes (Eichwald, 1841)

The taxonomic status of *D. haemobaphes* and related forms is in need of serious revision. The most common view (e.g. by Carausu et al. 1955; Karaman 1974; Stock 1974), was that the typical form of *D. haemobaphes* inhabits only the Caspian Sea, whereas its freshwater form, described by Martynov (1919) as "morpha *fluviatilis*", and most often treated as a subspecies, occurs in rivers and lakes. However, Straskraba (1962) and Barnard & Barnard (1983) have even raised the two forms, *D. h. haemobaphes* and *D. h. fluviatilis,* to the species rank, and this view was tentatively accepted by Jazdzewski & Konopacka (1988). On the other hand Konopacka (1998), who discovered this newcomer in the Vistula in 1997, after studying its morphology and comparing with the drawings of Carausu et al. (1955) and of Mordukhaj-Boltovskoj et al. (1969), did not find clear differences between the two forms and has retained the name *D. haemobaphes* for the form from the Vistula.

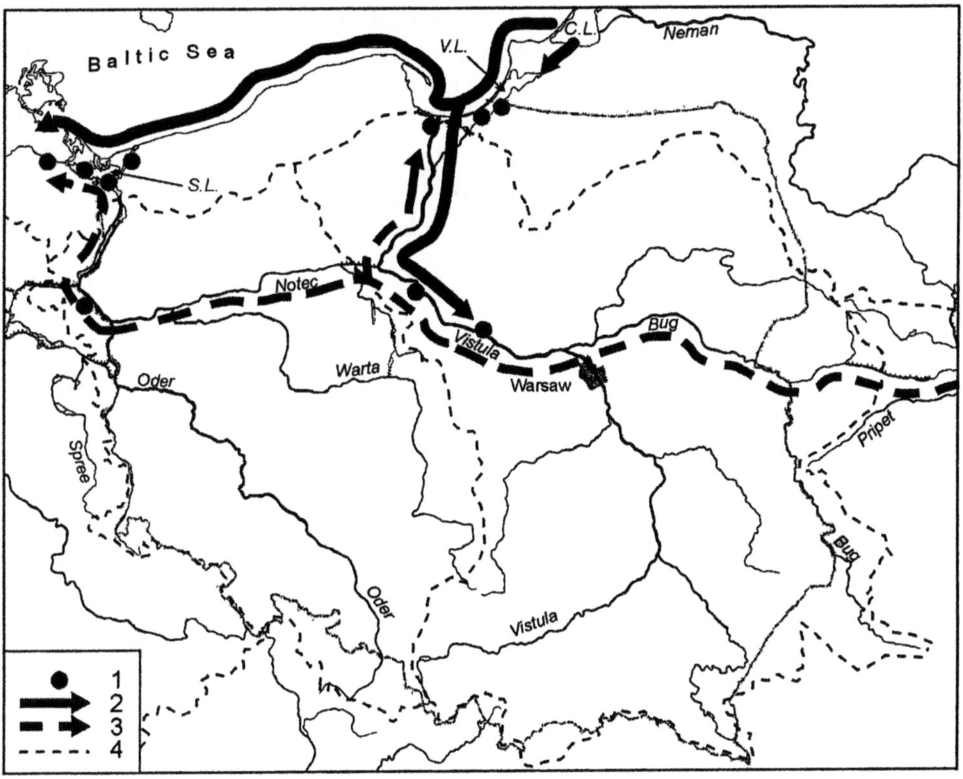

Figure 3. Distribution of *Pontogammarus robustoides* in Poland. (1) Literature and authors' records, (2) probable immigration route from Neman basin, (3) probable immigration route from Dnieper basin, (4) drainage basins boundaries.

Already Mordukhaj-Boltovskoj (1964) predicted that *D. haemobaphes* will soon penetrate another sea basins beyond the Ponto-Caspian basin. In the middle of 20th century the most upstream Danube localities of this species were some 1,700 km away from its mouth (Straskraba 1962), but recently this species has, moving upstream the Danube, crossed this river system reaching the North Sea basin by entering the Rhine drainage system (Schleuter et al. 1994; Nesemann et al. 1995; Leuchs & Schleuter 1996).

As mentioned above, *D. haemobaphes* was quite recently found as a new species of the Polish fauna and for the first time in the Baltic Sea basin (Konopacka 1998). Sampling along lower and middle Vistula flow in 1998 and 1999 revealed that the species is a dominant gammarid in this river, in lower course co-occurring with the less numerous *P. robustoides* and the scarce *C. ischnus*; in the middle course it was the only gammarid species. We have discovered also that *D. haemobaphes* is present and sometimes abundant in Bug (joining the Vistula with the Dnieper) and Notec (connecting Vistula with Oder) (Jazdzewski & Konopacka 2000) (Fig. 4). In Notec river *D. haemobaphes* co-occurs with the very abundant *Gammarus roeselii* and with the less numerous *G. varsoviensis*. In Vistula and Bug *C. curvispinum* was also usually a co-occuring amphipod species. Recently the present authors have found *D. haemobaphes* in the Vistula Lagoon and Gruszka (2000) and Müller et al. (2001) reported *D. haemobaphes* from the lower Oder. The abundant occurrence of *D. haemobaphes* in Vistula indicates that the species is at present in an active expansion phase; it has been recently noted also in the upper Volga system as far north as near Moscow in the Moskva river (Lvova et al. 1996) and Viatka river near Kirov (Tarasov 1996).

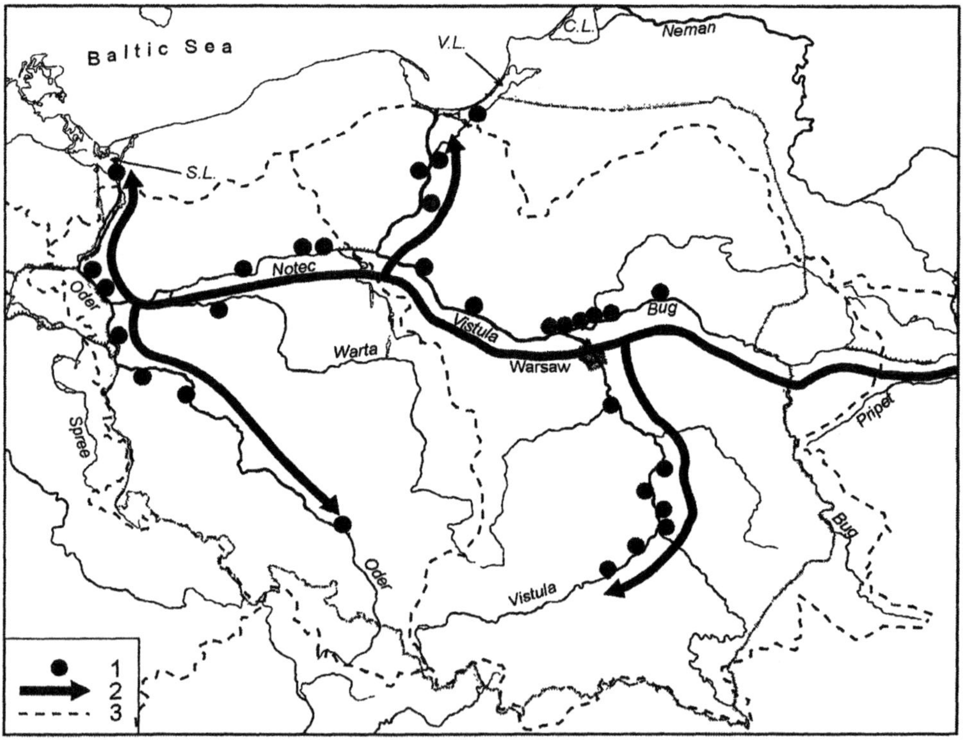

Figure 4. Distribution of *Dikerogammarus haemobaphes* in Poland. (1) Literature and author's records, (2) probable immigration route from Dnieper basin, (3) drainage basins boundaries.

2.2.6 Dikerogammarus villosus (Sowinsky, 1894)

This is a quite recent newcomer in Polish waters. The species was recorded in Oder in 2001 by German and Polish scientists (Müller et al., 2001; Jazdzewski & Konopacka unpubl. data). The species was collected in Oder only downstream from the canal connection of Oder with Havel - Spree systems (Fig. 5). The penetration of *D. villosus* into the Oder system is especially interesting since it first used the so-called southern corridor, i.e. Danube. In the upper reach of Danube, already in Germany, *D. villosus* was first recorded by Tittizer et al. (1994) in 1992, and soon it penetrated the Rhine river, even its lower reach via the Main-Danube canal (Bij de Vaate & Klink 1995). From the Rhine *D. villosus* started it's further, now eastward directed invasion using the Mittelland-canal joining the Weser, Elbe and Oder systems (Grabow et al. 1998; Zettler 1999; Rudolph 2000).

Now the species has reached the lower Oder flow, occurring there in common with other alien gammarids, like *D. haemobaphes*, *P. robustoides* and *G. tigrinus*, an invader with North-American origin (Jazdzewski & Konopacka unpubl. data; Müller et al. 2001).

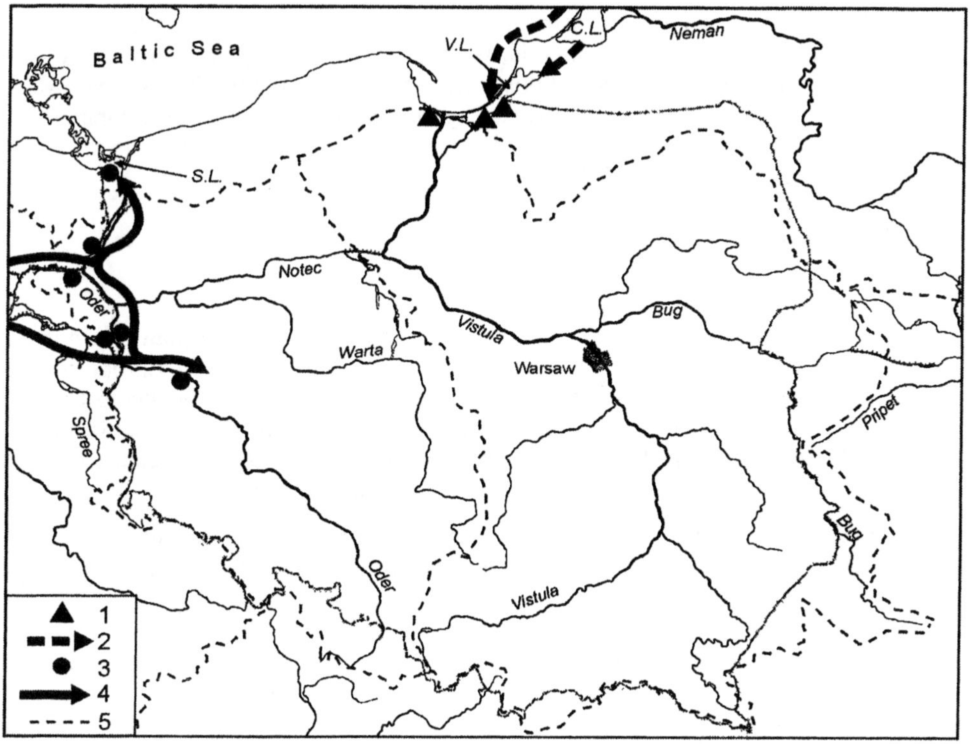

Figure 5. Distribution of *Dikerogammarus villosus* and *Obesogammarus crassus* in Poland. (1) author's records of *O. crassus*, (2) probable immigration route from the Curonian Lagoon, (3) literature and authors' records of *D. villosus*, (4) probable immigration route from Oder-Havel Kanal and Oder-Spree Kanal, (5) drainage basins boundaries.

2.2.7 Obesogammarus crassus (G.0. Sars, 1894)

This species has most recently been discovered in Polish waters. It was recorded in the Vistula Lagoon and the Dead Vistula (Konopacka & Jazdzewski, in press) (Fig. 5). It was placed in the genus *Pontogammarus* by Sowinsky (1904), however the revision by Stock (1974) has brought a new generic name, *Obesogammarus,* for a group of some 10 species inhabiting fresh and brackish waters of the Ponto-Caspian region (Stock et al. 1998), including *O. crassus*. Original distribution areas of *O. crassus* encompassed coastal Caspian Sea and lower courses of rivers emptying to this water body; in Volga the species entered as far as to Volgograd (Mordukhaj-Boltovskoj 1979). In the Black Sea area *O. crassus* occurred originally in brackish lagoons and coastal lakes entering upstream the lower courses of Kuban, Don, Dnieper, Ingulec, Boh, Dniester and Danube; in Danube it was noted up to the Morava river mouth (Lubjanov 1953; Carausu et al. 1955; Russev 1959, 1966; Mordukhaj-Boltovskoj 1960; Kaneva-Abadzhieva 1965; Dedju 1967, 1980; Birshtejn & Romanova 1968; Mordukhaj-Boltovskoj et al. 1969; Ioffe 1973; Jazdzewski 1980). *O. crassus*, like *P. robustoides*, was transplanted in the early 1960s to the Kaunas artificial lake on the Neman river in Lithuania and from there, after acclimatization, it was transferred to the Curonian Lagoon by humans, but also entered this basin by natural downstream migration (Gasjunas 1972). In Curonian Lagoon the species is a common gammarid, whereas in Vistula Lagoon it was noted only in 1999. Most probably it has entered this water body *via* the Pregel river system, which is connected to both Curonian and Vistula Lagoons. However, due to its comparatively high euryhalinity *O. crassus* could have dispersed southwestwards also along the Baltic Sea shores (average salinity 7 PSU).

2.3 MOLLUSCA

2.3.1 Dreissena polymorpha (Pallas, 1771)

The most important molluscan invader of Polish waters of Ponto-Caspian origin is the zebra mussel *D. polymorpha*. This species had a wide distribution in Europe in preglacial or at least in interglacial periods, its area has shrunken to the Ponto-Caspian basin during last Würm/Vistulian glaciation. From that region a new colonization of central and northern Europe begun, efficiently enhanced in 18th and 19th centuries by the construction of numerous canals joining different drainage systems.

In Poland this species lives in lakes, dam reservoirs and large rivers, as well as in brackish lagoons connected to the Baltic Sea, but not in the Baltic itself (Piechocki & Dyduch-Falniowska 1993). Especially dense population development *D. polymorpha* has attained in the Vistula Lagoon (3-5 PSU, Zmudzinski 1957) and in the Szczecin Lagoon (0.5-1.5 PSU, > 100,000 ind m^{-2}; Wiktor 1969). In northern Poland the species is very common, especially in the Mazurian and Pomeranian lakelands, in the southern half of the country its occurrence is practically restricted to the large rivers, Oder and Vistula, and to the artificial dam-lakes (Piechocki & Dyduch-Falniowska 1993, Fig. 61). Summaries of the extensive works on *D. polymorpha* in Poland can be found in the papers by Stanczykowska (1977, 1997).

Also in Poland the zebra mussel is still spreading into new habitats such as an artificial lake built on Pilica, left affluent of the Vistula. The reservoir was filled in 1973 and first zebra mussels were found there in 1997 (Abraszewska-Kowalczyk et al. 1999).

2.3.2 Litoglyphus naticoides (C. Pfeiffer, 1828)

Another well-known molluscan immigrant from the Ponto-Caspian basin is the freshwater and brackish-water snail *L. naticoides*. Its spreading westwards is also recognized as an effect of canal constructions in the last centuries. In Poland this species has the same distribution as zebra mussel, i.e. in large and medium-sized lowland rivers, in Szczecin and Vistula Lagoons, and in some lakes of the Mazurian Lakeland (Piechocki 1979). *L. naticoides* was widely recorded from Polish waters, but mainly from northern Poland and especially from large rivers like Vistula, Oder, Bug, Warta and Notec, due to the increasing pollution. However, it has recently become rare and has even disappeared from some of its former locations (Falniowski 1987).

2.4 PISCES

2.4.1 Neogobius melanostomus (Pallas, 1811)

First record of the Ponto-Caspian round goby in Polish waters was from the tip of Hel Peninsula (Gulf of Gdansk) in 1990 (Skora & Stolarski 1993). Skora (1996a) suggested, judging from the age of this first fish specimen (scale rings) that *N. melanostomus* might have inhabited the Gulf of Gdansk at least since 1987. The species increased its abundance, especially in the innermost, western bay of the Gulf of Gdansk, Bay of Puck, where it soon became one of the commonest fish in the coastal zone (Fig. 6), occupying habitats hitherto taken by autochthonous *Zoarces viviparus* and *Gobius niger* (Skora 1996b). Fears are expressed that this avalanche increase of *N. melanostomus* population will endanger local stocks of flounder (*Platichthys flesus*) because of the strong competition for food. On the other hand *N. melanostomus*, attaining the length of 20-25 cm, has already appeared in the local fish markets and begins to be appreciated by anglers (Kuczynski 1995; Skora & Stolarski 1995, 1996; Skora 1996b). Authors' unpublished observations from the Bay of Puck in 1998 and 1999 revealed a mass occurrence of this fish and at the same time a clear decrease of the hitherto abundant populations of littoral *Gammarus* populations, monitored for years by our laboratory. This could be a result of this enormous development of *N. melanostomus* population, although the diet of this fish in its original distribution area consists mostly of molluscs (Svetovidov 1969).

First suggestions by Skora & Stolarski (1993) were that the species entered the Baltic by means of ship ballast waters, however, other ways were later considered (Skora & Stolarski 1995, 1996). The first one, from the Black Sea via the canal joining Dnieper and Vistula systems (Pripet' - Bug rivers, as in the case of some above mentioned crustaceans) and the second, from the Caspian Sea, the Volga river and then through several canals joining the upper Volga with Onega and Ladoga lakes, and then the Gulf of Finland. The last possibility would be supported by the finding of *N. melanostomus* by Sokolov et al. (1989) in the Moskva river, a secondary affluent of the upper Volga, far north of the Caspian Sea.

Curiously enough, in nearly the same time (first record in 1990) *N. melanostomus* invaded North America (Jude et al. 1995) evidently with ship ballast water.

2.4.2 Neogobius gymnotrachelus(Kessler, 1857)
The Ponto-Caspian goad goby *N. gymnotrachelus* was recorded in the Bug river (Danilkiewicz 1996), quite near the entrance of the canal joining this river with Pripet', an affluent of Dnieper (Fig. 6). This species commonly inhabits large Ukrainian rivers emptying to the Black Sea - Dnieper, Dniester, and Boh. Despite the very old connection of the Dnieper and Vistula systems (canal Pripet' - Bug built in 1784), *N. gymnotrachelus* was never noted earlier in Bug, whereas Danilkiewicz caught it in summer 1995 in such an abundance that the species was said to be the second in dominance of the whole Bug ichthyofauna.

In 2000, Kostrzewa & Grabowski (2001) recorded the species for the first time in Vistula in the artificial Wloclawski reservoir built in the middle section of the river. Soon after this the species was found there in high abundance, commonly inhabiting empty *Anodonta* shells on the shallow sandy and muddy bottoms (Grabowski & Kostrzewa, pers. comm.).

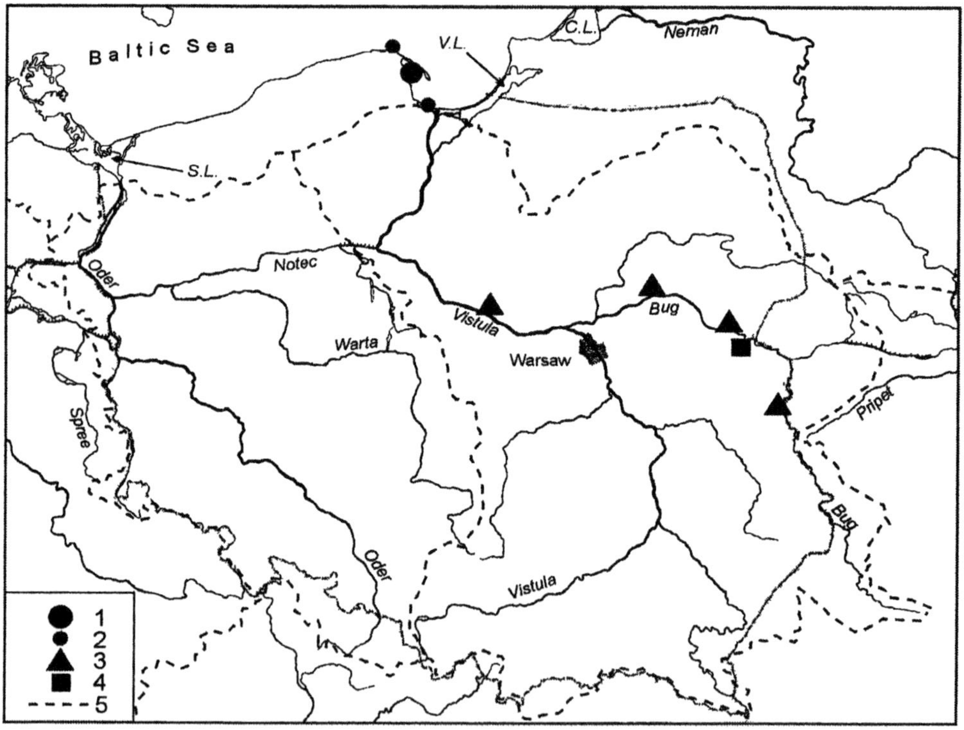

Figure 6. Records of *Neogobius* species in Polish waters. (1) mass occurrence of *N. melanostomus*, (2) single records of *N. melanostomus*, (3) record of *N. gymno-trachelus*, (4) record of *N. fluviatilis*, (5) drainage basin boundaries.

Once more one has an impression of a quite recent and rapid invasion. Danilkiewicz (1996), considering the "natural" entrance of the species into Vistula system, does not exclude the possibility of its introduction by the aquarists, especially when taking into account the recent fast growing, practically uncontrolled import of wild animals from the former Soviet Union countries to Poland.

2.4.3 Neogobius fluviatilis (Pallas, 1811)

This is the most recent Ponto-Caspian fish invader in the Baltic Sea drainage system. *N. fluviatilis* (monkey goby) was first noted in 1997 by Danilkiewicz (1998) in the Bug river near the Polish-Belorussian border downstream of the inflow of the river Muchawiec. This is a right side affluent of Bug connected by the canal with Pripet' i.e. the Dnieper system. Original distribution range of *N. fluviatilis* includes lower and middle reaches of such Ponto-Caspian rivers as Kuban, Don, Dnieper, Boh, Dniester and Danube. It was also recorded in middle reaches of Dnieper affluents - Desna and Pripet' (Berg 1962; Danilkiewicz 1998) from where it evidently entered the Bug river, i.e. the Baltic drainage system. Ichthyofauna of Bug, downstream of its canal connection with Dnieper system, was thoroughly investigated in the period 1971-1985 and in 1995, after the discovery of *N. gymnotrachelus*. Therefore, the record of *N. fluviatilis* only in 1997 would prove its very recent appearance in this river (Danilkiewicz 1998).

3 Conclusions

Except of the cladoceran, *Cercopagis pengoi*, probably introduced to the Baltic Sea with ballast water (Leppäkoski & Olenin 2000), the principal vector for the invasion of Polish fresh and brackishwaters by the remaining Ponto-Caspian invertebrates and fishes, was the removal of natural barriers between European drainage systems, i.e. canal constructions (Haas 1937; Jazdzewski 1980; Kinzelbach 1992). In the case of *P. robustoides* and *O. crassus* the evident vector acting prior to the "natural" invasion or ballast water introduction of these species to Polish Vistula and Szczecin Lagoons, was their intentional introduction in the Neman system in Lithuania in the 1960s (Gasjunas 1965, 1968).

Among the species discussed here we have to do with at least three "waves" of invasions. The oldest invaders have probably entered the Vistula and Oder systems soon after the opening of the 18th century canals joining these systems with the Black Sea catchment basin (Dnieper river), i.e. nearly 200 years ago. These species were probably *Cordylophora caspia* and *Dreissena polymorpha*, easily using boat hull fouling for their travel along canals and rivers.

Two amphipod species, *Chelicorophium curvispinum* and *Chaetogammarus ischnus*, which were recorded outside the Ponto-Caspian basin for the first time in the beginning of 20th century, may belong to the second, younger "wave". *C. curvispinum* attaches its tubes to hard substrates, therefore, boat hulls were a good medium enhancing far westward transport, whereas *C. ischnus* is known as a species commonly associated with zebra mussel clumps (Köhn & Waterstraat 1990).

The third wave of Ponto-Caspian newcomers into the Vistula and Oder drainage areas is evidently a rather recent one. Central European fresh- and brackish water basins of the

Baltic catchment area are recently monitored by the biologists thoroughly enough to assume that the delay in first records of the species of hydro-macrofauna is not longer than 2-5 years. The first records of all these species (*C. pengoi*, *D. haemobaphes*, *D. villosus*, *P. robustoides*, *O. crassus*, *N. melanostomus*, *N. gymnotrachelus* and *N. fluviatilis*) are all from the years 1988-2001, i.e. mostly in the last decade (Skora & Stolarski 1993; Rudolph 1997; Konopacka 1998; Danilkiewicz 1996, 1998; Gruszka 1999; Wawrzyniak-Wydrowska & Gruszka 2001; Konopacka & Jazdzewski, in press).

The invasion of Ponto-Caspian species in European freshwaters, and via freshwaters, into brackish coastal waters of the Baltic and North Seas, is related to their typically oligohaline preferences. Most of these species originally do not inhabit open waters of the Black Sea (salinity 17-18 PSU), but they live in estuaries and lagoons of the Black and Azov Seas of much lower salinity (mostly 0.5-5 PSU) as a relict fauna of the Sarmatian or Pontian Age. The salinity of both semi-enclosed Baltic and Black Seas has relative proportions of major ionic constituents nearly identical to the proportions present in oceanic waters, only of lower total concentration. The fully enclosed Caspian Sea (maximal salinity 12-13 PSU in the central and southern parts, 2-10 PSU in the northern part), has a very peculiar ionic content. Its water is especially rich in sulphates (some 25% of the whole salinity) and much richer in calcium and carbonate ions than oceanic waters, and its ion proportions are somewhat similar to freshwater conditions (Alekin 1956; Pearse & Gunter 1957).

The endemic Ponto-Caspian crustacean fauna is probably of freshwater origin. At present, various species exhibit various grades of euryhalinity. In the Caspian Sea itself over 60 endemic malacostracan species were recorded, of them some 15 species (mostly amphipods) penetrated Volga at different distances , being the most euryhaline taxa (Mordukhaj-Boltovskoj & Dzjuban 1976). Among these 15 taxa are all amphipod species mentioned in this article, and *Dreissena polymorpha.*

The species of the so-called Ponto-Caspian complex (see Mordukhaj-Boltovskoj 1964) do not occur, in general, in the open (central) Baltic Sea, where the surface salinity is 7-8 PSU, or at least do not compete with native fauna. At the moment the exceptions seem to be *Cercopagis pengoi* and especially *Neogobius melanostomus*, which do occur in "full" central Baltic salinity. Species of the Ponto-Caspian complex, such as *Chelicorophium curvispinum, Chaetogammarus ischnus, Pontogammarus robustoides, Dikerogammarus haemobaphes, D. villosus, Dreissena polymorpha,* and *Litoglyphus naticoides* are found only in freshwaters or in oligohaline lagoons, like Vistula Lagoon (salinity 3-5 PSU) or Szczecin Lagoon (0.5-1.5 PSU) and only there may they successfully compete with native fauna, the same is true for Curonian Lagoon, Gulf of Riga and Gulf of Finland.

An interesting question arises - what is the reason to this rather recent massive invasion of Ponto-Caspian species in central and western Europe? May be one of the answers is the increasing ionic content of large European rivers in last decades, caused by the industrial and agricultural pollution. This rise in the "salinity" of such rivers like Dnieper, Vistula, Oder or Danube could finally reach the "critical point" allowing several species of oligohaline preferences to start their rather quick conquest of new basins. Obviously the increasing transport is also responsible, however, when looking for a "trigger" of these invasions and attaining high abundances in whole river flows in comparatively short period, the present authors would rather favour the earlier mentioned hypothesis.

In the hydrofauna of the Vistula and Oder systems several other species of primarily Ponto-Caspian or at least of Pontic distribution range are noted. However, for various reasons these species should not be recognized as invasive ones. In this group the crayfish *Astacus leptodactylus* Eschscholz, 1823, the lamprey *Eudontomyzon mariae* (Berg, 1931), and some fish species, like *Gobio albipinnatus* Lukasch, 1933, *G. kessleri* Dybowski, 1862, *Barbus petenyi* Heckel, 1847, *B. cyclolepis waleckii* Rolik, 1970, *Sabanejewia aurata* (De Filippi, 1865), *Abramis sapa* (Pallas, 1814) and *Umbra krameri* Walbaum, 1792 could be recognised.

A. leptodactylus, a crayfish originally of Ponto-Caspian and Aral basins distribution range (Albrecht 1982; Banarescu 1991-95), was successively transplanted westwards for nearly two centuries to ponds and lakes, because of its commercial value as exquisite food. Consequently, it was patchily gaining new territories in central and western Europe (Kossakowski 1966; Noel 1992; Pöckl 1999). However, this crayfish appeared to be not resistant to the crayfish plague caused by *Aphanomyces astaci* Schikora and, during the 19th century, after the competitive and expansive American crayfish *Orconectes limosus* (Rafinesque, 1817) was introduced to European waters, the number of *A. leptodactylus* localities successively decreased. In Poland the species occurred in scattered localities in the whole country (Lenkowa 1962), but in last decades, due to the pollution and crayfish plague, *A. leptodactylus*, outcompeted by *O. limosus*, becomes a more and more rare species (Krzywosz et al. 1994; Bialokoz et al. 1996).

On the other hand the above-mentioned lamprey and fish species, which were recorded mainly in the Vistula basin, occur there at the northeastern borders of their natural distribution range and cannot be regarded as invasive species (Rembiszewski & Rolik 1975; Rolik & Rembiszewski 1987; Banarescu 1991-95). Witkowski (1989, 1996) recognizes all these taxa as native ones except *U. krameri*, of which several specimens were recorded as accidentally introduced alien fish species (Eysymont & Sloniewski 1969) and has probably not yet established any permanent population.

Studies on the ecological impact of alien species of Ponto-Caspian origin upon the native fauna in the Vistula and Oder systems are still not undertaken. Quantitative studies on the fish and invertebrates diets are urgently needed to estimate this impact. The present paper shows merely the qualitative aspect of these invasions, indicating only their possible routes and dates.

The present ecological status of alien species of Ponto-Caspian origin is diversified. *D. polymorpha* had formerly an important impact upon the ecosystems of many lakes in northern Poland as well as in the Vistula and Szczecin Lagoons (Wiktor 1969; Stanczykowska 1977, 1997; Wisniewski 1990; Stanczykowska & Lewandowski 1993). Mass occurrence of zebra mussel created thick shell deposits along the shores of these waterbodies. At present, however, despite some new records of *D. polymorpha* in some regions of Poland and information on its fouling effect upon the hydrotechnical installations (Abraszewska-Kowalczyk et al. 1999), formerly rich populations of zebra mussel in our lakes and lagoons have decreased recently (e.g., Lewandowski et al. 1997; Stanczykowska & Stoczkowski 1997; Stanczykowska et al. 1997). *Chelicorophium cur-*

vispinum occurs in masses in lower and middle reaches of Oder and Vistula, and less abundantly in large affluents of these rivers, such as Bug, Warta and Notec, whereas *Chaetogammarus ischnus*, although it has penetrated some new localities rather recently (Jazdzewski & Konopacka 1990), is becoming quite rare. In last samplings in the Vistula *C. ischnus* occurred only in low abundances, an evident consequence of the competition with the fresh Ponto-Caspian newcomers, especially *Dikerogammarus haemobaphes*, the species of similar ecological preferences (lito- and phytophilous).

At present the most successful gammarid invaders are *Dikerogammarus haemobaphes* and *Pontogammarus robustoides*. The first is an absolute dominant gammarid in the lower Vistula, except of the Wloclawek dam reservoir where *P. robustoides* gains the dominance in stagnant water conditions. The latter species, together with *O. crassus*, being more tolerant to brackish-water conditions, are now very abundant in the oligohaline Vistula Lagoon, locally competing with success, especially at the southern shores, with native *Gammarus duebeni* and another alien species *G. tigrinus* (Jazdzewski & Konopacka unpubl. data).

Wawrzyniak-Wydrowska & Gruszka (2001) reported similarly very high population densities of *P. robustoides* from the oligohaline Szczecin Lagoon. In gammarid populations of the lower flow of the Oder, *D. haemobaphes* co-occurred with *G. tigrinus* and *D. villosus* and it was a dominant gammarid in the samples taken by Müller et al. (2001).

An impact of mass occurrence of *Neogobius melanostomus* upon the local native fauna in the shallow and sheltered Baltic Bay of Puck, was mentioned above. Recent data (Grabowski & Kostrzewa, pers. comm.) indicate that also the population of *N. gymnotrachelus* has a tendency to increase rapidly in the artificial Wloclawek reservoir, built on the lower Vistula flow.

Ecological success and impact of the remaining Ponto-Caspian newcomers, *C. pengoi*, *D. villosus*, *O. crassus* and *N. fluviatilis*, is not yet sufficiently estimated, although the clogging of fishing nets by the water flea *C. pengoi* was already reported by Hornatkiewicz-Zbik (1999, 2000). It is obvious, however, that the situation can change quickly and the thorough monitoring of the Vistula and Oder basins is of utmost importance. Especially interesting seems to be the case of *D. villosus*, the species whose immigration route into Polish waters was evidently the most complicated (upstream the Danube, downstream the Rhine and eastwards through the German Mittelland canal) but also probably the most rapid one. The data by Van der Velde et al. (2000) indicate that this species can quickly outcompete earlier invaders.

THE CASPIAN-VOLGA-BALTIC INVASION CORRIDOR

YURY V. SLYNKO[1*], LIUDMILA G. KORNEVA[1], IRINA K. RIVIER[1], VLADIMIR G. PAPCHENKOV[1], GRIGORY H. SCHERBINA[1], MARINA I. ORLOVA[2] & THOMAS W. THERRIAULT[3]

[1]*Institute of Biology of Inland Waters, Russian Academy of Science, Borok, Russia*

[2]*Zoological Institute, Russian Academy of Science, St. Petersburg, Russia*

[3]*Great Lakes Institute for Environmental Research, University of Windsor, Windsor, Canada*

[*]Corresponding author *syv@ibiw.yaroslavl.ru*

Abstract

The north-south transfer of species in the Volga River basin is not new, but the scale and nature of invasions changed along the Volga-Baltic corridor following transformation of the Volga River from a riverine environment to one of a series of cascading reservoirs. Southward penetration of northern species was facilitated by the formation of a cold-water hypolimnion in the Volga reservoirs. Following reservoir impoundment, 106 invasive species have been found in the Volga River basin, a process that occurred over two different time periods. The first period of invasions occurred between 1940 and 1970, and involved many northern species (77% of total species) moving downstream by passive dispersal. The second period of invasions is still on going and involves invasions by many Ponto-Caspian species (51% of total species) while new invasions by northern species has decreased substantially (7% of total species). The proportion of exotic species (i.e., invaders originating from basins not adjacent to the Volga basin) increased from 7% during the first period to 41% during the second period. Since the late 1970s, water temperatures in the Volga basin have continued to increase and it is postulated that many invasions during the second period are related to global climatic change.

1 Introduction: the formation of the European part of the Russian intercontinental waterway

We define an aquatic invasion corridor as a system of waterways connecting previously geographically isolated river and sea basins, thereby allowing the active or passive dispersal of aquatic species beyond their historical, natural ranges. Also, available human-mediated vectors play a catalytic role in species dispersal along these invasion corridors. The Ponto-Volga-Baltic invasion corridor was formed along Europe's largest meridional river, the Volga River. The hydrology of the Volga has remained unchanged since the end of the Valday glaciation period (approximately 10,000 BP) and the entire catchment area is part of the Caspian basin (Obidientova 1977). The Volga River basin is divided into three main parts, the Upper, Middle and Lower Volga. The Volga-Ahtuba flood plain and the delta are considered separate parts of the Lower Volga (The Volga and its Life 1978). Berg (1962) attributed the entire basin to the Ponto-Caspian-Aral province of the Mediterranean sub-region while the basin itself belongs to the Palaeo-Arctic region. However, according to Starobogatov (1970) the entire Volga basin belongs to the European-Siberian sub-region and the Lower and Middle Volga are included in the Volga-Ural province, while the Upper Volga is included in the Baltic

E. Leppäkoski et al. (eds.), Invasive Aquatic Species of Europe, 399–411.

province. Indigenous biota of the Volga was formed by the end of Valday glacial period. Until recently, the Volga fauna consisted primarily of freshwater Palaeo-Arctic species with a unknown number of Ponto-Caspian species found in the Lower Volga.

Even before the building of dams, the Volga River was an important transportation route in Russia. This fact greatly contributed to the uniting of territories from the Baltic to Black and Caspian Seas and the creation of a united Russia. Since the time of Peter the Great, attempts have been made to build a waterway connecting the Volga River and the Baltic, White, Caspian, Azov and Black Seas. Between 1703 and 1709 the Vyshnevolotsk Waterway was built to connect one Volga River tributary, the Tvertsa River, with the Msta River. Connections were also established with Ilmen and Ladoga Lakes and the Baltic Sea. In 1718 Vyshnevolotsk Reservoir was impounded, however this waterway never developed into a significant European transportation route. In 1810 the Mariinskaya Waterway was built, connecting the Volga River and the Baltic Sea through the Neva, Sheksna, Kovzha and Vytegra Rivers and through Ladoga and Onega Lakes. In 1811 another waterway, the Tikhvinskaya, was created to connect the Volga River with the Baltic Sea via Rivers Neva, Mologa and Ladoga Lake. In 1829 the Severo-Dvinskiy Canal was opened connecting the Volga River with the White Sea basin via the Sheksna, Sukhona, Northern Dvina Rivers and Lake Kubenskoye. However, due to considerable seasonal and yearly water level fluctuations, regular ship traffic was not possible in this system until the 1940s.

Full regulation of the Volga River flow was not possible until reconstruction resulted in the formation of the United Deep Waterway System of canals that was completed by the former USSR in the 1930s. New inter- and intra-basin canals and a cascade of reservoirs were constructed to maintain a constant depth of 3.5 m, a depth necessary for ship traffic. As a result, the speed of traffic, and consequently the amount of transported cargo, increased. In addition, the White-Baltic and Moscow Canals were built in 1933 and 1937 respectively. The latter formed a circular intra-basin waterway in Central Russia by connecting the upper reaches of the Volga and Oka Rivers. In 1952 the Volga-Don Canal was opened providing a direct connection between the Volga-Caspian Sea basin and the Azov and Baltic Seas.

In 1964 the Volga-Baltic Waterway was fully reconstructed. A series of reservoirs along the Volga River were impounded including Uglich (1940), Rybinsk (1947), Verkhnevolzhskoye (1947), Gorki (1957), Kuybyshev (the largest; 1957), Volgograd (1960), Saratov (1968) and Cheboksary (1981) Reservoirs. Impoundment of these reservoirs resulted in full regulation of the entire river channel except for the Volga-Ahtuba flood plain (The Volga and its Life 1978). In addition, Tsimlyansk (on the River Don), Sheksna, and Vytegra Reservoirs, the system of Kama River Reservoirs and several smaller man-made lakes were impounded to maintain the Moscow Canal.

As a result of this development, the Volga River became the largest transcontinental waterway in Russia. The river is 3,530 km long and its catchment area is in total 1,360 km^2. There are 12 large and more than 300 medium and small reservoirs resulting in a total surface area of 25,660 km^2 and a total volume of 180.5 km^3, of which 85 km^3 is considered usable (Avakyan & Shirokov 1994). The Caspian-Volga-Baltic route is responsible for more than 70% of all cargo transported by the Russian river fleet. With respect to cargo transportation, Volga River ship traffic is second only to the St. Lawrence Seaway (> 300 million tons per year by the mid 1990s; The Problem of Territorial Re-distribution... 1985). About 40% of the fleet transporting goods along the Volga River from the Ponto-Caspian region to the Baltic and White Seas belongs to the "sea-river" class. In 2000 an agreement was signed between Russia, India, Iran and Oman creating a united transportation corridor, termed the "North-South" corridor, in order to provide a transportation route between the Persian Gulf and Indian Ocean to the Baltic Sea (Enactment of the Government of Russian Federation 2000). A primary focus of this global venture is the Volga-Baltic Waterway, a waterway that is expected to handle 50 times more cargo than at present. In 2001 this transportation corridor entered the first stage of active development.

2 Early stage of biological invasions

Since the creation of the canal and the intensified ship traffic during the 18th and 19th centuries, increased invasions by several Caspian species have been noted. This is particularly true for species that prefer solid substrates and either foul ship hulls or are able to actively migrate in the littoral zone. For example, in the late 19th century the bivalve mollusc *Dreissena polymorpha* and the crayfish *Astacus leptodactylus* were found in the Northern Dvina River (Starobogatov & Andreeva 1994) and in the Baltic Sea (Nowak 1951). During the early 20th century, a mysid, *Paramysis ullskyi* and three amphipod species (*Pontogammarus sarsi*, *Dikerogammarus haemobaphes* and *Corophium curvispinum*) were found near the mouth of the Mologa River, a tributary of the Volga River (The Volga and its Life 1978). However, regulation of river flow between the 1940s and 1960s resulted in an even more pronounced invasion pattern.

3 Recent "North-South" invasions

The early invasion sequence consisted of plant and animal species of northwestern limnophilic origins primarily due to the impoundment of the Rybinsk Reservoir, which started in the 1940s. The Volga-Baltic and Severo-Dvinsky Waterways provided major corridors for these invasions (The Volga and its Life 1978). The predominant donor waterbodies were large lakes in the Pskov, Leningrad, Vologda and Archangelsk regions (including Lakes Ladoga and Onega), and estuaries of large rivers in the Baltic and White Sea basins. Successful southward invasions by northern limnophilic species is determined primarily by environmental conditions including deceleration of river flow, formation of large lacustrine water bodies along the river channel (i.e. Rybinsk, Ivankovo and Kuybyshev Reservoirs), pronounced thermal stratification of water masses and existence of a cold water hypolimnion. Therefore passively moving pelagic limnophilic species of northern origin (algae, zooplankton and pelagic fish) have dominated among invaders between the 1940s and 1960s.

3.1 PHYTOPLANKTON AND AQUATIC VASCULAR PLANTS

Between the late 1950s and early 1960s phytoplankton species of the genus *Stephanodiscus* (*S. binderanus*, *S. hantzschii* and *S. minutulus*) increased in abundance from north to south and now dominate the spring-summer algal community of the Volga River (Korneva 1999).

Anthropogenic transformation of river discharge led to an increase in the proportion of limnophilic species, including *S. binderanus* (Korneva 2001), a cosmopolitan indicator of highly eutrophic waters. The abundance of this species in Ivankovo, Rybinsk and Gorki Reservoirs was several orders of magnitude higher than in the unregulated part of the Volga River even during the period after its first appearance (1955-1957). In the Upper Volga reservoirs (Ivankovo, Uglich, Rybinsk and Gorki), the maximum abundance of this species reached 5,880,000 cells l^{-1} and attained a biomass of 2.78 g m^{-3} in 1989-1991 (unpubl. data). In the Middle Volga reservoirs (Kuybyshev and Cheboksary) maximum abundance reached 704,000 cells l^{-1} and biomass reached 1.67 g m^{-3}. In contrast, maximum abundance was 62,000 cells l^{-1} and biomass was 0.109 g m^{-3} in Saratov and Volgograd Reservoirs and unregulated parts of the river. This finding supports the hypothesis that the southward decreasing abundance of *S. binderanus,* first identified during the 1950s. is still apparent. By the end of the 1960s, after all main Volga-Kama reservoirs were impounded, the proportion of small-sized species including *S. hantzschii* and *S. minutulus* increased. These species are cosmopolitan, alkaliphilic and tolerant to high concentrations of readily degradable organic matter (α-mesosaprobes) (Korneva & Genkal 2000). In 1969-1972 the average number of *S. hantzschii* (syn: *S. tenuis*) decreased from 250,000 to 81,000 cells l^{-1} in the Upper Volga, to 65,000 cells l^{-1} in the Middle Volga and to 0 cells l^{-1} in the Lower Volga. Among higher aquatic plants only a northern species, *Potamogeton wolffgangii,* spread far to the south, into Kuybyshev Reservoir and the Kama River mouth (Papchenkov 1997).

3.2 ZOOPLANKTON

The Volga River zooplankton was enriched by 18 species of northern invaders: *Heterocope appendiculata*, *Eudiaptomus gracilis*, *E. graciloides*, *Eurytemora lacustris*, *Cyclops kolensis*, *Limnosida frontosa*, *Daphnia cristata*, *D. longiremis*, *Bythotrephes longimanus*, *Synchaeta verrucosa*, *S. lakowitziana*, *Conochiloides natans*, *Keratella hiemalis*, *Bosmina coregoni*, *B. longispina*, *B. kessleri*, *B. obtusirostris*, and *B. crassicornis* (The Volga and its Life 1978). Impoundment of Uglich and Ivankovo Reservoirs facilitated invasion by planktonic crustaceans belonging to the central limnofaunistic complex. The proportion of "northern invaders" at these sites was relatively low: *B. longimanus* and *H. appendiculata* were represented by single specimens and *C. kolensis* was rare. In the cascades' northernmost reservoir (Rybinsk), conditions were more favourable for the establishment of the northwestern limnofaunistic complex. Between 1946-1948 (i.e., about 5 years after impoundment) the regular inflow of northern invaders to this lacustrine waterbody allowed the development of large populations of *D. cristata*, *B. longispina*, *B. longimanus*, *E. gracilis*, *E. graciloides*, and *H. appendiculata* (The Rybinsk reservoir 1972). These new zooplankton species of northern origins formed the base of a new food chain for planktivorous fish (The Volga and its Life 1978; Rivier 1993).

In the Volga River cascade, zooplankton of northern origins formed two complexes: the winter complex that actively develops under ice cover and during early spring (8 forms of genus *Notholca*, *Synchaeta verrucosa*, *S. lakowitziana*, *Conochiloides natans*, *Keratella hiemalis*, *Cyclops kolensis*, *Daphnia longiremis*, *D. cristata*, etc.) and the summer stenolimnophilic complex that develops following spring turnover (*Heterocope appen-*

diculata, *Limnosida frontosa*, *Bosmina longispina*, *Bythotrephes longimanus*, etc). In the Upper Volga, especially in Rybinsk Reservoir, the winter complex dominates from December until June, while in Volgograd Reservoir; the duration of the winter complex is considerably shorter (Rivier 1986).

3.3 FISH SPECIES

Two fish species belonging to the Arctic fish complex, the cisco *Coregonus albula* and the smelt *Osmerus eperlanus*, naturalized in Volga River soon after the impoundment of Rybinsk Reservoir (Ivanova 1982). Prior to reservoir impoundment both species occurred only as relict populations in some lakes in the Upper Volga basin (Lakes Seliger, Peno, Volgo, and Plescheevo). Both species appeared in Rybinsk Reservoir from Lake Beloye and from there they spread widely through Rybinsk, Uglich and Ivankovo Reservoirs. From the Upper Volga basin both species moved downstream along the Volga River and by the late 1970s smelt had reached Saratov Reservoir and cisco had reached Volgograd Reservoir (Yakovleva 1975; Nebol'sina 1975). Until 1995-1998 smelt dominated the pelagic fish complex of the three main reservoirs in the Upper Volga basin and was an important food item for pelagic predators including zander, Volga zander, asp, perch and burbot (Ivanova 1982). In 1989, in Beloye Lake (River Kem' mouth) the nine-spined stickleback, *Pungitius pungitius* was identified as a new invader. In 2001 single specimens of this species were found in Rybinsk Reservoir (River Ukhra mouth). In 2001 the river flounder (*Platichthys flesus*) was identified in Gorki Reservoir. Probably this species will penetrate the Upper Volga reservoirs from the north via the Volga-Baltic corridor.

4 Recent "South-North" invasions

River flow and powerful rapids in the Samara Luka region limited the upstream expansion of Ponto-Caspian species. Impoundment of lower reservoirs in the Volga cascade and construction of locks led to the removal of rapids as natural barriers for species dispersal to upstream locations and transformed the Volga River into a deep-water waterway. The presence of a continuous waterway in conjunction with the opening of the Volga-Don Canal facilitated increased shipping activities. Stabilization of reservoir ecosystems and intensification of ship traffic increased the rate of invasions by Ponto-Caspian species to the White and Baltic Seas basins. Turning the Volga River and other large rivers within the basin into a cascade of limnic, interconnected waterbodies resulted in the formation conditions similar to those during late Pleistocene Khvalyn transgression of the Caspian Sea, when many Ponto-Caspian species were found at latitudes comparable to Moscow (Mordukhai-Boltovskoi 1960). Removal of natural hydrochemical barriers such as flood plain terraces led to increased mineralisation (Tyuryukanov et al. 1996) and nutrient inputs to the Volga Reservoir (The Volga and its Life 1978) and has facilitated the dispersal and establishment of Ponto-Caspian species upstream in the Volga River. Global warming is considered to be another important factor facilitating the northward range expansion of southern species. For example, in Rybinsk Reservoir the average water temperature has increased since the 1960s from 12 °C to 15 °C (unpubl. data). Since the late 1970s there has been both qualitative and quantitative increases in Ponto-Caspian species (i.e., the "southern" flow of invaders and successful naturalization of some of the exotic subtropical and tropical species became obvious (Fig. 1, 2). In contrast, the "northern" flow of invasions by euryhaline zoobenthos and benthophilic fish species of the estuarine complexes of the Caspian, Azov and Black Seas have dominated for the last four decades.

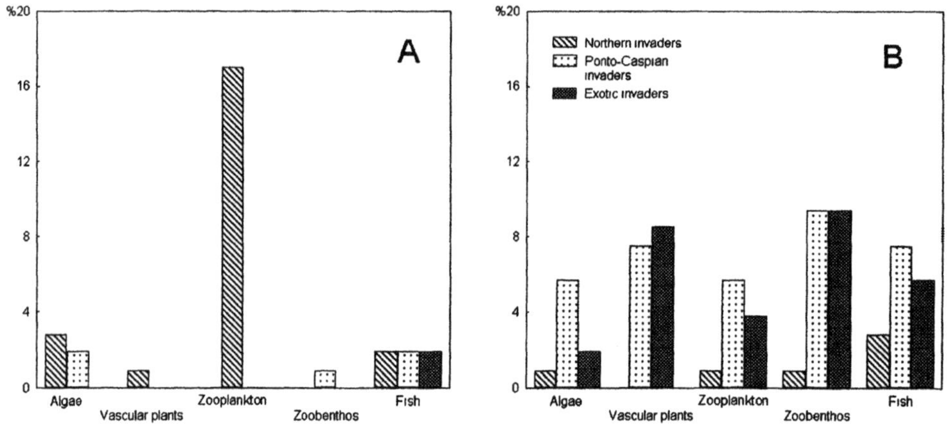

Figure 1. Relative shares of nonindigenous animal and plant species in the Volga River basin in the 1940s-1970s (A) and 1970s-2000s (B).

4.1 PHYTOPLANKTON

By the late 1960s, a new euryhaline alga, *Sceletonema subsalsum,* was found along the entire Volga River (Genkal & Kuzmin 1980) but was most abundant at downstream locations. Maximum abundance and biomass of this species is usually observed during the summer-spring period at water temperatures between 10 and 22 °C. This cosmopolitan species prefers stagnant waters with high concentrations of organic matter (Korneva 2001). Mass development of *S. subsalsum* was noted in 1954-1964 during the summer period in the near-mouth and central parts of the Northern Caspian Sea. This species is a common component of the algal communities in the Caspian and Azov Seas (Makarova 1969). In the 1960s it spread northward from the brackish seas to the Volga River reservoirs and it successfully conquered its own ecological niche. At the same time, a Caspian Sea species, *Thalassiosira incerta,* rapidly spread through the Volga River reservoirs. This species is euryhaline and eurythermal and its abundance decreases from south to north. Currently, this species is found in all reservoirs of the Volga cascade (Genkal 1992). However, it is numerous only in the Lower and Middle Volga reservoirs (unpubl. data for 1989-1991). In the late 1960s, other brackish water species of the genus *Thalassiosira* (*T. pseudonana, T. guillardii* and *T. weissflogii*) appeared in algal communities of the Volga River (The Volga and its Life 1978). These species are scarce in reservoirs and can be identified only by using electron microscopy. Since the mid 1980s the distribution of the brackish water Caspian species *Actinocyclus normanii* was noted in the Volga reservoirs (Korneva 2001). Its appearance is explained by climatic changes that have resulted in increased Volga River discharge, increased Caspian Sea levels, decreased salinity in its northern part and changes in water ionic composition.

Further dispersal of *S. subsalsum T. incerta* and *A. nordmanni* northward is now restricted north of Kuybyshev Reservoir.

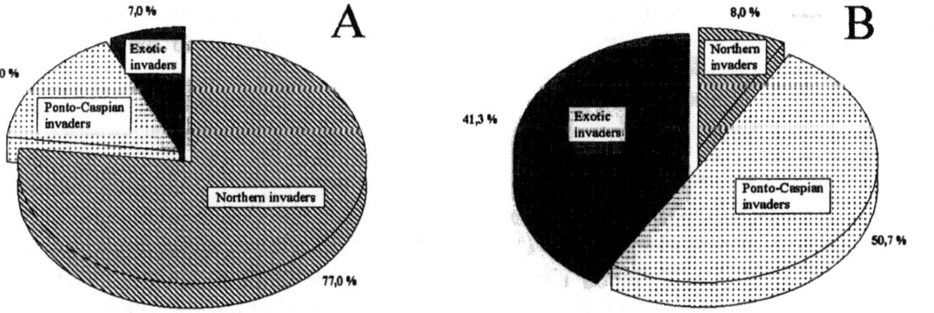

Figure 2. Relative shares of northern, Ponto-Caspian and exotic invaders in the Volga River basin in the 1940s-1970s (A) and 1970s-2000s (B).

4.2 AQUATIC VASCULAR PLANTS

Since the 1970s-1980s the northward spread of nonindigenous vascular plants became more intense. Among these invasive plants are such typical southern hydrophytes as *Lemna gibba*, *Vallisneria spiralis* and *Zannichellia palustris*, helophytes *Alisma gramineum*, *Bolboschoenus koshewnikowii*, *Phragmites altissimus*, *Scirpus tabernaemontanii*, and *Typha laxmannii*. Except for *P. altissimus* and *T. laxmannii* all these species are already present in the Upper Volga (e.g., *A. gramineum* became dominant in Rybinsk Reservoir). Duckweed, *L. gibba,* is numerous in the Yaroslavl part of the Volga basin although 10-15 years ago it was rare. Relatively rapid northward spread along small rivers and ponds also has been demonstrated by *Z. palustris*. *B. koshewnikowii* and *S. tabernaenontanii*. *V. spiralis,* native to the Volga delta, was reported from cooling ponds of thermal electric power plants; now it can be found far to the north, into the Yaroslavl region (Lisitsyna et al. 1993). *T. laxmannii*, previously inhabited salt marshes, meadows and banks in the southern region of the Volga River, has since appeared near the city of Kazan (Papchenkov 1985) and the Ponto-Caspian species *P. altissimus* has been found near the city of Ulyanovsk.

4.3 ZOOPLANKTON

Ponto-Caspian species are a relatively minor component of zooplankton communities of the Volga River. Their distribution northwards takes place slowly and such species have only been found in Kuybyshev Reservoir. In mid summer, *Heterocope caspia* is numerous in this reservoir. Two species from different limnofaunistic provinces, *H. appendiculata* and *H. caspia* also inhabit this reservoir (The Kuybyshev Reservoir 1983). Single specimens of *Cornigerius m. maeoticus* were reported in 1991 and 1995 (A. F. Timokhina pers. comm.). The Caspian species *Calanipeda aquae-dulcis* now inhabits the Lower Volga, including Volgograd Reservoir (Vyushkova & Gurova 1968). In addition, two cercopagid species also were found in this reservoir: *Cercopagis pengoi* from the Sea of Azov and the confamiliar "northern invader" *Bythotrephes longimanus*. Both species co-occur and have similar feeding types, but their trophic relationships remain to be determined. Volgograd and Tsimlyansk Reservoirs are inhabited by two podonid subspecies of Azov Sea origin: *Podonevadne trigona ovum* and *Cornigerius m. maeoticus* (Vyushkova 1971). In the Caspian Sea other subspecies are found including *P. t. trigona* and *C. m. hircus* suggesting that polyphemoids of the Sea of Azov origin have invaded the Volga Reservoir cascade.

4.4 ZOOBENTHOS

In addition to two mussel species, *Dreissena polymorpha* and *D. bugensis,* natural expansion of two oligochaetes, *Potamothrix vejdovskyi* and *P. heusheri* were reported. These species first appeared in the Upper Volga reservoirs in 1969 (Semernoy 1974). The isopod *Jaera sarsi* moved along the Kama River upstream to the Belaya River mouth (Lyakhov & Mordukhay-Boltovskoy 1973). The Caspian gastropod *Theodoxus pallasi* was found between the cities of Volgograd and Saratov, at the southernmost extent of its range (The Volga and its Life 1978). The Ponto-Azovian estuarine gastropod *Lithoglyphus naticoides* invaded the Volga delta in 1971 (Biserova 1996) and has since been observed in Gorki and Kuybyshev Reservoirs (unpubl. data). Several amphipods typical of the Lower Volga River and the Volga Delta expanded their ranges upstream including *Pontogammarus robustoides* (found near Volgograd), and *P. crassus* and *Stenogammarus deminutus*, both found between Volgograd and Saratov (Belavskaya & Vyushkova 1970). Recently, up to 14 Ponto-Caspian gammarid species inhabited the Kuybyshev Reservoir (Pirogov et al. 1990). Gammarids (13 species) have also penetrated the Kama cascade of reservoirs. Many of these benthic macroinvertebrate species have reached high abundance and biomass and play a significant role in the functioning of Volga reservoir ecosystems.

The zebra mussel *D. polymorpha* was repeatedly found in the Upper Volga basin during the last century. However, its mass development was noted in this system only following impoundment of the Rybinsk Reservoir (Ovchinnikova 1954). Found for the first time in 1954, this species occupied the entire waterbody by 1968 and by 1980-1990 *Dreissena*-dominated communities occupied about 25% of the bottom area in the deep-water part of Rybinsk Reservoir. In the early 1980s average biomass of *D. polymorpha* reached 594 g m^{-2} and by the early 1990s it had increased to 1,341 g m^2. The zebra mussel biomass in Rybinsk Reservoir was in total 732,623 tons in 1990. First reports of zebra mussels in Ivankovo Reservoir were made in 1953 (Fenyuk 1959). In the other Volga reservoirs, the invasion of this species was faster: by the second year after its first appearance it became dominant in many reservoirs (The Volga and its Life 1978; The Kuybyshev reservoir 1983). By the early 1990s, *D. polymorpha* dominated benthic communities of canal sections of reservoirs and were often found in deep water habitats of the limnic parts of these reservoirs (Scherbina et al. 1997). Being a powerful filter-feeder zebra mussels intercept considerable organic matter preventing its sedimentation in the profundal zones of lakes and river canals. Metabolic by-products of the mussels (agglutinates and feces) are a primary food source for many detritivores. The mussels play an important role as prey items for molluscivorous fish. The roach is the most active consumer of zebra mussels in reservoirs and the growth rate and condition factor of this species increase considerably when feeding on zebra mussels.

Another freshwater dreissenid, the quagga mussel *Dreissena bugensis*, was until recently only found in European estuaries, reservoirs and man-made canals of the Black Sea basin (Mills et al. 1996). In the Volga basin (Kuybyshev Reservoir), this species was found for the first time in 1992 (Antonov 1993). Since 1994 established populations of the quagga mussel were identified in the Volga delta and in low-brackish, shallow waters of the Northern Caspian Sea. Presumably, invasion of this species took place in the 1980s by ship traffic from Black Sea estuaries via the Volga-Don waterway (Orlova et al., 1998). The quagga mussel was found for the first time in Rybinsk Reservoir during 1997 (Orlova et al. 2000).

A group of Ponto-Caspian species that were unable to reach upstream locations of the Middle Volga River, were able to "jump" over the Upper Volga and the Volga-Baltic system of canals directly to the eastern Gulf of Finland, the terminal site of this invasion corridor based on ship traffic. Either within ballast water or on ship hulls, the hydroid polyp *Cordylophora caspia* and the oligochaetes *Paranais frici, Potamothrix vejdovskyi,* and *P. heusheri* penetrated the Neva Estuary and the eastern Gulf of Finland and are now established (Panov et al. 1999). *D. polymorpha* and the fish hook water flea *Cercopagis pengoi* became dominant species in the eastern Gulf of Finland in the 1990s. The Chinese mitten crab *Eriocheir sinensis* was noted during the 1990s in the Volga River delta. In 2001 adult specimens were found for the first time in Cheboksary and Rybinsk Reservoirs (unpubl. data), suggesting crab invasions in Volga River reservoirs originating from both southern (River Don and Volga River delta) and northern locations (Gulf of Finland).

4.5 FISH SPECIES

The increasing rate of invasions is evident for some Ponto-Caspian fish species. In the mid 1960s, the pipefish *Syngnathus nigrolineatus* penetrated the Volga basin through the Volga-Don Canal. From Tsimlyansk Reservoir this species expanded its range both down- and upstream in the Volga River. It has since naturalized in all Lower Volga reservoirs, the Volga-Ahtuba flood plain, the Volga Delta in the coastal part of the Northern Caspian Sea, and in Kuybyshev Reservoir (Sharonov 1971). Several Caspian species, including several species of gobies, two species of shad and the small southern stickleback *Pungitius platygaster* have move northwards. The Caspian shad *Alosa caspia* was noted in the Saratov Reservoir, the goby *Benthophilus stellatus* in the Gorki Reservoir, two other goby species, *Neogobius iljini* and *N. melanostomus* and the small southern stickleback have reached the Rybinsk Reservoir (Slynko et al. 2000). The Caspian kilka *Clupeonella cultriventris* has shown the most striking rate and intensity of range expansion, first noted in the late 1950s this species had naturalized in Kuybyshev Reservoir by the mid-1960s. In the mid-1970s, it was found in the Gorki Reservoir and in the early 1990s it was found in the Rybinsk Reservoir. Kilka became the dominant species of pelagic planktivorous fish assemblages in almost all reservoirs of the Volga cascade, including the Rybinsk Reservoir. The northward expansion of the kilka is still continuing. In 2000 established populations were identified in Ivankovo Reservoir and in 2001 it dominated fish communities from Sheksna Reservoir to Beloye Lake.

Some fish species have expanded their ranges beyond the Volga reservoir system. For example, in 1994 in Ivankovo Reservoir we found the Caspian goby *Neogobius melanostomus,* while this species is absent from downstream reservoirs (Uglich and Rybinsk). This finding suggests that this species moved from the Middle Volga through the River Oka - River Moscow - Moscow Canal system. The same pattern has been observed for the bitterling *Rhodeus sericeus* which moved beyond its original range downstream to the Volga River and its delta (The Astrakhan Reserve 1991) and upstream to the Ivankovo Reservoir (Slynko et al. 2000). Detection in 2001 in Kuybishev Reservoir of a river flounder, supports the hypothesis that invasions in the Volga reservoir system occur from both northern and southern locations.

5 Invasions of thermophilic species

In addition to species invasions of waterbodies adjacent to the Ponto-Volga-Baltic corridor, the invasion by species originating from tropical and subtropical, eutrophic and highly mineralised waters has increased, mainly due to the discharge of large volumes of thermal effluents. Factors facilitating the distribution of termophilic species accidentally or intentionally acclimatized in different regions of the corridor have also increased. During 1989-1991 in Kuybyshev and Cheboksary Reservoirs, two new *Thalassiosira* species (*T. faurii* and *T. gessneri*) were identified (Genkal & Korneva 2002). The first is known from Central African lakes while the second is known from North American rivers. In 1990, *Hemiaulus* sp., a marine species was also identified in the Kuybyshev Reservoir. Several species of thermophilic vascular plants have spread in the Volga basin: *Potamogeton biformis* (originating from Kazakhstan and spread upstream to the Oka River mouth), *Butomus junceus* (from the southern part of West Siberia) (Lisitsyna & Papchenkov 2000), *Persicaria hypanica* (from the Ukraine and now spreading upstream through the Volga to Rybinsk Reservoir). Also, the North American invaders *Elodea canadensis* and *Epilobium adenocaulon* have developed large abundances in this system (Skvortsov 1995). During the last decade rapid range expansion was noted for *Bidens frondosa*, native to North America, and is currently approaching Rybinsk Reservoir. Other rapid invaders include *Zizania latifolia*, a Far Eastern species introduced in the European part of Russia in 1934 (Matveev & Solovyeva 1997). During the 1950s-1970s numerous attempts were made to introduce this species to waterbodies of game territories, shallow parts of newly impounded reservoirs and ponds. These attempts were usually unsuccessful, resulting in a lack of further interest. However, in the 1980s-1990s, this species exhibited "self-expansion" from regions of its intentional introductions. In Ivankovo, Uglich and Gorki Reservoirs this species has spread over large areas and is now replacing aboriginal species. Recently, several attempts have been made to intentionally introduce the tropical species *Pistia stratioides* and *Eichhornia crassipes*. However, it is unlikely these introductions will be successful despite reports of *P. stratioides* in 1989 in one water body in the Volga delta and in 1991 in waters around the city of Astrakhan (Barmin & Kuzmina 1993).

Since the early 1970s exotic zooplankton species have been reported in the Upper Volga River including *Brachionus* sp. (typical of polluted waters), *Lecane bulla*, *Hexartra mira*, *Keratella tropica, Moina* sp. etc. Dramatic increases in the number of *Moina micrura* was noted in Ivankovo Reservoir during the summer of 1973 when temperatures were well above average and water levels were low. Increased abundance of the rotiferan *Brachionus* was also noted in the Rybinsk Reservoir during hot, low water years (1972, 1973, 1989), when the impact of anthropogenic pollution became especially pronounced.

6 Intentional introductions

Between 1948 (when intentional introductions were started) and 1963 (when introductions drastically increased), 66 species of benthic macroinvertebrates have been intentionally introduced to waters of the former USSR including the Volga basin (Karpevich 1975). The Caspian mysids *Paramysis lacustris*, *P. intermedia*, *P. ullskyi* and *P. baeri* were introduced extensively into Kuybyshev, Volgograd and Rybinsk Reservoirs. In Rybinsk Reservoir the mysids did not survive the first winter after introduction and further introductions were stopped. The shrimp *Palaemon modestus*, native to the Far East, was intentionally introduced into the Rybinsk Reservoir but following two failed attempted introductions, one in 1957 and the other in 1959 no additional attempts were made (The Volga and its Life 1978). Between 1958-1965 more than 15.7 million mysids were introduced into the Kuybyshev Reservoir and more than 14.4 million were introduced into the Volgograd Reservoir (Ioffe 1968). None of the introduced mysid species became dominant among higher crustaceans in either Kuybyshev or Volgograd Reservoir. Currently, the amphipods *Dikerogammarus haemobaphes* and *Pontogammarus obesus* dominate reservoir benthic communities.

In addition to mysids, in 1960 two polychaete species, *Hypania invalida* and *Hypaniola kowalewskyi*, in total 15,400 specimens were introduced to Volgograd Reservoir from the Don delta (Ioffe 1968). The latter species was never found again. It should be noted that there were no intentional introductions of *H. invalida* into other Volga reservoirs. This species appeared in the Lower Volga basin benthos in 1968 (S. V. Danilova pers. comm.). In 1977 *H. invalida* was found for the first time in two reservoirs (Saratov and Kuybyshev) in densities up to 1,000 ind m^{-2} at 25-30 m depth (Dzuban & Slobodchikov 1980). At that time, in Saratov Reservoir its numbers were low (20-90 ind m^{-2}), suggesting a recent appearance in the waterbody. However, by 1979 the number of *Hypania* had increased to 1,200-1,400 ind m^{-2}, reaching a level typical for the silty sediments of the deep-water zone of Kuybyshev Reservoir. In 1993, *Hypania* was found in the Moskva River (Lvova et al. 1996) and in 1994 in the Uglich Reservoir (Scherbina et al. 1997). Thus, between 1971-1994 *H. invalida* invaded a major part of the Volga from the Volgograd to the Ivankovo Reservoir and, via the canal, appeared in the Moskva River. Individual body weight and fecundity of this species increased significantly along its northward invasion pathway (Scherbina 2001). The dwelling tubes are placed either between shells in zebra mussel clusters or above the bottom surface. This makes the worm readily available for many benthivorous fish. In 1955-1956 70,000 specimens of the bivalve mollusc *Monodacna colorata*, naturally inhabiting the Caspian and Black Seas were introduced into Veselovsky Reservoir in the Don basin. In 1965-1970, 1.6 million specimens were released into Kuybyshev Reservoir, where it is now established. In 1967 *M. colorata* was found in the Saratov Reservoir.

In 1962-1965, the Baikalian gammarids *Microropus possolskyi* and *Gmelinoides fasciatus* were intentionally introduced in the Gorki Reservoir. *G. fasciatus* naturalized successfully and inhabits many waterbodies in north-western Russia. In 1986 it was found for the first time in the Rybinsk Reservoir and in 1994 in the Beloye Lake (Scherbina et al. 1997). In its natural range this species prefers shallow (2-3 m) waters overgrown by vegetation, where it lives at depths of 6-10 m in *Dreissena*-dominated reservoir communities (The Kuybyshev reservoir 1983). The greatest abundance and biomass of *G. fasciatus* (54,000 ind m^{-2} and 160 g m^{-2}, respectively) are known from Lake Ladoga (Panov 1996). Since first appearing in Upper Volga River reservoirs, *G. fasciatus* has become a primary prey species for young perch, ruffe, silver bream, roach and other benthivorous fish. To date, none of the intentionally introduced mysid species have established in the Volga River reservoirs, including Gorki. The Baikalian gammarid *G. fasciatus* and the Caspian polychaete *H. invalida* have spread considerably through the Volga basin since the mid-1980s. The amphipod *G. fasciatus* expanded its range due to high mobility as an adult, while the settled worm (lacking a mobile larval stage) expanded its range due to bottom trawling in zebra mussel clusters (Scherbina 2001). Intentional introduction of *G. fasciatus* to Karelian lakes in the 1970s resulted in "self-introduction" to Lake Ladoga in the 1990s and has since been reported from the Neva river estuary during 1997 (Alimov et al. 1998). Recently, *G. fasciatus* has invaded all ecosystems in the northern part of the Ponto-Volga-Baltic corridor including the Kuybyshev Reservoir downstream.

Although more than 20 species of fish were intentionally introduced in the Volga basin, a majority of these introductions were not successful (Kudersky 2001). Ecological conditions in the Volga-Baltic corridor were favourable only for species such as the common carp *Cyprinus carpio*, eel *Anguilla anguilla* and grass carp *Ctenopharyngodon idella*. In addition, some species introduced accidentally have also succeeded including Amur sleeper *Perccottus glenii*, *Pseudorasbora parva* and the guppy *Poecilia reticulata*. After the intentional introduction in 1965 to the Volga delta, the grass carp invaded upstream rivers and as early as the late 1960s it had successfully naturalized in the Kuybyshev Reservoir.

In the 1960s, *A. anguilla* was intentionally introduced in Lake Seliger and since the 1970s it has been found along the entire Volga River and desalinated parts of the Caspian Sea. Historically, there have been several attempts to intentionally introduce carp in the Middle and Upper Volga River reservoirs but until the 1990s these attempts were unsuccessful. During the 1990s, carp have formed sustainable populations in Gorki and Rybinsk Reservoirs where they exhibit both increased abundance and expanded ranges in the Upper Volga (Slynko et al. 2001). *P. glenii* was introduced accidentally in small water bodies; and has expanded its range in the Middle and Lower Volga basins (Evlanov et al. 2000) including Ivankovo Reservoir. The invasion of this species as well as of the guppy is facilitated by the abundance of thermal effluents in the Volga basin and continuing uncontrolled introduction of game fish and unwanted aquarium species. Similar reasons for the sleeper expansion are given for the Leningrad region (Alimov et al. 1998). A fast expansion of *P. parva* became evident in the Don River basin, including Tsimlyansk Reservoir during the 1990s.

7 Conclusions

The major factor determining species invasions along the Volga River is shipping activities along this continuous deep-water system connecting the Ponto-Caspian region and northwest Russia. Intended introductions and accidental releases, as well as secondary natural and human-mediated dispersal of introduced organisms, also contribute significantly to xenodiversity along the Ponto-Caspian - Volga-Baltic Waterway. Reconstruction of paleolimnological conditions of the Khvalyn and Novo-Caspian transgressions, have changed riverine conditions to limnetic ones. The disappearance of natural hydrological and chemical barriers by the transformation of the Volga River into a cascade of reservoirs has facilitated the establishment of invasive species along the entire corridor. As a result the "two-way" trans-European Volga-Baltic invasion corridor became a new habitat for both species of northern origins and those of Ponto-Caspian origins. Since the late 1970s, the influence of global warming and thermal pollution from power plants, as well as eutrophication, increased along the Volga River and facilitated the establishment of thermophilic and mesosaprobic species. The expected intensification of shipping traffic along the United Deep Waterway System and the development of this transportation route into the Persian region, will create an invasion corridor for predominantly euryhaline and stenolimnophilic organisms from Atlantic and Indo-Pacific coastal regions via this new transcontinental route. Continued global warming and the recent Caspian Sea transgression itself, suggest further acceleration of species invasions along the Ponto-Volga-Caspian corridor.

It is false to say that every species that could have been introduced would be in our waters by now. The timeline of introductions does not suggest a saturation level. The chance of a species to become introduced, established and then to become a serious problem for the environment or economy at the recipient coast is small. But, one single introduced species may be able to cause severe ecological change and economic damage, and this might be the next species about to arrive.

Impacts

ECOLOGICAL IMPACT OF PONTO-CASPIAN INVADERS IN THE BALTIC SEA, EUROPEAN INLAND WATERS AND THE GREAT LAKES: AN INTER-ECOSYSTEM COMPARISON

HENN OJAVEER[1*], ERKKI LEPPÄKOSKI[2], SERGEJ OLENIN[3] & ANTHONY RICCIARDI[4]
[1]*Estonian Marine Institute, Tallinn, Estonia*
[2]*Department of Biology, Åbo Akademi University, Turku, Finland*
[3]*Coastal Research and Planning Institute, Klaipeda University, Klaipeda, Lithuania*
[4]*Redpath Museum, McGill University, Montreal, Canada*
[*]Corresponding author *henn@sea.ee*

Abstract
We review the ecological impacts of Ponto-Caspian invaders in the Baltic Sea, selected lakes and rivers in inland Europe, and the North American Great Lakes. Each of these regions has been invaded multiple times in recent decades by Ponto-Caspian invertebrates and fishes. In attempt to identify predictable patterns, we compared the magnitude and direction of the impacts of these invaders on ecosystem components (phytoplankton, zooplankton, benthos and fish) and processes within each region. Most introductions appear to have increased local species richness rather than replace native species. Their impact on functional (rather than taxonomic) diversity, however, is of ecological importance. Ponto-Caspian organisms have become dominant members of various trophic levels (herbivores, detritivores and consumers) in benthic and pelagic food webs. They have apparently caused large-scale ecosystem impacts in the Great Lakes, whereas their role in Baltic Sea ecosystem is less pronounced, more spatially limited and, in part, camouflaged by other long-term ecological changes such as eutrophication. In all three regions, Ponto-Caspian invaders have altered multiple abiotic and biotic components and energy flow of ecosystems. Owing to the paucity of studies using pre- and post-invasion datasets, we cannot yet produce robust predictions for most of these invaders.

1 Introduction

During the past 200 years, biological invasions associated with human activities have resulted in largescale mixing of previously isolated biotas. Species endemic to Ponto-Caspian basins (Black, Caspian and Azov seas and the adjacent rivers emptying into these seas) have become established in inland Europe, the Baltic Sea and, most recently, the North American Great Lakes. In Europe, the Caspian and Ponto-Azovian fauna have spread in different manners and at different geological periods to Turkey, the Aral ba-

E. Leppäkoski et al. (eds.), Invasive Aquatic Species of Europe, 412–425.

sin, the Euphrates river, European river systems, and the Adriatic Sea (Bacescu 1966). This expansion was facilitated by two major activities. Firstly, the construction of numerous canals and reservoirs on Ponto-Caspian rivers allowed organisms to disperse to Central and West European river systems through previously disconnected waterways by active migration, attachment to hulls and barge traffic, or transport in ship ballast tanks (Jazdzewski 1980). Secondly, Ponto-Caspian crustaceans were transplanted as food to stimulate fish production in European lakes and reservoirs during the 20th century; within an immense former USSR program of transplantations of Peracarida, beginning about 1950, more than 30 amphipod species from the "Caspian complex" were used for acclimatization purposes (Gasiunas 1964). Consequently, a biased selection of over 40 Ponto-Caspian species have expanded their ranges from the southeast European source pool into central and western Europe (Jazdzewski 1980; Kinzelbach 1995; Olenin & Leppäkoski 1999). The colonization of European ports by these species led to their immigration to North America by transoceanic shipping. Most nonindigenous species discovered in the Great Lakes since the mid-1980s are Ponto-Caspian endemics and more are expected to arrive in the future (Ricciardi & MacIsaac 2000).

There are many documented examples of invasions altering aquatic communities and ecosystems (e.g., Van den Brink et al. 1993a; Strayer et al. 1999). One approach to develop useful predictions about their impacts is to compare the effects of multiple, spatially independent invasions of a given species (e.g. Grosholz & Ruiz 1996; Ricciardi et al. 1997, 1998). This empirical approach tests whether a species has similar, predictable impacts in different regions. Here we apply this approach to a group of Ponto-Caspian species invading inland Europe, the Baltic Sea, and the North American Great Lakes, these areas representing a model in which three previously isolated biotas have become exposed to each other through human-mediated introduction of species.

2 Similarities and differences between regions

The three basins compared in this review exhibit many of striking similarities, but also evident differences (Table 1). They are isolated from the ocean by physical and ecological barriers. Geologically and biologically, they are young formations, and natural changes are still taking place when their ecosystems are subjected to maturation processes (Leppäkoski & Mihnea 1996). Similar threats to the aquatic environment are present in these areas, e.g., increasing urbanisation, pollution and eutrophication due to heavy industries, intense agriculture and aquaculture, overexploitation of fish stocks, and abundant sea traffic and port activities.

3 Invasive species of Ponto-Caspian origin in the three areas compared

Ponto-Caspian species contribute significantly to animal life in most of West European rivers (Table 2). Tittizer (1996) listed 31 nonindigenous invertebrate species from German rivers, among them 11 species of Ponto-Caspian origin. Amongst macroinvertebrate species in German rivers, their number has been estimated at 6-14%, with the rivers Rhine, Neckar and Odra being the most invaded (> 12%); in lowland rivers and canals close to the North Sea coast their prevalence increases to 20-30% (oligochaetes and chironomid larvae excluded; Tittizer 1996). Among the 29 nonindigenous fish species recorded in the rivers Elbe and Rhine, there are five species originating from the

Black Sea basin via the river Danube (Lelek 1996). Bacescu (1966) reported 16 Caspian invertebrate species from the Danube, 1,000 km from the Black Sea coast, some of which are recent invaders in the West European rivers and/or the Baltic Sea (e.g., the flatworm *Palaeodendrocoelum romanodanubialis,* polychaete *Hypania invalida,* amphipods *Pontogammarus obesus, Corophium curvispinum* and mysid *Limnomysis benedeni*) whereas some other species (mysids *Paramysis lacustris* and *P. intermedia*) were restricted to the lower 200 km part of the river.

Table 1. Basic characteristics for the Black Sea (together with its tributaries, the core donor area for invasive Ponto-Caspian species), the Baltic Sea, and the Laurentian Great Lakes. Data for Baltic and Black Seas from Leppäkoski & Mihnea (1996), those for the Great Lakes from Beeton (1965), Vollenweider et al. (1974) and Charlton (1991).

	Black Sea	Baltic Sea	Great Lakes
Latitude (°N)	41-46	54-66	41-49
Surface area, km^2	423,500	412,600	244,160
Drainage area, km^2	2,405,000	1,729,000	521,830
Age[a], yr	9,000	7,500	< 10,000
Salinity[b], PSU			surface:
- surface	17-18	6-7	L. Ontario, 0.20
- bottom	22	10-13	L. Erie, 0.17
Primary production[b], g C m^{-2} yr^{-1}	150-200	140-160	L. Superior, < 90 L. Huron, 100 L. Michigan, 130 L. Ontario, 180 L. Erie, 160-310
Changes in water transparency in the 1900s, m	from > 18 to > 8	from > 9 to > 6	L. Erie (central): from 5 (1960s) to > 8 (1990s)

a) for the Black and Baltic Seas, last penetration of salty waters, b) central parts, offshore waters.

In the Baltic Sea, 22 species of Ponto-Caspian origin have been recorded; of them over 12 have been able to establish permanently (Baltic Sea Alien Species Database 2001). Their proportion diminishes westward: *Dreissena polymorpha* is the only species belonging to this group in Danish coastal waters (Knudsen 1989), only *Cordylophora caspia* has been reported from the North Sea coast of Germany, and *D. polymorpha, C. caspia*, and *Lithoglyphus naticoides* have been reported from brackish waters of the Netherlands (Reise et al. 1999).

In the Great Lakes, 15 Ponto-Caspian species are known to have been established, 13 of these having invaded during the past two decades; in fact, most of the animals that have invaded the Great Lakes since 1985 are Ponto-Caspian in origin (Ricciardi & MacIsaac 2000; Ricciardi 2001). Of the species listed in the Table 2, better-studied invaders such as *D. polymorpha, Echinogammarus ischnus, Hemimysis anomala, Cercopagis pengoi, C. caspia, C. curvispinum,* and *Neogobius melanostomus* were selected for further analysis.

4 Effects on abiotic and biotic environment

Nonindigenous Ponto-Caspian species have affected both abiotic environment and biotic components in the recipient areas studied. In this impact evaluation, available relevant literature on the selected exotic species has been used.

Table 2. Species of Ponto-Caspian origin recorded in European inland waters, the Baltic Sea and the Great Lakes and the date when first observed (? – undocumented).

Taxon	Inland Europe	First record	Baltic Sea	First record	Great Lakes	First record	Reference
Hydrozoa							
Cordylophora caspia	?	?	Curonian and Vistula Lagoons	1800s	Lake Erie	1956	Davis 1957, BMB WG NEMO 2001
Maeotias marginata			Väinameri	1999			Väinölä & Oulasvirta 2001
Turbellaria							
Dendrocoelum romanodanubiale	River Main	?					Schleuter & Schleuter 1998
Oligochaeta							
Paranais frici			Neva Bay	1995			BMB WG NEMO 2001
Potamothrix heuscheri			Neva Bay	1995			BMB WG NEMO 2001
P. vejdovskyi			Neva Estuary	1995			BMB WG NEMO 2001
Polychaeta							
Hypania invalida	Danube, Rhine Rivers	? 1995					Tittizer 1996
Crustacea, Cladocera							
Cercopagis pengoi	Reservoirs and canals on Dniepr and Don Rivers		Gulf of Riga	1992	Lake Ontario	1998	Mordukhai-Boltovskoi & Rivier 1987; MacIsaac et al. 1999; BMB WG NEMO 2001
Crustacea, Amphipoda							
Corophium curvispinum	Vistula, Odra, Elbe, Severn Lower Rhine Rivers	1913 1923 <1935 1987	Curonian, Vistula and Szczecin lagoons	1920s (possibly in the 1850s)			Van den Brink et al. 1989; BMB WG NEMO 2001
C. mucronatum					Lake St Clair	1997	Grigirovich & MacIsaac 1999
Echinogammarus ischnus	Vistula, Ems, Rhine Rivers	1928 <1978 1989	Curonian Lagoon	1962	Detroit River	1995	Gasiunas 1964; Van den Brink et al. 1993b; Witt et al. 1997; BMB WG NEMO 2001
Chaetogammarus warpachowskyi	Main and Rhine Rivers Lithuanian Lakes	 1960s	Curonian lagoon	1962			Gasiunas 1964; Jazdzewski 1980
Obesogammarus crassus	Lithuanian Lakes	1960s	Curonian lagoon	1962			Gasiunas 1964; Jazdzewski 1980
Pontogammarus robustoides	Lithuanian Lakes	1960	Curonian lagoon	1962			Gasiunas 1964; Jazdzewski 1980
Dikerogammarus villosus	Danube, Main, Rhine Rivers	 1994 1995					Bij de Vaate & Klink 1995; Tittizer 1996; Giesen 1998;
D. haemobaphes	Danube Main, Rhine, Moscow Rivers	 1993 1994					Jazdzewski 1980; Lvova et al. 1996; Tittizer 1996; Giesen 1998;
Crustacea, Mysidacea							
Hemimysis anomala	Lithuanian Lakes, Main River	1960s 1997	Curonian lagoon	1962			Gasiunas 1964; Schleuter & Schleuter 1998; Ketelaars et al. 1999

Taxon	Inland Europe	First record	Baltic Sea	First record	Great Lakes	First record	Reference
Limnomysis benedeni	Danube River Lithuanian Lakes	? 1960s	Curonian lagoon	1962			Gasiunas 1964; Wittman 1995
Paramysis lacustris	Lithuanian Lakes	1960s	Curonian lagoon	1962			Gasiunas 1964
Crustacea, Isopoda							
Jaera istri	Danube, Main, Rhine Rivers	1995					Tittizer 1996
Bivalvia							
Dreissena polymorpha	Elbe, Rhine Seine, Loire, Odra Rivers	<1828 <1860	Curonian and Vistula lagoons	<1824	Lake St Clair	1988	Hebert et al. 1989; BMB WG NEMO 2001
D. bugensis	Volga River	Late 1980s			Lake Ontario	1989	Mills et al. 1993; Orlova et al. 2000
Mollusca, Gastropoda							
Lithoglyphus naticoides	Rhine River	1800s	Curonian and Vistula Lagoons	1800s			Tittizer 1996
Pisces							
Neogobius melanostomus	Moscow River	mid 1980s	Gulf of Gdansk	1990	St. Clair River (Lake Huron outflow)	1990	Sokolov et al. 1989; Jude et al. 1992; Skora & Stolarski 1993;
Proterorhinus marmoratus					St. Clair River (Lake Huron outflow)	1990	Jude et al. 1992
A. stellatus	?	?	Gulf of Finland, Gulf of Bothnia	1999			S. Anatsky, pers. comm.
Hucho hucho	Rhine, Elbe Rhone Rivers	1957-1960					Lelek 1996
Huso huso	Rhine River	?	Gulf of Riga	1962			Lelek 1996; BMB WG NEMO 2001
Abramis sapa	Rhine River	?					Lelek 1996
Umbra krameri	Rhine River	?					Lelek 1996

4.1 ABIOTIC COMPONENTS

Among the Ponto-Caspian invaders, it is *D. polymorpha* that has been primarily responsible for water transparency alterations. Even if populations of zebra mussels store huge amounts of organic matter, energy and calcium carbonate, their filtration ability is likely to be the most important impact at the ecosystem level. Water transparency in Lake St. Clair and western Lake Erie doubled within a few years after zebra mussel invasion (Holland 1993; Griffiths 1993). Also, turbidity declined by 60% within four years of invasion in Saginaw Bay (Skubinna et al. 1995). In Europe, water transparency increased after the invasion of *D. polymorpha* from 2 to 4 m in the Lake Lukomskoe, Belarus (Lyakhnovich et al. 1983) and in Lake Como, Italy (Binelli et al. 1997; Table 3). The clearing capacity of zebra mussel populations in the Dutch lakes IJsselmeer and Markermeer was sufficient to filter both lakes at least once or twice a month, the filtration rate being regulated by the content of suspended matter: the filtration rate shows an inverse exponential relationship with the total dry matter content of the water (Reeders et al. 1989). In Mikolajskie lake (Poland), *D. polymorpha* became the main filter-feeding species. A zebra mussel population, estimated at 236 tons, was able to filter 53,

Unionidae 0.9, and Sphaeriidae 600,000 m^3 of water during the 6 months season. The estimated quantity of seston filtered by *D. polymorpha* was 160 tons, whereas unionids filtered 2.5 tons and sphaerids about 2 tons dry weight, respectively (Stanczykowska 1975). Here consumption of seston amounts to almost 9% of pelagic primary production, while faeces reach 13% of the annual sedimentation of tripton (Stanczykowska et al. 1975). In the Szczecin Lagoon (Polish coast), the *D. polymorpha* population was estimated to consume 53 tons of seston h^{-1} (Stanczykowska 1977). Similar effects were caused by *C. curvispinum*, whose colonization lower River Rhine coincided with a decrease in total suspended matter due to increase in overall filtration capacity of the benthic community (Van den Brink et al. 1993a).

Invasions have altered the cycling of nutrients in some of study basins. Multifold increases in concentrations of total phosphorus, ammonia, nitrate and nitrite in the surface waters of Lake Erie were observed following invasion by *Dreissena* spp. (Strayer 1999; Makarewicz et al. 2000). Mellina et al. (1995) showed by mass balance modelling and lab experiments that zebra mussels have uncoupled the phosphorus-chlorophyll relationship in Lake Erie and Lake St. Clair. Total phosphorus, phosphate and nitrate values decreased after the invasion of *D. polymorpha* in Lake Como, Italy (Binelli et al. 1997). *Dreissena* beds, similarly to those formed by *Mytilus edulis* in marine habitats, release huge amounts of inorganic nutrients into the water column directly or via the microorganisms and meiofauna that process the feces and pseudofeces.

Zebra mussels are a primary food source for waterfowl in the lower Great Lakes and transfer organochlorine contaminants to these predators (Mazak et al. 1997). Because zebra mussels have relatively high contaminant concentrations in their tissues (Brieger & Hunter 1993), and because round gobies (*Neogobius melanostomus*) feed predominantly on zebra mussels in the Great Lakes (Ray & Corkum 1997), they may transfer PCBs and other organic contaminants to their predators and further lead to increased exposure of humans to contaminants (Strayer 1999). However, this hypothesis, originally proposed by Jude et al. (1995), has not been tested. Nevertheless, zebra mussels are altering pathways and environmental fate of contaminants in Lake Erie. They also increase sedimentation of PCB's and cadmium by up to 10-fold (Dobson & Mackie 1998). *D. polymorpha* is likely altering bioaccumulation pathways of heavy metals and organic contaminants in the Rhine-Meuse basin via waterfowl (e.g. *Aythya farina* and *A. fuligula*) which prey on the mussels (Hendriks et al. 1998). Similar studies are lacking in the Baltic Sea.

Shell deposits of *Dreissena* have changed former soft bottoms (sand or silt) into shell gravel, and create patches of hard substrate for sessile species on sites in the coastal lagoons of the Baltic Sea (Olenin & Leppäkoski 1999) and in the Great Lakes (Garton et al. 1998). The amphipod *C. curvispinum* builds networks of mud tubes on firm substrata; it has transformed hard substrata in the lower Rhine River by fixing up to 4 cm of mud on stones (Van den Brink et al. 1993a). *C. caspia* forms dense "bushes" - colonies on submerged trees, hydrotechnical constructions, piers, etc., and traps particulate material on its stolons, and creates microhabitats for several associated species in, e.g., the Curonian Lagoon (Olenin & Leppäkoski 1999).

4.2 BIOTIC COMPONENTS

Most invasions by Ponto-Caspian species have not resulted in native species loss to date, and so they have generally increased local species richness (alpha diversity) in European inland waters (Jazdzewski 1980; Kinzelbach 1995) and in the Great Lakes (e.g., Stewart et al. 1998a,b). However, several native species in the Great Lakes and inland Europe have declined locally following invasion (see below). A more subtle long-term consequence of invasions is the increasing homogenization of aquatic ecosystems (e.g., Rahel 2000). Ponto-Caspian species have become dominant members of several food webs in inland Europe (Kinzelbach 1995), the Baltic Sea (Olenin & Leppäkoski 1999), and the Great Lakes (Stewart et al. 1998a; MacIsaac et al. 1999; Johannsson et al. 2000). Thus, variation in species composition between regions (beta diversity) has been reduced. Very little attention has been devoted to assessing impact of species of Ponto-Caspian origin on native biota and renewable resources in aquatic ecosystems, with the exception of *D. polymorpha* in European inland waters and in the Great Lakes. Even these assessments have been based on a retrospective rather than predictive approach. Ecosystems have assimilated nonindigenous Ponto-Caspian species to a certain extent; in fact, most Ponto-Caspian species appear to be ecologically benign in their new areas of occurrence. There are examples, however, of large-scale effects on structural and functional diversity with prominent food-web impacts in the most heavily affected recipient areas for the most successful and aggressive invaders in the three basins considered in this review.

4.2.1 Bacterioplankton

Selective removal of large bacteria (< 0.9μm) by D. *polymorpha* was observed at eutrophic stations in Saginaw Bay, Lake Huron. However, *D. polymorpha* favoured, probably indirectly, the development of heterotrophs by nutrient excretion at the oligotrophic environment (Cotner et al. 1995). *D. polymorpha* also appears to have caused 'blooms' of *Microcystis aeruginosa* in Lake Erie and Saginaw bay owing to selective grazing (Vanderploeg et al. 1996).

4.2.2 Phytoplankton and primary production

Zebra mussels have caused dramatic declines (30-90%) in phytoplankton biomass in western Lake Erie and Saginaw Bay (Lake Huron): chlorophyll *a* concentrations fell by approximately 60% in both of these habitats after the establishment of dense mussel beds (reviewed by MacIsaac 1996; Strayer 1999). Planktonic diatom populations in western Lake Erie declined by almost 90% within a few years following mussel invasion (Holland 1993). In Europe a few observations are available. Significant reduction in chl *a* values (from 18-13.3 μg l^{-1} to 8.6 μg l^{-1}), possibly as a result of *H. anomala* predation on algae were recorded in Biesbosch reservoir, the Netherlands (Ketelaars et al. 1999). Chl *a* values also decreased after the invasion of *D. polymorpha* in Lake Como, Italy (Binelli et al. 1997) and *H. anomala* to Biesbosch reservoir, the Netherlands (Ketelaars et al. 1999). In the Baltic Sea, such direct observations are lacking. However, Kotta et al. (1998) suggested that at sites of high abundances of *D. polymorpha* limitation of phytoplankton production may occur in the Gulf of Riga.

4.2.3 Benthic algae

Biomass of benthic algae has increased and composition has changed (from dominance by diatoms to filamentous green algae) in Saginaw Bay, presumably as a result of increased light penetration caused by zebra mussel filtration (Lowe & Pilsbury 1995; Strayer 1999).

4.2.4 Zooplankton

Dramatic effects on zooplankton were evident after the invasion of *Hemimysis* invasion to Biesbosch reservoir, the Netherlands: *Daphnia* spp., *Bosmina* spp., Ostracoda and Rotifera declined sharply in abundance. This was explained by predation by *H. anomala* (Ketelaars et al. 1999). Similarly, significantly lower population abundances of a small-bodied cladoceran *Bosmina* sp., a probable prey of *C. pengoi*, were recorded after invasion of this predatory cladoceran to the Gulf of Riga (Ojaveer et al. 2000). Considerable spatial differences were recorded in the zooplankton community of the Lake Ontario in the first year of invasion of the predatory *C. pengoi*: in the eastern part where *C. pengoi* was present, the cladoceran community was dominated by *Daphnia retrocurva* whereas in the western part where *C. pengoi* was absent, *Bosmina longirostris* prevailed. However, besides invertebrate predation, these differences could also be due to differences in population timing (Barbiero et al. 2000). No predation rate data exist to directly determine the impact of *Cercopagis* invasion on zooplankton or phytoplankton in the Great Lakes. However, based on the relative sizes of *Cercopagis* and *Bythotrephes*, and on *Bythotrephes* apparent selectivity for small zooplankton prey, it may be expected that *Cercopagis* will impact primarily small crustaceans and/or rotifers (Grigorovich et al. 1998).

The disappearance of the invertebrate predators *Leptodora kindtii* and *Bythotrephes longimanus* after invasion of *Hemimysis anomala* to Biesbosch reservoir, the Netherlands, appears to be linked to competition for food with this mysid species (Ketelaars et al. 1999). Small zooplankton (e.g. rotifers and copepods) declined by 74% in western Lake Erie at the same time that zebra mussel populations increased to densities up to 300,000 ind m^{-2} (MacIsaac et al. 1995). Declines in zooplankton are related to both ingestion by zebra mussels and the loss of phytoplankton stocks because of zebra mussel filtration. *Dreissena* veligers now contribute 10% to 25% of zooplanktonic production in Lake Erie (Johannsson et al. 2000). A similar decline was observed in the Hudson River (Atlantic coast of North America). After zebra mussels became abundant in the river in 1992, phytoplankton biomass was reduced by 80-90% and zooplankton biomass by more than 70% (Pace et al. 1998). The invasion of *D. polymorpha* in Lake Lukomskoe similarly caused a significant decrease in total abundance of filter-feeding zooplankton (Karataev & Burlakova 1995).

4.2.5 Zoobenthos

Benthic communities in Lake St. Clair and western Lake Erie have increased in species richness following zebra mussel invasion (Griffiths 1993; Stewart et al. 1998a), at least initially, although one study showed that diversity on rocky substrata in Lake Ontario returned to pre-invasion levels after several years (Haynes et al. 1999). The diversity of native mussel (unionid) communities has declined rapidly following zebra mussel invasion in the Great Lakes and St. Lawrence River (Schloesser & Nalepa 1994; Ricciardi et al. 1996, 1998). This is attributable primarily to the effects of fouling on native mussels but also to food competition, reducing food concentrations to levels too low to fuel the unionid and sphaeriid populations; in fact, the zebra mussel has had effects on all major taxa of freshwater molluscs in North America (Strayer 1999). Similar high mortalities have rarely been reported from European lakes invaded by zebra mussels; distinct unfavourable effects of overgrowth with *Dreissena* on the body growth and body weight of Unionidae were not observed in the lakes of northern Poland (Lewandowski 1976). Ricciardi et al. (1998) offered three reasons for the difference in documented impacts of *Dreissena* on native mussels in both regions: (i) very few European studies have examined changes in mussel communities following invasion; (ii) densities and infestation levels of *Dreissena* on native mussels are at least 10-fold lower in Europe compared with North America; and (iii) central European freshwater unionid mussel fauna (1 species *Margaritifera*, ca 10 species Unionidae (e.g. Haas 1969)) is depauperate compared with the North American fauna (~300 spp. described) and has had evolutionary experience with *Dreissena*, whose distribution expanded into northern and central Europe prior to the last glaciation before retreating to Ponto-Caspian basins. Thus, European species may have already experienced selection pressures to adapt to fouling by *Dreissena*.

Densities of macroinvertebrates on rocky substrata increased 2-8 fold following the development of zebra mussel densities of 1,500-4,000 ind m^{-2} in the upper St. Lawrence River. Field experiments using artificial substrata and manipulated zebra mussel densities demonstrated that the change in abundance was due to mussel colonization through enhancing populations of deposit-feeding organisms (primarily amphipods), small gastropods, and predatory invertebrates (leeches, flatworms, water mites), and by displacing net-spinning caddis flies (Ricciardi et al. 1997). The associated fauna in zebra mussel aggregations is stimulated, in part, by feces of the mussels. A population of *D. polymorpha* at 2,200 ind m^{-2} has been estimated to produce feces 3,520 mg·dw 24 h m^{-2} (Stanczykowska 1975). Madenjian's (1995) model for western Lake Erie predicts that 50,000 ind m^{-2} in the basin produce 1.4 megatons of pseudofeces (from ingesting 6.4 megatons of phytoplankton) over six months. Deposit-feeders, especially amphipods and chironomid larvae, have also increased in response to zebra mussel colonization in Lake St Clair and Lake Erie (Griffiths 1993; Stewart et al. 1998a). Following zebra mussel invasion, fresh shell-free benthic biomass increased by a factor of 50 in western Lake Erie, by a factor of 7 in the central Lake Erie, and by a factor of 40 in eastern Lake Erie; dreissenid mussels themselves account for over 95% of the benthic biomass (Johannsson et al. 2000). Huge interannual variations in density and biomass are characteristic of *D. polymorpha* populations. In the Mikolajskie lake, its biomass varied within 0-1,800 g shell-free wwt m^{-2} within a couple of years (Stanczykowska 1975).

D. polymorpha was the dominant species in Polish dimictic lakes and in the least fertile (100-300 μg·P l^{-1}) polymictic lakes where it accounted for 40-70% of mollusc abundance (Stanczykowska 1984). In the Baltic Sea, distribution areas of *D. polymorpha* and the native filter-feeder *M. trossulus* normally do not overlap. However, in some areas in the eastern Gulf of Riga these species co-occur (J. Kotta, pers. comm.).

In many cases the invasive species represent a new function (e.g., feeding strategy) or an entirely new trophic level and will, thus, restructure the community into which they were introduced. Non-native nektobenthic species (e.g., mysids and amphipods) swim actively and spend part of their time in the water column, and may also dwell within or on the bottom. They form dense populations in the Curonian Lagoon, and serve as important links in energy transfer between pelagic and benthic subsystems. *D. polymorpha* is known to have caused increases in local density and diversity of other benthic invertebrates in lakes and rivers (see above) and coastal lagoons (Olenin & Leppäkoski 1999). The mussels also offer refuge from large predators for associated species. Predator-avoidance behaviour of the snail *Physella heterostropha*, in combination with the structurally complex habitat provided by *Dreissena* beds, can reduce predation of fish on the snails (Stewart et al. 1999). Likewise, the amphipod *E. ischnus* may be displacing native *Gammarus fasciatus* in Lake Erie (Dermott et al. 1998), in part owing to use of *Dreissena* refuges from fish predation by the former species (H. MacIsaac, pers. comm.).

Reduction in macroinvertebrate species richness and substantial decrease in abundance of several macrozoobenthic taxa (Chironomidae, *Hydropsyche contubernalis, Asellus aquaticus*, *Gammarus tigrinus*, *Potamopyrgus antipodarum* and *D. polymorpha*,) was observed concomitantly with the enormous abundance increase of *C. curvispinum* in the Lower Rhine. The proposed mechanism is competition for food (Van den Brink et al. 1993a; Strayer 1999) and the transformation of substrate with muddy encrustations that prevent settlement of zebra mussel larvae and other fauna that prefer bare hard substrates (Rajagopal et al. 1999; Van der Velde et al. 1999). Dick & Platvoet (2000) showed that *Dikerogammarus villosus* is rapidly eliminating a native European amphipod *Gammarus duebeni* and an introduced amphipod *G. tigrinus* from parts of the Netherlands (particularly the Markermeer and the IJsselmeer). Microcosm experiments demonstrate that predation by *D. villosus* on both species occurs over a range of water conductivities. Both newly-moulted and intermoult animals are preyed upon. Recently, Kelleher et al. (1999) proved that invasions of exotic amphipods have had positive impact in the lower Rhine River by restoring predator-prey relationships, since the breakdown in its trophic structure in 1986. The same study demonstrated that *C. curvispinum*, together with other invasive crustaceans, play a significant role in energy mediation and trophic interactions in the lower Rhine River ecosystem.

4.2.6 Fish and birds

Populations of native benthic fishes, mottled sculpins *Cottus bairdi* and longperch *Percina caprodes*, in the St. Clair River declined sharply after the arrival of the round goby *Neogobius melanostomus*, apparently due to competition for food and space (Jude et al. 1995). Round gobies can penetrate interstitial spaces in coarse sediments to obtain food, and therefore may become predators of lake trout eggs; they have been shown to prey on lake trout eggs in the laboratory (Chotkowski & Marsden 1999). In the Gulf of Gdansk (Baltic Sea), competition for food and displacement of several native demersal fishes (e.g., flatfishes, eelpout *Zoarces viviparus*, and black goby *Gobius niger*) by invading *N. melanostomus* has been suggested to take place (Skora & Stolarski 1993, K.E. Skora & J. Rzeznik, pers. comm.). In addition, the same authors claimed that high abundances of the round goby may cause relaxation of predation pressure on several native fish in the Baltic Sea (e.g., sandeel *Ammodytes tobianus* and sprat *Sprattus sprat-*

tus) by being more favourable food for most abundant piscivores than the above-named species.

Increased water transparency following zebra mussel invasion has promoted prolific growth of macrophytes in Lake St. Clair (MacIsaac 1996) and Saginaw Bay, Lake Huron (Skubinna et al. 1995). In Lake St. Clair, this provoked a shift in fish communities: walleye *Stizostedion vitreum*, which are adapted to turbid conditions, were replaced by fish that are favoured by weed beds (northern pike *Esox lucius*, muskellunge *E. masquinongy*, and bass *Micropterus* spp.) (MacIsaac 1996). In addition to the habitat change, *Dreissena* can control walley abundance through the food-web: according to individual-based model prediction, shunting of energy from pelagic to benthic pathways by *Dreissena* results in elimination of walleye high-recruitment years in Oneida Lake that led to 30% reduction in adult walleye abundance (Rutherford et al. 1999). In Europe, adult individuals of *D. polymorpha* are directly incorporated into the food-web by being eaten by several species of fish (whitefish, eel, carp, bream, white bream, orfe, sturgeon and flounder), by crayfish *Orconectes limosus* (native to North America), and by waterfowl such as coot *Fulica atra*, tufted duck *Aythya fulicula*, common pochard *A. ferina*, mallard *Anas platyrhynchos*, gadwell *A. strepera*, red-crested pochard *Netta rufina*, eider *Somateria mollissima*, goldeneye *Bucephala clangula*, goosander *Mergus merganser*, merganser *M. serrator*, moorhen *Gallinula chloropus*, gulls *Larus* spp., sheld-duck *Tadorna tadorna*, and dabchick *Podiceps ruficollis* (Stanczykowska 1977).

By consuming *C. pengoi*, the diet composition of Baltic herring *Clupea harengus membras*, smelt *Osmerus eperlanus*, three-spined stickleback *Gasterosteus aculeatus*, and nine-spined stickleback *Pungitius pungitius* has changed after invasion of the cladoceran. The introduction of *C. pengoi* to the Baltic Sea may prove beneficial to fisheries production if it enhances utilization and transfer of mesozooplankton biomass (e.g., *Bosmina*) to planktivorous fishes (Ojaveer et al. 2000). Exotic amphipods are the most important prey taxa for dominant benthivorous fish in the lover Rhine River. Although their importance for fish has been constant over the last decade (Kelleher et al. 1999), significant changes in diet composition of certain fish species (perch and eel) and dietary overlap between them has occurred. These changes appear to be due to alterations in food supply caused by invasion of the amphipod *C. curvispinum* (Kelleher et al. 1998). Yellow perch *Perca flavescens* kept in enclosures with and without zebra mussels in an experimental pond (using fish and sediment from Lake St. Clair) showed differences in diet and growth rate (Thayer et al. 1997): growth rates of adult perch were higher in the presence of zebra mussels, presumably because of increased densities of invertebrate prey.

5 Comparative analysis of ecosystem impacts

When comparing the ecological changes caused by invasive species, we should distinguish between types and scales of impact. The types of impact are the same or very similar in all three regions compared. Invasive Ponto-Caspian species modify abiotic and biotic conditions for other species, they alter the pathways of energy transfer and influence cycling of contaminants within new ecosystems, thus, acting as habitat and ecosystem engineers (Jones et al. 1994; Stewart et al. 1998b, 1999; Olenin & Leppäkoski 1999; Table 3).

The magnitude and geographical dimension of the ecological impacts of Ponto-Caspian invasions in most cases are quite different between the inland European freshwater bodies, Baltic Sea and Great Lakes. The most prominent and larger-scale changes, both in the pelagic and benthic ecosystems, have recently taken place in the Great Lakes (Table 4). This is mainly due to the invasion of the zebra mussel, which has had multiple ecosystem impacts both in habitat modifications and at all trophic level. Another invader, *C. pengoi* was discovered in the Great Lakes only in 1998 and has not been present long enough for its impact to be measured.

Ecological impacts of the Ponto-Caspian species in European inland waters are locally substantial but certainly less pronounced at the regional scale. The Baltic Sea, the only brackish-water habitat amongst the studied water-bodies, has seemingly suffered by Ponto-Caspian species invasions the least with the highest impact to benthic communities; the impact of *C. pengoi* on the pelagic subsystem remains to be assessed in more detail. The fact that the Baltic Sea has suffered relatively the least could partly be due to geographic location of the basin: it is situated in a zone of colder climate than the remaining regions studied. This fact may be important for development of populations of Ponto-Caspian species originating from considerably warmer climatic conditions.

In addition, it should be stressed that although ecological effects of similar Ponto-Caspian species were generally unidirectional in the three areas compared, different species were often responsible for major ecosystem changes in these three geographically remote regions. Due to large differences in invasion histories of the same species in the three regions, quite often the ecological effects are not directly comparable either because of lack of relevant studies in the past or of different population abundance (and consequently, magnitude of impact) of each species in different systems.

Table 3. Examples of habitat engineering and ecosystem changes caused by Ponto-Caspian invasive species in the Baltic Sea, inland European freshwater bodies and North American Great Lakes (criteria largely adopted from Stewart et al. 1998b, 1999; Olenin & Leppäkoski 1999). [1) modifies rocky bottom or sediment substrate, 2) provides refuges from predators and currents, 3) traps and accumulates POM in interstitial microhabitats, 4) increases water clarity (= lowers concentrations of POM and suspended solids), 5) affects macrophyte canopy, 6) redirects energy from pelagic to benthic subsystems or *vice versa*, 7) provides additional prey to planktivorous and/or benthivorous fish, 8) provides food for waterfowl, 9) excludes competing species, 10) increases soluble (bioavailable) nutrients (N, P), 11) influences cycling of organic and/or metal contaminants, 12) ? – undocumented].

Species/function	1	2	3	4	5	6	7	8	9	10	11
C. caspia	x	x	x								
C. pengoi							x				
P. robustoides						X	x				
E. ischnus	x					x	x		x		x?
C. curvispinum	x	x	x	x	x	x	x		x		x?
H. anomala						x	x				
D. polymorpha	x	x	x	x	x	x	x	x	x	x	x
N. melanostomus						x	x	x	x		x

Another very important difference is the effort expended to quantity effects of exotic species in the three regions. Experimental studies on invasive Ponto-Caspian species are very recent (performed mainly during the 1990s) in all three areas compared in this review, but most advanced in North America. Therefore, even though the effects of introduced species appear to be strongest in the Great Lakes, it is possible that this reflects, in part, effort devoted to the topic. Available evidence indicates that all three regions have been moderately to strongly altered by invasions over very short periods of time.

Several ecosystem processes (e.g., nutrient and contaminant cycling, energy flow), involving multiple trophic levels, are affected. In general, the largest community changes were recorded for zooplankton and zoobenthos. However, effects on diversity have not been well studied for the majority of these invasions; although most invasions appear to add to local (alpha) diversity, in some cases diversity declines have occurred at this scale. As the same small pool of common dominant species become established (e.g., *N. melanostomus*, *D. polymorpha*, *C. pengoi*) in different areas, we can expect variation in diversity across regions (i.e. beta diversity) to decline. Some evidence suggests that these co-evolved invaders may be facilitating each other's population growth and spread (Ricciardi 2001). Therefore, we can predict that further ecosystem disruption is likely as more Ponto-Caspian invaders expand their ranges into these regions.

Table 4. Documented effects of Ponto-Caspian invasions on affected ecosystems in inland Europe, Baltic Sea, and the North American Great Lakes. Effect strength is indicated as strong (***), moderate (**), or weak/none (*),? – undocumented.

	Inland Europe	Baltic Sea	Great Lakes
Abiotic components			
Transparency	***	*	***
Nutrient cycling	**	**	**
Contaminant cycling	*?	?	**
Benthic substrate	***	**	***
Biotic components			
Phytoplankton	**	?	***
Zooplankton	***	**	***
Fish	?	*	**
Zoobenthos	***	***	***
Periphyton	*	?	**
Macrophytes	?	?	**

Based on generally similar patterns of ecological effects of Ponto-Caspian invaders in the different regions studied and previous evidence from inland Europe, future introductions by *C. curvispinum* and *H. anomala* to the Great Lakes ecosystems could be predicted to cause significant impacts; these should include, amongst others, severe changes in the increase in filtration capacity of these ecosystems and further transformations in phytoplankton, zooplankton and zoobenthos communities, including probable declines in zebra mussel densities (see Van den Brink et al. 1993a; Kelleher et al. 1998; Ketelaars et al. 1999).

Our review revealed substantial gaps in the knowledge of the ecological effects of Ponto-Caspian invaders in the various invaded regions. We lack basic impact studies on important ecosystem processes and/or trophic levels (e.g., changes in native species richness, phytoplankton/primary productivity, contaminant cycling, fisheries), even for the most abundant and well-recognized invasive species. Moreover, only a few studies examine the cascading effects of invasions in aquatic food webs. However, this knowledge is of high priority importance for proper ecosystem management. The scarcity of predictive information demands that systematic and well-coordinated research be initiated to provide better scientific advice for the management of aquatic invaders, but aquatic ecosystem especially.

FAST TURNOVER IN DOMINANCE OF EXOTIC SPECIES IN THE RHINE RIVER DETERMINES BIODIVERSITY AND ECOSYSTEM FUNCTION: AN AFFAIR BETWEEN AMPHIPODS AND MUSSELS

GUIDO HAAS[1*], MATTHIAS BRUNKE[2] & BRUNO STREIT[1]
[1]*Abt. Ökologie und Evolution, Zoologisches Institut der Johann Wolfgang Goethe-Universität Frankfurt am Main, Frankfurt am Main, Germany*
[2]*Institut für Gewässerökologie und Binnenfischerei Berlin, Germany*
[*]Corresponding author *haas-limno@t-online.de*

Abstract
A high turnover in macroinvertebrate composition took place between 1993 and 2001 in the Upper and Middle Rhine because of successful invasion of exotic invertebrates. At the end of the 1980s, the Ponto-Caspian crustacean *Corophium curvispinum* and the Asiatic clam *Corbicula* spp. invaded and colonized rapidly the Rhine. In the middle of the 1990s, another Ponto-Caspian crustacean, *Dikerogammarus villosus,* colonized predominantly the littoral riprap. Today, the predatory *D. villosus* dominates the macroinvertebrate fauna throughout the navigable sector of the Rhine. Before its invasion, *C. curvispinum, Gammarus tigrinus* and *Echinogammarus ischnus* were the co-occurring nonindigenous amphipods. Since 1996, all gammarid species declined, except *D. villosus,* which nowadays occurs in high densities of more than 3,000 ind m^{-2}. *D. villosus* interfered into the competition for space between the two filter-feeders *C. curvispinum* and *Dreissena polymorpha.* The impact of these exotics on biodiversity is species-specific: *Dreissena* provides new substrate for other invertebrates, *Corophium* reduces available habitat areas by transferring coarse and hard surfaces to fine and muddy areas, and *Dikerogammarus* exerts a strong predatory effect, in particular on *C. curvispinum.* The exotic mass invaders *C. curvispinum, D. polymorpha* and *Corbicula* spp. strongly influence ecosystem functions by linking pelagic and benthic processes by their intense filter feeding activity and thereby enhance the capacity for self-purification of the river system.

1 Introduction

The existence and impact of exotic species was recognized for large rivers in Germany since 1950 (Thienemann 1950; Kureck 1992) and for smaller rivers in eastern Germany, such as the Spree river, since about 1900. In the last two decades the invertebrate community of the Lower, Middle and Upper Rhine was increasingly exposed to exotic species (Van der Velde et al. 2000; Haas 2001). Among the invertebrate invaders in the Upper and the Middle Rhine, the most frequent exotic species refer to crustaceans and molluscs (Kinzelbach 1995). The exotic species may contribute to only small portions (10 to 15%) of total species richness in regulated rivers such as the Rhine, however, they can dominate the total abundance and biomass by more than 80% (Tittizer et al. 2000; Haas 2001). The most important donor-region for invaders of Central Europe is the Ponto-Caspian basin.

The catchment of the Rhine is the central part of one of the most dense networks of navigation waterways of canals, regulated river channels and seaports (Eckoldt 1998): navigation routes connect the North Sea via Rhine and Danube with the Black Sea, which in turn is connected by canals and rivers to the Caspian Sea (the Pontocaspis)

E. Leppäkoski et al. (eds.), Invasive Aquatic Species of Europe, 426–432.

since 1952. A further navigation route to the Black Sea and Caspian Sea is the German Mittelland-Elbe-Vistula-Pripyat'-Bug-Dnieper canal system (Jazdzewski 1980). Thus two biogeographical regions are connected, the Western European region and the Pontocaspis. The two navigation networks, which also are connected by canals, are the transport pathways for exotic aquatic organisms. Thereby two distributional modes can be distinguished: (i) a local release of animals by exchange of ballast water and (ii) a spread by migrations of animals along the waterway network, which often follows the artificial release.

The Rhine appears to be very susceptible to invasions for several reasons. Its position within the navigation networks between Eastern and Western Europe and overseas promotes frequent releases and enables spread of exotic species. The indigenous fauna is depauperated because of extreme pollution and accidental contaminations from 1960 to 1986. The river is still eutrophic and receives treated sewage effluent from numerous plants. The river morphology is strongly degraded because of straightening, reduction of channel networks to one main channel, disconnection from flood plains, and embankments along the whole shore. In consequence benthic habitats were monotonized to extreme areas like instable degrading sediments, consolidated gravel sheets, and stable and coarse grained riprap.

Nowadays the artificial riprap forms one of the most important habitats in terms of area, longitudinal connectivity, stability during floods, protection against ship-induced waves and enhanced sediment movement. In this study, the invertebrate fauna of riprap down to water depth of 6 m was investigated during a period of eight years to investigate turnover of exotic species.

2 Study site and methods

The faunal investigations were conducted at 16 stations along the Upper and Middle Rhine from 1993 to 2001 (Fig. 1).

The samples were taken by the aid of the laboratory ship "Argus" (HLUG Wiesbaden) allowing an water-level independent sampling. A polyp grab crane was used to remove boulders from the riprap. Surface areas between 0.3 m^2 and 0.5 m^2 were brushed carefully under water. The water containing animals and organic matter was drained through a 0.5 mm mesh. Animals were removed alive by the aid of a stereomicroscope and transferred into alcohol.

Physical and chemical variables were measured onboard from 1993 to 1996 and analyzed by the laboratory of the HLUG Wiesbaden. Water-quality stations located at the cities of Worms and Mainz provided also chemical and physical data. The water quality improved throughout the period from 1993 to 1999 as indicated by a reduced mean specific conductance from 1,070 $\mu s\ cm^{-1}$ in 1993 to 643 $\mu S\ cm^{-1}$ in 1999. The average oxygen saturation showed no trend and ranged between 101 and 112% in this period.

3 Results and discussion

Three successional phases in the development of benthic communities in the German part of the Rhine following the extreme toxic and organic contamination from 1960 to

1975 can be delineated. Except the first phase, each phase was characterized by a mass development of exotic crustaceans and molluscs.

(i) From 1970 to 1986, the aquatic community was species poor and still in an early stage of recovery. Only sewage resistant taxa such as *Erpobdella octoculata* (Hirudinea), *Asellus aquaticus* (Isopoda), *Sphaerotilus natans* (Bacteriophyta), *Radix ovata* (Gastropoda), sponges, chironomids and oligochaetes occurred (Wendling 1993), because of remaining organic pollution. An initial development of euryocous indigenous species such as *Hydropsyche contubernalis* (Trichoptera) was recorded (Wendling 1993) and the exotic zebra mussel *Dreissena polymorpha* colonized most hard substratum. But the Sandoz spill in Basel on November 1986, in conjunction with the emission of nearly 30 tons of toxic chemicals (insecticides, fungicides and herbicides, Van Dijk et al. 1995) into the Rhine (followed by a spread downstream), reset the recovering process. In 1987 faunal densities in the benthic habitat were still close to zero (Hartog et al. 1992). Yet *Dreissena* was able to quickly recolonise the Rhine following the Sandoz spill because of import of pelagic larvae from unaffected sites.

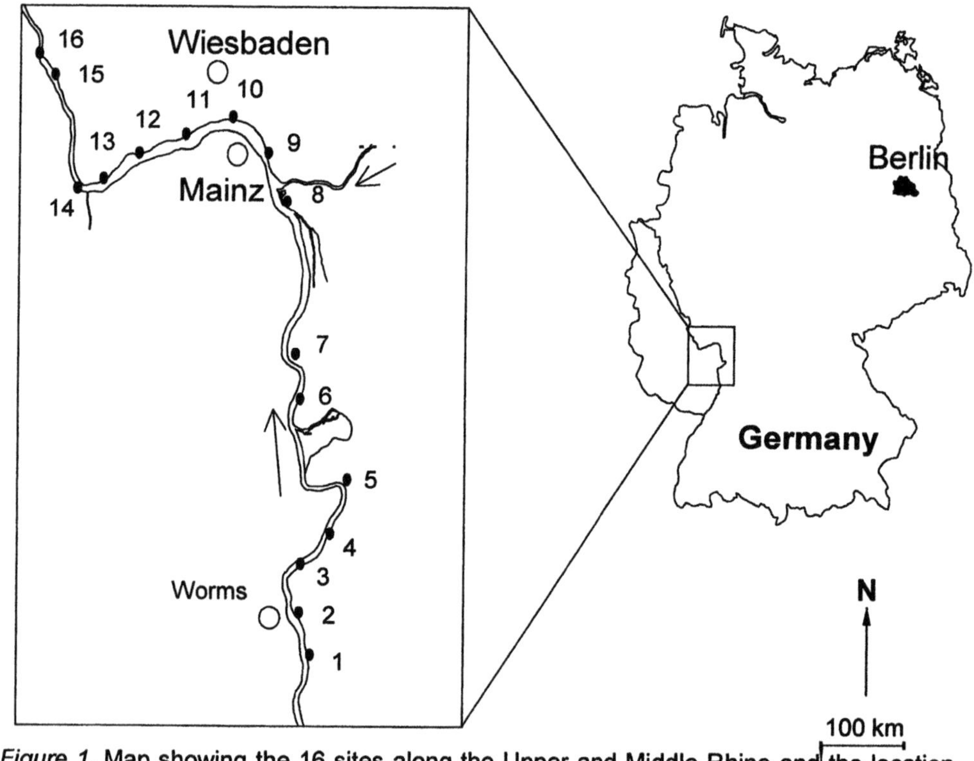

Figure 1. Map showing the 16 sites along the Upper and Middle Rhine and the location within Germany.

(ii) In 1987 and 1988 the exotic *Corophium curvispinum* (Amphipoda) and two *Corbicula* species (*C. fluminea* and *C. fluminalis*; Bivalvia) invaded the Rhine. Already in

1989, the population density of *C. curvispinum* in the Middle and Lower Rhine was so high that the hard substratum of riprap was overgrown due to the species engineering activity. The animals constructed extensive mats of dense silty tubes that covered any available hard surfaces. Stony substratum was covered totally by very fine matter, which was removed by the animals from suspension in the water column. In consequence, other epilithic invertebrates were negatively affected by this muddy layer. In particular the sessile zebra mussel, *D. polymorpha* was the most impaired species by a lost competition for space. The zebra mussel population collapsed because (i) adult shells were rapidly overgrown by *C. curvispinum* and (ii) the muddy tubes inhibited the development of new *Dreissena* patches since their planktonic larvae can only settle on hard and stable surfaces to mature (Haas 1993; Van der Velde et al. 1994; Tittizer & Krebs 1996; Van der Velde et al. 1998).

The Asiatic clam *Corbicula* is restricted to the gravely and sandy river bottom aside the riprap, because of lacking sticking structures like e.g. byssus threads. *C. fluminea* and *C. fluminalis* reached mean densities of 470 and 220 ind m^{-2}, respectively, within the interstices of sand and gravel. Disregarding the sparsely populated river bottom in the shipping canal, the mean population densities reached between 1,200 and 1,800 ind m^{-2} corresponding to 1,300 and 1,900 g m^{-2} fresh weight biomass. The maximum value was 3,149 ind m^{-2} (corresponds to 6,700 g m^{-2} fresh weight biomass) (Meister 1997).

(iii) A new phase started with the appearance of *Dikerogammarus villosus* in 1995. The species invaded the Rhine by passing the Rhine-Main-Danube canal and colonized the riprap of the river section below the confluence with the Main (Haas & Streit 1998). Densities of 50 to 100 ind m^{-2} were already detected for 1995. In 1996, *D. villosus* also spread into upstream sections and reached average densities of about 500 ind m^{-2} (Fig. 3). In 2000 and 2001 maximum densities of 3,000 ind m^{-2} were recorded. Juvenile specimens smaller than 5 mm were not counted, so that the actual abundance of *D. villosus* might be much higher.

In the middle of 1997 and early 1998 three further exotic invertebrate species reached the Rhine River which originated from the Danube and the Pontocaspis, respectively: *Jaera sarsi* (Isopoda), *Dendrocoelum romanodanubiale* (Turbellaria) and *Hypania invalida* (Polychaeta). In 1999, *J. sarsi* spread along the Upper and Middle Rhine and reached maximum densities of more than 4,000 ind m^{-2}. Recent samples from 2001 showed a moderate decline of the species numbers of *J. sarsi* (Fig. 3).

Since 1996, population densities of *Corophium curvispinum* dropped from more than 10,000 to 500 ind m^{-2} (Fig. 2). Gut content analyses showed that *D. villosus* captured in particular *C. curvispinum*. So *D. villosus* interfered into the interspecific competition for space between the two filter-feeders *C. curvispinum* and *Dreissena polymorpha*. Unoccupied muddy tubes of *C. curvispinum* were soon degraded. Consequently the population of the zebra mussel recovered and reached a high level of density.

Subsequent to the appearance of *D. villosus*, two further gammarid species in the Upper Rhine, *Gammarus tigrinus* and *Echinogammarus ischnus*, declined significantly; *G. tigrinus* disappeared finally in 1999 (Fig. 3). The recent arrival and mass development of *D. villosus* in coincidence with strong changes in abundances of other native or exotic species support the theory that predatory invaders have the most powerful effect on the species community (Moyle & Light 1996; Mooney & Cleland 2001). However, animals that actively change their environment like *C. curvispinum* exert strong impact on the availability of microhabitats for other species.

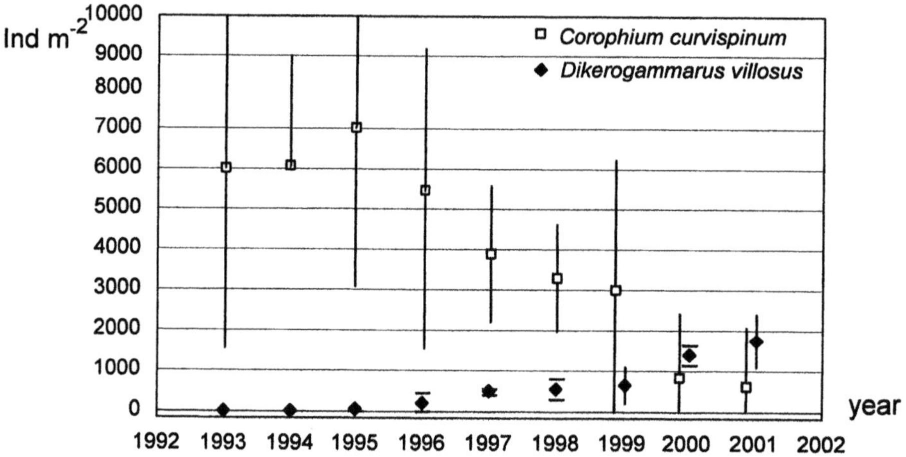

Figure 2. Average annual densities of *Dikerogammarus villosus* and *Corophium curvispinum* in the Upper Rhine from 1993 till 2001. Bars represent standard deviations.

In coincidence with these faunal changes, further decreases in abundances of characteristic species in the Rhine such as *Heptagenia sulphurea, Ephoron virgo* (both Ephemeroptera), *Aphelocheirus aestivalis* (Heteroptera) and turbellarian species (*Dendrocoelum lacteum, Dugesia lugubris* and *Dugesia tigrina*) were recorded (Haas 2001). The influence on these species by *D. villosus* cannot be excluded, however, it is unresolved. For example, Kureck et al. 2001 found a significant decrease of the ephemeropteran *Ephoron virgo* in presence of *D. villosus* because of predation. The large body size (~25 mm) of *D. villosus* relative to other amphipods in the Upper and Middle Rhine explains in part its successful predatory behaviour (Dick & Platvoet 2000). The largest individuals are in general the most aggressive and superior competitors (MacNeil et al. 1997). Presently *D. villosus* distributes into other large navigable rivers (e.g. Moselle and the Rhône; Devin et al. 2001). Because of its broad salinity and temperature tolerances (Muskó 1992; Bij de Vaate & Klink 1995), *D. villosus* is identified as a potential invader of waterways worldwide possibly resulting in a cosmopolitan distribution (Dick & Platvoet 2000; Devin et al. 2001).

Today, in the Upper und Middle Rhine the competition for space between the two exotic filter-feeders *Corophium curvispinum* and *Dreissena polymorpha* has lost its importance for the species composition. Therefore the abundance of *D. polymorpha* and other macroinvertebrate taxa, that depend on solid surfaces such as gastropods (e.g. *Ancylus fluviatilis, Bithynia tentaculata, Potamopyrgus antipodarum*) will probably increase because of a growing availability of hard substratum in the littoral riprap following the decline of *C. curvispinum*. Other native species also profit from the enhanced availability of hard substrata due to the disappearance of muddy-covered stony surfaces. Trichopteran species such as *Psychomyia pusilla* showed a strong increase in abundance and extended their range over the whole studied Upper und Middle Rhine. It builds up sandy-muddy tubes close to the water surface, thereby evading predation pressure by *D. villosus*. Other sessile species such as *Spongilla fragilis* (Spongillidae) and *Plumatella emarginata* (Bryozoa) are probably also positively affected.

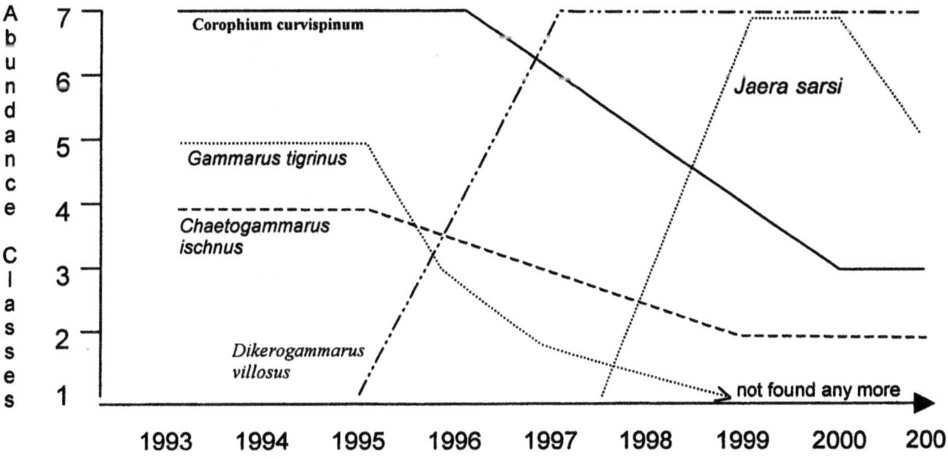

Figure 3. Schematic overview of changes in abundance classes of four exotic amphipod and one isopod species (*Jaera sarsi*) in the Upper and Middle Rhine (1993 to 2001). Abundance classes are: (1) single proof, 2) rare (2-10 m^{-2}), (3) moderate (11-50 m^{-2}), (4) regular (51-200 m^{-2}), (5) common (201-500 m^{-2}), (6) frequent (501-1,000 m^{-2}), (7) abundant (> 1,000 m^{-2}).

4 Conclusions

In the last 15 years, there was a major shift in the macroinvertebrate fauna in the Upper and Middle Rhine. Exotic crustaceans and molluscs arrived mainly from the Ponto-Caspian region and dominated macroinvertebrate assemblages in terms of abundance and biomass. Some exotic species eventually cause a reorganization of biotic interactions and changes in ecosystem functions and thus became keystone species in the new environment. The recently invaded *Dikerogammarus villosus* can be assessed as such a keystone species in that it affects abundances and distribution patterns of other amphipod species. The strong decline of *Corophium curvispinum* and the disappearance of *Gammarus tigrinus* are probably consequences of predation by the abundant *D. villosus*. Isotope analyses evidenced that *D. villosus* is on the same trophic level as some predatory fish species (Marguillier 1998).

Corbicula spp., *Dreissena polymorpha* and *Corophium curvispinum* strongly influence ecosystem functions by linking pelagic and benthic processes by their intense filter feeding activity (Basu & Pick 1997; Hakenkamp & Palmer 1999). *Corbicula* spp. can reach the highest biomasses and filtration rates among molluscs (Strayer et al. 1999), which was reflected by their high biomass in the Rhine. The high filtration rates and biomasses of exotic filter feeders indicate a strong retention and transformation of suspended organic particles (e.g. phytoplankton). As a result the mass development of

Corbicula spp. and the changes in population densities of *Corophium* and *Dreissena* in the Rhine River affect ecosystem processes and enhance the capacity for self-purification of the river system.

On a global scale, the pool of exotic species in the Rhine serves as a source for biological invasions to North America. This distributional pathway is discussed for the introduction of *Dreissena polymorpha* (Kinzelbach 1995) into the Great Lakes at the end of the 1980s and for *Echinogammarus ischnus* into the Great Lakes in the early 1990s (Ricciardi & MacIsaac 2000). Especially the central traffic position of the Rhine in Europe in combination with its abiotic conditions, e.g. in terms of temperature and salinity, emphasizes its significance as a recipient water body for xenofaunistic components (the Ponto-Caspian basin is clearly identified as a donor region) and as a basis for the pool of exotic Eurasian species to invade North America.

The use of large rivers for shipping and as transport pathways for pollutants strongly interfered in the indigenous aquatic biodiversity. The altered physical, chemical and hydrographic habitat conditions favour generalist species. The fauna of the navigable Rhine will probably further be dominated by exotic macroinvertebrate species, maintaining the uniform characteristics of invertebrate communities in shipping lanes. Under the perspective of the widespread occurrence of effective invaders like *Dreissena* and *Corbicula*, these exotics do not contribute to native biodiversity, however, they may play a important role for ecosystem functions like retention of organic material, which were once maintained by indigenous taxa such as extinct or endangered unionid mussels. Ecosystem engineers like *Dreissena* and *Corophium* increase the biocomplexity (Colwell 1998): *Dreissena* colonies provide a spatially complex habitat for smaller invertebrates, which also may profit from excretions by the mussels. The extensive muddy layers built by *Corophium* not only visualize the amount of retained particulate matter, they are probably also a site of extensive microbial transformations of organic matter. The activity of these exotics intensifies energy flow through freshwater food webs and nutrient cycling and thus the capacity for self-purification. Such ecosystem services render them not only to be exotic invaders endangering native biodiversity but also to be keystone species. Biological invasions will continue to occur and we are still far away from making sound risk assessments of present invaders and predictions of future invaders.

IMPACTS OF ZEBRA MUSSELS ON AQUATIC COMMUNITIES AND THEIR ROLE AS ECOSYSTEM ENGINEERS

ALEXANDER Y. KARATAYEV[1*], LYUBOV E. BURLAKOVA[1] &
DIANNA K. PADILLA[2]
[1]*Department of Biology, Stephen F. Austin State University, Nacogdoches, TX, USA*
[2]*Department of Ecology and Evolution, State University of New York at Stony Brook, Stony Brook, NY, USA*
[*]Corresponding author *akaratayev@sfasu.edu*

Abstract

Zebra mussels (*Dreissena polymorpha*) are not only an extremely aggressive invasive species, often dominating water bodies they invade, they are also very effective ecosystem engineers, altering the environments they invade. They are effective engineers, altering both ecosystem structure and function. They change existing and provide new habitat for other organisms, affect trophic interactions and the availability of foods for both pelagic species and other benthic species, and they affect the rates of other ecosystem processes including mineralization of nutrients, oxygen availability and sedimentation rates. These physical impacts on the environment feedback directly to other species that interact with or are impacted by zebra mussels, or indirectly through food chains, disturbance, succession, or other longer-term community and ecosystem processes.

1 Introduction

To date, *Dreissena polymorpha* has been the most aggressive freshwater invader worldwide. Dreissenids are the only freshwater bivalves that attach to hard substrates in high densities and have a planktonic larval stage. This life history facilitates their abilities as invaders, and allows them to become enormously abundant when introduced into a new water body. Once introduced their populations can grow rapidly, and the total biomass of a population can exceed 10 times that of all other native benthic invertebrates (Sokolova et al. 1980a; Karatayev et al. 1994a; Sinitsyna & Protasov 1994). The zebra mussel is frequently competitively dominant over native benthic fauna, and can impact all components of the freshwater ecosystem, especially benthic animals (Sokolova et al. 1980b; Karatayev et al. 1994a; Karatayev & Burlakova 1995a; Karatayev et al. 1997, and others). Most of the impacts of zebra mussels in freshwater systems are a direct result of their functioning as ecosystem engineers.

1.1 WHAT IS AN ECOSYSTEM ENGINEER?

The definition, role and consequences of species as ecosystem engineers were formalized in two papers by Jones et al. (1994, 1997). They defined ecosystem engineers as species that "directly or indirectly control the availability of resources to other organisms by causing physical state changes in biotic or abiotic materials" (Jones et al. 1994, 1997). Engineering organisms can cause physical modification of the environment, and influence the maintenance or creation of habitats. Their ecological effects on other species can occur because of physical state changes caused, either directly or indirectly, by the engineer, and because engineers affect the control and use of resources by other species.

E. Leppäkoski et al. (eds.), Invasive Aquatic Species of Europe, 433–446.

Many different species have been identified as ecosystem engineers in all types of habitats, and their impacts on systems can vary tremendously. There are many examples of ecosystem engineers from aquatic habitats. These include (but are not limited to) fishes in rivers (Flecker 1996), insects in streams and pools (Wotton et al. 1998), crayfish in riffles and pools (Statzner et al. 2000), angiosperms in marshes (Bruno 2000), and especially bivalves (Dame 1993, 1996).

1.2 BIVALVES AS ECOSYSTEM ENGINEERS

The role of bivalves in and impacts on aquatic ecosystems has long been recognized, however, most of this work has focused on marine bivalves (reviewed in Dame 1993, 1996). In part this is because the ecosystem effects of species are most obvious when they are dense or extremely abundant, as is often seen with marine bivalves that form dense beds or reefs. Dame (1996) identified four main ecosystem level roles of bivalves: trophic, nutrient cycling, structural, and monitors and indicators. Suspension feeding by bivalves will have large impacts on the plankton as well as suspended sediments. The reduction of filtering and suspension feeding bivalves will greatly alter many aspects of the planktonic community and affect benthic-pelagic coupling. Bivalves affect nutrient cycling by consuming particulate and dissolved organic matter and excreting inorganic nutrients. Because of their relatively large body sizes and rapid processing rates, they can have large impacts on nutrient cycling. Structurally, bivalves impact both the physical and biological environment. Physically, bivalves create structure with their shells and can move or stabilize sediments. Biologically they affect community structure (both in the water column and on the benthos) and can influence community stability, diversity and interspecies links. Marine bivalves have also been important indicators of environmental stress and their shells provide important records of environmental changes over both the short and long term.

Most of the research on the role of bivalves as ecosystem engineers has focused on native species, in environments where they are abundant and clearly play important roles. Attention to the natural role of bivalves in these communities has been particularly focused on areas where native species are over harvested or are lost due to disease or human disturbance, resulting in dramatic changes in ecosystems as a result of the loss of these important engineering species. The alternative situation that has recently gained attention is when an invasive bivalve causes dramatic changes in an environment, such as has been seen when the mat forming bivalve, *Musculista senhousia*, invaded near-shore habitats in North America (Crooks & Khim 1999).

2 Zebra mussels as ecosystem engineers

Zebra mussels have all of the properties of marine ecosystem engineers. Just by their presence, they change the characteristics of biotic and abiotic environments in aquatic systems. Other changes are caused by the activities of zebra mussels, especially their feeding and filtering. The changes caused by the presence or activities of zebra mussels can be systems wide, or have just local impacts. Zebra mussels and their activities will impact multiple levels in aquatic systems, affecting species composition, species interactions, community structure, and ecosystem properties. These changes or impacts may be direct or indirect in action (Fig. 1).

2.1 CHANGES TO THE ENVIRONMENT CAUSED BY ZEBRA MUSSEL STRUCTURES

Zebra mussels are macroinvertebrates with hard, calcium carbonate shells. This hard shell surface increases the substratum available for other species that live attached to or associated with hard substrata, including other zebra mussels.

Zebra mussels attach to hard substrata and each other with byssal proteinacious threads, creating complex three-dimensional structures. This complex creates habitat and refuge for a variety of species, which would otherwise not be common or present as epifauna in a soft bottom habitat.

Their shells do not decompose or dissolve readily in lakes that can support zebra mussel populations. Therefore the shells and shell fragments of dead animals can accumulate on the bottom, creating reef-like structures. These reefs provide additional hard substrata for animals and create a hard-substratum habitat in an otherwise soft sediment environment.

2.2 CHANGES TO THE ENVIRONMENT CAUSED BY ZEBRA MUSSEL FUNCTIONING

Zebra mussels not only cause changes to the environment due to their physical presence, they also have large impacts as a direct result of their biological activity.

2.2.1 Water clarity

Zebra mussels constantly filter the water for both feeding and respiration. Water currents are created by the beating of cilia on their ctenidia (gills), and water is almost constantly circulated through their siphons, and across their gills. As water is moved across their gills, particulates are constantly removed. Zebra mussels are extremely efficient at removing particulates from water, and seem to remove all particles larger than 0.4 μm in an unselective manner (Roditi et al. 1996). However, they are very selective in which of these particles they consume (Baker et al. 2000). Filtered particles are either consumed or bound in mucus, preventing immediate resuspension. This efficient filtering by large numbers of zebra mussels will greatly increase water clarity because both inorganic particulates and plankton will be removed.

Increasing water clarity has several impacts on aquatic ecosystems. Depth of the photic zone is extended, total lake volume available for photosynthesis by phytoplankton is greatly enlarged, and macrophytes, whose depth distribution is frequently limited by light, can live at greater depths and cover more of the bottom area. In addition, more light will reach more areas of the bottom of a water body, enhancing primary productivity over greater areas of the bottom.

2.2.2 Deposition on the bottom

When zebra mussels are in a water body, they build a direct connection between the planktonic portion of the water body and the benthos (benthic-pelagic coupling) and greatly enhance the rates of deposition of both organic and inorganic material on the bottom. Filtered particles (Section 2.2.1) are sorted, and either consumed, or rejected. Rejected particles are bound in mucus and expelled as pseudofeces, deposited onto the bottom. In general, this will result in greatly increased organic content of sediments, and provides and enhanced food subsidy for benthic deposit feeders. In addition, sedimentation rates can increase several orders of magnitude.

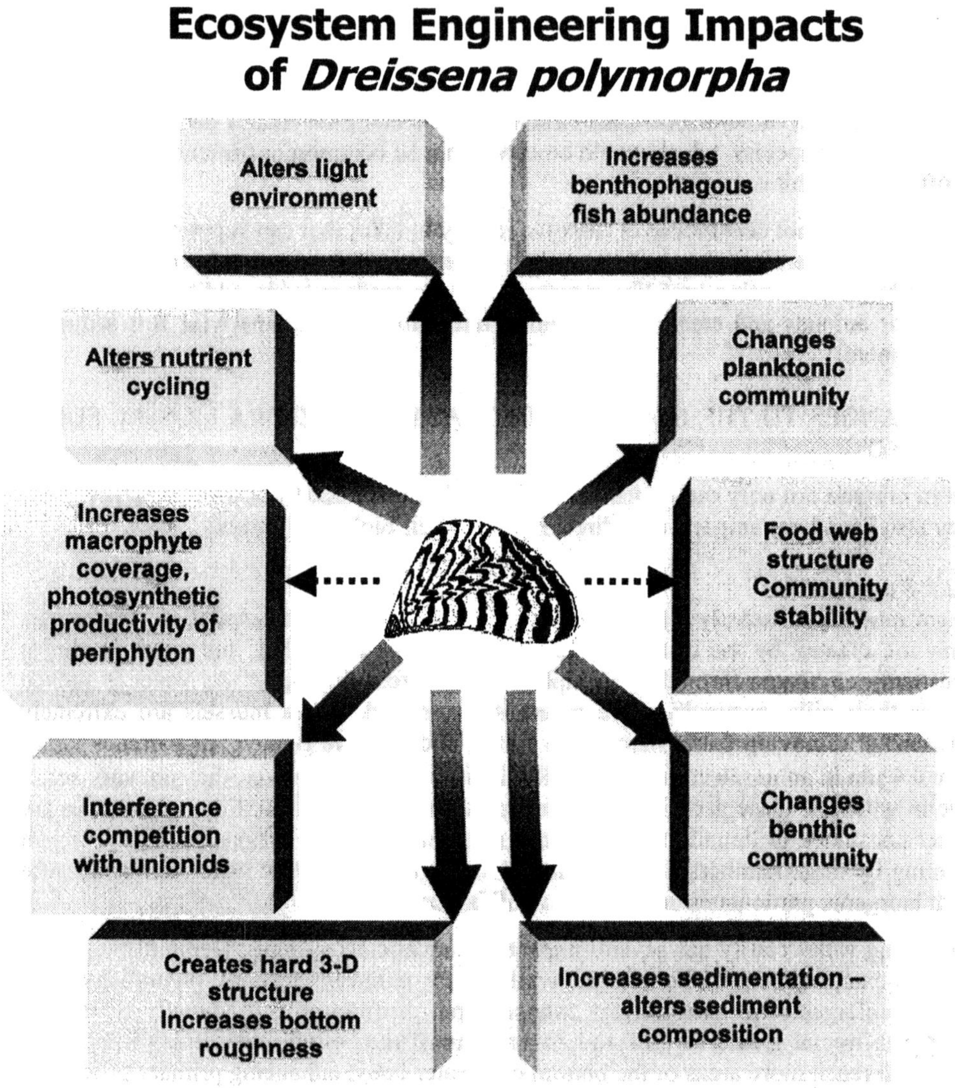

Figure 1. Direct (solid arrows) and indirect (dashed arrows) impacts of *Dreissena* on invaded aquatic ecosystems.

2.2.3 Mineralization and chemical cycles

Zebra mussel physiological activities can have a large impact on cycling of elements in the ecosystem. Through respiration they can have a large impact on oxygen levels, particularly in systems where photosynthesis may be limited by light (Effler & Siegfried

1994; Caraco et al. 2000). Because zebra mussels can be enormously abundant, their high rates of consumption and excretion can affect phosphorus and nitrogen cycles. In addition, zebra mussels secrete hard calcium carbonate shells. These shells are very stable, and will remain long after the animal has died.

2.2.4 Biotic community changes

Zebra mussels can cause a variety of changes in biotic communities as well. Selective feeding and removal of phytoplankton can directly cause changes in the composition of the plankton. Zebra mussels are limited in the size of particles they effectively remove from the plankton, and these limitations appear to be size based rather than quality based, and result from mechanical or functional limitations. Zebra mussels filter particles from 0.4 - 750 μm (Ten Winkel & Davids 1982; Mikheev et al. 1994; Roditi et al. 1996) with some reports to 1200 μm (Horgan & Mills 1997). Changes in substrata availability and the presence of refuges caused by zebra mussel structures (Section 2.1) create habitat for species that would otherwise be infrequent or absent in the environment. Increases in the organic composition of sediments creates a food base for benthic feeding species, again increasing their abundance relative to environments without zebra mussels (Karatayev et al. 1983, 1994a; Karatayev & Burlakova 1992, 1995a; Botts et al. 1996; Stewart et al. 1998).

In addition to these direct impacts on biotic communities, the activities of zebra mussels can have indirect impacts on systems. By increasing light, species composition and community structure can change. Selective removal or selective impact on different species can alter competitive interactions among species in the plankton. Changes in nutrient abundance and cycling can affect community diversity and stability. And, changes in community composition and dynamics can alter food webs and species not directly impacted by zebra mussels.

3 Species and system responses to changes in the environment caused by zebra mussels

The long history of zebra mussel invasion of freshwaters across eastern and western Europe has provided biologists with many systems where the direct and indirect impacts of zebra mussel ecosystem engineering can be measured. Much of the work has been done in regions that are part of the former Soviet Union, Eastern and Western Europe (refs. in Limanova 1964, 1978; Stanczykowska 1977, 1997; Schloesser et al. 1994; Karatayev et al. 1997, 1998). Complimentary work has been done in North America where zebra mussels have invaded lakes and rivers since the mid-1980s. In general, the North American research supports the findings in Europe.

3.1 SPECIES COMPOSITIONAL CHANGES

Dreissena provides food and shelter for many benthic invertebrates, and they increase density and biomass in zebra mussel beds, while others (mainly filter feeders) may decrease or disappear from the community (Sebestyen 1937; Sokolova et al. 1980b; Karatayev et al. 1983; Karatayev & Lyakhnovich 1990; Karatayev & Burlakova 1992; Stewart et al. 1998, 1999).

After *Dreissena* invade water body, they dominate the benthos, forming dense populations of very effective filter feeders. This causes a radical shift in the benthic trophic structure (Lvova-Kachanova & Izvekova 1978; Sokolova et al. 1980a,b; Karatayev & Burlakova 1992; Karatayev et al. 1994a, 1997). Native filter feeders are out-competed by *D. polymorpha*, and decrease in abundance. Simultaneously, animals that are deposit feeders greatly increase in abundance (Karatayev & Burlakova 1992; Karatayev et al. 1994a).

Historically, prior to the *D. polymorpha* invasion, the benthic community of Lukomskoe Lake (Belarus) was dominated by species considered to be detritus gatherers (45% of total benthic biomass) and detritus filterers (41%) (Karatayev & Burlakova 1992; Karatayev et al. 1997). After *D. polymorpha* invaded in late 1960s the relative abundance of detritus gatherers increased to 70% of the total biomass (excluding zebra mussel biomass), due to the increased food supply provided by zebra mussel pseudofeces and feces. The relative abundance of predators and omnivores in the community also increased, and the proportion of native detritus filterers decreased 8-fold. Excluding *D. polymorpha*, the benthic community was dominated by animals using food from sediments (Karatayev & Burlakova 1992; Karatayev et al. 1994a). The same patterns have been observed in 4 other water bodies studied in different regions across the former Soviet Union (Karatayev et al. 1994a).

Shelters created by zebra mussel shells have a positive impact on species abundance, providing refuges from predation and other disturbances. Shell habitats created by zebra mussels have been shown to be the primary mechanism for increased abundances of snails and gammarid amphipods (Botts et al. 1996; Stewart et al. 1998). Stewart et al. (1999) suggested that elevated invertebrate densities in natural *Dreissena* beds are caused by reduced predation rates on these organisms, as well as invertebrates immigrating to *Dreissena* beds from other habitats to avoid predators. They found an 8-fold increase in snail survivorship in the presence of *Dreissena* shells.

Dreissena have also been shown to have positive effect on isopods (*Asellus aquaticus*, Wolnomiejski 1970; Karatayev & Lyakhnovich 1990), larval chironomids (*Microtendipes* gr. *chloris* and *Limnochironomus* gr. *nervosus*, Wolnomiejski 1970), leeches (*Helobdella stagnalis*, Wolnomiejski 1970), snails (*Lymnaea lagotis*, Karatayev et al. 1983; *Amnicola*, reviewed in Strayer 1999), amphipods (*Gammarus lacustris*, Karatayev et al. 1983; Karatayev & Lyakhnovich 1990), and oligochaetes (*Aulodrilus limnobius*, *Psammoryctides albicola*, *Limnodrilus hoffmeisteri*, Afanasiev 1987). Botts et al. (1996) found that in Lake Erie the densities of amphipods, chironomids, oligochaetes, turbellarians, and hydrozoans were significantly higher in sand with zebra mussel druses than in bare sand. However, not all species are positively affected by the presence of zebra mussels. Negative effects of zebra mussels were found for native suspension feeders: chironomid larvae, *Glyptotendipes paripes* (Sokolova et al. 1980a,b), *Stictochironomus psammophilus* (Karatayev et al. 1983) and sphaeriid bivalves (Strayer et al. 1998).

Impacts of zebra mussels on unionid bivalves are of special concern because they are presently endangered or threatened in many environments where they were previously in high abundance. High densities of zebra mussels seem to have only negative affects on unionids (reviewed by Karatayev et al. 1997; Burlakova et al. 2000). By attaching to unionids, zebra mussels can make it more difficult for them to burrow and move through sediment. They can weight down their host unionid, resulting in burial in very soft sediments, can increase drag and the likelihood of dislodgement by water motion for species living near shore, prevent opening unionid valves for respiration, feeding and reproduction, or preventing the closing of the valves. *D. polymorpha* may directly compete with unionid bivalves for food, available space, and cause unionids physiological stress and symptoms of starvation, including declines in glycogen content, declines in organic matter contents in guts, and increased shell deformities.

The impact of zebra mussels on unionids host depends on: (i) *D. polymorpha* density (Lewandowski 1976; Ricciardi et al. 1995; Burlakova et al. 2000); (ii) time since invasion by *D. polymorpha* (Sebestyen 1937; Karatayev et al. 1997; Burlakova et al. 2000); (iii) characteristics of bottom substrata (Karatayev 1983; Burlakova et al. 2000); (iv) unionid species (Lewandowski 1976; Arter 1989; Haag et al. 1993); and (v) unionid sex (Haag et al. 1993).

Mass moralities of unionids caused by *D. polymorpha* overgrowth are most common during the initial stages of colonization by zebra mussels a new water body, when mussel populations are growing rapidly. This type of mass mortality is well documented both in European (Sebestyen 1937; Karatayev & Burlakova 1995b, Burlakova et al. 2000) and North American water bodies (Haag et al. 1993). However, *D. polymorpha* invasion did not result in complete elimination of unionid bivalves in any European water bodies. Moreover, after initial peaks in zebra mussel abundance, *D. polymorpha* coexist with unionid bivalves (Lewandowski 1976; Karatayev 1983; Miroshnichenko et al. 1984; Miroshnichenko 1987; Ponyi 1992; Nichols & Amberg 1999; Burlakova et al. 2000). Soft sediments provide refuges for unionids. By burrowing in silt, unionids can clean incrusted mussels from their shells (Arter 1989; Nichols & Amberg 1999; Burlakova et al. 2000). Water-level fluctuations and waves may also help to remove attached zebra mussels from unionid shells (Schloesser et al. 1997).

3.2 COMMUNITY CHANGES

As a rule, when *D. polymorpha* invade new freshwaters, they become the dominant benthic species in terms of biomass, with 10-50 times the total mass of all other benthic invertebrates combined (Sokolova et al. 1980a; Lyakhnovich et al. 1988; Protasov & Afanasiev 1990; Karatayev 1992; Karatayev et al. 1994a; Karatayev & Burlakova 1995a). In addition, *Dreissena* aggregations or druses have a dramatic direct effect on benthic communities, and a totally different community forms in their presence (Karatayev et al. 1997). The presence of individual *D. polymorpha* on the bottom, however, does not change the qualitative or quantitative composition of the benthic community (Karatayev 1988; Karatayev et al. 1994a).

After *D. polymorpha* invaded Lukomskoe Lake (Belarus), only 26 of the previous 49 taxa of littoral benthic animals were part of both the sandy littoral community and the community found in zebra mussel druses (Karatayev 1983; Karatayev et al. 1983). After invasion, the sandy littoral community was very similar to pre-invasion, dominated by chironomids and oligochaetes, most of which were small and live within the sediment. The druse community was composed of larger animals such as snails, amphipods, isopods, trichopterans, and leeches. The dominant species in the sand habitat was the chironomid *Stictochironomus psammophilus*, however, only a single individual of this species was found in druses. The snail *Lymnaea lagotis*, and the amphipod *Gammarus lacustris* were dominant in zebra mussel druses, but only a single *L. lagotis,* and no *G. lacustris* were found in sandy areas. Similar patterns were found for subdominant taxa in these two habitat types (Karatayev 1988; Karatayev et al. 1983, 1994a). Invertebrate biomass (excluding zebra mussels) was 8 times greater in druses even though densities were 1.5 times lower than in sandy sediments, thus the community was dominated by larger species. Therefore, druses are responsible for the creation of a new community of bottom invertebrates, not generally found in sandy sediments. Large changes in species composition and density/biomass of benthic invertebrates have been found in all systems where zebra mussels have invaded (Wiktor 1969; Kharchenko & Protasov 1981; Stewart et al. 1998).

Indirect effects, such as increasing light penetration, also have large effects on community composition once zebra mussels invade a water body. Increased light penetration positively affects macrophytes and benthic algae beds (Lyakhnovich et al. 1988; Reeders & bij de Vaate 1990; Skubinna et al. 1995; Lowe & Pillsbury 1995). Vascular macrophytes beds positively affect abundance and diversity of benthic invertebrates, providing them additional food and substrate. In contrast, thick beds of filamentous algae could decrease abundance of macroinvertebrates (Haynes et al. 1999).

3.3 CHANGES IN INTERSPECIES INTERACTIONS

When zebra mussels are added to freshwater communities, they consume phytoplankton and therefore will compete with zooplankton for microalgal foods. Decreases in phytoplankton and concomitant decreases in zooplankton abundance have been seen in a variety of water bodies (Lyakhnovich et al. 1983, 1988; Mitrakhovich et al. 1983; Reeders et al. 1993; Karatayev et al. 1994a; Karatayev & Burlakova 1995a; Pace et al. 1998, and others). Also, as indicated above, zebra mussels compete for planktonic food with benthic suspension feeders, especially unionid bivalves.

In addition to competition for food, zebra mussels can compete for space on hard substrata, and convert sandy substrata to hard substrata, displacing the typical sandy community (Wiktor 1969; Kharchenko & Protasov 1981; Karatayev et al. 1983, 1994a, 1997; Stewart et al. 1998). There is also a suggestion that zebra mussels may compete for benthic space with fish by covering areas used by fish for nests and rearing their young (Marsden 1997).

In addition to these direct effects on species interactions, zebra mussels have indirect effects on species interactions through their effects on nutrient availability and turnover. Inedible algae may increase in abundance through competitive release of nutrients due to the grazing by zebra mussels reducing other algal populations, or by increased availability of limited nutrients through zebra mussel excretion (Arnott & Vanni 1996).

The abundance of benthic feeding fishes is usually enhanced by the presence of zebra mussels because of the increase in density and mean body size of benthic invertebrates (Lyakhnovich et al. 1988; Karatayev 1992; Karatayev & Burlakova 1995a). Planktivorous fish, however, may be negatively impacted if the presence of zebra mussels is associated with decreased zooplankton abundance, or if increased water clarity increases predation rates on larval fishes (Francis et al. 1996). Piscivorous fish may be positively affected because of an increase in benthic feeding fish.

3.4 CHANGES IN FOODWEBS AND ENERGY FLOW

To our knowledge the only studies on the impact of *D. polymorpha* on food webs and the energy flow through trophic levels have been conducted in Lukomskoe Lake, Belarus (Karatayev 1992; Karatayev & Burlakova 1995a). Before the invasion of zebra mussels, the total primary production in this lake was 2,596 kcal m^{-2} (phytoplankton 98%, macrophytes 2%). The secondary production (nonpredatory zooplankton + nonpredatory zoobenthos) was 3.7% of total primary production. Fish production was 0.15% of the total primary production (Karatayev & Burlakova 1995a) (Table 1).

Table 1. Biomass (B, kcal m^{-2}) and production (P, kcal m^{-2} year^{-1}) of Lukomskoe Lake prior to and following the invasion of zebra mussels (modified from Karatayev & Burlakova 1995a).

Trophic level	Pre-invasion		10 years after invasion		20 years after invasion	
	B	P	B	P	B	P
Phytoplankton	50.9	2,544.5	12.5	624.5	16.1	805.0
Macrophytes	40.9	51.1	132.8	166.0	81.3	101.6
Total primary production	91.8	2,595.6	145.3	790.5	97.5	906.6
Nonpredatory zooplankton	6.40	93.8	0.80	14.8	3.91	68.1
Nonpredatory zoobenthos	0.38	1.4	3.22	12.2	6.44	24.5
Zebra mussels	0.00	0.0	27.80	16.7	40.00	24.0
Nonpredatory zooplankton + nonpredatory zoobenthos	6.78	95.2	31.82	43.7	50.35	116.6
Predatory zooplankton	1.24	13.2	0.23	5.4	0.99	17.2
Predatory zoobenthos	0.02	0.1	0.28	1.0	0.56	2.0
Predatory zooplankton + predatory zoobenthos	1.26	13.3	0.51	6.4	1.55	19.2
Benthivorous+planktivorous fish	8.75	3.5	17.00	6.8	21.25	8.5
Piscivorous fish	1.25	0.5	3.00	1.2	1.25	0.5

Ten years after the zebra mussel invasion in Lukomskoe Lake, macrophyte production had increased 3.3 times, and phytoplankton production decreased more than 4 times (Table 1). Total primary production decreased more than 3 times. Although secondary production declined from 95 to 44 kcal m^{-2}, the relative role of benthic invertebrates dramatically increased from 3 to 77% of the total production (nonpredatory zooplankton + zoobenthos). Fish production more than doubled. The conversion of primary production to higher trophic levels increased for nonpredatory invertebrates from 3.7% (before zebra mussels invasion) to 5.5% (after invasion), and for fish from 0.15% (before invasion) to 1% (after invasion) of the total primary production. This high rate of fish production is typical of commercial fish ponds, but is much higher than most natural lake communities (Bullion & Winberg 1981).

Twenty years after the initial invasion of the Lukomskoe Lake, the zebra mussel population density declined and, in terms of biomass, became relatively stable. Total primary production had increased 13%, and the contribution of macrophytes to the total decreased from 20% to 11% (Table 1). The production of nonpredatory zooplankton and zoobenthos more than doubled. Fish production also increased slightly, and remained approximately at 1% of primary production (Karatayev 1992; Karatayev & Burlakova 1995a). Therefore, *Dreissena* became the major consumer of primary productivity, and the most important conduit of energy fixed through photosynthesis by phytoplankton to higher trophic levels in the ecosystem (Table 1).

3.5 SYSTEMWIDE EFFECTS

In most freshwater ecosystems, the benthic community and the pelagic community are considered functionally separate. Dynamics in planktonic systems are often studied completely without consideration of the benthos. In most freshwaters, this distinction is probably not a bad approximation, because in the water column primary productivity is driven by external nutrient inputs, and zooplankton feed primarily on phytoplankton. Larger nekton are often planktivorous, although in Europe, there are many benthic feeding fishes. The major link between the benthos and the pelagic system in water bodies without *D. polymorpha* has been through predation by nektonic feeders, fishes, on benthic invertebrate production. Benthic production is driven by the slow rain of suspended organic material to the bottom and to a small extent by the filtering activity of suspension feeders such as unionids. In the benthos, most species are considered to feed on detritus or other benthic organisms. Thus, the typical benthic freshwater system is considered to be detritus dominated, rather than relying on large amounts of primary productivity or direct links to planktonic processes. And, in general, the benthos are not capable of controlling processes or dynamics in the planktonic system.

However, zebra mussels are functionally different than most benthic invertebrates in freshwater. Although they have large impacts on the structure and function of the benthos (Section 3.1, 3.2), they also have a large direct impact on the planktonic community. They filter large volumes of water and transport this material removed from the water column to the benthos (Lvova 1979; Lvova et al. 1980; Wiktor 1969; Shevtsova and Kharchenko 1981; Protasov et al. 1983; Stanczykowska 1977; Reeders et al. 1989; Karatayev & Burlakova 1995a,b; Karatayev et al. 1997). Therefore, they provide a direct link between processes in the plankton and those in the benthos and by their deposition of pseudofeces and feces, provide a direct conduit for primary productivity in the water column to the benthos to a much larger extent than any other process. In addition, they are capable of controlling pelagic processes by removal of particulate matter and increasing water transparency and the volume of the photic zone. They impact phytoplankton standing stock, and thus can influence planktonic trophic interactions (Lyakhnovich et al. 1983, 1988; Mitrakhovich et al. 1983; Shevtsova et al. 1986; Karatayev 1992; Reeders et al. 1993; Karatayev et al. 1994a; Karatayev & Burlakova 1995a). In addition, phytoplankton productivity will influence zebra mussel growth and reproduction, producing the possibility of direct feedback between the benthos and the pelagic system that were not present prior to their invasion.

D. polymorpha shift materials form the pelagic to the benthos by transporting suspended matter including detritus, phytoplankton, bacterioplankton, and small zooplankton from the water column to the benthic community (Lvova 1980; Karatayev 1988, 1992; Lyakhnovich et al. 1983, 1988; Kharchenko & Lyashenko 1985; Karatayev & Burlakova 1992, 1995a; Reeders et al. 1993). A small portion of filtered material is metabolised and used for *D. polymorpha* growth and the rest is available to other benthic organisms.

The movement of large amounts of seston from the plankton to the benthos induced changes in all aspects of freshwater ecosystems after the invasion of *D. polymorpha* (Lyakhnovich et al. 1983, 1988; Mitrakhovich et al. 1983, Reeders & Bij de Vaate 1990; Karatayev & Burlakova 1992, 1995a; Reeders et al. 1993; Karatayev et al. 1997). For example, in Lukomskoe Lake shortly after *D. polymorpha* invasion (late 1960s) water transparency in summer increased from 1.8 to 4 m, and seston concentrations decreased 3-fold (Lyakhnovich et al. 1983; Karatayev & Burlakova 1995a). The amount of dissolved organic mater in the water column after *D. polymorpha* invasion also decreased. Increased water transparency resulted in an expansion of macrophyte cover (from 6 to 30% of total lake area) mainly due to an increase in the depth at which macrophytes can grow (from 2.5 to 5 m). After the invasion of zebra mussels, the biomass of phytoplankton and zooplankton declined more than 10 times. In contrast, the biomass of benthic invertebrates increased more than 10 times. Fish productivity doubled, and the composition of the commercial catch in the Lukomskoe Lake is now characterized by benthophagous fishes which feed mainly on zebra mussels including roach, rudd, white bream and bream (Karatayev 1983, 1988, 1992; Lyakhnovich et al. 1983; Mitrakhovich et al. 1983; Lyakhnovich et al. 1988; Karatayev et al. 1994b; Karatayev & Burlakova 1995a).

By the mid-1980s, when *D. polymorpha* abundance in Lukomskoe Lake declined after its initial invasion (late 1960s), summer transparency decreased to 3 m, but remained above pre-invasion levels (1.8-2.0 m). Similar patterns were found for both phytoplankton and zooplankton; their biomass decreased when zebra mussels initially reached very high population density, but increased after *D. polymorpha* densities declined. Again, they did not return to their pre-invasion abundance. The extent of macrophytes also decreased from 30% to 20% of the lake surface area, but still remains higher than pre-invasion levels (6%) (Karatayev 1992; Karatayev & Burlakova 1995a).

D. polymorpha was associated with similar changes in the Narochanskie lake system (mesotrophic Naroch Lake, eutrophic Myastro Lake and highly eutrophic Batorino Lake). *D. polymorpha* invaded in the mid-1980s, after approximately 40 years of research on these lakes (Ostapenya et al. 1993, 1994a,b). Shortly after the invasion, water transparency increased 1.3-2.4 times, seston was reduced 2.3-6.9 times, and chlorophyll concentration decreased 2.7-6.9 times. Organic carbon content, BOD_5, primary production, respiration and biomass of phytoplankton also decreased (Ostapenya et al. 1993, 1994a,b). After zebra mussels invaded, highly eutrophic Lake Batorino became eutrophic (Ostapenya et al. 1994b), and eutrophic Lake Myastro became slightly eutrophic (Ostapenya et al. 1994a). Thus, *D. polymorpha* can be used to control the negative affects of anthropogenic eutrophication including increased phytoplankton abundance and decreased water clarity (Karatayev 1983, 1992). Some west-European scientists also have proposed using *D. polymorpha* as biofilters to decrease the effects of anthropogenic eutrophication in lakes (Reeders et al. 1989, 1993; Reeders & bij de Vaate 1990; Noordhuis et al. 1992).

Similar changes subsequent to the recent appearance of *Dreissena* have reported in the North American lakes. Following *D. polymorpha* invasion, water transparency (Leach 1993; Fahnenstiel et al. 1995a; Caraco et al. 1997), benthic algal abundance (Lowe & Pillsbury 1995), and macrophyte beds (Skubinna et al. 1995) have increased. At the same time, turbidity has decreased (Skubinna et al. 1995), as has chlorophyll, and phytoplankton abundance and production (Leach 1993; Fahnenstiel et al. 1995a,b; Caraco et al. 1997). The density of native benthic animals increased and was accompanied with changes in benthic community structure (Stewart & Haynes 1994; Wisenden & Bailey 1995; Botts et al. 1996). In all cases, patterns of the effects of *D. polymorpha* are similar to those found in European freshwaters.

Based on data from long-term studies in Europe and recent data from North America, we suggest the following generalizations of aquatic ecosystem response to *D. polymorpha* invasion.

(i) Water transparency increases 1.5-2 or more times (Stanczykowska 1968, 1977; Lvova 1979; Karatayev 1983, 1992; Fahnenstiel et al. 1995a).

(ii) The amount of seston in the water column decreases 1.5-10 times (Lvova et al. 1980; Kharchenko & Lyashenko 1985; Leach 1993; Reeders et al. 1993; Ostapenya et al. 1994a,b; Karatayev & Burlakova 1995a).

(iii) The amount of organic matter in the water column decreases (Kharchenko & Lyashenko 1985; Ostapenya et al. 1994a,b; Karatayev & Burlakova 1995a).

(iv) The rate of biogeochemical conversion from organic to inorganic matter increases (Kharchenko & Lyashenko 1985; Shevtsova 1989).

(v) Biochemical oxygen demand in the water column (BOD_5) decreases up to 1.5 times (Kharchenko & Lyashenko 1985; Ostapenya et al. 1994a,b)

(vi) The total density and biomass of phytoplankton decreases 1.5-4 times, and total chlorophyll decreases (Lyakhnovich et al. 1988; Karatayev 1992; Reeders et al. 1993; Ostapenya et al. 1994a,b; Karatayev & Burlakova 1995a; Caraco et al. 1997). However, the clearance rates and pseudofeces production of *Dreissena* differs for different algal species consumed (Mikheev et al. 1994, Berg et al. 1996; Ten Winkel & Davids 1982), therefore, the net effect will depend on algal community structure as well.

(vii) There are considerably less data on the effects of zebra mussels on bacterioplankton. Effects may depend on the trophic status of the water body and the bacterioplankton community structure. Zebra mussels can effectively reduce large (>0.9 μm) bacteria, while smaller size bacteria could escape predation. Simultaneously, high rates of mussel nutrient excretion may facilitate growth of the small bacteria under low nutrient conditions (Cotner et al. 1995). Small bacteria can also benefit from *Dreissena* grazing on flagellated protozoans, their major predators (Findlay et al. 1998). Ostapenya et al. (1994a,b) found somewhat increased abundance of bacterioplankton after the introduction of zebra mussels.

(viii) With increased transparency, the photic zone for macrophytes greatly increases. This results in a much greater portion of the lake bottom covered with macrophytes and greater total macrophyte production (Lyakhnovich et al. 1988; Reeders & bij de Vaate 1990; Skubinna et al. 1995).

(ix) With increased light reaching the bottom, periphyton and benthic algae increase in both standing stock and primary productivity as has been seen in North America (Lowe & Pillsbury 1995).

(x) Zooplankton abundance decreases (Lyakhnovich et al. 1983; Mitrakhovich et al. 1983; Mitrakhovich 1984; Shevtsova et al. 1986; Karatayev et al. 1994a; Pace et al. 1998). Microzooplankton (rotifers, protozoans, copepod nauplii) may be affected more than larger crustaceans (MacIsaac et al. 1995; Pace et al. 1998). Macrophyte beds associated with clearer water may provide increased spatial refuge for large cladocerans (Mayer et al. 2000). After zebra mussels established in Lukomskoe Lake, the structure of zooplankton community changed - the former dominant complex of species typical for eutrophic lakes shifted to that typical for mesotrophic lakes (Mitrakhovich et al. 1983; Mitrakhovich 1984).

(xi) Zoobenthos change in abundance and in taxonomic and trophic structure as discussed above (Wiktor 1969; Kharchenko & Protasov 1981; Karatayev 1983, 1992; Karatayev et al. 1983; Karatayev & Burlakova 1992, 1995a; Botts et al. 1996; Stewart et al. 1998).

(xii) The effect on fish may be direct or indirect. The direction and intensity of these effects are dependent on the feeding method of the majority of the fish in a water body. However, the complexity of food web interactions, and because different life stages of fishes (and zebra mussels) can have different trophic roles, precise predictions are impossible to make. *Dreissena* planktonic larvae can compose up to 70% of zooplankton density in summer months and are readily consumed by many species of fish (Karatayev et al. 1994b, Molloy et al. 1997). In general, we may expect an enhancement of all benthic feeding fishes, even those that do not feed on *Dreissena*, as *Dreissena* increases the biomass of other benthic invertebrates (Kharchenko & Protasov 1981; Karatayev 1983, 1992; Lyakhnovich et al. 1983, 1988; Karatayev & Burlakova 1992, 1995a; Stewart & Haynes 1994, and others). Direct or indirect negative effects could be expected on planktivorous fish through competition with larvae for zooplankton, by increasing fish predation on larvae via increased water clarity (Francis et al. 1996). However, to date, no negative changes in the abundance of planktivorous fish have been found (Trometer & Busch 1999; Mayer et al. 2000).

4 Heterogeneity of impacts of zebra mussels

Although many generalizations can be made about the impacts of zebra mussels and their functioning on freshwater species and ecosystems, it is important to recognize that impacts will not be uniform across a given water body, and impacts and their consequences are likely to change over time.

4.1 LOCAL VERSUS SYSTEM WIDE EFFECTS/IMPACTS

The impacts of zebra mussels, or any other biological agent, are likely to be most intense close to the animals, depend on local and total abundance and densities, and the size of the water body being affected. Zebra mussels are benthic, and will be restricted to a more or less two-dimensional surface. However, the water column above the benthos is a three-dimensional habitat, and the amount of three-dimensional habitat affected by zebra mussels will depend on the size and bathymetry of the water body. Therefore, we should expect heterogeneity of impacts due to local effects and conditions, and this hinders our ability to predict system or lake-wide long-term impacts at the present time.

Zebra mussels will have a local impact by creating structure, and providing food and shelter for benthic species. However, this effect will not necessarily be important at distances away from druses. Similarly depending on water mixing rates, lake morphology, and turnover rates, the effects of zebra mussels clearing and filtering water will have very different effects (Reed-Andersen et al. 2000) and may be very local in deep water lakes (Ackerman et al. 2001).

Because species composition, food web links, and biotic interactions as well as physical characteristic including nutrient loading can vary tremendously among water bodies, it is presently difficult to predict more than general patterns expected impacts of zebra mussel invasion.

4.2 SHORT-TERM VERSUS LONG-TERM EFFECTS

It is well known that immediately after invasion, populations of *D. polymorpha* grow to very high densities, but due to density-dependent processes total sustainable biomass declines as the system changes. Ultimately densities well below the maximum achieved persist (Sebestyen 1937; Zhadin 1946; Lvova 1980; Karatayev 1983; Karatayev & Burlakova 1995a). However, all populations of zebra mussels do not stabilize and can vary widely over time (Ramcharan et al. 1992; Stanczykowska & Lewandowski 1993).

Shortly after initial invasion, when population levels are climbing and high, *D. polymorpha* will have their largest and most obvious effects on communities, and most of the impacts will be direct effects (Section 3; Karatayev et al. 1997). The effects of *D. polymorpha* on the aquatic communities after the initial stages of invasion are much less predictable, and much more likely to be caused by indirect effects through changes in the ecosystem. These changes may be much more influenced by local conditions that short term direct effects.

At present, we need much more work on whole system impacts of zebra mussel invasion, and especially the mechanistic basis of proposed impacts. Like all areas of science, there is a persistent danger in finding correlations and assuming a causal relationship. Natural systems vary tremendously, and spurious correlations are not uncommon. Therefore, before a given change in a system is attributed to zebra mussel invasion, we need to know the natural variability among systems, and the mechanisms by which zebra mussels cause such impacts. We also need to determine the likelihood that zebra mussel impacts are not strictly linear, but may be very non-linear, for example responding in a threshold fashion, resulting in alternate states in communities and systems rather than simple changes as a function of zebra mussel density. These same problems are faced by all scientists who study environmental impacts, especially those potentially linked to human activity. Use of scientific and statistical methodology developed to test for environmental impacts may prove to be especially useful for assessing and determining the impacts of zebra mussels and other aquatic invaders.

Risk Assessment

HAZARD ANALYSIS OF AQUATIC SPECIES INVASIONS

STEPHAN GOLLASCH
GoConsult, Hamburg, Germany
sgollasch@aol.com

Abstract
Anthropogenically supported invasions of nonindigenous species (NIS) in aquatic ecosystems seem to increase worldwide. Many aspects of invasions remain nearly unpredictable. Among them, unfortunately, are the most wanted answers to: Which species will invade, when will it arrive, where will the species invade and what impact may be caused by this new invader? Until today these questions can be answered only on a theoretical or broad scale. Accordingly an indication of habitats at risk can be given only on a limited base. We know that certain areas such as estuaries and areas with high input of NIS, such as ports, waterways and aquaculture sites, are high-risk areas for further human mediated species introductions. Matching salinity and climate conditions in donor and recipient region enable a first but incomplete estimation of the species' potential to survive in new habitats. Taking into account the voyage duration of a ship (short term voyages support the survival rate of specimens in the ballast tank) the picture comes clearer, but still is far from prediction. The need for an improved risk assessment becomes clear as one introduced species can cause severe damage to economy or environment.

1 Introduction

Scientists have tried to mathematically assess the invasion rate of nonindigenous species (NIS) since a very long time. Darwin (1900) estimated that 5% and Williamson (1989) that 10% of the (intentionally and accidentally) introduced species may form self-sustaining populations for at least several generations. It was estimated that approximately 10% of all established NIS will (periodically) occur in high or massive densities. This 10's rule was mainly based on introductions to terrestrial habitats (Holdgate 1986; Simberloff 1986, 1989; Williamson & Brown 1986). Williamson (1996) revised the 10's rule pointing out that "10" actually ranges between 5 and 20. It is not clear if this rule can be applied for aquatic ecosystems.

The number of NIS recorded in different coastal waters is significantly different, varying from less than 25 introduced species to more than 200 in other areas (Ben-Tuvia 1953; Rubinoff 1968; Ben-Eliahu 1972; Walford & Wicklung 1973; Farnham 1980; Krapp & Sconfietti 1983; Leppäkoski 1984, 1994; Knudsen 1989; Zibrowius 1991; Boudouresque et al. 1992; Utting & Spencer 1992; Galil 1994; Jansson 1994; Carlton &

E. Leppäkoski et al. (eds.), Invasive Aquatic Species of Europe, 447–455.

Cohen 1995; Gollasch 1996; Eno et al. 1997; Minchin 1997; Ruiz et al. 1997; Zaitsev & Mamaev 1997; Lodge et al. 1998; see regional reviews in this volume). It seems obvious that certain communities are more open for introductions than others.

The increasing utilisation of the coastal zone for aquaculture enforced the need to minimise unwanted impacts caused by NIS. Former studies indicate that shipping is one of the major vector of species invasion and further that each vessel entering coastal waters has the potential to introduce an unwanted NIS. Each introduced species can cause severe harm to the environment and economy.

In order to identify the potential of a vessel to introduce unwanted species the ballast water needs to be sampled and analysed. However, not all vessels calling for a port can be sampled as their number in certain areas is huge. Therefore a risk assessment or decision support system is needed to concentrate on target vessels containing ballast water and species therein of critical origin. A selection of critical vessels enables an effective application of ballast water treatment and management measures.

Several methods have been developed in order to identify and/or quantify the risks of future species introductions. The risk assessment methods follow the steps to evaluate:
(i) the probability that the introduced species will survive in its new environment
(ii) the probability that the introduced species will establish a self-reproducing population and
(iii) the probability that the introduced species will cause harm (Hayes 1997).

2 Review of risk assessment methods

2.1 AUSTRALIAN APPROACH

A structured approach to decision making concerning the risk posed by individual vessels is highly desirable for the effective administration of any countries' ballast water management regime. Critical factors can be taken into account concerning the potential risk posed by any vessel's voyage and as a consequence the action required of an individual vessel on a local specific basis (MEPC40/INF.7; Hayes 1997; see Hewitt and Hayes below in this chapter).

2.1.1 Decision support system

As a possible way to minimise the risk of introducing NIS with ballast water, Australia has proposed a Decision Support System to evaluate the risk posed by each incoming vessel. The risk assessment component takes into account such criteria as the port of uptake of the ballast water (climate and species composition), the treatment of the ballast water en-route, the tolerance of the species which could have been taken onboard with ballast water and transported to the area of planned discharge and the estimated survival rates of the species in the ballast water during the voyage.

The estimation of the survival rate is based on results achieved through sampling a ballast tank before departure as well as immediately after the ballast water uptake, and further during the voyage. Other aspects are the length of the journey and the daytime of the ballast water uptake. The daytime is of importance due to the daily migration of species in the water column. Several studies showed that with increasing time in the ballast tank the number of species and specimens decreased dramatically (MEPC40/INF.7; Hayes 1997).

The Decision Support System performs a tank-by-tank risk assessment based on information supplied by the ship's master, and allows international vessels to determine before they arrive in Australian waters if their ballast water poses a risk of introducing exotic marine pests. Tanks identified as carrying high-risk ballast water will require treatment and/or management by a method acceptable to the Australian Quarantine and Inspection Service (AQIS), including:

(i) exchange of ballast water at sea through sequential exchange, flow-through or dilution
(ii) non-discharge of high risk ballast tanks
(iii) tank-to-tank transfer, preventing discharge of high risk tanks' ballast water, and
(iv) comparable treatment options as they are developed.

To ensure accurate reporting by ships' masters of their uptake and ballast water management arrangements, AQIS will be verifying information and will advise masters about the records that must be maintained.

2.1.2 Target species - a black list

Lists of target species (recognised as harmful and unwanted) compiled by scientists and authorities are in preparation in different countries. The main purpose of the Australian target species list is to minimise the spread of these species within the country by preventing secondary introductions in domestic trade and to minimise future (multiple) introductions of these species by international shipping. At present Australia's target species are: *Balanus eburneus, Callinectes sapidus, Charybdis japonica, Hemigrapsus penicillatus, H. sanguineus, Lepeophtheirus salmonis, Limulus polyphemus, Nippoleucon hinumensis, Pseudodiaptomus marinus, Rhithropanopeus harrisi, Tortanus dextrilobatus, Hydroides dianthus, H. dirampha, Marenzelleria viridis, Pileolaria berkeleyana, Liza ramada, Neogobious melanostomus, Pagrus major, Siganus rivulatus, Blackfordia virginica, Maeotias inexspectata, Crepidula fornicata, Ensis directus, Limnoperna fortunei, Mya arenaria, Perna perna, P. viridis, Petricolaria pholadiformis, Philine* spp. *(auriformis?), Chaetoceros concavicornis, C. convolutus, Dinophysis norvegica, Pfiesteria shumwayae, Pseudo-nitzschia seriata, Pyrodinium bahamense (*var. *compressum)* and *Womersleyella setacea.* The list will be modified from time to time as additional information becomes available (Paterson 1996; Hayes pers. comm.).

A particular interest was voiced with regard to gaining a greater understanding of donor regions as an important aspect of risk assessment. It was noted that obtaining such information must be a two-way process: donor ports should attempt to provide information on species of concern that departing vessels may take on in ballast, whereas receiving ports should also seek to obtain such information from the donor ports. It is critical to maintain updated information through continuing port surveys, since the non-native biota of many ports and harbours are unknown or under constant change, due to plankton blooms, new invasions, water quality changes, and other factors.

2.1.3 New regulation

AQIS implemented the ballast water risk assessment approach on July 1st 2001. It applies to all vessels engaged in international voyages that carry ballast water or intend to discharge ballast in Australian waters (www.aqis.gov.au/docs/bulletin/ ab1200_13.htm). This new regulation received widespread support from various stakeholders in government and the shipping industry. Australia's current requirements for ballast water management include mandatory aspects, reporting by vessels as to whether they undertook ballast water exchange at sea, providing access to safe onboard sampling points, and the requirement to discharge sediment only on land. The mandatory aspects of the new arrangements include:

(i) accurate reporting to AQIS regarding ballast water arrangements under the Quarantine Act 1908. Vessels that do not carry ballast water are still required to undertake mandatory reporting to AQIS,
(ii) if required, undertaking exchange and/or other treatment and management options as directed by AQIS before discharging ballast water in Australia,
(iii) re-submission and/or updating of ballast water information provided when details for the voyage have altered,
(iv) access to safe onboard ballast sampling points,
(v) disposal of sediment on land resulting from ballast tank and/or hold cleaning in accordance with AQIS requirements, and
(vi) no discharge of ballast water within Australian waters without written permission from a quarantine officer.

After the new arrangements were implemented in July 2001, vessels in international voyages intending to discharge ballast in Australian waters are asked to manage their ballast water en route by:

(i) using the Australian Ballast Water Decision Support System by lodging ballast water information at the 'last port of call' or en route, and determining tanks posing a high risk to intended ports of discharge, and/or
(ii) undertaking an independent treatment procedure of ballast water (exchange or other comparable method accepted by AQIS) before entering Australian waters.

2.2 USA APPROACH

2.2.1 Target species

In the USA the target species list is entitled "America's Least Wanted" and focuses on those non-native species that are introduced and are threatening natural environments. Aquatic and terrestrial species are listed: zebra mussel, purple loosestrife, flathead catfish, tamarisk, rosy wolfsnail, leafy spurge, green crab, *Hydrilla*, balsam wooly adelgid, miconia, Chinese tallow and brown tree snake. It is stated that further spread of this already introduced species should be avoided (Nature Conservancy 1998).

2.2.2 Matching climate

The obvious alternative to a species specific approach (see above) is that proposed by the International Council for the Exploration of the Sea (ICES) Working Group on Introductions and Transfers of Marine Organisms (WGITMO) 1996 report: the environmental matching assessment between recipient and donor ports. Environmental matching alone will not provide a progressive assessment and will necessarily become more conservative with additional information from successful developments. The environmental match between port regions will not necessarily represent the biological capabilities of the species. In contrast a species-specific approach will become less conservative as information becomes available. However, species are more likely to become established in environments that are similar to those of their origin. Therefore, if the port of loading and port of discharge are ecologically comparable the risk of a species introduction is relatively high.

Table 1. Colonisation probability of NIS, according to matching salinity in donor and recipient region, after Carlton (1985).

	DONOR region		
RECIPIENT region	Fresh water	Brackish water	Salt water
Freshwater	high	medium	low
Brackish water	medium	high	high
Salt water	low	high	high

2.3 MATCHING CLIMATE AND SALINITY METHOD, GERMANY

During the German shipping study (1992 – 1996) all NIS sampled from the ballast water, tank sediments and ship hulls were characterised by an estimated probability of establishment in German waters. The potential for an establishment was estimated in accordance with the scheme developed by Carlton (1985) comparing the salinity tolerance of the species and the salinity conditions of the receiving waters (see above). In addition a comparable scheme structure was employed to take into account the climate in the area of origin (donor area) and the potential recipient area.

Table 2. Colonisation probability of NIS, according to matching climate in donor and recipient area, after Gollasch (1996)

	DONOR region			
RECIPIENT region	Arctic& Antarctic	Cold-temperate	Warm-temperate	Tropics
Arctic & Antarctic	high	medium	low	low
Cold-temperate	medium	high	medium	low
Warm-temperate	low	medium	high	medium
Tropics	low	low	medium	high

During the German shipping study all species native to cold-temperate climate areas were quoted in category “establishment highly probable”. As the number of species and specimens decreased with an increasing duration in the ballast tank species native to cold-temperate areas of the northern hemisphere of the Atlantic Ocean (North American east coast and the upwelling area off western Africa) were quoted as high risk species due to the comparable short duration of the ships voyage and matching climates. 32 NIS were quoted in this category. Among others, the decapod *Hemigrapsus penicillatus*, native to cold-temperate areas of Japan, was listed in this category. *H. penicillatus* is believed to be one of the most recent macroinvertebrate invaders to European waters (first record from the Atlantic coast of France in 1994). The establishment of *H. penicillatus* indicates that the applied model gives a useful first estimation on the probability of establishment and therefore indicates the importance of matching climates in donor and recipient areas for the colonisation of new habitats.

2.4 RISK ASSESSMENT OF NORDIC COUNTRIES

The Nordic initiative on "Risk Assessment of Marine Alien Species in Nordic Waters" (Gollasch & Leppäkoski 1999) studied the application of risk assessment models to Nordic Countries. A semi-quantitative model (low - medium - high risk) was compiled and applied to vectors of introduction and target organisms. The tentative list of parameters for ballast water introductions includes: vessel ballasting characteristics, ballast water treatment, characteristics of donor and receiving ports or geographical areas, voyage route and duration, and relevant biological information for the key/target species. Information on the key/target species should include environmental requirements such as temperature, salinity, and light/energy requirements during different stages of the life cycle (including resting stages), habitat requirements, and known biotic interactions.

In addition to the biological component it was concluded that ports in marine and brackish water areas are more open for introductions due to the fact that most NIS in Nordic Countries were introduced by shipping. Further, the ballast water released originates mostly from ports located in estuaries or coastal waters. It should be noted that species are transported between freshwater ports by shipping as well, but in a much lesser dimension. As example ballast water transported between the two freshwater ports St. Petersburg (Russia) and Hamburg (Germany) has the potential to introduce freshwater species not being able to colonise both ports by natural means due to the "salinity barrier" between both ports (Baltic and North Seas).

2.5 EXCEPTIONS

It has to be taken into account that all general rules or models have their exceptions and cannot be applied for all habitats and/or species:

2.5.1 Climate and salinity

Matching temperatures in the area of origin and new habitats do not take into account the potential of species to tolerate or adapt to temperatures uncommon to its native range. A well-known example is the ship-boring mussel *Teredo navalis* (often called ship worm because its wormlike habitus), what is believed to be of tropical or warm-temperate origin. It was introduced to northern Europe and other continents with wooden sailing vessels. Nowadays the species causes damages to wooden man-made installations from cold temperate to tropical climates. The first documented record in Europe was during a mass occurrence of the species resulting in damages to high tide protection installations, quays and wharves along the coasts of the Netherlands, Germany and Denmark in the 1730s. The species was often found in the western Baltic Sea due to secondary introductions by ships or salt-water inflows from the North Sea. Until the early 1990s no self-reproducing population was observed in the Baltic Sea. In the late 1990s larvae of the shipworm were found off the eastern German Baltic coast, indicating a self-reproducing population. The tropical marine species *Teredo navalis* was surprisingly able to adapt to cold climates and to lower salinities. None of the risk assessment models based on matching environmental regimes alone would have quoted this species on the list of target species for the introduction into cold-temperate and brackish waters. Another example is the establishment of the tropic green alga *Caulerpa taxifolia* in the Mediterranean, where it surprisingly can survive winter temperatures down to 7-10°C (Wallentinus pers. comm.).

2.5.2 "Ecological niche"/partly empty niche

The diatom *Odontella (=Biddulphia) sinensis*, native to tropical waters, was first recorded in the Danish waters of the North Sea in 1903. It was assumed that the species was introduced by a ship (Ostenfeld 1908). Many native species (benthic and planktonic species) of the genus *Odontella* occurred at that time and today in the North Sea. Nevertheless *O. sinensis* spread rapidly through European waters and established a self-reproducing population. It is believed that the reason for its successful establishment was an empty niche, related to the tendency of the species to bloom as late as in November. But even during blooming periods other native species occur in higher numbers. Long-term investigations showed that the population growth of other phytoplankton species was depressed by high populations of *Odontella sinensis*. Therefore, no empty niche but a partly empty niche was available. The role of vacant or partly vacant ecological niches and the community structure should be included in assessments to manage future invasions (Williamson 1996).

2.5.3 No empty niche available

NIS can invade in areas were no empty niche is available when more resistant to e.g. pollution or due to a higher reproduction rate than native species. The zebra mussel *Dreissena polymorpha* invaded e.g. the freshwater areas of the Baltic Sea and North American Great Lakes. The ecological niche of this species is characterised as fresh water filter feeder. Both systems have native freshwater filter feeders belonging to the family Unionidae. The native mussels were regionally driven extinct by the invader. As result the non-availability of an empty niche cannot generally be considered as an excluding factor for further species introductions.

3 Summary of actions to reduce the risk

3.1 INTENTIONAL INTRODUCTIONS (THE ICES "CODE OF PRACTICE")

The International Council for the Exploration of the Sea (ICES) was founded in 1902. Within its 19 member countries today, ICES has the task to encourage member countries to conduct investigations and research on the sea and co-ordinate research interests, especially in regard to living resources. The Working Group on Introductions and Transfers of Marine Organisms developed the ICES Code of Practice dealing with e.g. quarantine measures to prevent the introduction of non-target organisms when importing living material for aquaculture purposes. ICES adopted the "Code of Practice" in October 1973. This code has been modified several times (most recent update in 1995) and was translated into most of the ICES member languages.

3.2 UNINTENTIONAL INTRODUCTIONS (IMO ASSEMBLY RESOLUTION A. 868(20))

The International Maritime Organization (IMO) Marine Environment Protection Committee (MEPC) had a specific interest in the field of unwanted introduced species by ballast water as demonstrated in 1973 when the International Conference on Marine Pollution adopted Resolution 18, drawing attention to the transport of aquatic organisms and pathogens around the world in ships' ballast tanks.

Australia was the first country to bring the ballast water problem into focus and has played a key part in proposing the development of control mechanisms for the release of the ballast water in the early 1990s when MEPC formed a working group to consider research information and solutions proposed by Member States of the IMO and by non-governmental organisations. The working group concluded that voluntary guidelines were the appropriate first step in addressing this problem. MEPC adopted guidelines by resolution in 1991 and in 1993 these were adopted by the IMO Assembly under resolution A.774 (18) entitled "International Guidelines for Preventing the Introduction of Unwanted Aquatic Organisms and Pathogens from Ships Ballast Water and Sediment Discharges". In 1997 the IMO Assembly adopted Resolution A.868 (20) "Guidelines for the Control and Management of Ship's Ballast Water to Minimise the Transfer of Harmful Aquatic Organisms and Pathogens".

This IMO Assembly Resolution is an extremely important step towards the development of provisions in addressing this global problem. IMO has put forward these guidelines to limit the movement of organisms by ballast water world-wide which include the informing of ships on areas where ballast water uptake should be avoided due to the presence of harmful algal blooms and known unwanted contaminants, precautionary procedures when taken on ballast water in shallow areas, ballasting with freshwater, discharging ballast water and sediments to on-shore facilities (if available) and the exchange of ballast water at sea.

The IMO Assembly Resolution A.868 (20) recommends an exchange of ballast water in open oceans as far as possible from the coast wherever possible. Compared with coastal waters, deep ocean waters contain less organisms and species occurring in open ocean waters very often are not able to survive in coastal zones and vice versa. Where open-ocean exchange is not possible, requirements developed within regional agreements may be applicable, particularly in areas within 200 nautical miles from the coast. It was recommended not to ballast e.g. at night (bottom living organisms may migrate towards the water surface being more likely taken in), in shallow areas where the number of organisms is higher and during e.g. algal blooms. It was further noted that the appropriateness of risk assessment and risk management approaches relevant to ballast water management should be reviewed in the future, with consideration of the recent implementations of mandatory risk assessment procedures by some jurisdictions.

4 Summary

It is impossible to indicate whether an ecological system will be open or resistant to biological invasions (Williamson 1996) and that human mediated accidental introductions are believed to be unpredictable (Zaitsev & Mamaev 1997). During controversial scientific discussions it is agreed upon that in general all ecological communities are invasible.

Serious predictions on the risk to receive further invasions and their potential impact on native economy and ecology need long-term surveys (Zaitsev & Mamaev 1997; Lodge et al. 1998). Until today it was not possible to exactly predict which organism will survive and establish in new habitats. The reason for this unsatisfactory situation is the enormously high number of factors that need to be taken into account (e.g. minimum individual numbers of founder population, climate, salinity, habitat structure, food availability and predators) (Acil Economics 1994; Hayes 1997). Furthermore, it is difficult to predict the time of future species introductions.

Some characteristics of invading species are known to be in common in successful invaders. The limiting factor for most of the successful invaders is their flexibility in regard to temperature and salinity tolerance, habitat selection and food. Hence species from similar latitude, i.e. similar climate, and similar water bodies in respect of salinity will have a greater chance for establishment once introduced.

To survive in a recipient area is substantial, but only the first step to a sustainable population. Secondly the newly introduced specimens need to reproduce effectively enough to form a successful founder population. Additionally the ability to survive the introducing process, especially the conditions transporting vectors provide, and the time of transportation are limiting factors (Gollasch 1996).

As previous studies have shown all habitats worldwide are in principle open for introductions and each invader has a potential to severely threaten the biodiversity and economics in the area it invaded. Therefore, effective and environmentally sound treatment measures are needed to reduce the number of new invaders.

RISK ASSESSMENT OF MARINE BIOLOGICAL INVASIONS

CHAD L. HEWITT[1*] & KEITH R. HAYES[2]
[1]*CSIRO Marine Research, Centre for Research on Introduced Marine Pests, Marmion, Western Australia, Australia*
[2]*CSIRO Marine Research, Centre for Research on Introduced Marine Pests, Hobart, Tasmania, Australia*
[*]Corresponding author *chad.hewitt@csiro.au*

Abstract
Risk assessment has proven a useful tool in managing potential hazards in business and industry. Environmental hazards however, have proven to be less tractable to quantitative assessment. Here we discuss the application of risk assessments in the context of marine biological invasions and provide a summary of the progress to date in the Australian context. Risk assessment for marine biological invasions is being applied to aid in the management of ballast water using a target species approach assessed in discrete elements of the ballast water cycle (uptake, transit survival, and discharge). Risk assessments for marine bioinvasions provide another tool to aid managers in controlling and reducing the rate of current invasions. In contrast to a blanket ban on ballast water imports (i.e. mandatory ballast water exchange at sea), a risk-assessed approach can provide a mechanism to better understand the invasion problem and determine better operating procedures, high-risk routes or system failures (e.g. false reporting of ballast exchange, incomplete or inadequate exchange). Similarly, risk assessment can aid in the process of management strategy evaluation to determine what components of the invasion process are more susceptible to control, or more apt to fail under specific management regimes.

1 Introduction

Risk is defined as the likelihood of an undesired event occurring as a result of some behaviour or action (including no action). Risk assessment is the means by which the frequency and consequences of such events are determined, and should properly be accompanied by an expression of any uncertainty in the assessment process.

The consequences of the undesired event in question are usually adverse (i.e. one does not usually consider the risk of winning the lottery) and are expressed in terms of explicitly defined assessment endpoints. Assessment endpoints are simply an expression of the values that one is trying to protect by undertaking the risk assessment procedure. Thus assessment endpoints distinguish environmental risk assessment (ecological endpoints) from human health risk assessment (human fatality or injury endpoints).

The risk associated with ship's ballast water and sediment discharges can be defined as the likelihood of an undesired event occurring as a result of these actions. Recognition that the interpretation of this definition is entirely dependent on the endpoint of the assessment is paramount. If the endpoint is the establishment of an exotic species in a new locality, then the risk is expressed in terms of the likelihood of establishment. If the endpoint is environmental (or ecological) damage (notwithstanding the additional problems of definition that the assessor may face), then the risk must be defined as the likelihood of environmental damage arising as a result of the introduction and establishment of an exotic species.

E. Leppäkoski et al. (eds.), Invasive Aquatic Species of Europe, 456–466.

Notice how the definition of risk is sensitive to the assessment endpoint (which in itself is simply an expression of value). In the first part of the paragraph above, risk was expressed in terms of the establishment of an exotic species – there is thus an implicit assumption that the establishment of any exotic species in a new locality is an undesired event. This is equivalent to an expression of environmental value that wishes to preserve 'natural' or existing species assemblages. By contrast, in the second part of the paragraph, risk is defined in terms of environmental damage. In this definition the establishment of a new species *per se* does not constitute the undesired event to be avoided – one is merely concerned with the subsequent environmental damage that could arise as a result of the establishment of a new species. Thus if an assessor could guarantee that a particular exotic species would have no adverse effect on the environment, then under this definition, there would be no risk.

Of course one can never guarantee that an exotic species, once established, will have no adverse environmental effect, and thus in practice there is always the potential risk of environmental damage arising as a result of the introduction of an exotic species to a new locality. The point, however, should be clear: an individual's understanding and perception of risk is defined as a function of their environmental values.

This intrinsically subjective perception and understanding of risk is the very reason why decisions regarding the acceptability of risk are correctly taken out of the risk assessment process and are made part of a much wider socio-economic and political debate. Thus risk cannot be defined in terms of what is acceptable or unacceptable during the risk assessment process. Rather the risk must be assessed, the uncertainty associated with this assessment communicated, and then the acceptability, or otherwise, of the risk gauged by a wider audience of stakeholders.

2 Ecological risk assessment and biological invasions

Bio-invasion risk assessments are used by various regulatory agencies to manage imports or translocations of one kind or another (Hayes, in review). Most bioinvasion risk assessments are based on, or mirror, the Office Internationale des Epizooties (OIE) "chain of events" framework. This framework views bio-invasion as the culmination of a series of steps, each of which must be successfully negotiated by the invading species, and to which a probability of success can be assigned. The overall probability of success - i.e. invasion – is the product of the probabilities assigned to each step in the event chain.

The OIE framework is simple, effective and approaches bioinvasions from the perspective of the invading species. The framework's efficacy can be improved, however, by incorporating basic tenets of the Quantitative Risk assessment (QRA) paradigm. The QRA paradigm consists of five steps: hazard identification, frequency assessment, consequence assessment, risk calculation and uncertainty analysis. It was originally developed for complex industrial systems, but its techniques can be applied equally well to complex ecological systems.

An alternative approach is environmental matching which seeks to use the environmental similarity between two regions (e.g. between the donor and recipient ports) as a surrogate of measure of bio-invasion risk. For example, the Queensland Ports Corporation ballast water risk assessment (Hilliard et al. 1997) measures the environmental similarity of 40 environmental variables. Carlton et al. (1995) developed a similar, but much simpler, approach for the Great Lakes of North America based purely on comparing the salinity of ballast water to that of the recipient port.

Bio-invasion risk assessments based on environmental comparison avoid the question; "which species will be the next invader?" They are therefore well suited to a translocated habitat or environment containing a multitude of species (e.g. ballast water, the soil in the root ball of a plant, a consignment of Siberian timber, etc.). This approach assumes that the likelihood of invasion is directly proportional to the biophysical similarity between donor and recipient environments. As a first step in the risk assessment process this is a useful approach, but as a stand-alone assessment it is seriously flawed (but see discussion in the previous chapter). Notwithstanding the difficulties of measuring biophysical similarity, this approach is virtually impossible to improve empirically (Hayes & Hewitt 2000), and is not conservative for species with broader environmental tolerances than their current range. The water hyacinth, *Eichhornia crassipes*, is a spectacular example of the inability of environmental-matching to predict the future range of an invading exotic (Mack 1996). *E. crassipes* is a hydrophyte native to the Amazon basin, South America. It has large, beautiful purple and violet flowers and is therefore a popular ornamental plant for ponds. It was deliberately introduced into North America in the late 19th century and quickly spread through rivers in sub-tropical Florida and now persists as far north as the central valley in Northern California (37°45' N). Clearly climate matching would have failed to predict the full extent of this plant's invasive range (Mack 1996).

3 What risk evaluations have occurred for bioinvasions?

Initially, an evaluation of the likelihood for invasion to occur in a given area requires an assessment of the various hazards associated with vectors, trade routes and species. These three hazard assessments will allow for a qualitative evaluation of the likely risks posed to an environment on the basis of past activities.

Hazard can be defined as a situation that in particular circumstances could lead to harm (The Royal Society 1983) or alternatively considered as a substances or activity's propensity for risk. Hazard is often perceived as solely a function of a substance's intrinsic properties but, as emphasised in the definition above, it is more usefully conceptualised as a function of both the intrinsic properties of a substance and circumstance.

A substance's intrinsically hazardous properties can often only be realised under a very specific set of circumstances. Any expression of hazard should therefore properly acknowledge both the intrinsic properties and the circumstances required in order for harm to be realised. The measure of the likelihood of these circumstances and the magnitude of the subsequent harm is a measure of risk. Accordingly hazard assessments must address both a substance's intrinsic properties and the circumstances required for the manifestation of harm as a result of these properties. This is particularly true for introduced species because the likelihood of an introduced organism becoming established, and the effects that follow, depends on the characteristics of the organism (its intrinsic properties) and the environment into which it is introduced (the circumstances).

Much of past invasion theory, however, has tended to focus on only one half of this equation, for example expressing the species hazard solely in terms of characteristics that make a species invasive. Indeed as noted by Parker & Kareiva (1996), the idea of using life history traits alone to predict invasiveness has been considered as a basis for US federal policy on the importation of exotic species, and for assessing the risks associated with Genetically Engineered Organisms.

Some of the most sophisticated invasion hazard assessments, using for example discriminant analysis (Rejmanek 1996) or correspondence analysis (Richardson et al. 1990), serve only to enforce the point by concluding that chance interactions can confound invasion predictions based on life history characteristics alone. In this manner they further underscore the idiosyncrasies inherent in biological invasions, and as Orr (1995) correctly concludes, these chance idiosyncrasies cannot be predicted in advance by general statements based only on the biology of the organism. This not to say that life history based hazard assessment approaches are not useful, they do undoubtedly hold some predictive ability but rather they will never be able to provide a complete description of invasion hazard or invasion risk.

4 Identification of vector hazards

At least 13 anthropogenic vectors are, or have been, responsible for spreading marine organisms beyond natural bio-geographic boundaries (Table 1). The dominant vectors vary over time and within biogeographical region. In Australia the dominant vectors are hull fouling, accidental release associated with mariculture (predominately oysters) and ballast water (Hewitt et al. 1999), with hull fouling associated with the highest percentage of introduced species (Thresher et al. 1999; Hewitt 2002). In New Zealand hull fouling is the most important vector - 69% of the 159 introduced marine organisms arrived via hull fouling, 6% arrived via ballast water (and ballast sediment), whilst 21% could have arrived via either mechanism (Cranfield et al. 1998). In San Francisco Bay, hull fouling and ballast water are the most important vectors, followed by mariculture of the Atlantic oyster, Crassostrea virginica, and the Pacific (Japanese) oyster, Crassostrea gigas, (Cohen & Carlton 1995). In Britain, oyster mariculture is the most important identifiable source of nonindigenous species (NIS) for the top 50 species, followed by hull fouling and ballast water (Eno et al. 1997). Oyster farming has introduced at least 14 nonindigenous marine plants into the Mediterranean, rating second only to the Suez Canal as a vector in this region (Ribera & Boudouresque 1995).

Mass oyster introductions around the world have, by and large, ceased. In their place, hull fouling and ballast water have emerged as the leading vectors for NIS introductions. The international community is working hard to managing invasion pathways associated with international shipping, particularly ballast water, but has largely ignored the role of other vessels and vessel-related pathways. It is becoming increasingly clear that small craft, including ocean-going yachts, recreational and commercial fishing vessels (both displacement and trailerable), are also responsible for introducing NIS to new sites; and contributing to their subsequent spread.

Table 1. Anthropogenic vectors responsible for the spread of marine organisms around the globe.

Anthropogenic vectors
Ships: accidental with vessel fouling (including boring into wooden hulls)
Ships: accidental with solid ballast (rocks, sand, etc)
Ships: accidental with ballast water
Fisheries: deliberate translocations of fish and shellfish to establish or support a new fishery
Fisheries: accidental with deliberate translocations of fish and shellfish (particularly oysters)
Fisheries: accidental with seaweed packing for bait and fishery products
Plant introductions: deliberate translocation of plant species (e.g. for erosion control)
Plant introductions: accidental with deliberate plant translocations
Biocontrol: deliberate translocation as a biocontrol agent
Biocontrol: accidental translocation with deliberate biocontrol release
Canals: natural range expansion through man-made canals
Individual release: deliberate and accidental release by individuals (e.g. aquarium discards)
Scientific release: deliberate and accidental release as a result of research activities

5 Identification of trade route hazards

An evaluation of hazard associated with trade routes will be primarily based on two factors: frequency or strength and the likelihood of transporting species that can survive in the recipient environment. Trade route strength can simply be measured as the number of ship visits (for hull fouling) or a function of the tonnage of ballast water discharged from a port (or region).

The second component of determining the hazard posed by various trade routes is to determine the likelihood that species arriving from that location will survive in the recipient port/region. Frequently, this has been determined based on the similarity between the donor port/region and the recipient port. This approach assumes that the success of repeated inoculation is heavily dependent on the biophysical similarities between the donor ports and the recipient port (Hilliard & Raaymakers 1997).

Theoretically, the strengths of trade routes can be weighted by the likelihood of survival to provide a simplistic hazard assessment of trade routes. Difficulty arises however, with the determination of environmental similarity. Environmental-match assessments identify similarities or differences between two locations according to the criteria selected. In an evaluation of 12 Queensland (Australia) ports, Hilliard et al. (1997) selected 40 environmental characteristics on which to perform a similarity analysis with the primary trading donor ports. The environmental characteristics they selected included attributes that were not explicitly useful in determining invasion potential (e.g., distance to nearest coral reef, distance to river) and had no relevance to species-specific environmental tolerances. Consequently potentially hazardous routes that lie well within the tolerance range of an individual species but are between two ports that differ in the selected environmental parameters were not identified (for example, Fig. 1)

Similarly, the temporal and spatial scales of environmental information from donor and recipient ports will define the similarity/dissimilarity match. Temporal factors will affect the comparison between high and low latitude ports and between Northern and Southern Hemisphere ports. The most conservative stance would be to evaluate similarity on the basis of maxima and minima over multiple years rather than individual months. Alternately, microenvironments in ports could confound match predictions. For example, the discharge into a port of power-station cooling water could raise the water temperature at a site, making it capable of supporting species that could not survive elsewhere in the port (Carlton 1992).

Consequently, either species-specific tolerances must be used to derive the match between donor and recipient regions, or a sufficient surrogate for species must be used. Target species identification (discussed below) is problematic and likely to vary between regions according to socio-political forces. Alternately, the use of bioprovinces as surrogates for species has been adopted by the CRIMP risk assessment. Bioprovinces represent significant and cohesive faunal and floral assemblages with 60-80% turnover at the edges. The environmental range (e.g., temperature and salinity maxima and minima) for the bioprovince can be used as a conservative estimator of the environmental tolerances of the species within the province (Hewitt et al. in prep).

6 Identification of species hazards

Several countries have adopted a 'target' species approach to assessing risks or for determining where available resources for biosecurity should be directed. In most circumstances, our ability to identify the current extent of invasions in marine environments is poor. First, the taxonomic knowledge of native species is lacking in most marine environments and bioprovinces. Recognising species from other provincial regions around the world becomes an added difficulty. Consequently, a majority of specimens collected in field surveys remain unidentified (see Hewitt 2002). Second, the establishment of sampling programmes to collect and identify invaders is still in its infancy, with few countries participating in such efforts.

Australia established a National Introduced Species Baseline Port Survey Programme in 1996. Thus far, 25 ports have been evaluated, with the intent to have all 72 international ports of first call evaluated using a single suite of sampling protocols and design specifications (see Hewitt & Martin 1996, 2001). These protocols have been adopted on a trial basis by the International Maritime Organisation (IMO) for the Global Ballast Water Programme in which six demonstration sites in developing countries or economies in transition will be sampled. These sites are: Sepetiba Bay, Brazil; Dalian, China; Mumbai (Bombay), India; Khark Island, Iran; Saldanha Bay, South Africa; and Odessa, Ukraine. Similarly, the New Zealand Ministry of Fisheries Marine Biosecurity Group has funded a National Port Survey Project for port evaluations and has put forth the Australian protocols as the basis for developing a sampling strategy.

While these sampling programmes will develop lists and distributions of recognised introduced species, the development of target lists for management activities is more problematic. Target lists can be generated for a variety of reasons: (i) recognised pest species that form the basis of a 'black list' approach; (ii) target species that act as surrogates for unrecognised potential invaders; or (iii) predicted next pests. The deductive approach is intrinsic to the development of a black list in most quarantine efforts. While the black list has numerous critics, targeted species may be a necessary first step in the development of marine bio-invasion management practices. Terrestrial pest management and quarantine rely heavily on the deductive identification of species of concern.

These targeted species are then used to determine quarantine inspection and containment strategies. Hayes & Sliwa (in review) chose the following selection criteria to identify potential marine pest that been introduced via hull fouling or ballast water:

(i) it has been reported in a shipping vector or has a ship-mediated invasion history; AND,
(ii) the vector still exists; AND,
(iii) it has been responsible for environmental and/or economic harm; AND,
(iv) it is exotic to Australia or present in Australia but subject to official control (i.e. listed, restricted or otherwise legislated by an authorised national authority).

These criteria are transparent and consistent with international standards. There are two reasons why it is important to generate species lists in this manner. First most biological risk assessment and risk management techniques are based on hazardous species (Hayes 1997). Almost all of Australia's (and New Zealand's) biosecurity initiatives are explicitly species-specific, as evidenced by the numerous lists of "notifiable" or "harmful" organisms and pathogens in aquatic (e.g. Humphrey 1995; Forrest et al. 1997; Arthington et al. 1999; Champion & Clayton 2000; Kailola 2000) and terrestrial (e.g. Humphries et al. 1991; Morley 1993) environments. Second, if these lists are to be used for quarantine purposes, and thereby potentially limit trading practices, they must be generated in an objective and scientific manner. As a signatory to the World Trade Organisation Agreement on the Application of Sanitary and Phytosanitary Measures (the SPS Agreement), Australia has a number of international obligations in this regard, including:

(i) quarantine measures to be based on science;
(ii) transparency - open and consultative decision-making on quarantine matters; and,
(iii) risk assessment - restrictions based on actual risk, assessed as objectively as possible (Doyle et al. 1996).

These obligations necessarily apply to the way in which notifiable or harmful species are identified from the thousands of potential invaders around the world.

Alternatively, an inductive approach whereby the determinants of invasion success and likelihood to become a pest species are derived from biological principles. Several attempts have been made to determine the suite of characters that make for an 'ideal' invader (Kareiva 1999; Parsons 1983; Ritte & Safriel 1977; Safriel & Ritte 1980; Willan 1987). These lists of life-history attributes have largely failed due to numerous exceptions to the rule (Crawley 1987; 1989; Lodge 1993a,b; Simberloff 1989). In order for the inductive approach to succeed, additional attention must be given to the synergies between the characteristics of invading species, the effect of the transport vector acting as a filter for species characteristics, and the effect of the recipient environment.

Recently, the first co-operative "risk assessment" study for the Baltic Sea area was completed (Gollasch & Leppäkoski 1999). This study developed hazard assessments for several recognised invading species using a deductive approach. Based on previous invasion history, and expert opinion from invasion researchers, a list of 'next pests' was drawn up and life history characteristics collated from the literature. Similarly, hazard assessments of five northwest European harbours along the salinity gradient from St. Petersburg to Bergen were developed (Gollasch & Leppäkoski 1999). These ports were evaluated for trading profiles, environmental characteristics and floral and faunal baseline information.

7 Risk assessment for bioinvasions

As previously stated, risk assessment is the means by which the frequency and consequences of undesired events are determined. Clearly hazard evaluations are a necessary first step in this process, however, these evaluations alone cannot adequately determine risk. Hayes (1997) reviewed ecological risk assessment methodologies in relation to marine bioinvasions and concluded that the development of a complete analysis of the risks posed via the invasion process has yet to occur.

Two previous marine bio-invasion "risk assessments" were identified: the bio-economic risk assessment commissioned by the Australian Quarantine and Inspection Service (AQIS) in 1993 (AQIS 1994) and the Ports Corporation of Queensland ballast water risk assessment (Hayes et al. 1996). ACIL Economics and Policy Ltd. primarily focussed on assessing the cost of toxic dinoflagellate introductions to the Australian economy. While this assessment followed the OIE evaluation process (Hayes 1997), the ultimate deliverable was fraught with significant uncertainties due to the endpoint selection.

The Port Corporation of Queensland (PCQ) ballast water risk assessment was described as a first-pass semi-quantitative risk assessment in which five stages were undertaken to determine overall risk. The trading partner ports to 12 Queensland ports were identified (stage 1) and ranked according to ballast water discharge (stage 2). Sub-samples of the identified overseas trading partner ports were selected for collation of environmental data and a subsequent similarity matrix of environmental overlap determined (stage 3). Target species were identified from the sub-sample of overseas trading partner ports that could possibly establish in Queensland waters (stage 4). Finally, an assessment of the establishment risk based on the first four stages (stage 5).

The PCQ assessment relies primarily on a hazard assessment of trading partners fundamentally driven by an assessment of environmental similarity. The primary assumption is that ports with similar environments will likely be able to support similar species. While this is likely to be true, the assessment of risk must necessarily undertake a broader perspective. As previously discussed, numerous instances can be identified in which a species' environmental tolerance far exceeds a single location (e.g. port), resulting in a species' ability to survive in two ports that have no environmental overlap (see Fig. 1).

In addition, environmental similarity driven assessments typically will not account for the carry-over of ballast water from one port to another. Most ballast water reporting currently requires only the Last Port of Call (LPOC) resulting in a flawed assessment when ballast water drawn from previous ports is carried over. In a study of American shipping patterns, Carlton et al. (1995) found that a vessel's LPOC was a poor predictor of ballast source: 53% of all vessels (and 66% of container vessels) contained ballast water that did not originate from the LPOC.

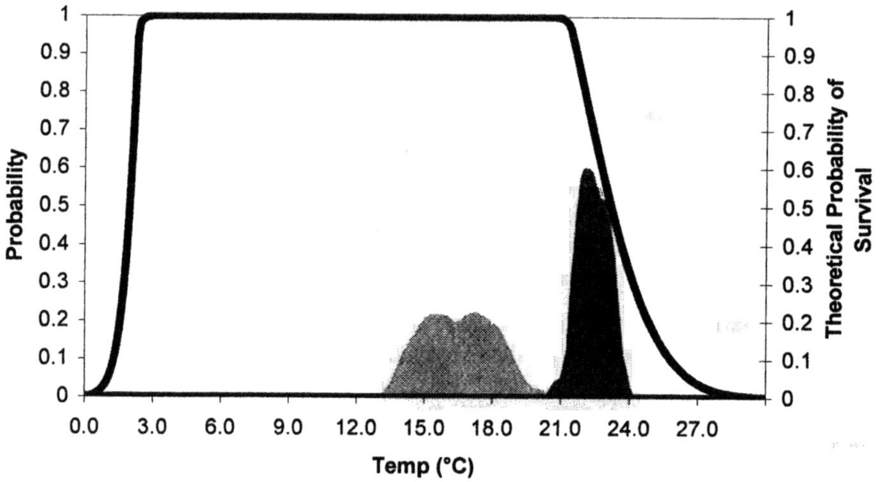

Figure 1. Example of uni-dimensional (temperature) environmental matching. Hobart (light grey) and Sydney (dark grey), Australia, have no environmental overlap, however the environmental tolerance (–) of the northern Pacific seastar, *Asterias amurensis*, (theoretical probability distribution; literature based thermal tolerance range is 1.1° to 29°C) overlaps both port environments.

Despite these problems, environmental-matching has merit when based on species-specific environmental tolerances, rather than port-to-port comparisons. The risk assessment developed by CRIMP advocates a combined approach (Hayes & Hewitt 1998 unpubl. ms).

A more rigorous quantitative approach to the ballast water invasion cycle requires a species-specific approach, however extant risk assessment models do not adequately address the steps of the invasion cycle. Hazard identification is the key component of any risk assessment process. Hazards that are not identified in the early stages of the assessment will not be taken into account at later stages, leading ultimately to an under-estimate of risk (increased Type II error).

Numerous methods for identifying hazards are available including the collation of expert 'heuristics'; the use of fault tree analyses, and the use of hazard and operability analyses. These methods are well suited to vector-based hazard identification (see above), however in order to develop a species-specific analysis it is assumed that the criteria for identifying target species (the species hazard) are accepted and defined in advance. The protection afforded by a species-specific assessment can be improved by selecting target species that are representative of broad guilds or that are among the most robust relative to the determinants of ballast water transport (e.g., species with wide environmental tolerances, pelagic larvae or resistant, resting life-stages).

The CRIMP risk assessment evaluates invasion risk using the endpoint of inoculation into the recipient port. The assessment is based on target species, however incorporates a non-target component whereby the environmental extremes of biological provinces are used as surrogates for species. The need for this becomes apparent if the risk assessment is to have wider applicability in situations where no target species can be readily agreed.

The invasion cycle can be broken into multiple steps: vector infection, transport survival, discharge or inoculation survival, establishment and/or dispersal, and impacts. The uncertainty associated with each step increases with each step, i.e. from infection to impacts. Quantitative probabilistic techniques become increasingly inappropriate as one moves from low to high uncertainty. It may not be necessary to quantify all of the steps in the invasion sequence. For species that are a priori pests (i.e. accepted as target species through another selection process) with a well-documented impact history, quantified estimates of inoculation (i.e. all those steps up to and including survival in the recipient area), which are relatively certain, may be sufficient from a risk manager's perspective.

8 Endpoint selection

The advantages of quantitative risk assessments are enormous – risk estimates relative to acceptance criteria become meaningful; risk management strategies can be compared; and cost-benefit analyses conducted. To achieve this, however, an earlier endpoint in the ballast water invasion cycle (such as inoculation survival) must be used in the assessment. Note also that the assessment framework remains a suitable platform to address the probability of establishment and subsequent economic/environmental impact when these components can be estimated with reasonable accuracy and scientific rigor.

Clearly endpoint selection becomes a critical component in the development of a risk management strategy. As previously discussed, risk assessment from a quarantine perspective results in a barrier control perspective. This in itself determines desirable endpoints; inoculation is deemed to be a breakdown of the barrier and consequently is the earliest point at which an evaluation of risk becomes meaningful. In order for inoculation risk to be determined, the likelihood of short-term survival (e.g. 48 hrs) becomes necessary. In contrast, other perspectives might select endpoints that accept inoculations or even establishments of new species, but would assess the risk relative to the impacts on native communities or economically sensitive operations such as mariculture activities.

9 Discussion

The prevention of bioinvasions occurs as a barrier control or quarantine response. As with all other quarantine management systems, risk assessments are merely tools that managers can use to create a filter rather than providing complete protection (Carlton et al. 1995). Risk assessment provides an estimate of the likelihood that a non-desirable event occurs. This in itself does not provide the manager an ability to determine the acceptability of an event. That decision requires an open and transparent discussion with a wider audience of stakeholders to determine the socio-political valuation of the level of required protection. This may not be equivalent for all species of concern. Similarly the level of protection may not be identical for all locations. For example, ports bordering on Marine Protected Areas or RAMSAR sites may desire higher levels of protection than ports without these 'sensitive' areas nearby.

At present three options for marine bioinvasions present themselves:
(i) do nothing, that is accept that whatever level of risk exists with current activities (e.g. ballast water discharge and hull fouling transport) it is acceptable;
(ii) control everything and adopt a zero tolerance perspective (no risk is acceptable); or
(iii) undertake a risk management strategy in which management is based on an assessment of the risks posed.

The last strategy has the ability to provide clear incentives for industry, while protecting the marine environment. In the absence of a knowledge base to adequately determine risk, the precautionary principle must be employed.

Currently, some nations (e.g. Australia) are experimenting with systems to allow a selective application of risk mitigation based upon voyage-specific risk assessments. Limited funds available to quarantine agencies can be more readily focussed on ships that pose a higher relative risk. It must be noted however, that managing vessels on a voyage-by-voyage basis required a large overhead in terms of reporting systems and a knowledge base to support the risk assessments.

Risk assessments for marine bioinvasions provide another tool to aid managers in controlling and reducing the rate of current invasions. In contrast to a blanket ban on ballast water imports (i.e. mandatory ballast water exchange at sea), a risk-assessed approach can provide a mechanism to better understand the invasion problem and determine better operating procedures, high-risk routes or system failures (e.g. false reporting of ballast exchange, incomplete or inadequate exchange). Similarly, risk assessment can aid in the process of management strategy evaluation to determine what components of the invasion process are more susceptible to control, or more apt to fail under specific management regimes.

Acknowledgments

We would like to acknowledge Ron Thresher and Nic Bax for useful discussions in the development of the ideas in this paper.

RISK BASED METHODOLOGY TO ASSESS INVASIVE AQUATIC SPECIES IN BALLAST WATER

GERD P. HAUGOM[1*], HANNA. L. BEHRENS[2] &
AAGE B. ANDERSEN[2]
[1]*Det Norske Veritas Research, Høvik, Norway*
[2]*Det Norske Veritas Maritime, Technical Solutions, Høvik, Norway*
[*]Corresponding author *Gerd.Petra.Haugom@dnv.com*

Abstract
This chapter describes how the methodology and principles commonly applied in risk assessment can be utilised to minimise the risk for transfer of invasive and harmful aquatic species. The risk-assessment approach is exemplified by focusing on the transfer vector that is commonly regarded as the most important - international and national transfer of invasive aquatic species in ships' ballast tanks. The transfer of species on ship exteriors, commonly called hull fouling, is also considered. This transfer vector is also thought to be of significant importance. The general structure of the risk- assessment methodology will be similar for these two transfer vectors, although the importance and relevance of certain variables will differ. The specific risk-assessment methodology described is based on EMBLA, a risk-based ballast water decision support system developed by Det Norske Veritas (DNV).

1 Introduction

The risk-assessment methodology described in this chapter is based upon the methodology applied in EMBLA, a risk-based ballast water decision support system developed by Det Norske Veritas (DNV). The main objectives of DNV, an international, independent foundation founded in 1864, are to safeguard life, property and the environment. One method to achieve this is the provision of professional services for managing risk. The work being undertaken by DNV in developing EMBLA and the Ballast Water Risk Assessment Methodology, which is and integrated part of EMBLA, address several of these issues.

EMBLA is a tool custom built for identification and assessment of ballast water management options that are environmentally safe without compromising the safety of vessels, trades or personnel. EMBLA has been developed to accommodate the increasing demands of international port authorities and ship owners to compile, assess and verify environmental information and data provided by vessels and the increasing requirement to manage and utilise such information efficiently. EMBLA integrates biological and shipping knowledge in a structured risk assessment methodology.

A main advantage with this risk based approach is the cost and time savings, thereby the efforts can be focused on the vessels and trades that represent the highest risks regarding invasion of potentially harmful aquatic species.

2 Why use a risk based approach to prevent introduction of invasive species?

Based on the view that any introduction of invasive or non-native species should be avoided, many have argued for a strict rule-based approach as means of regulating and controlling the use of ballast water. For a rule-based approach to be successful in terms

E. Leppäkoski et al. (eds.), Invasive Aquatic Species of Europe, 467–476.

of reducing the potential transfer of invasive or non-native species considerable bureaucracy and very comprehensive control mechanisms would be required.

It may be claimed that rule-based systems in general lack the flexibility required to allow case-specific assessments in order to optimise the efficiency of the work to control the risks of invasive species. Rule-based systems will therefore generate costs that benefit neither the environment nor the shipping industry, governments or society in general. The costs associated with the implementation of risk-controlling measures, which may prove inefficient or unnecessary for the specific vessel or location, is the main disadvantage of a rule-based system. Another disadvantage is the significant time lag to incorporate new knowledge into a rule-set. A well-structured risk-based approach would be dynamic and flexible and could therefore significantly reduce costs and increase the efficiency and reliability of ballast water management.

3 Quantitative versus qualitative approaches to risk assessment

Risk assessments of invasive aquatic species must assess and interpret large quantities of data. A qualitative (risk) assessment will typically be more "open" to subjective interpretations and value judgements than a quantitative assessment. A comparison of different options and an assessment of whether the risk level is acceptable will therefore be more difficult for qualitative assessments.

The main aim of quantitative risk-assessment methods is to present all relevant impacts in a form that can be aggregated for interpretation by decision-makers (Wathern 1984, 1988). Quantification enforces clear views and prevents an extensive use of vague terms like: may, might, can, probably, could, shall and should. It enforces clear priorities and decisions and enables compilation of large data quantities, particularly for comparison of alternatives or comparison of actual risk with acceptable risk.

Numerical results offer a good basis for clear result presentation, and can overcome subjective risk-perception problems. This is commonly seen as the most serious obstacle of risk management (WHO CEMP 1992). It is widely known that low probability/high consequence events tend to be overestimated, while high probability/low consequence events tends to be underestimated.

4 The challenge

The main challenge when aiming to undertake a detailed risk assessment for invasive aquatic species is the lack of relevant comprehensive data sources structured on a format suitable for risk assessment and decision support. Until very recently no historical databases enabling risk quantification had been gathered for ballast water management strategies (Hayes 1998).

It should be acknowledged that the relevant ecological properties of potentially invasive aquatic species are not fully understood, and further that interactions both between species and between species and their physical environment (e.g. stress, salinity and temperature tolerances) are numerous and complex. This is also the case for the complex processes that thrive in both natural and human-induced ecological changes. A consequence of this is that databases, even if comprehensive data had been gathered, cannot be expected to cover all data required on a format suitable for a complete risk assessment.

A key success factor for risk assessment of invasive aquatic species is the prioritisation of the factors that require assessment, based on a combination of relevant historical data, general biological knowledge and expert judgement. Preferably, databases for risk assessment of invasive aquatic species should be purpose built - or at least modified to suit this purpose - as databases built to serve other purposes cannot be expected to contain the required data on a format and structure suitable for risk assessment. Work to compile such data in databases and integrate the data in a risk assessment structure is in progress. A comprehensive database of environmental and other port-specific data required in the Hazard Screening process (described in this chapter) for major European and international ports are for example under development in DNV (unpubl.). Other interested parties are gathering data regarding nonindigenous aquatic species, for example the Nonindigenous Aquatic Species (NAS 2001) information resource providing information of nonindigenous species reported from the waters of the United States.

5 Risk assessment structure

Risk in relation to invasive aquatic species can be defined as the likelihood of undesired/unwanted invasive species establishing and causing biological, economical, safety-related or social damage in areas where the species did not occur naturally. A complete risk assessment of the potential risks due to alien invasive aquatic species that might be introduced to Europe will have to identify and assess all relevant vectors for the transfer of aquatic species, and all relevant species. To perform such a risk assessment, several parameters have to be defined, such as categories of consequences and risk-acceptance criteria for the assessment.

Studies of the geographical patterns of typical "ballast-dependent" trades have revealed that not all trades represent the same element of risk for unwanted transfer of invasive aquatic species. The risk of invasive aquatic species being introduced depends on numerous biological, environmental and vessel-specific parameters. "Safe" and "high risk" ballast voyages can be identified by systematic assessments of the bioinvasion risks associated with actual ballast voyages. Hence, the principles of biogeographical zones, regions and provinces (Ekman 1953; Briggs 1974) can be combined with those commonly adopted in risk assessments, in order to safeguard against the transfer of harmful aquatic organisms. By assessing the voyages that might represent an unacceptable risk for unwanted biological invaders, requirements to preventive or risk-reducing measures can also be identified and optimised.

In the following, a risk-assessment methodology is outlined by focusing on the transfer vector that is commonly regarded as one of the two most important, the transfer of species in ships' ballast tanks (Hayes & Hewitt 1998). A similar methodology could in principal be used for hull fouling, which is generally considered to be the other main transfer vector.

The risk-assessment framework can be used as a tool to determine which phase(s) of the ballast water transfer process that are critical to the risk level and where it is most efficient to introduce risk control measures to reduce the risk of introduction of the hazard species. [Species known to be invasive/nonindigenous and known to have caused harm (economical, ecological, safety, or social-occupational/health) in areas where the species do not naturally occur.]

Hence the cost-efficiency of risk-reducing measures can be optimised for individual vessels or ports (depending on the scope of the actual risk assessment).

The main steps in a well-structured risk assessment can be outlined as follows:
(i) Hazard screening – identification of relevant hazard species and the relevant risk factors for species transfer.
(ii) Analyse hazards – assess the probability/likelihood of the identified hazards/hazard species and the factors that affect these probabilities.
(iii) Consequence (impact) assessment to determine risk level.
(iv) Compare actual risk with acceptable risk
(v) Assess the use of risk-reducing measures
The structure of the risk assessment framework is illustrated in Fig. 1 below.

5.1 HAZARD SCREENING

The purpose of the hazard screening is to identify all significant and relevant risk factors. The main risk factors include:
(i) Origin (and quantity) of ballast water, and where it will be discharged (and in what quantity).
(ii) Toxic algae bloom status/history of the location of origin of the ballast water.
(iii) Biogeographical compatibility of donor and recipient ports (of ballasting and de-ballasting locations if different).
(iv) Identification of relevant hazard species. This may be derived based on for example local target lists, records of relevant historical species, or detailed biological knowledge of the relevant ports.
(v) Environmental compatibility between recipient port and the identified target list species or historical species.

For simplicity, the relevant invasive aquatic species that are identified as a result of the hazard screening are called "hazard species" in the following.

5.2 ANALYSIS OF HAZARDS

For an assessment of the risks associated with ballast water in shipping, the hazard analysis consists of three defined steps, according to the ballast water transfer sequence:
(i) Ballasting
(ii) Transfer
(iii) De-ballasting
The hazard analysis assesses the probability of survival in viable populations for all the hazard species identified in the hazard screening (and their relevant life stages) through these three phases.

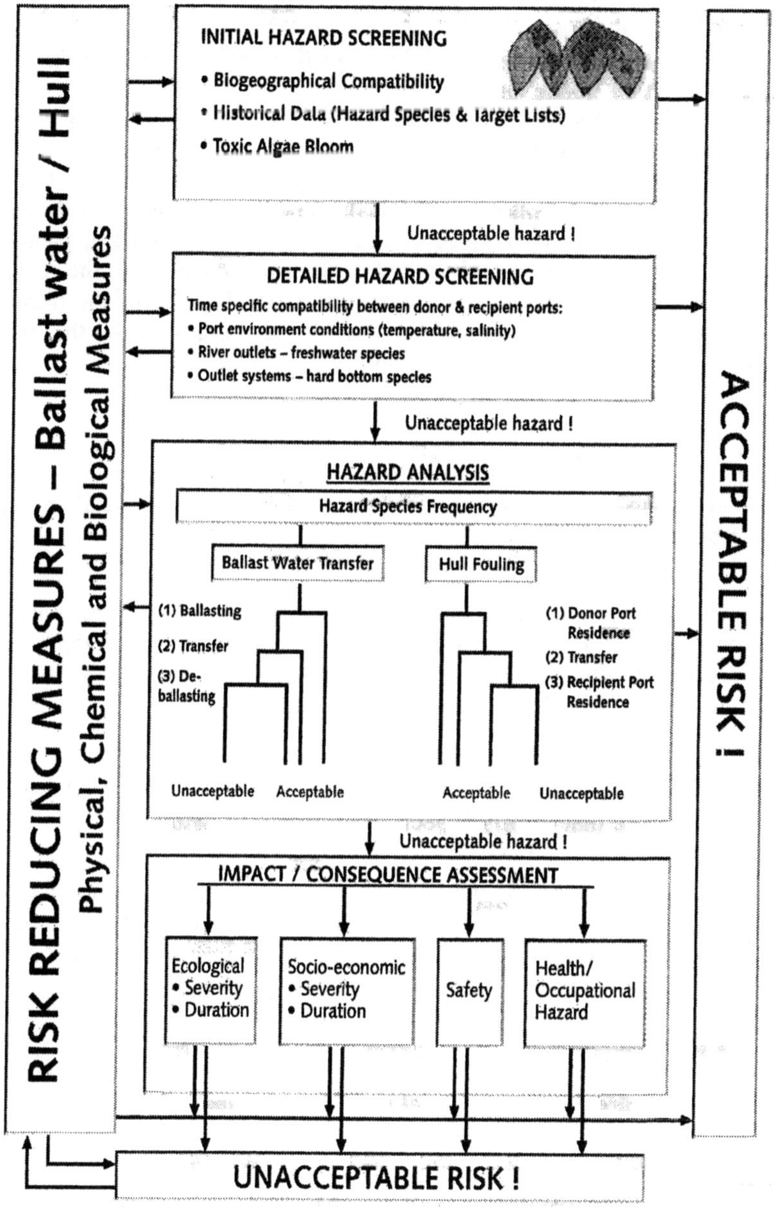

Figure 1. Structure of risk assessment framework.

The Ballasting phase assesses the probability of the hazard species being present in viable quantities in the ballast water when/if taken on board the vessel. These probabilities depend on several factors, including the biology of the assessed organisms (e.g. migration patterns, life stage, habitus), the environmental conditions in the ballasting port (e.g. traffic and pollution level), and technical ballasting conditions. Examples of the latter are depth of intake of ballast water and potential presence and efficacy of filters. The Transfer phase assesses the probabilities of the hazard species surviving in viable populations in the ballast tanks during the voyage. This is a function of the biology of the species (e.g. temperature and darkness tolerance and nutrition requirements) and factors that affect the conditions in the ballast tank (e.g. duration of the voyage and temperature and oxygen changes during this time). The De-ballasting phase assesses the probability of the survival and establishment of the hazard species in the recipient port. These probabilities are a function of factors such as the biology of the hazard species (including life-cycle stages), and the environmental conditions in the port of de-ballasting (e.g. biological sensitivity and stability of local fauna).

5.3 IMPACT ASSESSMENT

Three categories of impacts are discussed in this section: (i) Ecological, (ii) Economical, and (iii) Safety. Although not discussed here, an assessment of social impacts might also be undertaken.

5.3.1 Ecological impact

Ecological impacts include the disturbances of coastal and marine resources such as changes in pre-existing biota and loss of biodiversity caused by the hazard species. The severity of ecological impacts can be categorised in different ways. One method which categorises the impacts into three categories, is outlined below:

"High" ecological impact. This implies the invasive spread of a hazard organism, and that the uncontrolled spread of this organism leads to the loss of ecological diversity or stability; For example that a native species are threatened with extinction, thereby irreversibly altering the whole balance of the exposed ecosystem.

"Medium" ecological impact. Medium ecological impact implies that the hazard organism has a high rate of spread, but it does not have the potential to dominate the local ecosystem and will therefore have a limited effect on the ecosystem balance.

"Low" ecological impact. Low ecological impact implies that the hazard organism does not have the potential to dominate the local ecosystem. It can co-exist in balance with the local ecosystem without affecting the dynamic structure and balance of the system to any noticeable degree. Low ecological impact can generally be defined as acceptable from an ecological viewpoint.

The duration of the impact are categorised as:

(i) *"Seasonal" damage.* Seasonal damage is non-permanent impacts implying that the ecosystem damage is reversible.

(ii) *"Permanent" damage.* Permanent damage has been defined as damage that will last for at least 1 year and is considered non-reversible.

Further refinement of the impact categories can be carried out, but the above categories are considered sufficient for most practical purposes.

Whether the ecological risk is considered acceptable or unacceptable depends upon the acceptance criteria. These can vary depending on local ecological acceptance criteria, and are expected to be stricter for example for Marine Protected areas and for areas that are considered as particularly sensitive.

5.3.2 Economical impact

Economical impacts include disturbances causing commercial harm to coastal and marine resources. Economical losses for fish farms and commercial fisheries and the economical consequences caused by organisms relating to clogging of tubes and pipelines are two examples of economical impacts. Further details of economical impacts are given in chapter 7. The severity of economical impacts can be categorised in different ways. A method that categorises the impacts into three categories is outlined below:

"High" economical impact. High economical impact implies permanent and serious economical damage to existing industries and their economic basis. Examples of this are impacts that will be detrimental for fish farming or local fisheries on a permanent basis, and impacts which cause permanent clogging of important cooling water pipes. The substantial economical problems encountered in the Great Lakes area is a relevant example.

"Medium" economical impact. Medium economical impact implies limited permanent damage, limited damage of a contained or long duration, or more serious damage of seasonal (short) character.

"Low" economical impact. Low economical impact will typically be economical impact(s) that are considered acceptable based on the local acceptance criteria.

There may be cases where the economical impact of an introduced species can be financially beneficial, that is if the species can be potentially harvested or farmed. If required, an assessment of positive economical impacts can be included, although environmental risk assessments and environmental impact assessments tend to focus primarily on negative impacts.

As the suggested risk assessment approach is species-specific (analysing a set of identified hazard species), potentially beneficial species will only be assessed if they are considered a potential "hazard" either by the recipient area or by another authority setting the criteria for selection of hazard species.

The duration of the impact will be categorised as:

(i) *"Seasonal" or "Immediate".* Seasonal or immediate are non-permanent impacts, i.e. the economical damage can be recovered.
(ii) *"Permanent".* Permanent impacts are non-reversible impacts. A permanent impact does not have to be immediate.

Further refinements of the categories are possible, but the above categories are considered sufficient for most practical purposes. Whether the economical risk is considered acceptable or unacceptable depends upon local conditions and local acceptance criteria.

There is no automatic link between high ecological and high economical impacts. A seasonal ecological damage can for example cause permanent economical damage on existing facilities and be detrimental to fish farms.

5.3.3 Safety

Safety related impacts include safety of the vessel, and the safety of people that might be affected. This has particular relevance regarding an assessment of risk-reducing measures, as some of these are known to have potential safety implications for the vessel. Safety issues include possible health and occupational hazards for people that might be exposed to invasive aquatic species, for example viruses and bacteria. There are several reasons for including potential safety impacts in ballast water risk assessments. Some risk-controlling measures, e.g. mid-ocean exchange of ballast water, are known to have potentially serious safety impacts for vessels. Safety issues can also be related to the potential side effects of applied treatment methods. These include, but are not necessary limited to, residues of chemicals in the ballast water after treatment and safety-related hazards linked to the treatment method. Safety impacts are usually measured in potential for loss of life (PLL) and/or injury. This type of measure is widely used, for example to determine risk levels in the offshore industry. Ballast water safety acceptance criteria could therefore be linked to established acceptance criteria for the vessels. Vessel-specific acceptance criteria might however be impractical and unfeasible.

6 Risk-reducing measures

Risk-reducing measures in the form of preventive and technical measures are usually only assessed when the risk level is higher than that which is considered acceptable or when some regulatory framework requires such measures to be implemented. The introduction of feasible risk-reducing measures can reduce the risk level considerably, and reduce unacceptable risk to an acceptably low level.

Examples of risk-reducing measures are improved ballasting procedures, and various methods or combination of methods of treatment of ballast water. Ballast water treatment methods are usually categorised as chemical, biological, mechanical, or physical.

Preventive measures are expected to have different effects on different species and are therefore likely to have specific areas of application. Assessment standards or norms for such measures are therefore required to assure efficient application of preventive measures.

The modular risk assessment approach adopted enables case-specific comparison of the risk levels associated with different risk-reducing measures. Usually, the effect of a risk-reducing measure, or a combination of measures, will be determined by a reassessment starting from the Hazard Analysis module. This is because most risk-reducing measures will have some effect during one of the three ballast water transfer phases, ballasting, transfer or de-ballasting. In some cases, where measures like altering the time or location of the ballasting or de-ballasting operation might be considered, the reassessment may start right at the beginning of the hazard screening and even affect the case-specific list of hazard species. The assessment of risk-reducing measures will in all cases be species-specific, enabling the effect onto each identified hazard species to be determined.

7 Risk acceptance criteria

The risk assessment has to be put in a relevant context to enable determination as to whether the risk level is acceptable or not. The most transparent and only verifiable way of doing this is to compare the actual risk level with some clear and well-defined acceptance criteria. Acceptance criteria can be quantitative, semi-quantitative or qualitative; they can be biological, purely risk-based, or based on an agreed legal framework. This may be in the form of country-specific regulations or general requirements to vessels' ballast water management. Some local regulations and guidelines have been developed, for example by Australia and the United States.

The acceptable risk will also be stricter for areas that are acknowledged to be particularly sensitive. Examples of such are identified Marine Protected Areas (MPA) according to the Jakarta Convention and Special Areas (SA) according to MARPOL.

As no ballast water treatment method (except that of avoiding all ballasting operations) can eliminate the considered risks, the establishment of risk acceptance criteria is essential in the development of standards for assessing ballast water treatment methods (risk-reducing measures). Such acceptance criteria may be related to the mortality rate following treatment e.g. concentration of living representative species from some defined groups. An introduced or proposed treatment method can then be assigned a treatment impact profile with reference to application details (e.g. dose-response relationships) for the representative species (Andersen et al. 2001; Haugom et al. 2001).

Obtaining agreement within the relevant interest groups regarding the acceptance criteria can be a significant challenge, as views will often differ. Objective definitions of acceptable risk will not always be the obvious solution for all parties. Acceptance criteria can be based on legislative frameworks among the relevant authorities or international organisations. Regarding invasive aquatic species, such criteria are under development in a Ballast Water Working Group under the GloBallast programme (a cooperative initiative between IMO, UNDP and GEF). A recent GloBallast workshop (GloBallast 2001) considered methodological approaches to develop standards for ballast water treatment methods. As such standards are being developed, the relevant issues should be integrated into the assessment framework for risk-acceptance criteria.

7.1 BIOLOGICAL ACCEPTANCE CRITERIA

Biological acceptance criteria are used to determine which species are considered a potential hazard by the recipient port/authorities, for example by applying the target species approach. Factors influencing the biological acceptance criteria include:

(i) The local authorities.
(ii) Local regulations and legislations.
(iii) International agreements.
(iv) Ecological status/vulnerability of the recipient area, for example if the recipient area is within or in the vicinity of a Marine Protected Area or a Special Area.

(v) Ecological status of the donor area, for example if there is a history of or current occurrence of toxic algae blooms.

7.2 RISK-BASED ACCEPTANCE CRITERIA

Risk-based acceptance criteria can be quantitative or qualitative and can be expressed in a variety of ways, for example as "Acceptable risk for introduction of unwanted defined hazard species." The acceptable risk can be detailed and expressed as:
(i) Acceptable risk level per vessel de-ballasting in an area,
(ii) Acceptable risk level per ton foreign ballast water de-ballasted in the area,
(iii) Acceptable concentration or quantity of hazard species.

The local port authorities can decide the maximum total annual risk for permanent introduction of the defined unwanted species that is considered acceptable. A maximum "average acceptable risk per vessel" that wish to de-ballast in the port, can then be determined based on expected traffic data and the total expected discharged volume of ballast water in that specific port.

NATURE CONSERVATION IMPLICATIONS OF MARINE BIOLOGICAL INVASIONS

CLARE ENO[1*] & JOHN P. HAMER[2]
[1]*Countryside Council for Wales, Gwynedd, Wales, UK*
[2]*School of Ocean Sciences, University of Wales, Bangor, Anglesey, UK*
[*]Corresponding author *c.eno@ccw.gov.uk*

Abstract
Invasive species pose a serious threat to nature conservation. Despite there being much recent legislature to protect natural biodiversity, measures to prevent the introduction of non-native marine species are minimal. Unless applied immediately, control methods are largely ineffective and spread is inevitable. Nonetheless it is important to protect habitats and species of nature conservation interest and prevention is better than cure.

1 Introduction - Assessing relative importance and threats

A species, habitat, community (biotope) or site may have nature conservation interest by virtue of its diversity, rarity, naturalness, fragility, representativeness etc. (see Mitchell 1987). Its importance on a local, national, regional or global scale will depend upon how it matches up to these criteria. Anything that detracts from the reason for its importance may be considered a threat to nature conservation interests. The aims and objectives of nature conservation are to maintain and restore biodiversity by the protection or restoration of habitats, species, and biotopes. Nature conservation goes beyond ecology in being strategic and frequently involves active management which often focuses on the protection of species or habitats that are considered to be rare, threatened or at risk from human activities. There are different ways of achieving the conservation of species or habitats. A central mechanism is the designation and protection of areas that contain the species or habitats to be protected (see below).

Many non-native species are innocuous, present in low abundance and/or of limited distribution. However, occasionally, species are invasive and can have significant effects on their surroundings. The seriousness of an invasion will be site specific and also depend upon the nature of the invading species and the fragility of the marine community affected. When an introduced species becomes invasive it can be particularly detrimental to nature conservation interests. The impact of invasive species is recognised around the world as one of the most significant threats to the conservation of natural biodiversity, and IUCN (1999) considers that on a global scale they may be even more damaging than habitat loss. The most heavily affected areas around the world include oceanic islands, large lakes, coastal inlets and estuaries. This is probably on account of their relative isolation. Other areas tend to be contiguous with terrestrial or oceanic expanses and thereby exposed to a range of species over historic time. However, as a consequence of invasive species, hundreds or even thousands of species have gone extinct in the past few centuries on islands and many more will be threatened with extinction (IUCN 1999).

Terrestrial invasions are well documented, and although the public may appreciate their potential threat to biodiversity, many people regard some with approval and even affection.

E. Leppäkoski et al. (eds.), Invasive Aquatic Species of Europe, 477–483.

In Britain, the (invasive) grey squirrel is a favourite animal in urban parks and suburban gardens where few other mammals are seen, while sycamore and rhododendron are plants which many would think of as characteristic and desirable species in both town and country areas where they occur (JNCC 2001). The use of non-native species in farming, aquaculture and for recreational purposes has increased during this century e.g. in Britain (Bullock et al. 1997). Farmed strains of native species, such as salmon, can bring commercial returns, yet their escape into the wild can cause irreversible degradation to genetically distinct stocks. As a consequence, there are a number of competing perspectives that challenge or support views held by nature conservationists. Regardless, there are dedicated programmes for managing terrestrial and freshwater invasives, especially within sites protected for their nature conservation interest (JNCC 2000), whereas there are few similar examples for marine sites. Environmental pollution and habitat destruction provide new conditions in which non-native species can thrive, for example the invasion of the Black Sea by *Mnemiopsis leidyi* (GESAMP 1997); global climate change may further favour the spread and establishment of invasive species (IUCN 1999).

2 Legislation - relevance to marine nature conservation

In the last two or three decades, most of the major statutes relevant to nature conservation have called for control of introduced species. The extent of coverage gives an indication of the seriousness with which the potential threat of non-native species is viewed from an international, regional and UK standpoint, (see Council of Europe 1996). However, despite such commitments, at present few countries have developed the comprehensive legal and institutional systems that are capable of responding effectively to the threat of non-native species (IUCN 1999). Legal provisions that cover marine introductions appear in several global treaties as well as in an increasing number of regional conventions. However, there is no law that imposes a total prohibition in this field (De Klemm 1994). The usual method is that deliberate introductions become subject to a permit requirement, which is usually embodied in national legislation. UK law (Box 1) was some of the first and strictest relating to the introduction of non-native species. However, the deterrents are only applicable to deliberate releases and measures are taken to ensure they do not escape into the wild and have not as yet attempted to minimise the possibility of incidental releases e.g. as a consequence of trade.

Britain can be used to illustrate how the statutes apply in practice. A network of sites and an array of species have or will have full protection under national and European law in fulfilment of the Habitats Directive and Birds Directive (Natura 2000 sites). Additionally, under the Convention on Biological Diversity, species and habitats have been identified as priority or otherwise and afforded Action Plans. These are mainly voluntary, but have considerable backing and may involve actions to preclude the introduction or spread of non-native species. Under the Countryside and Rights of Way Act 2000, the Secretary of State (England) or the National Assembly for Wales now has a duty to take, or promote the taking by others, of reasonable steps to further the conservation of organisms and habitats which are of principal importance to conserving biological diversity as defined by the 1992 Convention.

Box 1. Nature conservation legislation of relevance to marine biological invasions.

The most recent and far-reaching international convention covering this subject is the Convention on Biological Diversity of 5 June 1992 that requires its Contracting Parties, as far as possible and appropriate, '*to prevent the introduction of, control or eradicate those alien species which threaten ecosystems, habitats or species*' (Article 8 h).

The Bonn Convention (1979) on the Conservation of Migratory Species of Wild Animals stipulates provision for '*the protection of (the) habitats (of migratory species) from disturbances, including strict control of the introduction of, or control of already introduced exotic species detrimental to the migratory species.*'

The Bern Convention on the Conservation of European Wildlife and Natural Habitats (1982), Article 11.2, requires '*each Contracting Party to strictly control the introduction of non-native species.*'

Some of the Protocols to the Regional Seas Conventions, concluded under the auspices of the United Nations Environment Programme specifically mention the question of introductions, including the Geneva Protocol (1982) on Specially Protected Areas in the Mediterranean; the Protocol concerning Protected Areas and Wild Fauna and Flora in the Eastern African Region (Nairobi 1985); and the Protocol on Specially Protected Areas and Wildlife in the Caribbean Region (Kingston 1990). The United Nations Convention on the Law of the Sea (UNCLOS) also makes strong recommendation on this point.

Regional instruments:

In relation to the Benelux Convention on the Conservation of Nature and Protection of the Landscape (1982) (which applies to Belgium, Netherlands and Luxemburg) a Decision was adopted in 1983 which requires all the governments to take steps to prohibit the introduction into the wild of non-native animal species.

In the European Union, two recent Directives are central to nature conservation and both refer to introduced species. Article 22b of the EC Directive on the Conservation of Natural Habitats and of Wild Flora and Fauna (1992), or 'Habitats' Directive, requires Member States to "*ensure that the deliberate introduction into the wild of any species which is not native to their territory is regulated so as not to prejudice natural habitats within their natural range or the wild native fauna and flora and, if they consider it necessary, prohibit such introduction.*"

The Directive on Wild Birds (1979) Article 11 states that "*Member States shall see that any introduction of species of bird which does not occur naturally in the wild state in the European territory of the member States shall not prejudice the local fauna and flora.*"

The Habitats' Directive refers only to deliberate introductions, whereas the 'Birds' Directive appears to be concerned also with accidental introductions, but only the introduction of birds (Council of Europe 1996).

The Wildlife and Countryside Act 1981 which applies to Great Britain and territorial waters out to 12 nautical miles, states that it is an offence to release (or allow to escape) into the wild any kind of non-native animal, except under licence. In addition, an introduced species of animal, once it is established in the wild, and any plant, established or not, known to have detrimental effects on the environment, may be listed in Schedule 9 of the Wildlife and Countryside Act 1981. It is illegal to release or allow to escape into the wild any species so listed, and action can be taken to remove such species from the wild. This piece of national legislation is one of the clearest for dealing with introductions.

3 Threats and consequential effects on nature conservation interests

Non-native species, especially invasive species, pose a threat to native biodiversity because they have the potential to alter the natural state of an ecosystem into which they are introduced. Such changes may consequently affect nature conservation interests. This may occur through the displacement of native species through competition, predation, smothering and alteration of habitats or the introduction of diseases and parasites and the alteration of the genetic pool through hybridisation. Table 1 provides examples of the effects of marine non-native species upon nature conservation interests in Europe (see also Eno 1996).

4 Response to non-native species which pose a threat to nature conservation interests

The control of invasive species in the terrestrial environment is carried out by destroying the invasive species themselves, and taking action to prevent their spread. In the marine environment species may disperse or secondarily spread (e.g. to Britain and Ireland) by different vectors to those involved with their primary introduction to Europe). This makes it harder to control the spread of species once introduced into European waters. Furthermore, once established, controlling an introduction in the marine environment is virtually impossible although a number non-native species have died out through natural causes.

A number of eradication programmes have been tried for non-native marine species. Most successful have been reactive programmes aimed at eradicating newly recorded species before they become properly established. Indications suggest that the removal of *Caulerpa* from California appears to have been successful, mainly due to the rapidity of action. A team working from the Centre for Research on Introduced Marine Pests (CRIMP) are on constant alert to eradicate any newly discovered marine introductions in Australian waters. More frequently, removal programmes serve to raise public awareness of the presence of non-native species and highlight the importance of measures to prevent further introductions. More realistically, control measures have to be adopted that aim to control the spread of a species and prevent it from becoming a dominant species in sites of nature conservation importance (see for example measures adopted to control spread of the fungal crayfish plague by restricting movement of crayfish between fresh water bodies).

Table 1. Effects of selected marine non-native species in European waters on nature conservation interests.

Effects on nature conservation interests	Non-native species
Displacement of native species	*Caulerpa taxifolia, Elminius modestus, Sargassum muticum, Styela clava, Crepidula fornicata, Hemigrapsus penicillatus*
Introduction of new pest and parasites which affect native species	*Crepidula fornicata, Anguillicola crassus*
Habitat alteration	*Caulerpa taxifolia, Crepidula fornicata, Spartina anglica, Eriocheir sinensis*
Changes in food webs, particularly through dietary competition and predation	Most non-native fauna
Degradation of integrity of gene pool through hybridization	*Spartina anglica*
Associated effect of commercial harvesting	*Mercenaria mercenaria*

On occasions, biological control methods have been employed in terrestrial situations. These sometimes involve introducing parasites or predators of the invasive species from its natural geographical range. There are considerable ecological risks associated with such measures, even more so in the sea due to its contiguous nature. Guidelines have been developed by the International Council for the Exploration of the Sea (ICES 1997). Britain and Ireland are islands. However, because of their proximity to other coastal areas and the nature of the marine environment, they need to be considered in a regional context. It is worth remembering that not many marine introductions have become established in British waters and only a small proportion of those cause problems to the environment or sea users.

5 Case history

A consideration of the spread of *Sargassum muticum* throughout the United Kingdom serves to highlight the process of introduction, establishment, spread and secondary introduction and ultimately the initiation of control measures to minimise the impact of this invasive species upon marine nature conservation interests. *Sargassum muticum* is widespread in Europe and has been reported from the Mediterranean and along the North Sea and Atlantic coasts of Portugal, Spain, France, Belgium, The Netherlands, Denmark, southern Norway and Sweden (Eno et al. 1997).

Sargassum has successfully invaded numerous sites of nature conservation importance in the United Kingdom (Davidson 1999). *Sargassum* was first recorded attached (as opposed to free-floating) in southern England in 1971 (Farnham et al. 1973) where it had arrived from France. It is thought to have spread attached to small boats and through the drifting of unattached fertile adult plants, despite its primary introduction to France in association with oysters. Since its arrival in Britain, it has been recorded from Strangford Lough Marine Nature Reserve (MNR), the Fleet lagoon candidate Special Area of Conservation (cSAC), the Solent cSAC and Lundy MNR. Most recently the alga has been reported in the Menai Strait and Conwy Bay cSAC and discussions have begun regarding what control measures are required. In the 1970s, a variety of chemical, biological and physical methods were tried to control the spread of *Sargassum* (Farnham & Jones 1974). These were generally unsuccessful and discontinued, all except some mechanical cutting of weed in inlets. On account of nature conservation concerns, physical control measures have been considered and applied to limit the spread of *Sargassum* in Strangford Lough in Northern Ireland (Davidson 1999). However, because of the invasive nature of the plant and the fact that it is now widespread in the UK, the effectiveness of such methods will be limited. This serves to highlight the importance of preventing introductions and if they do occur, acting quickly by embracing the precautionary principle.

6 Discussion

Non-natives are not generally considered to be part of the natural environment and hence their nature conservation interest is limited. On occasions, species have become naturalised and may augment communities in some way. The movement of the invasive cord grass *Spartina anglica* has sometimes been associated with the spread of the native cord grass *Spartina maritima*, which can provide an interesting feature of nature conservation interest. Species can become habituated to new areas, and they can become integrated into the natural community as seen with *Sargassum* in Northern Ireland, which is now considered to have an associated fauna of interest. Occasionally the original geographic location of a species is unknown and it is only recorded from sites to which it has been introduced. Such is the case with the hydroid *Clavopsella navis*, which, worldwide, has only ever been recorded in four localities, two on ships and two in the vicinity of harbours (Eno et al. 1997). On the basis of the 'rarity' criterion some consider it to have a nature conservation interest. Marine invasions are irreversible which has severe implications to biodiversity that Biodiversity Action Plans alone cannot combat. Nature conservation should really aim to maintain (or restore) the natural state of an environment. Any introduction (whether an impact is quantifiable or not) should be considered as undesirable and should be avoided wherever possible.

The risks associated with an introduction not only need to consider the potential effect of that species on native ecosystems (Hayes 1998) but also effects of control methods and associated effects e.g. of harvesting a species of commercial value. Such harvesting has occurred with a variety of bivalve molluscan introductions. The risks posed by invading species can be considered as biosecurity threats, of threat to the ecological welfare or the well-being of humans, plants or animals of a country (IUCN 1999).

It is often perceived there are different thresholds triggering potential human response (according to the perceived 'seriousness' of an invasion) which are afforded different priorities on a national/global scale:
(i) Top priority: essentially a political, generally public health concern, e.g. Cholera, toxic dinoflagellates
(ii) Medium priority: threats to economic or commercial interests including fishing/aquaculture, industrial abstraction etc., e.g. in response to zebra mussels *Dreissena polymorpha* in the Great Lakes
(iii) Low priority: other threats to the environment and nature conservation interests (particularly in relation to important habitats, species and communities); ecology (biological studies)
(iv) Frequently perceived levels of priority are biased towards commercial interests, however, the environmental effects should not be undervalued.

7 Future considerations

The threat posed to natural communities needs to be pushed up the political agenda. Getting wildlife provisions to bite outside of protected areas has been a major hurdle. However, governments (UK/EU) are now moving towards a process of integration of environmental policies. These need to translate into actions to ensure threats posed by spread of invasive species can be minimised globally, not just in countries which have already suffered from such effects e.g. US/Canada/Australia. Furthermore, the number of major international and other statutes covering the issue of non-native species illustrates the severity of the problem - and the need to act collectively. Ways to avoid potential problems include:
(i) applying the precautionary principle - if in doubt do not allow an introduction
(ii) strict control of import of all live specimens
(iii) use of native instead of introduced species
(iv) adoption of regulations on regional basis
(v) attempt eradication *before* species establishes.

Furthermore, the following actions should be pursued:

(i) the ICES Code of Practice (ICES 1995) should be given legal force and become binding upon permit-issuing authorities

(ii) the IMO guidelines for deballasting in open water should become legally binding world-wide and additional measures may be needed to reduce the transport of species by other ship-associated vectors

(iii) IUCN (1999) guidelines for the prevention of biodiversity loss due to biological invasion need to be extended to include marine species and converted into actions

(iv) link vectors to pathways (at a local, national, regional and global level) so control mechanisms can be better targeted

(v) do not become complacent about the potentially tremendous threat posed by Genetically Modified Organisms.

Treatment Measures

PREVENTIVE TREATMENT AND CONTROL TECHNIQUES FOR BALLAST WATER

ALAN TAYLOR[1*], GEOFF RIGBY[2], STEPHAN GOLLASCH[3], MATTHIAS VOIGT[4], GUSTAAF HALLEGRAEFF[5], TRACY MCCOLLIN[6] & ANDERS JELMERT[7]

[1]*Alan H. Taylor & Associates Melbourne, Victoria, Australia*
[2]*Reninna Consulting Pty. Limited, Charleston N.S.W, Australia*
[3]*GoConsult, Hamburg, Germany*
[4] *Dr. Voigt-Consulting, Stolpe, Germany*
[5]*University of Tasmania, Hobart, Tasmania, Australia*
[6]*FRS Marine Laboratory Aberdeen, Aberdeen, Scotland*
[7]*Floedevigen Research Station, Norway*
[*]Corresponding author *aht@ahtaylor.com*

Abstract
It is apparent that no single or simple universal solution presently exists for shipboard treatment or management to prevent the transfer of viable non-native organisms in ballast water. Only a very limited number of the treatment options listed below has been shown to be 100% effective (and only for some specific organisms in some cases), environmentally sound, cost effective and safe during application. Several technologies (including the widely recognised forms of water exchange currently in use throughout the world) or a combination of technologies (tool box) may however, be at least partially effective and feasible in terms of economic and shipboard constraints. Currently heat treatment, mechanical removal of organisms in combination with UV treatment, and chemical treatment of ballast water are considered the most promising approaches. However, concerns have been expressed regarding residual environmental polluting components, health and safety problems related to storage of chemicals and compatibility with cargo carried on board as well as direct and indirect handling of chemicals by crew members. Given the global nature of shipping and therefore the transport of non-native organisms in ballast water, the International Maritime Organization (IMO) and various countries, are considering the adoption of a Ballast Water Convention, which would include a technical effectiveness standard that would form the initial basis for acceptance of the various treatment options. The IMO Convention would aim to achieve a standard approach to ballast water management but in the meantime some countries, such as Australia, have introduced their own ballast water regulations. It is important that ongoing research and development aimed at developing and demonstrating existing and new treatment techniques be maintained in order that the IMO can incorporate the best available technology into the proposed convention. In summary, there is no current "stand-alone" treatment option, that covers all possible scenarios. However, a combination of methods could result in cost-effective management options.

E. Leppäkoski et al. (eds.), Invasive Aquatic Species of Europe, 484–507.

1 Introduction

The amount of ballast water transported annually on a world-wide basis has been estimated by scientists and engineers to be around 10 billion tonnes (Rigby & Taylor 1995) highlighting the global dimension of the problem to be controlled. It was further estimated that some 7,000 taxa are being transported in international shipping each day (Carlton pers. comm.). One reason for this great diversity of organisms arises from the three different "habitats" inside ballast water tanks: (i) tank walls, (ii) ballast water, and (iii) the sediment. Because of the diversity in ship design and improved technology (e.g. double hulls, higher economical cruising speeds), the survival rates of some species carried in ballast tanks have increased, and consequently, many introductions of nonindigenous organisms in new locations have occurred in recent years highlighting the need for effective ballast water management.

The control of ballast water introductions is likely to be based on a quarantine approach. This approach does not intend to provide an absolute barrier to prevent the introduction of unwanted nonindigenous species (Carlton et al. 1995) but rather aims for a significant reduction in risk. It has become clear that no single treatment process is likely to universally achieve the required inactivation, kill or total removal of all unwanted organisms in ballast water. Although a single process may be appropriate in some circumstances, a two stage treatment approach may be necessary to achieve the required level of effectiveness in other cases. Two stage treatment may comprise some form of mechanical removal of organisms followed by a physical or chemical treatment method. Comprehensive overviews on physical and chemical treatment options have been compiled, however, the depth of the information available on the individual treatment options varies greatly. While some techniques have been tested in the laboratory only with single or a few species, others have followed a long and costly test procedure involving numerous species and/or large scale trails in land based facilities or onboard ships (Fig. 1).

The mention of trade names and producers in this contribution does not necessarily imply an endorsement of the product by the authors but is mentioned as an example of the relevant technology.

2 Research initiatives on ballast water treatment

Since 1995 some fifteen research initiatives on ballast water treatment have been completed. Project objectives included the testing of several treatment options with regard to their effectiveness, environmentally soundness and practicability on board of vessels. The treatment options considered include ballast water exchange, filtration, heat treatment, coagulation/flocculation, pH adjustment, chemical treatment (ozone, glutaraldehyde, oxygen deprivation, hydrogen peroxide based formulations and chlorine) and UV. A further 14 projects with comparable objectives are underway and will be completed by 2003 (GloBallast Programme 2001). At this stage the various methods of treatment that have been put forward have been frequently referred to as a "tool box" from which the most effective (in achieving the desired level of effectiveness), practical (safe and easy to apply, not compromising deck safety or operational requirements, not damaging existing ship installations such as pumps, ballast tank integrity, ballast tank coating, isolators and sealing rings), cost effective and environmentally sound combination should be selected. Several of the options are straightforward statements of best practice but in many circumstances the choices available to a ship's Master may be limited.

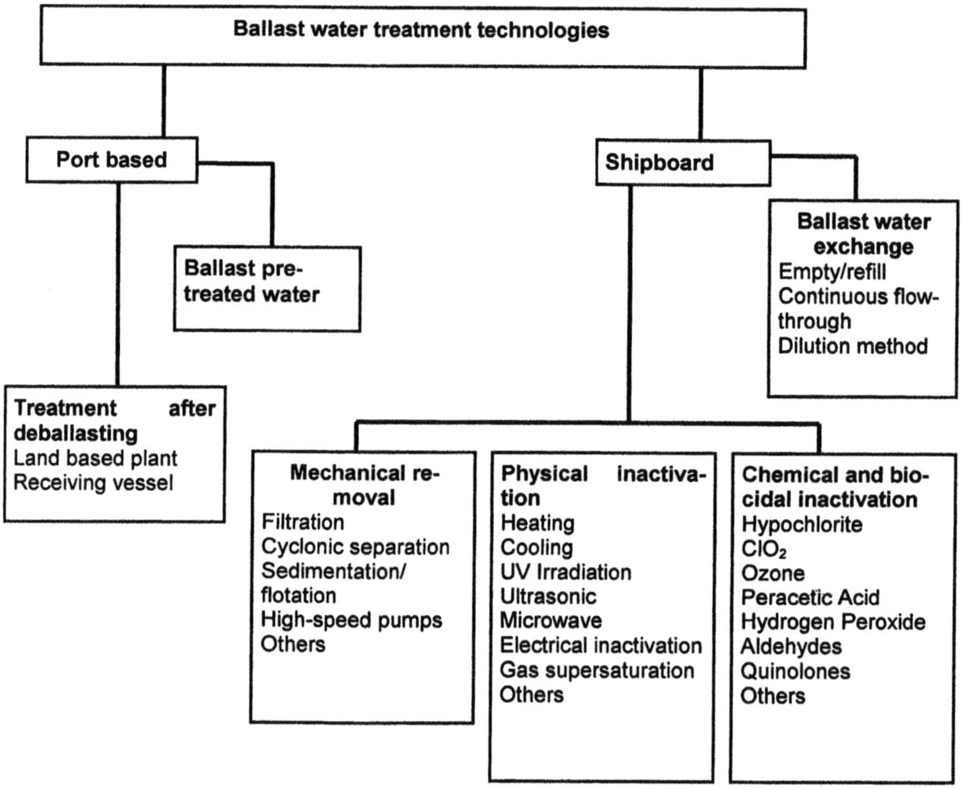

Figure 1. Port based and shipboard ballast water treatment options.

3 IMO Assembly Resolution

To date, international guidelines have been adopted as the IMO Assembly Resolution A.868(20) (see below). The IMO has not generally promoted regionally different systems, emphasising that a universal global approach is preferred to solve the ballast water problem. However it has been realised that some local restrictions may be appropriate to manage or control a particular organism of concern. Further, using different management options and treatment techniques could result in unwanted regional restrictive practices, restraints of trade and competitive advantages. Unfortunately, some concerned countries have already implemented voluntary and mandatory guidelines on ballast water, mainly (at least initially) based on the concept of ballast water exchange.

The International Maritime Organization's Marine Environment Protection Committee (MEPC) has had a specific interest in the field of unwanted aquatic species introductions by ballast water since 1973 when the International Conference on Marine Pollution adopted Resolution 18, drawing attention to the transport of aquatic organisms and pathogens around the world in ships′ ballast tanks. In the late 1980s, the MEPC formed a working group to consider research information and solutions proposed by Member States of the IMO and by non-governmental organizations. The working group concluded that voluntary guidelines were the appropriate first step in addressing this problem. MEPC adopted guidelines by resolution in 1991 and in 1993 these were adopted by the IMO Assembly under Resolution A.774 (18) entitled "International Guidelines for Preventing the Introduction of Unwanted Aquatic Organisms and Pathogens from Ships Ballast Water and Sediment Discharges". This was then replaced in 1997 by the IMO Assembly Resolution A.868 (20) "Guidelines for the Control and Management of Ship's Ballast Water to Minimize the Transfer of Harmful Aquatic Organisms and Pathogens". The IMO Assembly Resolution A.868 (20) includes a recommendation that an exchange of ballast water is carried out in open oceans as far as possible from the shore. The mid-ocean exchange of ballast water is currently the only readily available method that can be used in order to minimize the risk of transfer of unwanted organisms on existing vessels. Compared with coastal waters, deep ocean waters are generally expected to contain fewer organisms and, in addition, species occurring in open ocean waters are generally not able to survive in coastal zones and vice versa. Where open ocean exchange is not possible, requirements developed within regional agreements may be applicable, particularly in areas within 200 nautical miles from shore.

It is noted that no form of ballast exchange should be undertaken unless it is included in the ship's Ballast Water Management Plan and approved by the ship's Classification Society via the ship's "Trim and Stability" booklet. It is always the responsibility of the ship's Master to ensure that any operation carried out at sea is done so in a safe manner. In addition to the exchange of ballast water at sea the guidelines include reference to ballast water management practices that would reduce the risk of introducing non-native species such as:

(i) ballast water uptake should be avoided in the presence of harmful algal blooms and known unwanted contaminants (e.g. Cholera disease outbreaks),
(ii) precautionary procedures when taking on ballast water in shallow areas,
(iii) discharging ballast water and sediments to on-shore facilities (if available) and
(iv) avoiding ballast water uptakes at night as many zooplankton organisms migrate towards the water surface in darkness.

The MEPC Ballast Water Working Group is currently working towards a set of legally binding regulations in the form of a stand alone IMO Convention. The working group is scheduled to present the final draft of the guidelines to the Committee in 2002 and a diplomatic conference for its adoption is planned for 2003.

The following list of management options and treatment techniques to reduce the risk of species translocations (not necessarily in order of preference or effectiveness) provides an overview of options and does not claim to be fully comprehensive.

3.1 BALLAST WATER EXCHANGE

Ballast water exchange was originally developed as a method to be used by vessels on trans-oceanic journeys. The basis is that the water that was loaded in the port is exchanged for deep oceanic water that will contain fewer organisms that are unlikely to survive in coastal waters. The exchange process was also recommended for when a vessel was travelling between two fresh water ports as the increase in salinity would kill any freshwater organisms remaining in the tanks and the oceanic species would not survive in freshwaters. There are currently three methods by which a mid-ocean exchange of ballast water may be achieved:

3.1.1 Option 1 Empty/refill (=reballasting)

An empty-refill is as it sounds; the ballast tanks are emptied of port water and then refilled with oceanic water. Stripping pumps or eductors should be used wherever possible to minimise the amount of originally ballasted water remaining in the tanks. Trials have shown that for deballasting undertaken in such a manner at least 95% of the original water can be replaced (Table 1) (Rigby 1994; Rigby & Taylor 2001; Miller 1998; Wonham et al. 1996). However, it has to be noted that the exchange of 95% of the volume of the ballast water may not be equivalent to the exchange of 95% of the organisms in ballast tanks as these are not necessarily equally distributed in the ballast water, but may accumulate at the bottom and tank walls (GloBallast Programme 2001). On many ships this method may result in unacceptable bending moments or shear stresses (Rigby & Hallegraeff 1994; Karaminas 2000), but potentially could be 100% effective at removing all the original ballast water on some vessels. In practice, many woodchip carriers that claimed to have undergone reballasting still had sediments present in the tanks that included toxic dinoflagellate cysts (Hallegraeff & Bolch 1991, 1992). (Details on safety aspects further below).

3.1.2 Option 2 Continuous flow-through of ballast water (=ballast exchange)

A continuous flow through system allows continuous sea-to-sea circulation of ballast water while the ballast tanks remain filled, i.e. sea water is pumped into the ballast tanks while the tank is simultaneously overflowed from the top of the tank. Where the flow-through method is employed, at least three times the tank volume should be pumped through the tank (on some vessels this has been shown to correspond to a replacement of approximately 95% of the original water). Some pipework modifications may be necessary on some ships to enable this option to be utilised safely and effectively (Taylor & Rigby 2001).

In contrast to deballasting in high seas during bad weather using the empty/refill technique, the continuous flow through system does not impose excessive bending moments or shearing forces and minimises stability problems. However, Rigby & Hallegraeff (1993, 1994) demonstrated that by emptying certain ballast tanks on the bulk carrier *Iron Whyalla* the still water bending moment may be much higher than the maximum allowable value. This fact, in combination with the high number of organisms in the remaining water bodies in the ballast tank after emptying (option 1, see above), made the flow through option more favourable. However, future research should be carried out to confirm this view which is based on a limited number of trials.

3.1.3 Option 3 Dilution Method

The dilution method is a further development of the continuous flow through technique. After the installation of additional pipework on the vessel continuous ballasting from the top of the tanks via one pipe system and at the same time continuous deballasting by a second pipe system at the bottom of the tank may be carried out (IMO MEPC 38/13/2 1996; Villac et al. 2000). Mathematical modelling of the effectiveness of this method was carried out resulting in a comparable effectiveness as the ballast tank flushing for three times the ballast tank volume (Armstrong et al. 1999).

3.2 EFFECTIVENESS OF THE BALLAST WATER EXCHANGE

3.2.1 Water replacement efficiency

The efficiency of replacement of the original water will depend on the design of the ship's ballast tank, safety requirements, sea conditions, quantity of water pumped and the pumping system design. Water replacement efficiencies have recently been reviewed by Rigby & Taylor 2001 and Rigby 2001 (Table 1). Trials have shown that three times volumetric exchange of ballast water result in approximately 95% removal of viable algal cells and approximately 60% removal of zooplankton organisms. However, as the inoculation size to introduce a new algal species is about 1,000 cells a removal of 95% of these organisms would not eliminate the risk of species invasions as millions of specimens remain in the water (GloBallast Programme 2001). The efficiency of water exchange using the continuous flushing option is lower if the ship is not at sea, as the mixing is less efficient compared to when the vessel is moving as at sea (Figure 2; Rigby & Hallegraeff 1994).

Table 1. Efficiency of water exchange for various ocean exchange options.

Mode of Exchange	% Water exchanged	Reference
Continuous flushing, 0.5 tank volume	39.3	Rigby & Hallegraeff 1994
Continuous flushing, 1 tank volume	63.2	Rigby & Hallegraeff 1994
Continuous flushing, 2 tank volumes	86.5	Rigby & Hallegraeff 1994
Continuous flushing, 3 tank volumes	95.0	Rigby & Hallegraeff 1994
Continuous flushing, 4 tank volumes	98.2	Rigby & Hallegraeff 1994
Continuous flushing, 3 tank volumes	AA90, 99	Taylor & Bruce 2000
Empty/refill (20 tonnes residual per db tank)	99.6	Calculated
(calculated for 50 tonnes residual water)	99.2	
Non-dedicated tanks empty/refill	100	Rigby 1994
Empty/refill	95.0	Miller 1998
Dilution/flushing	90.0	IMO MEPC 38/13/2 1996, 1998
Dilution/flushing	86-96	Villac et al. 2001
Sequential empty/refill	>99	Wonham et al. 1996
Continuous flushing/heating	>99	Rigby et al. 1997
Ocean exchange for salinity increase of brackish water;		
(ocean 35 PSU, brackish 15 PSU)	75	
(ocean 35 PSU, brackish 5 PSU)	83	

3.2.2 Organism removal efficiency

The efficiency of removal of organisms as distinct from the original ballast water, is a complex issue, which will be affected by the nature and behaviour of organism in the tanks, the design of tanks, mixing within the tanks and the types and behaviour of sediments (Table 2). For example some fast swimming zooplankton may remain after several tank volumes have been replaced, as indicated by recent sampling studies on the *Iron Whyalla* (Sutton et al. 1998).

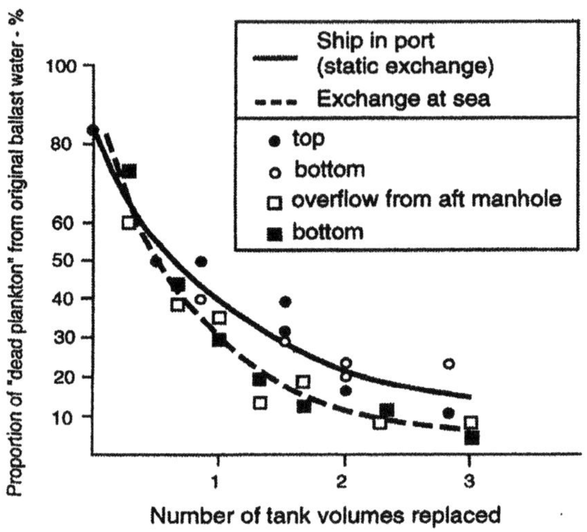

Figure 2. Continuous flushing ballast exchange efficiencies on the *Iron Whyalla* (Rigby & Hallegraeff 1994).

Shipboard based microscopic examination of organism removal during ballast exchange and heating trials on the *Iron Whyalla*, showed that in addition to the effects of heating, the efficiency of removal of phytoplankton contained in the water after heat treatment was similar or higher than for water exchange (Rigby & Hallegraeff, 1994; Rigby et al. 1997). Flushing trials by Ruiz & Hines (1997) in wing tanks on the *S/R Long Beach* and the *S/R Benecia* showed a 60% water exchange on the basis of salinity and less than 90% on the basis of coastal plankton communities. In another trial, the exchange of 3 tank volumes resulted in water exchange efficiencies of 70-100% on the basis of salinity and greater than 95% on the basis of coastal organisms.

In a recent trial involving the Dilution Method of continuous flushing on the oil carrier M/V *Lavras*, a water exchange efficiency of 90% was achieved with a phytoplankton exchange of 96%. Chlorophyll 'a' exchange was estimated as 86% (Villac et al. 2001).

From the flow through ballast exchange trials carried out by Taylor & Bruce (2000), it was concluded that there was retention of phyto- and zooplankton from the source port for the *Spirit of Vision* trials and that biological efficacy was at variance with the tracer dilution efficiencies. Approximately 50% of one of the taxa, the diatom *Thalassiosira* spp., appeared to settle in the bottom layer of the tank after exchange. However, the trials on another vessel, the *Iver Stream,* indicated that the flow through dilution method was relatively effective at reducing the planktonic organisms originally ballasted in the source port. A 90-100% reduction in the means of depth stratified counts of source port indicator taxa was achieved.

Many other studies showing survival of organisms have been reported following ballast exchange in some form or other, however in many cases there has been no quantitative assessment of replacement of the originally ballasted water and researchers have relied on the word-of-mouth comments of the ship's officers. Consequently, it is difficult to accurately compare differences between water exchange and biological exchange. For example, Locke et al. (1993), using freshwater zooplankton and salinity as indicators (for ships originating from fresh and brackish water ports), estimated that for 24 vessels entering the Great Lakes region, the efficiency of zooplankton exchange was 67% and for water exchange was 86%.

Table 2. Efficiency of organism removal for various ocean exchange options.

Mode of Exchange	% organisms removal	Reference
continuous flushing	>95 (phytoplankton)	Rigby & Hallegraeff 1994; Rigby et al. 1997
continuous flushing	<90 (coastal organisms)	Ruiz & Hines 1997
empty/refill, 3 tank volumes	>95 (coastal organisms)	Ruiz & Hines 1997
continuous flushing	96 (phytoplankton)	IMO MEPC 1998
flow through	90-100 (selected taxa)	Taylor & Bruce 2000
empty/refill, 1 tank volume	67 (plankton), estimated	Locke et al. 1993
empty/refill, 1 tank volume	48 (phytoplankton)	Dickman & Zhang 1999
dilution	86-96 (phytoplankton)	Villac et al. 2001

Harvey et al. (1999) reported that although complete exchanges had been reported for vessels entering the Great Lakes, the presence of coastal species suggested that the exchanges were incomplete.

During a scientific study in the framework of the European Concerted Action "Introductions with ships" (1997-1999) the container vessel *Pusan Senator* was accompanied in 1999 on a voyage from Kaohsiung (Taiwan) to Hamburg (Germany) to investigate the zooplankton present in ballast tanks over time (Fig. 3). In addition, the ballast water was exchanged enroute using (three times) the empty/refill method. A comparison of the zooplankton community before and after exchange revealed that the number of species increased considerably after water exchange but in contrast the number of individuals showed a continuously decreasing trend. Some species survived the entire 23 day voyage. It was concluded that a mid-ocean exchange does not necessarily result in a complete exchange of the taxa in the ballast water and reduces the density of zooplankton individuals inside the tank on a limited scale.

In examining the effectiveness of open ocean reballasting (empty and refill) in reducing the number of diatoms and dinoflagellates in ballast water for 14 container ships travelling from Oakland, California to Hong Kong, Zhang and Dickman (1999) reported the Master estimated that 95 to 99% of the original water was removed and their analyses showed that 87% of the diatoms and harmful dinoflagellates had been removed. In a similar study (Dickman & Zhang 1999) for 5 container ships travelling from Manzanillo, mid ocean reballasting resulted in 48% removal of diatoms and dinoflagellates (again with the Master estimating 95-99% water replacement). The differences between the two studies was attributed to the fact that the latter study involved much older ships, which did not have such efficient ballast water exchange systems whereas the first study involved ships that had only been in use for approximately 1 year and had more modern ballast water exchange designs. However in the absence of any quantitative measurements on water exchange it is difficult to make a direct comparison.

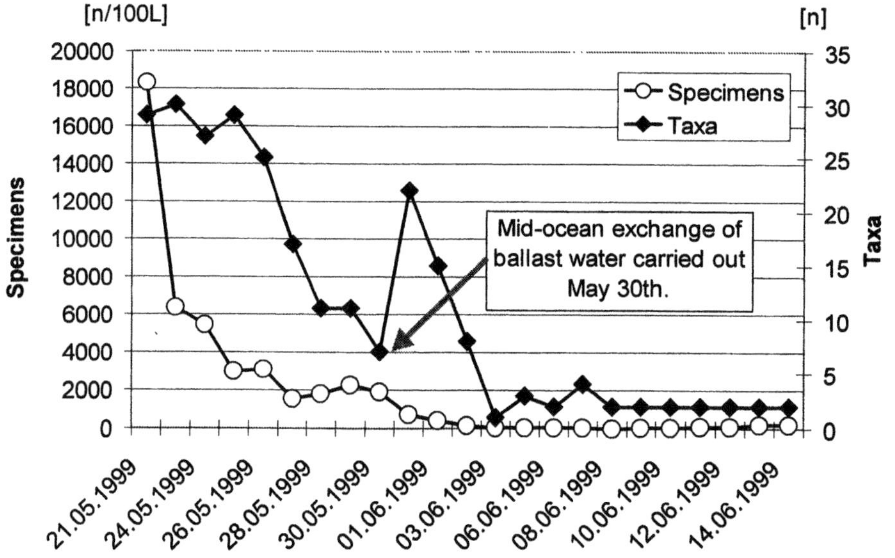

Figure 3. Number of zooplankton taxa and individuals from the ballast water of the portside side tank during a ships voyage from Kaohsiung to Hamburg. Origin of the ballast water investigated: Hong Kong. After sampling on May 30th a mid-ocean exchange of the ballast water was undertaken. All samples were taken using plankton net with a mesh size of 55 µm.

The above limited observations and the diversity of results concerning differences between water exchange or replacement efficiency and removal/replacement of organisms show that it is not considered realistic to seek to establish a generalised definitive relationship between the two. In addition to natural mortality in the tanks, tank geometry and ballast hydraulic design, type of exchange operation and other operational features specific to each ship and the types of organisms present will have a marked effect on their behaviour within the tanks.

3.3 SIGNIFICANCE OF SEDIMENTS

The significance and role of sediments in relation to ballast exchange has been poorly researched. Analyses of sediments removed from tanks have identified the presence of harmful organisms. For example, Hallegraeff & Bolch (1992) have reported the presence of toxic dinoflagellate resting spores in various types of sediments. It is feasible for sediments to remain in ships for many years (interdocking interval). However, in many ships, sediment is confined to areas well away from locations where the main flow of water occurs and hence may not play a significant role on organism dispersal. Work currently in progress is aimed at examining some of these effects in more detail (Rigby & Hallegraeff 2001).

3.4 LIMITATIONS OF BALLAST EXCHANGE

Whilst ballast exchange is currently the primary option available to owners, operators and ship's Masters for treatment of ballast water, there are some limitations and potential disadvantages as far as the possibility of organism translocation is concerned.

The precise location of ballast water exchange needs to be carefully chosen. In studies by Macdonald & Davidson (1998) and Forbes & Hallegraeff (2001) mid ocean exchange of ballast water resulted in an increase in the species diversity of phytoplankton in ballast tanks. Macdonald & Davidson (1998) found that although exchange of ballast water in regional seas may reduce the risk from polluted European harbour waters, it may result in transport of potentially harmful phytoplankton species from the North Sea or Irish Sea to UK coastal areas, where conditions are likely to be sufficiently similar for survival. Details of the type of mid ocean exchange were not reported in detail in this study, but it was reported that the diversity of diatoms and dinoflagellates increased in 69% and 85% of cases, and abundance increased in 31% and 85% of cases respectively. A more detailed follow on study using one of the vessels from the preliminary study is being carried out by McCollin et al. (2001). This study involves taking samples before, during and after an exchange process. The vessel can carry out either an empty-refill or a flow through exchange process. The project is currently in the initial stages but preliminary results would seem to indicate that the exchange process in regional seas is not as effective as in deep oceanic waters.

In a recent study Forbes & Hallegraeff (2001) reported that 80% of woodchip ships operating between Japan and Triabunna in Tasmania reballast in coastal waters off Papua New Guinea and bring a new viable tropical/cosmopolitan inoculum, mixed with remnants of old Japanese plankton into Triabunna. Monitoring of woodchip vessels arriving in Triabunna, which claim to have exchanged 100%, indicated that 80% of ships still contain up to 30 culturable diatom species (including potentially toxic *Pseudo-nitzschia*).

Forbes & Hallegraeff (2001) noted that there may be the problem of ships carrying tropical plankton directly into North Australian tropical ports (such as Port Hedland and Gladstone), without undertaking any exchange and relying on the natural ballast water mortality to take its toll. Here it would be preferable to exchange the water. Scientific studies have shown that even after 116 days living zooplankton can be found in ballast tanks (Gollasch 1996) and under certain circumstances zooplankton species reproduce in ballast tanks (Lenz et al. 2000). Further Forbes & Hallegraeff (2001) showed that the exchange of ballast water could increase the species diversity of ballast tanks, especially in many domestic shipping routes, where no deep water exchange zones occur.

These observations illustrate that although mid ocean exchange is currently the most widely practised treatment option, caution needs to be exercised in assuming that use of this option will always reduce the risk of translocating undesirable organisms, especially if the exchange water is compatible with the receiving water. In utilising exchange on relatively long international voyages, care will need to be exercised in the future in choosing areas for exchange where organisms in the ocean water are not likely to be more of a problem than those present in the original ballast water. Similar caution needs to be taken for shorter international and coastal voyages where suitable reballasting zones may be limited or insufficient time available for exchange in an appropriate area.

It is also relevant to re-iterate that exchange of water has been adopted as an initial attempt to minimise the risks of discharging harmful organisms in ballast water and that research aimed at developing and demonstrating more effective treatment options should be encouraged and continued, so that improved systems can be utilised in the future.

Even if it is assumed that the efficiency of removal of organisms is the same as the water replacement efficiency in ocean exchange, it is important to realise that large numbers of harmful organisms may still be present in the water discharged into the receiving port. This is especially true when ballasting occurs during an algal bloom in the ballasting region. This point can be illustrated by considering an algal bloom of the toxic dinoflagellate, *Gymnodinium catenatum*. Typical cell densities during a bloom might be 10^5/L. This means that in a Cape Size carrier containing 55,000 m^3 ballast water, the number of cells present would be $5.5x10^{12}$. Typically, it can be assumed that about 1% of these cells could produce cysts, and hence the number of residual cysts in the water remaining on the ship after flushing with three tank volumes (assuming an exchange efficiency of 95%) would be $2.75x10^9$. This number of cysts would obviously represent a very serious threat when discharged into a receiving port, since a viable inoculum of several hundred cysts might be expected to be sufficient for a new introduction of this species (Hallegraeff 1998). Even if 99.9% of the original water were replaced during the exchange, the number of cysts still remaining would be 5.5×10^7 (Rigby & Taylor 2001).

3.5 SAFETY ASPECTS

As long as safety permits, open–ocean ballast water exchange is currently the best option available as a first step in order to minimize the number of species and individuals unintentionally introduced with ballast water until more effective treatment options are approved. However, ballast water exchange can pose serious concerns and every eventuality must be taken into account when deciding whether it would be safe to exchange ballast at sea.

A study of the "Ship Operational and Safety Aspects of the Ballast Water Exchange at Sea" was carried out by Woodward et al. (1992) who concluded that ballast/deballast operations may be carried out safely if wave heights were below a maximum value. Using hydrostatic data hull bending moments and stabilities are investigated to find the tank-emptying operations representing the maximum safety. At-sea analysis for hull bending moment, shear and rate of slamming was carried out using both linear and non-linear analysis. From the small sample of three ships (a dry bulk carrier, a tanker and a container ship) it appears that the critical wave height lies between 10 and 20 feet. The sample is too small to support a more definite conclusion on the maximum safe wave height. However, a complete reballasting at sea using the empty/refill method can be unsafe even in good weather conditions owing to stability problems. A sequential operation of continuous flushing of tanks with ocean water would be an alternative workable option as it reduces stability problems (Rigby & Hallegraeff 1993).

Karaminas (2000) has analysed the use of sequential exchange (empty-refill) on 26 existing ships of various types, configurations and sizes. This study showed that many of the ships did not have sufficient strength to undertake this type of exchange safely, or in some cases, only under light or minimum ballast conditions. The use of a diagonal sequential method could be an effective method for reducing the still water bending moments and shear forces to within permissible levels. It is essential that any of the ocean exchange options must only be undertaken if safe conditions can be maintained. This decision always rests with the Master of the ship.

In addition the IMO sub-committee on Ship Design and Equipment prepared guidelines on safety aspects of ballast water exchange taking into consideration:

(i) crew safety,
(ii) structural integrity as well as stability of ships and in particular stated to avoid over and under pressurisation of ballast tanks,
(iii) free surface effects on stability and sloshing loads in tanks that may be slack,
(iv) admissible weather conditions,
(v) maintenance of adequate intact stability in accordance with an approved trim and stability booklet,
(vi) permissible sea-going strength limits of shear forces and bending moments in accordance with an approved loading manual,
(vii) torsional forces,
(viii) minimum/maximum forward and aft draughts, and
(ix) wave induced hull vibration.

It was suggested that the ballast water management plan should include designated control personnel responsible for the ballast water exchange and crew training for familiarisation. Furthermore, it was noted that a need exists to evaluate the safety of the long term aspects of ballast water exchange, taking into account relevant safety matters, including safety of crews and ships, ship´s position, weather condition, ballast system inspection and maintenance, machinery performance and availability (IMO MEPC39/7, IMO MEPC39/7/1, IMO MEPC39/7/4, IMO MEPC39/8).

From the above it can be summarised that every ship is different, even sister ships, in relation to the safety of ballast water exchange carried out by the sequential or flow through methods. There are too many variables involved to make any specific conclusions suitable for all ships at all times by any regulating authority. If the owners or operators of a ship select ballast water exchange as their preferred option for ballast water management, the safety of the ship and crew always resides with the master under international law (SOLAS Convention). In addition, as the movement of ballast water at sea, within the hull envelope, of any vessel is considered a "CRITICAL SAFETY PROCESS" as it can affect the trim, stability, the bending moment or the shear forces acting upon the hull of the ship as well as the possibility of under or over pressurising the ballast water tanks, hence the safety of the ship and its crew. Being a critical safety process it is also a requirement under the ISM Code to have a procedure to cover any movement of ballast water at sea. As the movement of ballast water can have an affect on the trim, stability, bending moment or shear forces acting upon the hull of the ship, the operation requires approval by the ship's Classification Society and by the Flag State Administration.

4 Mechanical removal of species in ballast water

Mechanical technologies are based on particle-size or specific weight to separate or remove organisms from the ballast water. To its advantage mechanical treatment does not generally produce any unwanted environmental side effect.

4.1 FILTRATION

Filtration of ballast water is one of the most environmentally sound methods. The overall advantage of this method is its use during ballast water intake. The backwashing (filter cleaning) material may be returned immediately to the region from which the ballast water was taken without any treatment.

Filter systems under consideration include self-cleaning backwashing filter systems, microfiltration and granular filtration. Woven mesh filters, made from synthetic fibres, are available as automatically self-cleaning units. They could potentially be retro-fitted to existing ships or incorporated into the design of new vessels. Automatic cleaning of the filters can be programmed either for specific time intervals or at specific pressure differences across the filters. This would involve stainless steel brush and suction scanner filter mechanisms for 'coarse' and 'fine' filters. It is claimed that the 'coarse' filter would remove most of the larger zooplankton whilst the second in-line 'fine' filter would remove most of the smaller zooplankton and much of the large and medium sized phytoplankton. In early experiments filter capacities of 500-1,000 m^3 per hour were achieved at the various mesh sizes tested. As pump capacities in smaller vessels, e.g. container ships and cruise liners, are below this figure filtration would not slow down the ballasting or de-ballasting operations. However, resizing of the pumps may be required to cope with the increased filter resistance. Alternatively, with no modification, a reduction in pump capacity and concomitant increase in ballasting time could result.

Recent investigations included test filters with screen sizes of 25, 50 and 100 μm and a flow rate of about 5,000 m^3 per hour. The fine screens were protected from larger floating objects and organisms by the additional use of 5 mm screens prior to the water entry into the system. Filters reached a 92-99% percent removal of larger zooplankton and 74-94% removal of smaller zooplankton and phytoplankton (Taylor & Rigby 2001; GloBallast Programme 2001). Bacteria attached to larger organisms and suspended solids were significantly reduced using the smaller filter (25 μm). However, total bacteria remain largely unaffected by filtration. It was suggested that the effectiveness of filter systems could be increased by the use of an additional technique such as heat or UV as secondary treatment (Cangelosi et. al. in print; Rigby & Taylor 2001).

The first full scale investigation of a ballast water treatment system, other than mid-ocean exchange and one form of heat treatment, is a filtration experiment currently in use on the cruise liner M/S Regal Princess (P&O Princess Cruises, United States) with a total ballast tank capacity of 4,213 m^3.

4.2 CYCLONIC SEPARATION

Cyclonic separation has been proposed as a relatively simple and inexpensive way of removing larger particles and organisms from ballast water. Water and particulates enter a separation unit tangentially, thus inducing a circular flow. The water is then drawn through tangential slots and accelerated into the separation chamber. Centrifugal action ejects particles heavier than the water to the perimeter of the separation chamber. Suspended solids gently drop along the perimeter and end up in the calm collection chamber of the separator. In addition to the removal of organisms, a large fraction of the sediment usually deposited in the ballast water tanks could be removed. Solids may be periodically purged or continuously bled from the separator. A flow rate of up to about 3,000 m^3 per hour has been achieved. The pressure drop across this separator is relatively low (by comparison with filtration), so that the pumps already present on the ships would generally be sufficient. It seems feasible that this system could be used on board ships by incorporating the separator in a ballast tank recirculation system. However, cyclonic separation of organisms with a specific gravity similar to that of sea water (such as jellyfish, chaetognaths and phytoplankton) is likely to be limited (Armstrong 1997).

To increase the overall efficacy a UV-light unit could be applied as secondary treatment after the cyclonic separation.

4.3 SEDIMENTATION AND FLOTATION

Other separation treatment processes such as sedimentation and flotation have also been considered in terms of their application to ballast water treatment. The former essentially increases the settling of material from the water column under gravitational forces, with or without the use of coagulant chemicals to assist sedimentation. The latter also entails the use of coagulants and the injection of fine air bubbles into a 'flotation tank'. The bubbles attach themselves to coagulated organisms or particulates and float them to the surface. Efficacy data are not available (Carlton et al. 1995).

4.4 PUMPING VELOCITY

The use of a high velocity ballast water pumps during ballast water intake and discharge could minimize the survival rate of some macro-organisms due to mechanical damage (Woodward 1990; Carlton et al. 1995). The installation of additional units in order to create high velocity jets of water in ballast tanks or pipework would involve high costs. Only limited data is available on the efficacy of this method (Carlton et al. 1995), although some recent trials using a pipe section fitted with internal mixing elements reported some progress (GloBallast Programme 2001). However, the usual maximum velocity in ballast water systems should not exceed 3 m s^{-1}, because of erosion of the internal surfaces of the ballast pipes especially in the areas of 'bends' in the piping system. High velocity treatment techniques within the ballast system are extremely difficult to install and it is expensive to replace ballast water piping in service what makes this option less favourable.

5 Physical inactivation of species in ballast water

Although a number of physical techniques have been investigated, the most promising options currently available are heat treatment and Ultra Violet (UV) irradiation. Physical treatment options for ballast water focus on the removal of organisms or on changing the physical properties or hydrodynamic characteristics of the water to kill / inactivate the taxa present.

5.1 HEAT TREATMENT

Temperatures around 40 °C will kill or inactivate many organisms of concern frequently found in ballast water. Heating of ballast water to temperatures of this order has been proposed as an appropriate means to kill most toxic marine organisms (AQIS 1993). For example, exposure to temperatures of 36 to 38 °C over a period of 2 to 6 h was sufficient to kill zebra mussels in pipes. More recent data showed that 35 to 38 °C for a period of 4 to 5 h effectively killed vegetative cells of toxic and non-toxic dinoflagellates (Hallegraeff et al. 1997). With many organisms, the temperature required will generally be lower for longer periods of heating (Bolch & Hallegraeff 1993, 1994). In a laboratory study Bolch and Hallegraeff (1993) demonstrated that short-term (30 to 90 s) exposure to temperatures above 40 °C were effective in killing cysts of the dinoflagellates *Gymnodinium catenatum* and *Alexandrium tamarense*, whereas temperatures as low as 35 to 38 °C were sufficient after 4 h heating. Time temperature relationships required to kill a number of model organisms have been examined by Mountfort et al. (1999, 2000). These laboratory findings were subsequently confirmed in full scale shipboard trials, where the ship's pipework was modified to enable waste heat from the main engine cooling circuit to heat the water in one of the ballast tanks by flushing with the heated water (Rigby et al. 1997, 1999). In these trials, onboard microscopic observation of heated water, showed that none of the zooplankton (mainly chaetognaths and copepods) and only very limited original phytoplankton (mainly dinoflagellates) survived the heat treatment. The original organisms were reduced to flocculent amorphous detritus. Subsequent culturing efforts on the heated ballast tank samples only produced growth of some small (5μm) diatoms and colourless ciliates. Although no toxic dinoflagellate cysts were present in the tanks, based on earlier laboratory experiments, it is probable that these would have been effectively killed by the temperatures achieved during the heating trial, since essentially all of the water reached 37-38°C.

Heating of ballast water as described above also has the added advantage that organisms contained in sediments would also be subjected to these temperatures (in fact higher temperatures are experienced at the bottom of the tanks where the ballast water is pumped into the base of the tanks).

This form of heating may not be appropriate for short (domestic) voyages or where heat losses to the ocean are high (for example where sea temperatures are low).

In addition to the heating/flushing option, a number of other options (involving recirculation of water from the ballast tanks) utilising alternative heat exchanger designs and/or additional heat or steam from the main engine or exhaust system are possible (Sobol et al. 1995; Thornton 2000; Mountfort et al. 1999, 2000; Rigby & Taylor 2001). The suitability, cost and practicality of these alternatives will depend on the ship's heat balance, length of voyage, engine cooling system design, availability of steam and the additional equipment involved.

Although some concerns have been expressed about possible effects of elevated temperatures on tank internal coatings (Armstrong 1997), these effects are not likely to be significant given the temperatures involved and time that the tank surfaces are exposed (Rigby et al. 1997).

Concerns that such mild heating of ballast water could stimulate the growth of pathogenic bacteria such as *Vibrio cholerae* have not been substantiated by simulated laboratory culture experiments (Desmarchelier & Wong 1998).

5.2 COOLING TREATMENT

A reduction of the temperature of the ballast water near the freezing point requires e.g. a cooling unit, additional pipework and power, not to mention the safety aspects relating to the ships hull and tanks that could occasion if the water were to freeze. Further studies to evaluate the feasibility should take into account the safety aspects, likely costs, temperature related impact on the pipework and ballast tanks as well as trials on the treatment effectiveness on ballast water organisms.

5.3 ULTRAVIOLET IRRADIATION

UV irradiation is commonly used for sterilising large amounts of potable or wastewater and for the purification in aquaculture and fisheries (Carlton et al. 1995). UV irradiation operates by causing photochemical reactions of biological components such as nucleic acid (DNA and RNA) and proteins. The lower UV wavelengths are generally more effective. However, irradiation at these wavelengths shows a lower transmission in water. Due to a higher concentration of inorganic solutes, the transmission in seawater is slightly less than in freshwater. It may further be affected by the organic load, suspended solids or air bubbles. The effectiveness of UV treatment further depends largely upon the size and morphology of organisms. Viruses require similar dosages to bacteria. Algae require larger dosages than bacteria due to their size and pigmentation.

The biological effectiveness of UV treatment is not necessarily a simple function of irradiance and exposure time. Experiments with a phytoplankton species showed that a short exposure at high irradiance was found to be more effective than long exposure at low irradiance (Cullen & Lesser 1991). Confirmation by Hallegraeff et al. (1997) and Montani et al. (1995) have shown that the germination of cysts of *Alexandrium*, *Gymnodinium*, *Protoperidinium*, *Scrippsiella* and *Gyrodinium* occurred after exposure to UV radiation. However, it has to be noted that many cyst walls are impermeable to UV. It has been suggested that organisms have repair mechanisms which may enable them to recover from the UV treatment provided their exposure does not exceed a certain time. Similar observations were made with bacteria and other phytoplankton species, demonstrating the importance of specifying exposure time as well as irradiance level.

In-line flow treatment would appear feasible and the most practical option for retro-fitting a UV treatment system on ships. Treatment could take place at the time of ballasting and/or de-ballasting. This method is likely to be practicable and environmentally sound (no toxic side effects) and no adverse effects in pipework pumps, sealing rings or coating are known (Müller 1995; Müller & Reynolds 1995). No specific health, safety or environmental concerns appeared to be associated with the use of UV systems on board ship, however, the possibility exists that UV radiation might cause mutation of genetic material in the organisms treated. Capital and running costs for suitable systems are likely to be significant (Müller 1995; Müller & Reynolds 1995; Rigby & Taylor 2001).

Disadvantages of UV treatment include the possibility that some smaller organisms could pass the UV unit in the shadow of larger organisms or suspended solids without any treatment (Armstrong 1997), the reduced penetration of UV irradiation in turbid ballast waters, what is a major limitation for the use in ballast water treatment (Rigby et al. 1993) and the recovery of the phytoplankton following exposure to UV irradiation. On the basis of the experimental results, UV appears promising for the treatment of some organisms in ballast water and its effectiveness may be increased by initial filtering of ballast water to avoid shadowing effects of larger organisms and suspended solids.

5.4 ULTRASONICS

The use of ultrasonics for controlling hull fouling dates back to the 1950s, however, demonstration of its potential application for ballast water treatment purposes has not been investigated to any great extent (Subklew 1963; Müller 1995; Müller & Reynolds 1995). Ultrasound is thought to be mediated through various responses that may be fatal to marine organisms. These are heat generation, pressure wave deflections, cavitation and possibly the degassing effect of ultrasound causing removal of much of the oxygen. Higher frequencies, warmer temperatures and lower concentrations of dissolved matter have been found to increase the effect of ultrasound pulses. Plankton mortality has also been observed in the presence of ultrasound and is considered in part to be attributable to the cavitation process (Armstrong 1997).

The implementation of ultrasonics would require the installation of in-line transducers because ultrasound is unlikely to penetrate sediments. With respect to health and safety aspects, problems may arise with noise from some transducer types. There may also be some as yet unknown implications for the ship's structural integrity and health of crews following repeated exposure to ultrasound. It has also been suggested that the cavitation process could cause damage to tank coatings or structures (Müller 1995; Müller & Reynolds 1995). The application of sonic disruption as treatment to the ballast water has so far not been tested.

5.5 MICROWAVE

Beside the use of ultrasonics the application of microwaves (wavelength 0.1-1 nm) for ballast water treatment has been listed. The application of microwaves is relatively new but has been developed to treat waste waters (AQIS 1993). Microwaves have not been tested to treat ballast water organisms.

5.6 ELECTRICAL REMOVAL OF SPECIES IN BALLAST WATER

Ballast water treatment with electrical currents may cause serious damage to macro-organisms (Woodward 1990). The inactivation of dinoflagellate cysts had been achieved by the use of an electric shock during the exposure to 100V for 5 sec (Montani et al. 1995). Hallegraeff et al. (1997) claim that this mortality is due to generation of heat and reactive oxygen species.

The installation of an electronic unit in the sea chest of the vessel near the ballast water intake may provide a deterrent for the entry of some macro-organisms. Comparable units have been used in intake areas of power plants to prevent the unwanted impact of macro-organisms but have not been implemented onboard ships.

5.7 ELECTROCHEMICAL BALLAST WATER TREATMENT

An electrochemical technology using large-scale Boron-Doped Diamond (BDD) electrodes (electrolysis cell) was recently tested in laboratory trials. The self-cleaning BDD electrodes show significant technical and economic advantages due to their specific production capacity of disinfectant and oxidant substances. Completed treatment trials showed that the system has an outstanding production of chemical oxidizing agents such as strong hydroxyl radicals. In addition it may also support the production of other oxidizing agents like chlorine, persulfates, ozone and hydrogen peroxide in many kinds of water. The disinfectants produced by the system are of very high activity (i.e. more than four times higher than a typical chemical dosage at similar concentrations with regard to total chlorine). The flow rate of 7,000 m^3 ballast water in two days resulted in a total energy consumption <420 kWh (pumping included). The system needs little extra maintenance and energy compared to other treatment techniques, has a compact design (enabling retrofitting onboard existing ships) and is highly effective to treat microorganisms being of major concern in unwanted ballast water introductions (e.g. disease agents and pathogens) (Pupunat et al. 2001, CSEM, Switzerland and Tanksystem, Switzerland pers. comm.).

5.8 GAS SUPER-SATURATION

As a coarse definition gas supersaturation (GSS) is a condition where the water contains more dissolved gas than it normally will at a given depth and barometric pressure. When GSS is established, it will tend to move towards equilibrium.

Several groups of organisms have been found to be sensitive to GSS and when exposed to high levels of GSS, they may suffer from acute Gas Bubble Trauma (Colt 1986). Acute gas bubble trauma means formation of bubbles in tissues and the vascular system, which often leads to occlusions of vessels, emphysema, and haemorrhages (Weitkamp & Katz 1980). Jelmert (1999) has suggested that this susceptibility could be utilized as the target for an environmentally friendly ballast water treatment, where air (approximately 80% N_2) or N_2 are injected into the ballast water.

While only multicellular organisms are believed to be directly susceptible, some additional effects can be hypothesised. The use of N_2 will reduce the available O_2 content and impair the respiration of all aerobic organisms. The injection of large amounts of gas by a dedicated injection system will produce a high number of small bubbles. These represent a large hydrophobic area sifting through the water column. Inorganic particles, as well as bacteria and other organisms with hydrophobic surfaces will be transported to the surface. Organisms with cavities or structures where small bubbles may be entrapped will also be transported to the surface where an increased bacterial activity could increase the breakdown of injured or impaired organisms. The method is under testing and evaluation. However, at high levels of supersaturation lethal treatment has been demonstrated (Jelmert pers. comm.).

6 Chemical and biocidal inactivation of species in ballast water

A large number of chemical disinfectants are commercially available that have been used successfully for many years in land-based potable and wastewater treatment applications. Target organisms include protozoa, vegetative and resting cells of algae, bacteria and viruses. Within the general treatment efficacy requirements, that each chemical treatment of ballast water needs to meet, it should be environmentally sound and any residual chemical should be fully biodegradable to avoid accumulation of toxic substances in the remaining ballast water and/or sediments inside the ballast water tanks.

The costs involved in the use of some chemicals, operating and material costs are considered to be prohibitive (Müller 1995; Müller & Reynolds 1995; Rigby et al. 1993; Rigby & Taylor 2001). Beside the costs, the storage and use of chemicals on board could be impracticable.

With some chemicals several tonnes would be required to treat the large amounts of ballast water on large ships such as bulk liquid and ore carriers being fully loaded with ballast water. In addition, both inorganic and organic biocides would present a range of health and safety problems related to the storage and handling of chemicals, their compatibility with cargoes carried onboard ships, as well as those related to the direct and indirect handling of chemicals by ships' crews.

Most of the traditional biocides produce by-products which are likely to be environmentally unacceptable and/or might require specialist operator skills. Suitable dispersal mechanisms would also need to be addressed, in particular mechanisms for penetrating the sediment layer. The injection during ballast water uptake seems to be an appropriate option to apply and dose the chemical. In cases where ballast water in cargo holds would be treated with chemicals, these tanks would have to be cleaned intensively before cargo could be loaded in the same tanks (Carlton et al. 1995). Although there is extensive testing ongoing for possible residues, the long-term accumulating effects are unknown.

Biocides suggested for the use to treat ballast water include ozone, hydrogen peroxide, potassium permanganate, chlorination (chlorine dioxide, chloramines and sodium/calcium hypochlorite), ozonation, electrolytically generated metal ions (copper and silver ions), oxygen deprivation (de-oxygenation) using e.g. reducing agents, such as sulphur dioxide or sodium sulphite, coagulants (aluminium and ferric sulphate, ferric and poly aluminium chloride as well as cationic polymers), pH and salinity adjustment, antifouling paints as ballast tank coatings, organic biocides (3-trifluoromethyl-4-nitophenol, bromine, formaldehyde, glutaraldehyde, isothiazolone, quaternary ammonia compounds, sodium amines, EDTA, peracetic acid, bisulfite, iodine (periodate), use of bacterial pyrogens (endotoxins) and organic algacides containing isothiazolone) (Ridgway & Safarik 1991; Bolch & Hallegraeff 1994; Voigt 2000; Fuchs (Degussa) pers. comm.; GloBallast Programme 2001).

The most recently proposed treatment technique to inactivate or kill microoganisms in ballast water using chemicals is an advanced oxidation technology. This patented technology consists of a combination of ozone (O_3), two UV systems with different wavelength spectra and catalysts. The suggested ozonolytic/photolytic/photocatalytic redox processes are stated to operated simultaneously in a closed system consisting of a reactor and a control panel. The reactor is made of titanium and houses low-pressure UV lamps and catalysts placed close to the lamps. A large number of radicals are generated within the reactor. These hydroxyl radicals are aggressive and can break down virtually any organic compound to carbon dioxide and water. They are reported to be short lived (a few nanoseconds). This technique could be considered as an effective secondary treatment after particles and large organisms are removed by e.g. filter systems. The first prototype of the system will have a flow rate of several hundred m^3 per hour, while systems with further increased capacity are under development. Performance or cost data are not available at present (Patrick Dahl, pers. comm. BenRad Marine Technology AB, Stockholm Sweden). Possible corrosion problems (production of aggressive hydroxyl radicals) need to be considered unless the reactor is well separated from the main piping system.

One advantage of most chemicals is the comparatively easy application to the ballast water when it is taken on board. However, one of the shortcomings is that with every (even with partly) exchange of ballast water, the entire volume of water in the ballast tank has to be treated. Otherwise the dilution of the substances with the remaining water would result in insufficient dosing of the chemicals for the newly added ballast water. This limits the applicability of stand-alone chemical treatments to those types of ships, that exchange the ballast water completely in one lot (e.g. bulk carriers and oil tankers). In contrast, container vessels frequently exchange only part of their ballast water, or transfer part of the ballast water between ballast tanks when loading or unloading in the ports of call. This problem can be overcome if all the ballast water on board is treated to the desired level, however when ballast water is loaded on top of existing ballast water it would require the whole tank to be retreated with the chemical.

7 Constant volume of ballast water

Another option is the carriage of a constant volume of ballast water onboard without any discharges or uptakes of additional water. This option seems to be applicable to a very limited number of vessels. Ships which usually carry very little amounts of ballast water as e.g. cruise liners, could minimize their ballast water discharge to a minimum or even could prevent any discharge by pumping the ballast on board from one tank into another.

8 Alternating salinities in ballast water and area of discharge

Wherever possible, alternating salinities of ballast water and area of discharge for ballast water could be used. Firstly, to discharge marine ballast water in freshwater areas (e.g. the North American Great Lakes, freshwater ports) and secondly, to discharge freshwater ballast in marine ports could help to minimize the survival of organisms after discharge. It is believed that most of the freshwater organisms cannot survive marine conditions and vice versa. Knowing that there are exceptions to this rule, this option could be used to minimize the risk but cannot exclude further species introductions. However, there are many trading routes in the world where ships do not have the opportunity to take freshwater ballast onboard. The use of an onboard desalination unit for this purpose is probably extremely time consuming and associated with enormous need of energy.

9 Fresh or treated water ballasting

Providing ships with treated or fresh water appears to be a useful option in unique circumstances (Carlton et al. 1995; Rigby & Taylor 2001). An International Seminar on Fresh Water Ballasting in 1983 discussed the use of fresh water ballast for oil carriers. It was suggested that oil carriers could load fresh water instead of oceanic ballast water before cruising back to the oil exporting country without any cargo. Many of the oil exporting countries are located in arid or semi-arid climates where rainfall is scarce. The agriculture of some countries could benefit from this fresh water imported in segregated ballast tanks. The applicability of this option is limited to very special circumstances on certain trading routes and the availability of freshwater at donor ports.

10 Reception facilities

The possibility of land-based reception facilities for ballast water has not been ruled out for the treatment of smaller volumes of ballast water. The reception facility would need to be placed in certain areas of ports taken into account quarantine regulations. Land-based facilities could provide an acceptable means of control, but appear to have very high costs involved e.g. for pipework of large diameter in larger ports. Furthermore ships would need to be equipped with appropriate piping systems to connect the ballast water outlet with the ports ballast water pipeline. Many ports do not have any land areas available to house such storage facilities.

In order to reduce costs for the pipework an especially designed tanker (ballast water treatment vessel) could act as reception facility. The ballast water to be discharged could be pumped to this vessel located along side the discharging vessel (AQIS 1993; Taylor & Rigby 2001).

11 Cost effectiveness of various treatment options

In addition to the technical effectiveness of the various treatment options, the cost effectiveness will play an important role in the selection and long-term viability of a particular treatment option (Rigby & Taylor 2001; Table 3).

Exchange of ocean water in its simplest form (with no additional equipment) provides the most cost effective option (1.48-2.25^c/m^3). These costs are reduced by approximately 50% (for the empty/refill option) if gravity ballasting can be accomplished. The capital costs associated with additional equipment (that could be required in some cases for safe or effective operation) can result in an increase up to approximately 18.67^c/m^3. The heating /flushing process provides the next most cost effective option at 3.35^c/m^3. Use of a Hi Tech system (Thornton 2000) involving recirculation, higher temperatures and additional heat exchange equipment has been estimated at = 5.42^c/m^3. A range of representative costs for a selection of ships and other potential treatment options are also summarised in Table 3.

Costs for the container ship are relatively high compared to the other ships (where additional capital equipment is required) since the quantity of water treated is quite low. It may be possible to reduce some of these capital costs by reducing the capacity of the new equipment. However this aspect would need to be considered as part of the development of the Ballast Water Management Plan to allow the optimum overall outcome to be achieved.

Table 3. Indicative comparative ballast water treatment costs. [Costs have been estimated for three different types of ship classes; a Capesize Bulk Carrier, an LNG Carrier and a Container Vessel after Rigby & Taylor (2001). Capital costs have been based on the use of a Capital Recovery Factor (incorporating an 8% interest rate over a 10 year period) of 0.149 and a voyage schedule generally involving ballasting 12 times per year. Operating costs have been based on the use of additional fuel and additional maintenance associated with operating the equipment. The estimated costs are based on the volume of ballast water on board the ship, except in the case of the container vessel where a cost is indicated for the quantity of ballast water actually replaced or treated, as container ships do not usually completely discharge their ballast water in port]. Efficiency details regarding removal or inactivation of species see text.

Treatment option	Costs per m^3 [US$]
Exchange of ocean water in its simplest form (with no additional equipment).	1.48-2.25[c]
Costs are reduced by approximately 50% (for the empty/refill option) if gravity ballasting can be accomplished	
Exchange of ocean water including capital costs associated with additional equipment (that could be required in some cases for safe or effective operation)	up to 18.67[c]
Heating/flushing process	3.35[c]
Using recycled process water	4.16[c]
Heating/flushing process using a Hi Tech system (Thornton 2000) involving recirculation, higher temperatures and additional heat exchange equipment	5.42[c]
Hydrocyclones	6.49-26.33[c]
Continuous backflushing filtration (with a relatively high capital cost component)	7.07-19.31[c]
UV irradiation	9.80-31.39[c]
Chemical treatment (based on operating cost alone)	14.46[c]-$24.10
UV combined with hydrocyclone	16.28-57.71[c]
UV combined with filtration	16.87–50.69[c]
Land based treatment	20.48[c]-$8.31
Dedicated treatment ship	32.53[c]
Use of fresh water	50.00[c]–72.29[c]

The relatively high costs of some options (resulting from the equipment capital costs) will probably mean that preference will be given to those involving little or no capital. However standards which take into account biological effectiveness will ultimately have an influence on the most appropriate choice. It is important to note that the shipping industry has currently generally accepted the costs of ballast water exchange (as specified in the current IMO Guidelines) as being reasonable. Treatment technologies involving significantly higher costs will have a direct impact on freight rates. Further, it has to be noted that the most cost effective method is not necessarily at the same time the most effective method regarding its removal or inactivation capacity of species in ballast water.

The capital cost accounts for a large proportion of the overall cost of retrofitting equipment to existing ships for some treatment options. This situation will be less of an issue in relation to a new ship where new designs can be readily included and the additional capital may be very minor compared to the total cost of the ship.

12 Improved designs to enhance better ballast water management and treatment on new and existing ships

The adoption of many of the above treatment or management options will require the retrofitting or modification of existing pipework or equipment to permit the new procedures to be put into practice in a safe, technically effective, environmentally acceptable, practical and cost effective way. For many ships this will involve substantial costs.

For new ships the cost of incorporating new designs and installation of new equipment will represent a very minor additional cost. Consequently it is important that adequate consideration be given to these concepts at the new ship design stage.

The following design concepts have been suggested to facilitate and enhance better ballast water management and treatment. Several of these are specifically related to facilitating the implementation and operation of the various forms of ocean exchange, water drainage and minimisation of sediment accumulation as well as water sampling required for research, monitoring and compliance testing (more specific details have been presented by Taylor & Rigby 2001):

12.1 NEW AND EXISTING SHIPS

(i) fitting of tanker hatches, where possible, as an alternative to manholes to allow more ready access to tanks would be beneficial. It is also suggested that the tank immediately below the opening be kept free of obstructions that may impede lowering of sampling nets and other equipment

(ii) fitting of quick-release couplings (such as Kamlock Coupling Caps) to sounding pipes or sampling pipes would be beneficial as well as their specific location within ballast tanks to enhance sampling

(iii) modifications to standard sounding pipes to allow for better sampling by incorporating a number of holes located circumferentially down the length of the pipe (say 25 mm diameter, 1 m apart) to allow for relatively free flow of water in the tank to be represented and to make sampling at a particular position more effective

(iv) new ships' design attention to providing access to tanks (especially where access is not normally required) to enhance sampling of sediments

(v) if ocean exchange using the empty/refill option is selected, consideration should be given at the design stage to have sufficient strength built into the hull girder to allow this operation to be undertaken safely

(vi) if ocean exchange using the continuous flushing option is selected, new ship designs should examine options to allow ballast water overflow to take place in a safe and convenient manner; examples include
- doubling the number air pipes
- installation of Tanker Hatches; and
installation of internal overflow pipes to avoid water flowing over the deck

(vii) if ocean exchange using the dilution method is selected, consideration can be given to the installation of deck pipes to load ballast water through the top of the ballast tanks (IMO MEPC 38/13/2 1996) or alternative internal piping arrangements (Armstrong 1997)

(viii) if heating and heating /flushing is selected as the treatment option (Rigby & Taylor 2001) it is recommended that appropriate pumps and piping be included in new ship designs. Heating could also be enhanced by designing the ballast tanks so that there is provision for one tank to be empty at any one time and that each tank is strengthened to allow ballast water to be pumped sequentially through a heating system and then into an empty tank. Under these conditions, treatment time would be minimised and the problem of mixing of partially treated water with treated water would be avoided. This arrangement would also require an additional pump for circulation and pipework to permit water to be transferred in/out of all tanks

12.2 BALLAST TANK DESIGN FOR NEW SHIPS

(i) install additional drain holes in longitudinals and intercostals

(ii) install larger drain holes in floors at intersection of longitudinals and intercostals; and

(iii) install larger drain holes in horizontal and vertical longitudinals, corner gussets, panting stringers and intercostals where they butt up against watertight bulkheads to stop “hang up” of sediment and particles. This will allow better flow into the ballast water suction and stripping heads located in the last aft bays of the tanks. Naval Architects should undertake flow calculations (CFD) to determine the size of the drain holes to confirm that there is sufficient flow of water to the suction heads to match the capacity of the discharge of the ballast system. This will ensure that the maximum amount of water and sediment is discharged with the minimum ”hang up” or retention of sediment in the tanks.

(iv) where possible ”Hat Box” suctions be installed in tanks and ballast water holds

(v) ballast tanks should be installed with Butterworth type tank washing systems to minimise the retention of sediments. This suggestion is good for existing and new ships and in addition at the design stage consideration should be given to the enhanced manual removal of sediments at dry dockings.

Acknowledgements

The approval of the Department of Agriculture, Fisheries and Forestry - Australia to include some information from the Ballast Water Research Series Reports No. 12 and 13 is acknowledged with thanks.

THE BIOLOGICAL EFFICACY OF OPEN OCEAN EXCHANGE – IMPLICATIONS FOR BALLAST WATER MANAGEMENT

BELLA S. GALIL[1*] & NORBERT HÜLSMANN[2]
[1]*National Institute of Oceanography, Israel Oceanographic and Limnological Research, Haifa, Israel*
[2]*Institute of Zoology/Protozoology, Free University Berlin, Berlin, Germany*
[*]Corresponding author *bella@ocean.org.il*

Abstract
The urgent need to control ballast-mediated bioinvasions prompted the maritime industry and the legislators to adopt open-ocean ballast water exchange without rigorously testing its effectiveness in terms of eliminating ballast-entrained biota. A review of recent studies of biota entrained in ballast water and sediments following open-ocean exchange raised questions as to the reliability of the procedure as an effective control measure.

1 Introduction

Shipping is considered the largest single vector for the movement of nonindigenous marine species across the globe (Ruiz et al. 1997). Ballast-mediated bioinvasions into freshwater, estuarine and marine habitats have altered the structure of host ecosystems and have caused significant economic losses (Carlton & Geller 1993). The volume of ballast taken and discharged daily is estimated in the tens of millions of tons (Hay & Tanis 1998). Open ocean exchange (OOE) of ballast water is at present the single widely practiced procedure relied upon by management to reduce the risk of ballast-mediated bioinvasions. Indeed, the International Maritime Organization (IMO), in its resolution A.868(20) adopted November 1997, recognized ballast water exchange to be appropriate in the short term as an operational quarantine measure. The premise for advocating OOE is that it replaces the entrained coastal species with oceanic plankton species that are ill adapted for survival in near-shore environments. Moreover, where harbours are riverine or estuarine, the osmotic stress of salinity change following OOE is perceived to act as a biocide (Smith et al.1999). The urgent need to control ballast-mediated invasions following the damages caused by the Eurasian zebra and quagga mussels (*Dreissena polymorpha, D. bugensis*) in the North American Great Lakes and the Mississippi river system, the North Pacific toxic dinoflagellates, seastar (*Asterias amurensis*), *Undaria* seaweed in Australia, and the American comb jelly (*Mnemiopsis leidyi*) in the Black Sea, prompted the maritime industry and the legislators to adopt open-ocean ballast water exchange without rigorously testing its effectiveness in terms of eliminating ballast-entrained biota.

2 Effectiveness of ballast water exchange

Rigby & Hallegraeff (1994) who studied the effectiveness of ballast water exchange aboard a bulk carrier using tracer dye, found that 37%, 13%, and 5% of the original ballast water remain after exchanging one, two, and three tank volumes respectively. These amounts were later readjusted upward (AQIS Ballast Water Exchange verifica-

E. Leppäkoski et al. (eds.), Invasive Aquatic Species of Europe, 508–510.

tion Workshop 1998). Hay & Tanis (1998) reported that in two experiments using water tracer dye in 115 m^3 ballast tank, dilution rates during a three volumetric exchange were "significantly lower than the predicted 95%". The assumption underlying the above experiments that ballast-entrained biota will be diluted at the same rate as water was not borne out. Rigby & Hallegraeff's (1994) assertion that "The proportion of living organisms surviving ocean exchange will be only a fraction of the above [5%]" is wholly inferred from the dye experiment and the dead plankton retrieved. After mid-ocean empty-refill ballasting in a coal carrier, the plankton assemblage represented less than 2% of the density but nearly 40% of the taxa initially ballasted (Wonham et al. 2001). Following open-ocean empty-refill ballasting in 14 newly-built container vessels, Zhang & Dickman (1999) reported that on arrival 15 harmful diatom and dinoflagellate species were found in non-reballasted tanks, whereas eight species were found in a reballasted tank, though their abundance was 87% lower than in non-reballasted tanks, and blamed it on improper deballasting. In older vessels biotic efficacy of open ocean reballasting is nearly halved, with only 48% removal of diatoms and dinoflagellates (Dickman & Zhang 1999). A study of ballast exchange aboard oil-tankers, utilizing Rhodamine dye and fluorescent microspheres, showed that though empty-refill is more efficient than flow-through in removing inert particles, efficiency of biotic removal varies significantly among voyages and taxa (Smith et al. 2001). A study of ballast exchange aboard a container vessel found that though tracer dye dilution efficiency was greater than 90%, much of the entrained phyto- and zooplankton were retained (Taylor & Bruce 2000). Recently Rigby & Taylor (2001) affirmed that "In some cases the process of exchange may present an even worse scenario than discharging the residual originally ballasted organisms". En-route exchange within a regional sea clearly defeats its purported intent: Macdonald and Davidson (1998) found that the diversity of diatoms and dinoflagellates increased in 69% and 85% of cases, and abundance increased in 31% and 85% of cases, following ballast water exchange in the North Sea.

Uptake and retention of sediment in ballast tanks varies with tank location and configuration, position of ballast water intakes, ports visited, and ballasting procedures. Some vessels accumulate tens of tonnes of sediments that contain their own resident biota (Gollasch 1996; Hay & Tanis 1998). Rigby et al. (1993) reported that 14 of the 32 vessels that exchanged ballast water in mid-ocean contained "significant amounts of sediments, including dinoflagellates cysts". In fact, in an earlier manuscript Rigby and Hallegraeff (1992) observed that "even after exchanging 3 tank volumes..... up to 25% of plankton sediment was still retained". Heterotrophic protist communities, some of great diversity, were identified in all sediment samples collected from container vessel ballast tanks (Galil & Hülsmann 1997). It has been postulated that at least part of the biota that avoids dilution during OOE remains in the residual water and sediment. Wonham et al. (2001) attribute some of the faunal increase following open ocean empty-refill exchange to coastal organisms being stirred up from the bottom of the tank following exchange. OOE may also provide animals retained in sediments or water residue with fresh supplies of oxygen and food. Already Williams et al. (1988) noted that OOE "may not been quite so effective with benthic taxa, unless the sediments deposited in the tanks are removed at the same time as the water is exchanged".

OOE is believed to act most efficaciously when the salinity differential is greatest. Yet, Locke et al. (1993) found living rotifers and cyclopoid calanoid copepods in eight of the 24 sampled vessels entering the North American Great Lakes and originating in fresh or brackish water ports that reported saltwater ballast exchange. Hülsmann & Galil (2001) demonstrate that marine protists are capable of surviving over four weeks in fresh water, and suffer no ill effects when restored to seawater. As protists form a major component of marine microbial food webs and may have significant impacts on total food web structure (Sherr & Sherr 1988) their survival may be instrumental in supporting complex ballast-entrained food webs.

Very few studies have directly measured the efficacy of OOE, and those few were biased by the taxa sampled and vessel type. All but three studies were carried out aboard bulk carriers that tend to exchange all or most their ballast in one operation, rather than on container vessels, which have a complex ballast history; and most sampled only plankton, though many successful invaders are benthic species. Yet, even those studies raised questions as to the reliability of OOE as an effective control measure: "It is widely agreed that the current exchange practice, with the limitations of current ship design, is inadequate" (Reeves 1999).

3 Conclusion

Following reservations raised concerning the effectiveness of removal of harmful organisms, and ship safety constraints (Karaminas 2000) relating to OOE, the Ballast Water Working Group, Marine Environment Protection Committee (MEPC) of IMO agreed that "Ballast Water Exchange should be regarded as an interim solution and that the aim is to produce safe and more effective alternative ballast water treatment options that will replace Ballast Water Exchange" (MEPC 46/3 2000).

Acknowledgements

This work was supported by a grant from the Ministry of the Environment, marine & coastal environment division, Israel.

FILTRATION AS A BALLAST WATER TREATMENT MEASURE

ALLEGRA CANGELOSI
Northeast-Midwest Institute, Washington DC, USA
acangelo@nemw.org

Abstract
Filtration has been proposed as a ballast treatment in both shipboard and shore-based applications. Operated at the port of origin, filtration can dramatically reduce densities of larger plankton and other taxa in the ballast intake stream without generating chemical or thermal residuals. It also could enhance or complement performance of biocidal treatment systems. Pilot scale tests suggest that filtration is already operationally and economically feasible for some classes of ships, including many of concern in the global translocation of organisms, such as container and passenger ships. Applications to larger ships with higher flow-rates could become feasible with design improvements to today's filter technology. This chapter describes current uses of filtration on ships, types of filtration proposed for the ballast treatment application, performance evaluations of filtration as a ballast treatment, filtration as a component in compound treatment systems, pore size considerations, and the merits of filtration versus cyclonic separation as a primary treatment method.

1 Introduction

Filtration of ballast water is a straightforward means of reducing transfers of aquatic organisms by ships. It has been proposed as both a stand-alone treatment and the primary component of compound treatment systems (Laughton et al. 1992; AQIS 1993; Carlton et al. 1995; National Research Council 1996; Oemcke 1999). Pilot (340 $m^3 h^{-1}$) and bench-scale demonstrations of filtration as a potential ballast treatment have shown it to be extremely effective at removing zooplankton and some forms of phytoplankton from harbor water (Cangelosi et al. 2001; Galil 2001). Mechanical tests at the pilot scale suggest that filtration can be both operationally and economically feasible for at least some ship applications (Parsons & Harkins 2000).

Filtration offers the advantage of producing no residuals such as waste heat or chemical by-products. As a shipboard treatment, filtration would be best employed during ballast uptake, removing and returning matter entrained in the intake stream to the source harbor prior to the ship's departure. In this way, filtration could prevent altogether the movement of many near coastal organisms across the open ocean – formerly a natural barrier to transoceanic dispersal – as well as the accumulation of sediments in ballast systems. As a component of a port-based treatment system, filtration could offer a contingency treatment strategy for vessels unable to affect ballast water exchange due to safety concerns, or otherwise treat their ballast water through shipboard technology. If operated at the discharge point, ballast filtration would require proper disposal of filtered matter.

Clearly, filters alone cannot prevent all ballast-mediated organism transfers. Operational and space requirements of filters increase as mesh size decreases. Accordingly, these requirements restrict the lower bound of effective removal size in a shipboard applica-

E. Leppäkoski et al. (eds.), Invasive Aquatic Species of Europe, 511–519.

tion; some organisms will always be able to pass through a shipboard filtration system. Still, the potential window of effectiveness of shipboard ballast filtration subsumes a large proportion of the taxa of known concern, including fish, benthic and epibenthic organisms, and many forms of plankton. Moreover, in both the shipboard and port-based applications, filtration can improve performance of secondary treatment systems such as ultraviolet radiation and chemical biocides through reducing particulate matter that may consume or interfere with these treatments. Filtration also reduces the treatment burden on secondary treatment by largely removing many zooplankton and phytoplankton taxa from the intake stream

Size requirements also restrict the upper bound flow-rates at which a shipboard filtration system can function effectively, and filter vendors are actively exploring ways to make their systems more compact. However, the full range of flow requirements of many classes of ships of concern in global movements of organisms – including container ships, small tankers, cruise ships and some St. Lawrence Seaway-sized bulk cargo carriers – already can be accommodated by today's filter designs.

2 Filters, ships and ballast water

Filters have a long history of shipboard use. Manually cleaned cartridge strainers cleanse fuel oil and lubrication oil for diesel engines and generators. Filters also remove water from ship air compressors. It is not uncommon for these filters to have a nominal wire mesh pore size of 30 μm – low enough to remove most zooplankton from an intake stream. These shipboard uses of filters, however, have involved relatively low flow-rates (100-150 $m^3 h^{-1}$) compared to the flow requirements of ballast water treatment. Still, these uses attest to the fact that filters can operate successfully in the shipboard environment, and that they have a legitimate place in the suite of potential ballast water treatments (R. Harkins pers. comm.).

Filtration comprises several distinct technologies, which vary fundamentally in their approach to removing particles and self-cleaning. These differences in turn imply a variety of pump capacities, back pressure demands, numbers of moving parts, and structural materials among other features. Filter designs which commercial vendors have proposed for testing in the Great Lakes Ballast Technology Demonstration Project (the Project), and those proposed for approval under the California State ballast regulations provide some insight into leading ballast treatment filter technologies. Each approach offers intriguing advantages and warrants testing at the pilot- and full-scales.

The filtration systems proposed for testing by the Project were all envisioned for installation in the engine or pump room of the ship in-line with the ballast pump to treat water upon intake. Each design was self-cleaning and balanced in different ways the variables of unit size, mechanical complexity, and demands on ship operational systems to maximize biological efficiency and minimize diverted energy, intake flow and routine maintenance required for cleaning and operation. The designs offered included:

(i) A cylindrical mesh screen filter that self-cleans automatically (triggered by a drop in flow pressure) through periodic backflushing of a small volume of the ballast intake flow and spiral suction removal of the filter cake for discharge back into the source harbor (Amiad Filtration Systems - www.amiadusa.com).

(ii) A depth filter consisting of stacked discs a centimeter or two in width with microscopic grooves and ridges. Water flows laterally through the stacked discs and particles tumble and become entrained. The mechanism self-cleans (triggered by a drop in flow pressure) by releasing pressure on the discs such that they separate using the pressure head of the ship's cooling water or fire main systems. (Arkal – www.arkal-filters.com).
(iii) A wedge wire strainer, which automatically self-cleans by direct contact of the filter membrane with a rotating wire brush and backflushing of the debris (Helland – www.hellandstrainer.com).

An externally mounted shipboard filtration concept also has been proposed to the State of California Lands Commissions. According to this concept, a filter suspended from the ship's hull strains intake into or discharge from the ship's ballast system and is retracted for storage on the ship deck during the voyage. This concept, if viable, could be quite useful for ships for which retrofitting a system into the engine room is overly expensive or impracticable. It could also provide a useful shore- or barge-based back-up system in ports.

3 Effectiveness testing of filtration as ballast treatment

Intensive testing of filter systems as potential ballast treatment systems is now underway primarily in the United States (Cangelosi et al 2001, Parsons & Harkins 2000, 2001; Waite 2001) and Singapore (J. Matheichal pers. comm.). The Great Lakes Ballast Technology Demonstration Project (the Project) was the first to undertake comprehensive biological and operational tests on commercially available filter systems in the United States, and has the longest running test program. The Project filtration trials have provided valuable early benchmarks on operational and biological performance of various filter types and sizes.

3.1 M/V ALGONORTH TRIALS

The first tests took place in 1997 at a flow rate of 340 $m^3 h^{-1}$ on board an operating commercial bulk cargo vessel (*M/V Algonorth*) at various locations in the Great Lakes/St. Lawrence Seaway System. The Project performed these shipboard trials using a deck-mounted automatic back-flush screen filter (ABSF) designed by Ontario Hydro Technologies, Inc. Two filter units (a 250 μm pre-filter and a 25, 50, 100, or 150 μm polishing filter) were installed in series on the ship's deck along with a diesel pump piped to draw water either from the ballast tanks or the sea. Biological trials utilized matched control and treatment upper wing tanks equipped with cable trolleys for direct tank sampling using identical plankton net transects.

Operational assessments revealed that the 250 μm prescreen was not necessary to enhance polishing filter performance even at the finest polishing screen pore sizes, and that all polishing filters performed well enough to warrant further evaluation (Parsons & Harkins 2000). In terms of biological effectiveness, each polishing filter mesh size tested significantly reduced zooplankton density relative to the controls. The smaller size screens appear to have performed better than the larger screen sizes though the effect was not statistically significant due to variation of the ambient species assemblages present in the various source waters. The sizes of the organisms in the control samples did not differ across trials or treatment sets. Fig. 1 shows the percent zooplankton removed by the various screen sizes tested.

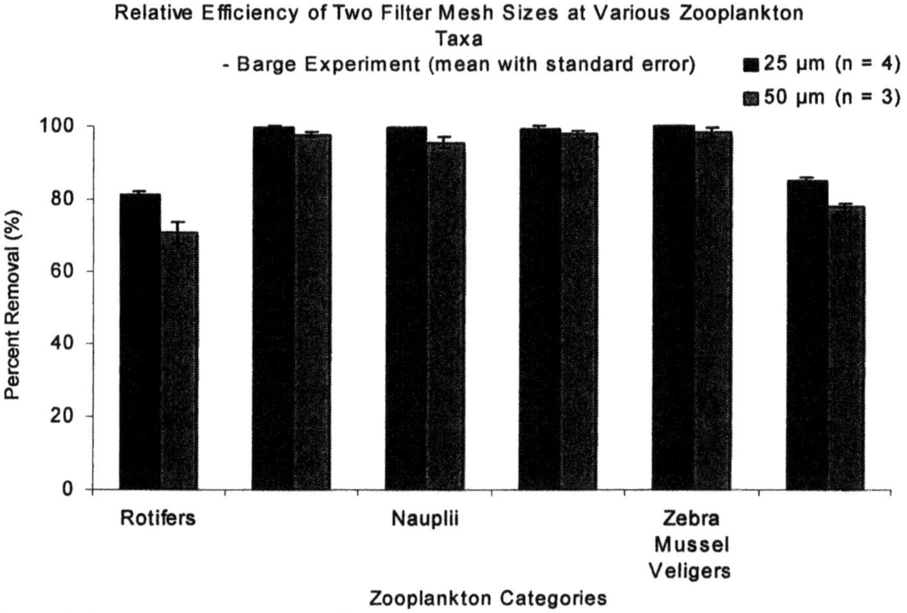

Figure 1. Zooplankton removal efficiency by various filter screen sizes.

3.2 GREAT LAKES BARGE TRIALS

From 1998 until the present, Project equipment trials have continued upon a stationary barge platform. These trials also were conducted at a flow-rate of 340 $m^3 h^{-1}$, and took place at two locations in Lake Superior with sharply contrasting physical, chemical and biological characteristics. As part of these barge-based trials, the Project also has evaluated depth filtration, cyclonic separation and ultraviolet radiation as prospective treatment system components.

The Project tests of the ABSF system under harbor conditions onboard the stationary barge (Fig. 2) were more rigorous and controlled than the shipboard. The effectiveness of 25 µm ABSF was compared with 50 µm ABSF at a single site in Lake Superior (Duluth/Superior Harbor). The barge experimental platform for these tests comprised the 340 $m^3 h^{-1}$ diesel pump, ABSF and three identical catchment tanks of 700 L each. The 250 µm prefilter was replaced with an intake strainer with 48 mm pore size.

Samples were collected from in-line taps located upstream and downstream of the treatment system, concentrated with a 20 µm net, and analyzed immediately using Acusizer Particle Sizing System. The mechanical tests showed the commercially available ABSF at 50 µm to be more operationally efficient and better suited to shipboard application than at 25 µm. However, both screens showed strong performance removing about 90% of all particles above 50 μm, and the 25 μm screen removing about 85% of all particles above 25 μm (Parsons & Harkins 2000).

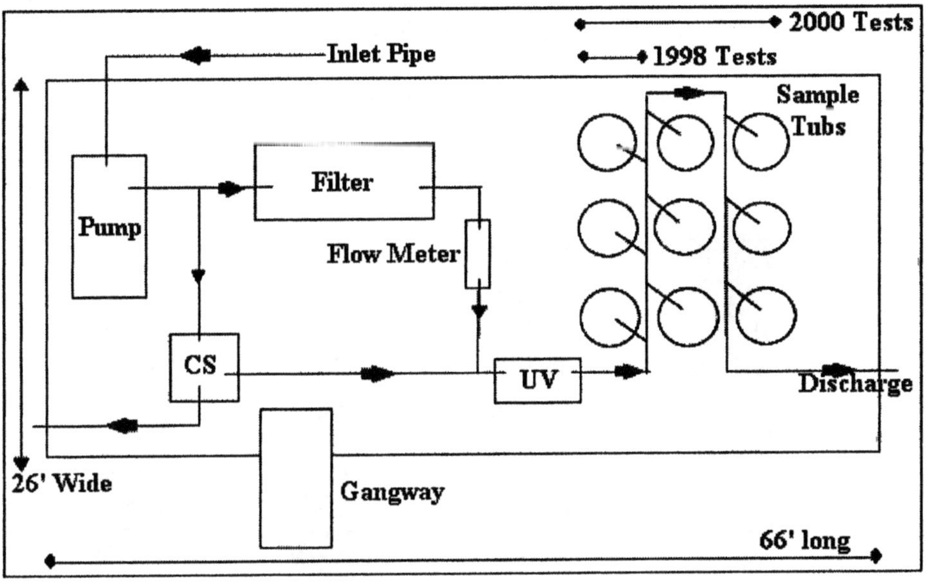

Figure 2. Great Lakes Ballast Technology Demonstration Project barge-based experimental platform.

Biological samples were collected from the triplicate catchment tubs, which were filled consecutively with treated or untreated water. Whole water samples were taken for bacteria counts, and all of the water in the catchment tubs was drained through bottom outlets into a 20 μm plankton net and concentrated to 1 L for plankton and attached bacteria analysis. The objectives, methods, and findings of the biological experiment are detailed in Cangelosi et al. (2001). Figures 3 and 4 show percent removal relative to controls of ambient zooplankton and phytoplankton taxa by the two filter sizes.

The Project tests also showed that the filters did not have the effect of increasing relative to controls the number of smaller particles in the discharge stream through break-up of algal filaments or colonies. Fig. 5 shows the distribution of algal filament sizes in treatment and control samples.

4 Filtration as compound system component

Most documented biological invasions are of larger zooplankton, benthos, fouling organisms and fish (Waite & Kazumi 2001). However, several studies (Hallegraeff & Bolch 1991; Carlton & Geller 1993; McCarthy & Khambaty 1994; Knight et al. 1999; Ruiz et al. 2000) highlight the threat that microzooplankton, phytoplankton, even bacteria and viruses pose. Red tide, for example, caused by toxic dinoflagellates, is a well-known public health threat. It can be transported via ballast water in the form of cysts of 20-40 μm diameter. Scientists suspect that toxic dinoflagellates can reproduce to high densities from only a few propagules (Hallegraeff & Bolch 1991). Some zooplankton are also quite small (10 μm) and capable of asexual reproduction. Though very little is known about the ecological changes that may result from ship-mediated transfers of free-standing bacteria, if microecological communities vary greatly from one region of the world to the next, these changes could be profound.

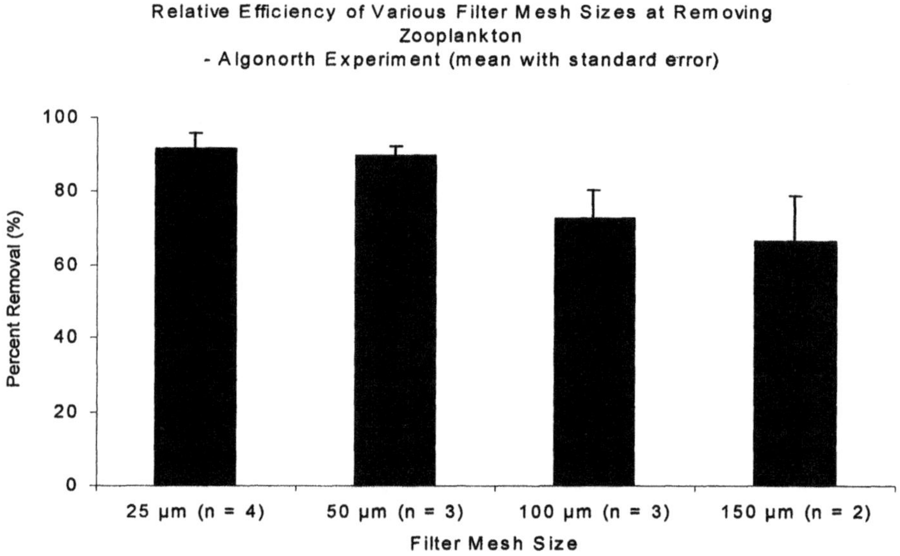

Figure 3. Efficiency of removal of ambient zooplankton taxa by 25 and 50 μm filtration.

Filtration by itself is inadequate to treat against the threats posed by very small (i.e. < 50 μm) freestanding organisms. Indeed, there is no known single treatment that can effectively guard against ballast-transfers of every form of aquatic life that is not accompanied by untenable economic, environmental or safety trade-offs. Yet filtration's special forte — the ability to exclude larger aquatic life forms (i.e. > 50 or 100 μm in width) and sediment particles from ballast intake with no environmental residuals - is an extremely valuable one because these taxa are indeed notorious culprits of ballast-mediated biological invasions. Efforts are therefore underway to make filtration technology more compact and practical for application to many classes of ships, and to combine filtration with a secondary treatment, such as ultraviolet radiation (UV) or an environmentally sound biocide, to optimize biological and operational effectiveness.

In 2000 and 2001, the Project tested UV as a possible secondary treatment with 40 μm prefiltration, and experiments with UV and 100 μm depth prefiltration are underway. The experiments involving UV radiation measure effects of treatment on phytoplankton growth and zooplankton mortality and reproduction, and bacteria and virus viability. The results so far show that UV contributes significantly to system effectiveness by significantly reducing culturable bacteria, viruses and phytoplankton. UV alone, however, did not reduce zooplankton to the extent possible with filtration as a pretreatment. This provides empirical evidence for the advantages of a compound treatment system in which zooplankton removal is achieved in the primary stage (Cangelosi et al. 2001).

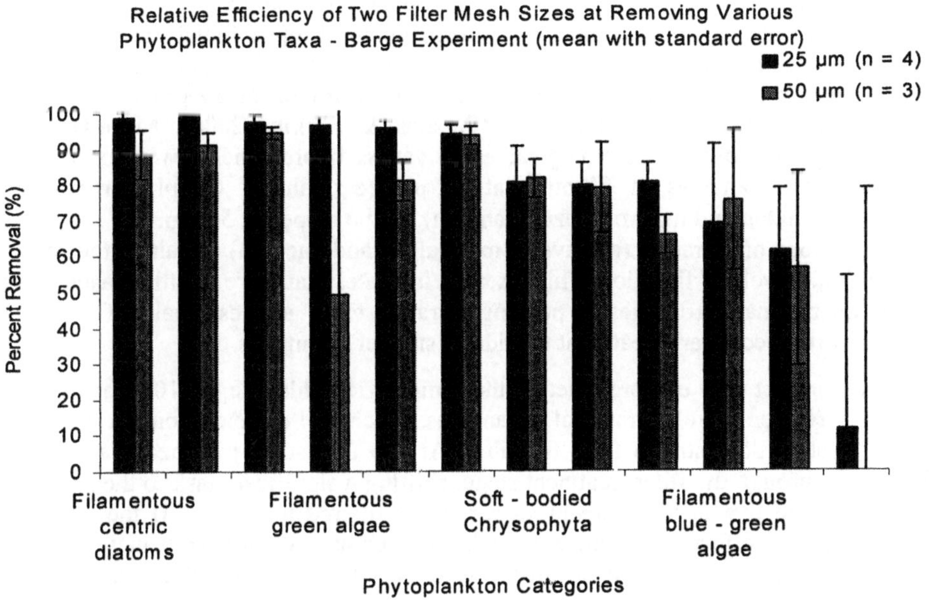

Figure 4. Efficiency of removal of ambient phytoplankton taxa by 25 and 50 µm filtration.

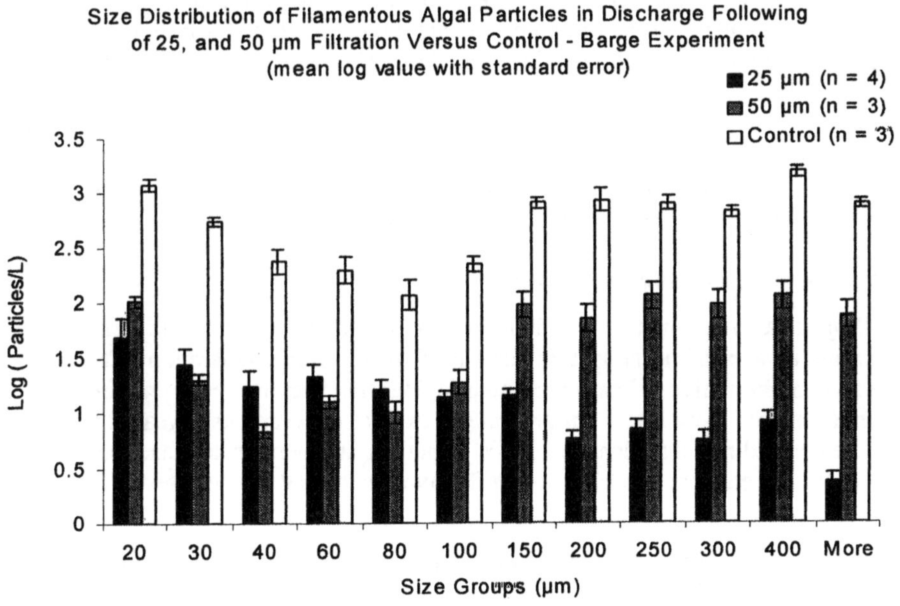

Figure 5. Size distribution of algal filaments with and without filtration at 25 and 50 µm

5 Filter pore size considerations

Clearly, the finer the filtration, the greater the operational challenges associated with it, while biological performance improves. However, the Project found that these trends are not necessarily linear, helping to define the level of filtration that is most efficient. The Project 's filter trials showed that the screen filtration at 50 µm is much more operationally efficient than at 25 µm (Parsons & Harkins 2000). Meanwhile, the biological tests showed that biological effectiveness improvements were only subtle; most of the advantages of filtration at 25 µm (e.g. almost complete removal of macrozooplankton and most microzooplankton) can be reaped at 50 µm.
The limitations of filtration (relative to microalgae and bacteria) remained the roughly same at both levels of filtration. This research indicates that there is little reason to suffer the operational difficulties of pushing filtration to 25 µm, especially if it is to be coupled with a secondary treatment to address smaller organisms.

On-going project tests explore whether the same logic holds true for 100 µm filtration, which offers even more operational advantages. Much will depend upon the strength of the UV system downstream from the filter. At 100 µm, some macrozooplankton will likely pass through the filter treatment stage, shifting a significant task to the secondary treatment stage that is not there in source water filtered to 50 µm. If the secondary treatment is less effective on larger organisms (such as UV), this shift may create the need for significantly higher doses of the secondary treatment to achieve target reductions. This need may outweigh the operational advantages of coarser filtration for some classes of ships.

Ultimately, the treatment vendor and the ship owner will weigh these advantages and disadvantages to arrive at the optimum treatment combinations for each class and make of ship. It is important to note that the type of filtration will likely confound direct comparisons of filter effectiveness based on pore sizes alone. Depth filtration at 100 µm may well remove a wider range of zooplankton than screen filtration at 100 µm because organisms tumble through the depth filter exposing both their long and short dimensions to the filter pores. Organisms orient to the flow of the uniplanar screen filters, with high odds of presenting their narrowest dimension to the screen (Cangelosi et al. in prep).

6 Filtration versus cyclonic separation

The Project also evaluated a commercially available cyclonic separator (Hyde-Optimarin) as a possible substitute for filtration. Cyclonic separation (CS) has been offered as an alternative primary treatment to filtration that has fewer moving parts and can handle higher flow rates. Unfortunately, Project experiments revealed that the CS system tested did not significantly reduce organism numbers or increase mortality. The CS also did not enhance the biological effectiveness of UV in these tests (Cangelosi et al. 2001).

A pilot-scale study of the same system reported only minimal removal of certain organisms by CS (Jelmert 1999). The system, rated to 100 µm, also had a low overall particle removal efficiency. It removed only 30.5% of all particles 100 µm and greater in size. In comparison, screen and disc filtration removed over 90% of all particles above the ratings tested (50 µm and 100 µm, respectively) (Parsons & Harkins in prep.) (Fig. 5).

At present, CS may contribute to treatment through performing a mechanical role in helping to protect the UV system from larger damaging particles over time.

It also may reduce particle loadings in those circumstances in which large numbers of heavier particles are entrained in the ballast intake. Until CS technology applications to ships improve, however, CS should not be considered as a substitute for filtration for particle or organism removal.

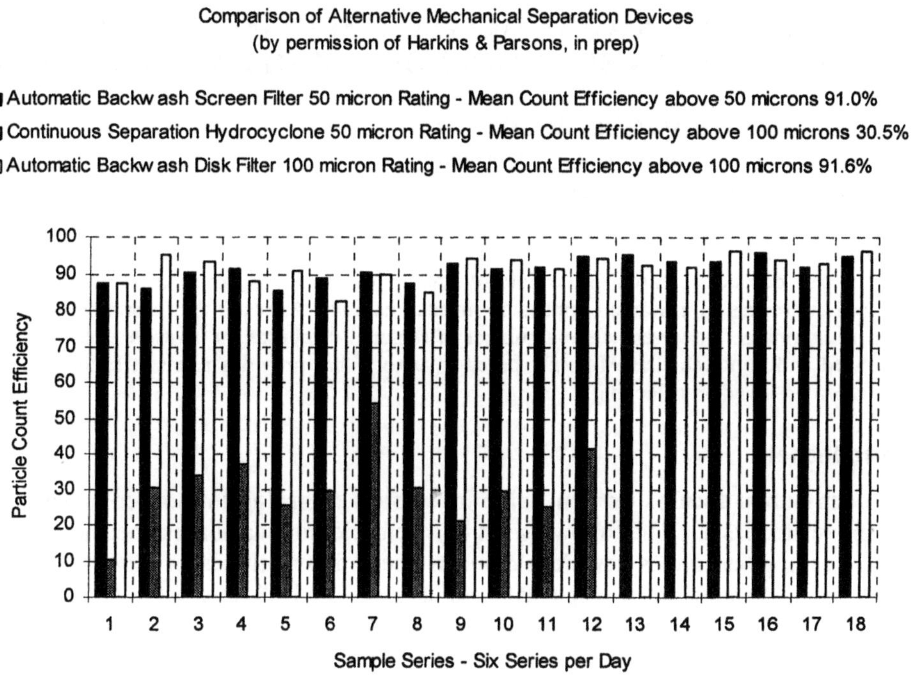

Figure 6. Mean particle removal efficiencies of a cyclonic separator and two filtration technologies at 340 $m^3 h^{-1}$.

Acknowledgements

I wish to thank the funders of the Great Lakes Ballast Technology Demonstration Project, including the Great Lakes Protection Fund, the U. S. Environmental Protection Agency, the National Oceanic and Atmospheric Administration, and the Legislative Commission on Minnesota Resources; my co-principal investigator Rick Harkins of the Lake Carriers' Association; members of the biological team, including Donald Reid, Mary Balcer, and Xenqing Gao; our private sector collaborators including the owners, officer and crew of the M/V Algonorth, the Seaway Port Authority of Duluth and the DM & IR iron ore terminal in Two Harbors, MN. I would also like to thank Nicole Mays for her editorial and technical support.

Databases

DATABASES ON AQUATIC ALIEN SPECIES: NORTH AND MEDITERRANEAN SEAS AND NON-EUROPEAN INITIATIVES

STEPHAN GOLLASCH
GoConsult, Hamburg, Germany
sgollasch@aol.com

Abstract
Over the past decades research in the field of aquatic invasions has expanded rapidly due to an increased awareness of unwanted impacts in invaded habitats. As result the information dissemination is enormous. However, while a number of regional databases on invasive species exist, a centralized and comprehensive Internet based inventory that provides relevant species-specific information on invaders is lacking.

1 Introduction

This contribution focuses on non-Baltic inventories regarding introduced aquatic species and related issues. The following list of inventories and databases does not claim to be fully comprehensive as new developments occur frequently, especially on the Internet. A brief introduction to each database is given and all Internet references were checked between August 27th and September 5th 2001.

2 Europe (excluding the Baltic Sea)

2.1 NORTH SEA, SCOTTISH AND BRITISH COASTS

(i) Reise K, Gollasch S & Wolff WJ (1999) Introduced marine species of the North Sea coasts. Helgoländer Meeresuntersuchungen 52: 219-234. This account lists about 80 nonindigenous species that form self-sustaining populations in the North Sea.
(ii) Welch D, Carss DN, Gornall J, Manchester SJ, Marquiss M, Preston CD, Telfer MG, Arnold H & Holbrook J (2001) An audit of alien species in Scotland. Scottish Natural Heritage Review No 139. The report lists 988 species considered alien and gives basic information about their origin, vector of introduction, habitats and impact.
(iii) Eno NC (1996) Non-native marine species in British waters: effects and controls. Aquatic Conservation: Marine & Freshwater Ecosystems 6: 215-228.

E. Leppäkoski et al. (eds.), Invasive Aquatic Species of Europe, 520–524.

(iv) Eno NC, Clark RA & Sanderson WG (eds) (1997) Non-native marine species in British waters: a review and directory (Directory of Non-native Marine Species in British Waters). Joint Nature Conservation Committee, 152 pp. The Internet based directory based on the above mentioned publications includes new records since its publication (http://www.jncc.gov.uk/marine/dns/).

2.2 MEDITERRANEAN REGION

(i) The CIESM Atlas of Exotic Species is the first attempt to provide a comprehensive summary of recent marine immigrants in the Mediterranean Sea. Individual species pages present illustrations, diagnostic features, biological information, references and a distribution map (http://www.ciesm.org).
(ii) *Caulerpa taxifolia* Database located at Laboratoire Evironnement Marin Littoral (LEML), Université de Nice-Sophia Antipolis. The site provides photographs, distributional maps, results of research initiatives and impacts (http://www.unice.fr/LEML select www.caulerpa.org http://www.com.univ-mrs.fr/basecaul).
(iii) Alien crayfish species, Results of the workshop: "The Introduction of Alien Species of Crayfish in Europe. How to make the best of a bad situation? University of Florence, Italy, September 24-27, 1997 Museo "La Specola" (http://www.unifi.it/unifi/dbag/workshop/Welcome.html).

2.3 EUROPEAN NETWORK ON AQUATIC INVASIVE SPECIES RESEARCHERS (ERNAIS)

ERNAIS is a joint initiative of the author and Vadim Panov (Russian Academy of Sciences, Zoological Institute, St. Petersburg, Russia) Chairman of the International Association of Theoretical and Applied Limnology (SIL) Working Group on Aquatic Invasive Species (WG AIS). Key objectives of ERNAIS: (a) to create a network facilitating a close cooperation and information exchange between invasion biologists European-wide, (b) the development of an international database on aquatic alien species and (c) to document the impact of invaders. At present more than 58 scientists from 21 European countries join this initiative (http://www.zin.ru/projects/invasions/gaas/ernaismn.htm).

3 North America

(i) NAS – Nonindigenous Aquatic Species (U.S.)
This site focuses on biogeographic accounts of nonindigenous aquatic species and includes scientific reports, online queries, data sets, regional contact lists, and general information of nonindigenous algae, plants, vertebrate and invertebrate aquatic species. The geographical coverage is the USA (http://nas.er.usgs.gov, http://www.fcsc.usgs. gov/R4finalreport.pdf).
(ii) US Great Lakes
Since the 1800s, more than 140 exotic aquatic organisms including, algae vascular plants, invertebrates and fish became established in the Great Lakes (http://www.great-lakes.net/envt/flora-fauna/invasive/invasive.html).
(iii) Woods Hole Oceanographic Institution lists a page on harmful algae at http://habserv1.whoi.edu/hab/.

(iv) National Database of Marine and Estuarine Invasions (U.S.) Smithsonian Environmental Research Center (SERC) has developed and maintains a database of marine and estuarine invasions in the U.S. This database compiles a detailed invasion history of approximately 500 different species of plants, fish, invertebrates, and algae that have invaded coastal states of the North America. The database identifies which species are invading, as well as when, where, and how they invaded; it also summarizes any existing information on the ecological and economic impacts of each invader. SERC is now working with the U.S. Geological Survey (USGS) which maintains a parallel database for primarily freshwater invasions (see NAS above), to functionally link the marine database (SERC) and freshwater database (USGS) - creating access to information in both databases through a single website.

(v) Aquatic Invasion Research Directory (AIRD)
The Smithsonian Environmental Research Center (SERC) Edgewater, MD, USA developed an international database of researchers and research studies (Aquatic Invasions Research Directory (AIRD)). The scope of the Directory includes (a) ecology of aquatic invasions: vectors, impacts, risk assessment and response, (b) ecology of ballast water, (c) prevention and treatment measures, and (d) policy and management details (http://invasions.si.edu/aird.htm).

(vi) Gulf of Mexico
Inventory of nonindigenous species in the Gulf of Mexico and research needs for these species as a cooperative program of the Gulf of Mexico Program and the Gulf Coast Research Laboratory Museum. (http://museum.ims.usm.edu/, http://lionfish.ims.usm.edu/~musweb/invaders.html, http://lionfish.ims.usm.edu/~musweb/nis/nistoc.html, and http://www.gmpo.gov/).

(vii) Sea Grant Nonindigenous Species (SGNIS)
The SGNIS Site of the National Sea Grant College Program, produced by the Great Lakes Sea Grant Network contains a comprehensive collection of research publications and education materials on zebra mussels and other aquatic nuisance species (http://www.sgnis.org, http://www.ansc.purdue.edu/sgnis).

(viii) Invasivespecies.gov
This site provides details of (a) the impacts of invasive species and the Federal Government's response, (b) selected species profiles, (c) links to agencies and organizations dealing with invasive species, and (d) databases (Invasive Species Databases, Expertise Databases, General Databases, Terrestrial Plant Databases, Terrestrial Animal Databases, Aquatic Plant Databases, Aquatic Animal Databases, Microbial Databases and Regional Databases) (http://www.invasivespecies.gov/index.shtml).

(ix) San Francisco Bay
The San Francisco Estuary is globally seen one of most invaded aquatic ecosystems with more than 200 introduced species. Cohen AN & Carlton JT (1995) Nonindigenous Aquatic Species in a United States Estuary: A case Study of the Biological Invasions of the San Francisco Bay and Delta. Report for the United States Fish and Wildlife Service, Washington D.C. and the National Sea grant

College Program, Connecticut Sea Grant, 200 pp (http://nas.er.usgs.gov/publications/sfinvade.htm).

(x) Zebra Mussel links
The key task of the Western Zebra Mussel Task Force (WZMTF) is to prevent and/or slow the spread of zebra mussels into western North America (http://www.usbr.gov/zebra/wzmtf.html. This site provides links to various related sites).

4 Australia and New Zealand

(i) The Centre for Research on Introduced Marine Pests (CRIMP) is Australia's national centre for research on the impacts and management of introduced marine species. A database of nonindigenous species established in Australian waters was compiled and contains more than 75 marine invertebrates, algae, plants, and fishes (Furlani DM (1996) A Guide to the Introduced Marine Species in Australian Waters. CRIMP, Technical Report 5, unbound file of approx. 200 pp).

(ii) CRIMP Technical Reports and Information Sheets on marine introductions and detailed reports on selected invaders (Invasive Marine Pests Database) (http://www.ml.CSIRO.au/~spinks/CRIMP/index.html).

(iii) Weeds in New Zealand. Inventory of terrestrial and aquatic plants in New Zealand (http://www.boprc.govt.nz/www/green/weeds.htm).

5 Global databases on ballast water and introduced aquatic species

(i) UNESCO-IOC Global Directory of Marine (and Freshwater) Professionals (GLODIR) is a database, containing information on individuals involved in all aspects of aquatic research and management including contact details, description of activities and citations of most important and/or most recent papers, *inter alia* data related to the invasive species problem (http://ioc2.unesco.org/glodir).

(ii) Global Invasive Species Programme (GISP). The database was developed by IUCN's Invasive Species Specialist Group (ISSG) as a contribution to Global Invasive Species Programme (GISP). GISP provides details on the top 100 invasive terrestrial and aquatic species worldwide. The detailed reports aim to illustrate the range of impacts caused by biological invasion (http://www.iucn.org, http://iucn.org/themes/ssc/pubs/policy/invasivesEng.htm, http://indaba.iucn.org/external/themes/ssc/programs/ invasives/issg http://www.issg.org/database/welcome).

(iii) Food and Agriculture Organization (FAO) Database on Introductions of Aquatic Species (DIAS). This site is an important source of information on benefits and risks of species introductions. It was initiated by R. Welcomme in the early 1980s. The database contains about 3,150 records (http://www.fao.org/waicent/faoinfo/fishery/statist/fisoft/dias/index.htm).

(iv) Intergovernmental Oceanographic Commission (IOC) Harmful Algal Blooms (HAB) Programme. HAB are of major concern as these species severely impact e.g. aquaculture activities resulting in loss of harvest of affected industries. The IOC Harmful Algae Bloom Expert Directory HABDIR is available at http://ioc.unesco.org/iocweb/default.htm, http://ioc.unesco.org/hab/FINAL.PDF, http://ioc.unesco.org/hab/data1.htm.

(v) The Centre for Research on Introduced Marine Pests (CRIMP) and the Smithsonian Environmental Research Center (SERC) have developed a standardized format for their databases on marine invasions of Australia and the U.S., respectively. Each database includes detailed information on hundreds of species. These two databases are now being linked, as separate database nodes, in the first step toward creating a distributed international database. When fully operational, data on exotic species can be obtained from both databases (residing in Australia and the U.S.) through a single web-based query at one location. The long-term goal is to establish additional nodes in multiple countries, providing information access and exchange on an international scale.

INTERNET DATABASE ON ALIEN SPECIES IN THE BALTIC SEA

SERGEJ OLENIN[1*], ERKKI LEPPÄKOSKI[2] & DARIUS DAUNYS[1]
[1]*Coastal Research and Planning Institute, Klaipeda University, Lithuania*
[2]*Department of Biology, Åbo Akademi University, Turku/Åbo, Finland*
[*]Corresponding author *serg@gmf.ku.lt*

Abstract

An Internet Database on aquatic alien species in the Baltic Sea area was developed as an initiative of the Baltic Marine Biologists' Working Group on Nonindigenous Estuarine and Marine Organisms in 1997; in 2000 a new concept of the online Database appeared with support received from the Baltic Marine Environment Protection Commission (HELCOM). This regional project encourages the exchange of data within the Baltic Sea area, providing a competent system regarding biological invasions, vectors of introduction, spread of alien species and their impacts on environment and economy. The Baltic Database represents an important regional node for a future global information system on invasive species.

1 Introduction

In the beginning of the 21st century, the Baltic Sea is known to host about 100 aquatic alien species. The first attempt to systematise information on these species was done nearly half a century ago, when Nikolaev (1951) and Segerstråle (1957) independently made their overviews on human-mediated introductions of aquatic plants and animals into the Baltic Sea. At that time, these species were not considered to pose a threat to the environment of the Baltic Sea. It took more than two decades before the first recognition of the introduction of nonindigenous species as an important negative anthropogenic impact to the Baltic Sea ecosystem (Leppäkoski 1984). One decade later, in spite of the appearance of comprehensive reviews (Leppäkoski 1994; Jansson 1994), the scientific literature on the alien species in the Baltic Sea area still remained rather scarce and scattered with notes often published in local languages only.

In 1994, the Baltic Marine Biologists (BMB) decided to establish a new working group on Nonindigenous Estuarine and Marine Organisms (WG NEMO). BMB was founded in the late 1960s with the aim to promote studies on the biological diversity, structure, function and sustainable management of the ecosystems of the Baltic Sea Area (the Baltic Sea, Belt Seas, the Sound and Kattegat) (BMB 2001). During three decades of its existence BMB has established more than 30 working groups dealing with different aspects of biology and ecology of the Baltic Sea. Establishment of the WG NEMO by this non-governmental scientific organisation became a remarkable event, indicating a growing academic interest to the problem of invasive species. One of the key aims of the WG NEMO was to collect and summarise information on nonindigenous aquatic plants and animals in the Baltic Sea. Relevant data on non-native species were compiled since than and the first version of the Database (at that time called Inventory) of Baltic alien species appeared on the Internet in 1997 as a result of the Working Group activity (Baltic Sea Alien Species Database 2001 2001). It was developed by a team of biologists and programmers from the Coastal Research and Planning Institute, Klaipeda University, Lithuania and Department of Biology, Åbo Akademi University, Finland.

E. Leppäkoski et al. (eds.), Invasive Aquatic Species of Europe, 525–528.

Physically it is located on a server of Klaipeda University (http://www.ku.lt/nemo/mainnemo.htm).

In its initial phase, the Database included ca. 80 species previously found in the Baltic Sea and adjacent waters. A selection of 12 species were considered in greater detail in the form of case histories providing information on species taxonomy, identification features, area of origin, vector of introduction, distribution history in the Baltic, abiotic preferences, life cycle and ecological and economic impacts. By the same time, the invasive species problem gradually became an important issue in HELCOM, an intergovernmental organisation responsible for Baltic Marine Environment Protection (HELCOM 2001). In 2000, HELCOM funded further work on the Database. This paper presents the most important features of its new version.

2 Goals and structure of the Database

The goals of the new Database were formulated as follows: (i) to provide a qualified reference system on alien species for the Baltic Sea area, available online for environmental managers, researchers, students and all concerned; (ii) to update the information on the Baltic Sea alien species, their biology, vectors of introduction, spread, impacts on environment and economy through the online questionnaire involving the data input from research institutions and responsible environmental authorities; and (iii) to encourage the exchange of data among different geographical regions and thereby to serve a node in the Global Information System for Invasive Species. During the year 2000, a new concept of the Database was developed. It is now an interactive user-friendly tool, which includes several information retrieving options: "Database Search", "Baltic Regions" and "Species Directory" (Fig. 1). The "Species Directory", being the backbone of the Database, is largely based on the former "Species Inventory". It contains individual species entries. An entry includes the complete taxonomy of a species and available comments, complementing and specifying the Database features (year of introduction, ecological impact, etc., see below). For some species additional data on synonyms and common names, taxonomic information and relevant drawings, life history, rate and means of spread, useful links to other Internet sources, etc. are also included. Information on each species is standardised according to eight major features (Table 1).

The Database Search tool is a direct way to retrieve information according to the major features. It allows the retrieval of data by a single feature (i.e. by "Taxon") or by combined features (i.e. "Taxon" and "Origin" and "Ecological impact"), including multiple selections of items within any feature. A list of species, retrieved according to the selected criteria, is linked to relevant individual entries on species and references. Using the "Baltic Regions" option allows retrieval of geographically related information. The Baltic Sea is subdivided into several regions according to their physical-geographical characteristics, major invasive corridors and available information on alien species. It is planned that in the future the Database will include more Baltic regions, as new data will appear. Baltic regions might be selected from the list or an interactive map. Each region window contains a summary on environmental conditions and invasion history as well as data retrieved from the Database: the total number of alien species found in this region, lists of introduction vectors involved, donor areas, ecological impacts and impacts on uses/resources as well as the numbers of species relevant to each of these features. Further, one can click on the numbers and retrieve a list of relevant species and features as in "Database Search". The "References" contains bibliographic descriptions of all information sources used in the Database. "Research Network" contains information on the WG NEMO activity, study areas and expertise of the WG members and links to partner web sites.

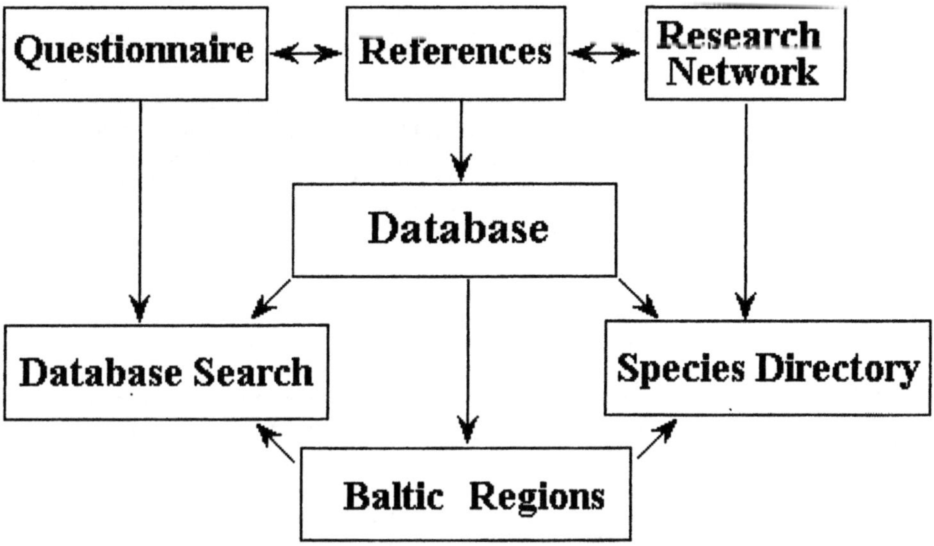

Figure 1. Structure of the Database on Alien Species of the Baltic Sea.

3 Data acquisition and control

The information in the Database originates from the members of the WG NEMO and other researchers involved in aquatic alien species studies in the Baltic Sea area, published papers, environmental reports, "grey literature" and Internet sites. Presently this information is being checked and prepared for publication on the Internet by editors. An online Questionnaire is currently being developed, which will allow gaining data directly from monitoring institutions and academic researchers working in the Baltic Sea area. Currently the WG NEMO advises on reporting formats on invasive alien species for a new early warning system on "abnormal" events being developed by HELCOM.

4 Further development

The Baltic Sea has received alien species from North America, the Ponto-Caspian region, Southeast Asia and other parts of the world. In turn, the Baltic may also be a donor area, i.e. its species may be exported over the environmental barriers to other geographical regions. Thus, inter-regional cooperation in gathering and disseminating information on alien species is of vital importance. In this context, the Baltic Database is seen as an important regional node in a global information exchange system on biological invasions. The number of similar Internet databases being developed by national and international institutions in other parts of the world is also growing (Ricciardi et al.

2000; Weidema 2000 and references therein). Therefore we support the idea that a future Global Information System for Invasive Species (Ricciardi et al. 2000) needs closer coordination and cooperation at national and international levels.

Table 1. Division of information by major categories and relevant features in the Database on Alien Species in the Baltic Sea.

Feature	Items within a feature and comments
Taxon	Phylum/division or class that an alien plant or animal species belongs to
Ecofunctional group	Phytoplankton (auto-, hetero- or mixotrophs)*; Zooplankton (phytophagous, predacious); Macrophytes; Invertebrate parasites; Benthic macrofauna (suspension- and deposit-feeders, omnivores, nektobenthic organisms or wood-borers)*; Fishes (phytophagous, planktivorous, benthophagous or predacious)*; Birds; Mammals
Origin	Known or probable donor area
Vector	Intentional introduction for stocking or ornamental purposes; Unintentional introduction by shipping (incl. ballast water of ships, boat hulls, rafts and flotsam moved via inland canals) or associated with aquaculture; Unknown - none of the above vectors has been proven
Time of introduction	Year of the first observation in the Baltic Sea; Approximate decade or century of the introduction; Unknown time of introduction
Status	Data on species establishment in the Baltic Sea or adjacent freshwater bodies: Established, Not established; Status unknown)
Ecological impact	Known or likely impact on native biodiversity and ecosystem functioning, including: Competition for food and/or space; Habitat change; Food-prey for native species; Predation on native species; Herbivory; Hybridisation with native populations; Parasitism; Toxicity - excretion of phytotoxins; Community dominance - quantitative changes in community structure; Benthic-pelagic interaction – introducing new linkages between benthic and pelagic environment; Bioaccumulation - storage of toxic substances
Impacts on uses/resources	Known or likely impact on uses of the sea/natural resources and human health, including: Aquaculture; Aquatic Transport; Fisheries; Tourism; Water Abstractions; Water Quality

* Further subdivision within a given ecofunctional group

Epilogue

TOWARD UNDERSTANDING PATTERNS OF COASTAL MARINE INVASIONS: A PROSPECTUS

GREGORY M. RUIZ[1*] & CHAD L. HEWITT[2]
[1]*Smithsonian Environmental Research Center, Edgewater, Maryland 21037 USA*
[2]*CSIRO Marine Research, Centre for Research on Introduced Marine Pests, Marmion, Western Australia, AUSTRALIA*
[*]Corresponding author *ruiz@serc.si.edu*

Abstract
Understanding invasion patterns and processes depends greatly upon empirical measures. Although observation and theory have resulted in a conceptual framework for invasion ecology, the empirical data needed to rigorously test many key hypotheses and develop robust predictions lag far behind. This gap is especially conspicuous for marine systems, existing both in the quality and quantity of descriptive data. At the present time, most analyses that evaluate patterns of invasion or test specific hypotheses derive data from the existing literature, which is extremely uneven in space and time. This "by-catch" approach to data collection can result in biases, creating *apparent* patterns that must be viewed with caution. Quantitative field surveys, which employ standardized and repeatable measures, can be used to remove such bias. We call for a concerted international effort to conduct quantitative surveys, designed explicitly to test a variety of hypotheses and to produce the high-quality empirical data that is now lacking. Without this information base, many fundamental questions in marine invasion ecology will remain unresolved, limiting advances for basic science as well as its ability to guide effective management and policy.

1 Introduction

Biological invasions are common in coastal marine and estuarine habitats. Invasions, the establishment of species outside of their historical range, can result from either natural dispersal or transfer by human activities (e.g., Carlton 1989; Vermeij 1991). However, the human transfer of species has greatly increased the background rate of invasions, and the rate of human-mediated invasions appears to be increasing for many geographic regions (e.g., Cohen & Carlton 1998; Hewitt et al. 1999; Reise et al. 1999; Ruiz et al. 2000; Hewitt 2002). As a result, hundreds of nonindigenous species (NIS) are now known to be established in marine systems, especially coastal bays and estuaries, throughout the world (e.g., Carlton 1979; Por 1978; Pollard & Hutchings 1990a,b; Asakura 1992; Coles et al. 1997, 1999a; Eno et al. 1997; Cranfield et al. 1998; Gollasch & Leppäkoski 1999; Hewitt et al. 1999; Reise et al. 1999).

E. Leppäkoski et al. (eds.), Invasive Aquatic Species of Europe, 529–547.

Importantly, not only have many NIS become established, but communities are often dominated by NIS in terms of number of organisms and biomass (e.g., Cohen & Carlton 1995; Reise et al. 1999). In some cases, it is clear that invasions have caused dramatic shifts in food webs, chemical cycling, and disease outbreaks (e.g., Cloern 1996; Strayer et al. 1999; Burreson et al. 2000; Grosholz et al. 2000). Although the impacts of most invasions remain unexplored (Ruiz et al. 1999), there is no doubt that biological invasions have become a major force of ecological change operating on a global scale.

Invasion ecology has truly emerged as a discipline over the past few decades, gaining a great deal of momentum (e.g., Williamson 1999). This emergence results in large part from the conspicuous and strong effects of invasions across a broad range of ecological processes. There is also a great deal of public concern about invasions, driven both by observed ecological changes as well as associated economic impacts, such as fishery losses. We have therefore witnessed explosive growth in both basic and applied research areas of invasion biology. Furthermore, a variety of guidelines and regulations are being implemented throughout the world to minimize the risk of future invasions.

With increased attention to invasion ecology, it is evident that many fundamental gaps remain in our knowledge about invasion patterns and processes, limiting advances in both the basic and applied science (e.g., Williamson 1996, 1999; Mack 2000). Recent reviews (Carlton 1996a; Vermeij 1996) have outlined some of these gaps, in areas that are pivotal to our understanding and management of marine invasions, providing an "agenda" or "road map" for the invasion ecology. Both reviews provide a compelling menu of questions that deserve attention and underscore the need for quantitative field data. We concur with the priorities outlined in each review, and with the need for particular types of data (now lacking), although little guidance was provided as to specific methods needed to address these priorities.

In this chapter, we review some of the patterns and predictions that have been of great interest in invasion ecology, providing our perspectives on approach and methods necessary to address these questions, especially in marine systems. We focus most attention on collection of descriptive field data needed to test a broad range of hypotheses, exploring merits and constraints associated with particular methods. Finally, we outline a strategy to measure patterns and rates of invasion and suggest formation of a distributed international network for this purpose, allowing coordinated implementation and information-exchange on the global scale at which invasions operate.

2 Hypotheses

Several themes have emerged for invasion ecology that are of central importance for basic and applied research, management, and policy concerning invasions. These themes appear repeatedly in the literature and are common to invasion ecology of freshwater, marine, and terrestrial habitats (e.g., Elton 1958; Drake et al. 1989; Lodge 1993a; Carlton 1996a; Vermeij 1996; Williamson 1996, 1999; Mack et al. 2000). Perhaps most frequent among these are two general questions, in various permutations, that include: "What species characteristics influence invasion success?" and, "What makes a system susceptible to invasion?"

These themes have led to a variety of hypotheses in invasion ecology that have become generally accepted, despite limited empirical support, especially in marine systems. For example, we frequently encounter statements that:

(i) particular life-history characteristics or biological attributes increase invasion success,
(ii) disturbance or low diversity increases invasion success, and
(iii) low latitude systems are less invasible than temperate systems (perhaps inversely related to diversity).

To some extent these concepts have become part of the conventional wisdom or dogma surrounding invasions, both in the scientific and popular literature. Some theoretical support exists for each of these statements, but empirical support is often equivocal. Moreover, few tests of these concepts exist in marine systems, especially in nature or at relatively large spatial scales (e.g., Carlton 1996; Vermeij 1996; Ruiz et al. 1999).

Here, we review briefly some of these central concepts in invasion ecology and support for them. A comprehensive review for each of these areas (individually or combined) is clearly beyond the scope of this chapter. Our intent is merely to illustrate various dimensions of these concepts, including complexities associated with the existing data and their impact upon interpretation.

2.1 CHARACTERISTICS OF INVADERS

Many attempts have been made to identify the critical life history characteristics leading to successful introductions, as a management strategy to aid in quarantine and control efforts (Hayes 1998; Hayes & Hewitt 1998; Karieva 1996; Willan 1987), to identify potential bio-control species (DeBach 1974; Beddington et al. 1978) or for development of ecological theory (e.g., Ritte & Safriel 1977; Safriel & Ritte 1980, 1983; Parson 1983; Crawley 1987; Carlton 1996a; Mack 1996). The primary focus of attributes has relied heavily on the parallels drawn between the theoretical predictions of MacArthur & Wilson (1963, 1967) concerning colonists, the examination of fugitive species (e.g. White & Pickett 1985), and the *r*- selected character traits of Southwood (1977).

2.1.1 Terrestrial lessons

Baker (1965) was among the first to review characteristics of an 'ideal' weed, listing those traits that influence invasion success (Table 1). However, this approach to identify *a priori* potential invaders, or species of particular concern, has been elusive: invaders do not exhibit all listed traits, nor is it clear how many or which attributes most influence invasion success. These criteria have been variously adapted to examine invasive plants in multiple geographic regions (e.g. Bazzaz 1986; Rejmánek 1989; Rejmánek & Richardson 1996). Exceptions to all of the criteria make the list a qualitative but not wholly predictive management tool (Joenje 1987; Williamson 1999).

This ideal weed list has also been adapted for animals with mixed results (Mayr 1965; Ehrlich 1986; Kitching 1986; Lodge 1993a). Despite successfully defining life history attributes that are often associated with *r*-strategists, these lists are typically post-hoc endeavors, often with contradictory results. Erhlich (1986) suggested that a successful colonist should be larger than most relatives, whereas Kitching (1986) suggested that the classic invader should be small in size. Similarly, Mayr (1965) proposed that successful avian invaders should have good dispersal capabilities, but Veltman et al. (1996) found that only one life history characteristic was significantly correlated with avian invasion success - migration - whereby migratory birds tended to fail. Rejmanek (1996) provided a synthetic analysis of seed plant invasiveness, demonstrating that no single character suite is necessarily sufficient to predict invasion success (see also Williamson & Fitter 1996, Williamson 1999).

Table 1: Baker's (1965) characteristics of an 'ideal' weed.

Characteristic
No special requirements for germination;
Discontinuous germination (self-controlled) and great longevity of seed;
Rapid seedling growth;
Spends a short time in vegetative condition prior to flowering (precocious);
Maintains a continuous seed production as long as growing conditions permit;
Self-compatible, but not obligatory self-pollinated or apomictic;
Cross-pollinated by a non-obligatory pollinator or by wind;
Very high seed output in favourable conditions;
Seed production and has a high tolerance of a wide range of conditions (climactic and edaphic);
Special adaptations for short and long distance dispersal;
Vigorous vegetative reproduction as perennial;
Has brittleness at lower nodes, or of the rhizomes as perennial;
Demonstrates an ability to regenerate from severed portion of rootstock;
Has the ability to compete by special means (e.g., rosette formation, choking growth, exocrine production, etc...)

2.1.2 Freshwater lessons

The characteristics developed for weeds in terrestrial systems have been adapted to aquatic freshwater environments by numerous workers (Arthington & Mitchell 1986; Lodge 1993a,b). Arthington & Mitchell (1986) provided an evaluation of the requirements for invasion by aquatic plants. They noted that all determinants of success appeared to be related to dispersal and reproduction such as:

(i) a reliance on vegetative reproduction (often as the only method),
(ii) association with human transport mechanisms, and
(iii) rapid reproduction rates; they also noted that dispersal by vegetative propagules was important in sexually sterile plants.

Ricciardi & Rasmussen (1998) presented a modified version of the attribute list and described a mechanism for identifying potential invaders, based upon a step-wise evaluation of active transport vectors, donor regions defined by those vectors, and species attributes. This approach may enhance the explanatory power of attributes to characterize invasions, as it controls for vector and source region, both of which may greatly influence invasion success (Vermeij 1991; Carlton 1996a).

2.1.3 Summary

Across environments and taxa, those species that exhibit the now well-known r-selected characteristics are *theoretically* predicted to be the most successful colonists of new environments. There is little empirical support for this paradigm at present (Crawley 1987, 1989; Simberloff 1989; Lodge 1993a,b) and many conflicting proposed attributes.

Indeed, when the attributes associated with successful invaders are compared with non-invasive congeners the traits are not significantly different; in many instances non-invasive species have the same suite of 'invader characteristics' (Paula & Eston 1987; Trowbridge 1995, 1998). Two alternative interpretations have been offered: either the non-invading congeners are 'likely' to become invaders (Paula & Eston 1987; Carlton 1996a); or the life history characteristics are insufficient predictors of invasion tendency (Parson 1987; Trowbridge 1995, 1998). More recently, Roy et al. (2001) have demonstrated that larger sized marine bivalves have successfully extended range limits between the Pleistocene and present more frequently than smaller species in the north-eastern Pacific. They suggest that successful nonindigenous marine bivalves follow a similar pattern, however Miller et al. (2002) provided an alternative interpretation.

2.2 CHARACTERISTICS OF INVASIBLE COMMUNITIES

Elton (1958) did not explicitly list ideal characteristics of an invader but rather suggested the recipient community as the primary determinant of invasion success. This has clearly emerged as a major theme in invasion ecology, with many permutations, including the effects of disturbance (from any combination of abiotic and biotic interactions) on invasion success.

A broad range of hypotheses focus on the potential importance of invasion resistance, emerging from characteristics of the recipient environment that affect survival and establishment of new species. Londsdale (1999) has illustrated this concept with the equation: $E = IS$, where E is the number of successful invasions, resulting from the product of number of exotic species introduced (I) and the survival rate of these species (S). Resistance is an independent attribute of the recipient environment that influences invasion success, through differential effects (among sites or times) on survivorship. Resistance therefore arises from variation in S, and can be defined by measuring residuals from the relationships between I and S (Williamson 1996; Londsdale 1999), resulting from some combination of abiotic and biotic factors.

Although we expect variation in resistance (or invasibility) among sites to be extensive, explaining a great deal of the observed variation, predictive ability in this area remains limited. Strong support exists for abiotic and biotic resistance (e.g., Case 1990; Pimm 1991; Rejmanek 1989; Baltz & Moyle 1993; Symstad 2000; Stachowicz et al. 1999; Tilman 1997, 1999; Blackburn & Duncan 2001; see also Ruiz et al. 2000 for review). However, there are also some exceptions, making generalizations premature (Robinson et al. 1995; Levine 2000). Thus, much of the data must be viewed as idiosyncratic at the present time (Williamson 1999), and caution is required in interpreting invasion risk as a function of particular environmental characteristics (for recent efforts see Hillard et al. 1997; Gollasch & Leppäkoski 1999; Hayes & Hewitt 2000).

Altered or increased disturbance regimes have often been associated with increased introductions, possibly through reduction in biotic resistance, including often those species considered to have r-selected traits (Rejmánek 1989; Moyle & Light 1996; Rejmánek & Richardson 1996). While there is some support for a relationship between disturbance and invasion (e.g., Hobbs 1989; Pimm 1989; Hobbs & Huenneke 1992; Horvitz 1997), there are also many cases that undermine a general outcome (Lawton &

Brown 1986; Stohlgren et al. 1999). Furthermore, even where a correlation between invasion and disturbance is perceived, many confounding factors often exist that limit interpretation of a causal relationship. For example, many marine introductions are located in habitats associated with significant alterations by human activity, yet these are also centers of human activities that may simply deliver more propagules to "disturbed" sites (Ruiz et al. 1999, 2000). In addition, invasions have continued to occur in more pristine, undisturbed environments (Carlton 1987, 1989).

The alteration of ecosystem properties due to previous introductions, operating perhaps as a disturbance agent, may also affect the ability for new invaders to colonize a new region (Drake 1989; Nichols et al. 1990; Vitousek et al. 1997; Simberloff & VonHolle 1999). Strong evidence for such interactions exists in studies of native community reassembly. Both theoretical (e.g., Post & Pimm 1983; Case 1990, 1991; Drake 1990; Nee 1990) and empirical (e.g., Dickerson & Robinson 1986; Pickett & White 1985; Gray et al. 1987; Case & Bolger 1991) work suggests that the sequence of species additions can have significant impact on the success or failure of subsequent colonists. For example, Drake found that differing assembly sequences of the same community affected the success of a species to become established (Drake 1990, 1991). However, the extent to which such interactions influence the outcome of invasions in the natural environment remains largely untested.

2.3 LATITUDINAL GRADIENTS IN INVASIONS

Relatively few low latitude introductions have been detected in terrestrial, freshwater or marine systems (Drake et al. 1989; Sax 2001; Hewitt 2002) resulting in the hypothesis that temperate systems are more invadable than tropical ones. Higher 'biotic pressure,' associated with increased diversity of native tropical communities, is hypothesized to confer increased resistance to invasions (*sensu* Elton 1958). Alternatively, tropical versus temperate latitudes have significantly different historical patterns of trade and development, suggesting that the opportunity for introductions has not been equal between regions.

Sax (2001) demonstrated that for introduced, well-known, terrestrial and freshwater taxa (plants, birds, mammals, freshwater fishes) a latitudinal gradient in exotic diversity is nonlinear with a strong peak between 25° and 30° latitude. In the tropics (below 25°), no significant change in species numbers was apparent, however, above 30° a linear reduction in exotic diversity with latitude correlated with a concomitant increase in latitudinal range (with the exception of South American mammals). Sax (2001) concluded that above the subtropics, naturalized species in both hemispheres "interact with their environments, sorting into the patterns of distribution and diversity for the same reasons that native species do."

Recent evaluations of invasions in tropical marine systems (e.g., Coles et al. 1997; 1999a,b; Hewitt 2002) have provided evidence that low latitudes are certainly susceptible to invasions. High latitude invaders that achieve spectacular population size are well known, such as the northern Pacific seastar *Asterias amurensis* invasion in southeastern Australia (Buttermore et al. 1994; Talman et al. 1999) and the Asian clam *Potamocorbula amurensis* into San Francisco Bay, California (Carlton et al. 1990). Similar examples are now emerging for tropical systems. Coles et al. (1997, 1999a,b) identified over 96 introduced species in the Hawaiian Islands, including the Caribbean barnacle *Chthamalus proteus,* which now occupies the high intertidal environment of several Hawaiian islands (Southward et al. 1998). Similarly, the black striped "mussel" *Mytilopsis sallei* has been introduced throughout Indo-Pacific port systems, including those in India, Indonesia, Hong Kong, Singapore (Morton 1981, 1989) and most recently, the Port of Darwin, Australia (Bax 1999; Willan et al. 2000).

Despite these documented invasions of tropical marine systems, the strong perception that lower latitudes are less susceptible to invasions may be in part a result of the reduced sampling effort and increased native biodiversity of which we have a poor taxonomic grasp (Hewitt 2002). Ultimately, evaluations of latitudinal gradients in invasion success suffer from differential sampling intensity and taxonomic knowledge.

3 Hypothesis testing

A variety of approaches have been used to test hypotheses about patterns of invasion and the underlying mechanisms that may generate particular patterns. Briefly, these approaches rely on the same tools used in other areas of ecology. Theoretical models have been used successfully to explore the potential interaction of various demographic characteristics, behaviors, environmental conditions, and species interactions that can influence invasion processes and patterns (e.g., Williamson 1989; Case 1990; Moyle & Light 1996; Tilman 1997, 1999). Laboratory and field experiments have been used frequently to manipulate particular parameters, under controlled conditions, and test for predicted outcomes on the level of populations, communities, or ecosystem function (Robinson & Dickerson 1984; Robinson et al. 1995; Robinson & Edgemon 1988; Schoener & Spiller 1995; Stohlgren et al. 1999; Tilman 1999; Levine 2000; Symstad 2000). Descriptive field-based measures have also been used often to describe specific patterns of invasion and test for correlation or association with particular factors (e.g., Simberloff 1989; Schmitt & Osenberg 1996; Ruiz et al. 1999).

Each approach has its strengths and weaknesses. Controlled experiments and theoretical models explore explicitly the effect of one or more variables, while controlling for many others, and provide arguably the best resolution. Although these methods have produced useful insights, they can sometimes fail to predict actual outcomes and patterns in nature. This disparity can result when such methods address only a limited range of spatial or temporal scales at which ecological processes occur (Diamond 1986), or when they fail to include key factors or forces, including especially indirect effects and interactions (e.g., Menge 1995). Furthermore, the effect of interactions among a range of environmental and biological characteristics may be especially difficult to predict for NIS, as they are colonizing novel communities outside their historical ranges (e.g., Grosholz & Ruiz 1996; Ruiz et al. 1999).

In our view, the best approach to test hypotheses about invasion patterns, and the effects of particular factors on invasion patterns, requires a combination of methods, including: (i) descriptive measures that demonstrate both the existence of patterns and correlation to the respective factor in field communities and (ii) experimental results that demonstrate the role of that factor in creating observed patterns.

Descriptive field measures are necessary both to accurately depict patterns of invasion, integrating across multiple scales and interactions, and to test whether hypothetical relationships – as predicted by theory or demonstrated in laboratory experiments – actually exist in nature. Although descriptive data cannot often control for many confounding effects necessary to demonstrate cause-effect relationships or underlying mechanisms, they serve a critical role in testing whether predicted patterns (that would result from a particular cause or mechanism) are evident. Thus, descriptive field data are necessary but not sufficient to demonstrate a causal relationship between dependent and independent variables.

The inferences that can be drawn from any approach depend upon the explicit design under which data are collected, especially including quality, quantity, as well as spatial and temporal scale (Yoccoz et al. 2001). In this article, we wish to focus particular attention on methods used for descriptive field measures, exploring some of the present constraints and opportunities in this area. We draw heavily from our joint knowledge of methods used in the United States and Australia, highlighting extensive surveys that are now underway in each country. Although many of the examples are specific to these two countries, the issues apply generally to Europe and elsewhere. Moreover, we suggest field surveys should operate on a synergistic and coordinated manner on a global scale, rather than a national scale, providing: (i) greater statistical power to test hypotheses and (ii) greater information sharing to develop, implement, and evaluate management strategies.

3.1 DESCRIPTIVE FIELD MEASURES

Descriptive field measures can test whether particular factors (e.g., life-history traits, donor region, recipient region, biotic resistance, disturbance regime) may explain patterns of invasion that are observed in nature. Carlton (1996a) and Vermeij (1996) have both suggested this approach as a critical test for many hypotheses. For example, successful versus unsuccessful invaders can be compared from a particular source (or donor) region to test for association between invasion success and specific character traits. The number of invasions from a single source region to different recipient regions can also be measured to test for effect of recipient region (biotic or abiotic characteristics), or disturbance regime, on invasion success.

In addition to testing specific *a priori* hypotheses, including those relevant to specific management objectives, field measures can provide important *a posteriori* insights about invasion patterns and rates. For example, such measures have been used to assess the rate of invasions as well as the relative importance of particular transfer mechanisms, source regions, or taxonomic groups to spatial and temporal invasion patterns (e.g., Cohen and Carlton 1995; Hewitt et al. 1999; Reise et al. 1999; Ruiz et al. 2000; Hewitt 2002).

In all cases, the success of these analyses is predicated both upon the availability of data on patterns of invasion as well as the quality of measures (Ruiz et al. 2000; Yoccoz et al. 2001). Unfortunately, the quantity and quality of existing data is rather limited at the present time (Carlton 1996a; Vermeij 1996). Invasions have not been assessed in many regions. Furthermore, where assessed, there may remain some significant gaps (e.g., the percent of established NIS that have been detected, or the habitats and taxonomic groups examined may differ among locations), and this could have a significant effect on the outcome of comparisons. For this reason Carlton (1996a) has suggested the need for extensive surveys. Here, we evaluate further various methods to achieve this goal.

3.1.1 Methods

Two types of methods have been used to identify field patterns of invasion: Records Syntheses and Field Surveys (hereafter synthesis and survey methods, respectively). The synthesis method involves a compilation of existing information from one or more sources, including published literature and reports, museum collections, unpublished records from various sources (e.g., monitoring or research programs, personal communications, theses). In contrast, the survey method employs direct field observations, using any one of a variety of field survey techniques.

Although both methods produce useful information about invasions, the data that result from synthesis and survey methods differ in many important ways, affecting the inferences that can be drawn and the rigor with which hypotheses can be tested (Table 2; see also Ruiz et al. 2000). The synthesis method provides a summary of the current state of knowledge about *detected* invasions for one or more coastal regions. A key advantage of the synthesis method is the relative speed with which data can be gathered and analyzed, across multiple spatial and temporal scales, indicating which species are known to be invaders to a particular region as well as their respective invasion histories (e.g., date of arrival, transfer mechanism or vector, pattern of spread) and ecological or economic impacts. The synthesis method may often provide a useful coarse measure of invasion patterns for large, conspicuous taxa in well-studied bays and estuaries.

Table 2. Relative value of each characteristic associated with the three methods of analysis: Synthesis, Visual Scan Field Survey (VSS), and Quantitative Sampling Field Survey (QSS).

Characteristic	Analysis Method		
	Record Synthesis	Field Survey: VSS	Field Survey: QSS
Effort	Low	Moderate - High	High
Amount of existing data			
Spatial	High	Moderate	Low
Temporal	High	Low	Low
Taxonomic	High	Moderate	Low
Potential biases	High	Low - Moderate	Low
Repeatability	Low	Low - Moderate	High

Analysis and interpretation of invasion patterns based upon synthesis methods requires a great deal of caution, because the quality and quantity of existing data on biotic communities is extremely uneven among geographic locations, habitats, taxonomic groups, and time periods. It is important to recognize that data used by the synthesis method are not from studies designed explicitly to detect invasions. Instead, these data are effectively "by-catch", gleaned from a broad range of collection methods and intensities, habitat types, and taxonomic expertise in space and time. As a result, patterns that emerge from synthesis methods are prone to strong biases in search effort. The potential importance of such bias is especially apparent when comparing the extent of data and effort for tropical versus temperate estuaries, taxonomic groups with organisms of large versus small body size (e.g., bivalve molluscs versus protists), or for recent versus historical periods. However, even patterns for well-studied regions may exhibit strong spatial and temporal variation in search effort (and thus detection of established species) by taxa and habitat, creating a mismatch between actual versus perceived invasion patterns.

Synthesis methods are often used as a proxy for analysis of spatial and temporal patterns of invasion. However, the potential biases associated with this approach are rarely addressed and may lead to erroneous conclusions about invasion patterns (Ruiz et al. 2000). Despite some attempts to standardize for search effort using the synthesis method (e.g., Cohen & Carlton 1998; Coles et al. 1999a, Hewitt et al. 1999), these analyses have not successfully controlled for many of the independent variables known to influence detection (e.g., Rosenzweig 1995; Hayek & Buzas 1997). Thus, synthesis methods can quickly describe the patterns of *detected* invasions, or *apparent* patterns, that often require further measures to test the validity of these patterns or possibly to groundtruth data quality.

In contrast, survey methods can control for search effort in analysis of invasion patterns, removing many of the potential biases that exist with synthesis methods. A wide range of survey methods have been used in coastal marine communities, for measurement of invasion patterns as well as other community attributes (e.g., Cohen & Carlton 1995; Eno et al. 1997; Hewitt et al. 1999; Reise et al. 1999, Coles & Eldregde 2002). We consider these methods to fall into two broad categories, which differ in some characteristics: Visual Scan Surveys (VSS) and Quantitative Sampling Surveys (QSS).

VSS are perhaps the most common survey method, typically involve collections made at one or more sites through visual searching, and result in a cumulative species list per site (e.g., Cohen et al. 1998; Lambert & Lambert 1998; Hewitt et al. 1999; Hines & Ruiz 2000). For many taxonomic groups, VSS may provide a relatively comprehensive species list for a site, but the measures may not be easily repeated or standardized across space or time. These surveys are often performed using a "rapid assessment" strategy – a brief visit to each site for focal collections – and produce largely qualitative data. Although search time may be standardized per site for VSS, the search path is haphazard and not standardized. This approach could result in substantially different effort in terms of spatial dispersion, area, and number of organisms or taxonomic groups examined across sites, surveys, or times. Another source of variation likely exists in the participants, who vary among surveys and in taxonomic expertise. Thus, unless the same taxonomic specialists are present and searching on each survey, the effort applied to search and analysis among taxonomic groups may vary greatly among surveys.

By comparison, QSS employ a spatially explicit sampling design, comprised of sampling units that are standardized by area (as well as depth and volume), to create cumulative species list within and among sites. Examples of such sampling units include core samples of soft-sediment communities, scrape or photographic samples of hard substrate communities, settling plates or other materials as collectors, plankton net tows, and traps (e.g., Hewitt & Martin 1996, 2001). QSS have several advantages over VSS. First, the sampling effort is easily repeatable in space and time. Although both methods are dependent upon taxonomic expertise, the collecting effort itself is standardized with

QSS such that anyone could repeat the survey. Furthermore, additional information is preserved in collections by QSS that are not often available with VSS, including spatial dispersion and abundance of the respective species as well as covariance in distribution among species.

Contemporary surveys of either type (VSS or QSS) do not exist for most geographic regions, habitats, and taxonomic groups, yet survey data are critical to advance our understanding about key patterns and mechanisms of invasion. Standardized, unbiased data that can result from surveys are required to test hypotheses about spatial or temporal patterns of invasion, characteristics of species or recipient systems that may promote invasions, or effects of propagule supply or management strategies on the establishment of new invading populations. The current paucity of such high-quality data is a major obstacle to invasion ecology, limiting both the basic and applied science.

There is currently great interest in detection and monitoring of NIS in coastal ecosystems, both to estimate patterns or rates of invasion and to guide particular management actions, for which we advocate the use of surveys designed explicitly with these goals. Analysis by surveys has some distinct disadvantages compared to synthesis methods:

(i) collection of data via surveys is much more resource and time-intensive;
(ii) survey data are not immediately available and require a significant effort to obtain, especially across multiple locations, habitats, and taxonomic groups; and
(iii) analysis of temporal patterns via survey methods requires repeated measures over time and does often not permit "backcasting" or retrospective analysis of historical patterns (where historical surveys do not exist).

However, the alternative approach, using synthesis methods, provides data of such uneven search effort and quality that distinction between real patterns of invasion versus bias is problematic (e.g., Ruiz et al. 2000). Thus, in our view, only surveys can provide confidence in the data quality and therefore interpretation of invasion patterns (e.g., see also Jackson & Johnson 2001 for discussion of a similar issue in the paleontological record).

Below, we outline key elements to the design and implementation of field surveys, providing examples of programs to survey marine invasions in Australia and the United States. We emphasize the paramount importance of standardized, repeatable measures as the fundamental building blocks for analysis of invasion patterns. We intend this review to be illustrative rather than prescriptive; importantly, there are multiple measures that can achieve the same goals, and we have discussed only a subset. Further, although we do not wish to discount the value of synthesis methods or the use of VSS for particular purposes, we emphasize QSS as the best approach to avoid unwanted bias, providing the highest degree of resolution and repeatability.

3.1.2 Design of field surveys

Field surveys include many elements that must be addressed explicitly in the process of design, implementation, and analysis. Thus far, as a central theme, we have focused exclusively on the general approach used to collect invasion records, emphasizing quantitative survey methods (QSS). There are clearly many aspects of survey design to consider. In addition, other elements that require attention, having important consequences for the possible analyses and interpretation, include:

(i) taxonomic identification,
(ii) reference material,
(iii) geographical information,
(iv) information management, and
(v) environmental characteristics.

We discuss briefly each of these elements below.

Field survey. Surveys are characterized by the specific method of sampling as well as the spatial and temporal scales of sampling, number of samples, and variables measured in each sample. For example, hard substrate communities can be sampled by destructive removal of all organisms in a specific area, by deployment and retrieval of settling plates, or by photographic methods. Samples may be restricted to a particular port or region, a particular habitat type (e.g., depth, substrate, wave exposure), a specific spatial dispersion (random, stratified random, uniform, haphazard), and particular time periods (e.g., seasons or years). Finally, the number of samples collected, method of examination (e.g., use of microscope, inclusion of all taxa), the type of analysis (species present, abundance, size structure) may also vary.

Each of these attributes defines functionally the "search effort" and can greatly influence the number or type of species detected (e.g., Rosenzweig 1996; Hayek & Buzas 1997). Although search effort should be standardized across space or time, when comparing spatial or temporal patterns (respectively), the most appropriate design will depend upon the goals (e.g., see recent review by Yoccoz et al. 2001, with emphasis on detection and spatial variation in field surveys).

Taxonomy & reference material. The quality of data that result from surveys depends greatly upon taxonomic identification and knowledge. There are at least two different issues associated with taxonomic knowledge that impact data quality and analysis of invasions. First, taxonomic expertise is clearly critical to the correct identification of species, as many organisms may go undetected by the untrained observer. Such under-detection can occur even for those with good working knowledge of a local biota who may be unaware of species from other regions that are similar in appearance.

Second, historical knowledge of the distribution of many species or particular regions may be poorly resolved, causing many species to go undetected as NIS. For example, many species that are now cosmopolitan may have originally been restricted to a particular geographic region prior to dispersal by human activities. In addition, the biota for some regions, such as eastern North America or many tropical regions, were poorly described prior to extensive human movement, creating the potential for many early invasions to be undetected.

Although the issue of taxonomic identification may appear easily addressed, there are now many groups of organisms for which taxonomic experts are rare or non-existent. Resources and expertise in taxonomy have eroded throughout much of the world in the past decades (National Research Council 1995). As a result, identification of all species may not be immediately available for each survey. This underscores the importance of establishing reference or voucher collections, which can play a critical role in providing comparative material for both confirmation of known species and resolution of unknown species. Ideally, such reference collections would include both morphological and genetic vouchers, as molecular tools may be particularly effective in resolving taxonomy in some cases (e.g., Geller 1996).

The status of many species as exotic or native remains problematic and difficult to resolve. To acknowledge this problem, Carlton (1996b) advocates the term "cryptogenic" to describe species of uncertain origin (not clearly native or exotic at a specific location). Thus, in addition to taxonomic identification of species, each should be classified among the three categories: native, non-native, or cryptogenic to a particular region. Using only two categories for analysis of invasion patterns may lead to erroneous conclusions, whereas including a category for cryptogenic species provides an explicit assessment of existing biogeographic knowledge for each species and region. Taxa for which species resolution is not possible should therefore be regarded as cryptogenic until either native or introduced status can be ascertained.

Geographic information and information management. The utility of surveys can be enhanced greatly by using existing tools to (i) reference all data in a geographically explicit fashion and (ii) manage all data in a referential database. Obtaining the latitude and longitude of all sites with a geographical positioning system (GPS), with any additional site descriptions, can be critical for many analyses of temporal or spatial patterns of invasion, especially using results from past surveys or across multiple, independent surveys. Furthermore, databases can provide a powerful tool both to manage survey data, making it possible to rapidly link or share data on a broad scale (Ruiz 1998), and to add further information for analyses (see below).

Environmental conditions. Environmental conditions may be responsible for, or shape, many invasion patterns. For example, current, temperature, salinity, or nutrient characteristics may play an important role in the invasibility of particular sites or the dominance of particular taxa (e.g., Baltz & Moyle 1993; Carlton 1996a; Hayes & Hewitt 2000; see also Ruiz et al. 1999, and references therein).

Although the primary focus of many surveys is detection of NIS or measurement of invasion patterns, it is desirable to measure environmental conditions associated with each survey site, and possibly each sample, during the survey period. This approach provides a rich source of information on local conditions, often not readily available at the appropriate scale, which can be used to test formally some specific hypotheses about invasion patterns and whether particular covariates explain a significant amount of observed variation.

3.1.3 Case histories: implementation of field surveys in Australia and the U.S.

Quantitative field surveys have been undertaken on a continental scale in both Australia and the United States. Both programs were designed explicitly to measure patterns of invasion and test a broad range of hypotheses about invasion patterns and mechanisms. While these survey programs are intended to measure current spatial and taxonomic patterns of invasion in each country, they were also designed to (i) develop a baseline for measurement of temporal patterns and (ii) provide a basis for comparison across continents.

Within each program, the use of standardized protocols for the quantitative surveys as well as associated elements (i.e., information management, geographically explicit data, and environmental conditions) has been a high priority. This focus on standardization is key to removing potential sources of bias and enhancing the utility of the current surveys (as discussed above). However, this approach is also intended to permit direct comparisons with measures in other geographic regions or future surveys (i.e., spatial replication and repeated measures, respectively), establishing quantitative data for comparisons across multiple geographic and temporal scales.

Although both survey programs have evolved somewhat independently, we are now developing a collaborative approach for comparisons between Australia and the U.S. Here, we discuss briefly the scope and features of each survey program. Some of the collaborative elements that are presently underway, as well as potential future directions, are discussed in the next section.

Australia – surveys of multiple habitats. Over the past 6 years, Hewitt (and others) have implemented an ambitious program of quantitative field surveys, or port baseline surveys, which are focused in bays throughout Australia with major commercial ports (Fig. 1a). These surveys were designed to sample the resident biota across a wide range of habitats (Hewitt & Martin 1996, 2001), including:

(i) hard substrate – photographic and comprehensive "scrape" samples for fixed areas of vertical surfaces that are at multiple depths;
(ii) soft-sediments – core samples of fixed area and depth;
(iii) mobile epibenthic biota (e.g., crabs and fish) – traps that are set for standard time periods.

Great care has been given not only to the spatial dispersion and replication of samples for the port surveys, but also other key elements (Hewitt & Martin 1996, 2001). More specifically, sample processing and preservation is standardized, information for each sample is geographically referenced, taxonomic identification is verified by taxonomic experts, specimens are maintained in a reference collection, and resulting data is managed with database tools that include environmental information. This level of detail was necessary in order to guarantee consistency in sampling intensity and quality as additional groups undertook port baseline surveys.

Results from the Australian port surveys are just now emerging. CRIMP (Hewitt et al.) has conducted or been involved with over 16 surveys around Australia and 21 additional port surveys have been conducted (or currently being implemented) by at least six external organisations using the methods of Hewitt and Martin (1996, 2001) (Fig. 1a). Hewitt et al. (1999) presented detailed results of the port survey for a single bay, Port Philip Bay. In addition, some initial analyses of invasion pattern across 4 temperate and 4 tropical bays found that 49.2% of NIS detected by the surveys occurred in the tropics (Hewitt 2002). Among these 8 bays, there was a significant increase in number of recognized invasions with latitude (Fig. 2; $F_{[1,6]} = 7.26$, $p = 0.036$, $r^2 = 0.555$; Hewitt 2002). Although this preliminary latitudinal pattern requires further analyses, including consideration of the possible importance of cryptogenic species in the result (as discussed above), it illustrates the use of survey data to test key hypotheses about important correlates to invasion pattern in space or time.

United States – surveys of fouling communities. A similar, broadscale program of quantitative field surveys has been implemented in the U.S., although this program focuses exclusively on fouling communities at the present time. In each of 16 different bays and estuaries (Fig. 1b), the epifaunal fouling community is being surveyed by Ruiz and colleagues using settling plates (14x14cm collectors made of polyvinyl chloride, or PVC), and boring organisms such as shipworm and gribbles are surveyed with wood blocks. At all bays, plates are deployed in a stratified, random design. As a minimum for each bay, 200 PVC plates are deployed during the summer in a standardized fashion: the high salinity zone (>20 PSU) is divided into 10 equal-area strata; 20 plates and 2 wood blocks are randomly distributed in a single block (1 km length) in each stratum; the plates are each located at the same depth (1 m below MLLW). After 3 months, the plates and blocks are collected, and all organisms present are identified to species.

Using this survey method, additional measures are underway at three core sites to assess the effect of time (season, year, and duration of deployment), depth, and salinity on species richness (i.e., number or percent of NIS detected). Three bays were selected as core sites, one on each coast of the U.S., including Chesapeake Bay, San Francisco Bay, and Tampa Bay (Fig. 1b). In addition, at each core site, results of this survey method are being compared to the port survey methods (for hard substrata) being used in Australia (Hewitt & Martin 1996, 2001).

The fouling community surveys began in 1999, and analyses are only just beginning. Despite clear differences in survey method between the U.S. fouling surveys and the Australian port surveys, the methods for taxonomic identification, geo-referenced information, reference collections, and data management are nearly identical. The parallel nature of these two survey programs stems from (i) our collaborative development of the fouling surveys and (ii) the similar goals that are driving both survey programs. More specifically, the data from fouling surveys will be used to test for correlation of invasions with latitude, vector activity, and various environmental and community characteristics. In addition, we expect to use the fouling surveys as baseline for analysis of temporal patterns of invasion through repeated measures, especially at the core sites.

4 Conclusions and Future Directions

Understanding of marine invasions that can lead to accurate predictions and effective management is greatly limited by the paucity of contemporary surveys, especially those that control adequately for search effort and provide confidence in data quality. Two types of critical gaps are evident in existing surveys. First, data on marine invasions is not presently available for many global regions. Second, where available, data on marine invasions are incomplete and of variable quality among sites, as many assessments are not contemporary and most of the existing data have been collected largely as "by catch" from a composite of methods. These gaps place significant constraints on interpretation of the existing data and, in particular, its value to basic and applied issues in invasion ecology.

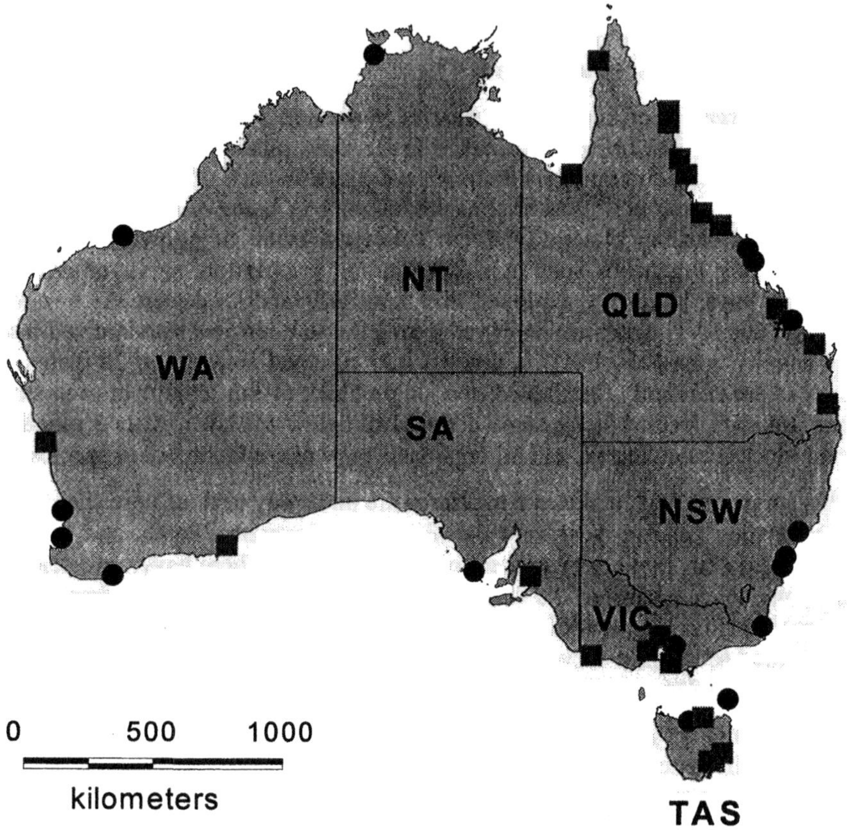

Figure 1a. Distribution of port surveys in Australia; surveys undertaken in whole or in part by CRIMP (●), or other organisations (■) including James Cook University, Central Queensland University and University of Queensland (Queensland), Marine and Freshwater Resources Institute (Victoria), Aquenal Pty Ltd (Tasmania), Geraldton Port Authority and Murdoch University (Western Australia).

Although numerous studies have used synthesis methods to summarize the invasions known for particular regions, underscoring the prevalence of NIS and some providing important insights (e.g., Cohen & Carlton 1995, 1998; Hewitt et al. 1999; Reise et al. 1999), the approach suffers from an uneven and largely undocumented search effort. More specifically, each synthesis study uses a unique assortment of data, gleaned from the cumulative historical records in each target region, creating conspicuous differences in the data quality (methods, taxonomic and habitat focus, taxonomic expertise) and quantity (amount and temporal distribution of research) among studies. As a result, it remains difficult to assess the degree of bias in patterns of invasion that emerge from synthesis methods, and to compare data or patterns across studies that use this method. In our view, there exists no adequate method to standardize these historical "by-catch" data, and to remove the many biases that may result, from synthesis methods.

We advocate the use of standardized, quantitative surveys to measure patterns of invasion. This approach removes many biases evident in the existing data and can be used to develop a high-quality information base necessary to

(i) test hypotheses about marine invasions,
(ii) develop robust predictions, and
(iii) guide management and policy decisions.

The survey programs described for Australia and the U.S. provide two examples of standardized, quantitative surveys that can be replicated in space and time. Although each program is focused primarily on a single country, some efforts are also underway to expand the geographic scope of each survey method and to develop common information management tools that can be used elsewhere. More specifically, the U.S. fouling community survey has been replicated at two sites is Australia, Swan River (Perth) in Western Australia and Sydney Harbor in New South Wales, and at least one of these locations will become a core site for temporal measures (as described above).

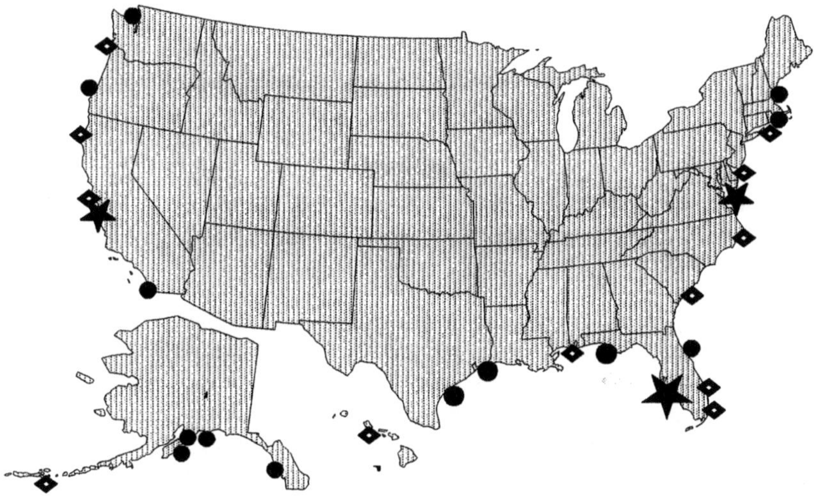

Figure 1b. Distribution of fouling community surveys in the U.S. Surveys undertaken by SERC through 2001 are shown as filled symbols (• baseline survey, ★ core sites; see text for explanation). Future surveys planned at additional sites shown as open symbols (◊). Alaska (to the left) and Hawaii (to the right) are shown at the bottom of the figure.

In addition, the Australian port survey programme is being replicated by the International Maritime Organization Global Ballast Water Management Programme (in conjunction with the Global Environment Facility and the United Nations Development Program) in six demonstration sites: Brazil, China, India, Iran, South Africa, and Ukraine (IMO 2001). Similarly, the New Zealand Ministry of Fisheries has embarked on a port survey programme, primarily using the Hewitt & Martin (1996, 2001) methods.

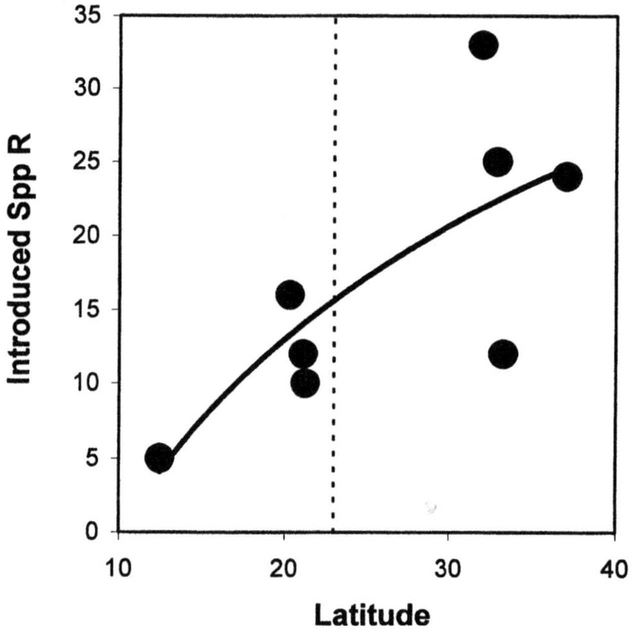

Figure 2. Correlation of introduced species richness and latitude ($y = 18.846Ln(x) - 43.466$; $r^2 = 0.555$) for Australian port surveys (data from Hewitt 2002). Dotted line represents Tropic of Capricorn.

In a collaborative effort, the Australian and U.S. survey programs have also developed a common database format for management of data on marine invasions, resulting from their respective surveys as well as other sources. This tool assures our respective data is collected and managed in similar fashion, and should facilitate comparisons and analyses across continents that we are now planning. We intend to make this database tool freely available to others who may wish to use it, creating opportunities for further collaborations and comparisons.

We call for a concerted international effort to implement standardized, quantitative field surveys that can provide high-quality and comparable data across the globe. This appears ambitious at first glance. However, many independent research programs that measure marine invasion patterns already exist in countries throughout the world. Furthermore, many countries are calling for development of "early-warning systems", "early-detection systems", and "rapid-response programs" that will require (by definition) some level of monitoring for new arrivals by standardized surveys.

A coordinated program of standardized surveys and data acquisition across countries could provide an effective and efficient approach to collect key data, needed to advance invasion ecology in marine systems. This strategy should be advanced at both the national and international level. There are many possible methods that could be adopted for standardized surveys, of which we have discussed only two as examples. A set of standard surveys and information management, implemented broadly, would pay tremendous dividends in two general ways:

(i) Replication of a standard survey would increase the low sample size that now exists on a country-by-country basis, improving greatly the resolution, power, and generality of analyses. Thus, data collected from throughout the world can provide important insights into invasion processes in Europe or any other global region.

(ii) Invasions in one global region are greatly relevant as potential future invasions to another, but no clear system exists for information exchange on new invasion, or their biological and environmental characteristics, that may be important for other regions. Concerted information management could serve this purpose.

To implement such a program, we suggest creation of a distributed international network. More specifically, research on invasions is conducted at a series of nodes that are in communication with each other. Anyone can participate in the network, which would develop and use a set of guidelines to include at least some standard protocols for field surveys and data management in their research program. Ideally, the network could both engage those groups or individuals involved presently in invasion research and encourage / facilitate participation by others. Control of the research and resulting data would remain with each participant for their respective component(s), and efforts for establishing cross-system comparisons or data-sharing would be negotiated among the participants. We have begun to develop this network approach to understanding marine invasion patterns, beginning with coordinated activities between Australia and the U.S., and we invite others to join us.

Acknowledgements

We thank S. Gollasch, E. Leppäkoski, and S. Olenin for the opportunity to contribute to this volume. Although the volume focuses primarily on invasion biology in European waters, the editors encouraged global perspective and participation, which we believe is so critical to the field. This article benefited from discussions with many of our colleagues, including especially J. Carlton, M. Campbell, J. Crooks, P. Fofonoff, J. Geller, S. Gollasch, E. Grosholz, A. Hines, K. Hayes, R. Martin, W. Miller, B. Steves, and R. Thresher. Our research in this area has been supported by funding to GMR (U.S. Fish & Wildlife Service, U.S. Department of Defense Legacy Program, National Sea Grant, Regional Citizens' Advisory Council of Prince William Sound, and Smithsonian Marine Station) and CLH (Agriculture, Fisheries and Forestry - Australia, Australian Quarantine and Inspection Service, Environment Australia, numerous port authorities and CSIRO Marine Research).

References

Abbasi SA, Nipaney PC & Schaumberg GD (1990) Bioenergy potential of eight common aquatic weeds. Biological Wastes 34: 359-366

Abrahamsson S (1973) The freshwater crayfish *Astacus astacus* in Sweden and the introduction of the American crayfish *Pacifastacus leniusculus*. Freshwater Crayfish 1: 27-40

Abrahamsson SAA (1966) Dynamics of an isolated population of the crayfish *Astacus astacus* Linné. Oikos 17: 96-107

Abraszewska-Kowalczyk A, Jobczyk I & Pazera E (1999) Invasion of zebra mussel *Dreissena polymorpha* (Pallas, 1771) in the Sulejowski Dam Reservoir and in the lower Pilica River. XV Krajowe Seminarium Malakologiczne, Lodz, 23-25.IX.1999, Wyd. UL, pp 3-5 (in Polish)

Achituv Y (1973) On the distribution and variability of the Indo-Pacific sea star *Asterina wega* (Echinodermata: Asteroidea) in the Mediterranean Sea. Marine Biology 18: 333-336

Achituv Y (1999) Recent changes in the distribution of two species of *Asterina* (Echinodermata: Asteroidea) along the Mediterranean shore of Israel. Israel Journal of Zoology 45: 121-125

ACIL Economics Ltd (1994) Bio-economic Risk Assessment of the Potential Introduction of Exotic Organisms through Ships' Ballast Water. In: AQIS (ed) Ballast Water Research Series 6. Australian Government Publishing Service, Canberra

Ackefors H (1999) The positive effects of established crayfish introductions in Europe. In: Gherardi F & Holdich DM (eds) Crayfish in Europe as Alien Species. How to Make the Best of a Bad Situation, pp 49-61. A.A. Balkema, Rotterdam, Brookfield

Ackerman JD, Loewen MR & Hamblin PF (2001) Benthic-pelagic coupling over a zebra mussel reef in Western Lake Erie. Limnology and Oceanography 46: 892-904

Adams RM, Remington JS, Steinberg J & Seibert J (1970) Tropical fish aquariums. A source of *Mycobacterium marinum* infections resembling sporotrichosis. Journal of the American Medical Association 211: 457-461

Admiraal W, Van der Velde G, Smit H & Cazemier WG (1993) The rivers Rhine and Meuse in the Netherlands: present state and signs of ecological recovery. Hydrobiologia 265: 97-128

Afanasiev SA (1987) The differences in oligochaete distribution in periphyton on substrate with different structure. In: Water Oligochaeta. Proceedings of the 6th All Union Symposium, pp 38-41. Riga, Latvia (in Russian)

Agerberg A (1993) Genetic and phenotypic variation in the freshwater crayfish *Pacifastacus leniusculus* (Dana). PhD Thesis, Acta Universitatis Upsaliensis, Faculty of Science 446, 28 pp

Airoldi L (1998) Roles of disturbance, sediment stress and substratum retention on spatial dominance in algal turf. Ecology 79: 2759-2770

Akkermann R (1975) Untersuchungen zur Ökologie und Populationsdynamik des Bisams (*Ondatra zibethicus* L.) an einem nordwestdeutchen Verlandungsee III. Verhalten und Populationsdynamik. Zeitschrift für Angewandte Zoologie 62: 281-338

Aladin NV & Plotnikov IS (2000) The danger of large-scale ecological catastrophe in the Caspian. The Caspian Bulletin 4: 112–126

Albrecht H (1982) The system of European crayfishes (Decapoda, Astacidae): proposals and assumptions. Mitteilungen aus dem Hamburgischen Zoologischen Museum und Institut 79: 187-210 (in German)

Alderman DJ (1996) Geographichal spread of bacterial and fungal disease of crustaceans. Revue Scientifique et Technique office International des Epizooties 15: 603-632

Alderman DJ & Polglase JL (1988) Pathogens, parasites and commensals. In: Holdich DM & Lowery RS (eds) Freshwater Crayfish: Biology, Management and Exploitation, pp 167-212. Croom Helm (Chapman & Hall), London

Alderman DJ, Holdich DM & Reeve ID (1991) Signal crayfish acting as vectors in crayfish plague. Aquaculture 86:3-6

Aleem AA (1948) The recent migration of certain Indo-Pacific algae from the Red Sea into the Mediterranean. New Phycologist 47: 88-94

Aleem AA (1950) Some new records of marine algae from the Mediterranean Sea. Meddelanden Göteborgs Botaniska Trädgrad 18: 275-288

Aleem AA (1993) The Marine Algae of Alexandria, Egypt. University of Alexandria. Egypte, 154 pp

Alekin OA (1956) Elements of Hydrochemistry. Wydawnictwo Geologiczne, Warszawa, 311 pp (in Polish)

Alekseev AP & Jakovenko VJ (1999) Acclimatized pink salmon in the White Sea. Rybnoe khozyaistvo 1: 49 (in Russian)

Alekseev AP& Kulachkova VG (2000) Pink salmon from the Far East in the basins of the White and Barents seas (the second, "the Magadan" period of acclimatization). In: Vidi-vselentsy v evropeyskikh moryakh Rossii, pp 249-257. Izdatelstvo Kolskogo Nauchnogo Tzentra Rossiyskoy Academii Nauk, Apatity (in Russian)

Alexandrov B & Zaitsev Yu (2000) Chronicle of exotic species introduction into the Black Sea. International Symposium "The Black Sea ecological problems" (31 October-5 November, Odessa, Ukraine). OCNTI, pp 14-19

Alimov AF (1981) Functional Ecology of Freshwater Bivalves. Nauka, Leningrad, 248 pp (in Russian)

Alimov AF, Alekseev AP & Kulachkova VG (1998) Directions and basic researches on The White Sea Project in 1995-1997. In: Problems of study, rational exploitation and conservation of natural resources of the White Sea. Proceedings of the International Conference, Archangel1998, St. Petersburg, pp 5-10 (in Russian)

Alimov AF, Panov VE, Krylov PI, Telesh IV, Bychenkov DE, Zimin VL, Maximov AA & Filatova LA (1998) The problem of anthropogenic introductions of nonindigenous organisms in the Gulf of Finland basin. In: Ecological situation in St. Petersburg and Leningrad Region in 1997, pp 243-248. St. Petersburg (in Russian)

Allen YC, Thompson BA & Ramcharan CW (1999) Growth and mortality rates of the zebra mussel, *Dreissena polymorpha*, in the lower Mississipi River. Canadian Journal of Fisheries and Aquatic Sciences 56: 748-759

Alongi G, Cormaci M, Furnari G & Giaccone G (1993) Prima segnalazione di *Caulerpa racemosa* (Chlorophyta, Caulerpales) per le coste italiane. Bollettino Accademia Gioenia Scienze Naturali Catania 26: 49-53

Amorim A & Dale B (1998) Distribution of cysts from toxic or potentially toxic dinoflagellates along the Portuguese coast. In: Reguera B, Blanco J, Fernández ML & Wyatt T (eds) Harmful Algae, pp 64-65. Xunta de Galicia and Intergovernmental Oceanographic Commission of UNESCO, Vigo and Paris

Andersen AB, Behrens HL & Haugom GP (2001) Methodological approach to develop standards for assessment of harmful aquatic organisms in ballast water. GloBallast Workshop, London, March 2001

Anderson LWJ & Keppner S (2001) *Caulerpa taxifolia*: marine algal invader provokes quick response in U.S. waters. Aquatic Nuisance Species (ANS) Digest 4(2): 21-23

Andersson J, Sandström O & Hansen HJM (1991) Elver (*Anguilla anguilla* L.) stockings in a Swedish thermal effluent - recaptures, growth and body condition. Journal of Applied Ichthyology 7: 78-89

Anninsky BE (1995) Seasonal dynamics of the chemical composition of organic matter in the ctenophore *Mnemiopsis leidyi* A.Agassiz in the Black Sea. Oceanology 35: 395-399

Anninsky BE, Romanova ZA, Abolmasova GI, Gucu AC & Kideys AE (1998) Ecological and physiological state of the ctenophore *Mnemiopsis leidyi* (Agassiz) in the Black Sea in autumn 1996. In: Ivanov L & Oguz T (eds) NATO TU-Black Sea Project: Ecosystem Modeling as a Management Tool for the Black Sea, Symposium on Scientific Results, pp 249-262. Kluwer Academic Publishers

Annoni D, Bianchi I, Girod A & Mariani M (1978) Introduction of *Dreissena polymorpha* (Pallas) (Mollusca, Bivalvia) in the coastal mollusc beds of Lake Garda (North Italy). Quaderni Civica Stazione Idrobiologica Milano 6: 77-84

Anonymous (1923) Trout release in The Netherlands. Fischerei Zeitung 26 (30): 251 (in German)

Anonymous (1982) Status (1980) of introduction of non indigenous marine species to north Atlantic waters. ICES Cooperative Research Report116: 1-87

Anonymous (1984) Relazione delle ricerche svolte sul banco di vongole veraci (*Venerupis decussata*) della Sacca di Goro, Istituto di Zoologia - Unversità di Ferrara, unpubl. report, 15 pp

Anonymous (1995) ICES Code of Practice on the Introductions and Transfers of Marine Organisms 1994. ICES, Copenhagen 12 pp

Anonymous (2000) Final Report of the Joint Government/Industry Working Group on Infectious Salmon Anaemia (ISA) January 2000, 7 pp

Anonymous (2000) The Large Aquaria Guide. Fishes, Plants and Utensils. Könemann, Kiel, 287 pp

Antipova TV, Denisenko NV & Semenov VN (1989) Distribution of benthic species and problems of biogeographic regionalization of North seas. In: Zhizn i sreda polyarnikh morey, pp 146-157. Nauka, Leningrad (in Russian)

Antonov PI (1993) On invasion of bivalve mollusc *Dreissena bugensis* (Andr.) into the Volga reservoirs. In: Ecological Problems of the Large River Basins, pp 52-53. Tolyatti (Abstract; in Russian)

Antonov PI (1996) The bivalve species *Dreissena bugensis* (Ant.) as new to the Volga fauna. In: Kutikova LA (ed) International Conference. Problems of Aquatic Ecology in Inland Waters and their Malakofauna, pp 9-10. Zoological Institute RAS, St-Petersburg (Abstract; in Russian)

Antsulevich A &Välipakka P (2000) *Cercopagis pengoi* - new important food object of the Baltic herring in the Gulf of Finland. International Revue of Hydrobiology 85: 609-619

Anttila CK, Daehler CC, Rank NE & Strong DR (1998) Greater male fitness of a rare invader (*Spartina alterniflora*, Poaceae) threatens a common native (*Spartina foliosa*) with hybridization. American Journal of Botany 85: 1597-1601

AQIS (1993) Shipping ballast water trials on the bulk carrier M.V. "Iron Whyalla". AQIS Ballast Water Research Series, Report 2, 123 pp

AQIS (1994) Bio-economic risk assessment of the potential introduction of exotic organisms through ship's ballast water. Ballast Water Research Series 6, Australian Government Publishing Service, Canberra

AquaSense (1998a) Ballast water. Overview of available data and estimation of possible risks. AquaSense, Amsterdam, Report 1162

AquaSense (1998b) Risk assessment of ballast water. Ecological and economic consequences, frequency analysis and measures. AquaSense, Amsterdam, Report 1290

Ardabieva AG, Tatarintseva TA & Teletskaya OV (2000) The species introducers of phytoplankton in the Caspian Sea. In: Species Introducers in the European Seas in Russia, Murmansk, pp 16-17 (in Russian)

André F, L'Hardy-Halos MT & Saldanha L (1982) Observations nouvelles sur la morphologie et la répartition géographique de trois Céramiales: *Ceramium cinnabarium*, *Mesothamnion caribaeum* et *Ctenosiphonia hypnoides*. Cryptogamie, Algologie 3: 3-20

Argyrou M, Demetropoulos A & Hadjichristophorou M (1999) Expansion of the macroalga *Caulerpa racemosa* and changes in soft bottom macrofaunal assemblages in Moni Bay, Cyprus. Oceanologica Acta 22: 517-528

Ariani AP & Serra V (1969) Sulla presenza di *Portunus pelagicus* (L.) in acque italiane con osservazioni sulla morfologia della specie (Crustacea Decapoda). Archivio Botanico Biogeografico Italiano 14: 187-206

Arkhiptseva NT, Baranov IV, Zabelina GM, Pokrovsky VV, Serebrova SA, Tereshenkov II & Tsybaleva GA (1977) Lakes of the North Vuoksa Basin. Proceedings of GosNIORKh 124: 83-134 (in Russian)

Armonies W (2001) What an introduced species can tell us about the spatial extension of benthic populations. Marine Ecology Progress Series 209: 289-294

Armonies W & Reise K (1999) On the population development of the introduced razor clam *Ensis americanus* near the island of Sylt (North Sea). Helgoländer Meeresuntersuchungen 52: 291-300

Armstrong G (1997) Ballast system design for flow-through exchange of ballast water. TransIMarE 109: 257-269

Armstrong G, Rose A & Holde A (1999) An analysis of flow-through ballast water exchange. TransIMarE 3(2): 59-70

Arnott DL & Vanni MJ (1996) Nitrogen and phosphorus recycling by the zebra mussel (*Dreissena polymorpha*) in the Western basin of Lake Erie. Canadian Journal of Fisheries and Aquatic Sciences 53: 646-659

Arrignon JCV, Gérard P, Krier A & Laurent PJ (1999) The situation in Belgium, France, and Luxemburg. In: Gherardi F & Holdich DM (eds) Crayfish in Europe as Alien Species. How to Make the Best of a Bad Situation, pp 129-140. A.A. Balkema, Rotterdam, Brookfield

Arter HE (1989) Effect of eutrophication on species composition and growth of freshwater mussels (Mollusca, Unionidae) in Lake Hallwil (Aargau, Switzerland). Aquatic Sciences 51: 87-99

Arthington AH & Mitchell DS (1986) Aquatic invading species. In: Groves RH & Burdon JJ (eds) Ecology of Biological Invasions: an Australian perspective, pp 34-53. Australian Academy of Science, Canberra, Australia

Arthington AH, Kailola PJ, Woodland DJ & Zalucki JM (1999) Baseline environmental data relevant to an evaluation of quarantine risk potentially associated with the importation to Australia of ornamental finfish. Report to the Australian Quarantine and Inspection Service, Agriculture, Fisheries and Forestry (Canberra, Australia)

Artimo A (1960) The dispersal and acclimatization of the muskrat, *Ondatra zibethicus* (L.), in Finland. Papers on Game Research 21: 1-101

Artuz I (1991) Ctenophore invasion in the Marmara Sea. Cumhuriyet Bilim Teknik Dergisi 9, 571: 9-11 (in Turkish)

Asakura A (1992) Recent introductions of marine benthos into Tokyo Bay (review): process of invasion into an urban ecosystem with discussion on the factors inducing their successful introduction. Journal of the Natural History Museum and Institute, Chiba 2(1): 1-14

Athanasiadis A (1990) Evolutionary biogeography of the North Atlantic antithamnioid algae. In: Garbary DJ & South GR (eds) Evolutionary Biogeography of the Marine Algae of the North Atlantic, pp 219-240. Springer-Verlag, Berlin

Atkins SM, Jones AM & Garwood PR (1987) The ecology and reproductive cycle of a population of *Marenzelleria viridis* (Annelida: Polychaeta: Spionidae) in the Tay Estuary. Proceedings of the Royal Society of Edinburgh 92B: 311-322

Atlas of the Arctic (1985) Moscow, 204 pp (in Russian)

Augener H (1913) Polychaeten von Franz-Joseph Land II. Zoologischer Anzeiger 41: 253-273

Australian Quarantine and Inspection Service (AQIS) (1993) Ballast Water Treatment for the Removal of Marine Organisms. Ballast Water Research Series Report 1. Australian Quarantine and Inspection Service (AQIS), Canberra, Australia

Avakyan AB & Shirokov VM (1994) Rational Use and Protection of Water Resources. Victor Press, Ecaterinburg, 320 pp (in Russian)

Avinski V (1997) *Cercopagis pengoi* - a new species in the eastern Gulf of Finland ecosystem. In: Sarkkula J (ed) Proceedings of the Final Seminar of the Gulf of Finland Year 1996, March 17-18, pp 247-256. The Finnish Environment Institute, Helsinki, Finland

Avinski VA & Telesh IV (1999) Composition and numerical characteristics of zooplankton. In: Rumyantsev VA & Drabkova VG (eds) The Gulf of Finland in the Conditions of Anthropogenic Stress, pp 161-189. St. Petersburg (in Russian)

Avise JC (2000) Phylogeography. Harvard University Press, Cambridge, Massachusetts, 447 pp

Baback EV (1983) Pliocene and Quaternary Dreissenids from Euxinian Basin. Nauka, Moskow, 105 pp (in Russian)

Babkov I & Golikov AN (1984) Hydrobiocomplexes of the White Sea. Leningrad. 102 pp (in Russian)

Bacescu M (1966) Die kaspische Reliktfauna im ponto-asowschen Becken und in anderen Gewässern. Kieler Meeresforschung 22: 176-188

Bachrach AJ (1972) Psychological Research. An Introduction. Random House, New York, 153 pp

Baglinièri JL & Maisse G (1999) Biology and Ecology of the Brown and Sea Trout. Springer/Praxis Publishing, Chicester, UK, IV 286 pp

Baker HG (1965) Characteristics and modes of origin of weeds. In: Baker HG & Stebbins GL (eds) The Genetics of Colonizing Species, pp 147-172. Academic, New York, NY

Baker LD & Reeve MR (1974) Laboratory culture of the lobate ctenophore *Mnemiopsis mccradyi* with notes on feeding and fecundity. Marine Biology 96: 57-62

Baker S (1999) Eradication of *Myocastor coypus* in United Kingdom. Workshop on the Control and Eradication of Non-native Terrestrial Vertebrates, Malta, pp 83-87. Environmental Encounters, Council of Europe Publishing

Baker SJ & Clarke CN (1988) Cage trapping coypus (*Myocastor coypus*) on baited rafts. Journal of Applied Ecology 25: 41-48

Baker SM, Levinton JS & Evan WJ (2000) Particle transport in the zebra mussel, *Dreissena polymorpha* (Pallas). Biological Bulletin (Woods Hole) 199: 116-125

Ballesteros E & Rodríguez-Prieto C (1996) Presència *d'Asparagopsis taxiformis* (Delile) Trevisan a Balears. Bolletin Societat Història Natural Balears 39: 135-138

Balon EK (1995) Origin and domestication of wild carp *Cyprinus carpio*: from Roman gourmets to the swimming flowers. Aquaculture 129: 3-48

Balss H (1927) Bericht über die Crustacea Decapoda (Natantia und Anomura). Zoological Results of the Cambridge Expedition to the Suez Canal, 1924. XIV. Transactions of the Zoological Society of London 22: 221-227

Baltic Marine Environment Bibliography (1970 -) http://otatrip.hut.fi/vtt/baltic/search.html (2001-11-05)

Baltic Sea Alien Species Database (2001) Olenin S, Leppäkoski E & Daunys D (eds) Alien Species Directory. www.ku.lt/nemo/mainnemo.htm

Baltic Sea Alien Species Database (2002) Olenin S, Leppäkoski E & Daunys D (eds) Alien Species Directory. www.ku.lt/nemo/mainnemo.htm

Baltz DM (1991) Introduced fishes in marine systems and inland seas. Biological Conservation 56:151-177

Baltz DM & Moyle PB (1993) Invasion resistance to introduced species by a native assemblage of California stream fishes. Ecological Applications 3: 246-255

Banarescu P (1991-1995) Zoogeography of Fresh Waters. Aula-Verlag, vol. 1-3, 1617 pp

Banoub MW (1963) Survey of the Blue Crab *Callinectes sapidus* (Rath.) in Lake Edku in 1960. Alexandria Institute of Hydrobiology. Notes and Memoirs 69: 1-18

Barash A & Danin Z (1973) The Indo-Pacific species of Mollusca in the Mediterranean and notes on a collection from the Suez Canal. Israel Journal of Zoology 21: 301-374

Barash A & Danin Z (1977) Additions to the knowledge of Indo-Pacific Mollusca in the Mediterranean. Conchiglie 13: 85-116

Barash A & Danin Z (1982) Contribution to the knowledge of mollusca in the Bardawil Lagoon. Bollettino Malacologico 18: 107-128

Barash A & Danin Z (1992) Fauna Palaestina. Mollusca 1: Annotated List of Mediterranean Molluscs of Israel and Sinai. Israel Academy of Science and Humanities, Jerusalem. 405 pp

Barbiero RP, Little RE & Tuchman ML (2000) Results from the USEPA's biological open water surveillance programm of the Laurentian Great Lakes: III. crustacean zooplankton. Journal of Great Lakes Research 27: 167-184

Barbour MG & Rodman JE (1970) Saga of the west coast sea-rockets: *Cakile edentula* ssp. *californica* and *C. maritima*. Rhodora 72: 370-386

Barmin AN & Kuzmina EV (1993) *Pistia stratioides* L. (Araceae) in the Astrakhan city waterbodies. Proceedings of the 3rd Conference on Aquatic Plants of Inland Waters and the Quality of their Waters, pp 25-26. Petrozavodsk (in Russian)

Barnard JL & Barnard CM (1983) Freshwater Amphipoda of the World. Hayfield Association, Mt. Vernon, Virginia, 830 pp

Barreto R, Charudattan R, Pomella A & Hanada R (2000) Biological control of neotropical aquatic weeds with fungi. Crop Protection 19: 697-703

Barton NH & Gale KS (1993) Genetic analysis of hybrid zones. In: Harrison RG (ed) Hybrid Zones and the Evolutionary Process, pp 13-45. Oxford University Press, New York

Bartsch I & Kuhlenkamp R (2000) The marine macroalgae of Helgoland (North Sea): an annotated list of records between 1845 and 1999. Helgoland Marine Research 54: 160-189

Baskin Y (1998) Winners and losers in a changing world. Global changes may promote invasions and alter the fate of invasive species. BioScience 48: 788-792

Baster J (1759) Natuurkundige Uitspanningen, behelzende eenige waarnemingen over sommige zee-planten en zee-insecten, benevens derzelver zaadhuisjes en eijernesten, eerste stukje. Te Haarlem bij J. Bosch

Bastrop R, Jürss K & Sturmbauer C (1998) Cryptic species in a marine polychaete and their independent introduction from North America to Europe. Molecular Biology and Evolution 15: 97-103

Bastrop R, Röhner M & Jürss K (1995) Are there two species of the polychaete genus *Marenzelleria* in Europe? Marine Biology 121: 509-516

Bastrop R, Röhner M, Sturmbauer C & Jürss K (1997) Where did *Marenzelleria* spp. (Polychaeta: Spionidae) in Europe come from? Aquatic Ecology 31: 119-136

Basu BK & Pick FR (1997) Phytoplankton and zooplankton development in a lowland, temperate river. Journal of Plankton Research 19: 237-253

Bax N (1999) Eradicating a dreissenid from Australia. Dreissena! 10: 1-5

Bazikalova AY (1945) Amphipoda of Lake Baikal. Proceedings of Baikal Limnological Station 1. 440 pp (in Russian)

Bazzaz FA (1986) Life history of colonising plants: some demographic, genetic, and physiological features. In: Mooney HA & Drake JA (eds) Ecology of Biological Invasions of North America and Hawaii, Ecological Studies, pp 96-110. Springer-Verlag, New York

Bazzocchi P (1985) Prima segnalazione di *Strombus (Conomurex) decorus raybaudii* Nicolay & Romagna Manoja, 1983 per l'isola di Cipro. Bollettino Malacologico 21: 64

Beddington JR, Free CA & Lawton JH (1978) Characters of successful natural enemies in models of biological control of insect pests. Nature 273: 513-519

Bedulli D & Franchini DA (1978) *Dreissena polymorpha* (Pallas): primi rinvenimenti nel fiume Po e predazione su di essa da parte di *Rattus norvegicus* (Berk.). Quaderni Civica Stazione Idrobiologica Milano 6: 85-92

Beeton AM (1965) Eutrophication of the St Lawrence Great Lakes. Limnology and Oceanography 10: 240-254

Bekman MY (1962) Ecology and production of *Micruropus possolskyi* Sow. and *Gmelinoides fasciatus* Stebb. Proceedings of Limnological Institute 2(22): 141-155 (in Russian)

Bekman MY & Bazikalova AY (1951) Biology and production capability of some Baikalian and Siberian amphipods. Proceedings of thematic meetings 1: 61-67 (in Russian)

Bellan-Santini D, Arnaud PM, Bellan G & Verlaque M (1994) Résultats préliminaires sur la faune d'invertébrés du peuplement à Caulerpa taxifolia des côtes de Provence (Méditerranée Nord-occidentale). In: Boudouresque CF, Meinesz A & Gravez V (eds) First International Workshop on *Caulerpa taxifolia*, pp 365-369. GIS Posidonie publ., Marseille

Bellemo G, Curiel D, La Rocca B, Scattolin M & Marzocchi M (2001) First report of a filamentous species of *Desmarestia* (Desmarestiaceae, Fucophyceae) in the Lagoon of Venice (Italy, Mediterranean Sea). Botanica Marina 44: 541-545

Bellemo G, Curiel D, Marzocchi M, Luri M, Pavan M, Vecchiato N & Scattolin M (1999) Aggiornamento di nuove specie algali per la Laguna di Venezia. Lavori Società Veneta. Scienze Naturali 24: 55-66

Belmonte G (1998) *Pteriacartia*, a new genus of Acartiidae (Calanoida, Diaptomidae) for *Acartia josephinae* Crisafi, 1974. Journal of Marine Systems 15: 359-368

Belmonte G & Potenza D (2001) Biogeography of the family Acartiidae (Calanoida) in the Ponto-Mediterranean Province. Hydrobiologia 453/454:171-176

Belmonte G, Mazzocihi MG, Prusova I & Shadrin NV (1994) *Acartia tonsa*: a species new for the Black Sea fauna. Hydrobiologia 292/293: 9-15

Belsher T & Pommellec (1988) Expansion de l'algue d'origin japonaise *Sargassum muticum* (Yendo) Fensholt, sur les côtes françaises, de 1983 à 1987. Cahiers de Biologie Marine 29: 221-231

Belsher T, Cabioch J, L'Hardy-Halos MT, Castric-Fey A & Le Roux A (1997) Effets des introductions d'espèces: les algues macrophytes. In: Dauvin JC (ed) Les biocénoses marines et littorales françaises des côtes Atlantique, Manche et Mer du Nord, synthèse, menaces et perspectives. Laboratoire de Biologie des Invertébrés Marins et Malacologie–Service du Patrimoine naturel, Paris, pp 229-242

Belyavskaya LI & Vyushkova VP (1970) Bottom fauna of the Volgograd reservoir. Proceedings Saratov Branch of State Research Institute of the Lake and River Fisheries 10: 93-107 (in Russian)

Ben Maiz N (1986) Flore algale (Rhodophyta, Phaeophyceae, Chlorophyceae, Bryopsidophyceae) de l'étang de Thau (Hérault). PhD Thesis, Ecologie, Université d'Aix-Marseille II, 354 pp

Ben Maiz N, Boudouresque CF & Gerbal M (1986) Flore algale de l'étang de Thau: *Grateloupia doryphora* (Montagne) Howe et *G. filicina* (Wulfen) C. Agardh. Thalassographica 9: 39-49

Ben Maiz N, Boudouresque CF, Riouall R & Lauret M (1987) Flore algale de l'étang de Thau (France, Méditerranée): sur la présence d'une Rhodymeniale d'origine japonaise, *Chrysymenia wrightii* (Rhodophyta). Botanica Marina 30: 357-364

Ben Tuvia A (1953) New Erythrean fishes from the Mediterranean coast of Israel. Nature 172: 464-465

Ben Tuvia A (1970) Man-made changes in the eastern Mediterranean and their effect on the fishery resources. Journées ichthyologique, CIESM, Rome pp 179-185

Ben Tuvia A (1973) Man-made changes in the Eastern Mediterranean Sea and their effect on the fishery resources. Marine Biology 19: 197-203

Ben Tuvia A (1983) The Mediterranean Sea, B. Biological aspects. In: Kechum BN (ed) Estuaries and Enclosed Seas, pp 239-251. Elsevier

Ben Tuvia A (1985) The impact of the Lessepsian (Suez Canal) fish migration on the eastern Mediterranean ecosystem. In: Moraitou-Apostolopoulou M & Kiotosis V (eds) Mediterranean Marine Ecosystems, pp 367-375. Plenum Press, New York

Ben Yami M & Glaser T (1974) The invasion of *Saurida undosquamis* (Richardson) into the Levant Basin - an example of biological effect of interoceanic canals. Fishery Bulletin 72: 359-373

Ben-Eliahu MN (1972) Studies on the migration of the Polychaeata through the Suez Canal. 17th Congress International Zoology. Conseqeuénces biologiques des canaux inter-océans 3

Berezina NA, Khlebovich VV, Panov VE & Zaporozhec NV (2001) Salinity resistance of the amphipod *Gmelinoides fasciatus* (Stebb.), introduced into the Gulf of Finland (Baltic Sea) basin. Reports of the Academy of Sciences 379(3): 414-416 (in Russian)

Berg D, Garton DW, MacIsaac HJ, Panov V & Telesh I (2002) Identification of a source and erosion of founder effects in an exotic species. Freshwater Biology 47: 275-282

Berg DJ, Fisher SW & Landrum PF (1996) Clearance and processing of algal particles by Zebra mussels (*Dreissena polymorpha*). Journal of Great Lakes Research 22: 779-788

Berg LS (1928) About origin of northern elements in the Caspian fauna. Proceedings of the Academy of Science of the USSR 14(3) (in Russian)

Berg LS (1948) Fish of freshwaters and adjacent countries. Part 2. Moscow-Leningrad. 466 pp (in Russian)

Berg LS (1962) Distribution of freshwater fishes of Europe. In: Nikolskij GV & Obruchev DV (eds) Akad. L.S. Berg, Izbrannye trudy 5: 236-319 (in Russian)

Berg LS (1962) On the origin of northern elements in Caspian fauna. In: Select. General Biology. Biogeography and Palaeoichthyology vol 5, pp 143-148. USSR Academy of Science, Moscow-Leningrad (in Russian)

Berger VJa, Bryazgin VF & Naumov AD (1999) Chinese mitten crab *Eriocheir sinensis* as a new element of the East Fennoscandian Fauna. In: Biologocal Basis of the Study, Management and Protection of Flora, Fauna and the Soil Cover in Eastern Fennoscandia. Petrozavodsk, p 260 (in Russian)

Berman T, Azov Y, Schneller Y, Walline P & Townsend DW (1986) Extent, transparency, and phytoplankton distribution of the neritic waters overlying the Israeli coastal shelf. Oceanologica Acta 9: 439-447

Berthe F & Thébault A (2000) EU directives for the control of diseases affecting bivalve molluscs: current state and prospects. Verona Meeting 'Towards the year 2000: what changes in aquaculture?'

Berthold G (1882) Über die Vertheilung der Algen im Golf von Neapel, nebst einem Verzeichniss der bisheer daselbst beobacheten Arten. Mittheilungen der Zoologischen Station zu Neapel 3: 393-536

Besiktepe S, Ozsoy E, Abdullatif FM & Oguz T (2000) Hydrography and circulation of the Marmara Sea. In: Ozturk B, Kadioglu M & Ozturk H (eds) Marmara Sea 2000 Symposium, pp 314-326. TUDAV Publication, Istanbul (in Turkish)

Best EPH & Bakker JP (eds) (1993) Netherlands-Wetlands. Developments in Hydrobiology 88: 1-328

Bevanger K & Henriksen G (1995) The distributional history and present status of the American mink (*Mustela vison* Shreber, 1777) in Norway. Annales Zoologici Fennici 32: 11-14

Bialokoz W, Chybowski L & Krzywosz T (1996) Endangered crayfish species in Suwalki voyvodeship. Zoologica Poloniae 41(suppl.): 87-91 (in Polish)

Bianchi CN & Morri C (2000) Marine biodiversity of the Mediterranean Sea: situation, problems and prospects for future research. Marine Pollution Bulletin 40: 367-376

Bianchi I, Girod A & Mariani M (1974) Densità, strutture di popolazione e distribuzione di *Dreissena polymorpha* nel bacino idrografico del Benaco. Archiv für Molluskenkunde 104: 97-105

Bianchi I, Girod A & Mariani M (1976) Ritrovamento di *Dreissena polymorpha* (Pallas) nel lago di Valvestino (Brescia). Natura Bresciana Annali Museo Civico Storia Naturale Brescia 13: 115-116

Bianco PG (1991) Nuove forme alloctone nel Lago di Campotosto (Abruzzo): *Padogobius martensi* (Gunther, 1861) (Pisces, Gobiidae), *Cobitis taenia* (Linnaeus, 1758) (Pisces, Cobitidae) e *Piscicola geometra* (Linnaeus, 1758 (Hirudinea, Piscicolidae). Atti 4° Convegno Nazionale Associazione Italiana Ittiologi Acque Dolci, Riva del Garda, pp 245-252

Bianco PG (1998) Freshwater fish transfers in Italy: history, local changes in fish fauna and a prediction on the future of native populations. In: Cowx IG (ed) Stocking and Introduction of Fish, pp 167-185. Fishing News Books. MPG Books Ltd., Bodmin, Cornwall

Bibliotti M & Abdelahad (1990) *Halophila stipulacea* (Forssk.) Aschers. (Hydrocharitaceae): espèce nouvelle pour l'Italie. Posidonia Newsletter 3: 23-26

Bick A (1995) Entwicklungsbedingte Veränderungen von morphologischen Merkmalen bei *Marenzelleria viridis* (Verrill), (Spionidae) und *Manayunkia aestuarina* (Bourne), (Sabellidae). Mitteilungen des hamburgischen zoologischen Museums und Instituts 92: 41-53

Bick A & Burckhardt R (1989) Erstnachweis von *Marenzelleria viridis* (Polychaeta, Spionidae) für den Ostseeraum, mit einem Bestimmungsschlüssel der Spioniden der Ostsee. Mitteilungen aus dem Zoologischen Museum Berlin 65: 237-247

Bick A & Zettler ML (1997) On the identity and distribution of two species of *Marenzelleria* (Polychaeta, Spionidae) in Europe and North America. Aquatic Ecology 31: 137-148

Bidoux C & Magne F (1989) Étude de quelques Acrochaetiales (Rodophyta) devant être rapportées au genre *Rhodothamniella*. Cryptogamie, Algologie 10: 33-55

Bij de Vaate A & Klink A (1995) *Dikerogammarus villosus* Sowinsky (Crustacea: Gammaridae) a new immigrant in the Dutch part of the lower Rhine. Lauterbornia 20: 51-54

Bij de Vaate A, Jazdzewski K, Ketelaars HAM, Gollasch S & Van der Velde G (submitted) Geographical patterns in range extension of Ponto-Caspian macroinvertebrate species in Europe. Canadian Journal of Fisheries and Aquatic Sciences

Bilecik N (1990) Distribution of sea snail "*Rapana venosa* (V.)" in Turkish coast of Black Sea and its impact on Black Sea fisheries. Directorate of Fisheries Research Institute Publication, Bodrum, 32 pp (in Turkish)

Billard C (1992) *Fibrocapsa japonica* (Rhaphidophyceae), planktonic marine alga reported for the first time in France. Cryptogamie Algologie 13: 225-231 (in French)

Binder E (1965) Un mollusque envahissant, la *Dreissena polymorpha*. Musee de Geneve 54: 2-4

Binelli A, Provini A & Galassi S (1997) Trophic modifications in Lake Como (N. Italy) caused by the zebra mussel (*Dreissena polymorpha*) Water, Air and Soil Pollution 99: 633-640

Binney WG & Gould AA (1870) Report on the Invertebrata of Massachusetts. Comprising the Mollusca. Wright & Potter, Boston. 524 pp

Biochino GI (1994) Polymorphism and geographical variability. In: Starobogatov YaI (ed) *Dreissena* (*Dreissena polymorpha* (Pall.) Bivalvia, Dreissenidae). Systematics, Ecology and Practical Importance, pp 56-67 Series "Fauna species of Russia and neighbouring countries". Moscow (in Russian)

Biro P (1997) Temporal variation in Lake Balaton and its fish populations. Ecology of Freshwater Fish 6: 196-216

Birshtejn JA & Romanova NN (1968) Order Amphipoda. In: Birshtejn JA et al. (eds) Atlas of the Invertebrates of the Caspian Sea. Izdatelstvo "Pishchevaja Promyshlennost", Moskva, 241-269 (in Russian)

Bischof K, Kräbs G, Hanelt D & Wiencke C (2000) Photosynthetic characteristics and mycosporine-like amino acids under UV radiation: a competitive advantage of *Mastocarpus stellatus* over *Chondrus crispus* at the Helgoland shoreline. Helgoland Marine Research 54: 47-52

Biserova LI (1996) Parasites of the molluscan invader *Lithoglyphus naticoides* (C. Pfr.) from the delta of the Volga River. In: Problems of Hydrobiology of the inner Waters and their Malacofauna, pp 12-13. St.Petersburg (Abstract; in Russian)

Bishara N (1976) Contributions to the biology of penaeid prawns in Lake Manzalah, Egypt. I. Growth studies and length-weight relationship. Aquaculture 8: 337-349

Bitar G & Zibrowius H (1997) Scleractinian corals from Lebanon, Eastern Mediterranean, including a non-lessepsian invading species (Cnidaria: Scleractinia). Scientia Marina 61: 227-231

Blackburn SI, Bolch CJS, Haskard KA & Hallegraeff GM (2001) Reproductive compatibility among four global populations of the toxic dinoflagellate *Gymnodinium catenatum* (Dinophyceae). Phycologia 40: 78-87

Blackburn TM & Duncan RP (2001) Determinant of establishment success in introduced birds. Nature 414: 195-197

Blanc G (1997) Introduced pathogens in European aquatic ecosystems: theoretical aspects and realities. Bulletin Francais de la Pêche de la Pisciculture 344-5: 489-515

Blanchard M (1996) Spread of the slipper limpet *Crepidula fornicata* (L., 1758) in Europe. Current state and consequences. Scientia Marina 61(Suppl. 2): 109-118

BMB (2001) The Baltic Marine Biologists. www.smf.su.se/bmb/ (2001-07-23)

BMB WG NEMO (2001) Alien Species Database. Olenin S, Leppäkoski E & Daunys D (eds) Alien Species Directory. www.ku.lt/nemo/mainnemo.htm

Boaden PJS (1995) The adventive seaweed *Sargassum muticum* (Yendo) Fensholt in Strangford Lough, Northern Ireland. Irish Naturalists' Journal 25: 111-113

Boalch GT (1994) The introduction of non-indigenous marine species to Europe: planktonic species. In: Boudouresque CF, Briand F & Nolan C (eds) Introduced Species in European Coastal Waters. European Commission Ecosystems Research Report 8: 1-111

Boalch GT & Harbour DS (1977) Unusual diatom off the coast of south west England and its effect on fishing. Nature 269: 687-688

Board PA (1970) Some observations on the tunnelling of shipworms. Journal of Zoology 161: 193-201

Bochert A, Richard D & Bochert R (1997) *Marenzelleria* cf. *viridis* and the sulphide regime. Aquatic Ecology 31: 223-231

Bochert R (1997) *Marenzelleria viridis* (Polychaeta: Spionidae): a review of its reproduction. Aquatic Ecology 31: 163-175

Bochert R & Bick A (1995) Reproduction and larval development of *Marenzelleria viridis* (Polychaeta: Spionidae). Marine Biology 123: 763-773

Bogdanova DP & Konsoulov A (1993) On the distribution of the new ctenophora species *Mnemia mccradyi* in the Black Sea along the Bulgarian coastline in the summer of 1990. Comptes rendus de l'Academie Bulgare des Sciences 46: 71-74

Bogi C & Khaizallah NH (1987) Nota su alcuni molluschi di provenienza indo-pacifica raccolti nella Baia di Jounieh (Libano). Notiziario del CISMA 10: 54-60

Boileau MG & Hebert PDN (1993) Genetics of the zebra mussel (*Dreissena polymorpha*) in populations from the Great Lakes region and Europe. In: Nalepa TF & Schloesser DW (eds) Zebra Mussels: Biology, Impacts, and Control, pp 227-238. Lewis Publishers, Boca Raton

Bokn T & Lein TE (1978) Long-term changes in fucoid association of the inner Oslofjord, Norway. Norwegian Journal of Botany 25: 9-14

Bolch CJS (2000) Evolution and biogeography of the *Gymnodinium catenatum* complex: A multidisciplinary approach. 9th International Conference on Harmful Algal Blooms, Tasmania, Australia, 7-11 February 2000, p 8. University of Tasmania, Hobart (Abstract)

Bolch CJS & Hallegraeff GM (1993) Chemical and physical treatment options to kill toxic dinoflagellate cysts in ships' ballast water. Journal of Marine Environment and Engineering 1: 23-29

Bolch CJS & Hallegraeff GM (1994) Ballast water as a vector for the dispersal of toxic dinoflagellates. In: Proceedings of the Conference & Workshop Nonindigenous Estuarine and Marine Organisms (NEMO), Seattle, Washington, April 1993, pp 63-67

Bolch CJS, Blackburn SI, Hallegraeff GM & Vaillancourt RE (1999) Genetic variation among strains of the toxic dinoflagellate *Gymnodinium catenatum* (Dinophyceae). Journal of Phycology 35: 356-367

Boni L, Ceredi A, Guerrini F, Milandri A, Pistocchi R, Poletti R & Pompei M (2000) Toxic *Gonyaulax grindley* Reincke in the north-western Adriatic Sea (Italy). 9th International Conference on Harmful Algal Blooms, Tasmania, Australia, 7-11 February 2000, p 87. Univ. of Tasmania, Hobart (Abstract)

Boni L, Mancini L, Milandri A, Poletti R, Pompei M & Viviani R (1992) First cases of diarrhoetic shellfish poisoning in the Northern Adriatic Sea. In: Vollenweider RA, Marchetti R & Viviani R (eds) Marine Coastal Eutrophication, pp 419-426 Science of Total Environment Suppl. Elsevier Science, Amsterdam

Boni L, Pompei M & Reti M (1986) Maree colorate e fioriture algali lungo la costa dell'Emilia-Romagna dal 1982 al 1985 con particolare riguardo alla comparsa di *Protogonyaulax tamarensis*. Nova Thalassia 8: 237-245

Borodich ND (1979) Baikalian amphipod *Gmelinoides fasciatus* (Stebbing) (Amphipoda, Gammaridae) in the Kubyshevskoe Reservoir. Zoological Journal 58: 920-921 (in Russian)

Bosch van den F, Hengeveld R & Metz JAJ (1992) Analysing the velocity of animal range expansions. Journal of Biogeography 19: 135-150

Bossard CC, Randall JM & Hoshovksy MC (eds) (2000) Invasive Plants of California's Wildlands. University of California Press, Berkeley, 360 pp

Boström C & Bonsdorff E (2000) Zoobenthic community establishment and habitat complexity - the importance of seagrass shoot-density, morphology and physical disturbance for faunal recruitment. Marine Ecology Progress Series 205: 123-138

Botnen HD, Evensen D & Johannessen PJ (2000) Ballast water, a paradise for stowaways - results of the Sture Project. IFM-rapport 2

Botts PS, Patterson BA & Schloesser DW (1996) Zebra mussel effects on benthic invertebrates: physical or biotic? Journal of the North American Benthological Society 15: 179-184

Boudouresque CF (1994) Les espèces introduites dans les eaux côtières d'Europe et de Méditerranée: Etat de la question et conséquences. In: Boudouresque CF, Briand F & Nolan C (eds) Introduced Species in European Coastal Waters. CEC Ecosystem Research Report 8: 8-27

Boudouresque CF & Ribera MA (1994) Les introductions d'espèces végétales et animales en milieu marin - conséquences écologiques et économiques et problèmes législatifs. In: Boudouresque CF, Meinesz A & Gravez V (eds) First International Workshop on *Caulerpa taxifolia*, pp 29-102. GIS Posidonie, Marseille

Boudouresque C-F & Ribera MA (1995) Introduced marine plants, with special reference to macroalgae: Mechanisms and impact. Progress in Phycological Research 11: 187-268

Boudouresque CF & Verlaque M (2001) Biological pollution in the Mediterranean Sea: invasive versus introduced macrophytes. Marine Pollution Bulletin 44: 32-38

Boudouresque CF, Ballesteros E, Cinelli F, Henocque Y, Meinesz A, Pesando D, Pietra F, Ribera MA & Tripaldi G (1996) Synthèse des résultats du programme CCE-LIFE "Expansion de l'algue verte tropicale *Caulerpa taxifolia* en Méditerranée". In: Ribera MA, Ballesteros E, Boudouresque CF, Gómez A & Gravez V (eds) Second International Workshop on *Caulerpa taxifolia*, pp 11-57. Publicacions Universitat de Barcelona

Boudouresque CF, Gerbal M & Knoepffler-Péguy M (1985) L'algue japonaise *Undaria pinnatifida* (Phaeophyceae, Laminariales) en Méditerranée. Phycologia 24: 364-366

Boudouresque CF, Meinesz A, Verlaque M & Knoepffler-Peguy M (1992) The expansion of the tropical alga *Caulerpa taxifolia* (Chlorophyta) in the Mediterranean. Journées phycologiques 13: 144-145

Boudry P, Heurtebise S, Collet B, Cornette F & Gerard A (1998) Differentiation between populations of the Portuguese oyster, *Crassostrea angulata* (Lamarck) and the Pacific oyster, *Crassostrea gigas* (Thunberg), revealed by mtDNA RFLP analysis. Journal of experimental marine Biology and Ecology 226: 279-291

Bour W (1990) The fishery resources of Pacific island countries. Part 3: *Trochus*. FAO Fisheries Technical Paper 272(3): 1-89

Bousfield EL & Hoover PM (1997) The amphipod superfamily Corophioidea on the Pacific coast of North America. Part V. Family Corophiidae. Corophiinae new subfamily. Systematic and distributional ecology. Amphipacifica 2(3): 67-139

Bower SM, McGladdery SE & Price IM (1994) Synopsis of infectious diseases and parasites of commercially exploited shellfish. Annual Review of Fish Diseases 4: 1-199

Bowman MF & Bailey RC (1998) Upper pH tolerance limit of the zebra mussel (*Dreissena polymorpha*). Canadian Journal of Zoology 76: 2119-2123

Boycott AE & Oldham C (1938) The mollusca of the Brecon and Newport and other detached or isolated canals. Proceedings of the Malacological Society of London 23: 33-50

Boyle PJ (1988) Marine wood biodeterioration and woodboring crustaceans. In: Thompson M-F, Sarojiani R & Nagabhushanam R (eds) Marine Biodeterioration: Advanced Techniques Applicable to the Indian Ocean, pp 167-188. A. A. Balkema, Rotterdam

Braarud T (1950) The immigration of *Colpomenia peregrina* in Norwegian waters. Blyttia 8: 125-126

Braarud T & Heimdal BR (1970) Brown water off the Norwegian coast in autumn 1966. Nytt magasin for botanikk17: 91-97

Braarud T & Nygaard I (1967) Phytoplankton I-III. In: The Oslofjord and its pollution problems. Norwegian Institute of Water Research 4: 1-171 (in Norwegian)

Brattegard T & Holthe T (eds) 1997) Distribution of marine, benthic macroorganisms in Norway. A tabulated catalogue. Preliminary edition. DN 1997-1. Directorate for Nature Management, Trondheim, Norway. 409 pp

Bravo I & Ramilo I (1999) Distribution of microreticulate dinoflagellate cysts from the Galician and Portuguese coast. Scientia Marina 63: 45-50

Bravo I, Reguera B, Martinez A & Fraga S (1990) First report of *Gymnodinium catenatum* Graham on the Spanish Mediterranean coast. In: Graneli E, Sundström B, Edler L & Anderson DM (eds) Procedings of the Fourth International Conference Toxic Marine Phytoplankton, Lund, Sweden, pp 449-452. Elsevier, New York

Breber P (1980) Annual gonadal cycle in the carpet-shell clam *Venerupis decussata* in Venice Lagoon, Italy. Proceedings of the National Shellfisheries Association 70: 31-35

Breber P (1985a) L'introduzione e l'allevamento in Italia dell'arsella del Pacifico, *Tapes semidecussatus* Reeve (Bivalvia; Veneridae). Oebalia NS 11: 675-680

Breber P (1985b) On-growing of the carpet-shell clam (*Tapes decussatus* (L.)): two years' experience in Venice Lagoon. Aquaculture 44: 51-56

Breber P (1991) Experience with Manila clam (*Ruditapes philippinarum* Adams & Reeve) syn. (*Tapes semidecussatus* Reeve) on-growing in Italy. Rivista Italiana di Acquacoltura 26: 11-17

Breber P (1992) An account of the acclimatisation of the Manila clam, *Ruditapes philippinarum* (Adams & Reeve) syn. *Tapes semidecussatus* Reeve (Mollusca; Bivalviae; Venerinae), in Italian waters. Lavori della Società Italiana di Malacologia 24: 47-52

Breber P (1996) L'Allevamento della Vongola Verace in Italia. CLEUP, Padova, 157 pp

Breeman AM, Meulenhoff EJS & Guiry MD (1988) Life history regulation and phenology of the red alga *Bonnemaisonia hamifera*. Helgoländer Meeresuntersuchungen 42. 535-551

Brenchley GA & Carlton JT (1983) Competitive displacement of native mud snails by introduced periwinkles in the New England intertidal zone. Biological Bulletin 165: 543-558

Breton G, Faasse M, Noël P & Vincent T (2002) A new alien crab in Europe: *Hemigrapsus sanguineus* (Decapoda: Brachyura Grapsidae). Journal of Crustacean Biology 22: 184–189

Brieger G & Hunter RD (1993) Uptake and depuration of PCB 77, PCB 169, and hexachlorobenzene by zebra mussels (*Dreissena polymorpha*). Ecotoxicology and Environmental Safety 26: 153-165

Briggs JC (1974) Marine Zoogeography. McGraw-Hill, New York. 475 pp

Bright C (1998) Life out of Bounds. Bioinvasion in a Borderless World. Norton & Co, New York. 287 pp

Brinck P (1988) The restoration of the crayfish production in a plaque stricken country. Journal of Aquatic Production 2: 53-60

Brouwer E & Den Hartog C (1996) *Crassula helmsii* (Kirk) Cockayne, een adventief op droogvallende, zandige oevers. Gorteria 22: 149-152

Bruijs MCM, Kelleher B, Van der Velde G & Bij de Vaate A (2001) Oxygen consumption, temperature and salinity tolerance of the invasive amphipod *Dikerogammarus villosus*: indicators for further dispersal via ballast water transport. Archiv für Hydrobiologie 152: 633-646

Bruno J (2000) Facilitation of cobble beach plant communities through habitat modification by *Spartina alterniflora*. Ecology 81: 1179-1192

Bryazgin VF et al. (1981) Animals and Plants of the Barents Sea. Zhivotniye I rasteniya Barentseva morya. Izdatelstvo Kolskogo Filiala Academii Nauk SSSR, Apatity. 188 pp (in Russian)

Bubinas A (1979) Fish feeding on the acclimatized mysid of the Caspian relict type *Mesomysis kowalewskyi* Czern. (*Paramysis lacustris*). LTSR MA Darbai, Serija B 4(88): 89-96 (in Russian)

Buchmann K, Mellergaard S & Køie M (1987) *Pseudodactylogyrus* infections in eel: a review. Diseases of Aquatic Organisms 3: 51-57

Buia MC, Petrocelli A & Saracino OD (1998) *Caulerpa racemosa* spread in the Mediterranean Sea: first record in the Gulf of Taranto. Biologia Marina Mediterranea 5: 527-529

Bullion VV & Winberg GG (1981) Correlation between primary production and fish productivity of water bodies. In: Winberg GG (ed) Principles of Research of Freshwater Ecosystems, pp 5-10. Nauka Press, Leningrad (in Russian)

Bullock JM, Hodder KH, Manchester SJ & Stevenson MJ (1997) Review of information, policy and legislation on species translocation. JNCC Report 261, JNCC, Peterborough

Bulnheim HP (1977) Effects of inbreeding on the relative fitness of the amphipod *Gammarus duebenii* Liljeborg, 1852. Crustaceana Suppl 4: 3-14

Burckhardt R, Schumann R & Bochert R (1997) Feeding biology of the pelagic larvae of *Marenzelleria* cf. *viridis* (Polychaeta: Spionidae) from the Baltic Sea. Aquatic Ecology 31: 149-162

Burlakova LE, Karatayev AY & Padilla DK (2000) The impact of *Dreissena polymorpha* (Pallas) invasion on unionid bivalves. International Review of Hydrobiology 85: 529-541

Burrel VG & Van Engel WA (1976) Predation by and distribution of a ctenophore, *Mnemiopsis leidyi* (A. Agassiz) in the York River estuary. Estuarine and Coastal Marine Science 4: 235-242

Burreson EM, Stokes NA & Friedman CS (2000) Increased virulence in an introduced pathogen: *Haplosporidium nelsoni* (MSX) in the eastern oysters *Crassostrea virginica*. Journal of Aquatic Animal Health 12: 1-8

Butler IV MJ & Stein RA (1985) An analysis of the mechanisms governing species replacement in crayfish. Oecologia 66: 168-177

Buttermore RE, Turner E & Morrice MG (1994) The introduced northern Pacific seastar *Asterias amurensis* in Tasmania. Memoirs of the Queensland Museum 36: 21-25

Cabioch J, Castric Fey A, L'Hardy Halos MT & Rio A (1997) *Grateloupia doryphora* and *Grateloupia filicina* var. *luxurians* (Rhodophyta, Halymeniaceae) from the coasts of Brittany (France). Cryptogamie Algologie 18: 117-137 (in French)

Cabioch L, Gentil F, Glaçon R & Retière C (1977) Le macrobenthos des fonds meubles de la Manche: distribution générale et écologie. In: Keegan BF, Ceidigh PO & Boaden PJS (eds) Biology of Benthic Organisms, pp 115-128. Pergamon Press

Cabot D (1999) Ireland. The New Naturalist. Harper Collins, London

Cabrini M, Chiurco R, Cok S, Martecchini E, Pecchiar I & Ganis P (1996) The occurrence of *Alexandrium* spp. along the Adriatic coasts. In: Yasumoto T, Oshima Y, Fukuyo Y (eds) Harmful and Toxic Algal Blooms, pp 82-84. Intergovernemental Oceanographic Commission of UNESCO, Paris

Cachia C (1981) Notes on some uncommon species of molluscs from the Maltese Islands. Bollettino Malacologico 17: 291-294

Caddy JF & Griffiths RC (1990) A perspective on recent fishery-related events in the Black Sea. General Fisheries Council for the Mediterranean Studies and Reviews 63: 43-71

Calloway BC & Turner RD (1988) Brooding in the Teredinidae (Mollusca: Bivalvia). In: Thompson M-F, Sarojiani R & Nagabhushanam R (eds) Marine Biodeterioration: Advanced Techniques Applicable to the Indian Ocean, pp 215-226. AA Balkema, Rotterdam

Calman WT (1927) Appendix to the report on the Crustacea Decapoda (Brachyura). Zoological Results of the Cambridge Expedition to the Suez Canal, 1924. XIII. Transactions of the Zoological Society of London 22: 217

Campbell SJ (1999) Occurrence of *Codium fragile* subsp. *tomentosoides* (Chlorophyta: Bryopsidales) in marine embayments of southeastern Australia. Journal of Phycology 35: 938-940

Campos NH & Türkay M (1989) On a record of *Charybdis helleri* from the Caribbean coast of Colombia. Senckenbergiana Maritima 20: 119-123

Cangelosi A, Balcer M, Blatchely ER III, Dawson R, Gao X, Huq A, Knight IT, Mays NL, McGregor B, Reid D, Sturtevant R, Swan C, Taverna J, Wells C & Wright D (2001) Experimental studies comparing biological effectiveness of commercially available ballast treatment systems. Northeast-Midwest Institute, Washington DC, USA

Cangelosi A, Balcer M, Dawson R, Knight IT, Mays NL, Reid D, Swan C, Taverna J & Wright D (in prep) Biological effectiveness of depth filtration and ultraviolet radiation as a compound ballast treatment system

Cangelosi A, Knight I, Balcer M, Gao X, Huq A, McGreevy JA, McGregor B, Reid D, Sturtevant R & Carlton JT (in press) Bioeffectiveness of automatic backwash filtration as a ballast water treatment measure

Caraco NF, Cole JJ, Findlay SEG, Fischer DT, Lampman GG, Pace ML & Strayer DL (2000) Dissolved oxygen declines in the Hudson River associated with the invasion of the zebra mussel (*Dreissena polymorpha*). Environmental Science & Technology 34: 1204-1210

Caraco NF, Cole JJ, Raymond PA, Strayer DL, Pace ML, Findlay SEG & Fischer DT (1997) Zebra mussel invasion in a large, turbid river: phytoplankton response to increased grazing. Ecology (Washington D C) 78: 588-602

Carausu S, Dobreanu E & Manolache C (1955) Amphipoda of Brackish and Fresh Waters. Fauna Republici populare Romine, Crustacea. 409 pp (in Romanian)

Carlton JT & Geller JB (1993) Ecological roulette: The global transport of nonindigenous marine organisms. Science 261: 78-82

Carlton JT (1979) History, Biogeography and Ecology of the Introduced Marine and Estuarine Invertebrates of the Pacific Coast of North America. PhD Thesis, University of California, Davis, 904 pp

Carlton JT (1985) Transoceanic and interoceanic dispersal of coastal marine organisms: the biology of ballast water. Oceanography and Marine Biology Annual Review 23: 313-371

Carlton JT (1987) Patterns of transoceanic marine biological invasions in the Pacific Ocean. Bulletin of Marine Science 41: 452-465

Carlton JT (1989) Man's role in changing the face of the ocean: Biological invasions and implications for conservation of near-shore environments. Conservation Biology 3: 265-273

Carlton JT (1992) Introduced marine and estuarine molluscs of North America: An end-of-the-20th-century perspective. Journal of Shellfish Research 11: 489-505

Carlton JT (1994) Biological invasions and biodiversity in the sea: the ecological and human impacts of nonindigenous marine and estuarine organisms. In: Proceedings of Conference and Workshop Nonindigenous Estuarine and Marine Organisms, pp 5-11. U.S. Department of Commerce, NOAA, Washington

Carlton JT (1996) Biological invasions and cryptogenic species. Ecology 77: 1653-1655

Carlton JT (1996) Marine bioinvasions: The alteration of marine ecosystems by nonindigenous species. Oceanography 9: 1-7

Carlton JT (1996) Pattern, process, and prediction in marine invasion ecology. Biological Conservation 78: 97-106

Carlton JT (1999) Molluscan invasions in marine and estuarine communities. Malacologia 41: 439-454

Carlton JT (1999) The scale and ecological consequences of biological invasions in the world's oceans. In: Sandlund OT, Schei PJ & Viken Å (eds) Invasive Species and Biodiversity Management, pp 195-212. Kluwer Academic Publishers, Dordrecht

Carlton JT (2000) Quo Vadimus Exotica Oceanica? Marine bioinvasion ecology in the twenty-first century. In: Pederson J (ed) Marine Bioinvasions. Proceedings of the First National Conference, January 24-27, 1999, pp 6-23. Massachusetts Institute of Technology, Cambridge, MA

Carlton JT (2001) Introduced species in U.S. coastal waters: environmental impacts and management priorities. Pew Oceans Commission, Arlington, Virginia, 28 pp

Carlton JT & Ruckelshaus MH (1997) Nonindigenous marine invertebrates and algae. In: Simberloff D, Schmitz DC & Brown TC (eds) Strangers in Paradise. Impact and Management of Non-Indigenous Species in Florida, pp 187-201. Island Press, Washington, D.C and Covelo CA

Carlton JT & Scanlon JA (1985) Progression and dispersal of an introduced alga: *Codium fragile* ssp. *tomentosoides* (Chlorophyta) on the Atlantic coast of North America. Botanica Marina 28: 155-165

Carlton JT, Reid DM & van Leeuwen H (1995) Shipping study: The role of shipping in the introduction of non-indigenous aquatic organisms to the coastal waters of the United States (others than the Great Lakes) and an analysis of control options. National Sea Grant College Programme/CT Sea Grant Project R/ES 6, 213 pp

Carlton JT, Thompson JK, Schemel LE & Nichols FH (1990) Remarkable invasion of San Francisco Bay (California, USA) by the Asian clam *Potamocorbula amurensis* 1. Introduction and dispersal. Marine Ecology Progress Series 66: 81-94

Caron DA & Finlay B (1994) Protozoan links in food webs. In: Hausmann K & Hülsmann N (eds) Progress in Protozoology. Proceedings of the IX International Congress of Protozoology, pp 125-133. G. Fischer, Stuttgart.

Carqueija CRG & Gouvea EP (1996) The occurrence in coastal Brazil of a Portunidae (Crustacea, Decapoda), originating from the Indo-Pacific and the Mediterranean. Nauplius 4: 105-112

Carrada GC, Casotti R, Modigh M & Saggiomo V (1991) Presence of *Gymnodinium catenatum* (Dinophyceae) in a coastal Mediterranean lagoon. Journal of Plankton Research 13: 229-238

Cary PR & Weerts PGJ (1992) Growth and nutrient composition of *Azolla pinnata* R. Brown and *Azolla filiculoides* Lamarck as affected by water temperature, nitrogen and phosphorus supply, light intensity and pH. Aquatic Botany 43: 163-180

Case TJ (1990) Invasion resistance arises in strongly interacting species-rich model competition communities. Proceedings of the National Academy of Science 87: 9610-9614

Case TJ (1991) Invasion resistance, species build-up and community collapse in metapopulation models with interspecies competition. Biological Journal of the Linnean Society 42: 239-266

Case TJ & Bolger DT (1991) The role of introduced species in shaping the distribution and abundance of island reptiles. Evolutionary Ecology 5: 272-290

Casper SJ & Krausch H-D (1980) Pteridophyta und Anthophyta. 1. Lycopodiaceae - Orchidaceae. Süsswasserflora von Mitteleuropa 23. Gustav Fischer Verlag, Stuttgart, 403 pp

Casper SJ & Krausch H-D (1981) Pteridophyta und Anthophyta. 2. Sauraceae - Asteraceae. Süsswasserflora von Mitteleuropa 24. Gustav Fischer Verlag, Stuttgart, 533 pp

Castel J, Dauvin JC & Glémarec M (1997) Les conditions générales en Atlantique, Manche et Mer du Nord. In: Dauvin JC (ed) Les biocénoses marines et littorales françaises des côtes Atlantiques, Manche et Mer du Nord, synthèse, menaces et perspectives. Laboratoire de Biologie des Invertébrés marins et Malacologie. Service du Patrimoine naturel, Paris, pp 6-16

Castric-Fey A, Beaupoil C, Bouchain J, Pradier E & L'Hardy-Halos MT (1999a) The introduced alga *Undaria pinnatifida* (Laminariales, Aliraceae) in the rocky shore ecosystem of the St Malo area: morphology and growth of the sporophyte. Botanica Marina 42: 71-82

Castric-Fey A, Beaupoil C, Bouchain J, Pradier E & L'Hardy-Halos MT (1999b) The introduced alga *Undaria pinnatifida* (Laminariales, Aliraceae) in the rocky shore ecosystem of the St Malo area: growth rate and longevity of the sporophyte. Botanica Marina 42: 83-96

Catalano E & Zava B (1993) Sulla presenza di *Stephanolepis diaspros* Fr. Brunn. nelle acque italiane (Osteichthyes, Monacanthidae). Supplemento Ricerche Biologia Selvaggina 21: 379-382

Cavaliere A & Berdar G (1975) Presenza di *Callinectes sapidus* nello stretto di Messina (Crustacea, Decapoda, Portunidae). Bollettino di Pesca Piscicoltura e Idrobiologia 30: 315-322

CBD (1992) The Convention on Biological Diversity, UN Conference on Environment and Development, Rio de Janeiro. www.biodiv.org

Ceccherelli G & Cinelli F (1999a) The role of vegetative fragmentation in dispersal of the invasive alga *Caulerpa taxifolia* in the Mediterranean. Marine Ecology Progress Series 182: 299-303

Ceccherelli G & Cinelli F (1999b) A pilot study of nutrient enriched sediments in a *Cymodocea nodosa* bed invaded by the introduced alga *Caulerpa taxifolia*. Botanica Marina 42: 409-417

Ceccherelli G & Cinelli F (1999c) Effects of *Posidonia oceanica* canopy on *Caulerpa taxifolia* size in a north-western Mediterranean bay. Journal of Experimental Marine Biology and Ecology 240: 19-36

Ceccherelli G & Piazzi L (2001) Dispersal of *Caulerpa racemosa* fragments in the Mediterranean: Lack of detachment time effect on establishment. Botanica Marina 44: 209-213

Ceccherelli G, Piazzi L & Cinelli F (2000) Response of the non-indigenous *Caulerpa racemosa* (Forsskål) J. Agardh to the native seagrass *Posidonia oceanica* (L.) Delile: effect of density of shoots and orientation of edges of meadows. Journal of Experimental Marine Biology and Ecology 243: 227-240

Cecere E (1989) Un'alga della famiglia delle Soliariaceae (Rhodophyta, Gigartinales) nel Mar Piccolo di Taranto. In: Camera di Commercio I.A.A. Lecce (ed) Convegno Nazionale; Alghe e loro utilizzazione, pp 177-178. Salentina, Galatina

Cecere E (1990) Sulla presenza nel Golfo di Taranto di una specie nuova per il Mediterraneo: *Solieria filiformis* (Kützing) Gabrielson. Oebalia 16 (2 suppl.): 629-631

Cecere E, Petrocelli A & Saracino D (2000) *Undaria pinnatifida* (Fucophyceae, Laminariales) spread in the central Mediterranean: its ocurrence in the Mar Piccolo of Taranto (Ionian Sea, southern Italy). Criptogamie, Algologie 21: 305-309

Cederwall H, Jermakovs V & Lagzdins G (1999) Long-term changes in the soft-bottom macrofauna of the Gulf of Riga. ICES Journal of Marine Science Suppl. 56: 41-48

Cellot B, Mouillot F & Henry CP (1998) Flood drift and propagule bank of aquatic macrophytes in a riverine wetland. Journal of Vegetation Science 9: 631-640

Celona A (2000) First record of a tiger shark *Galeocerdo cuvier* (Peron & Le Sueur, 1822) in the Italian waters. Annales, annali di studi istriani e mediterranei. Sectio Historiae naturalis 10: 207-210

Center TD, Dray FA, Jubinsky GP & Grodowitz MJ (1999) Biological control of water hyacinth under conditions of maintenance management: Can herbicides and insects be integrated? Environmental Management 23: 241-256

Cerenius L & Söderhäll K (1992) Crayfish diseases and crayfish as vectors for important diseases. Finnish Fisheries Research 14: 125-133

Cesari P & Pellizzato M (1985) Molluschi pervenuti in Laguna di Venezia per apporti volontari o casuali. Acclimazione di *Saccostrea commercialis* (Iredale & Roughely, 1933) e di *Tapes philippinarum* (Adams & Reeve, 1850). Bollettino Malacologico italiano 21: 237-274

Chamberlin RV (1920) The polychaeta collected by the Canadian Arctic expedition, 1913-1918. Report of the Canadian Arctic Expedition 1913-1918, 9B: 1-41B

Champion PD & Clayton JS (2000) Border control for potential aquatic weeds. Stage 1. Weed risk model. Science for Conservation 141, Department of Conservation, Wellington, New Zealand

Chaouachi B, Dos Santos A & Ben Hassine OK (1998) Occurrence of the shrimp *Metapenaeus monoceros* (Fabricius, 1798) in the Gulf of Gabes: a first record for Tunisian coasts. IV International Crustacean Congress, Amsterdam (Abstract)

Chapman AS (1999) From introduced species to invader: what determines variation in the success of *Codium fragile* ssp. *tomentosoides* (Chlorophyta) in the North Atlantic Ocean. Helgoländer Meeresuntersuchungen 52: 277-289

Chapman JW (2000) Climate effects on the geography of nonindigenous peracaridan crustacean introductions in estuaries. In: Pederson J (ed) Marine Bioinvasions. Proceedings of the First National Conference, January 24-27, 1999, pp 66-80. Massachusetts Institute of Technology, Cambridge, MA

Chapman JW & Dorman JA (1975) Diagnosis, systematics, and notes on *Grandidierella japonica* (Amphipoda: Gammaridea) and its introduction to the Pacific coast of the United States. Bulletin of the Southern California Academy of Sciences 74: 104-108

Charlton MN (1991) Lake Erie water quality in 1990: Restoration or resilience? In: Proceedings of the 34th Conference of the International Association for Great Lakes Research, Buffalo, NY (USA), 2-6 June 1991, p 121

Chauvaud L, Jean F, Ragueneau O & Thouzeau G (2000) Long-term variation of the Bay of Brest ecosystem: benthic-pelagic coupling revisited. Marine Ecology Progress Series 200: 35-48

Cheney DP, MacDonald BF & Elston RA (2000) Summer mortality of Pacific oysters *Crassostrea gigas* (Thunberg): initial findings on multiple environmental stressors in Puget Sound, Washington, 1998. Journal of Shellfish Research 19: 353-359

Cheng L (1989) Factors limiting the distribution of *Halobates* species. In: Ryland JS & Tyler PA (eds) Reproduction, Genetics and Distribution of Marine Organisms, pp 357-362. Olsen and Olsen, Fredensborg, Denmark

Chevalier A (1923) Note sur les *Spartina* de la flore française. Bulletin Societé Botanique de France 70: 54-63

Chew KK (1990) Global bivalve shellfish introductions. World Aquaculture 21: 9-24

Chikwenhere GP (1994) Biological control of water lettuce in various impoundments of Zimbabwe. Journal of Aquatic Plant Management 32: 27-29

Chilton C (1911) Notes on the dispersal of marine crustacea by means of ships. Transactions of the New Zealand Institute 43: 131-133

Chotkowski MA & Marsden JE (1999) Round goby and mottled sculpin predation on lake trout eggs and fry: field predictions from laboratory experiments. Journal of Great Lakes Research 25: 26-35

Christiansen ME (1977) The Chinese mitten crab found for the first time in Norway. Fauna (Oslo) 30: 134-138

Chuykov YuS (1994) The Zooplankton of the North fore-Caspian and the North Caspian. The Astrakhan Committee on Ecology and Natural Resources

CIESM Atlas of Exotic species in the Mediterranean Sea. www.ciesm.org/atlas

Cigarría J & Elston R (1997) Independent introduction of *Bonamia ostreae*, a parasite of *Ostrea edulis*, to Spain. Diseases of Aquatic Organisms 29: 157-158

Cinelli F (1985) On the biogeography of the benthic algae of the Mediterranean. In: Moraitou-Apostolopoulou M & Kiortsis V (eds) Mediterranean marine ecosystems, pp 49-56. Plenum Press, New York

Cinelli F & Sartoni G (1969) *Acrothamnion* J. Ag. (Rhodophyta, Ceramiaceae): genere algale nuovo per il mare Mediterraneo. Pubblicazioni della Stazione Zoologica di Napoli 37: 567-574

Cinelli F, Salghetti U & Serena F (1984) Nota sull'areale di *Acrothamnion preissii* (Sonder) Wollaston nell'Alto Tirreno. Quaderni del Museo di Storia naturale di Livorno 5: 57-60

Clark PFC, Rainbow PS, Robbins RS, Smith B, Yeomans WE, Thomas M & Dobson G (1998) The alien Chinese mitten-crab, *Eriocheir sinensis* (Crustacea: Decapoda: Brachyura), in the Thames catchment. Journal of the marine biological Association of the United Kingdom 78: 1215-1221

Clarke D (1952) The infestation of a waterworks by *Dreissena polymorpha*, a freshwater mussel. Journal of the Institution of Water Engineers 6: 370-379

Clavell A & Polo L (1998) Estudio comparativo entre *Apoglossum gregarium* (Dawson) Wynne y *Apoglossum ruscifolium* (Turner) J.Agardh (Delesseriacea, Rhodophyta), en la Costa Brava (Girona). Algas 3-4

Clayton MN, Wiencke C & Kloser H (1997) New records of temperate and sub Antarctic marine benthic macroalgae from Antarctica. Polar Biology 17: 141-149

Clinton WJ (1999) Executive Order 13112. Invasive Species. The White House, Washington, D.C.

Clode D & Macdonald DW (1995) Evidence for food competition between mink (*Mustela vison*) and otter (*Lutra lutra*) on Scottish islands. Journal of Zoology 237: 435-444

Cloern JE (1996) Phytoplankton bloom dynamics in coastal ecosystems: a review with some general lessons from sustained investigations of San Francisco, California. Reviews of Geophysics 34: 127-168

Coan EV, Scott PV & Bernard FR (2000) Bivalve Seashells of Western North America. Santa Barbara Museum of Natural History Monographs 2, 764 pp

Cobb JS & Harlin MM (1980) Mute swan (*Cygnus olor*) feeding and territoriality affects diversity and density of rooted aquatic vegetation. American Zoologist 20: 882

Cofrancesco AF (1998) Overview and future direction of biological control technology. Journal of Aquatic Plant Management 36: 49-53

Cohen A, Mills C, Berry H, Wonham M, Bingham B, Bookheim B, Carlton J, Chapman J, Cordell J, Harris L, Klinger T, Kohn A, Lambert C, Lambert G,

Li K, Secord D & Toft J (1998) Puget Sound Expedition: A Rapid Assessment Survey of Non-Indigenous Species in the Shallow Waters of Puget Sound. Olympia, WA: Washington State Department of Natural Resources & United States Fish and Wildlife Service. 37 pp
Cohen AN & Carlton JT (1995) Biological study: Non-Indigenous aquatic species in a United States estuary: a case study of the biological invasions of the San Francisco Bay and Delta. US Fisheries and Wildlife and National Sea Grant College Program Report PB96-166525, Springfield, Virginia, USA, 273 pp
Cohen AN & Carlton JT (1998) Accelerating invasion rate in a highly invaded estuary. Science 279: 555-558
Cohen AN, Carlton JT & Fountain MC (1995) Introduction, dispersal and potential impacts of the green crab *Carcinus maenas* in San Francisco Bay, California. Marine Biology 122: 225-237
Coles SL & Eldredge LG (2002) Nonindigenous species introductions on coral reefs: a need for information. Pacific Science 56: 191-209
Coles SL, DeFelice RC, Eldredge LG & Carlton JT (1997) Biodiversity of Marine Communities in Pearl Harbor, Oahu, Hawaii with Observations on Introduced Exotic Species. Bishop Museum Technical Report 10, Bishop Museum, Honolulu, Hawaii
Coles SL, DeFelice RC, Eldredge LG, Godwin LS, Pyle RL & Suzimoto A (1999b) Nonindigenous marine species introductions in the harbors of the south and west shores of Oahu, Hawaii. Bishop Museum Technical Report 15. Honolulu, Hawaii
Coles, SL, DeFelice RC, Eldredge LG & Carlton JT (1999a) Historical and recent introductions of non-indigenous marine species into Pearl Harbor, Oahu, Hawaiian Islands. Marine Biology 135: 147-158
Colt J (1986) Gas supersaturation - impact on the design and operation of aquatic systems. Aquacultural Engineering 5: 49-85
Colwell R (1998) Balancing the biocomplexity of the planet's life system: A twenty-first century task for science. BioScience 48: 786-787
Comps M (1969) Observations relatives à l'affection branchiale des huîtres portugaises *Crassostrea angulata*. Revue des Travaux Institut des Pêches maritimes 33: 151-160
Comps M, Grizel H, Tige G & Duthoit J-L (1975) A new parasite of the digestive gland in marine molluscs. Comptes Rendus de l'Academie des Sciences 281: 179-181 (in French)
Comps M, Tigé G & Grizel H (1980) Etudes ultrastructurales sur un protiste parasite de l'huître plate *Ostrea edulis*. Comptes Rendus de l'Académie des Sciences 209D: 83-384
Confortini I, Maio G, Marconato E, Salviati S, Dell'Antonio L, Bertolo A & Perini V (1994) Prima segnalazione di *Blicca bjoerkna* (Linnaeus, 1758) in Italia (Pisces, Cyprinidae). Atti 5° Convegno Nazionale Associazione Italiana Ittiologi Acque Dolci, Montecchio Maggiore, pp 375-378
Conrad V (1991) Experiments on the re-establishment of sea trout (*Salmo trutta trutta* L.) in Rheinland-Pfalz. Fischökologie Aktuelle 4: 11-13 (in German)
Conroy DA (1975) An evaluation of the present state of the world trade in ornamental fish. Rome, FAO Fisheries Technical Paper 146: 1-128
Cook CDK (1990) Aquatic Plant Book. SPB Academic Publishing, The Hague, 228 pp
Cordell JRR & Morrison SM (1996) The invasive Asian copepod *Pseudodiaptomus inopinus* in Oregon, Washington, and British Columbia estuaries. Estuaries 19: 629-638
Cormaci M & Furnari G (1988) *Antithamnionella elegans* (Berthold) Cormaci & Furnari (Ceramiaceae, Rhodophyta) and related species, with the description of two new varieties. Phycologia 27: 340-346
Cormaci M, Duro A & Furnari G (1982) Considerazioni sugli elementi fitogeografici della flora algale della Sicilia. Naturalista siciliano 4: 7-14
Cormaci M, Furnari G & Serio D (1991) First record of the austral species *Plocamium* (Gigartinales, Rhodophyta) from the Mediterranean Sea. Cryptogamie, Algologie 12: 235-244
Cormaci M, Furnari G, Alongi G, Dinaro R & Pizzuto F (1992) On the occurrence in Sicily of three Florideophyceae new to the Mediterranean Sea. Botanica Marina 35: 447-449
Cosson J (1999) On the progressive disappearance of *Laminaria digitata* on the coasts of Calvados (France). Cryptogamie Algologie 20: 35-42 (in French)
Costa ML, Santos MC & Carrapico F (1999) Biomass characterization of *Azolla filiculoides* grown in natural ecosystems and wastewater. Hydrobiologia 415: 323-327
Costanza R, d'Arge R, de Groot R, Farber S, Grasso M, Hannon B, Naeem S, Limburg K, Pa-ruelo J, O'Neill RV, Raskin R, Sutton P & Van den Belt M (1997) The value of the world's ecosystem services and natural capital. Nature 387: 253-260
Costello MJ, Emblow CS & Picton BE (1996) Long term trends in the discovery of marine species new to science which occur in Britain and Ireland. Journal of the marine biological Association of the United Kingdom 76: 255-257
Cotner JB, Gardner WS, Johnson JR, Sada RH, Cavaletto JF & Heath RT (1995) Effects of zebra mussels (*Dreissena polymorpha*) on bacterioplankton: evidence for both size-selective consumption and growth stimulation. Journal of Great Lakes Research 21: 517-528
Couch JA, Farley CA & Rosenfield A (1966) Sporulation of *Minchinia nelsoni* (Haplosporida, Haplosporiidae) in *Crassostrea virginica* (Gmelin). Science 153: 1529-1531
Council of Europe (1996) Introductions of non-native organisms into the natural environment. Standing Committee of the Convention on the Conservation of European Wildlife and Natural Habitats (Bern Convention). Nature and environment 73
Cowx IG (1997) Introduction of fish species into European freshwaters: economic successes or ecological disasters? Bulletin Francais de la Pêche de la Pisciculture 344-5: 57-77
Cowx IG, Bergot F & Vigneux E (1996) L'introduction d'escapes de poissons dans les eaux douces europeennes: succes economiques ou desasters ecologiques? Bulletin Francais de la Peche et de la Pisciculture 344-345: 57-77
Cox EF & Allen JA (1999) Stand structure and productivity of the introduced *Rhizophora mangle* in Hawaii. Estuaries 22: 276-284
Cranfield HJ, Gordon DP, Willan RC, Marshall BA, Battershill CN, Francis MP, Nelson WA, Glasby CJ & Read GB (1998) Adventive marine species in New Zealand. NIWA Technical Report 34, National Institute of Water and Atmospheric Research, Wellington, New Zealand, 48 pp
Crawford GI (1935) *Corophium curvispinum* G.O. Sars var. *devium* Wundsch, in England. Nature 136: 685
Crawley MJ (1987) What makes a community invasible? In: Gray AJ, Crawley MJ & Edwards PJ (eds) Colonization, Succession, and Stability, pp 429-453. Blackwell Scientific, Oxford, UK
Crawley MJ (1989) Chance and timing in biological invasions. In: Drake JA, Mooney HA, diCastri F, Groves RH, Kruger FJ, Rejmanek M & Williamson M (eds) Biological Invasions: A Global Perspective, SCOPE 37, pp 407-423. John Wiley & Sons, New York
Creese RG, Hooker SH, De Luca S & Wharton Y (1997) The ecology and environmental impact of the introduced Asian date mussel, *Musculista senhousia*. New Zealand Journal of Marine and Freshwater Research 31: 225-236
Cremades J (1993) National report for Spain. In: Report of the Working Group of Introductions and Transfers of Marine Organisms, Aberdeen, 26-28 April 1993, p 45. ICES CM 1993/F:3
Crisafi P (1974) Inquinamento e speciazione. *Acartia josephinae* e *A. enzoi* (Copepoda Calanoida), specie nuove del Mare Mediterraneo. Bollettino Pesca Piscicoltura e Idrobiologia 29: 5-10
Crisp DJ (1958) The spread of *Elminius modestus* Darwin in north-west Europe. Journal of the marine biological Association of the United Kingdom 37: 483-520
Cristescu M, Hebert PDN, Witt J, MacIsaac HJ & Grigorovich I (2001) An invasion history for *Cercopagis pengoi* based on mitochondrial gene sequences. Limnology and Oceanography 46: 224-229
Critchley AT (1983) The establishment and increase of *Sargassum muticum* (Yendo) Fensholt population within the Solent area of Southern Britain. II: An investigation of the increase in canopy cover of the alga at low water. Botanica Marina 26: 547-552
Critchley AT, Farnham WF & Morell LS (1986) An account of the attempted control of an introduced marine alga, *Sargassum muticum*, in southern England. Biological Conservation 35: 313-332
Critchley AT, Farnham WF, Yoshida T & Norton TA (1990) A bibliography of the invasive alga *Sargassum muticum* (Yendo) Fensholt (Fucales; Sargassaceae). Botanica Marina 33: 551-562
Cronk QC & Fuller JL (1995) Plant Invaders: The Threat to Natural Ecosystems. Chapman and Hall, London 255 pp
Crooks JA & Khim HS (1999) Architectural vs. biological effects of a habitat-altering, exotic mussel, *Musculista senhousia*. Journal of Experimental Marine Biology and Ecology 240: 53-75
Crosby AW (1972) The Columbian Exchange. Biological and Cultural Consequences of 1492. Greenwood Press. 268 pp
Crosby AW (1986) Ecological Imperialism. The Biological Expansion of Europe 900-1900. Cambridge University Press. 368 pp
Cukerzis JM (1988) *Astacus astacus* in Europe. In: Holdich DM & Lowery RS (eds) Freshwater Crayfish: Biology, Management and Exploitation, pp 309-340. Croom Helm (Chapman & Hall), London
Cullen JJ & Lesser MP (1991) Inhibition of photosynthesis by ultraviolet radiation as a function of dose and dosage rates - results for a marine diatom. Marine Biology 111: 183-190
Cullinane JP & Whelan PM (1981) Ecology distribution and seasonality of *Cryptonemia hibernica* Guiry et Irvine on the south coast of Ireland. In: Levring T (ed) 10th International Seaweed Symposium, pp 259-264. Walter de Gruyter & Co., Berlin
Culver CS & Kuris AM (2000) The apparent eradication of a locally established introduced marine pest. Biological Invasions 2: 245-253
Curiel D, Bellemo G & Marzocchi M (1996) New records of marine algae in the Lagoon of Venice. Giornale Botanico Italiano 130: 352
Curiel D, Bellemo G, Iuri M, Marzocchi M & Scattolin M (1999a) First record of the genus *Sorocarpus* Pringsheim (Fucophyceae, Ectocarpaceae) in the Mediterranean. Botanica Marina 42: 7-10
Curiel D, Bellemo G, La Rocca B, Scattolin M & Marzocchi M (2002) First report of *Polysiphonia morrowii* Harvey (Ceramiales, Rhodophyta) in the Mediterranean. Botanica Marina 45: 66-70
Curiel D, Bellemo G, Marzocchi M, Scattolin M & Parisi G (1998) Distribution of introduced Japanese macroalgae *Undaria pinnatifida*, *Sargassum muticum* (Phaeophyta) and *Antithamnion pectinatum* (Rhodophyta) in the Lagoon of Venice. Hydrobiologia 385: 17-22
Curini-Galletti M (1988) Notes and Tidings. La Conchiglia 232-233: 14
Currie DR, McArthur MA & Cohen BF (2000) Reproduction and distribution of the invasive European fanworm *Sabella spallanzanii* (Polychaeta: Sabellidae) in Port Phillip Bay, Victoria, Australia. Marine Biology 136: 645-656
Cvetkov LP & Marinov TM (1986) Faunistic enrichment of the Black Sea and changes in its benthic ecosystems. Hidrobiologija (Sofia) 27: 3-21
d'Hondt JL & Occhipinti Ambrogi A (1985) *Tricellaria inopinata* n.sp., un nouveau Bryozoaire Cheilostome de la faune méditerranéenne. P.S.Z.N.I. Marine Ecology 6: 35-46
Daehler CC (2001) Two ways to be an invader, but one is more suitable for ecology. Bulletin of the Ecological Society of America 82: 101-102
Daehler CC & Strong DR (1996) Status, prediction and prevention of introduced cordgrass *Spartina* spp. invasions in Pacific estuaries, USA. Biological Conservation 78: 51-58

Dagg M (1977) Some effects of patchy food environments on copepods. Limnology and Oceanography 22: 99-107

Dahl E (1946) The Amphipoda of the Sound 1. Terrestrial Amphipoda. Lunds Universitets Årsskrift 42(6): 1-53

Dame RF (1993) Bivalve Filter Feeders in Estuarine and Coastal Ecosystem Processes. Springer-Verlag, Berlin, 579 pp

Dame RF (1996) Ecology of Marine Bivalves. An Ecosystem Approach. CRC Press, Boca Raton, FL, 254 pp

Damgaard J (2001) Genetic variation among, and phylogeny of sea skaters (Gerridae: *Halobates*). www.zmuc.dk/EntoWeb/staff/jd2.htm

Damgaard J, Andersen NM, Cheng L & Sperling FAH (2000) Phylogeny of sea skaters, *Halobates* Eschscholtz (Hemiptera: Gerridae), based on mtDNA sequence and morphology. Zoological Journal of the Linnean Society 130. 511-526

Dando PR & Southward AJ (1980) A new species of *Chthamalus* (Crustacea: Cirripedia) characterized by enzyme electrophoresis and shell morphology: with a revision of other species of *Chthamalus* from the western shores of the Atlantic Ocean. Journal of the Marine Biological Association of the United Kingdom 60: 787-831

Danell K (1977a) Dispersal and distribution of the muskrat (*Ondatra zibethica* L.) in Sweden. Swedish Wildlife Research 10: 1-26

Danell K (1977b) Short-term plant succession following the colonization of a northern Swedish lake by the muskrat, *Ondatra zibethica*. Journal of Applied Ecology 14: 933-947

Danell K (1978) Population dynamics of the muskrat in a shallow Swedish lake. Journal of Animal Ecology 47: 697-709

Danell K (1979) Reduction of aquatic vegetation following the colonization of a northern Swedish lake by the muskrat, *Ondatra zibethica*. Oecologia (Berlin) 38: 101-106

Danell K (1985) Population fluctuation of the muskrat in coastal Nothem Sweden. Acta Theriologica 30: 219-227

Danell K (1996) Introductions of aquatic rodents: lessons of the muskrat *Ondatra zibethicus* invasion. Wildlife Biology 2: 213-220

Danilkiewicz Z (1996) Racer (goad) goby *Neogobius gymnotrachelus* (Kessler, 1857) (Perciformes, Gobiidae) - a new species in the fish fauna of the Baltic Sea basin. Komunikaty Rybackie 2/96: 27-29 (in Polish)

Danilkiewicz Z (1998) The monkey goby, *Neogobius fluviatilis* (Pallas, 1811), Perciformes, Gobiidae - another Pontic element in the ichthyofauna of the Baltic Sea basin. Fragmenta faunistica 41: 269-277 (in Polish)

Danilov P (1995) Canadian and European beavers in Russian northwest. The Third Nordic Beaver Symposium 15.-17.9.1992 Mekrijärvi, pp 10-16. Finnish Game and Fisheries Research Institute, Helsinki

Darwin C (1900) Origin of Species by Means of Natural Selection or the Preservation of Favoured Races in the Struggle for Life. John Murray, London, 703 pp

Daunys D, Schiedek D & Olenin S (2000) Species strategy near its boundary: the *Marenzelleria* cf. *viridis* (Polychaeta, Spionidae) case in the south-eastern Baltic Sea. International Review of Hydrobiology 85: 639-651

Daunys D, Zettler M & Gollasch S (1999) *Marenzelleria viridis*. In: Gollasch S, Minchin D, Rosenthal H & Voigt M (eds) Exotics Across the Ocean. Case Histories on Introduced Species, pp 61-67 Logos Verlag, Berlin

Davenport KE (1996) Characteristics of the current international trade in ornamental fish, with special reference to the European Union. Revue Scientifique & Technique Office International Epizootie 15: 435-443

Davidson DM (1999) *Sargassum muticum* in Strangford Lough, 1985-1998; A review of the introduction and colonization of Strangford Lough MNR and cSAC by the invasive brown alga *Sargassum muticum*. Report to the Environment & Heritage Service, DoE (NI), 85 pp

Davis AR, Roberts DE & Cummins SP (1997) Rapid invasion of a sponge-dominated deep-reef by *Caulerpa scalpelliformis* (Chlorophyta) in Botany Bay, New South Wales. Australian Journal of Ecology 22: 146-150

Davis CC (1957) *Cordylophora lacustris* Allman from Chagrin Harbor, Ohio. Limnology and Oceanography 2: 158-159

Davis MA & Thompson K (2000) Eight ways to be a colonizer; two ways to be an invader: a proposed nomenclature scheme for invasion ecology. Bulletin of the Ecological Society of America 81: 226-230

Davis MA & Thompson K (2001) Invasion terminology: should ecologists define their terms differently than others? No, not if we want to be of any help! Bulletin of the Ecological Society of America 82: 206

Davis MA, Grime JP & Thompson K (2000) Fluctuating resources in plant communities: a general theory of invasibility Journal of Ecology 88: 528-534

Davis N, Villablanca FX & Roderick GK (1999a) Determining the source of individuals: multilocus genotyping in nonequilibrium population genetics. Trends in Ecology and Evolution 14: 17-21

Davis N, Villablanca FX & Roderick GK (1999b) Bioinvasions of the medfly *Ceratitis capitata*: source estimation using DNA sequences at multiple intron loci. Genetics 153· 351-360

Dawson AG (1992) Ice Age Earth: Late Quaternary Geology and Climate. Routlege, London

Dawson FH (1990) The spread of *Crassula helmsii* (Kirk) Cockayne in Britain. BES Industrial Ecology Group proceedings of a meeting on the biology and control of invasive plants. University of Wales, College of Cardiff, 20-21 September 1990

Dawson FH (1996) *Crassula helmsii*: Attempts at elimination using herbicides. Hydrobiologia 340: 241-245

De Angelis DL & Gross LJ (1992) Individual based models and approaches in ecology. Chapman and Hall, New York

De Beaufort LF (1954) Veranderingen in de flora en fauna van de Zuiderzee (thans IJsselmeer) na de afsluiting in 1932. Nederlandse Dierkundige Vereniging, C. de Boer Jr, Den Helder, 359 pp

De Casabianca ML, Laugier T & Posada F (1995) Petroliferous wastewaters treatment with water hyacinths (Raffinerie de Provence, France): Experimental statement. Waste Management 15: 651-655

De Jonge JA (1990) Het Zeeuwse goud. Geschiedenis van de mossel- en oestercultuur. De Bataafsche Leeuw, Amsterdam. 144 pp

De Klem C (1994) The introduction of exotic species and the law. In: Boudouresque CF, Briand F & Nolan C (eds) Introduced species in European coastal waters. European Commission Ecosystems Research Report 8: 85-92

De Mars H & Bouman A (2002) Grote waternavel ook een bedreiging voor zwak gebufferde wateren. De Levende Natuur 103: 22-25

De Masi F & Gargiulo GM (1982) *Grateloupia doryphora* (Montagne) Howe (Rhodophyta, Cryptonemiales) en Mediterranée. Allionia 25: 105-108

de Montaudoüin X & Sauriau PG (1999) The proliferating Gastropoda *Crepidula fornicata* may stimulate macrozoobenthic diversity. Journal of the Marine Biological Association of the United Kingdom 79: 1069-1077

de Montaudoüin X & Sauriau PG (2000) Contribution to a synopsis of marine species richness in the Pertuis Charentais Sea with new insights in soft-bottom macrofauna of the Marennes-Oléron Bay. Cahiers de Biologie Marine 41: 181-222

de Montaudoüin X, Audemard C & Labourg PJ (1999) Does the slipper limpet (*Crepidula fornicata L.*) impair oyster growth and zoobenthos biodiversity? A revisited hypothesis. Journal of Experimental Marine Biology and Ecology 235: 105-124

de Montaudoüin X, Labarraque D, Giraud K & Bachelet G (2001) Why does the introduced gastropod *Crepidula fornicata* fail to invade Arcachon Bay (France)? Journal of the Marine Biological Association of the United Kingdom 81: 97-104

De Nie HW (1996) Atlas van de Nederlandse zoetwatervissen. Media Publishing, Doetichem, 151 pp

De Waal LC, Child LE, Wade PM & Brock JH (eds) (1994) Ecology and Management of Invasive Riverside Plants. J. Wiley & Sons, Chicester, 217 pp

De Wreede RE (1983) *Sargassum muticum* (Fucales, Phaeophyta): regrowth and interaction with *Rhodomela larix* (Ceramiales, Rhodophyta). Phycologia 22: 153-160

DeBach P (1974) Biological Control by Natural Enemies. Cambridge University Press, Cambridge, UK

Dedju II (1967) Amphipods and Mysids of the Dniester and Prut Drainage Systems. Izdatelstvo "Nauka", Moskva, 171 pp (in Russian)

Dedyu II (1980) Amphipods of Fresh and Salt Waters of the South-west Part of the USSR. Kishinev, 222 pp (in Russian)

Dehus P, Phillipson S, Bohl E, Oidtmann B, Keller M & Lechleiter S (1999) German conservation strategies for native crayfish species with regard to alien species. In: Gherardi F & Holdich DM (eds) Crayfish in Europe as Alien Species. How to Make the Best of a Bad Situation, pp 149-159. A.A. Balkema, Rotterdam, Brookfield

Delacoste M, Baran P, Lascaux JM, Abad N, Besson JP, Bergot F & Vigneux E (1996) Bilan des introductions de salmonides dans les lacs et ruisseaux d'altitude des Hautes-Pyrenees. Bulletin Francais de la Peche et de la Pisciculture 344-345: 205-219

Delile A (1813) Florae Aegyptiacae illustratio. In: Description de l'Egypte ou recueil des observations et des recherches qui ont été faites en Egipte pendant l'expédition de l'armée française (1798-1801). Histoire Naturelle 2: 145-320

Delmastro GB & Balma GAC (1990) Sulla comparsa del ciprinide transalpino *Rutilus rutilus* (L. 1758) nelle acque piemontesi (Osteichthyes, Cyprinidae). Atti 3° Convegno Nazionale Associazione italiana Ittiologi Acque Dolci, Perugia. Rivista Idrobiologia 29: 163-168

Demel K (1933) List of invertebrates and fishes from the Polish Baltic Sea. Fragmenta faunistica Musei Zoologici Polonici 2(13): 121-136 (in Polish)

Demir M (1977) On the presence of *Arca (Scapharca) amygdalum* Philippi, 1847 in the harbour of Izmir, Turkey. Istanbul Universitesi Fen Fakultesi Mecmuasi B 42: 197-202

den Hartog C (1997) Is *Sargassum muticum* a threat to eelgrass beds? Aquatic Botany 58: 37-41

Denker C (1907) Een nieuw schelpdier voor onze fauna. De Levende Natuur 12: 215

Dermott R, Witt J, Young M & Gonzáles M (1998) Distribution of the Ponto-Caspian amphipod *Echinogammarus ischnus* the Great Lakes and replacement of native *Gammarus fasciatus*. Journal of Great Lakes Research 24: 442-452

Desmarchelier P & Wong F (1998) The potential for *Vibrio cholerae* to translocate and establish in Australian waters. AQIS Ballast Water Research Series, Report 10, 45 pp

Detenbeck NE, Galatowitsch, SM, Atkinson J & Ball H (1999) Evaluating perturbation and developing restoration strategies for inland waters in the Great Lakes basin. Wetlands 19: 789-820

Devin S, Beisel JN, Bachmann V & Moreteau JC (2001) *Dikerogammarus villosus* (Amphipoda: Gammaridae): another invasive species newly established in the Mosell River and French hydrosystems. Annales de Limnologie 37: 21-27

Dexbach NK (1935) *Dreissena polymorpha* - Verbreitung im europäischen Teile der USSR und die sie bedingenden Faktoren. Verhandlungen der Internationale Vereinigung für Theoretische und Angewandte Limnologie 7: 432-438

di Castri F (1989) History of biological invasions with special emphasis on the Old World. In: Drake JA, Mooney HA, diCastri F, Groves RH, Kruger FJ, Rejmanek M & Williamson M (eds) Biological Invasions: A Global Perspective, pp 1-30. SCOPE 37, John Wiley & Sons, New York

di Castri F (1990) On invading species and invaded ecosystems: the interplay of historical chance and biological necessity. In: di Castri F, Hansen AJ &

Debussche M (eds) Biological Invasions in Europe and the Mediterranean Basin, pp 3-15. Kluwer Academic Publishers, Dordrecht
Di Geronimo I & Fredj G (1987) Les fonds à *Errina aspera* et *Pachylasma giganteum*. In: Barrier P, Di Geronimo I & Montenat C (eds) Le Détroit de Messine (Italie), évolution tectono-sédimentaire récente (Pliocène et Quaternaire) et environnement actuel. Document et Travaux Institut Géologique Albert de Lapparent *(IGAL)* 11: 243-247
Di Geronimo I (1971) Prima segnalazione sulle costa italiane di *Brachiodontes variabilis* (Krauss). Bollettino delle sedute dell' Accademia Gioenia di scienze naturali in Catania 10: 847-852
Di Natale A (1982) Extra-Mediterranean species of Mollusca along the southern Italian coasts. Malacologia 22: 571-580
Diamond J (1986) Overview: laboratory experiments, field experiments, and natural experiments. In: Diamond J & Case TJ (eds) Community Ecology, pp 3-23. Harper & Row, New York
Diapoulis A, Verlaque M, Huve P & Boudouresque CF (1985) Sur la présence d'une espèce du genre *Sarconema* (Solieriaceae, Rhodophyta) en Grèce. Vie et Milieu 35: 57-59
DIAS (2001) Database on Introductions of Aquatic Species. www.fao.org/fi/statist/fisoft/dias/index.htm (2001-08-14)
Dick JTA & Platvoet D (2000) Invading predatory crustacean *Dikerogammarus villosus* eliminates both native and exotic species. Proceedings of the Royal Society of London B 267: 977-983
Dick JTA, Elwood RW & Irvine DE (1990) Displacement of the native Irish freshwater amphipod *Gammarus duebeni* by the introduced *Gammarus pulex*. Irish Naturalist Journal 23: 313 – 316
Dickerson JE & Robinson JV (1986) The controlled assembly of microcosmic communities: the selective extinction hypothesis. Oecologia 71: 12-17
Dickman M & Zhang F (1999) Mid-ocean exchange of container vessel ballast water. 1. effects of vessel type in the transport of diatoms and dinoflagellates from Manzanillo, Mexico, to Hong Kong, China. Marine Ecology Progress Series 176: 253-262
DIE (1998) Fisheries Statistics. State Institute of Statistics, Prime Ministry of Republic of Turkey, Ankara, 34 pp (in Turkish)
Dietz TH, Lessard D, Silverman H & Lynn JW (1994) Osmoregulation in *Dreissena polymorpha*: the importance of Na, Cl, K and particularly Mg. Biological Bulletin 187: 76-83
Dijkema R (1992) The risk of provoking toxic dinoflagellate blooms in the Dutch coastal waters through immersion of imported bivalves, originating from red tide areas. ICES CM 1992/K:48 Ref. E, 15 pp
Dijkema R (1993) Molluscan fisheries and culture in the Netherlands. NOAA Technical Report NMFS 129: 115-135
Dimitrakis KG (1989) Shells from Greek Seas. 67 pp (in Greek)
Dobson AP & May RM (1986) Patterns of invasions by pathogens and parasites. In: Mooney HA & Drake JA (eds) Ecology of Biological Invasions of North America and Hawaii, pp 56-79. Springer Verlag, New York
Dobson EP & Mackie GL (1998) Increased deposition of organic matter, polychlorinated biphenyls, and cadmium by zebra mussels (*Dreissena polymorpha*) in western Lake Erie. Canadian Journal of Fisheries and Aquatic Sciences 55: 1131-1139
Dobzhansky T (1964) How do the genetic loads affect the fitness of their carriers in *Drosophila* populations? American Naturalist 98: 151-166
Domm SR, McCauley RW, Knott E & Ackerman JD (1993) Physiological and taxonomical separation of two dreissenid mussels in the Laurentian Great Lakes. Canadian Journal of Fisheries and Aquatic Sciences 50: 2294-2297
Doude van Troostwijk WJ (1976) The muskrat (*Ondatra zibethicus* L.) in the Netherlands, its ecological aspects and their consequences for man. Rijksinstituut voor Natuurbeheer, Arnheim, Verhandeling 7: 1-136
Dowidar NM & Ramadan SE (1976) Family Penaeidae from the Mediterranean waters of Egypt. Thalassia Jugoslavica 8: 121-126
Doyle KA, Beers PT & Wilson DW (1996) Quarantine of aquatic animals in Australia. Revue scientifique et technique de l'Office international des épizooties 15: 659-673
Drake IA, Mooney HA, DiCastri F, Groves RH, Kruger FJ, Rejmanek M & Williamson M (eds) (1989) Biological Invasions, A Global Perspective. SCOPE 37, 525 pp. John Wiley & Sons, New York
Drake JA (1989) Communities as assembled structures: do rules govern pattern? Trends in Ecology and Evolution 5: 159-164
Drake JA (1990) The mechanics of community assembly and succession. Journal of theoretical Biology 147: 213-234
Drake JA (1991) Community-assembly mechanics and the structure of an experimental species ensemble. American Naturalist 137: 1-26
Drinkwaard AC (1999) History of cupped oysters in European coastal waters. Aquaculture Europe 24: 7-11
Drinkwaard AC (1999) Introductions and developments of oysters in the North Sea area: a review. Helgoländer Meeresuntersuchungen 52: 301-308
Duda TF (1994) Genetic population structure of the recently introduced Asian clam, *Potamocorbula amurensis*, in San Francisco Bay. Marine Biology 119: 235-241
Dukes JS & Mooney HA (1999) Does global change increase the success of biological invaders? Trends in Ecology and Evolution 14: 135-139
Dumont H (1995) Ecocide in the Caspian Sea. Nature 337: 673–674
Dumont HJ (1998) The Caspian Lake: History, biota, structure and function. Limnology and Oceanography 43: 44-52
Dunstone N (1993) The Mink. T & AD Poyser, London, 232 pp
Duris Z, Jurasz W, Kublakova M, Varecha D (2000) Ponto-Caspian invading predatory water-flea *Cercopagis pengoi* in the Gulf of Gdansk, Poland (Crustacea, Cladocera). Acta Facultatis Rerum Naturalium Universitatis Ostraviensis (Biologica-Ecologica) 192: 51-56
Dyrynda P, Fairall VRR, Occhipinti Ambrogi A & d'Hondt JL (2000) The distribution and origins of *Tricellaria inopinata* (d'Hondt & Occhipinti Ambrogi, 1985), an invasive bryozoan new to the Atlantic. Journal of Natural History 34: 1993-2006
Dzuban NA & Slobodchikov NB (1980) *Hypania invalida* (Grube, 1860) in the Volga reservoirs and the hydrobiological monitoring. Hydrobiology Journal 16: 56-59 (in Russian)
Ebenhard T (1988) Introduced birds and mammals and their ecological effects. Swedish Wildlife Research 13: 1-107
Eckoldt M (ed) (1998) Flüsse und Kanäle. Die Geschichte der deutschen Binnenwasserstraßen. DLV Verlag, Braunschweig
Edgar GJ (1997) Australian Marine Life. Reed Books, Kew, Victoria, Australia, 544 pp
Edlund MB, Taylor CM, Schelske CL & Stoermer EF (2000) *Thalassiosira baltica* (Grunow) Ostenfeld (Bacillariophyta), a new exotic species in the Great Lakes. Canadian Journal of Fisheries and Aquatic Sciences 57: 610-615
Edsman L & Jonsson A (1996) The effect of size, antennal injury, ownership, and ownership duration on fighting success in male signal crayfish, *Pacifastacus leniusculus* (Dana). Nordic Journal of Freshwater Research 72: 80-87
Edwards C (1976) A study in erratic distribution the occurrence of the medusa *Gonionemus* in relation to the distribution of oysters. Advances in Marine Biology 14: 251-284
Edwards M, John AWG, Johns DG & Reid PC (2001) Case history and persistence of the non-indigenous diatom *Coscinodiscus wailesii* in the north-east Atlantic. Journal of the Marine Biological Association of the United Kingdom 81: 207-211
EEA (1999) State and pressures of the marine and coastal Mediterranean environment, Environmental Assessment Series 5: 1-137
Effler SW & Siegfried C (1994) Zebra mussel (*Dreissena polymorpha*) populations in the Seneca River, New York: impact on oxygen resources. Environmental Science & Technology 28: 2216-2221
Ehrhold A, Blanchard M, Auffret JP & Garlan T (1998) Conséquences de la prolifération de la crépidule (*Crepidula fornicata*) sur l'évolution sédimentaire de la baie du Mont-Saint-Michel (Manche, France). Comptes Rendus de l'Académie des Sciences, Sciences de la terre et des planètes 327: 583-588
Ehrlich PR (1986) Which animal will invade? In: Mooney HA & Drake JA (eds) Ecology of Biological Invasions of North America and Hawaii, Ecological Studies, pp 79-95. Springer-Verlag, New York
Ekman S (1916) Systematische und tiergeographische Bemerkungen über einiger glacialmarine Relikte des Kaspischem Meeres. Zoologischer Anzeiger 47
Ekman S (1953) Zoogeography of the Sea. Sidgwick & Jackson Ltd., London
Elbrächter M (1999) Exotic flagellates of coastal North Sea waters. Helgoländer Meeresuntersuchungen 52: 235-242
Eldredge LG (1987) Coral reef alien species. In: Salvat B (ed) Human Impacts on Coral Reefs: Facts and Recommendations, pp 215-228. Antenne Museum E.P.H.E., French Polynesia
Eldredge LG (1994) Introductions of commercially significant aquatic organisms to the Pacific Islands. South Pacific Commission (Noumea, New Caledonia), Inshore Fisheries Research Project, Technical Document 7, 127 pp
Elston RA, Farley CA & Kent ML (1986) Occurrence and significance of bonamiasis in European flat oyster *Ostrea edulis* in North America. Diseases of Aquatic Organisms 2: 49-54
Elton CS (1958) The Ecology of Invasions by Animals and Plants. Methuen, London. 181 pp
Enactment of the Government of the Russian Federation "On signing of the agreement on the international transport corridor "North-South" (2000). Chairman of the Government of the Russian Federation M. M. Kasyanov, Moscow (Sept. 13, 2000)
Engel S (1995) Eurasian watermilfoil as a fishery management tool. Fisheries 20: 20-27
Eno NC (1996) Non-native species in British waters: effects and controls. Aquatic Conservation: Marine and Freshwater Ecosystems 6: 215-228
Eno NC (1998a) Non-native marine species in British waters: ballast water introductions. ICES Co-operative Research Report 224: 43-49
Eno NC (1998b) The introduction to British waters of non-native marine molluscs and their implications for nature conservation interests. Journal of Conchology Special Publication 2: 287-294
Eno NC, Clark RA & Sanderson WG (1997) Non-native Species in British Waters: a Review and Directory. Joint Nature Conservation Committee, Peterborough, 152 pp
Enserink M (1999) Biological invaders sweep in. Science 285: 1834-1836
Environment Agency (1999) Freshwater crayfish in Britain and Ireland. Environment Agency, Bristol
Ermala A (1996) European and Canadian beaver. In: Linden H, Hario M & Wikman M (eds) On the tracks of game, pp 30-33. Finnish Game and Fisheries Research Institute, Helsinki (in Finnish)
Ermala A, Helminen M & Lahti S (1989) Some aspects of the occurrence, abundance and future of the Finnish beaver population. Suomen Riista 35: 108-118 (in Finnish, English summary)
Ermala A, Lahti S & Vikberg P (1999) Beaver populations continue to increase. Metsästäjä 4/99: 28-31 (in Finnish)
ERNAIS (2001) European Research Network on Aquatic Invasive Species. www.zin.ru/projects/invasions/gaas/ernaismn.htm
Essink K (1994) Foreign species in the Wadden Sea, do they cause problems? Wadden Sea Newsletter 1: 9-11
Essink K (1998) Dispersal and development of *Marenzelleria* spp. (Polychaeta, Spionidae) populations in NW Europe and The Netherlands. Helgoländer Meeresuntersuchungen 52: 367-372
Essink K & Dekker R (2000) Invasion ecology of *Marenzelleria* cf. *wireni* (Polychaeta: Spionidae) in the Dutch Wadden Sea. ICES Council Meeting U:04: 13pp

Evans DW & Mathews MA (1999) *Anguillicola crassus* (Nematoda, Dracunculoidea); first documented record of this swimbladder parasite of eels in Ireland. Journal of Fish Biology 55: 665-668

Evlanov IA, Kozlovskiy SV & Rosenberg GS (2000) Current state of fisheries management of Middle Volga. Tolyatti, 24 pp (in Russian)

Eysymont L & Sloniewski B (1969) On entirely incidental findings of mudfish (*Umbra krameri* Walbaum, 1792) in Poland (Osteichthyes, Esociformes). Przeglad Zoologiczny 13(3): 215-219 (in Polish)

Fabbri R & Landi L (1999) Nuove segnalazioni di molluschi, crostacei e pesci esotici in Emilia-Romagna e prima segnalazione di *Corbicula fluminea* (O.F. Muller, 1774) in Italia. Quaderni di Studi e Notizie di Storia Naturale della Romagna 12: 9-20

Fabricius E (1983) Canada goose in Sweden. Statens Naturvårdsverk Report 1678: 1- 85 (in Swedish, English summary)

Fahnenstiel GL, Bridgeman TB, Lang GA, McCormick MJ & Nalepa TF (1995b) Phytoplankton productivity in Saginaw Bay, Lake Huron: effects of zebra mussel (*Dreissena polymorpha*) colonization. Journal of Great Lakes Research 21: 465-475

Fahnenstiel GL, Lang GA, Nalepa TF & Johengen TH (1995a) Effects of zebra mussel (*Dreissena polymorpha*) colonization on water quality parameters in Saginaw Bay, Lake Huron. Journal of Great Lakes Research 21: 435-448

Falniowski A (1987) Hydrobioidea of Poland (Prosobranchia: Gastropoda). Folia Malacologica 1: 1-122

Famà P, Olsen JL, Stam WT & Procaccini G (2000) High levels of intra- and inter-individual polymorphism in the rDNA ITS1 of *Caulerpa racemosa* (Chlorophyta). European Journal of Phycology 35: 349-356

FAO (1997) FAO Database on Introduced Aquatic Species. www.fao.org/waicent/faoinfo/fishery/statist/fisoft/dias/mainpage.htm (2001-08-10)

Farabegoli A, Ferrari I, Manzoni C & Pugnetti A (1989) Prima segnalazione nel Mar Adriatico del copepode calanoide *Acartia tonsa* Dana. Nova Thalassia 10: 207-208

Farnham WF (1980) Studies on aliens in the marine flora of southern England. In: Price JH, Irvine DEG & Farnham WF (eds) The Shore Environment 2: Ecosystems., pp 875-914. Academic Press, London

Farnham WF (1994) Introduction of marine benthic algae into Atlantic European waters. In: Boudouresque CF, Briand H & Nolan C (eds) Introduced Species in European Coastal Waters. CEC Ecosystem Research Report 8: 32-36

Farnham WF (1997) Invasive species on the English Channel and Atlantic coasts. In: Dynamics of Invasive Marine Species: Application to the Expansion of *Caulerpa taxifolia* in the Mediterranean. Actes de Colloques Septembre 1997, Académie des Sciences, pp 15-35. Lavoiser Technique & Documentation, Paris

Farnham WF & Critchley AT (1997) Especes invasives sur les côtes de la Manche et de l'Atlantique. In: Dynamique d'especes marine invasives: application a l'expansion de *Caulerpa taxifolia* en Méditerranée, Proceedings International Seminar Academic Sciences Edition, pp 15-35. TecDoc, Paris

Farnham WF & Irvine LM (1973) The addition of a foliose species of *Grateloupia* in the British marine flora. British Phycological Journal 8: 208-209

Farnham WF & Jones EBG (1974) The eradication of the seaweed *Sargassum muticum* from Britain. Biological Conservation 6: 57-58

Farnham WF, Fletcher RL & Irvine LM (1973) Attached *Sargassum* found in Britain. Nature 243. 231-232

Faunce CH & Lorenz JJ (2000) Reproductive biology of the introduced Mayan cichlid, *Cichlasoma urophthalmus*, within an estuarine mangrove habitat of southern Florida. Environmental Biology of Fishes 58: 215-225

Fauvel P (1923) Un nouveau serpulien d'eau saumâtre, *Mercierella enigmatica*. Bulletin de la Société Zoologique de France 47: 424-430

Fava G & Meggiato L (1995) Genetica biochimica e caratteri morfologici in popolazioni alto adriatiche di *Ruditapes philippinarum* e *R. decussatus*. Technical Report, Ricerche e Sperimentazioni, Programma Integrato Mediterraneo per le zone lagunari dell'Adriatico settentrionale, pp 203-209

Fedyakov VV (1984) White Sea's molluscs. Objective laws of distribution. Leningrad. 128 pp (in Russian)

Feldmann J & Magne F (1954) Excursion phycologique sur la côte des Albères. Notices botaniques et itinéraires commentés. VIII Congrès International Botanique 18: 33-43

Fenyuk VF (1959) The bottom fauna of the Ivankovo and Uglich reservoirs. Proceedings Institute of Biology of Reservoirs 1(4): 139-160 (in Russian)

Filipiak J, Trzebiatowski R & Sadowski J (1993) Determination of optimum protein demand for African catfish (*Clarias gariepinus*) of different size reared in cooling water. Zeszyty Naukowe Akademii Rolniczej w Sczcecinie 156: 23-32 (in Polish)

Findlay S, Pace ML & Fischer DT (1998) Response of heterotropic planktonic bacteria to the zebra mussel invasion of the tidal freshwater Hudson River. Microbial Ecology 36: 131-140

Finenko FA, Romanova ZA & Abolmasova GI (2000) The ctenophore *Beroe ovata* is a recent invader to the Black Sea. Ecologiya morya 50: 21-25 (in Russian)

Finenko GA, Anninsky BE, Romanova ZA, Abolmasova GI & Kideys AE (2001) Chemical composition, respiration and feeding rates of the new alien ctenophore, *Beroe ovata*, in the Black Sea. Hydrobiologia 451: 177-186

Finnish Game and Fisheries Research Institute (1992) Crayfish management and stocking. Finnish Fisheries Research 14: 1-159

Firkins I & Holdich DM (1993) Thermal studies with three species of freshwater crayfish. Freshwater Crayfish 9: 241-248

Fitzpatrick JF (1995) The Eurasian Far-eastern crawfishes: a preliminary review. Freshwater Crayfish 8: 1-11

Fjälling A & Fürst M (1988) The development of a fishery for the crayfish *Pacifastacus leniusculus* in Sweden 1960-86. Freshwater Crayfish 7: 223-230

Flassch JP & Leborgne Y (1992) Introduction in Europe, from 1972 to 1980, of the Japanese Manila clam (*Tapes philippinarum*) and the effects on aquaculture production and natural settlement. ICES Marine Science Symposia 194: 92-96

Flecker AS (1996) Ecosystem engineering by a dominant detritivore in a diverse tropical stream. Ecology 77: 1845-1854

Fleminger A & Kramer SH (1988) Recent introduction of an Asian estuarine copepod, *Pseudodiaptomus marinus* (Copepoda: Calanoida), into southern California embayments. Marine Biology 98: 535-541

Fletcher BL & Farrel P (1999) Introduced brown algae in the North Atlantic, with particular respect to *Undaria pinnatifida* (Harvey) Suringar. Helgoländer Meeresuntersuchungen 52: 259-275

Fletcher RL & Manfredi C (1995) The occurrence of *Undaria pinnatifida* (Phaeophyta: Laminariales) on the south coast of England. Botanica Marina 38: 355-358

Forbes E & Hallegraeff GM (2001) Transport of potentially toxic *Pseudo-nitzschia* diatom species via ballast water. In: John J (ed) Proceedings of the 15th International Diatom Symposium, Perth 1998. Koeltz Scientific Publishers (in press)

Forrest B, Taylor M & Hay C (1997) Foreign marine species in New Zealand: Towards a risk assessment and management model. Cawthron Report 424, Cawthron Institute, Nelson, New Zealand

Forslund P & Larson K (1991) Breeding range expansion of barnacle goose in the Baltic area. Ardea 79: 343-346

Forti A (1928) Nuove segnalazioni di alghe passivamente transportate a traverso al canale di Suez, poi naturalizzate. Nuovo Giornale Botanico Italiano 34: 1443-1451

Relini M & Torchia G (1992) Prima segnalazione di *Caulerpa taxifolia* (Vahl) C. Agardh in acque italiane. Doriana, suppl. Annali Museo civico di Storia naturale "G. Doria" Genova 279: 1-4

Fosså SA & Nilsen AJ (1996) The Modern Coral Reef Aquarium 1. Birgit Schmettkamp Verlag, Berlin, 367 pp

Foster J & Slater FM (1995) A global review of crayfish predation with observations on the possible loss of *Austropotamobius pallipes* in the Welsh Wye due to crayfish plague. Freshwater Crayfish 8: 589-613

Fox HM (1924) The migration of a Red Sea Crab through the Suez Canal. Nature 93: 714

Fox HM (1927) Introduction. Zoological Results of the Cambridge Expedition to the Suez Canal, 1924. Transactions of the Zoological Society of London 22: 1-89

Francis JT, Robillard SR & Marsden JE (1996) Yellow perch management in Lake Michigan: a multi-jurisdictional challenge. Fisheries (Bethesda) 21: 18-20

Franke H-D, Gutow L & Janke M (1999) The recent arrival of the oceanic isopod *Idotea metallica* Bosc off Helgoland (German Bight, North Sea): an indication of a warming trend in the North Sea? Helgoländer Meeresuntersuchungen 52: 347-357

Fredj G (1974) Stockage et exploitation des données en écologie marine. C. Considérations biogéographiques sur le peuplement benthique de la Méditerranée. Mémoires de l'Institut Océanographique (Monaco) 7: 1-88

Fritsch C (1895) Über die Auffindung einer marinen Hydrocharidae im Mittelmeer. Verhandlungen Zoologist-Botanischen Gesellschaftamten Wien 45: 104-106

Fritzsche D (1995) Leistungsanalytische und resistenzökologische Untersuchungen zur Emanzipation des Polychaeten *Marenzelleria viridis* (Verrill 1873) gegenüber den Faktoren Salinität und Sauerstoffpartialdruck. Dissertation. Universität Rostock, 106 pp

Froese R & Pauly D (eds) (2001) FishBase. www.fishbase.org (2001-08-09)

Froglia C & Speranza S (1993) First records of *Dyspanopeus sayi* (Smith, 1869) in the Mediterranean Sea (Crustacea: Decapoda: Xanthidae). Quaderni Istituto Ricerca Pesca Marittima 5: 163-166

Froglia C, Galil B, Noel P & Turkay M (1998) Alien species in the Mediterranean Sea. Crustacea: Decapoda & Stomatopoda. Working document, 35th Congress CIESM, Dubrovnik

Fukuyo Y, Ikegami T & Murase T (1995) Unwanted aquatic organisms in ballast tank. Report of the ballast water by treatment using main engine water cooling circuit and findings of the on-board research. ICES Annual Science Conference, 12 pp

Fuller PL, Nico LG & Williams JD (1999) Nonindigenous fishes introduced into inland waters of the United States. American Fisheries Society Special Publication 27, 613 pp

Fumari G & Scammacca B (1971) Présence de *Lophocladia lallemandii* (Mont.) Schmitz aux environs de Catane (Sicile orientale). Revue d'Algologie 10: 161-163

Fumari G, Cormaci M & Serio D (1999) Catalogue of the benthic marine macroalgae of the Italian cost of the Adriatic Sea. Bocconea 12: 5-214

Fürst M (1977) Introduction of *Pacifastacus leniusculus* (Dana) into Sweden: methods, results and management. Freshwater Crayfish 3: 229-247

Fürst M (1984) Introduction of the North American crayfish (*Pacifastacus leniusculus* Dana) into Sweden. In: European Inland Fisheries Advisory Commission, Symposium on Stock Enhancement in the Management of Freshwater Fisheries 2. EIFAC Technical Paper 42 Suppl 2: 400-404

Fürst M (1995) On the recovery of *Astacus astacus* L. populations after an epizootic of the crayfish plague (*Aphanomyces astaci* Shikora). Freshwater Crayfish 8: 565-576

Fürst M & Boström U (1978) Frequency of visible symptoms of crayfish plague in populations of *Pacifastacus leniusculus* Dana). Information Sötvattenslaboratoriet, Drottningholm 1: 1-24

Galil B (1994) Lessepsian migration - biological invasion of the Mediterranean. In: Boudouresque CF, Briand F & Nolan C (eds) Introduced species in European Coastal Waters. European Commission Ecosystems Research Report 8: 63-66

Galil BS (1986) Red Sea decapods along the Mediterranean coast of Israel: ecology and distribution. In: Dubinsky Z & Steinberger Y (eds) Environmental Quality and Ecosystem Stability, vol. IIIA/B, pp 179-183. Ramat Gan, Bar-Ilan University Press
Galil BS (1993) Lessepsian migration: New findings on the foremost anthropogenic change in the Levant Basin fauna. In: Della Croce NFR (ed) Symposium Mediterranean Seas 2000: 307-318
Galil BS (1994) Lessepsian migration - Biological invasion of the Mediterranean. In: Boudouresque CF, Briand F & Nolan C (eds) Introduced species in European coastal waters. European Commission, Ecosystem Research Report 8: 63-66
Galil BS (1997) Two Lessepsian migrant decapods new to the coast of Israel. Crustaceana 70: 111-114
Galil BS (2000) A sea under siege - alien species in the Mediterranean. Biological Invasions 2: 177-186
Galil BS (2001) Ballast Water Treatment: Arkal Pilot Study. National Institute of Oceanography, Haifa, Israel
Galil BS & Golani D (1990) Two new migrant decapods from the eastern Mediterranean. Crustaceana 58: 229-236
Galil BS & Hülsmann N (1997) Current research activities on ballast water and introduced marine species. Report of the SGBWS, La Tremblade, France, Annex 4: 28-30
Galil BS & Innocenti G (1999) Notes on the population structure of the portunid crab *Charybdis longicollis* Leene, parasitized by the rhizocephalan *Heterosaccus dollfusi* Boschma, off the Mediterranean coast of Israel. Bulletin of Marine Science 64: 233-241
Galil BS & Lewinsohn C (1981) Macrobenthic communities of the Eastern Mediterranean continental shelf. Marine Ecology PSZNI 2: 343-352
Galil BS & Lützen J (1995) Biological observations on *Heterosaccus dollfusi* Boschma (Cirripedia: Rhizocephala), a parasite of *Charybdis longicollis* Leene (Decapoda: Brachyura), a lessepsian migrant to the Mediterranean. Journal of Crustacean Biology 15: 659-670
Galil BS & Lützen J (1998) Jeopardy: host and parasite lessepsian migrants from the Mediterranean coast of Israel. Journal of Natural History 32: 1549-1551
Galil BS, Spanier E & Ferguson WW (1990) The Scyphomedusae of the Mediterranean coast of Israel, including two Lessepsian migrants new to the Mediterranean. Zoologische Mededelingen (Leiden) 64: 95-105
Galkin Yul (1998) Long-term changes in the distribution of molluscs in the Barents Sea related to the climate. Berichte zur Polarforschung 287: 100-143
Gallager SM, Barlow RB & Mann R (1981) Non-invasive, long-term recording of shipworm boring activity. Biological Bulletin of the Marine Biological Laboratory Woods Hole 161: 312
Gandolfi G, Zerunian S, Torricelli P & Marconato A (1991) I pesci delle acque interne italiane. Istituto Poligrafico e Zecca dello Stato, Roma, 616 pp
Garbary DJ, Vandermeulen H & Kim KY (1997) *Codium fragile* ssp. *tomentosoides* (Chlorophyta) invades the Gulf of St Lawrence, Atlantic Canada. Botanica Marina 40: 537-540
Garcia-Meunier P, Martel C, Pigeot J, Chevalier G, Blanchard G, Goulletquer P, Robert S & Sauriau PG (2002) Recent invasion of the Japanese oyster drill, *Ocinebrellus inornatus* (Récluz, 1851) along the French Atlantic coasts. evidence for specific molecular markers. Aquatic Living Resources (in press)
Garfunkel Z & Bartov Y (1977) The tectonics of the Suez Rift. Geological Survey of Israel 71: 1-44
Gargiulo MG, De Masi F & Tripodi G (1992) *Sargassum muticum* (Yendo) Fensholt (Phaeophyta, Fucales) is spreading in the Lagoon of Venice (Northern Adriatic Sea) Giornale Botanico Italiano 126: 259
Garton DW & Haag WR (1991) Heterozygosity, shell length, and metabolism in the European mussel, *Dreissena polymorpha*, from a recently established population in Lake Erie. Comparative Biochemistry and Physiology 99A: 45-48
Garton DW, Kennedy GW, Gannon JE, Mackey SD, Fuller JA & Liebenthal SL (1998) Zebra mussels invade Lake Erie muds. Nature 393: 27-28
Gasiunaite Z & Didziulis V (2000) Ponto-Caspian invader *Cercopagis pengoi* (Ostroumov, 1891) in Lithuanian coastal waters. Sea & Environment (Klaipeda) 4: 97-101
Gasiunas I (1963) The acclimatization of fodder crustaceans (of the Caspian relict type) into the reservoir of Kaunas Hydroelectric power station and the possibility of their transference into other water bodies. Lietuvos TSR MA Darbai, Serija C 1(30): 79-85 (in Russian)
Gasiunas I (1964) Acclimatisation of forage crustaceans into the Kaunas water power plant reservoir and possibility of their migration into other waters of Lithuania. Trudy Akademii Nauk Litovskoi. SSR, Serija B 1 (in Russian)
Gasiunas I (1968) A mysid *Hemimysis anomala* Sars acclimatization in the reservoir of the Kaunas HEPS. Lietuvos TSR MA Darbai, Serija C 3(47): 71-73 (in Russian)
Gasiunas I (1972a) Enrichment of fodder basis of water bodies of Lithuania by acclimatized crustaceans-like organisms from the Caspian Sea complex. In:Virbickas J (ed) On the Breeding of Fish and Crustaceans in the Water Bodies of Lithuania. Vilnius: 57-68 (in Russian)
Gasiunas I (1972b) Bentic fauna of Lithuanian waters. Vilnius, 64 pp (in Lithuanian)
Gasjunas II (1965) On the results of acclimatization of the Caspian complex invertebrates in Lithuanian waters. Zoologicheskij Zhurnal 44: 340-343 (in Russian)
Gasjunas II (1968) Acclimatization of higher Crustacea of the Caspian complex in Lithuanian lakes. Limnologija 3(1): 42-48 (in Russian)
Gebhardt H (1996) Ecological and economic consequences of introduction of exotic wildlife (birds and mammals) in Germany. Wildlife Biology 2: 205-211
Geldiay R & Balik S (1988) Freshwater Fishes of Turkey. The Aegean University Publication, Bornova, Izmir, 519 pp (in Turkish)
Geller JB (1996) Molecular approaches to the study of marine biological invasions. In: Ferraris J & Palumbi S (eds) Molecular approaches to zoology, pp 119-132
Geller JB, Walton ED, Grosholz ED & Ruiz GM (1997) Cryptic invasions of the crab *Carcinus* detected by molecular phylogeography. Molecular Ecology 6: 901-906
Genkal SI (1992) The Atlas of Diatom Algae of the Volga River Plankton. St. Petersburg, 128 pp (in Russian)
Genkal SI & Korneva LG (2002) New findings of diatom algae (Centrophyceae) from the Volga reservoirs. Algologia (in press; in Russian)
Genkal SI & Kuzmin GV (1980) On taxonomy and biology of little known freshwater species *Sceletonema* Grev. (Bacillariophyta). Hydrobiology Journal 16: 25-30 (in Russian)
Genovesi P (2000) Guidelines for eradication of terrestrial vertebrates: a European contribution to the invasive alien species issue. Convention on the Conservation of European Wildlife and Natural Habitats. T-PVS 2000 (65) 31 pp
Gerasimova OV, Kuzmin SA & Oganesyan SA (1996) Studies of the red king crab in the Barents Sea. Ribnoye khozyaystvo 2: 34-36 (in Russian)
Gerbal M, Ben Maiz N & Boudouresque CF (1985) Les peuplements à *Sargassum muticum* de l'étang de Thau: données préliminaures sur la flore algale. Congrès national de Sociétés Savantes 110: 241-254
Gerell R (1967) Dispersal and acclimatization of the mink (*Mustela vison* Schreb.) in Sweden. Viltrevy 4: 1-38
Gerell R (1985) Habitat selection and nest predation in a common eider population in southern Sweden. Ornis Scandinavica 16: 129-139
Gershwin L-A (2001) Systematics and biogeography of the jellyfish *Aurelia labiata* (Cnidaria: Scyphozoa). Biological Bulletin 201: 104-119
Gertmeier R & Romig T (1998) Die Süßwasserfische Europas. Kosmos, Stuttgart. 367 pp
GESAMP (1997) Opportunistic settlers and the problem of the ctenophore *Mnemiopsis leidyi* invasion in the Black Sea. GESAMP Reports and Studies 58: 1-84
Gherardi F & Holdich DM (1999) (eds) Crayfish in Europe as Alien Species. How to Make the Best of a Bad Situation. A.A. Balkema, Rotterdam, Brookfield, 299 pp
Gherardi F, Baldaccini GN, Ercolini P, Barbaresi S, De Luise G, Mazzoni D & Mori M (1999) The situation in Italy. In: Gherardi F & Holdich DM (eds) Crayfish in Europe as Alien Species. How to Make the Best of a Bad Situation, pp 107-128. A.A. Balkema, Rotterdam, Brookfield
Ghisotti F (1971a) Molluschi del genere *Crassostrea* nell'Alto Adriatico. Conchiglie 7: 113-124
Ghisotti F (1971b) Molluschi del genere *Crassostrea* nell'Alto Adriatico (nota aggiuntiva). Conchiglie 7: 152
Ghisotti F (1973) *Scapharca* cfr. *cornea* (Reeve), ospite nuova del Mediterraneo. Conchiglie 9: 68
Giaccone G (1978) Revisione della flora marina del mare Adriatico. Annuario del World Wildlife Fund (suppl.), Parco Marino Miramare (Trieste) 6(19): 1-118
Giaccone G & Di Martino V (1995) Le Caulerpe in Mediterraneo: un ritorno del vecchio bacino Tetide verso il dominio Indo-Pacifico. Biologica marina meditteranea 4: 388-392
Giaccone G & Geraci RM (1989) Biogeografia delle alghe del Mediterraneo. Anales Jardin Botánico Madrid 46: 27-34
Giacobbe MG, Oliva F, Laferla R, Puglisi A, Crisafi E & Maimone G (1995) Potentially toxic dinoflagellates in Mediterranean waters (Sicily) and related hydrobiological conditions. Aquatic Microbial Ecology 9: 63-68
Giacobbe S & Spano' N (2001) *Pilumnus inermis* (Decapoda, Brachyura) in the Straits of Messina and the southern Tyrrhenian Sea (Mediterranean Sea): distribution and some aspects of its ecology. Crustaceana 74: 659-672
Gibson FA (1961) Gaffkaemia in stored lobsters. ICES Shellfish Committee 58
Giesen S (1998) Taxonomische Stellung und populationsgenetische Struktur von Arten der gebietsfremden Amphipoden-Gattung *Dikerogammarus* (Stebbing 1899) in Main, Main-Donau-Kanal und Donau.
Giordani-Soika A (1951) Il *Neptunus pelagicus* (L.) nell'alto Adriatico. Natura 42: 18-20
Gittenberger E, Janssen AW, Kuijper WJ, Kuiper JGJ, Meijer T, Van der Velde G & De Vries JN (1998) De Nederlandse zoetwatermollusken. Recente en fossiele weekdieren uit zoet en brak water. Nationaal Natuurhistorisch Museum Naturalis, KNNV uitgeverij & Stichting EIS. Nederlandse Fauna 2: 1-288
Giusti F & Oppi E (1972) *Dreissena polymorpha* (Pallas) nuovamente in Italia (Bivalvia, Dreissenidae). Memorie Museo civico di Storia naturale Verona 20: 45-49
Glémarec M (1978) Distribution bathymétrique et latitudinale des bivalves du Golfe de Gascogne. Haliotis 9: 23-32
Glémarec M (1979) Les fluctuations temporelles des peuplements benthiques liées aux fluctuations climatiques. Oceanologica Acta 2: 365-371
Glémarec M (1991) Bathymetric and latitudinal distribution of Onuphid Polychaeta in the Bay of Biscay (Northeast Atlantic). Ophelia Suppl. 5: 547-554
GloBallast (2001) Workshop Report. A "possible options" paper. 1st International Ballast Water Treatment Standards Workshop, IMO London 28-30 March 2001
Golani D (1994) Niche separation between colonizing and indigenous goatfish (Mullidae) along the Mediterranean coast of Israel. Journal of Fish Biology 45: 503-513
Golani D & Ben Tuvia A (1995) Lessepsian migration and the Mediterranean fisheries of Israel. In: Armantrout NB (ed) Condition of the World's Aquatic Habitats. Proceedings of the World Fisheries Congress. Theme 1, pp 279-289. Science Publishers, Lebanon, New Hampshire

Gollasch S (1996) Untersuchungen des Arteintrages durch den internationalen Schiffsverkehr unter besonderer Berücksichtigung nichtheimischer Arten. PhD Thesis, Dr. Kovac, Hamburg, 314 pp

Gollasch S (1997) Removal of barriers to the effective implementation of ballast water control and management measures in developing countries. Report for the International Maritime Organization, London, 188 pp

Gollasch S (1999) *Eriocheir sinensis*. In: Gollasch S, Minchin D, Rosenthal H & Voigt M (eds) Exotics Across the Ocean. Case Histories on Introduced Species, pp 55-60. Logos Verlag, Berlin

Gollasch S (1999) The Asian decapod *Hemigrapsus penicillatus* (de Haan, 1835) (Grapsidae, Decapoda) introduced in European waters: status quo and future perspective. Helgoländer Meeresuntersuchungen 52: 359-336

Gollasch S & Leppäkoski E (1999) Initial Risk Assessment of Alien Species in Nordic Coastal Waters. Nord 1999:8. Nordic Council of Ministers, Copenhagen. 244 pp

Gollasch S & Mecke R (1996) Introduced organisms. In: Lozan JL, Lampe R, Matthäus W, Rachor E, Rumohr H & von Westernhagen H (eds) Warning Signals from the Baltic Sea, pp 146-150. Parey Buchverlag, Berlin (in German)

Gollasch S & Riemann-Zümeck K (1996) Transoceanic dispersal of benthic macrofauna: *Haliplanella luciae* (Verrill, 1898) (Anthozoa, Actiniaria) found on a ship's hull in a shipyard dock in Hamburg Harbour, Germany. Helgoländer Meeresuntersuchungen 50: 253-258

Gollasch S, Lenz J, Andres H-G & Dammer M (2000a) Survival of tropical ballast water organisms during a cruise from the Indian Ocean to the North Sea. Journal of Plankton Research 22: 923-937

Gollasch S, Macdonald E, Belson S, Botnen H, Christensen JT, Hamer JP, Houvenaghel G, Jelmert A, Lucas I, Masson D, McCollin T, Olenin S, Persson A, Wetsteyn LPMJ, Wallentinus I & Wittling T. Life in ballast tanks - results of European research initiatives (in review)

Gollasch S, Rosenthal H, Botnen H, Laing I, Leppäkoski E, Macdonald E, Minchin D, Nauke M, Olenin S, Utting S, Voigt M & Wallentinus I (2000b) Fluctuations rates of zooplankton taxa in ballast water during short-term and long-term ocean-going voyages. International Revue of Hydrobiology 85: 597-608

Gómez Garreta A (2001) Flora phycologica iberica 1. Fucales. Universidad de Murcia. Spain, 192 pp

Gómez Garreta A, Gallardo T, Ribera MA, Cormaci M, Furnari G, Giaccone G & Boudouresque CF (2001) Checklist of Mediterranean Seaweeds. III Rhodophyceae Rabenh. 1. Ceramiales Oltm. Botanica Marina 44: 425-460

Gómez Garreta A, Ribera MA & Seoane Camba JA (1979) Nuevas citas para la flora algológica de Baleares. Acta Botanica Malacitana 5: 29-38

Gómez O & Martinez-Iglesias JC (1990) Reciente hallazgo de la especie indopacifica *Charybdis helleri* (A.Milne Edwards, 1867) (Crustacea: Decapoda: Portunidae) en aguas cubanas. Caribbean Journal of Science 26(1-2): 70-72

Gomoiu M-T & Porumb I (1969) *Mya arenaria* L. - bivalve recently penetrated into the Black Sea. Rev. Roum. Zool. 14(3): 199-202

Gomoiu M-T & Skolka M (1996) Changements recents dans la biodiversité de la Mer Noire dus aux immigrants. In: Danube Delta - Black Sea System under Global Changes Impact, pp 49-65. Geo-Eco-Marina, Romanian Centre of Marine Geology and Geoecology 1: 34-47

Gorokhova E (1998) Zooplankton spatial distribution and potential predation by invertebrate zooplanktivores. 2nd BASYS Annual Science Conference, 23-25.09.1998, Stockholm, Sweden

Gorokhova E, Aladin N & Dumont HJ (2000) Further expansion of the genus *Cercopagis* (Crustacea, Branchiopoda, Onychopoda) in the Baltic Sea, with notes on the taxa present and their ecology. Hydrobiologia 429: 207-218

Gosliner TM (1995) Introduction and spread of *Philine auriformis* (Gastropoda: Opisthobranchia) from New Zealand to San Francisco Bay and Bodega Harbor. Marine Biology 122: 249-255

Gosling M (1989) Extinction to order. New Scientist 1654: 44-49

Gosling M & Baker S (1989) The eradication of muskrats and coypus from Britain. Biological Journal of the Linnean Society 38: 39-51

Gosling M, Baker SJ & Clarke CN (1988) An attempt to remove coypus (*Myocastor coypus*) from wetland habitat in East Anglia. Journal of Applied Ecology 25: 49-62

Gould AA (1841) Report on the Invertebrata of Massachusetts, Comprising the Mollusca, Crustacea, Annelida, and Radiata. Folsom, Wells, and Thurston, Cambridge. 373 pp

Goulletquer P & Héral M (1996) History, present conditions and future of the Molluscan fisheries of North America and Europe. Marine Molluscan production trends in France: from fisheries to aquaculture. Marine Fisheries Review, NOAA Technical Report NMFS 129 : 137-164

Goulletquer P, Soletchnik P, Le Moine O, Razet D, Geairon P, Faury N & Taillade S (1998) Summer mortality of the Pacific cupped oyster *Crassostrea gigas* in the Bay of Marennes Oleron (France). ICES Statutory Meeting 1998, Mariculture Committee CM 1998/CC:14, 21 pp

Grabow K, Eggers TO & A Martens (1998) *Dikerogammarus villosus* Sovinsky (Crustacea: Amphipoda) in navigable canals and rivers of northern Germany. Lauterbornia 33: 103-107 (in German)

Grall J, Chauvaud L, Thouzeau G, Fifas S, Glémarec M & Paulet YM (1996) Distribution de *Pecten maximus* (L.) et de ses principaux compétiteurs et prédateurs potentiels en rade de Brest. Comptes Rendus de l'Académie des Sciences, Sciences de la vie 319: 931-937

Granéli E (1987) Dinoflagellate blooms. Occurrence, causes and consequences in the marine environment. A review. Statens Naturvårdsverk Rapport 3293: 1-133 (in Swedish, English summary)

Grannum RK, Murfet NB, Ritz DA & Turner E (1996) The distribution and impact of the exotic seastar, *Asterias amurensis* (Lütken) in Tasmania. In: The Introduced Northern Pacific Seastar *Asterias amurensis* in Tasmania, pp 53-138. Australian Nature Conservation Agency

Gray AJ, Crawley MJ & Edwards PJ (1987) Colonization, Succession and Stability. Blackwell Scientific, Oxford, UK

Gray AJD, Marshall DF & Raybould AF (1991) A century of evolution in *Spartina anglica*. Advances in Ecological Research 21: 1-62

Greze VN (1951) Baikalian elements of fauna as acclimatization fund. Proceedings of All-Union Hydrobiological Society 2: 221-226 (in Russian)

Griffiths CL, Hockey PAR, Van Erkom Shurink C & Le Roux PJ (1992) Marine invasive aliens on South African shores: implications for community structure and trophic functioning. South African Journal of Marine Science 12: 713-722

Griffiths RW (1993) Effects of zebra mussels (*Dreissena polymorpha*) on the benthic fauna of Lake St. Clair. In: Nalepa TF & Schloesser DW (eds) Zebra Mussels: Biology, Impacts and Control, pp 415-437. Lewis Publishers, Ann Arbor, Michigan

Grigelis A (1980) Bentic glacial relict crustaceans found in glacial lakes of Lithuanian SSR during 1952-1978. Lietuvos TSR MA darbai, Serija C 2(90): 69-78 (in Russian)

Grigorovich IA & MacIsaac HJ (1999) First record of *Corophium mucronatum* Sars (Crustacea: Amphipoda) in the Great lakes. Journal of Great Lakes Research 25: 401-405

Grigorovich IA, Pashkova OV, Gromova YF & Van Overdijk CDA (1998) *Bythotrephes longimanus* in the Commonwealth of Independent States: variability, distribution and ecology. Hydrobiologia 379: 183-198

Grim J (1971) Tiefenverteilung der Dreikantmuschel *Dreissena polymorpha* (Pallas.) im Bodensee. AWE, Wasser Abwasser 112: 437-441

Grimsrud AS & Throndsen J (2000) Sand flagellates in the Oslofjord. Norske Havfoskeres Forening, Årsmøte 2000, Oslo (in Norwegian)

Grishin AN & Zolotarev PN (1988) Biological characteristics and stocks of *Rapana thomasiana* in the Black Sea. In: Proceedings of All-Union Conference "Resources and biological basis of the rational use of commercial invertebrates", pp 56-57. Vladivistok (in Russian)

Grizel H (1997) Les maladies des mollusques bivalves: risques et prévention. Revue Scientifique & Technique Office International des Epizooties 16: 161-171

Grizel H & Héral M (1991) Introduction into France of the Japanese oyster (*Crassostrea gigas*). Journal du Conseil International pour l'Exploration de la Mer 47: 399-403

Grodowitz MJ, Center TD, Cofrancesco AF & Freedman JE (1997) Release and establishment of *Hydrellia balciunasi* (Diptera: Ephydridae) for the biological control of the submersed aquatic plant *Hydrilla verticillata* (Hydrocharitaceae) in the United States. Biological Control 9: 15-23

Grofit E (1987) The trawl fishery in 1986. Fisheries and Fishbreeding in Israel 20: 36-42

Groombridge B (1993) 1994 IUCN Red List of Threatened Animals. World Conservation Monitoring Centre, Cambridge, UK

Groot SJ de (1985) Introductions of non-indigenous fish species for release and culture in the Netherlands. Aquaculture 46: 237-257

Groppali R (1999) Prima segnalazione di naturalizzazione del cobite di stagno *Misgurnus fossilis* (L.) in Italia. Atti 7° Convegno Nazionale Associazione Italiana Ittiologi Acque Dolci, Stazione della Carnia (UD), pp 227-228

Grosholz ED & Ruiz GM (1996) Predicting the impact of introduced species: lessons from the multiple invasions of the European green crab. Biological Conservation 78: 59-66

Grosholz ED, Ruiz GM, Dean CA, Shirley KA, Maron JL & Connors PG (2000) The impacts of a nonindigenous marine predator on multiple trophic levels. Ecology 81: 1206-1224

Gross EM, Meyer H & Schilling G (1996) Release and ecological impact of algicidal hydrolysable polyphenols in *Myriophyllum spicatum*. Phytochemistry 41: 133-138

Gruet Y & Baudet J (1997) Les introductions d'espèces d'invertébrés marins. In : Dauvin JC (ed) Les biocénoses marines et littorales françaises des côtes Atlantique, Manche et Mer du Nord : synthèse, menaces et perspectives. Laboratoire de Biologie des Invertébrés Marins et Malacologie-Service du Patrimoine naturel, Paris, pp 242-250

Gruet Y, Heral M & Robert J-M (1976) Premières observations sur l'introduction de la faune associée au naissain d'huîtres Japonaises *Crassostrea gigas* (Thunberg), importé sur la côte Atlantique Française. Cahiers de Biologie Marine 17: 173-184

Gruner H-E, Moritz M & Dunger W (1993) Wirbellose Tiere. In: Gruner H-E (ed) 4. Teil: Arthropoda (ohne Insecta) 1, Gustav Fischer, 4 Aufl., 1279 pp

Gruszka P (1991) *Marenzelleria viridis* (Verrill, 1873) (Polychaeta: Spionidae) - a new component of shallow water benthic community in the southern Baltic. Acta Ichthyologica et Piscatoria 21 Suppl: 57-65

Gruszka P (1999) The River Odra estuary as a gateway for alien species immigration to the Baltic Sea basin. Acta hydrochimica et hydrobiologica 27: 374-382

Gruszka P (2000) *Dikerogammarus haemobaphes* (Eichwald, 1841) (Crustacea: Amphipoda) - a new species in the Oder River. Materialy Zjazdowe, XVIII Zjazd Hydrobiologow Polskich, Bialystok, p 87 (Abstract; in Polish)

Gruvel A (1936) Contribution à l'étude de la Bionomie générale et de l'exploitation de la faune du Canal de Suez. Mémoires présentés l'Institut d'Égypte 29: 1-255

Guan R & Wiles PR (1996) Growth, density and biomass of crayfish, *Pacifastacus leniusculus*, in a British lowland river. Aquatic Living Resources 9: 265-272

Guan RZ (1994) Burrowing behaviour of signal crayfish, *Pacifastacus leniusculus* (Dana), in the river Great Ouse, England. Freshwater Forum 4. 155-168

Gücü AC & Bingel F (1994) Trawlable species assemblages on the continental shelf on the Northeastern Levant Sea (Mediterranean) with an emphasis on Lesseptian migration. Acta Adriatica 35: 83-100

Guerriero A, Meinesz A, D'Ambrosia M & Pietra F (1992) Isolation of toxic and potentially toxic sesqui- and monoterpenes from tropical green seaweed

Caulerpa taxifolia which has invaded the region of Cap Martin and Monaco. Helvetica Chimica Acta 75: 689
Guiry M (2001) A revised checklist of the seaweeds (benthic marine algae) of Britain and Ireland. http://seaweed.ucg-ie/check-list/Check-list.html (2001-07-24)
Guiry MD & Nic Dhonncha E (2001) AlgaeBase. www.algaebase.com (2001-06-08)
Gulliksen B (1971) A new record of *Gonionemus vertens* Agassiz in Norway. Det Kongelige Norske Videnskabers Selskab Skrifter 12: 4 pp
Gulliksen B, Palerud R, Brattegard T & Sneli J-A (eds) (1999) Distribution of marine benthic macroorganisms at Svalbard (including Bear Island) and Jan Mayen. Research Report 1999-4. Directorate for Nature Management, Trondheim, Norway. 148 pp
Gusinskaja SL & Zhdanova GA (1978) The distribution of northern and Caspian planktonic crustaceans in the River Dnieper reservoirs. Gidrobiologicheskij Zhurnal 14: 25-27 (in Russian)
Gutiérrez-Yurrita PJ, Martinez JM, Bravo-Utrera MÁ, Montes C, Ilhéu M & Bernardo JM (1999) The status of crayfish populations in Spain and Portugal. In: Gherardi F & Holdich DM (eds) Crayfish in Europe as Alien Species. How to Make the Best of a Bad Situation, pp 161-192. A.A. Balkema, Rotterdam, Brookfield
Haag WR & Garton DW (1995) Variation in genotype frequencies during the life history of the bivalve, *Dreissena polymorpha*. Evolution 49: 1284-1288
Haag WR, Berg DR, Garton DW & Farris JL (1993) Reduced survival and fitness in native bivalves in response to fouling by the introduced zebra mussel (*Dreissena polymorpha*) in Western Lake Erie. Canadian Journal of Fisheries and Aquatic Sciences 50: 13-19
Haahtela I (1963) Some new observations and remarks on the occurrence of the mitten crab, *Eriocheir sinensis* Milne-Edwards (Crustacea, Decapoda), in Finland. Aquilo Ser. Zoologica 1: 9-16
Haas F (1937) Bivalvia. Teil II. In: Bronn HG (ed) Klassen und Ordungen des Tierreichs. Leipzig, 678 pp
Haas G (1993) Untersuchung der Steinschüttungsfauna des hessischen Rheingebietes. Diplom-arbeit Albert-Ludwigs-Universität Freiburg
Haas G (2001) Entwicklung der Makro-Invertebratengemeinschaft im hessischen Rhein- und Untermainabschnitt in den Jahren 1993 bis 1999. Dissertation Johann Wolfgang Goethe-Universität Frankfurt am Main
Haas G & Streit B (1998) Die Einwanderung von *Dikerogammarus* sp. in den Rhein. Tagungs-bericht, Erweiterte Zusammenfassung der Jahrestagung der Deutschen Gesellschaft für Limnologie (DGL), Frankfurt am Main 1997, pp 154-158
Haas T (1969) Superfamilia Unionacea. Das Tierreich, Berlin
Habsburgo-Lorena AS (1979) Present situation of exotic species of crayfish introduced into Spanish continental waters. Freshwater Crayfish 4: 175-184
Habsburgo-Lorena AS (1983) Socioeconomic aspects of the crawfish industry in Spain. Freshwater Crayfish 5: 552-554
Hagemeijer WJM & Blair MJ (eds) (1997) The EBCC Atlas of European Breeding Birds: Their Distribution and Abundance. T & A Poyser, London. 903 pp
Hägg R (1948) Some notes on Quaternary and recent cirripedes. Geologiska föreningens i Stockholm förhandlingar 70: 346-349
Hagmeier A (1941) Die intensive Nutzung des nordfriesischen Wattenmeeres durch Austern- und Muschelkultur. Zeitschrift für Fischerei und deren Hilfswissenschaften 39:105-165
Hagmeier A & Kändler R (1927) Neue Untersuchungen im nordfriesischen Wattenmeer und auf den fiskalischen Austernbänken. Wissenschaftliche Meeresuntersuchungen (Abteilung Helgoland) 16: 1-90
Hahn A (1956) Fighting the shipworm. Die Küste 5: 49-71 (in German)
Hajdu S, Edler L, Olenina I & Witek B (2000) Spreading and establishment of the potentially toxic dinoflagellate *Prorocentrum minimum* in the Baltic Sea. International Review of Hydrobiology 85: 561-575
Hakenkamp CC & Palmer MA (1999) Introduced bivalves in freshwater ecosystems: the impact of *Corbicula* on organic matter dynamics in a sandy stream. Oecologia 119: 445-451
Halim Y (1990) On the potential migration of Indo-Pacific plankton through the Suez Canal. Bulletin de l'Institut Océanograhique, Monaco Special 7: 11-27
Halim Y & Messih MA (1999) The Suez Canal revisited. Changes in the Ascidian fauna of the canal. International conference on "Oceanography of the Eastern Mediterranean and Black Sea" 23-26/2/1999. EU Report, p 367
Hallegraeff GM (1993) A review of harmful algal blooms and their apparent global increase. Phycologia 32: 79-99
Hallegraeff GM (1998) Transport of toxic dinoflagellates via ships' ballast water: bioeconomic risk assessment and efficacy of possible ballast water management strategies. Marine Ecology Progress Series 168: 297-309
Hallegraeff GM & Bloch CJ (1991) Transport of toxic dinoflagellate cysts via ship's ballast water. Marine Pollution Bulletin 22: 27-30
Hallegraeff GM & Bolch CJ (1992) Transport of diatom and dinoflagellate resting spores in ships' ballast water: implications for plankton biogeography and aquaculture. Journal of Plankton Research 14: 1067-1084
Hallegraeff GM, Bolch CJ, Bryan J & Koerbin B (1990) Microalgal spores in ships' ballast water: A danger to aquaculture. In: Granéli E, Sundström B, Edler L & Anderson DM (eds) Toxic Marine Phytoplankton. Proceedings of the Fourth International Conference onToxic Marine Phytoplankton (26-30 June 1989), pp 475-480. Elsevier, New York
Hallegraeff GM, Valentine JP, Marshall JA & Bolch CJ (1997) Temperature tolerances of toxic dinoflagellate cysts: application to the treatment of ships' ballast water. Aquatic Ecology 31: 47-52
Hamel G (1926) Quelques algues rares ou nouvelles pour la flore méditerranéenne. Bulletin Museum Histoire national Paris 32: 6, 420
Hamel G (1930) Les Caulerpes méditerranéennes. Revue Algologique 5: 229-230
Hamel N (1852) Über das Project: Austern, wie auch Hummern, Seekrebse, Krabben und Miesmuscheln im Finnischen Meerbusen zu ziehen. Bulletin de la Classe Physico-Mathématique de L'Académie Impériale des Sciences de Saint-Pétersbourg 10(18-20): 308-318
Hamer JP, Lucas IAN & McCollin TA (2001) Harmful dinoflagellate resting cysts in ships' ballast tank sediments; potential for introduction into English and Welsh waters. Phycologia: 40: 246-255
Hamer JP, McCollin TA & Lucas IAN (1998) Transport of dinoflagellate cysts in ships' ballast tanks. Norges Teknisk-naturvitenskapelige Universitet Vitenskapsmuseet. Rapport Botanisk serie 1: 1-53
Hamer JP, McCollin TA & Lucas IAN (1999) Viability of decapod larvae in ships' ballast water. Marine Pollution Bulletin 38(Suppl 8): 646-647
Hamer JP, McCollin TA & Lucas IAN (2000) Dinoflagellate cysts in ballast tank sediments: Between tank variability. Marine Pollution Bulletin 40: 731-733
Hamon D (1996) Peuplements benthiques des fonds meubles: distribution de la faune benthique - distribution de la crépidule (*Crepidula fornicata*). In: Augris C & Hamon D (eds) Atlas thématique de l'environnement marin en baie de Saint-Brieuc. Brest, Ifremer, pp 46-71
Hänfling B & Brandl R (1998) Genetic differentiation of the bullhead *Cottus gobio* L. across watersheds in Central Europe: evidence for two taxa. Heredity 80: 110-117
Hansen G, Daugbjerg N & Henriksen P (2000) Comparative study of *Gymnodinium mikimotoi* and *Gymnodinium aureolum*, comb. nov. (=*Gyrodinium aureolum*) based on morphology, pigment composition, and molecular data. Journal of Phycololgy 36: 394-410
Harache Y & Novotny AJ (1976) Coho salmon farming in France. Marine Fisheries Review 38(8): 1-8
Hardeng G & Viker M (1997) New find of the Chinese mitten crab *Eriocheir sinensis* near Fredrikstad - yet another example of dispersion of introduced species. Natur i Østfold 16 (1-2): 47 (in Norwegian)
Hario M, Komu R, Muuronen P & Selin K (1986) Population trends among archipelago birds in Söderskär bird sanctuary 1963-86. Suomen Riista 33: 79-90 (in Finnish, English summary)
Härkönen S (1999) Forest damage caused by the Canadian beaver (*Castor canadensis*) in South Savo, Finland. Silva Fennica 33: 247-259
Harmelin-Vivien M, Harmelin JG & Francour P (1996) A 3-year study of the littoral fish fauna of sites colonized by *Caulerpa taxifolia* in the NW Mediterranean (Menton, France). In: Ribera MA, Ballesteros E, Boudouresque CF, Gómez A & Gravez V (eds) Second International Workshop on *Caulerpa taxifolia*, pp 391-397. Publicacions Universitat Barcelona
Haroun RJ & Prud'homme van Reine WF (1993) A biogeographical study of *Laurencia* and *Hypnea* species of the Macaronesian region. Courier Forschungsinstitut Senckenberg 159: 119-125
Harris ASD, Medlin LK, Lewis J & Jones KJ (1995) *Thalassiosira* species (Bacillariophyceae) from a Scottish sea-loch. European Journal of Botany 30: 117-131.
Harris LG & Tyrrell MC (2001) Changing community states in the Gulf of Maine: synergism between invaders, overfishing and climate change. Biological Invasions 3: 9-21
Harrison PG & Bigley RE (1982) The recent introduction of the seagrass *Zostera japonica* Aschers. and Graebn. to the Pacific coast of North America. Canadian Journal of Fisheries and Aquatic Sciences 39: 1642-1648
Hartl DL & Clark AG (1989) Principles of Population Genetics. Sinauer Ass., Inc. Sunderland, Massachusetts
Hartman G (1995) Patterns of spread of a reintroduced beaver *Castor fiber* population in Sweden. Wildlife Biology 1: 97-103
Hartog C den & van der Velde G (1987) Invasions by plants and animals into coastal, brackish and fresh water of the Netherlands. Proceeding Koninklijke Nederlands Academie Wetenshap C 90: 31-37
Hartog C den (1987) "Wasting disease" and other dynamic phenomena in *Zostera* beds. Aquatic Botany 27: 3-14
Hartog Den C, Van den Brink FWB & Van der Velde G (1992) Why was the invasion of the river Rhine by *Corophium curvispinum* and *Corbicula* species so successful? Journal of Natural History 26: 1121-1129
Harvey M, Gilbert M, Gauthier D & Reid DM (1999) A preliminary assessment of risks for the ballast water mediated introduction of nonindigenous marine organisms in the Estuary and Gulf of St. Lawrence. Canadian Technical Report of Fisheries and Aquatic Sciences 2268: 1-56
Hasle GR (1983) *Thalassiosira punctigera* (Castr.) comb. nov., a widely distributed marine planktonic diatom. Nordic Journal of Botany 3: 593-608
Hasle GR (1990) Diatoms in the Oslofjord and the Skagerrak. Species new to the area: Immigrants or overlooked in the past. Blyttia 48: 33-38 (in Norwegian, English abstract)
Haslin C & Pellegrini M (2001) Culture medium composition for optimal thallus regeneration in the red alga *Asparagopsis armata* Harvey (Rhodophyta, Bonnemaisoniaceae). Botanica Marina 44: 23-30
Hastein T & Gladhaug O (1973) The occurrence of the crayfish plaque and attempts to prevent further spread of the disease. In: Abrahamsson S (ed) Freshwater Crayfish, pp 182-184. First International Symposium on Freshwater Crayfish, Austria
Haugom GP, Kelley AE & Behrens HL (2001) The model group concept - a methodological approach for setting standards and comparing ballast water treatment performance. 1st International Conference on Ballast Water Management. Singapore November 2001 (Poster)
Hay CH (1990) The dispersal od sporophytes of *Undaria pinnatifida* by coastal shipping in New Zealand, and implications for further dispersal of *Undaria* in France. British Phycological Journal 25: 301-313
Hay CH & Tanis D (1998) Mid-ocean ballast water exchange: procedures, effectiveness, and verification. Cawthron Institute, Halifax, New Zealand Report 468: 1-66
Hayek L-AC & Buzas MA (1997) Surveying Natural Populations. Columbia University Press, New York
Hayes KR (1997) A review of ecological risk assessment methodologies. Centre for Research on Introduced Marine Pests, CSIRO Marine Research, Hobart, Tasmania, Australia. Technical Report 13, 116 pp

Hayes KR (1998) Ecological risk assessment for ballast water introductions: a suggested approach. ICES Journal of Marine Science 55: 201-212

Hayes KR & Hewitt CL (1998) Risk assessment framework for ballast water introductions. CRIMP Technical Report 14

Hayes KR & Hewitt CL (2000) Quantitative biological risk assessment of the ballast water vector: an Australian approach. In: Pederson J (ed) Marine Bioinvasions. Proceedings of the First National Conference, January 24-27, 1999, pp 66-80. Massachusetts Institute of Technology, Cambridge, MA

Hayes KR & Sliwa C (in review) Identifying Australia's next marine pests: A deductive approach. Marine and Freshwater Research

Hayes TM, Hilliard RW & Raaymakers S (1996) Ballast water risk assessment: 12 Queensland ports. EcoPorts Monograph Series 3, Ports Corporation of Queensland, Brisbane, Australia

Haynes JM, Stewart TW & Cook GE (1999) Benthic macroinvertebrate communities in Southwestern Lake Ontario following invasion of *Dreissena*: continuing change. Journal of Great Lakes Research 25: 828-838

Hayward PJ & Ryland JS (1990) The Marine Fauna of the British Isles and NW Europe. Clarendon Press, Oxford, 996 pp

Healand M & Beall E (1997) Studies on interspecific competition between the juveniles of coho salmon *Oncorhynchus kisutch*, and Atlantic salmon, *Salmo salar*, in freshwater. Bulletin Francais de la Pêche de la Pisciculture 344-5: 241-252

Hebert PDN, Muncaster BW & Mackie GL (1989) Ecological and genetic studies on *Dreissena polymorpha* (Pallas): a new mollusc in the Great Lakes. Canadian Journal of Fisheries and Aquatic Sciences 46: 405-409

Hebert PDN, Wilson CC & Lazar R (1991) demography and ecological impacts of the invading mollusc *Dreissena polymorpha*. Canadian Journal of Zoology 59: 405-409

Hedgpeth JW (1980) The problem of introduced species in management and mitigation. Helgoländer Meeresuntersuchungen 33: 662-673

Hedgpeth JW (1993) Foreign invaders. Science 261(Suppl 6): 34-35

Heilscher S (2000) Sustainability of the brown seaweed *Laminaria saccharina* for cultivation in the effluents of an Atlantic salmon ongrowth farm. A field study in Southeastern Norway. Institut für Biologie, Albert-Ludwigs-Universität, Freiburg im Breisgau. Diplomarbeit, 109 pp

Hein W (1911) A warning when obtaining crayfish for stocking. Allgemeine Fischerei Zeitung 36(8): 176-177 (in German)

Heinis F, Akkerman I, Essink K, Colijn F & Latuhihin MJ (1995) Biologische monitoring zoute rijkswateren, 1990-1993. Rijksinstituut voor Kust en Zee, rapport RIKZ-95.059

HELCOM (2001) The Baltic Marine Environment Protection Commission - Helsinki Commission. www.helcom.fi (2001-07-23)

Hendriks AJ, Pieters J & de Boer J (1998) Accumulation of metals, polycyclic (halogenated) aromatic hydrocarbons, and biocides in zebra mussel and eel from the Rhine and Meuse Rivers. Environmental Toxicology and Chemistry 17: 1885-1898

Hénocque Y (1997) Use of mechanical and chemical treatment methods, advantages and risks. In: Dynamics of Invasive Marine Species: Application to the Expansion of *Caulerpa taxifolia* in the Mediterranean. Actes de Colloques Septembre 1997, Académie des Sciences, pp 271-278. Lavoiser Technique & Documentation, Paris (in French)

Henriksen AZ, Sandvik A, Bergh K, Reiersen R & Natås O (1993) Serious case of gastroenteritis caused by domestic infection by *Vibrio cholerae* non-O1. Tidsskrift for den Norske Laegeforening 113: 3017-3018 (in Norwegian)

Henttonen P & Huner JV (1999) The introduction of alien species of crayfish in Europe: a historical introduction. In: Gherardi F & Holdich DM (eds) Crayfish in Europe as Alien Species. How to Make the Best of a Bad Situation, pp 13-22. A.A. Balkema, Rotterdam, Brookfield

Henttonen P, Huner JV & Lindquist OV (1984) Country report for Finland for the intersessional period of the European Inland Fisheries Advisory Commission 1982-1984. 48 pp

Heral M (1990) Traditional oyster culture in France In: Barnabé G (ed) Aquaculture, pp 342-387. Ellis Horwood, London

Herbst V (1982) Amphipods in salt polluted surface waters in lower Saxony. Gewässer und Abwässer 68/69: 35-40 (in German)

Herhaus KF (1978) First records of *Gammarus tigrinus* Sexton, 1939, and *Chaetogammarus ischnus* (Stebbing, 1906) (Crustacea, Amphipoda, Gammaridae) in the drainage basin of Ems and their range extension history. Natur und Heimat 38(3): 99-102 (in German)

Hersteinsson P (1999) Methods to eradicate the American mink (*Mustela vison*) in Iceland Workshop on the control and eradication of non-native terrestrial vertebrates, Malta, pp 25-30. Environmental Encounters, Council of Europe Publishing

Hessland I (1946) On the Quaternary *Mya* period in Europe. Arkiv för Zoologi 37A(8): 1-51

Hewitt CL (2002) The distribution and diversity of tropical Australian marine bioinvasions. Pacific Science 56: 213-222

Hewitt CL & Martin RB (1996) Port surveys for introduced marine species - Background considerations and sampling protocols. CRIMP Technical Report 4, Division of Fisheries, CSIRO, Hobart

Hewitt CL & Martin RB (2001) Revised protocols for baseline port surveys for introduced marine species - design considerations, sampling protocols and taxonomic sufficiency. CRIMP Technical Report 22. CSIRO Marine Research, Hobart

Hewitt CL, Campbell ML, ThresherRE & Martin RB (eds) (1999) Marine Biological Invasions of Port Phillip Bay, Victoria. CSIRO Center for Research on Introduced Marine Pests (CRIMP), Technical Report 20, 344 pp

Hewitt G (1993) Postglacial distribution and species substructure: lessons from pollen, insects and hybrid zones. In: Lees DR & Edwards D (eds) Evolutionary Patterns and Processes, pp 97-123. Linnean Society of London, Academic Press, London

Hilden O & Hario M (1993) The changing bird fauna of the archipelago. Forssa, 317 pp (in Finnish)

Hilge V (1980) Experiments made with the culture of the channel catfish (*Ictalurus punctatus*) in Germany. In: Experiences with new Native and Non-native Species in Aquaculture of the Federal Republic of Germany, pp 99-104. Arbeiten des Deutschen Fischerei Verbandes 30 (in German)

Hilliard RW, Walker S & Raaymakers S (1997) Ballast Water Risk assessment - 12 Queensland Ports: Stage 5 Report - Executive Summary & Synthesis of Stages 1-4. EcoPorts Monograph Series 14, Ports Corporation of Queensland, Brisbane,

Hindar K, Fleming IA, Jonsson N, Breistein J, Sægrov H, Karlsbakk E, Gammelsæter M & Dønnum BO (1996) Rainbow trout in Norway: occurrence, reproduction and establishment. NINA Oppdrags-melding 454: 1-32

Hines AH & Ruiz GM (2000) Biological invasions of cold-water coastal ecosystems: Ballast-mediated introductions in Port Valdez/Prince William Sound, Alaska. Final Report, Regional Citizens Advisory Council of Prince William Sound, 400 pp

His E (1977a) Observations préliminaires sur la présence de *Mytilicola orientalis* Mori (1935) chez *Crassostrea gigas* Thunberg dans le Bassin d'Arcachon. Bulletins de la Société Géologique de Normandie et des Amis du Muséum du Havre 64(2): 7-8

His E (1977b) Mytilicolides et myicolides parasites des lamellibranches d'interet commercial du Bassin d'Arcachon. Haliotis 8: 99-102

Hobbs HH (1988) Crayfish distribution, adaptive radiation and evolution. In: Holdich DM & Lowery RS (eds) Freshwater Crayfish: Biology, Management and Exploitation, pp 52-82. Croom Helm (Chapman & Hall), London

Hobbs III HH, Jass JP & Huner JV (1989) A review of global crayfish introductions with particular emphasis on two North American species (Decapoda, Cambaridae). Crustaceana 56: 299-316

Hobbs RJ (1989) The nature and effects of disturbance relative to invasions. In: Drake JA, Mooney HA, diCastri F, Groves RH, Kruger FJ, Rejmanek M & Williamson M (eds) Biological Invasions: A Global Perspective, SCOPE 37, pp 389-405. John Wiley & Sons, New York

Hobbs RJ & Huenneke LF (1992) Disturbance, diversity, and invasion: implications for conservation. Conservation Biology 6: 324-337

Hoek PPC (1882) Stand van het onderzoek de oester en haar cultures betreffende en het einde van het eerste onderzoeksjaar. V. Weekdieren. Tijdschrift Nederlandsche Dierkundige Vereeniging 6: 56-58

Hoek PPC (1911) Rapport over de schelpdiervisscherij en schelpdierteelt in de Noordelijke Zuiderzee. Verslag van den Staat der Nederlandsche Zeevisscherijen over 1910. Den Haag, 162 pp

Hoeksema BW (1983) Excavation patterns and spiculae dimensions of the boring sponge *Cliona celata* from the SW Netherlands. Senckenbergiana maritima 15: 55-85

Hoffmann M (1958) Die Bisamratte. Akademishe Verlagsgesellschaft Geest & Portig K.-G., Leipzig, 267 pp

Hogger JB (1988) Ecology, population biology and behaviour. In: Holdich DM & Lowery RS (eds) Freshwater Crayfish: Biology, Management and Exploitation, pp 114-144. Croom Helm (Chapman & Hall), London

Holcik J (1984) Some considerations on the role of introduced species of fish in the management of inland fisheries. EIFAC Techical Paper 42 Suppl. 2: 488-495

Holcík J (1991) Fish introductions in Europe with particular reference to its central and eastern part. Canadian Journal of Fisheries and Aquatic Sciences 48(Suppl.1): 13-23

Holdgate MW (1986) Summary and conclusions: characteristics and consequences of biological invasions. Philosophical Transactions of the Royal Society of London B 314: 733-742

Holdich DM (1988) The Sture Abrahamsson Lecture. The dangers of introducing alien animals with particular reference to crayfish. Freshwater Crayfish 7: xv-xxx

Holdich DM (1999) The negative effects of established crayfish introductions. In: Gherardi F & Holdich DM (eds) Crayfish in Europe as Alien Species. How to Make the Best of a Bad Situation, pp 31-47. A.A. Balkema, Rotterdam, Brookfield

Holdich DM & Gherardi F (1999) Native and alien crayfish in Europe: an introduction. In: Gherardi F & Holdich DM (1999) (eds) Crayfish in Europe as Alien Species. How to Make the Best of a Bad Situation, pp 3-9. A.A. Balkema, Rotterdam, Brookfield

Holdich DM & Lowery RS (1988) (eds) Freshwater Crayfish. Biology, Management and Exploitation. Croom Helm (Chapman & Hall), London, 498 pp

Holdich DM, Ackefors H, Gherardi F, Rogers WD & Skurdal J (1999a) Native and alien crayfish in Europe: some conclusions. In: Gherardi F & Holdich DM (eds) Crayfish in Europe as Alien Species. How to Make the Best of a Bad Situation, pp 281-292. A.A. Balkema, Rotterdam, Brookfield

Holdich DM, Gydemo R & Rogers WD (1999c) A review of possible methods for controlling nuisance populations of alien crayfish. In: Gherardi F & Holdich DM (eds) Crayfish in Europe as Alien Species. How to Make the Best of a Bad Situation, pp 245-270. A.A. Balkema, Rotterdam, Brookfield

Holdich DM, Harlogly MM & Firkins I (1997) Salinity adaptations of crayfish in British waters with particular reference to *Austropotamobius pallipes, Astacus leptodactylus* and *Pacifastacus leniusculus*. Estuarine, Coastal and Shelf Science 44: 147-154

Holdich DM, Rogers WD & Reynolds JD (1999b) Native and alien crayfish in the British Isles. In: Gherardi F & Holdich DM (eds) Crayfish in Europe as Alien Species. How to Make the Best of a Bad Situation, pp 221-235. A.A. Balkema, Rotterdam, Brookfield

Holland BS (2000) Genetics of marine bioinvasions. Hydrobiologia 420: 63-71

Holland RE (1993) Changes in planktonic diatoms and water transparency in Hatchery Bay, Bass Island area, western Lake Erie since the establishment of the zebra mussel. Journal of Great Lakes Research 19: 617-624

Holm L, Pancho JV, Herberger JP & Plucknett (1979) A Geographical Atlas of World Weeds. John Wiley & Sons, New York, 391 pp

Holmes JMC & Minchin D (1995) Two exotic copepods imported into Ireland with the Pacific oyster *Crassostrea gigas* (Thunberg). Irish Naturalists' Journal 25: 17-20

Holthuis LB (1961) Report on a collection of Crustacea Decapoda and Stomatopoda from Turkey and the Balkans. Zoologische Verhandelingen (Leiden) 47: 1-67

Holthuis LB & Gottlieb E (1955) The occurrence of the American Blue Crab, *Callinectes sapidus* Rath. in Israel waters. Bulletin of the Research Council of Israel 5B(2): 155-156

Holthuis LB & Gottlieb E (1958) An annotated list of the Decapod Crustacea of the Mediterranean coast of Israel, with an appendix listing the Decapoda of the Eastern Mediterranean. Bulletin of the Research Council of Israel 7B: 1-126

Honsell G (1993) First report of *Alexandrium minutum* in Northern Adriatic waters. In: Smayda TJ & Shimizu Y (eds) Toxic phytoplancton blooms in the sea, pp 127-132. Elsevier Science Publisher

Honsell G, Boni L, Cabrini M & Pompei M (1992) Toxic or potentially toxic dinoflagellate species from the Adriatic Sea. Science of the Total Environment 92: 107-114

Hooker JD & Harvey WH (1845) Algae antarcticae, being characters and descriptions of the hitherto unpublished species of algae, discovered in Lord Auckland's Group. London Journal of Botany 4: 249-276, 293-298

Hopkins CCE (2001) Actual and potential effects of introduced marine organisms in Norwegian waters, including Svalbard. Research Report 2001-1, 53 pp. Norwegian Directorate for Nature Management. http://english.dirnat.no/wbch3.exe?d=3539&toppgiff=

Horgan MJ & Mills EL (1997) Clearance rates and filtering activity of zebra mussel (*Dreissena polymorpha*): implications for freshwater lakes. Canadian Journal of Fisheries and Aquatic Sciences 54: 249-255

Hornatkiewicz-Zbik A (1999) Predatory cladocerans in the Vistula Firth. Komunikaty Rybackie 5: 10-11 (in Polish)

Hornatkiewicz-Zbik A (2000) *Cercopagis pengoi* (Ostroumov, 1891) in the Vistula Firth. Materialy Zjazdowe, XVIII Zjazd Hydrobiologow Polskich, Bialystok, p 97 (Abstract; in Polish)

Horvath TG, Lamberti GA, Lodge DM & Perry WL (1996) Zebra mussel dispersal in lake-stream systems: source-sink dynamics? Journal of the North American Benthological Society 15: 564-575

Horvitz CC (1997) The impact of natural disturbances. In: Simberloff D, Schmitz DC & Brown TC (eds) Strangers in Paradise: Impact and Management of Nonindigenous Species in Florida, pp 63-74. Island Press, Washington DC

Houghton DR & Millar RH (1950) Spread of *Styela mammiculata* Carlisle. Nature 185: 862

Houghton JT, Callander BA & Varney SK (eds) (1992) Climate Change 1992: The supplementary report to the IPCC scientific assessment. Intergovernmental Panel on Climate Change. Cambridge University Press, Cambridge, 200 pp

Howard AE (1994) The possibility of long distance transmission of *Bonamia* by fouling on boat hulls. Bulletin of the European Association of Fish Pathologists 14: 211-212

Hsu KJ, Montadert L, Bernoulli D, Cita MB, Erikson A, Garrison, RE, Kidd RB, Melieres F, Muller C & Write R (1978) History of the Mediterranean salinity crisis, Initial Report of DSDP 42/1: 1053-1078

Hughes B, Criado J, Dalany S, Gallo-Orsi U, Green A, Grussu M, Perennou C & Torres JA (1999) The status of the ruddy duck *(Oxyura jamaicensis)* in the western Palearctic: towards an action plan for eradication. Report by the Wildfowl & Wetlands Trust to the Council of Europe

Huiskes AHL (1979) Biological flora of the British Isles: *Ammophila arenaria* (L.) Link. Journal of Ecology 67: 363-382

Huisman EA (1981) Integration of hatchery, cage, and pond culture of common carp (*Cyprinus carpio* L.) and grass carp (*Ctenopharyngodon idella* Val.) in the Netherlands. Bio-Engineering Symposium for Fish Culture (FCS Publ. 1), pp 266-273

Huisman J & Weissing FJ (1999) Biodiversity of plankton by species oscillations and chaos. Nature 402: 407-410

Huisman J & Weissing FJ (2001) Biological conditions for oscillations and chaos generated by multispecies competition. Ecology 82: 2682-2695

Hülsmann N & Galil BS (2001) The effects of freshwater flushing on marine heterotrophic protists - implications for ballast water management. Marine Pollution Bulletin 42: 1082-1086

Humphrey JD (1995) Australian Quarantine Policies and Practices for Aquatic Animals and their Products: A review for the Scientific Working Party on Aquatic Animal Quarantine. Part 1: Review and Annexes. Bureau of Resource Sciences, Canberra, Australia. 264 pp

Humphries SE, Groves RH & Mitchell DS (1991) Plant invasions of Australian ecosystems: A status review and management directions. Australian National Parks and Wildlife Service, Canberra, Australia

Huner JV (1988) *Procambarus* in North America and elsewhere. In: Holdich DM & Lowery RS (eds) Freshwater Crayfish: Biology, Management and Exploitation, pp 239-262. Croom Helm (Chapman & Hall), London

Huner JV (1994) (ed) Freshwater crayfish aquaculture in North America, Europe and Australia. Families Astacidae, Cambaridae, and Parastacidae. Binghamton, New York: Haworth, 312 pp

Huner JV & Barr JE (1991) Red Swamp Crayfish: Biology and Exploitation. The Louisiana Sea Grant College Program, Lousiana State University, Baton Rouge, Louisiana, USA, 128 pp

Ibrahim KM, Nichols RA & Hewitt GM (1996) Spatial patterns of genetic variation generated by different forms of dispersal during range expansion. Heredity 77: 282-291

ICES (1982) Status (1980) of introductions of non-indigenous marine species to North Atlantic waters. ICES Cooperative Research Report 116: 1-87

ICES (1995) ICES Code of Practice on the Introductions and Transfers of Marine Organisms 1994. ICES Cooperative Research Report 204. 17 pp

ICES (1997) Report of the ICES Study group on marine biocontrol species; 13 pp

ICES/IOC/IMO SGBOSV (2001) Report of the 2001 Meeting of the ICES/IOC/IMO Study Group on Ballast Water and Other Ship Vectors, Barcelona, Spain, 44 pp

IMO (1995) International Maritime Organisation Guidelines on Ballast Water

IMO (1999) 21st Assembly Session, Resolution A.895(21) Anti-fouling systems used on ships. International Maritime Organization London, IMO News, 4/1999, 7 pp (www.imo.org/imo/news/499)

IMO (2001) GloBallast Programme Website, International Maritime Organization, http://globallast.imo.org

IMO MEPC 38/13/2 (1996) International Maritime Organisation, Marine Environmental Protection Committee, Report -Brazil

IMO MEPC39/7 (1997) Unwanted Aquatic Organisms in Ballast Water. Report of the Working Group on Ballast Water convened during MEPC 38

IMO MEPC39/7/4 (1997) Unwanted Aquatic Organisms in Ballast Water. Outcome of the 40th Session of the Sub-Committee on Ship Design and Equipment

IMO MEPC39/7/8 (1997) Unwanted Aquatic Organisms in Ballast Water. Draft MSC/MEPC Circular on Guidance on Safety Aspects of Ballast Water Exchange at Sea

Ingle RW (1963) Crustacea Stomatopoda from the Red Sea and the Gulf of Aden. In: Contributions to the knowledge of the Red Sea 26. Bulletin Sea Fisheries Research Station (Haifa) 33: 1-69

Insacco G & Zava B (1999) First record of the saddled snake eel *Pisodonophis semicinctus* (Richardson, 1848) in Italian waters (Osteichthyes, Ophichthidae). Atti Società italiana di Scienze naturali, Museo civico di Storia naturale Milano 140: 141-169

Ioffe CI (1973) Acclimatization resources of invertebrates in SSSR. Izvestija Gosudarstvennogo Nauchnogo Instituta Okeanografii i Rybnogo Khozjajstva 84: 18-68 (in Russian)

Ioffe TI (1968) Review of fulfilled works on acclimatization of the forage for fish invertebrates in reservoirs. News of the State Research Institute of Lake and River Fisheries 67: 7-29 (in Russian)

IUCN (1999) IUCN guidelines for the prevention of biodiversity loss due to biological invasions. Newsletter of the Species Survival Commission IUCN - The World Conservation Union 31: 28-42

IUCN (2000) IUCN Guidelines for the Prevention of Biodiversity Loss Caused by Alien Invasive Species. Species Survival Commission (SSC). Invasive Species Specialist Group (ISSG), Auckland, New Zealand, 14 pp (www.issg.org)

Ivanov VM & Semenova NN (2000) Parasitological consequences of animal introduction. Russian Journal of Ecology 31: 281-283

Ivanov VP, Kamakin AM, Ushivtzev VB, Shiganiva T, Zhukova O, Aladin N, Wilson SI, Harbison RG & Dumont H (2000) Invasion of the Caspian Sea by the comb jellyfish *Mnemiopsis leidyi* (Ctenophora). Biological Invasions 2: 255–258

Ivanova MN (1982) Populations variability of freshwater smelt. Rybinsk, 144 pp (in Russian)

Iyengar MOP & Ramanathan KR (1954) On a new species of *Halicystis* from South India. Journal of the Indian Botanical Society 33: 446-452

Jackson JBC & Johnson KG (2001) Measuring past biodiversity. Science 293: 2401-2404

Jacobsen N (1979) Aquaria Plants in Colour. Almqvist & Wiksell, Stockholm, 157 pp

Jalas J & Suominen J (eds) (1980) Atlas Florae Europeae. Distribution of Vascular Plants in Europe. 5. Chenopodiaceae to Basellaceae. The Committee for Mapping the Flora of Europe and Societas Biologica Fennica Vanamo, Helsinki, 119 pp

Jalas J & Suominen J (eds) (1994) Atlas Florae Europeae. Distribution of Vascular Plants in Europe. 10. Cruciferae (*Sisymbrium* to *Aubretia*). The Committee for Mapping the Flora of Europe and Societas Biologica Fennica Vanamo, Helsinki, 224 pp

Jalas J, Suominen J, Lampinen R & Kurtto A (eds) (1999) Atlas Florae Europeae. Distribution of Vascular Plants in Europe. 12. Resedaceae to Plantanaceae. The Committee for Mapping the Flora of Europe and Societas Biologica Fennica Vanamo, Helsinki, 250 pp

Janes R (1998a) Growth and survival of *Azolla filiculoides* in Britain. II. Sexual reproduction. New Phytologist 138: 377-384

Janes R (1998b) Growth and survival of *Azolla filiculoides* in Britain. I. Vegetative reproduction. New Phytologist 138: 367-375

Jansson K (1994) Alien species in the marine environment. Introductions to the Baltic Sea and the Swedish west coast. Swedish Environmental Protection Agency Report 4357: 1-68

Jansson K (2000) The marine environment. In: Weidema IR (ed) Introduced Species in the Nordic Countries, pp 43-86. Nord 2000:13. Nordic Council of Ministers, Copenhagen

Jarocki J & Demianowicz A (1931) On the occurrence of a Ponto-Caspian amphipod *Chaetogammarus tenellus* (G.O.Sars) in the Vistula River. Bulletin International de l'Academie Polonaise des Sciences et des Lettres (Cl. Math. Nat. B(II), 1931): 513-530 (in German)

Jaubert JM, Chisholm JRM, Ducrot D Ripley HT, Roy L & Passeron-Seitre G (1999) No deleterious alterations in *Posidonia* beds in the Bay of Menton (France) eight years after *Caulerpa taxifolia* colonization. Journal of Phycology 35: 1113-1119

Jazdzewski K (1975) Morphology, taxonomy and occurrence of gammarids from the genus *Gammarus* Fabr. and *Chaetogammarus* Mart. (Crustacea, Amphipoda) in Poland. Acta Universitatis Lodziensis, Lodz, 185 pp (in Polish)

Jazdzewski K (1980) Range extensions of some gammaridean species in European inland waters caused by human activity. Crustaceana Supplement 6: 84-107

Jazdzewski K & Konopacka A (1988) Notes on the gammaridean Amphipoda of the Dniester River basin and Eastern Carpathians. Crustaceana, Leiden, Suppl 13: 78-89

Jazdzewski K & Konopacka A (1990) New, interesting locality of the Ponto-Caspian gammarid *Echinogammarus ischnus* (Stebbing, 1898) (Crustacea, Amphipoda) in Poland. Przeglad Zoologiczny 34: 101-111 (in Polish)

Jazdzewski K & Konopacka A (1995) Malacostraca (excl. Oniscoidea). Katalog Fauny Polski, PWN, Warszawa, 13 (1), 165 pp (in Polish)

Jazdzewski K & Konopacka A (2000) Immigration history and present distribution of alien crustaceans in Polish waters. In: von Vaupel Klein JC & Schram FR (eds) The biodiversity crisis and Crustacea. Proceedings of 4th International Crustacean Congress, vol. 2, Brill, Leiden, Crustacean Issues 12: 55-64

Jeffries HP (1962) Succession of two *Acartia* species in estuaries. Limnology and Oceanography 7: 354-364

Jelmert A (1999) Process and apparatus for the control of undesirable organisms in a water system. Patent PCT NO 00/00385, filed 15 Nov. 2000

Jelmert A (1999) Testing the effectiveness of an integrated hydro cyclone water treatment. Optimarin Marketing AS, Norway (www.optimarin.com/test031.htm)

Jenkins R (2001) Churchill. A Biography. Farrar, Straus and Giroux, New York, 912 pp

Jennings DP & Williams JD (1992) Factors influencing the distribution of blackchin tilapia *Sarotherodon melanotheron* (Osteichthys: Cichlidae) in the Indian River system, Florida. Northeast Gulf Science 12: 111-117

JNCC (2000) The effectiveness of species protection legislation - advice to government. JNCC, Peterborough

JNCC (2001) Biological translocations: a conservation policy for Britain. Consultation draft. Joint Nature Conservation Committee, Peterborough

Joenje W (1987) The SCOPE programme on the ecology of biological invasions. an account of the Dutch contribution. In: Joenje W, Bakker K & Vlijm L (eds) The Ecology of Biological Invasions, pp 3-13. Proceedings of the Koninklijke Nederlandse Akadamie Wetenschappen C 90

Johannsson OE, Dermott R, Graham DM, Dahl JA, Millard ES, Myles DD & LeBlanc J (2000) Benthic and pelagic secondary production in Lake Erie after the invasion of *Dreissena* spp. with implications for fish production. Journal of Great Lakes Research 26: 31-54

Johnsen BO & Jensen AJ (1991) The *Gyrodactylus* story in Norway. Aquaculture 98: 289-302

Johnsen BO, Møkkelgjerd PI & Jensen AJ (1999) The parasite *Gyrodactylus salaris* on salmon in Norwegian watercourses, status report at the beginning of the year 2000. NINA Fagrapport 039: 1-39 (in Norwegian)

Johnson LE & Padilla DK (1996) Geographic spread of exotic species: ecological lessons and opportunities from the invasion of the zebra mussel *Dreissena polymorpha*. Biological Conservation 78: 23-33

Jónasson J, Stefansson SE, Gudnason A & Steinarsson A (1999) Genetic variation for survival and shell length of cultured red abalone (*Haliotus rufescens*) in Iceland. Journal of Shellfish Research 18: 621-625

Jones CG, Lawton JH & Shachak M (1994) Organisms as ecosystem engineers. Oikos 69: 373-386

Jones CG, Lawton JH & Shachak M (1997) Positive and negative effects of organisms as physical ecosystem engineers. Ecology 78: 1946-1957

Jones DS (1992) A review of Australian fouling barnacles. Asian Marine Biology 9: 89-100

Jonsell B (ed) (2001) Flora Nordica 2. Chenopodiaceae to Fumariaceae. The Bergius Foundation, The Royal Academy of Sciences, Stockholm, 430 pp

Jonsson A (1995) Life history differences between crayfish *Astacus astacus* and *Pacifastacus leniusculus* in embryonic and juvenile development, laboratory experiences. Freshwater Crayfish 8: 170-178

Josefsson M (1999) Introduction of alien species in Swedish lakes and rivers. Swedish Environmental Protection Agency Report 4941: 1-112 (in Swedish, English summary)

Jousson O, Pawlowski J, Zaninetti L, Meinesz A & Boudouresque CF (1998) Molecular evidence for the aquarium origin of the green alga *Caulerpa taxifolia* introduced to the Mediterranean Sea. Marine Ecology Progress Series 172: 275-280

Jousson O, Pawlowski J, Zaninetti L, Zechman FW, Din F, Di Guiseppe G, Woodfield R, Millar A & Meinesz A (2000) Invasive alga reaches California. Nature 408: 157-158

Jude DJ & Janssen J & Crawford G (1995) Ecology, distribution and impact of the newly introduced round and tubenose gobies on the biota of the St.Clair and Detroit Rivers. In: Munawar M, Edsall T & Leach J (eds) The Lake Huron ecosystem, ecology, fisheries and management, pp 447-460. SPB Academic Publishing

Jude DJ, Reider RH & Smith GR (1992) Establishment of Gobiidae in the Great lakes basin. Canadian Journal of Fisheries and Aquatic Sciences 49: 416-421

Jurd KM, Rogers DJ, Blunden G & Mclellan D (1995) Anticoagulant properties of sulfated polysaccharides and a proteoglycan from *Codium fragile* ssp. *atlanticum*. Journal of Applied Phycology 7: 339-345

Jürss K, Röhner M & Bastrop R (1999) Enzyme activities and allozyme polymorphism in two genetic types (or sibling species) of the genus *Marenzelleria* (Polychaeta: Spionidae) in Europe. Marine Biology 135: 489-496

Kailola PJ (2000) Development of an alert list for non-native freshwater fishes. Final report to Environment Australia, Patricia Kailola, Consultant (Newnham, Tasmania)

Kaneva-Abadzhieva V (1965) On the amphipod fauna of the Danube river at the Bulgarian bank. Izviestija na Zoologicheskija Institut i Muzej BAN 35: 165-178 (in Bulgarian)

Karaman G & Barnard JL (1979) Classificatory revisions in gammaridean Amphipoda (Crustacea). Part 1. Proceedings of the Biological Society of Washington 92(1): 106-165

Karaman G (1974) Crustacea, Amphipoda. Catalogus Faunae Jugoslaviae 3(3): 1-42

Karaman G (1977) Contribution to the knowledge of the Amphipoda. Revision of the *Echinogammarus* genera-complex (fam. Gammaridae). Arhiv Bioloshkih Nauka, Beograd 27(1-2): 69-93

Karaminas L (2000) An investigation of ballast water management methods with particular emphasis on the risks of the sequential method. Lloyd's Register of Shipping, 19 pp

Karatayev AY (1983) Ecology of *Dreissena polymorpha* Pallas and its effects on macrozoobenthos of the thermal power plant's cooling reservoir. PhD Thesis, Zoology Institute of Academy of Science Belorussian SSR, Minsk, Belarus 153 pp (in Russian)

Karatayev AY (1988) Ecology of Macroinvertebrates of Cooling Reservoirs of Belarus. Minsk, VINITI Press, Paper 875-B88 Dep. 178 pp (in Russian)

Karatayev AY (1992) The structure and function of benthic and periphyton's communities in cooling water reservoirs. PhD Thesis, Zoology Institute of the Academy of Science Republic Belarus, Minsk, Belarus 287 pp (in Russian)

Karatayev AY & Burlakova LE (1992) Changes in tropic structure of macrozoobenthos of an eutrophic lake, after invasion of *Dreissena polymorpha*. Biologiya Vnutrennikh Vod 93: 67-71 (in Russian)

Karataev AY & Burlakova LE (1995a) The role of *Dreissena* in lake ecosystems. Russian Journal of Ecology 26: 207-211

Karatayev AY & Burlakova LE (1995b) Present and further patterns in *Dreissena* population development in the Narochanskaya lakes system. Vestsi Akademii Navuk Belarusi. Seriya Biyologichnikh Navuk 3: 95-98 (in Belorussian)

Karatayev AY & Lyakhnovich VP (1990) Effect of *Dreissena polymorpha* Pallas on benthic crustaceans (*Gammarus lacustris* Sars, *Pallasea quadrispinosa* Sars and *Asellus aquaticus* L.) in Lukomskoe lake. In: Khmeleva NN et al. (eds) Species Within their Range: Biology, Ecology, and Productivity of Aquatic Invertebrates, pp 123-125. Navuka i Tekhnika Press, Minsk (in Russian)

Karatayev AY, Burlakova LE & Padilla DK (1997) The effects of *Dreissena polymorpha* (Pallas) invasion on aquatic communities in Eastern Europe. Journal of Shellfish Research 16: 187-203

Karatayev AY, Burlakova LE & Padilla DK (1998) Physical factors that limit the distribution and abundance of *Dreissena polymorpha* (Pall.). Journal of Shellfish Research 17: 1219-1235

Karatayev AY, Lyakhnovich VP, Afanasiev SA, Burlakova LE, Zakutsky VP, Lyakhov SM, Miroshnichenko MP, Moroz TG, Nekrasova MY, Skalskaya IA, Kharchenko TG & Protasov AA (1994a) The place of species in ecosystem. In: Starobogatov JI (ed) Freshwater Zebra Mussel *Dreissena polymorpha* (Pall.) (Bivalvia, Dreissenidae). Systematics, Ecology, Practical Meaning, pp 180-195. Nauka Press, Moscow (in Russian)

Karatayev AY, Mikheev VP, Afanasiev SA, Kirpichenko MY, Protasov AA, Shevtsova LV & Kharchenko TG (1994b) Practical uses and control on human made structures. In: Starobogatov JI (ed) Freshwater Zebra Mussel *Dreissena polymorpha* (Pall.) (Bivalvia, Dreissenidae). Systematics, Ecology, Practical Meaning, pp 206-221. Nauka Press, Moscow (in Russian)

Karatayev AY, Tishchikov GM & Karatayeva IV (1983) The specific community of benthic animals associated with *Dreissena polymorpha* Pallas. Biologiya Vnutrennikh Vod 61: 18-21 (in Russian)

Kareiva P (1996) Developing a predictive ecology for non-indigenous species and ecological invasions. Ecology 77: 1651-1652

Karlson B & Kuylenstierna M (2001) Checklist of phytoplankton in the Skagerrak-Kattegat (including heterotrophic protists). www.marbot.gu.se/SSS/SSShome.htm (2001-03-10)

Karlsson J (1995) Inventory of marine macroalgae in Østfold 1994: The area Heia-Torbjørnskjær. Report of the Tjärnö Marine Biological Laboratory, 21pp (in Norwegian)

Karlsson J & Loo L-O (1999) On the distribution and the continuous expansion of the Japanese seaweed - *Sargassum muticum* - in Sweden. Botanica Marina 42: 285-294

Karlsson J, Kuylenstierna M & Åberg P (1992) Contribution to the seaweed flora of Sweden: New or otherwise interesting records from the west coast. Acta Phytogeographica Suecica 78: 49-63

Karpevich AF (1952) Osmoregulatory capacities of *Dreissena* from the Northern Caspian Sea. Reports on biology, taxonomy and nutrition of fishes, chemistry of the Sea and net conservation 1: 127-131 (in Russian)

Karpevich AF (1955) Some data on morphological evolution in bivalve molluscs. Zoologicheskii Zhurnal 34 (1): 46-67 (in Russian)

Karpevich AF (1975) The theory and practice of acclimatization of aquatic organisms. Pischevaya promyshlennost, Moscow, 427 pp (in Russian)

Karpevich AF, Agapov VS & Magomedov GM (1991) Acclimatization and farming of the salmonid fishes-installers. 208 pp (in Russian)

Kauhala K (1996a) Distributional history of the American mink (*Mustela vison*) in Finland with special reference to the trends in otter (*Lutra lutra*). Annales Zoologici Fennici 33: 283-291

Kauhala K (1996b) Introduced carnivores in Europe with special reference to central and northern Europe. Wildlife Biology 2: 197-204

Kavaliauskiene J (1996) Algae of Lithuanian Lakes. Vilnius, 173 pp (in Lithuanian)

Kearney MP & Morton BS (1970) The distribution of *Dreissena polymorpha* (Pallas) in Britain. Journal of Conchology 27: 97-100

Keister JE, Houde ED & Breitburg DL (2000) Effects of bottom-layer hypoxia on abundances and depth distributions of organisms in Patuxent River, Chesapeake Bay. Marine Ecology Progress Series 205: 43-59

Keith P & Allardi J (1997) An assessment of freshwater fish introductions to France. Bulletin Francais de la Pêche de la Pisciculture 344-5: 181-191 (in French)

Keith P & Allardi J (1998) The introduced freshwater fish of France: status, impacts and management, pp 153-166. In: Cowx IG (ed) Stocking and

Introduction of Fish. Fishing News Books. MPG Books Ltd., Bodmin, Cornwall
Kelleher B, Bergers PJM, Van den Brink FWB, Giller PS, Van der Velde G & Bij de Vaate A (1998) Effects of exotic amphipod invasions on fish diet in the Lower Rhine. Archiv für Hydrobiologie 143: 363-382
Kelleher B, Van der Velde G, Giller PS & Bij de Vaate A (1999) Dominant role of exotic invertebrates, mainly Crustacea, in diets of fish in the lower Rhine River. In: Von Vaupel Klein JK & Scram FR (eds). The biodiversity crisis and Crustacea: Proceedings of the Fourth International crustacean congress, pp 35-46. A.A. Balkema, Rotterdam, Brookfield
Kelleher B, Van der Velde G, Giller PS & Bij de Vaate A (2000) Dominant role of exotic invertebrates, mainly Crustacea, in diets of fish in the lower Rhine. Crustacean Issues 12: 35-46
Keller C (1882) Die Fauna im Suez-Kanal und die Diffusion der Mediterranean und Erythräischen Thierwelt. Neue Denkschriften der Allgemeiner schweizerischen Gesellschaft für Naturwissenschaften, Zürich 28: 1-39
Kelly JM (1993) Ballast water and sediments as mechanisms for unwanted species introductions into Washington state. Journal of Shellfish Research 12: 405-410
Kennedy CR & Fitch DJ (1990) Colonization, larval survival and epidemiology of the nematode *Anguillicola crassus*, parasitic in the eel *Anguilla anguilla*, in Britain. Journal of Fish Biology 36: 117-131
Ketelaars HAM, Lambregts-van de Clundert FE, Carpentier CJ, Wagenvoort AJ & Hoogenboerem W (1999) Ecological effects of the mass occurrence of the Ponto-Caspian invader, *Hemimysis anomala* G. O. Sars, 1907 (Crustacea: Mysidacea), in a freshwater storage reservoir in the Netherlands, with notes on its autecology and new records. Hydrobiologia 394: 233-248
Kevrekidis K, Kevrekidis T & Psaltopoulou H (1996) First experiences on acclimatization and rearing of *Penaeus japonicus* Bate (Crustacea, Penaeidae), in Lagos lagoon, Aegean Sea. Oebalia 22: 91-98
Kharchenko TA (1995) *Dreissena*: Range, ecology, biofoulings. Gidrobiologicheskii Zhurnal 31(3): 3-20 (in Russian)
Kharchenko TG & Lyashenko AV (1985) Destruction of allochthonous organic matter in canals in the presence of *Dreissena*. Hydrobiological Journal 21: 93-98
Kharchenko TG & Protasov AA (1981) On consortia in water ecosystems. Gidrobiologicheskii Zhurnal 17: 15-20 (in Russian)
Kideys A & Niermann U (1994) Occurrence of *Mnemiopsis* along the Turkish coasts (from northeastern Mediterranean to Istanbul). ICES Journal of Marine Science 52: 423-427
Kideys AE (1994) Recent dramatic changes in the Black Sea ecosystem: The reason for the sharp decline in Turkish anchovy fisheries. Journal of Marine Systems 5: 171-181
Kideys AE (1996) Determination of respiratory energy loss of *Mnemiopsis* in the Black Sea. TÜBTAK Postdoctoral Project YDABÇAG-96/D, Ankara, Turkey, 33 pp (in Turkish)
Kideys AE & Gücü AC (1995) *Rhopilema nomadica*: a Lessepsian scyphomedusan new to the Mediterranean coast of Turkey. Israel Journal of Zoology 41: 615-617
Kideys AE & Romanova Z (2001) Distribution of gelatinous macrozooplankton in the southern Black Sea during 1996-1999. Marine Biology 139: 535-547
Kideys AE, Ghasemi S, Ghninejad D, Roohi A & Bagheri S (2001) Strategy for combating *Mnemiopsis* in the Caspian waters of Iran. A report prepared for the Caspian Environment Programme, Baku, Azerbaijan
Kideys AE, Gordina A, Bingel F & Niermann U (1999) The role of environmental conditions on the distribution of eggs and larvae of anchovy (*Engraulis encrasicolus* L.) in the Black Sea. ICES Journal of Marine Science 56: 58-64
Kideys AE, Kovalev AV, Shulman G, Gordina A & Bingel F (2000) A review of zooplankton investigations of the Black Sea over the last decade. Journal of Marine Systems 24: 355-371
Kieser D & Campell M (2000) National report for Canada. Report of the Working Group of Introductions and Transfers of Marine Organisms, Pärnu, Estonia, 27-29 March 2000, pp 19-22. ICES CM 2000/ACME:07, Ref.: E+F
Kilgour BM, Mackie GL, Baker MA & Keppel R (1994) Effect of salinity on the condition and survival of zebra Mussel (*Dreissena polymorpha*). Estuaries 17: 385-393
Kinne O & Rotthauwe HW (1952) Biological observations and investigations on the blood concentration in *Heteropanope tridentatus* Maitland (Decapoda). Kieler Meeresforschung 8: 212-216 (in German)
Kinzelbach R (1965) Die blaue Schwimmkrabbe (*Callinectes sapidus*), ein Neubürger im Mittelmeer. Natur und Museum 95: 293-296
Kinzelbach R (1992) The main features of the phylogeny and dispersal of the zebra mussel *Dreissena polymorpha* In: Neuman D & Jenner HA (eds) The Zebra Mussel *Dreissena polymorpha*, Ecology, Biological Monitoring and First Applications in Water Quality Management, pp 5-17. Gustav Fisher, Stuttgart
Kinzelbach R (1992) The main features of the phylogeny and dispersal of the zebra mussel *Dreissena polymorpha*. Limnologie aktuell 4: 5-17
Kinzelbach R (1995) Neozoans in European waters - Exemplifying the worldwide process of invasion and species mixing. Experientia 51: 526-538
Kinzie RA (1984) Aloha also means goodbye: a cryptogenic stomatopod in Hawaii. Pacific Science 38: 298-311
Kiørboe T, Møhlenberg F & Hamburger K (1985) Bioenergetics of the planktonic copepod Acartia tonsa: relation between feeding, egg production and respiration, and composition of specific dynamic action. Marine Ecology Progress Series 26: 85-97
Kirjavainen J & Westman K (1999) Natural history and development of the introduced signal crayfish, *Pacifastacus leniusculus*, in a small, isolated Finnish lake, from 1968 to 1993. Aquatic Living Resources 12: 387-401
Kitching RL (ed) (1986) The Ecology of Exotic Animals and Plants: Some Australian Case Histories. John Wiley & Sons, Brisbane, Australia
Klitin AK (1996) Fertility of the red king crab at the coasts of the Sakhalin and Southern Kuril Islands. Bulleten' Moskovskogo Obshchestva Estestvoispitateley Prirodi 101(6): 43-52 (in Russian)
Knight IT, Wells CS, Wiggins B, Russell H, Reynolds KA & Huq A (1999) Detection and enumeration of fecal indicators and pathogens in the ballast water of transoceanic cargo vessels entering the Great Lakes. General meeting of the American Society for Microbiology Chicago, IL. Abstract Q-71, p 546
Knoepffler M, Noailles MC, Boudouresque CF & Abelard C (1990) Phytobenthos des Pyrénés-Orientales: complément à l'inventaire. Présence d'espèces non-indigènes (*Sargassum* et *Undaria*). Bulletin Societé Zoologique France 113: 37-43
Knoepffler-Péguy M, Belsher T, Boudouresque CF & Lauret M (1985) *Sargassum muticum* begins to invade the Mediterranean. Aquatic Botany 23: 291-295
Knowlton N (1993) Sibling species in the sea. Annual Review of Ecology and Systematics 24: 189-216
Knudsen J (1989) Immigration of marine invertebrates to the Limfjord (Denmark) and the North Sea - Baltic transition area. In: Spanier E, Steinberger Y & Luria M (eds)Environmental Quality and Ecosystem Stability. Vol. IV-B, Environmental Quality, pp 135-145. ISEEQ Publication, Jerusalem
Kocak F, Ergen Z & Çinar ME (1999) Fouling organisms and their developments in a polluted and an unpolluted marina in the Aegean Sea (Turkey). Ophelia 50: 1-20
Kocatas A & Katagan T (1983) Crustacean fauna of Turkish coastal lagoons. Rapports et procès-verbaux des réunions, Commission internationale pour l'Exploration scientifique de la mer Méditerranée 28: 231-233
Kocatas A, Koray T, Kaya M, & Kara OF (1993) A review of the fishery resources and their environment in the Sea of Marmara. Studies and Reviews GFCM, FAO64: 87-143
Köhn J & Waterstraat A (1990) The amphipod fauna of Lake Kummerow (Mecklenburg, GDR) with reference to *Echinogammarus ischnus* Stebbing, 1899. Crustaceana 58: 74-82
Køie M (1991) Swimbladder nematodes (*Anguillicola* spp.) and gill monogeneans (*Pseudodactylogyrus* spp.) parasitic on the European eel (*Anguilla anguilla*). Journal du Conseil 47: 391-398
Köksal G (1988) *Astacus leptodactylus* in Europe. In: Holdich DM & Lowery RS (eds) Freshwater Crayfish: Biology, Management and Exploitation, pp 365-400. Croom Helm (Chapman & Hall), London
Kolar CS & Lodge DM (2000) Freshwater nonindigenous species: interactions with other global changes, pp 3-30. In: Mooney HA & Hobbs RJ (eds) Invasive Species in a Changing World. Island Press, Washington DC, Covelo, California
Kolar CS & Lodge DM (2001) Progress in invasion biology: predicting invaders. Trends in Ecology & Evolution 16: 199-204
Komarova TI (1991) Mysids (Mysidacea). Ukrainian Fauna 26, Higher Crustaceans N. 7. Kiev, 104 pp (in Russian)
Kondratiev GP (1962) About some peculiarities of filtration in *Dreissena polymorpha* Pallas. Trudy Saratovskogo otd. GosNIORKH 7: 280-283 (in Russian)
Konopacka A (1998) *Dikerogammarus haemobaphes* (Eichwald, 1841) (Crustacea, Amphipoda), a new species in Polish fauna and two other rare amphipods in the Vistula River. Przeglad Zoologiczny 42(3-4): 211-218 (in Polish)
Konopacka A & Jazdzewski K (in press) *Obesogammarus crassus* (G.O. Sars, 1894) - one more Ponto-Caspian species in Polish waters. Fragmenta faunistica
Konsoulov AS & Kamburska LT (1998) Ecological determination of the new Ctenophora - *Beroe ovata* invasion in the Black Sea. Oceanology (Bulgaria) 2: 195-198
Korneva LG (1999) The phytoplankton succession. In: The Ecology of the Rybinsk Reservoir Phytoplankton, pp 89-148. Tolyatti (in Russian)
Korneva LG (2001) Spatial and temporal distribution of invading diatoms in the Volga river basin reservoirs. U.S. Russia Invasive Species Workshop, pp 89-91. Borok
Korneva LG & Genkal SI (2000) Taxonomic composition and ecological-geographical characteristics of the Volga reservoir phytoplankton. In: The Catalogue of Plants and Animals of the Volga Basin Water Bodies, pp 5-112. Yaroslavl (in Russian)
Korringa P (1942) *Crepidula fornicata*'s invasion in Europe. Basteria 7: 12-23
Korringa P (1951a) *Crepidula fornicata* as an oyster-pest. Rapport & Procès-Verbaux Réunions Conseil Permanent International Exploration Mer 128: 55-59
Korringa P (1951b) The shell of *Ostrea edulis* as a habitat. Archives néerlandaises de zoologie 10: 32-152
Korringa P (1957) Water temperature and breeding throughout the geographical range of *Ostrea edulis*. L'Année Biologique 33: 1-15
Korringa P (1965) De Portugese oester, *Crassostrea angulata*, in Nederland. Basteria 29: 18-22
Korringa P (1976a) Farming the Flat Oysters of the Genus *Ostrea*. Elsevier, Amsterdam, 238 pp
Korringa P (1976b) Farming the Cupped Oysters of the Genus *Crassostrea*. Elsevier, Amsterdam, 224 pp
Kossakowski J (1966) Crayfishes. PWRiL, Warszawa, 292 pp (in Polish)
Kostrzewa J & Grabowski M (2001) Racer (goad) goby, *Neogobius gymnotrachelus* (Kessler, 1857) (Gobiidae, Perciformes) - a new fish species in the Vistula River. Przeglad Zoologiczny 45(1-2): 101-102 (in Polish)
Kotta J (2000) Impact of eutrophication and biological invasions on the structure and functions of benthic macrofauna. PhD Thesis, University of Tartu, 160 pp

Kotta J & Kotta I (1998) Distribution and invasion ecology of *Marenzelleria viridis* in the Estonian coastal waters. Proceedings of the Estonian Academy of Science Biology Ecology 47: 212-220

Kotta J, Orav H & Kotta I (1998) Distribution and filtration activity of the zebra mussel, *Dreissena polymorpha*, in the Gulf of Riga and in the Gulf of Finland. Proceedings of the Estonian Academy of Sciences Biology Ecology 47: 32-41

Koutnik MA & Padilla DK (1994) Predicting the spatial distribution of *Dreissena polymorpha* (zebra mussel) among inland lakes of Wisconsin: modeling with a GIS. Canadian Journal of Fisheries and Aquatic Sciences 51: 1189-1196

Koutsikopoulos C, Beillois P, Leroy C & Taillefer F (1998) Temporal trends and spatial structures of the sea surface temperature in the Bay of Biscay. Oceanologica Acta 21: 335-344

Koutsoubas D & Voultsiadou-Koukoura E (1991) The occurrence of *Rapana venosa* (Valenciennes, 1846) (Gastropoda, Thaididae) in the Aegean Sea. Bollettino Malacologico 26: 201-204

Kovalev A, Gubanova AD, Kideys AE, Melnikov VV, Niermann U, Ostrovskaya NA, Prusova IYu, Skryabin VA, Uysal Z & Zagorodnyaya JuA (1998a) Long-term changes in the biomass and composition of fodder zooplankton in coastal regions of the Black Sea during the period 1957-1996. In: Ivanov L & Oguz T (eds) NATO TU-Black Sea Project: Ecosystem Modeling as a Management Tool for the Black Sea, Symposium on Scientific Results, pp 209-220. Kluwer Academic Publishers

Kovalev A, Niermann U, Melnikov VV, Belokopytov V, Uysal Z, Kideys AE, Ünsal M & Altukhov D (1998b) Long-term changes in the Black Sea zooplankton: The role of natural and anthropogenic factors. In: Ivanov L & Oguz T (eds) NATO TU-Black Sea Project: Ecosystem Modeling as a Management Tool for the Black Sea, Symposium on Scientific Results, pp 221-234. Kluwer Academic Publishers

Kozhova OM & Izhboldina LA (1993) Spread of *Elodea canadensis* in Lake Baikal. Hydrobiologia 259: 203-211

Kraeplin K (1887) The German freshwater Bryozoa. A monography. Abhandlungen des Naturwissenschaftlichen Vereins Hamburg 10: 1-168 (in German)

Kramp PL (1922) Demonstration of *Gonionemus murbachi* from Kristianiafjord, Norway. Videnskabelige Meddelser fra Dansk Naturhistorisk Forening i Kobenhavn 74 XI

Krapp F & Sconfietti R (1983) *Ammothea hilgendorfi* (Böhm, 1879), an adventitious pycnogonid new for the Mediterranean Sea. Marine Ecology 4: 123-132

Krause W (1997) Charales (Charophyceae). Süsswasserflora von Mitteleuropa 18. Gustav Fischer Verlag, Stuttgart, 202 pp (in German)

Kremer P & Nixon S (1976) Distribution and abundance of the ctenophore, *Mnemiopsis leidyi* in Narragansett Bay. Estuarine and Coastal Marine Science 4: 627-639

Kristensen ES (1969) Attacks by *Teredo navalis* L. in inner Danish waters in relation to environmental factors. Videnskabelige Meddelelser Dansk Naturhistorisk Forening 132: 199-210

Kristensen ES (1979) Observations on growth and life cycle of the shipworm *Teredo navalis* L. (Bivalvia; Mollusca) in the Isefjord, Denmark. Ophelia 18: 235-242

Krom MD, Brenner S, Kress N, Neori A & Gordon LI (1992) Nutrient dynamics and new production in a warm-core eddy from the eastern Mediterranean Sea. Deep Sea Research 39: 467-480

Krylov PI, Bychenkov DE, Panov VE, Rodionova NV & Telesh IV (1999) Distribution and seasonal dynamics of the Ponto-Caspian invader *Cercopagis pengoi* (Crustacea, Cladocera) in the Neva Estuary (Gulf of Finland). Hydrobiologia 393: 227-232

Krzywosz T, Bialokoz W & Chybowski L (1994) Native crayfishes in the Suwalki district. Komunikaty Rybackie 1: 9-10 (in Polish)

Kube J, Zettler ML, Gosselck F, Ossig S & Powilleit M (1996) Distribution of *Marenzelleria viridis* (Polychaeta: Spionidae) in the southwestern Baltic Sea in 1993/94 - ten years after introduction. Sarsia 81: 131-142

Kuczynski J (1995) Round goby *Neogobius melanostomus* (Pallas, 1811) - an emigrant from the Ponto-Caspian basin in the Gulf of Gdansk. Bulletin of the Sea Fisheries Institute 2(135): 68-71 (in Polish)

Kudersky LA (2001) Fish acclimatization in water bodies of Russia: State and course of development. Voprosy Rybolovstva 2: 7-85 (in Russian)

Kuenen F (1999) Grote Watemavel: een nieuwe bedreiging voor Nederlandse ecosystemen? Scriptie 220. Department of Aquatic Ecology & Environmental Biology, University of Nijmegen: 1-28

Kuhn, SF (1996/97) *Rhizamoeba schnepfii* sp. nov., a naked amoeba feeding on marine diatoms (North Sea, German Bight). Archiv für Protistenkunde 147: 277-282

Kühn SF (1997a) Infection of *Coscinodiscus* spp. by the parasitoid nanoflagellate *Pirsonia diadema*: I. Selective infection behaviour for host species and individual host cells. Journal of Plankton Research 20: 443-454

Kühn SF (1997b) *Victoriniella multiformis*, gen. et spec. nov. (incerta sedis), a polymorphic parasitoid protist infecting the marine diatom *Coscinodiscus wailesii* Gran & Angst (North Sea, German Bight). Archiv für Protistenkunde 148: 115-123

Kühn SF, Drebes G & Schnepf E (1996) Five new species of the nanoflagellate *Pirsonia* in the German Bight, North Sea, feeding on planktic diatoms. Helgoländer Meeresuntersuchungen 50: 205-222

Kulida SV, Zelenkov VM, Yermakov SV & Timofeev VI (1998) Cultivation of trout smolts in brackish waters of the White Sea. In: Alimov AF (ed) Problems of study, rational exploitation and conservation of natural resources of the White Sea. Proceedings of the International Conference, Archangel, September 1998, St. Petersburg, pp 208-210 (in Russian)

Kureck A (1992) Neue Tiere im Rhein. Naturwissenschaften 79: 533-540

Kureck A, Bieg R & Ottenberg R (2001) Einfluß von Futtermenge und Neozoen auf Überlebensrate und Wachstum von *Ephoron virgo* (Ephemeroptera) im Rhein. Deutsche Gesellschaft für Limnologie, Tagungsbericht 2000 Eigenverlag, Tutzing

Kuris AM (1997) Conceptual framework for biocontrol of introduced marine pests. In: Thresher RE (ed) Proceedings of the First International Workshop on the Demography, Impacts and Management of Introduced Populations of the European Crab, *Carcinus maenas*. Centre for Research on Introduced Marine Pests Technical Report 11: 66-68

Kuris AM & Culver CS (1999) An introduced sabellid polychaete pest infesting cultured abalones and its potential spread to other California gastropods. Invertebrate Biology 118: 391-403

Kuzmin SA (1999) The red king crab. In: Obzor sostoyaniya syryevikh resursov Barentseva moray i Severnoy Atlantiki na 1999, pp 25-26. Izdatelstvo Polyarnogo Instituta Ribnogo Hozyaystva i Okeanografii, Murmansk (in Russian)

Kuzmin SA & Lobodenko SG (1997) Distribution and environmental conditions of the red king crab in the Barents Sea. Aquicultura: problemy I dostizheniya OI/VNIERH 8: 14-18 (in Russian)

Kuzmin SA, Olsen S & Gerasimova OV (1996) Barents Sea king crab (*Paralithodes camtschatica*): The transplantation experiments were successful In: Proceedings of the International Symposium on High Latitude Crabs: Biology, Management, and Economics. University of Alaska Sea Grant Report, pp 649-663

Kvasov DD (1975) The late Quaternary history of large lakes and internal seas of Eastern Europe. Nauka Publisher, Leningrad. 278 pp (in Russian)

Laabir M & Gentien P (1999) Survival of toxic dinoflagellates after gut passage in the Pacific oyster *Crassostrea gigas* Thunberg. Journal of Shellfish Research 18: 217-222

Laborel J (1992) Origine probable de la contamination du nouveau port des Lecques (Var) par *Caulerpa taxifolia*. Rapport préliminaire. Rapport du Laboratoire de Biologie marine et d'Ecologie du Benthos. Faculté des Sciences de Luminy. Université d'Aix-Marseille: 1-3

Lagzdins G & Pallo P (1994) *Marenzelleria viridis* (Verrill) (Polychaeta, Spionidae) - a new species for the Gulf of Riga. Proceedings of the Estonian Academy of Science Biology 43: 184-188

Laing I (1998) Ballast water discharges into coastal waters of England and Wales. ICES Cooperative Research Report 224: 36-42

Laing I (1999) *Coscinodiscus wailesii*. In: Gollasch S, Minchin D, Rosenthal H & Voigt M (eds) Exotics Across the Ocean. Case Histories on Introduced Species, pp 1-4. Logos Verlag, Berlin

Lamb HH (1977) Climate: Present, Past and Future, vol. 2. Methuen, London

Lambert CC & Lambert G (1998) Non-indigenous ascidians in southern California harbors and marinas. Marine Biology 130: 675-688

Lamprell KL (1998) Recent *Spondylus* species from the Middle East and adjacent regions, with the description of two new species. Vita Marina 45: 41-60

Lane CE (1959) Some aspects of the general biology of *Teredo*. In: Ray DL (ed) Marine Boring and Fouling Organisms, pp 137-144. University of Washington Press, Seattle

Langar H, Djellouli A, Ben Mustafa K & El Abed A (2000) Première signalisation de *Caulerpa taxifolia* (Vahl) C. Agardh en Tunisie. Bulletin Institut Scientifique et Technique de la Mer 27: 7-8

Lange CB, Hasle GR & Syvertsen E (1992) Seasonal cycle of diatoms in the Skagerrak, North Atlantic, with emphasis on the period 1980-1990. Sarsia 77: 173-187

Lardicci C & Castelli A (1986) *Desdemona ornata* Banse, 1957 (Polychaeta, Sabellidae, Fabricinae); new record in the Mediterranean Sea. Oebalia 13: 195-201

Larson A (1984) Neontological inferences of evolutionary pattern and process in the salamander family Plethodontidae. Evolutionary Biology 17: 119-217

Larsson U, Hajdu S, Andersson L & Edler L (1998) The free water mass. In: The Baltic Sea '97. Report on Environmental Monitoring, pp 28-33. SMF Stockholm University, Stockholm (in Swedish, English summary)

Laruelle F, Guillou J & Paulet YM (1994) Reproductive pattern of the clams, *Ruditapes decussatus* and *R. philippinarum* on intertidal flats in Brittany. Journal of the Marine Biological Association of the United Kingdom 74: 351-366

Lauckner G (1983) Diseases of marine Mollusca: Bivalvia. In: Kinne O (ed) Diseases of Marine Animals II, pp 477-961. Biologische Anstalt Helgoland, Hamburg

Laughton R, Moran T & Brown G (1992) A Review and Evaluation of Ballast Water Management Options to Reduce the Potential for the Introduction of Non-native Species to the Great Lakes. Pollutech Group of Companies, Inc., Ontario, Canada (www.pollutech.com)

Laurent PJ (1988) *Austropotamobius pallipes* and *A. torrentium*, with observations on their interactions with other species in Europe. In: Holdich DM & Lowery RS (eds) Freshwater Crayfish: Biology, Management and Exploitation, pp 341-364. Croom Helm (Chapman & Hall), London

Lauret M (1967) Morphologie, phénologie, répartition des *Polysiphonia* marins du littoral languedocien. I. Section *Oligosiphonia*. Naturalia Monspeliensia, Botanique 18: 345-373

Lauret M (1970) Morphologie, phénologie, répartition des *Polysiphonia* marins du littoral languedocien. II. Section *Polysiphonia*. Naturalia Monspeliensia, Botanique 21: 121-163

Lauret M, Riouall R & Dubois A (1985) L'acclimatation et la croissance de *Sargassum muticum* (Yendo) Fensholt (Phaeophyceae) dans l'étang de Thau. Actes du Congrès National des Sociétés Savantes 110: 223-231

Laursen KK (1971) Aquatic plants. Natur & Kultur, Stockholm, 132 pp (in Swedish)

Lavie B & Nevo E (1986) Genetic diversity of marine gastropods: contrasting strategies of *Cerithium rupestre* and *C. scabridum* in the Mediterranean Sea. Marine Ecology Progress Series 28: 99-103

Lawton JH & Brown KC (1986) The population and community ecology of invading insects. Philosophical Transactions of the Royal Society of London B 314: 607-617
Lazareva LP (1961) The effect of temperature and salinity on oxygen consumption by different sized ctenophore *Pleurobrachia pileus* O.F. Müller. Trudy Karadagskoj biologicheskoj stanstii, Kiev 17: 85-96
Lazauskiene L (1997) Efficiency assessment of fish-food basis enrichment with introduced crustaceans. Hydrobiological Society of Lithuania and Association of Fishery Enterprises "Lampetra", Report. Vilnius, 49 pp (in Lithuanian).
Lazzari G & Rinaldi E (1994) Alcune considerazioni sulla presenza di specie esotiche extra mediterranee nelle lagune salmastre del ravennate. Bollettino di Malacologia, Milano 30: 195-202
Le Borgne Y, Gras P, Comps M, Carresco G & Razet D (1983) Observations sur la reproduction des huîtres dans la Seudre (Bassin de Marennes Oléron) en 1972. ICES CM 1973/K:16
Leach JH (1993) Impacts of the zebra mussel (*Dreissena polymorpha*) on water quality and fish spawning reefs in Western Lake Erie. In: Nalepa TF & Schloesser DW (eds) Zebra Mussels: Biology, Impacts, and Control, pp 381-397. Lewis Publishers, Ann Arbor, Michigan
Lee CE & Bell MA (1999) Causes and consequences of recent freshwater invasions by saltwater animals. Trends in Ecology and Evolution 14: 284-288
Lein TE (1999) A newly immigrated red alga ('*Dasyphonia*', Dasyacea, Rhodophyta) to the Norwegian coast. Sarsia 84: 85-88
Lejeune A, Cagauan A & van Hove C (1999) *Azolla* research and development: Recent trends and priorities. Symbiosis 27: 333-351
Lelek A (1996) Die allochthonen und die beheimateten Fischarten unserer großen Flüße - Neozoen der Fischfauna. In: Gebhardt H, Kinzelbach R & Schmidt-Fischer S (eds) Gebietsfremde Tierarten. Auswirkungen auf einheimische Arten, Lebensgemeinschaften und Biotope pp 197-217. Ecomed, Landsberg
Lemaitre R (1995) *Charybdis helleri* (Milne Edwards, 1867), a nonindigenous portunid crab (Crustacea: Decapoda: Brachyura) discovered in the Indian River lagoon system of Florida. Proceedings of the Biological Society of Washington 108: 643-648
Lémée R, Pesando D, Durand-Clément M, Dubreuil A, Meinesz A, Guerriero A & Pietra F (1993) Preliminary survey of toxicity of the green alga *Caulerpa taxifolia* introduced into Mediterranean. Journal of Applied Phycology 5: 485-493
Lenders HJR, Aarts BGW, Strijbosch H & Van der Velde G (1998) The role of reference and target images in ecological recovery of river systems. Lines of thought in the Netherlands. In: Nienhuis PH, Leuven RSEW & Ragas AMJ (eds) New Concepts for Sustainable Management of River Basins, pp 35-52. Backhuys Publishers, Leiden
Lenkova A (1962) Studies on the causes of vanishing, protective means and restriction of the noble crayfish *Astacus astacus* (L.) in regard of the range extension of American crayfish *Cambarus affinis* Say. Ochrona Przyrody 28: 1-38 (in Polish)
Lenz J, Andres H-G, Gollasch S & Dammer M (2000) Einschleppung fremder Organismen in Nord- und Ostsee: Untersuchungen zum ökologischen Gefahrenpotential durch den Schiffsverkehr. UBA Project Water, Umweltbundesamt, Berlin 5, 273 pp
Leonardsson K (2001) Mjukbottenfauna. In: Wiklund K (ed) Bottniska viken 2000, pp 12-14. Umeå marina forskningcentrum, Umeå University, Sweden
Leppäkoski E & Mihnea PE (1996) Enclosed seas under man-induced change: A comparison between the Baltic and Black Seas. Ambio 25: 380-389
Leppäkoski E & Olenin S (2000a) Non-native species and rates of spread: lessons from the brackish Baltic Sea. Biological Invasions 2: 151-163
Leppäkoski E & Olenin S (2000b) Xenodiversity of the European brackish water seas: the North American contribution. In: Pederson J (ed) Marine Bioinvasions. Proceedings of the First National Conference, January 24-27, 1999, pp 107-119. Massachusetts Institute of Technology, Cambridge, MA
Leppäkoski E & Olenin S (2001) The meltdown of biogeographical peculiarities of the Baltic Sea: the interaction of natural and man-made processes. Ambio 30: 202-209
Leppäkoski E (1984) Introduced species in the Baltic Sea and its coastal ecosystems. Ophelia Suppl. 3: 123-135
Leppäkoski E (1994) Non-indigenous species in the Baltic Sea. In: Boudouresque CF, Briand F, Nolan C (eds) Introduced species in European coastal waters. CEC Ecosystem Research Report 8: 67-75
Leppäkoski E (1999) *Balanus improvisus*. In: Gollasch S, Minchin D, Rosenthal H & Voigt M (eds) Exotics Across the Ocean. Case Histories on Introduced Species, pp 49-54. Logos Verlag, Berlin
Leppäkoski E (2002) Harmful non-native species in the Baltic Sea - an ignored problem. In: Schernewski G & Schiewer U (eds) 'Baltic Coastal Ecosystems: Structure, Function and Coastal Zone Management. Central and Eastern European Development Studies, Springer Verlag (in press)
Leuchs H & Schleuter A (1996) *Dikerogammarus haemobaphes* (Eichwald, 1841), in Neckar, a small crustacean (Gammaridae) coming from the Danube. Lauterbornia 25: 139-141 (in German)
Lever C (1985) Naturalized Mammals of the World. Longman, London & New York, 487 pp
Lever C (1987) Naturalized Birds of the World. Longman, London & New York, 615 pp
Lever C (1998) Introduced fishes: an overview. In: Cowx IG (ed) Stocking and Introduction of Fish, pp 143-152. Fishing News Books. MPG Books Ltd., Bodmin, Cornwall
Levine JM (2000) Species diversity and biological invasions: relating local process to community pattern. Science 288: 852-854
Lewandowski K (1976) Unionidae as a substratum for *Dreissena polymorpha* Pallas. Polskie Archiwum Hydrobiologii 23: 409-420
Lewandowski K, Stoczkowski R & Stanczykowska A (1997) Distribution of *Dreissena polymorpha* (Pall.) in lakes of the Jorka river watershed. Polskie Archiwum Hydrobiologii 44(4): 431-443
Lewin J & Norris RE (1970) Surf-zone diatoms of the coasts of Washington and New Zealand (*Chaetoceros armatum* T. West and *Asterionella* spp.). Phycologia 9: 143-149
Lewinsohn Ch & Galil BS (1982) Notes on species of *Alpheus* (Crustacea, Decapoda) from the Mediterranean coast of Israel. Quaderni del Laboratorio di Tecnologia della Pesca, Ancona 3: 207-210
Lewinsohn Ch & Holthuis LB (1986) The Crustacea Decapoda of Cyprus. Zoologische Verhandelingen (Leiden) 230: 1-64
Lewinsohn Ch & Manning RB (1980) Stomatopod Crustacea from the Eastern Mediterranean. Smithsonian Contribution to Zoology 305: 1-22
Lewis J, Harris ASD, Jones KJ & Edmonds RL (1999) Long-term survival of marine planktonic diatoms and dinoflagellates in stored sediment samples. Journal of Plankton Research 21: 343-354
Lewis KM, Feder JL & Lamberti GA (2000a) Population genetics of the zebra mussel, *Dreissena polymorpha* (Pallas): local allozyme differentiation within midwestern lakes and streams. Canadian Journal of Fisheries and Aquatic Sciences 57: 637-643
Lewis KM, Feder JL, Horvath TG & Lamberti GA (2000b) Heterozygosity and fitness: no strong association in Great Lakes populations of the zebra mussel, *Dreissena polymorpha* (Pallas). Malacologia 42: 113-122
Liakhnovich VN, Karataev AYu, Liakhov SM, Andreev NI, Andreeva SI, Afanasiev SA, Diga, AK, Zakutsky VP, Zolotareva VI, Lvova AA, Nekrasova MYa, Osadchih VF, Pligin YuV, Protasov AA & Tishchikov GM (1994) Habitat parameters. In: Starobogatov YI (ed)Freshwater Zebra Mussel *Dreissena polymorpha* (Pall.) (Bivalvia, Dreissenidae) Taxonomy, Ecology and Practical Use, pp 109-119. Nauka, Moscow (in Russian)
Limanova NA (1964) *Dreissena* (Bibliography). In: Shtegman BK (ed) Biology and Control of *Dreissena*. Trudy Instituta Biologii Vnutrennikh Vod 7(10): 71-145 (in Russian) (Translated: Shtegman BK (ed) Biology and Control of *Dreissena*: A Collection of Papers. Jerusalem, Israel Program for Scientific Translations, 1968)
Limanova NA (1978) *Dreissena* (Dreissenidae. Lamellibranchia). Bibliography of Russian and Foreign Literature. Academiya Nauk SSSR. Biblioteka Instituta Vnutrebbikh Vod, 115 pp
Lindqvist OV (1988) Restoration of native European crayfish stocks. Freshwater Crayfish 7: 6-12
Lindqvist OV & Huner JV (1999) Life history characteristics of crayfish: What makes some of them good colonizers? In: Gherardi F & Holdich DM 1999 (eds) Crayfish in Europe as Alien Species. How to Make the Best of a Bad Situation, pp 23-30. A.A. Balkema, Rotterdam, Brookfield
Linnamies O (1956) On the occurrence and damages of beavers in Finland. Suomen Riista 10: 63-86 (in Finnish)
Lipkin Y (1972) Contribution to the knowledge of Suez canal migration. Marine algal and sea-grass flora of the Suez canal. Israel Journal of Zoology 21: 405-446
Lisitsyna LI & Papchenkov VG (2000) The Flora of the Water Bodies of Russia: Identification Keys for the Vascular Plants. Nauka Press, Moscow, 237 pp (in Russian)
Lisitsyna LI, Papchenkov VG & Artemenko VI (1993) The Flora of the Water Bodies of the Volga Basin. Identification Key Book of Flowering Plants. Hydrometeoizdat Press, St. Petersburg, 220 pp (in Russian)
Litvinchuk LF, Rivier IK & Panov VE (2001) Dynamics of abundance, structure of population and fecundity of Ponto-Caspian predaceous Cladoceran, *Cercopagis pengoi* (Ostroumov, 1891) (Polyphemoidea, Cercopagidae), in the Eastern Gulf of Finland, the Baltic Sea. Biology of Inland Waters 1: 57-62 (in Russian)
Locke A, Reid DM, van Leeuwen HC, Sprules WG & Carlton JT (1993) Ballast water exchange as a means of controlling dispersal of freshwater organisms by ships. Canadian Journal of Fisheries and Aquatic Sciences 50: 2086-2093
Locke A, Reid DM, Sprules WG, Carlton JT & van Leeuwen HC (1991) Effectiveness of mid-ocean exchange in controlling freshwater and coastal zooplankton in ballast water. Canadian Technical Report, Fisheries and Aquatic Sciences 1822: 1-93
Lodge D (1993) Species invasions and deletions: Community effects and responses to climate and habitat change. In: Kareiva PM, Kingsolver JG & Huey RB (eds) Biotic Interactions and Global Change, pp. 367-87. Sinauer Associates, Sunderland
Lodge DM (1993) Biological invasions: lessons for ecology. Trends in Ecology and Evolution 8: 133-37
Lodge DM & Hill AM (1994) Factors governing species composition, population size, and productivity of cool-water crayfishes. Nordic Journal of Freshwater Research 69: 111-136
Lodge DM, Stein RA, Brown KM, Covich AP, Brönmark C., Garvey JE & Klosiewski SP (1998) Predicting impact of freshwater exotic species on native biodiversity: Challenges in spatial scaling. Australian Journal of Ecology 23: 53-67
Löffler H (1996) Neozoen in der Fischfauna Baden-Württembergs - ein Überblick. In: Gebhardt H, Kinzelbach R & Schmidt-Fischer S (eds) Gebietsfremde Tierarten. Auswirkungen auf einheimische Arten, Lebensgemeinschaften und Biotope, pp 217-226. Ecomed, Landsberg
Lonsdale WM (1999) Global patterns of plant invasions and the concept of invasibility. Ecology 80: 1522-1536
Lonsdale WM & Lane AM (1994) Tourist vehicles as vectors of weed seeds in Kakadu National Park, Northern Australia. Biological Conservation 69: 277-283
Lovejoy C, Bowman JP & Hallegraeff GM (1998) Algicidal effects of a novel marine *Pseudoalteromonas* isolate (class Proteobacteria, gamma subdivision) on harmful algal bloom species of the genera *Chattonella*, *Gymnodin-*

ium, and *Heterosigma*. Applied and Environmental Microbiology 64: 2806-2813

Lowe RL & Pillsbury RW (1995) Shifts in benthic algal community structure and function following the appearance of zebra mussel (*Dreissena polymorpha*) in Saginaw Bay, Lake Huron. Journal of Great Lakes Research 21: 558-566

Lowery RS (1988) Growth, moulting and reproduction. In: Holdich DM & Lowery RS (eds) Freshwater Crayfish: Biology, Management and Exploitation, pp 83-113. Croom Helm (Chapman & Hall), London

Lowery RS & Holdich DM (1988) *Pacifastacus leniusculus* in North America and Europe, with details of the distribution of introduced and native crayfish species in Europe. In: Holdich DM & Lowery RS (eds) Freshwater Crayfish: Biology, Management and Exploitation, pp 281-308. Croom Helm (Chapman & Hall), London

Lozon JD & MacIsaac HJ (1997) Biological invasions: are they dependent on disturbance? Environmental Reviews 5: 131-144

Lu SY, Hsueh ML, Hsu TW & Chen TT (1999) A taxonomical note on *Rhizophora stylosa* in Taiwan. Taiwan Journal of Forest Science 14: 351-354

Lubjanov IP (1953) On the creation and the directed studies on the bottom fauna of small dam reservoirs of the SE Ukraine. Zoologicheskij Zhurnal 32(6): 1074-1083 (in Russian)

Lucas IAN, McCollin TA & Hamer JP (1999) Marine organisms in ships ballast water. Report to Report to Ministry for Agriculture, Fisheries and Food, UK

Lucy F (2000) Establishment of a monitoring survey for zebra mussels (*Dreissena polymorpha*) in Lough Allen and Lough Key. Proceedings of the 1999 Institutes of Technology Colloquium in Science and Computing, Letterkenny, Ireland

Ludyanskiy ML (1991) Are *Dreissena polymorpha* and *Dreissena bugensis* synonymous? New York Sea Grant 2: 2-3

Lugliè A, Giaccobe MG, Sannio A, Fiocca F & Sechi N (2001) First record of the dinoflagellate *Alexandrium catenella* (Whedon & Kofoid) Balech (Dinophyta), a potential producer of paralytic shellfish poisoning, in Italian waters (Sardinia, Tyrrhenian Sea) Abstracts of X OPTIMA Meeting, 13-19 September 2001, Palermo, p 237

Lumare F & Casolino G (1986) First record of *Penaeus japonicus* Bate, 1888 (Decapoda Natantia) along Italian coasts. Oebalia 13: 179-183

Lumare F & Palmegiano B (1980) Acclimatazione di *Penaeus japonicus* Bate nella Laguna di Lesina (Italia sud-orientale). Rivista italiana di Piscicultura e ittiopatologia 15: 53-58

Lüning K (1985) Meeresbotanik. Verbreitung, Ökophysiologie und Nutzung der marinen Makro-algen. Georg Thieme Verlag, Stuttgart, 375 pp

Lüning K (1990) Seaweeds, their Environment, Biogeography, and Ecophysiology. John Wiley, New York, 527 pp

Luther H (1979) *Chara connivens* in the Baltic Sea area. Annales Botanici Fennici 16: 141-150

Lvova AA & Makarova GE (1994) Gametogenesis, reproductive cycle. In: Starobogatov YI (ed) Freshwater Zebra mussel *Dreissena polymorpha* (Pall.) (Bivalvia, Dreissenidae) Taxonomy, Ecology and Practical Use, pp 138-149. Nauka, Moscow (in Russian)

Lvova AA (1979) The role of *Dreissena polymorpha* Pallas in the self-cleaning of Uchinskoe reservoir. In: Likharev IM (ed) Molluscs: Main results of their Study. 6th Meeting on the Investigation of Molluscs, pp 219-220. Nauka Press, Leningrad (Abstract; in Russian)

Lvova AA (1980) Ecology of zebra mussel (*Dreissena polymorpha polymorpha* (Pall.). Trudy Vsesoiuznogo Gidrobiologicheskogo Obshchestva 23: 101-119 (in Russian)

Lvova AA, Izvekova EI & Sokolova NY (1980) The role of benthic animals in the conversion of organic matter and in the processes of waterbody self-cleaning. Trudy Vsesoiuznogo Gidrobiologicheskogo Obshchestva 23: 171-177 (in Russian)

Lvova AA, Makarova GE., Alimov AF Karataev Ayu, Miroshnichenki MP, Zakutsky VP & Nekrasova MN (1994) Growth and production. In: Starobogatov YI (ed) Freshwater Zebra Mussel *Dreissena polymorpha* (Pall.) (Bivalvia, Dreissenidae) Taxonomy, Ecology and Practical Use pp 156-179. Nauka, Moscow (in Russian)

Lvova AA, Polyi AV & Sokolova NYu (1996) Ponto-Caspian immigrants in the Moscow River within Moscow city. Zoologicheskij Zhurnal 75: 1273-1274 (in Russian)

Lvova-Kachanova AA & Izvekova EI (1978) Zebra mussel and chironomids from Uchinskoe reservoir. In: Dunaeva TN et al. (eds) Plant and Animal Life in Moscow and its Environments, pp 119-121. Moscow University Press, Moscow (in Russian)

Lyakhnovich VP, Karataev AYu & Mitrakhovich PA (1983) Influence of *Dreissena polymorpha* Pallas on the ecosystem of a eutrophic lake. Biologija Vnutrennykh Vod 60: 25-28 (in Russian)

Lyakhnovich VP, Karatayev AY, Mitrakhovich PA, Guryanova LV & Vezhnovets GG (1988) Productivity and prospects for utilizing the ecosystem of Lake Lukoml, thermoelectric station cooling reservoir. Soviet Journal of Ecology 18: 255-259

Lyakhov SM & Mordukhay-Boltovskoy FD (1973) On the distribution of *Jaera sarsi* Valk (Crustacea, Isopoda) in the Volga basin. Information Bulletin of the Institute for Biology of Inland Waters 20: 29-33 (in Russian)

Maasdam R & Claassen THL (1998) Trends in water quality and algal growth in shallow Frisian lakes, The Netherlands. Water Science and Technology 37: 177-184

MacArthur RH & Wilson EO (1963) An equilibrium theory of insular zoogeography. Evolution 17: 373-87

MacArthur RH & Wilson EO (1967) The Theory of Island Biogeography. Princeton University Press, Princeton

MacCrimmon HR (1971) World distribution of rainbow trout (*Salmo gairdneri*). Journal of the Fisheries Research Board of Canada 28: 663-704

Macdonald E (1999) *Gyrodinium aureolum*. In: Gollasch S, Minchin D, Rosenthal H & Voigt M (eds) Exotics Across the Ocean. Case Histories on Introduced Species, pp 5-9. Logos Verlag, Berlin

Macdonald EM (1995) Dinoflagellate resting cysts and ballast water discharges in Scottish ports. ICES C.M. 1995/O: 10, 17 pp

Macdonald EM (1998) Dinoflagellate resting cysts and ballast water discharges in Scottish ports. ICES Cooperative Research Report 224: 24-35

Macdonald EM & Davidson RD (1998) Ballast water project. Fisheries Research Services Report 3/97. FRS Marine Laboratory, Aberdeen, 83 pp

Macdonald EM & Davidson RD (1998) The occurrence of harmful algae in ballast water discharges in Scottish ports and the effects of mid-water exchange in regional seas. In: Reguera B, Blanco J, Fernandez ML & Wyatt T (eds) Harmful Algae. Xunta de Galicia and Intergovernmental Oceanographic Commission of UNESCO, pp 220-223

Maciolek JA (1984) Exotic fishes in Hawaii and other islands of Oceania. In: Courtenay WR & Stauffer JR (eds) Distribution, Biology and Management of Exotic Fishes, pp 131-61. John Hopkins University Press, Baltimore

Maciolek NJ (1984) New records and species of *Marenzelleria* Mesnil and *Scolecolepides* Ehlers (Polychaeta; Spionidae) from northeastern north America. Linnean Society of New South Wales: 48-62

MacIsaac HJ (1996) Potential abiotic and biotic impacts of zebra mussels on the inland waters of North America. American Zoologist 36: 287-299

MacIsaac HJ, Grigorovich IA, Hoyle JA, Yan ND & Panov V (1999) Invasion of Lake Ontario by the Ponto-Caspian predatory cladoceran *Cercopagis pengoi*. Canadian Journal of Fisheries and Aquatic Sciences 56: 1-5

MacIsaac HJ, Lonnee CJ & Leach JH (1995) Suppression of microzooplankton by zebra mussels: importance of mussel size. Freshwater Biology 34: 379-387

Mack RN (1996) Predicting the identity and fate of plant invaders: Emergent and emerging approaches. Biological Conservation 78: 107-121

Mack RN, Simberloff D, Lonsdale WM, Evans G, Clout M & Bazzaz F (2000) Biotic invasions: causes, epidemiology, global consequences and control. Issues in Ecology 5: 1-22

MacNeil C, Dick JTA & Elwood RW (1997) The trophic ecology of freshwater *Gammarus* spp. (Crustacea: Amphipoda): problems and perspectives concerning the functional feeding group concept. Biological Review 72: 349-364

Madenjian CP (1995) Removal of algae by the zebra mussel (*Dreissena polymorpha*) population in western Lake Erie: a bioenergetics approach. Canadian Journal of Fisheries and Aquatic Sciences 52: 381-390

Maggs CA & Stegenga H (1999) Red algal exotics on North Sea coasts. Helgoländer Meeresuntersuchungen 52: 243-258

Maggs CA & Ward BA (1996) The genus *Pikea* (Dumontiaceae, Rhodophyta) in England and the North Pacific: Comparative morphological, life history, and molecular studies. Journal of Phycology 32: 176-193

Maggs CA, Douglas SE, Fenety J & Bird CJ (1992) A molecular and morphological analysis of the *Gymnogongrus devoniensis* (Rhodophyta) complex in the North Atlantic. Journal of Phycology 28: 214-232

Magne F (1992) *Goniotrichopsis* (Rhodophyceae, Porphyridiales) in Europe. Cryptogamie, Algologie 13: 109-112

Maitland PS (1969) A preliminary account of the mapping of the distribution of freshwater fish in the British Sea. Journal of Fish Biology 1(1): 45-58

Maitland PS (2000) Guide to Freshwater Fish of Britain and Europe. Octopus Publishing, London. 256 pp

Makarewicz JC, Bertram O & Lewis TW (2000) Chemistry of the offshore surface waters of Lake Erie: Pre- and Post-*Dreissena* introduction (1983-1993). Journal of Great Lakes Research 26: 82-93

Makarova IV (1969) Formation and relations of the Black, Azov and Caspian Sea phytoplankton. Botanical Journal 54: 389-398 (in Russian)

Makrushin AV (1985) The Anhydrobiosis of the Early Water Vertebrates: the preservation of vital capacity in dried state, 104 pp. Nauka, Moscow

Malonette-Rizzoli P & Hecht A (1988) Large-scale properties of the eastern Mediterranean: A review. Oceanologica Acta 11: 323-335

Malanotte-Rizzoli P, Manca BB, d'Alcala MR, Theocharis A, Brenner S, Budillon G & Özsoy E (1999) The Eastern Mediterranean in the 80s and in the 90s: the big transition in the intermediate and deep circulations. Dynamics of Atmospheres and Oceans 29: 365-395

Manabe T & Ishio S (1991) Bloom of *Coscinodiscus wailesii* and DO deficit of bottom water in Seto Inland Sea. Marine Pollution Bulletin 23: 181-184

Mandell GL, Bennett JE & Dolin R (eds) (1995) Principles and Practice of Infectious Diseases. pp 2087-2103. Churchill Livingstone Inc.

Mann R & Gallagher SM (1985) Growth, morphometry and biochemical composition of the wood boring Molluscs *Teredo navalis* L., *Bankia gouldi* (Bartsch), and *Nototeredo knoxi* (Bartsch) (Bivalvia: Teredinidae). Journal of Experimental Marine Biology and Ecology 85: 229-251

Mann R & Harding JM (2000) Invasion of the North American Atlantic coast by a large predatory Asian mollusc. Biological Invasions 2: 7-22

Mannonen A & Westman K (1998) Crayfish situation in Finland. In: Taugøl T (ed) Nordic-Baltic Workshop on Freshwater Crayfish Research and Management, May 22-26, 1998, Estonia. Østlandsforsking, Rapport 26/1998: 29-38

Mantelatto FLM & Dias LL (1999) Extension of the known distribution of *Charybdis hellerii* (A. Milne-Edwards, 1867) (Decapoda, Portunidae) along the western tropical South Atlantic. Crustaceana 72: 617- 620

Maran T & Henttonen H (1995) Why is the European mink (*Mustela lutreola*) disappearing? Annales Zoologici Fennici 32: 47-54

Maran T, Kruuk H, Macdonald DW & Polma M (1998) Diet of two species mink in Estonia: displacement of *Mustela lutreola* by *M. vison*. Journal of Zoology 245: 218-222

Marguillier S (1998) Stable isotopes ratios and food web structure of aquatic ecosystems. PhD Thesis, University of Brussels

Mariani M, Borsani G & Saronni G (1992) Distribution of *Dreissena polymorpha* (Pallas, 1711) in Northern Italy. In: Giusti F & Manganelli G (eds) Abstracts of 11th International Malacological Congress: 19-22

Marsden JE (1997) Common carp diet includes zebra mussels and lake trout eggs. Journal of Freshwater Ecology 12: 491-492

Marsden JE, Spiedle AP & May B (1996) Review of genetic studies of *Dreissena* spp. American Zoologist 36: 259-270

Marshall TC, Slate J, Kruuk LEB & Pemberton JM (1998) Statistical confidence for likelihood-based paternity inference in natural populations. Molecular Ecology 7: 639-655

Marston M & Villalard-Bohnsack M (2000) The use of molecular genetics to investigate the geographic origin and vector of an invasive red alga. In: Pederson J (ed) Marine Bioinvasions. Proceedings of the First National Conference, January 24-27, 1999, pp 244-250. Massachusetts Institute of Technology, Cambridge, MA

Martens A, Eggers IO & Grabow K (1999) First record of *Pontogammarus robustoides* (Grimm) in Mittelland-canal (Crustacea: Amphipoda). Lauterbornia 35: 39-42 (in German)

Martynov AB (1919) On the higher Crustacea in the region of Rostov over Don. Arbeiten der Naturforschende Gesellschaft der Donschen Universitat Rostov 1(3): 39-53 (in German)

Maslowski J (1992) Bottom macrofauna of the Szczecin Lagoon (northwestern Poland). Acta Hydrobiologia 34: 253-274

Masselin P, Amzil Z, Abadie E, Nézan E, Le Bec C, Carreras A, Chiantella C & Truquet P (2001) Paralytic shellfish poisoning on the French Mediterranean coast in autumn 1998: *Alexandrium* "*tamarense* complex" (Dinophyceae) as causative agent. In: Hallegraeff GM, Blackburn SI, Bolch CJ & Lewis RJ (eds) Proceeding of the Ninth International Conference Harmful Algal Blooms, Paris (in press)

Mather ME (1990) Mechanisms of a species replacement in a benthic stream community. PhD Thesis, The Ohio State University, 162 pp

Mathiasson S (1973) A moulting population of non-breeding Mute swans with special reference to flight-feather moult, feeling ecology and habitat selection. Wildfowl 24: 43-53

Matveev VI & Solovyeva VV (1977) Zyzania. Samara State Pedagogic University, Samara, 96 pp (in Russian)

May B & Marsden JE (1992) Genetic identification and implications of a second invasive species of dreissenid mussel in the Great Lakes. Canadian Journal of Fisheries and Aquatic Sciences 49: 1501-1506

Mayer CM, VanDeValk AJ, Forney JL, Rudstam LG & Mills EL (2000) Response of yellow perch (*Perca flavescens*) in Oneida Lake, New York, to the establishment of zebra mussels (*Dreissena polymorpha*). Canadian Journal of Fisheries and Aquatic Sciences 57: 742-754

Mayhew DF (1977) Reinterpretation of the extinct beaver *Trogontherium*. Philosophical Transactions of the Royal Society B 281: 407-438

Mayhoub H (1976) Recherches sur la végétation marine de la côte syrienne. Etude expérimentale sur la morphogenèse et le développement de quelques espèces peu connues. PhD Thesis, Université Caen, France, 286 pp

Mayhoub HJ (1989) Envahissement de la côte syrienne par une algue brune de mer Rouge. Damascus University Journal 5: 65-79

Mayhoub HJ (1990) Syrians Algae. A study on some new red algae in the Mediterranean. Damascus University Journal 6: 22-37

Mayr E (1965) The nature of colonising birds. In: Bakker HG & Stebbins GL (eds) The Genetics of Colonizing Species, pp 29-43. Academic, New York, NY

Mazak EJ, MacIsaac HJ, Servos MR & Hesslein R (1997) Influence of feeding habits on organochlorine contaminant accumulation in waterfowl on the Great Lakes. Ecological Applications 7: 1133-1143

McCarthy SA & Khambaty FM (1994) International dissemination of epidemic *Vibrio cholerae* by cargo ship ballast and other non-potable waters. Applied Environmental Microbiology 60: 2597-2601

McCarthy TK, Fitzgerald J & O'Connor W (1997) The occurrence of the zebra mussel *Dreissena polymorpha* (Pallas 1771), an introduced biofouling freshwater bivalve in Ireland. Irish Naturalists' Journal 25: 413-416

McCollin T, Macdonald EM, Dunn J, Hall C & Ware S (2001) Investigations into ballast water exchange in European regional seas. In: International Conference on Marine Bioinvasions, New Orleans April 9-11 2001, pp 94-95

McCollin TA, Hamer JP & Lucas IAN (1999) Marine organisms transported in ships' ballast water. Final Report to Report to Ministry for Agriculture, Fisheries and Food, UK

McDermott JJ (2000) Natural history and biology of the Asian shore crab *Hemigrapsus sanguineus* in the Western Atlantic: a review, with new information. In: Pederson J (ed) Marine Bioinvasions. Proceedings of the First National Conference, January 24-27, 1999, pp 193-199. Massachusetts Institute of Technology, Cambridge, MA

McGill DA (1961) A preliminary study of the oxygen and phosphate distribution in the Mediterranean Sea. Deep Sea Research 9: 259-269

McGill DA (1965) The relative supplies of phosphate, nitrate and silicate in the Mediterranean. Rapports et procès-verbaux des réunions, Commission internationale pour l'Exploration scientifique de la mer Méditerranée 18: 734-744

McLeod MJ (1986) Electrophoretic variation in North American *Corbicula*. American Malacological Bulletin 2: 125-132

McMinn A, Hallegraeff GM, Thompson P, Jenkinson AV & Heinjis H (1998) Cyst and radionucleotide evidence for the recent introduction of the toxic dinoflagellate *Gymnodinium catenatum* into Tasmanian waters. Marine Ecology Progress Series 161: 165-172

Medcof JC (1975) Living marine animals in a ships' ballast water. Proceedings National Shellfish Association 65: 54-55

Meesenburg H (1975) The colonisation of *Spartina* along Ho Bugt. Geografisk Tidskrift 71: 37-45 (in Danish)

Meijer T & Preece RC (2000) A review of the occurrence of *Corbicula* in the Pleistocene of North-West Europe. Geologie en Mijnbouw-NJG 79: 241-255

Meinesz A (1992) Modes de dissémination de l'algue *Caulerpa taxifolia* introduite en Méditerranée. Rapport Commission International Mer Méditerranée 33: 44

Meinesz A (1997) The use of Ascoglossan as a biological control of *Caulerpa taxifolia* in the Mediterranean Sea. In: Dynamics of Invasive Marine Species: Application to the Expansion of *Caulerpa taxifolia* in the Mediterranean. Actes de Colloques, Septembre 1997, Académie des Sciences, pp 291-300. Lavoiser Technique & Documentation, Paris (in French)

Meinesz A & Boudouresque CF (1996) Sur l'origine de *Caulerpa taxifolia* en Mediterranee. (On the origin of *Caulerpa taxifolia* in the Mediterranean Sea.). Comptes Rendus de l'Academie des Sciences, Sciences de la vie 319: 603-613

Meinesz A & Hesse B (1991) Introduction et invasion de l'algue tropicale *Caulerpa taxifolia* en Méditerranée nord-occidentale. Ocenologia Acta 14: 415-426

Meinesz A, Cottalorda JM, Chiaverini D, Cassar N & De Vaugelas J (eds) (1998) Suivie de l'invasion de l'algue tropicale *Caulerpa taxifolia* en Méditerranée: situation au 31 décembre 1997. LEML-UNSA publ. France, 238 pp

Meister A (1997) Lebenszyklus, Autökologie und Populationsökologie der Körbchenmuscheln *Corbicula fluminea* und *Corbicula fluminalis* (Bivalvia, Corbiculidae) im Inselrhein. Dissertation Technische Hochschule Darmstadt

Mekhanikova IV (1982) Life cycle of *Gmelinoides fasciatus* in Bratskoe Reservoir. In: Proceedings of All-Union Scientific Conference on Problems of Ecology of Baikal Area 3: 34 (in Russian)

Mellina E, Rasmussen JB & Mills EL (1995) Impact of zebra mussel (*Dreissena polymorpha*) on phosphorus cycling and chlorophyll in lakes. Canadian Journal of Fisheries and Aquatic Sciences 52: 2553-2573

Menard A, L'Hostis M, Leray G, Marchandeau S, Pascal M, Roudot N, Michel V & Chauvin A (2000) Inventory of wild rodents and lagomorphs as natural hosts of *Fasciola hepatica* on a farm located in a humid area in Loire Atlantique (France). Parasite 7: 77-82

Mendez Domingo C (1957) Sur l'existence du *Colpomenia peregrina* (Sauv.) Hamel dans la Méditerranée. Vie et Milieu 8: 92-98

Menge B (1995) Indirect effects in marine rocky intertidal interaction webs: patterns and importance. Ecological Monographs 65: 21-74

Menzel RW (1974) Portugese and Japanese oysters are the same species. Journal of the Fisheries Research Board of Canada 31: 453-456

Mianzan H (1999) Ctenophora. In: Boltovskoy D (ed) South Atlantic Zooplankton, pp 561-573. Backhuys Publishers, Leiden, The Netherlands

Micol M, Doncaster P, Jouventin P & Guedon G (1996) Demographie du ragondin. In: Jouventin P, Micol T, Verheyden C & Guedon G (eds) Le Ragondin. Biologie et methods de limitation des populations, pp 30-41. Association de Coordination Technique Agricole, Paris

Mienis HK (1984) *Strombus decorus persicus* found in Israel. Hawaiian Shell News 32: 4

Mienis HK (1999) *Strombus persicus* on the fishmarket of Yafo, Israel. De Kreukel 35: 112

Mienis HK (2001) Noord-Amerikaanse aquatische mollusken die voorkomen in het binnenwater van West Europa. Spirula/Correspondentie-blad van de Nederlandse Malacologische Vereniging 320: 54-55

Mienis HK, Galili E & Rapoport J (1993) The spiny oyster, *Spondylus spinosus*, a well established Indo-Pacific bivalve in the eastern Mediterranean off Israel (Mollusca, Bivalvia, Spondylidae). Zoology in the Middle East 9: 83-91

Mikheev VP (1964) About the rate of mortality in *Dreissena* under anoxic conditions. Trudy insituta Biologii Vnutrennikh vod 7(12): 76-80 (in Russian)

Mikheev VP (1967) Filter feeding in *Dreissena*. Trudy VNIIPRKh 15: 117-129 (in Russian)

Mikheev VP, Karatayev AY & Burlakova LE (1994) Feeding. In: Starobogatov JI (ed) Freshwater Zebra Mussel *Dreissena polymorpha* (Pall.) (Bivalvia, Dreissenidae). Systematics, Ecology, Practical Meaning, pp 120-137. Nauka, Moscow, Russia (in Russian)

Mikschi E, Wolfram G & Wais A (1996) Long-term changes in the fish community of Neusiedler See (Burgenland, Austria). In: Conservation of Endangered Freshwater Fish in Europe, pp 111-120. Birkhäuser Verlag, Basel

Millar RH (1960) The identity of the Ascidians *Styela mammiculata* Carlisle and *S. clava* Herdman. Journal of the Marine Biological Association of the UnitedKingdom 39: 509-511

Miller AW, Hewitt CL & Ruiz GM (2002) Invasion success: does size really matter. Ecology Letters 5: 159-162

Miller GC (1960) The taxonomy and certain biological aspects of the crayfish of Oregon and Washington. Thesis, Oregon State College, 216 pp

Miller W (1998) Ballast research at the Smithsonian Environmental Research Center. Paper presented at ICES/IMO/IOC Study Group Workshop on Ballast water and Sediments, The Hague, Netherlands, 23-24 March

Mills ED, Leach JH, Carlton JT & Secor CL (1993) Exotic species in the Great Lakes: A history of biotic crises and anthropogenic introductions. Journal of Great Lakes Research 19: 1-54

Mills L, Rosenberg G, Spiedle AP, Ludyanskiy M & Pligin Y (1996) A review of the biology and ecology of the quagga mussel (*Dreissena bugensis*), a second species of freshwater dreissenids introduced to North America. American Zoologist 36: 271-286

Minchin D (1996a) Desk assessment of discharges of shipping ballast water in Ireland. ICES CM 1996/E11, 13 pp

Minchin D (1996b) Management of the introduction and transfer of marine molluscs. Aquatic Conservation: Marine and Freshwater Ecosystems 6: 229-244

Minchin D (1997) The significance of ballast water in the introduction of exotic marine organisms to Cork Harbour, Ireland. Report of the SGBWS, La Tremblade, France, Annex 4: 35-36

Minchin D (1999) *Crepidula fornicata* (Linnaeus, 1758), Calyptraeidae, Gastropoda. In: Gollasch S, Minchin D, Rosenthal H & Voigt M (eds) Exotics Across the Ocean. Case Histories on Introduced Species. Logos Verlag, Berlin, Germany. 74 pp

Minchin D (2000a) Modes of introduction of exotic animals into Irish waters. In: Rushton BS (ed) Biodiversity: The Irish Dimension, pp 147-155. Royal Irish Academy, Dublin

Minchin D (2000b) Dispersal of zebra mussels in Ireland Verhandlungen der Internationalen Vereinigung für theoretische und angewandte Limnologie 27: 1576-1579

Minchin D (2001) Introduction of exotic species. In: Steele J, Thorpe S & Turekian K (eds) Encyclopedia of Ocean Sciences, pp 877-889. Academic Press, London

Minchin D & Duggan CB (1988) The distribution of the exotic ascidian, *Styela clava* Herdman, in Cork Harbour. Irish Naturalists' Journal 22: 388-393

Minchin D & Moriarty C (1998) Distribution of the zebra mussel in Ireland in 1997. Irish Naturalists' Journal 26: 38-42

Minchin D & Sheehan J (1995) The significance of ballast water in the introduction of exotic marine organisms to Cork Harbour, Ireland. ICES Cooperative Research Report 224: 12-23

Minchin D, Duggan CB & McGrath D (1987) *Calyptraea chinensis* (Mollusca: Gastropoda) on the west -coast of Ireland: a case of accidental introduction? Journal of Conchcology 32: 297-302

Minchin D, McGrath D & Duggan CB (1995) The slipper limpet, *Crepidula fornicata* (L.), in Irish waters, with a review of its occurrence in the North-Eastern Atlantic. Journal of Conchology 35: 247-254

Minelli A, Ruffo S & La Posta S (1995) Checklist delle specie della fauna Italiana. Fasc. 13-18. Edizioni Calderini, Bologna

Minkina NI & Pavlova EV (1995) Diurnal changes of the respiration rate in the comb jelly *Mnemiopsis leidyi* in the Black Sea. Oceanology 35: 222-225

Miroshnichenko MP (1987) Molluscs of the Tsimlyanskoe reservoir and their role in increasing fish productivity. Gosudarstvennyj Nauchno-Issledovatel'skij Institut Ozernogo i Rechnogo Rybnogo Khozyajstva 270: 61-70 (in Russian)

Miroshnichenko MP, Volvich LI & Skabichevskii BO (1984) Production and energetic balance of the most abundant molluscs in Tsimlyanskoe Reservoir. Sbornik nauchnykh trudov. Gosudarstvennyj Nauchno-Issledovatel'skij Institut Ozernogo i Rechnogo Rybnogo Khozyajstva 218: 19-30 (in Russian)

Mitchell R (1987) Conservation of marine benthic biocenoses in the North Sea and the Baltic. A framework for the establishment of a European network of marine protected areas in the North Sea and the Baltic. European Committee for the Conservation of Nature and Natural Resources. Nature and Environment Series 37

Mitchell-Jones AJ, Amori G, Bogdanovich W, Krystufek B, Reijnders PJH, Spitzenberger F, Stubbe M, Thissen JBM, Vohralik V & Zima J (1999) Atlas of European Mammals. The Academic Press, London, 496 pp

Mitrakhovich PA (1984) Dynamic and effect of different factors on zooplankton community of the thermal power plant's cooling reservoir. PhD Thesis, Zoology Institute of Academy of Science Belorussian SSR, Minsk, Belarus 115 pp (in Russian)

Mitrakhovich PA, Lyakhnovich VP & Boikova SA (1983) Zooplankton of lake Lukoml under the influence of thermal pollution. Hydrobiological Journal 19: 54-57

Mitretek Systems (2001) Toxins. www.mitretek.org/mission/envene/biological/agents/toxins.html (2001-09-03)

Mitskevich OI (1980) The production of *Gmelinoides fasciatus* Stebb., acclimated in Lake Otradnoe. Questions of acclimatization of feeding fishes and invertebrates. Suppl 25: 18-28 (in Russian)

Mitton JB (1997) Selection in Natural Populations. Oxford University Press, Oxford

Mizzan L (1993) Presence of the swimming crab of the genus *Callinectes* (Stimpson) (Decapoda, Portunidae) in the Venice lagoon (North Adriatic sea - Italy): first record of *Callinectes danae* Smith in European waters. Bollettino Museo civico di Storia naturale Venezia 42: 31-43

Mizzan L (1995) Notes on presence and diffusion of *Dyspanopeus sayi* (Smith, 1869) (Crustacea, Decapoda, Xanthidae) in the Venitian Lagoon. Bollettino Museo civico di Storia naturale Venezia 44: 121-129

Mizzan L & Zanella M (1996) First record of *Rhithropanopeus harrisii* (Gould,1841) (Crustacea, Xanthidae) in the Italian waters. Bollettino Museo civico di Storia naturale Venezia 46: 109-122

Mo T-A & Steien SH (1994) First observation of the eel swimbladder nematode *Anguillicola crassus* in Norway. Bulletin of the European Association of Fish Pathologists 14: 163-164

Mo T-A & Sterud E (1998) First observation of *Pseudodactylogyrus*spp. and other parasites in wild eel (*Anguilla anguilla*) in Norway. Bulletin of the Scandinavian Society for Parasitology 8: 1-5

Möbius K (1977) Die Auster und die Austernwirthschaft. Wiegandt, Hempel & Parey, Berlin. 126 pp

Modena M, Matricardi G, Vacchi M & Guidetti P (2000) Spreading of *Caulerpa racemosa* (Forsskål) J.Agardh (Bryopsidaceae, Chlorophyta) along the coasts of the Ligurian Sea. Cryptogamie, Algologie 21: 301-304

Moerdijk P (1986) Hoe staat het met de Japanse oester? Het Zeepaard 46: 170-171

Mojetta A & Ghisotti A (1996) Flore et Faune de la Méditerranée. Solar Publication, Paris, 318 pp

Molenaar H, Thibaut T & Meinesz A (2001) Experimental invasion dynamic of endemic *Posidonia oceanica* beds by introduced *Caulerpa taxifolia*. Phycologia 40 (4) Supplement: 14 (Abstract)

Moll F (1914) The borer (Genus *Teredo* Linné). Naturwissenschaftliche Zeitschrift für Forst- und Landwirtschaft 12: 505-564 (in German)

Mollion J (1984) Seaweed cultivation for phycocolloid in the Mediterranean. Hydrobiologia 204-205: 288-291

Molloy DP, Karatayev AY, Burlakova LE, Kurandina DP & Laruelle F (1997) Natural enemies of zebra mussels: predators, parasites and ecological competitors. Reviews in Fisheries Science 5: 27-97

Momot WT & Leering GM (1986) Aggressive interaction between *Pacifastacus leniusculus* and *Orconectes virilis* under laboratory conditions. Freshwater Crayfish 6: 87-93

Momot WT (1988) *Orconectes* in North-America and elsewhere. In: Holdich M & Lowery RS (eds) Freshwater Crayfish: Biology, Management and Exploitation. pp 262-282. Croom Helm (Chapman & Hall), London

Momot WT (1995) Redefining the role of crayfish in aquatic ecosystems. Reviews in Fisheries Science 3: 33-63

Monod Th (1932) Crustacés exotiques en Méditerranée. La Terre et la Vie Revue d'Histoire Naturelle Paris 2: 65-73

Montani S, Meksumpun C & Ishimi K (1995) Chemical and physical treatments for destruction of phytoflagellate cysts. Journal of Marine Biotechnology 2: 179-181

Monterosato TA (1878) Enumerazione e sinonimia delle conchiglie Mediterranee. Giornale di scienze naturali ed economiche di Palermo 13: 61-113

Monterosato TA (1899) Coquilles Marins de Chypre. Journal de Conchyliologie 47: 1-9

Montresor M, Sarno D & Zingone A (2000) Can we peacefully live together with harmful phytoplankton? The case of the Gulf of Naples. 9th International Conference on Harmful Algal Blooms, Tasmania, Australia, 7-11 February 2000, p 182. Univ. of Tasmania, Hobart (Abstract)

Montresor M, Zingone A & Sarno D (1998) Dinoflagellate cyst production at a coastal Mediterranean site. Journal of Plankton Research 20: 2291-2312

Mooney HA & Cleland EE (2001) The evolutionary impact of invasive species. Proceedings of the National Academy of Science 98: 5446-5451

Mooney HA & Hobbs RJ (eds) (2000) Invasive Species in a Changing World. Island Press, Washington, D.C., 457 pp

Morcos SA (1960) Die Verteilung des Salzgehaltes in Suez-Kanal. Kieler Meeresforschungen 16: 133-154

Morcos SA (1980) Seasonal changes in the Suez Canal following its opening in 1869: newly discovered hydrographic records of 1870-1872. In: Sears M & Merriman D (eds) Oceanography: The Past, pp 290-305. Springer-Verlag, Berlin

Mordukhai-Boltovskoi FD (1960) Caspian fauna in the Azov - Black Seas basin. Izdatelstvo Akademii Nauk SSSR, Moskva-Leningrad, 288 pp (in Russian)

Mordukhai-Boltovskoi FD (1964) Caspian fauna beyond the Caspian Sea. Internationale Revue der gesamten Hydrobiologie und Hydrographie 49: 139-176

Mordukhai-Boltovskoi FD (1974) Atlas of the Aral Sea Invertebrates. Pishchevaya promishlennost Moscow, 272 pp (in Russian)

Mordukhai-Boltovskoi FD & Chirkova EN (1971) About distribution of Baikalian amphipod *Gmelinoides fasciatus* (Stebbing) in Gorkovskoe Reservoir. Biology of Inland Waters Inform Bulletin 9: 38-42 (in Russian)

Mordukhai-Boltovskoi FD & Dzjuban (1976) Changes in composition and distribution of Volga River fauna caused by human activity. In: Biologicheskie produkcionnye processy v basejne Volgi. Institut Biologii Vnutrennykh Vodoemov AN SSSR, Izdatelstvo "Nauka", Leningrad: 67-82 (in Russian)

Mordukhai-Boltovskoi FD & Galinskij VL (1974) On the further distribution of the Caspian Polyphemoidea in the reservoirs of the Ponto-Caspian rivers. Informacionnyj Bjulletin Instituta Biologii Vnutrennykh Vodoemov AN SSSR 21: 40-44 (in Russian)

Mordukhai-Boltovskoi FD & Rivier IK (1987) The predatory cladocera (Onychopoda: Podonidae, Polyphemidae, Cercopagidae and Leptodorida) of the world fauna. Nauka, Leningrad. 184 pp (in Russian)

Mordukhai-Boltovskoi FD, Greze II & Vasilenko SV (1969) Amphipods - Amphipoda Latreille, 1816-1817. In: Vodjanickij VA (ed) Opredelitel fauny Chernogo i Azovskogo morej. Izdatelstvo "Naukova Dumka", Kiev: 440-524 (in Russian)

Moreira I, Ferreira T, Monteiro A, Catarino L & Vasconcelos T (1999) Aqautic weed and their management in Portugal: insights and the international context. Hydrobiologia 415: 229-234

Morello E & Solustri C (2001) First record of *Anadara demiri* (Piani, 1981) (Bivalvia: Arcidae) in Italian waters. Bollettino Malacologico 37: 231-234

Moriarty C & Fitzmaurice P (2000) Origin and diversity of freshwater fishes in Ireland. Verhandlungen der Internationalen Vereinigung für theoretische und angewandte Limnologie 27: 128-130

Morley RS (1993) A model for the assessment of the animal disease risks associated with the importation of animals and animal products. Revue scientifique et technique de l'Office international des épizooties 12: 1055-1092

Morrice MG (1995) The distribution and ecology of the introduced Northern Pacific seastar, *Asterias amurensis* (Lütken), in Tasmania. In: The introduced northern Pacific seastar *Asterias amurensis* in Tasmania, pp 1-50. Australian Nature Conservation Agency

Morris MJ, Wood AR & den Breeyen A (1999) Plant pathogens and biological control of weeds in South Africa: a review of projects and progress during the last decade. African Entomology: 129-137

Mortensen S (1993) On transfers and introductions of living organisms 2. Bivalves, examples of success and fiasco. Naturen 4/1993: 156-162 (in Norwegian)

Morton B (1981) The biology and functional morphology of *Mytilopsis sallei* (Recluz) (Bivalvia: Dreissenacea) fouling Visakhapatnam Harbour, Andhra Pradesh, India. Journal of Molluscan Studies 47: 25-42

Morton B (1989) Life history characteristics and sexual strategy of *Mytilopsis sallei* (Bivalvia: Dreissenacea), introduced into Hong Kong. Journal of Zoology London 219: 469-485

Morton BS (1971) Studies on the biology of *Dreissena polymorpha* Pall. Some aspects of filter-feeding and the effect of microorganisms upon the rate of filtration Proceedings of Malacological Society 39: 289-301

Morton BS (1979) Studies on the biology of *Dreissena polymorpha* (Pallas): Population dynamics. Proceedings of the Malacological Society of London 38: 471-482

Mountfort DO, Dodgshun T, Gibbs W & McCallin B (1999) Towards a feasible heat treatment system for ships' ballast water. Proceedings of Workshop on Ballast Water Problem - Where to from here? Ports Corporation of Qld. EcoPorts Monograph Series 19

Mountfort DO, Hay C, Taylor M, Buchanan S & Gibbs W (2000) Heat treatment of ship's ballast water: development of a model based on laboratory studies. Journal of Marine Environment and Engineering (in press)

Moyle PB & Light T (1996) Biological invasions of fresh water: empirical rules and assembly theory. Biological Conservation 78: 149-161

Moyle PB & Light T (1996) Fish invasions in California: do abiotic factors determine success? Ecology 77: 1666-1670

Muehlstein LK, Porter D & Short FT (1988) *Labyrinthula* sp., a marine slime mold producing the symptoms of wasting disease in eelgrass, *Zostera marina*. Marine Biology 99: 465-472

Muehlstein LK, Porter D & Short FT (1991) *Labyrinthula zosterae* sp. nov., the causative agent of wasting disease in eelgrass, *Zostera marina*. Mycologia 83: 180-191

Mueller KW & Hoffmann A (1999) Effect of freshwater immersion on attachment of the Japanese oyster drill, *Cerastostoma inornatum* (Récluz, 1851). Journal of Shellfish Research 18: 597-600

Müller J (1998) Genetic population structure of two cryptic *Gammarus fossarum* types across a contact zone. Journal of Evolutionary Biology 11: 79-101

Müller J (2000) Mitochondrial DNA variation and the evolutionary history of cryptic *Gammarus fossarum* types. Molecular Phylogenetics and Evolution 15: 260-268

Müller J & Schramm S (2001) Genetic evidence for a third *Dikerogammarus* invader of the upper Danube. Lauterbornia 41 (in press)

Müller J, Wöll S, Fuchs U & Seitz A (2001) Genetic interchange of *Dreissena polymorpha* populations across a canal. Heredity 86: 103-109

Müller JC (2001) Invasion history and genetic population structure of riverine macroinvertebrates - a review. Zoology: Analysis of Complex Systems (in press)

Müller K (1995) Disinfection of Ballast Water. A Review of Potential Options. Lloyd's Register, 29 pp

Müller K & Reynolds GL (1995) Disinfection of Ballast Water. A Review of Potential Options. Lloyd's Register, Report Summary, 38 pp

Müller O, Zettler ML & Gruszka P (2001) Distribution and status of *Dikerogammarus villosus* Sovinsky (Crustacea: Amphipoda) in the middle and the upper part of the River Oder and the adjacent waterways. Lauterbornia 41: 105-112 (in German)

Müller W (1984) Rearing of one-year-old silver carp (*Hypophthalmichthys molitrix*) in culture with carp (*Cyprinus carpio*) - Auswertung von Teichversuchen Fortschritte der Fischereiwis-senschaft 3: 115-120 (in German)

Munro ALS, Utting SD & Wallentinus I (eds) (1999) Status of introductions of non-indigenous marine species to North Atlantic waters. ICES Cooperative Research Report 231, 91 pp

Muravelko VM, Sparkovsky IA, Chinarina AD & Alexandrov DI (2000) Steelhead in the rivers of the East Murman. In: Species Introducers in the European Seas in Russia. Apatity, Publishing House of Kola Science Centre, pp 269-272 (in Russian)

Murugesan AG, Vijayalakshmi GS, Sukumaran N & Mariappan C (1995) Utilization of water hyacinth for oyster mushroom cultivation. Bioresource Technology 51: 97-98

Muskó IB (1992) Amphipoda species found in Lake Balaton since 1987. Miscnea. Zoology Hungary 7: 59-64

Mutlu E (1999) Distribution and abundance of ctenophores, and their zooplankton food in the Black Sea. II. *Mnemiopsis leidyi*. Marine Biology 135: 603-613

Myers GS (1940) An American cyprinodont fish *Jordanella floridae* reported from Borneo, with notes on the possible widespread introduction of foreign aquarium fishes. Copeia 1940(4): 267-268

Myrberget S (1987) Introduction of mammals and birds in Norway including Svalbard. Meddelelser fra Norsk Viltforskning 3(17): 1-26

Nagai S, Hori Y, Manabe T & Imai I (1995a) Morphology and rejuvenation of *Coscinodiscus wailesii* Gran (Bacillariophyceae) resting cells found in bottom sediments of Harima-Nada, Seto Inland Sea. Bulletin of the Japanese Society of Scientific Fisheries 61: 179-185

Nagai S, Hori Y, Manabe T & Imai I (1995b) Restoration of cell size by vegetative cell enlargement in *Coscinodiscus wailesii* (Bacillariophyceae). Phycologia 34: 533-535

Nair NB & Sraswathy M (1971) The biology of wood-boring teredinid molluscs. Advances in Marine Biology 9: 335-509

Nakai S, Inoue Y, Hosomi M & Murakami A (1999) Growth inhibition of blue-green algae by allelopathic effects of macrophyte. Water Science and Technology 39: 47-53

NAS (2001) Nonindigenous aquatic species information resource for the United States Geological Survey. http://nas.er.usgs.gov/ (2001-08-27)

Nastenko EV & Polishchuk LN (1999) The comb jelly *Beroe* (Ctenophora: Beroidea) in the Black Sea. Dopovidi Nac. Acad. Nauk Ukrainy 11: 159-161

Natili GL, Sola L, De Bonfils G & Gelosi E (1986) Prime osservazioni morfoecologiche su *Basilicthys bonariensis*, una specie ittica argentina introdotta nel lago di Nemi. Bollettino Zoologia 53: 99

National Research Council (1995) Understanding Marine Biodiversity: A Research Agenda for the Nation. National Academy Press, Washington, 128 pp

National Research Council (1996) Stemming the Tide: Controlling Introductions of Non-Indigenous Species by Ships' Ballast Water. National Academy Press, USA

Nature Conservancy (1998) America's least wanted: Alien species invasions of U.S. ecosystems (www.consci.tnc.org/library/pubs/dd/toc.html)

Naumenko EN & Polunina YuYu (2000) New cladocera species - *Cercopagis pengoi* (Ostroumov, 1891) (Crustacea) in the Vistula Lagoon of the Baltic Sea. ICES C.M./U: 14

Naumov DV (1960) Hydrozoans and hydromedusae of marine, brackish and freshwater basins of Soviet Union. Opredelitel fauny SSSR, Izdatelstvo Zoologicheskogo Instituta AN SSSR 70: 585 pp (in Russian)

Naylor E (1965) Biological effects of a heated effluent in docks at Swansea, S. Wales. Proceedings of the Linnean Society London 144: 253-268

Nebol'sina TS (1975) Volgograd Reservoir. Izvestiya GosNIORKh, 102: 130-148 (in Russian)

Nee S (1990) Community construction. Trends in Ecology and Evolution 5: 337-340

Neet C (1999) Gestion des neozoa et des especes terrestres exotiques en Suisse: un bilan des problemes et politiques actuelles. Workshop on the Control and Eradication of Non-native Terrestrial Vertebrates, Malta, pp 67-71. Environmental Encounters, Council of Europe Publishing

Neev D, Greenfield L & Hall JK (1985) Slice tectonics in the Eastern Mediterranean Basin. In: Stanley DJ & Wezel FC (eds) Geological Evolution of the Mediterranean Basin, pp 249-269. Springer Verlag, New York

Nehring S (1998) Establishment of thermophilic phytoplankton species in the North Sea - biological indicators of climatic changes? ICES Journal of Marine Science 55: 818-823

Nehring S (1998) Non-indigenous phytoplankton species in the North Sea: supposed region of origin and possible transport vector. Archive of Fishery and Marine Research 46: 181-194

Nehring S (1999) Oyster beds and *Sabellaria* reefs. In: De Jong F, Bakker JF, van Berkel CJM, Dankers NMJA, Dahl K, Gätje C, Marencic H & Potel P (eds) Wadden Sea Quality Status Report, pp 146-147. Common Wadden Sea Secretariat Wilhelmshaven, Wadden Sea Ecosystem 9

Nehring S (1999) Zur Rolle der Seeschifffahrt bei der Einschleppung fremder Tierarten (Makrozoobenthos) in deutsche Gewässer. In: Umweltaspekte der Seeschifffahrt, pp 101-110. GAUSS Forum, Bremen

Nehring S (2000a) On the occurrence of *Rhithropanopeus harrisii* (Gould, 1841) in German aquatic habitats: the successive spreading of an American neozoon (Crustacea: Decapoda: Panopeidae). Senckenbergiana maritima 30: 115-122 (in German)

Nehring S (2000b) Biodiversity and nature conservation in aquatic systems - on the status of introduced animal species. Wasser und Boden 51 (1/2): 23-26 (in German)

Nehring S (2001) After the TBT era: Alternative anti-fouling paints and their ecological risks. In: Kröncke I, Türkay M & Sündermann J (eds) Burning Issues of North Sea Ecology. Senckenbergiana maritima 31 (in press)

Nehring S & Leuchs H (1999) Neozoa (Makrozoobenthos) an der deutschen Nordseküste - Eine Übersicht. Bundesanstalt für Gewässerkunde, Koblenz, Bericht BfG-1200. 131 pp

Nehring S & Leuchs H (2000) Neozoan invertebrates in the brackish waters at the German North Sea coast. Lauterbornia 39: 73-116 (in German)

Nei M, Maruyama T & Chakraborty R (1975) The bottleneck effect and genetic variability in populations. Evolution 29: 1-10

Nekludov MN (1992) Some results of pink salmon acclimatization on European North. In: Scarlato OA (ed) Problems of study, rational exploitation and conservation of natural resources of the White Sea. Proceedings of the 5th regional meeting. Petrozavodsk, pp 237-239 (in Russian)

Nesemann H, Pöckl M & Wittmann KJ (1995) Distribution of epigean Malacostraca in the middle and upper Danube (Hungary, Austria, Germany). Miscellanea zoologica hungarica 10: 49-68

Nesler TP & Bergersen EP (1991) Mysids and their impacts on fisheries: an introduction to the 1988 mysid-fisheries symposium. In: Nesler TP & Bergersen EP (eds) Mysids in fisheries: hard lessons from headlong introductions. American Fisheries Society Symposium 9: 1-4

Neudecker T (1985) Untersuchungen zur Reifung, Geschlechtsumwandlung und künstlichen Vermehrung der Pazifischen Auster *Crassostrea gigas* in deutschen Gewässern. Veröffentlichungen Institut für Küsten- und Binnenfischerei Hamburg 88: 1-212

Nevessky EN, Medvedev VS & Kalinenko VV (1977) The White Sea. The sedimentogenesis and history of the development during Holocene. Moscow, Nauka Publisher, 236 pp (in Russian)

Nevesskaya LA (1976) About pedomorphosis as one of possible ways of species formation in bivalves. In: 50th anniversary of Soviet palaeontology and questions of taxonomy of ancient organisms, pp 48-52. Nauka, Leningrad (in Russian)

Ng PKL & Sivasothi N (eds) (2001) A Guide to Mangroves of Singapore. The Singapore Science Center (http://mangrove.nus edu sg/guidebooks/ contents .htm)

Ng PKL & Tan HH (1997) Freshwater fishes of Southeast Asia: potential for the aquarium fish trade and conservation issues. Aquarium Sciences and Conservation 1: 79-90

Nichols FH, Thompson JK & Schemel LE (1990) Remarkable invasion of San Francisco Bay (California, USA) by the Asian clam *Potamocorbula amurensis*. II. Displacement of a former community. Marine Ecology Progress Series 66: 95-101

Nichols SA (1991) The interaction between biology and management of aquatic macrophytes. Aquatic Botany 41: 225-252

Nichols SJ & Amberg J (1999) Co-existence of zebra mussels and freshwater unionids: population dynamics of *Leptodea fragilis* in a coastal wetland infested with zebra mussels. Canadian Journal of Zoology 77: 423-432

Nicolay K & Romagna Manoja L (1983) *Strombus (Conomurex) decorus raybaudii* n.spp. La Conchiglia 174-175: 20-21

Niemimaa J (1995) Activity patterns and home ranges of the American mink *Mustela vison* in the Finnish outer archipelago. Annales Zoologici Fennici 32: 117-122

Niermann U, Bingel F, Gorban A, Gordina AD, Gücü AC, Kideys AE, Konsulov A, Radu G, Subbotin AA & Zaika VE (1994) Distribution of anchovy eggs and larvae (*Engraulis encrasicolus* Cuv.) in the Black Sea in 1991 and 1992 in comparison to former surveys. ICES Journal of Marine Science 51: 395-406

Nijland R (2000) Present distribution of the crab *Hemigrapsus penicillatus* in The Netherlands. Het Zeepaard 60: 316-317 (in Dutch)

Nijssen H & de Groot SJ (1974) Catalogue of fish species of the Netherlands. Beaufortia 21:173-207

Nikolaev I (1951) On new introductions from distant areas in fauna and flora of the North and the Baltic Seas. Zoologicheskij Zhurnal 30: 556-561 (in Russian)

Nikolaev II (1963) New introductions in fauna and flora of the North and Baltic Seas. Zoologicheskij zhurnal 42(1): 20-27 (In Russian, English summary)

Nilova OI (1976) Some features of ecology and biology of *Gmelinoides fasciatus* Stebb., acclimatized in Lake Otradnoe in Leningrad Region. Proceedings of GosNIORKh 110: 10-15 (in Russian)

Nilsson C & Grelsson G (1995) The fragility of ecosystems: A review. Journal of Applied Ecology 32: 677-692

Nilsson N-A (1985) The niche concept and the introduction of exotics. Reports of the Institute of Freshwater Research Drottningholm 62: 128-135

Nizzamuddin M (1981) Contribution to the marine algae of Libya. Dictyotales. Bibliotheca Phycologica 54. J. Cramer. Vaduz, 122 pp

Noël P (1998) Preliminary results of a tagging experiment on the crab *Hemigrapsus penicillatus*, recently introduced in Bay of Biscay (France). 4th International Conference on Crustaceans, Amsterdam, 20-24/ 07/98

Noël P, Tardy E & D'Udekem D'Acoz C (1997) Will the crab *Hemigrapsus penicillatus* invade the coasts of Europe? Comptes Rendus de l'Académie des Sciences, Sciences de la Vie 320: 741-745

Noel PY (1992) Preliminary key for identification of Crustacea Decapoda of France. Collection Patrimoines Naturels 9: 146 pp (in French)

Noordhuis R (2001) Macrofauna in het IJsselmeergebied. De Levende Natuur 102: 173-178

Noordhuis RH, Reeders H & bij de Vaate A (1992) Filtration rate and pseudofeces production in zebra mussels and their application in water quality management. In Neumann D & Jenner HA (eds) The Zebra Mussel *Dreissena polymorpha*: Ecology, Biological Monitoring and First Applications in the Water Quality Management, pp 101-114. Gustav Fisher, Stuttgart

Nordström M, Högmander J, Nummelin J, Laine J, Laanetu N & Korpimäki E (2002) Variable responses of waterfowl breeding populations to long-term removal of introduced American mink. Ecography (in press)

Norman E (1977) The geographical distribution and the growth of the wood-boring molluscs *Teredo navalis* L., *Psiloteredo megotara* (Hanley) and *Xylophaga dorsalis* (Turton) on the Swedish west coast. Ophelia 16: 233-250

Norris RE (1989) Natalian *Botryocladia* (Rhodymeniales, Rhodophyceae), including description of a new, long axis-forming species. Botanica Marina 32: 131-148

North Sea Task Force (1993) North Sea Quality Status Report 1993. Oslo and Paris Commissions, London. Olsen & Olsen, Fredensborg, Denmark, 132 pp

Novaczek I & McLaclan J (1989) Investigations of the marine algae of Nova Scotia XVII. Vertical and geographic distribution of marine algae on rocky shores of the Maritime Provinces. Proceedings of Nova Scotian Institute of Science 38: 91-143

Nowak E (1971) The range expansion of animals and its cause. Zeszyty Naukowe 3: 1-255 (in Polish; Transl. by Smithsonian Inst. and Nat. Sc. Found., Wash. D.C. Foreign Sc. Publ. Dep. of the Nat. Center for Sc., Tech., and Economic Information, Warsaw, Poland, 1975: 1-163)

Nummelin J & Högmander J (1998) New method for controlling minks in the outer archipelago. Metsästäjä 1/98: 16-18 (in Finnish)

Nummi P (1996) Wildlife introductions to mammal-deficient areas: the Nordic countries. Wildlife Biology 2: 221-226

Nummi P (2001) Alien species in Finland. The Finnish Environment 466. Ministry of the Environment, Helsinki 40 pp

Nylund V & Westman K (1979) *Psorospermium haeckeli*, a parasite on the European crayfish, *Astacus astacus* found in Finland. In: Laurent PJ (ed) Freshwater Crayfish IV: International Symposium on Freshwater Crayfish, pp 385-390. Institut National de Recherche Argonomique, Thonon les Bains, France

Nylund V & Westman K (1992) Crayfish diseases and their control in Finland. Finnish Fisheries Research 14: 107-118

Nylund V & Westman K (1995) The crayfish mortality register as an aid in the control of crayfish diseases in Finland. Freshwater Crayfish 10: 363-373

Nylund V & Westman K (2000) The prevalence of crayfish plague (*Aphanomyces astaci*) in two signal crayfish (*Pacifastacus leniusculus*) populations in Finland. Journal of Crustacean Biology 20: 777-785

Nyström P & Strand JA (1996) Grazing by a native and a exotic crayfish on aquatic macrophytes. Freshwater Biology 36: 673-682

Nyström P (1999a) The effects of crayfish on interactions in freshwater benthic communities. PhD Thesis, Department of Ecology, Limnology, Lund University. 30 pp

Nyström P (1999b) Ecological impact of introduced and native crayfish on freshwater communities: European perspectives. In: Gherardi F & Holdich DM (1999) (eds) Crayfish in Europe as Alien Species. How to Make the Best of a Bad Situation, pp 63-85. A.A. Balkema, Rotterdam, Brookfield

Nyström P, Brönmark C & Granéli W (1999) Influence of an exotic and native crayfish species on a littoral benthic community. Oikos 85: 545-553

O'Brien SJ & Evermann JF (1988) Interactive influence of infectious disease and genetic diversity in natural populations. Trends in Ecology and Evolution 3: 254-259

O'Foighil D, Gaffney PM, Wilbur AE & Hilbisch TJ (1998) Mitochondrial cytochrome oxidase I gene sequences support an Asian origin for the Portuguese oyster *Crassostrea angulata*. Marine Biology 131: 497-503

O'Mahony JHT (1993) Phytoplankton species associated with imports of the Pacific oyster *Crassostrea gigas*, from France to Ireland. ICES C.M. 1993/F:26, 8 pp

O'Riordan RM (1996) The current status and distribution of the Australian barnacle *Elminius modestus* Darwin in Ireland. In: Keegan BF & O'Connor R (eds) Proceedings of the Irish Marine Science Symposium 1995, pp 207-218. Galway University Press, Galway

Öberg T (1997) The king crab creaps onwards. MiljöAktuellt (18 juni 1997): 19 (in Norwegian)

Obidientova OV (1977) Erosion Cycles and Formatting of Volga Valley. Nauka Press, Moscow, 240 pp (in Russian)

Occhipinti Ambrogi A & d'Hondt JL (1996) Introduzione di specie alloctone in laguna di Venezia: *Celleporella carolinensis* Ryland, 1979 (Bryozoa: Ascophora). Bollettino Museo civico di Storia naturale Venezia 46: 53-61

Occhipinti Ambrogi A (2000a) Biotic invasions in a Mediterranean Lagoon. Biological Invasions 2: 165-176

Occhipinti Ambrogi A (2000b) Recent developments of the history of Bryozoans in the lagoon of Venice: biodiversity and environmental stress. Proceedings of the 11th International Bryozoology Association Conference, pp 305-315. Allen Press, Lawrence Kansas

Occhipinti Ambrogi A (2001) National report for Italy. In: Report of the Working Group of Introductions and Transfers of Marine Organisms, Barcelona, Spain, 21-23 March 2001, pp 59-63. ICES CM 2001/ACME:08, Ref. E, F, Annex 4

Oceanography Atlas for the Summer Period (1998) Environmental Working Group. Joint U.S. Russian Atlas of the Arctic Ocean: CD

Oceanography Atlas for the Winter Period (1997) Environmental Working Group. Joint U.S. Russian Atlas of the Arctic Ocean: CD

Oemcke D (1999) The Treatment of Ships' Ballast Water. EcoPorts Monograph Series 18, Ports Corporation of Queensland, Australia (www.pcq.com.au)

Ofugi K, Satake M, McMahon T, Silke J, James KJ, Naoki H, Oshima Y & Yasumoto T (1999) Two analogs of azaspiracid isolated from mussels, *Mytilus edulis*, involved in human intoxication in Ireland. Natural Toxins 7: 99-102

Oganesyan SA, Kuzmin SA & Filina EA (1998) New data about reproduction function and breeding of the red king crab in the Barents Sea. Materialy otchetnoy sessii PINRO po itogam 1996-1997 gg. Izdatelstvo Polyarnogo Instituta Ribnogo Hozyaystva i Okeanografii, Murmansk, pp 180-188 (in Russian)

Ojaveer H & Lumberg A (1995) On the role of *Cercopagis* (*Cercopagis*) *pengoi* (Ostroumov) in Pärnu Bay and the NE part of the Gulf of Riga ecosystem. Proceedings of the Estonian Academy of Sciences Ecology 5: 20-25

Ojaveer H, Simm M, Lankov A & Lumberg A (2000) Consequences of invasion of a predatory cladoceran. ICES C.M. 2000/U: 16

Øksnebjerg B, Enzenross R & Enzenross L (1997) First record of Rhizocephala (Crustacea; Cirripedia) from Turkish waters, with notes on lessepsian migration. Stuttgarter Beiträge zur Naturkunde (Biologie) 557: 1-7

Olenin S & Chubarova S (1992) Recent introduction of the North American spionid polychaete *Marenzelleria viridis* (Verrill, 1873) in the coastal areas of the southeastern part of the Baltic area. International Symposium - Function of Coastal Ecosystems in Various Geographical Regions, Gdansk University, p 20

Olenin S & Leppäkoski E (1999) Non-native animals in the Baltic Sea: alteration of benthic habitats in coastal inlets and lagoons. Hydrobiologia 393: 233-243

Olenin S, Gollasch S, Jonushas S & Rimkute I (2000) En-route investigations of plankton in ballast water on a ships' voyage from the Baltic Sea to the open Atlantic coast of Europe. International Revue of Hydrobiology 85: 577-596

Olenin S, Olenina I, Daunys D & Gasiunaite Z (1999) The harbour profile of Klaipėda, Lithuania. In: Gollasch S & Leppäkoski E (eds) Initial risk assessment of alien species in Nordic coastal waters. pp 185-202. Nord 8. Nordic Council of Ministers, Copenhagen

Olenin S, Orlova M & Minchin D (1999) *Dreissena polymorpha* (Pallas, 1771). In: Gollasch S, Minchin D, Rosenthal H & Voigt M (eds) Exotics Across the Ocean, pp 37-42. Logos Verlag, Berlin

Oliver PG (1992) Bivalved Seashells of the Red Sea. Verlag Christa Hemmen & National Museum of Wales, Cardiff. 330 pp

Oliverio M (1995) The status of the living Mediterranean *Strombus*, or: What is a lessepsian migrant? Notiziario del CISMA 16: 35-40

Oliverio M, Gerosa G & Cocco M (1992) First record of *Pinctada radiata* (Bivalvia, Pteriidae) epibiont on the loggerhead sea turtle *Caretta caretta* (Chelonia, Cheloniidae). Bollettino Malacologico 28: 149-152

Olsvik PA (1996) The red king crab, *Paralithodes camtschatica* (Tilesius 1815), in the Barents Sea: Life history and future stock progress. Fauna 49(1): 20-33 (in Norwegian, English summary).

Opuszinski K (1978) Silver carp, *Hypophthalmichthys molitrix* (Val.), in carp ponds. 1. Fishery production and food relations. Ekologia Polska 27(1): 71-92

Oren OH (1957) Changes in the temperature of the Eastern Mediterranean Sea in relation to the catch of the Israel trawl fishery during the years 1954/55 and 1955/56. Bulletin de l'Institut Océanographique (Monaco) 1102: 1-12

Oren OH (1969) Oceanographic and biological influence of the Suez Canal, the Nile and the Aswan Dam on the Levant Basin. Progress in Oceanography 5: 161-167

Orlov YuI (1962) On acclimatization problem of the commercial crabs in the Barents Sea. Trudi Vesoyuznogo Gidrobiologicheskogo Obshchestva 12: 400-409 (in Russian)
Orlov YuI (1977) On implantation of the commercial crabs in the Barents Sea. Ribnoye khozyaystvo 9: 20-22 (in Russian)
Orlova MI (2000) The Volga-Caspian Basin as a region-donor and region recipient for anthropogenic introductions of the water invertebrates. In: Species-introducers in the European Seas in Russia. Murmansk 2000, pp 68-71 (in Russian)
Orlova MI, Khlebovich VV & Komendantov AYu (1998) Potential euryhalinity of *Dreissena polymorpha* (Pallas) and *Dreissena bugensis* (Andr.). Russian Journal of Aquatic Ecology 8: 17-28
Orlova MI, Starobogatov Y & Biochino G (2000). *Dreissena bugensis* (Andr.) range expansion in the Volga River and in the northern Caspian Sea: further invasion perspectives for the Baltic Sea region. ASLO Aquatic Science Meeting. American Society of Limnology and Oceanography, Copenhagen (Abstract)
Orr RL (1995) Generic non-indigenous aquatic organisms risk analysis review process (Draft). Aquatic Nuisance Species Task Force, United States Department of Agriculture
Orsi Relini L & Costa MR (1986) Cattura di un Marlin a Camogli: segnalazione di *Makaira indica* (Cuvier 1832) (Osteichthyes, Istiophoridae) nel Mediterraneo. Doriana, suppl. Annali Museo civico di Storia naturale "G. Doria" Genova 6 (259): 1-4
OSPAR (1997) Alien species in the marine environment: status and national activities in the OSPAR Convention area - revised report. Presented by Sweden. OSPAR Working Group in Impacts on the Marine Environment, Lisbon 7-10 October 1997. IMPACT 97/7/1-E
Ostapenya AP, Kovalev AA, Mikheeva TM, Babitsky VA, Zhukova TV, Kryuchkova NM, Kovalevskaya RZ, Kostyukevich SB, Inkina GA, Makarevich TA, Zhukov EP, Ikonnikov VF, Samusenko AM, Orlovsky AF, Rachevsky AN & Jakushko OF (1994a) Ecological Passport of Myastro Lake. Ekomir Press, Minsk 45 pp (in Russian)
Ostapenya AP, Kovalev AA, Mikheeva TM, Babitsky VA, Zhukova TV, Kryuchkova NM, Kovalevskaya RZ, Kostyukevich SB, Inkina GA, Makarevich TA, Zhukov EP, Ikonnikov VF, Samusenko AM, Orlovsky AF, Rachevsky AN & Jakushko OF (1994b) Ecological Passport of Batorino Lake. Ekomir Press, Minsk 45 pp (in Russian)
Ostapenya AP, Kovalev AA, Zhukova TV, Mikheeva TM, Kryuchkova NM, Babitsky VA, Kovalevskaya RZ, Kostyukevich SB, Inkina GA, Makarevich TA, Zhukov EP, Ikonnikov VF, Samusenko AM, Orlovsky AF, Rachevsky AN & Jakushko OF (1993) Ecological Passport of Naroch Lake. Ekomir Press, Minsk 95 pp (in Russian)
Ostenfeld CH (1908) On the immigration of *Biddulphia sinensis* Grev. and its occurence in the North Sea during 1903-1907. Meddelelser fra Kommissionen for Havundersogelser, Plankton 1(6): 1-25
Ostroumov AA (1892) On the occasion of American balanus in Seavastopol Bay. Vestn. estestvoispitateley (in Russian)
OTA (Office of Technology Assessment), U.S. Congress. 1993. Harmful Non-indigenous Species in the United States. OTA-F-565. U. S. Government Printing Office, Washington, D.C., 391 pp
Ovchinnikov IF (1954) The *Dreissena* of the Rybinsk reservoir. Proceedings III Ecolological Conference, pp 107-109. Kiev (in Russian)
Owens CS & Madsen JD (1995) Low-temperature limits of water hyacinth. Journal of Aquatic Plant Management 33: 63-68
Ozturk B (1998) Black Sea Biological Diversity, Turkey. Black Sea Environment Series 9. UN Publication, New York, 144 pp
Ozturk B (1999) Report of the 1999 Marmara Sea expedition. Submitted to the Fisheries Faculty, Istanbul University, 40 pp (in Turkish)
Ozturk B & Ozturk AA (1996) Biology of the Turkish Straits system. In: Briand F (ed) Dynamics of Mediterranean Straits and Channels. Bulletin of Oceanography Special 17: 205-221
Ozturk B & Ozturk AA (2000) Marine biodiversity and ship originated pollution in the Turkish Straits System (TSS). In: Proceedings of the Second International Conference on Oil Spills in the Mediterranean and Black Sea Regions, pp 203-212. The Bogazici University, Istanbul
Ozturk B, Ozturk AA & Algan N (2001) Ship originated pollution in the Turkish Straits System. In: Ozturk B & Algan N (eds) Proceedings of the International Symposium on the Problems of Regional Seas, pp 86-93. TUDAV Publication, Istanbul
Pace ML, Findlay SEG & Fischer D (1998) Effects of an invasive bivalve on the zooplankton community of the Hudson River. Freshwater Biology 39: 103-116
Padilla DK, Chotkowski MA & Buchan LAJ (1996) Predicting the spread of zebra mussels (*Dreissena polymorpha*) to inland waters using boater movement patterns. Global Ecology and Biogeography Letters 5: 353-359
Paetkau D, Calvert W, Stirling I & Strobeck C (1995) Microsatellite analysis of population structure in Canadian polar bears. Molecular Ecology 4: 347-354
Page RDM & Holmes EC (1998) Molecular Evolution: A phylogenetic approach. Blackwell Science, Oxford, 346 pp
Pallary P (1912) Catalogue des mollusques du littoral Méditerranéen de l'Egypte. Mémoires de l'Institut d'Egypte 7: 69-205
Palmer M (1994) Action plan for the conservation of the native crayfish *Austropotamobius pallipes* in the United Kingdom. JNCC Report No. 193, Joint Nature Conservation Committee, Peterborough
Panek FM (1987) Biology and ecology of carp. In: Cooper EL (ed) Carp in North America, pp 1-15. American Fisheries Society, Bethesda, Maryland
Panov VE & Berezina NA (2001) *Gmelinoides fasciatus* (Stebbing). www.zin.ru/projects/invasions/ gaas/gmefas.htm
Panov VE (1996) Establishment of the Baikalian endemic amphipod *Gmelinoides fasciatus* Stebb. in Lake Ladoga. Hydrobiologia 322: 187-192
Panov VE, Krylov PI & Telesh IV (1999) The St. Petersburg harbour profile. In: Gollasch S & Leppäkoski E (eds) Initial Risk Assessment of Alien Species in Nordic Coastal Waters, pp 225-244. Nord 8. Nordic Council of Ministers, Copenhagen
Panov VE, Timm T & Timm H (2000) Current status of an introduced Baikalian amphipod *Gmelinoides fasciatus* Stebbing, in the littoral communities of Lake Peipsi. Proceedings of the Estonian Academy of Sciences Biology Ecology 49 (1): 71-80
Papchenkov VG (1985) New and rare plant species of the autonomous republics of the Middle Volga. Botanical Journal 70: 1696-1697 (in Russian)
Papchenkov VG (1997) Notes on *Potamogeton gramineus* s. l. (Potamogetonaceae). Botanical Journal 82: 65-76 (in Russian)
Parker IM & Kareiva P (1996) Assessing the risks of invasion for genetically engineered plants: Acceptable evidence and reasonable doubt. Biological Conservation 78: 193–203
Parker IM, Simberloff D, Lonsdale WM, Goodell K, Wonham M, Kareiva PM, Williamson MH, von Holle B, Moyle PB, Byers JE & Goldwasser L (1999) Impact: toward a framework for understanding the ecological effects of invaders. Biological Invasions 1: 3-19
Parsons MG & Harkins RW (2000) The Great Lakes Ballast Technology Demonstration Project Mechanical Testing Program. Marine Technology 37(3): 129-140
Parsons MG & Harkins RW (in prep) Full-scale particle removalperformance of mechnical separation devices for the primary treatment of ballast water
Parsons PA (1983) The Evolutionary Biology of Colonizing Species. Cambridge University Press, London, UK
Parsons PA (1987) Features of colonizing animals: phenotypes and genotypes. In: Gray AJ, Crawley MJ & Edwards PJ (eds) Colonization, Succession and Stability, pp 133-154. Blackwell Scientific, Oxford, UK
Partensky F, Vaulot D, Couté A & Sournaria A (1988) Morphological and nuclear analysis of the bloom forming dinoflagellates *Gyrodinium* cf *aureolum* and *Gymnodinium nagasakiense*. Journal of Phycology 24: 408-415
Passamonti M (1996) Nuova segnalazione per le coste Tunisine di *Papyridea papyracea* (Gmelin, 1791) (Bivalvia: Cardidae). Bollettino Malacologico 32: 153-156
Paterson D (1996) Ballast water risk assessment (The use of the risk assessment and risk management technique as an aid to improved ballast water management). Ballast Water Working Group MEPC38, London, July 1996, 10 pp
Patzner RA (1998) The invasion of *Lophocladia* (Rhodomelaceae, Lophotaliae) at the northern coast of Ibiza (western Mediterranean Sea). Bolletí Societat Història Natural Balears 41: 75-80
Paula EJ & Eston VR (1987) Are there other *Sargassum* species potentially as invasive as *S. muticum*? Botanica Marina 30: 405-410
Pautsch F, Lawinski L & Turoboyski K (1969) On the ecololy of the crab *Rhithropanopeus harrisi* (Gould) (Xanthidae). Limnologica 7: 63-68 (in German)
Pavlova EV & Kemp RB (1999) Impact of anthropogenic load on the summer mesozooplankton in the Sevastopol Bay. In: Pavlova EV & Shadrin NV (eds) Sevastopol Aquatory and Coast: Ecosystem Processes and Services for the Human Society. Sevastopol, Aquavita 1999, pp 142-146 (in Russian)
Pearse AS & Gunter G (1957) Salinity. In: Hedgpeth JW (ed) Treatise on Marine Ecology and Paleoecology. Geological Society of America, Memoir 67: 129-157
Pelikan J, Svoboda J & Kvet J (1970) On some relations between the production of *Typha latifolia* and a muskrat population. Zoologicke Listy 19: 303-320
Pellizzato M (1990) Acclimazione della specie *Tapes philippinarum* e primi allevamenti in Italia. In: *Tapes philippinarum*, biologia e sperimentazione, pp 158-170. Ente Sviluppo Agricolo Veneto, Venezia, Italy
Pemberton RW (2000) Waterblommetjie *Aponogeton distachyos* (Aponogetonaceae), a recently domesticated aquatic food crop in Cape South Africa with unusual origins. Economic Botany 54: 144-149
Pérez R, Kaas R & Barbaroux O (1984) Culture expérimentale de l'algue *Undaria pinnatifida* sur les côtes de France. Science et Pêche 343: 3-15
Pérez R, Lee JY & Juge C (1981) Observations sur la biologie de l'algue japonaise *Undaria pinnatifida* (Harvey) Suringar introduite accidentellement dans l'étang de Thau. Science et Pêche 315: 1-12
Perrone C & Cecere C (1994) Two solieriacean algae new to the Mediterranean: *Agardhiella subulata* and *Solieria filiformis* (Rhodophyta, Gigartinales). Journal of Phycology 30: 98-108
Persoone G (1968) Ecologie des infusoires dans les salissures de substrats immergés dans un port du mer. I. Le film primaire et le recouvrement primaire. Protistologica 4 : 187-194
Persson A (2000) Raw cultures - a useful tool in search for organisms producing resting stages. In: 9th International Conference on Harmful Algal Blooms, 7-11 February 2000, Tasmania, Australia (Abstract)
Persson A (in press) The use of sediment slurry culture to search for organisms producing resting stages. In: Proceedings of the 9th International Conference on Harmful Algal Blooms, Tasmania, Australia, 7-11 February 2000, University of Tasmania, Hobart
Persson A, Godhe A & Karlson B (2000) Dinoflagellate cysts in recent sediments from the west coast of Sweden. Botanica Marina 43: 69 -79
Persson L-E (1990) The national Swedish environmental monitoring programme (PMK): Soft-bottom macrofauna monitoring off the south coast of Sweden - Annual Report 1990. Naturvårdsverket Rapport 3937: 5-12
Pervozvansky VYa & Bugaev VF (1992) Materials on ecology of the pike *Esox lucius* in the Ketert' River (basin of the White Sea). Voprosi ikhtiologii 32(1): 94-102 (in Russian)
Peters AF & Breeman AM (1992) Temperature responses of disjunct temperate brown algae indicate long-distance dispersal of microthalli across the tropics. Journal of Phycology 28: 428-438
Petersen HE (1918) Algae (excl. calcareous algae). Report of the Danish Oceanographic Expedition, 1908-1910, to Mediterranean and adjacent Seas, II, Biol. 2: 1-20

Petersen KS, Rasmussen KL, Heinemeier J & Rud N (1992) Clams before Columbus? Nature 359: 679

Piazzi L & Cinelli F (1999) Développement et dynamique saisonnière d'un peuplement méditerranéen de l'algue tropicale *Caulerpa racemosa* (Forsskål) J.Agardh. Cryptogamie, Algologie 20: 295-300

Piazzi L & Cinelli F (2000) Effects de l'expansion des Rhodophyceae introduites *Acrothamnion preissii* et *Womersleyella setacea* sur les communautés algales des rhizomes de *Posidonia oceanica* de Méditerranée occidentale. Cryptogamie, Algologie 21: 291-300

Piazzi L, Ceccherelli G & Cinelli F (1999) Interaction between the two invasive tropical algae *Caulerpa taxifolia* and *Caulerpa racemosa* in the Mediterranean. Abstracts 2nd European Phycological Congress, Montecatini, Italy, p 140

Piazzi L, Ceccerelli G & Cinelli F (2001) Threat to macroalgal diversity: effects of the introduced green alga *Caulerpa racemosa* in the Mediterranean. Marine Ecology Progress Series 210: 149-159

Picco P, Sakellariou D, Martini G, Margottini C, Manzella G, Nittis C, Souvermezoglou E, Panayotidis P & Henoque Y (1999) Natural characteristics. In: State and pressures of the marine and coastal Mediterranean environment. EEA Environmental Assessment Series 5: 30-46

Pichot Y, Comps M, Tigé G, Grizel H & Rabouin MA (1979) Recherche sur *Bonamia ostreae* gen.n., sp. n., parasite nouveau de l'huître plate *Ostrea edulis* L. Revue des Travaux Institut des Pêches maritimes 43: 131-140

Pickett STA & White PS (1985) The Ecology of Natural Disturbance and Patch Dynamics. Academic, Sydney, Australia

Picl K (1995) Fishes of Lakes and Rivers. Magna Books, Wigston, England. 223 pp

Piechocki A (1979) Molluscs (Mollusca). Gastropods (Gastropoda). Fauna slodkowodna Polski 7, PWN, Warszawa-Poznan, 187 pp (in Polish)

Piechocki A & Dyduch-Falniowska A (1993) Molluscs (Mollusca). Bivalves (Bivalvia). Fauna slodkowodna Polski, 7A, PWN, Warszawa, 204 pp (in Polish)

Pierce RW, Carlton JT, Carlton DA & Geller JB (1997) Ballast water as a vector for tintinnid transport. Marine Ecology Progress Series149: 295-297

Pieterse AH, de Lange L & Verhagen L (1981) A study on certain aspects of seed germination and growth of *Pistia stratiotes* L. Acta Botanica Neerlandica 30: 47-57

Pigeot J, Miramand P, Garcia-Meunier P, Guyot T & Séguignes M (2000) Présence d'un nouveau prédateur de l'huître creuse, *Ocinebrellus inornatus* (Récluz, 1851) dans le bassin conchylicole de Marennes-Oléron. Comptes Rendus de l'Académie des Sciences, Sciences de la Vie 323: 697-703

Pimm SL (1989) Theories of predicting success and impact of introduced species. In: Drake JA, Di Castri F, Groves RH, Kruger FJ & Mooney HA (eds) Biological_Invasions: a Global Perspective. SCOPE 37. John Wiley & Sons, New York

Pimm SL (1991) The Balance of Nature? University of Chicago Press, Chicago, Illinois

Pinkster S (1973) The *Echinogammarus berilloni* - group, a number of predominantly Iberian amphipod species (Crustacea). Bijdragen tot de Dierkunde 43 (1): 1-39

Pinkster S, Scheepmaker M, Platvoet D & Broodbaker N (1992) Drastic changes in the amphipod fauna (Crustacea) of Dutch inland waters during the last 25 years. Bijdragen tot de Dierkunde 61: 193-204

Pirogov VV, Filchakov VA, Zinchenko TD, Karpyuk MI & Edskiy LB (1990) New elements in the benthic fauna of the Volga-Caspian reservoir cascade. Zoological Journal 69: 138-142 (in Russian)

Pizzicori P, Castriota L, Marino G & Andaloro F (2000) *Seriola carpenteri*: a new immigrant in the Mediterranean from the Atlantic Ocean. Journal of Fish Biology 57: 1335-1338

Pöckl M (1999) Distribution of crayfish species in Austria with special reference to introduced species. Freshwater Crayfish 12: 733-750

Pollard DA & Hutchings PA (1990) A review of exotic marine organisms introduced to the Australian region. II. Invertebrates and algae. Asian Fisheries Science 3: 223-250

Ponder WF (1988) *Potamopyrgus antipodarum*: a molluscan coloniser of Europe and Australia. Journal of Molluscan Studies 54: 271-285

Ponyi JE (1992) The distribution and biomass of Unionidae (Mollusca, Bivalvia), and the production of Unio tumidus Retzius in Lake Balaton (Hungary). Archiv fur Hydrobiologie 125: 245-251

Por FD (1971) One hundred years of Suez Canal. A century of Lessepsian migration: retrospect and viewpoints. Systematic Zoology 20: 138-159

Por FD (1978) Lessepsian Migration. Springer-Verlag. Berlin, Heidelberg. 228 pp

Por FD (1989) The Legacy of Tethys. An Aquatic biogeography of the Levant. Kluwer, 214 pp

Por FD (1990) Lessepsian migrations. An appraisal and new data. Bulletin Institut Océanographique Monaco 7: 1-10

Portielje R & Roijackers RMM (1995) Primary succession of aquatic macrophytes in experimental ditches in relation to nutrient input. Aquatic Botany 50: 127-140

Post WM & Pimm SL (1983) Community assembly and food web stability. Mathematical Bioscience 64: 169-192

Preston CD & Croft JM (1997) Aquatic Plants in Britain and Ireland. Harley Books, Colchester, England. 365 pp

Preston G, Brown, G & van Wyk E (eds) (2000) Best management practices for preventing and controlling invasive alien species. Symposium Proceedings. The Working for Water Programme, Cape Town, South Africa, 316 pp

Pritchard JK, Stephens M & Donnelly P (2000) Inference of population structure using multilocus genotype data. Genetics 155: 945-959

Protasov AA & Afanasiev SA (1990) Principal types of *Dreissena* communities in periphyton. Hydrobiological Journal 26(4): 15-23

Protasov AA, Afanasiev SA & Ivanova OO (1983) The distribution and role of *Dreissena polymorpha* in the periphyton of Chernobyl water cooling reservoir. In: Molluscs: Systematics, Ecology and Patterns of Occurrence. 7th Meeting on the Investigation of Molluscs, pp 220-222. Nauka Press, Leningrad (Abstract; in Russian)

Pupunat L, Rychen P, Fontecha Camara MA, Haenni W, Klaus B & Rossi P (2001) Latest electrochemical water treatment technology for ballast water disinfection (biological inactivation). Proceedings of the 1st International Conference on Ballast Water Management, 1-2 November 2001, Singapore

Quayle DB (1969) Pacific oyster culture in British Columbia. Bulletin of the Fisheries Research Board of Canada 169: 1-192

Quayle DB (1996) Marine wood borers in British Columbia. Canadian Special Publication of Fisheries and Aquatic Sciences, 55 pp

Quéro JC, Du Buit MH & Vayne JJ (1998) Les observations de poissons tropicaux et le réchauffement des eaux dans l'Atlantique européen. Oceanologica Acta 21: 345-351

Quine W (1946) Truth by convention. In: Philosophical Essays for Alfred North Whitehead. Longmans Green, New York

Ragonese S, Giusto GB & Caruso JH (2001) Second record of the toadfish *Chaunax suttkusi* Caruso, 1989 in the Mediterranean Sea. Journal of Fish Biology 58: 291-294

Rahel FJ (2000) Homogenization of fish faunas across the United States. Science 288: 854-856

Rajagopal S, Van der Velde G & Bij de Vaate A (2000) Reproductive biology of the Asiatic clams *Corbicula fluminalis* and *Corbicula fluminea* in the river Rhine. Archiv für Hydrobiologie 149: 403-420

Rajagopal S, Van der Velde G, Paffen BGP, Van den Brink FWB & Bij de Vaate A (1999) Life history and reproductive biology of the invasive amphipod *Corophium curvispinum* (Crustacea: Amphipoda) in the Lower Rhine. Archiv für Hydrobiologie 144: 305-325

Ramacharan CW, Padilla DK & Dodson SI (1992) A multivariate model for predicting population fluctuations of *Dreissena polymorpha* in North American lakes. Canadian Journal of Fisheries and Aquatic Sciences 49: 150-158

Ramon E & Friedmann I (1965) The gametophyte of *Padina* in the Mediterranean. Proceedings of the International Seaweed Symposium 5: 183-195

Randall JE (1987) Introductions of marine fishes to the Hawaiian Islands. Bulletin of Marine Science 41: 490-502

Rannala B & Mountain JL (1997) Detecting immigration by using multilocus genotypes. Proceedings of the National Academy of Sciences, USA 94: 9197-9201

Ray WJ & Corkum LD (1997) Predation of zebra mussels by round gobies, *Neogobius melanostomus*. Environmental Biology of Fishes 50: 267-273

Raybaudi Massilia L (1983) A *Strombus* from the Mediterranean. La Conchiglia 15: 174-175

Raymond M & Rousset F (1995) GENEPOP (version 1.2): population genetics software for exact tests and ecumenicism. Journal of Heredity 86: 248-249

Rayss T (1941) Sur les Caulerpes de la côte palestinienne. Palestine Journal Botany 2: 103-124

Rayss T (1954) Les algues tropicales de la Méditerranée orientale et leur origine probable. Rapport 8ème Congrès International Botanique, Paris, 17: 148-149

Rayss T (1963) Sur la présence dans la Méditerranée orientale des algues tropicales de la famille des Solériacées. Lucrările Grădinii Botanice din Bucaresti 1961-1962: 91-108.

Razinkovas A (1990) Productional processes in the populations of mysids in water bodies of River Nemunas basin. Summary of PhD Thesis, Minsk (in Russian)

Redding T, Todd S & Midlen A (1997) The treatment of aquaculture wastewaters - A botanical approach. Journal of Environmental Management 50: 283-299

Reed-Andersen T, Carpenter SR, Padilla DK & Lathrop RC (2000) Predicted impact of zebra mussel (*Dreissena polymorpha*) invasion on water clarity in Lake Mendota. Canadian Journal of Fisheries and Aquatic Sciences 57: 1617-1626

Reeders HH & bij de Vaate A (1990) Zebra mussels (*Dreissena polymorpha*): a new perspective for water quality management. Hydrobiologia 200/201: 437-450

Reeders HH, bij de Vaate A & Noordhuis R (1993) Potential of the zebra mussel (*Dreissena polymorpha*) for water quality management. In: Nalepa TF & Schloesser DW (eds) Zebra Mussels: Biology, Impacts, and Control, pp 439-451. Lewis Publishers, Ann Arbor, Michigan

Reeders HH, bij de Vaate A & Slim FJ (1989) The filtration rate of *Dreissena polymorpha* (Bivalvia) in three Dutch lakes with reference to biological water quality management. Freshwater Biology 22: 133-141

Reeve MR, Walter MA & Ikeda T (1978) Laboratory studies of ingestion and food utilization in lobate and tentaculate ctenophores. Limnology and Oceanography 23: 740-751

Reeves E (1999) Exotics and public policy in the Great Lakes: the results of a workshop at the biennial Great Lakes Water Quality Forum, Milwaukee, Wisconsin, 23 and 26 September 1999. www.ijc.org/milwaukee/transcript/exotic/ballast.html

Reinbold T (1898) Meeresalgen von der Insel Rhodos. Hedwigia 37: 87-90

Reinhold M (1999) Displacement by inland vessels as a possibility of anthropochor dispersal in macroinvertebrates in relation to the faunal exchange Rhine/Main/Main-Danube canal/Danube. PhD thesis, Cuvillier Verlag, Göttingen, 146 pp (in German)

Reinhold M & Tittizer T (1998) *Limnomysis benedeni* Czerniavsky 1882 (Crustacea: Mysidacea), ein weiteres pontokaspishes Neozoon im Main-Donau-Kanal. Lauterbornia 33: 34-40

Reinhold M & Tittizer T (1999) Dispersal of invertebrates in cooling filter of ships. Wasser und Boden 51 (1/2): 61-66 (in German)

Reise K (1982) Long-term changes in the macrobenthic invertebrate fauna of the Wadden Sea: are polychaetes about to take over? Netherlands Journal of Sea Research 16: 29-36

Reise K (1990) Historische Veränderungen in der Ökologie des Wattenmeeres. Rheinisch-Westfälische Akademie der Wissenschaften, Vorträge N 382: 35-55

Reise K (1991) Ökologische Erforschung des Wattenmeeres. Biologie der Meere. Spektrum Akademischer Verlag, Heidelberg, pp 68-79

Reise K (1998) Pacific oysters invade mussel beds in the European Wadden Sea. Senckenbergiana maritima 28: 167-175

Reise K & Lackschewitz D (1998) Benthos des Wattenmeeres zwischen Sylt und Rømø. In: Gätje C & Reise K (eds) Ökosystem Wattenmeer, Austausch-, Transport- und Stoffumwandlungsprozesse, pp 55-64. Springer-Verlag, Berlin Heidelberg

Reise K, Gollasch S & Wolff WJ (1999) Introduced marine species of the North Sea coasts. Helgoländer Meeresuntersuchungen 52: 219-234

Rejmánek M (1989) Invasibility of plant communities. In: Drake JA, Mooney HA, diCastri F, Groves RH, Kruger FJ, Rejmanek M & Williamson M (eds) Biological Invasions. A Global Perspective, SCOPE 37, pp 369-388. John Wiley & Sons, New York

Rejmanek M (1996) A theory of seed plant invasiveness: the first sketch. Biological Conservation 78: 171–181

Rejmanek M (2000) Invasive plants: approaches and predictions. Australian Ecology 25: 497-506

Rejmánek M & Richardson DM (1996) What attributes make some plant species more invasive? Ecology 77: 1653-1655

Relini G, Relini M & Torchia G (1998) The role of fishing gears in spread of allochthonous species: the case of *Caulerpa taxifolia* in the Ligurian Sea. ICES Journal of Marine Science 57: 1421-1427

Relini M, Orsi L, Puccio V & Azzurro E (2000) The exotic crab *Percnon gibbesi* (H. Milne Edwards, 1853) (Decapoda, Grapsidae) in the Central Mediterranean. Scientia Marina 64: 337-340

Remane A (1934) The fauna of brackish waters. Verhandlungen der Deutschen Zoologischen Gesellschaft 36: 34-74 (in German)

Rembiszewski JM & Rolik H (1975) Cyclostomes and fishes. Katalog Fauny Polski 38, PWN, Warszawa, 249 pp (in Polish)

Renard E, Bachmann V, Cariou ML & Moreteau JC (2000) Morphological and molecular differentiation of invasive freshwater species of the genus *Corbicula* (Bivalvia, Corbiculidea) suggest the presence of three taxa in French rivers. Molecular Ecology 9: 2009-2016

Renault T (1996) Appearance and spread of diseases among bivalve molluscs in the northern hemisphere in relation to international trade. Revue Scientifique & Technique Office International des Epizooties 15: 556-561

Renault T, Stokes NA, Chollet B, Cochennec N, Berthe F, Gérard A & Burreson EM (2000) Haplosporidiosis in the Pacific oyster *Crassostrea gigas* from the French atlantic coast. Diseases of Aquatic organisms 42: 207-214

Reynoldson TB, Smith BD & Maitland PS (1981) A species of North American triclad (Paludicola; Turbellaria) new to Britain found in Loch Ness, Scotland. Journal of Zoology London 193: 531-539

Ribera MA (1994) Les macrophytes marins introduits en Méditerranée: biogéographie. In: Boudouresque CF, Briand F & Nolan C (eds) Introduced Species in European Coastal Waters. CEC Ecosystem Research Report 8: 37-43

Ribera MA (2001) Pathways of biological invasion of marine plants. GISP Program. Island Press (in press)

Ribera MA & Boudouresque CF (1995) Introduced marine plants with special reference to macroalgae: mechanisms and impact. Progress in Phycological Research 11: 187-268

Ribera MA, Gómez-Garreta A, Gallardo T, Cormaci M, Furnari G & Giaccone G (1992) Check-list of the Mediterranean Seaweeds. I. Fucophyceae (Warming, 1884). Botanica Marina 35: 109-130

Ricciardi A (2001) Facilitative interactions among aquatic invaders: is an "invasional meltdown" occurring in the Great Lakes? Canadian Journal of Fisheries and Aquatic Sciences 58: 2513-2525

Ricciardi A & McIsaac HJ (2000) Recent mass invasion of the North American Great Lakes by Ponto-Caspian species. Trends in Ecology and Evolution 15: 62-65

Ricciardi A & Rasmussen JB (1998) Predicting the identity and impact of future biological invaders: a priority for aquatic resource management. Canadian Journal of Fisheries and Aquatic Science 55: 1759-1765

Ricciardi A, Neves RJ & Rasmussen JB (1998) Impending extinctions of North American freshwater mussels (Unionoida) following the zebra mussel (*Dreissena polymorpha*) invasion. Journal of Animal Ecology 67: 613-619

Ricciardi A, Steiner WWM, Mack RN, Simberloff D (2000) Toward a global information system on invasive species. BioScience 50: 239-244

Ricciardi A, Whoriskey FG & Rasmussen JB (1995) Predicting the intensity and impact of *Dreissena* infestation on native unionid bivalves from *Dreissena* field density. Canadian Journal of Fisheries and Aquatic Sciences 52: 1449-1461

Ricciardi A, Whoriskey FG & Rasmussen JB (1996) Impact of the *Dreissena* invasion on native unionid bivalves in the upper St. Lawrence River. Canadian Journal of Fisheries and Aquatic Sciences 53: 1434-1444

Ricciardi A, Whoriskey FG & Rasmussen JB (1997) The role of the zebra mussel (*Dreissena polymorpha*) in structuring macroinvertebrate communities on hard substrata. Canadian Journal of Fisheries and Aquatic Sciences 54: 2596-2608

Richardson DM, Cowling RM, & Le Maitre DC (1990) Assessing the risk of invasive success in *Pinus* and *Banksia* in South African Mountain Fynbos. Journal of Vegetation Science 1: 629–642

Richardson DM, Pysek P, Rejmanek M, Barbour MG, Panetta FD & West CJ (2000) Naturalization and invasion of alien plants: concepts and definitions. Diversity and Distributions 6: 93-107

Richter S, Braband A, Aladin N & Scholtz G (2001) The phylogenetic relationship of "predatory water-fleas" (Cladocera: Onychopoda, Haplopoda) inferred from 12S rDNA. Molecular Phylogenetic and Evolution 19: 105-113

Rick HJ & Durselen CD (1995) Importance and abundance of the recently established species *Coscinodiscus wailesii* Gran and Angst in the German Bight. Helgoländer Meeresuntersuchungen 49: 355-374

Ricordi P (1993) Segnelazione di *Pinctada radiata* (Leach, 1814) per il Tirreno meridionale. Notizario S.I.M. 11: 41-42

Ridgway HF & Safarik J (1991) Biofouling of reverse osmosis membranes. In: Flemming HC & Geesey GG (eds) Biofouling and Biocorrosion in Industrial Water Systems, pp 81-112. Springer Verlag, Berlin

Rigatti Luchini S, Breber P, Strada R & Vidale R (1989) Osservazioni sulla diffusione territoriale di popolazioni bentoniche. In: Chieppa M (ed) Atti delle Giornate di Studio su Analisi Statistica di Dati Territoriali, pp 399-410. Società Italiana di Statistica, Bari

Rigby G & Hallegraeff G (1992) Ballast water exchange trials and marine plankton distribution on the MV "Iron Whyalla". Ballast water research series AQIS 2(2): 43-78

Rigby G & Taylor A (1995) Ballast water: its impacts can be managed. ICES Annual Science Conference, Aalborg, 8 pp

Rigby G & Taylor AH (2001) Ballast water treatment to minimize the risks of introducing nonindigenous marine organisms in Australian ports. Ballast Water Research Series, Department of Agriculture, Fisheries and Forestry, Australia 13: 1-93

Rigby GR (1994) Possible solutions to the ballast water problem. AQIS Ballast Water Symposium Proceedings, Canberra, 11-13 May

Rigby GR & Hallegraeff GM (1993) Ballast water exchange trial and behaviour of marine plankton on the M.V. "Iron Whyalla", in Japan and en-route to Canada. In: AQIS Ballast Water Research Series, Report 2, 123 pp

Rigby GR & Hallegraeff GM (1994) The transfer and control of harmful marine organisms in shipping ballast water: behaviour of marine plankton and ballast water exchange on the MV "Iron Whyalla". Journal of Marine Environment and Engineering 1: 91-110

Rigby GR, Hallegraeff GM & Sutton CA (1997) Ballast water heating and sampling trials on the BHP ship 'MV Iron Whyalla'. AQIS Ballast Water Research Series, Report 11

Rigby GR, Hallegraeff GM & Sutton CA (1999) Novel ballast water heating technique offers cost-effective treatment to reduce the risk of global transport of harmful marine organisms. Marine Ecology Progress Series 191: 289-293

Rigby GR, Steverson IG, Bolch CJ & Hallaegraeff GM (1993) The transfer and treatment of shipping ballast waters to reduce the dispersal of toxic marine dinoflagellates. In: Smayda TJ & Shimizu Y (eds) Toxic Phytoplankton Blooms in the Sea, pp 169-176. Elsevier

Rincé Y & Paulmier G (1986) Données nouvelles sur la distribution de la diatomée marine *Coscinodiscus wailesii* Gran & Angst (Bacillariophyceae). Phycologia 25(1): 73-79

Rindi F, Maltagliati F, Rossi F, Acunto S & Cinelli F (1999) Algal flora associated with a *Halophila stipulacea* (Forsskål) Ascherson (Hydrocharitaceae, Helobiae) stand in the western Mediterranean. Oceanologica Acta 22: 421-429

Riouall R (1985) Sur la présence dans l'etang de Thau (Hérault, France) de *Sphaerotrichia divaricata* (C. Agardh) Kylin et *Chorda filum* (L.) Stackhouse. Botanica Marina 27: 83-86

Riouall R, Guiry MD & Codomier L (1985) Introduction d'une espèce foliacée de *Grateloupia* dans la flore marine de l'etang de Thau (Hérault, France). Cryptogamie, Algologie 6: 91-98

Ritte U & Safriel UN (1977) The theory of island biogeography as a model for colonization in more complex systems. Proceedings of the 8th Scientific Meeting of the Israel Ecological Society, 1977, pp 288-294

Rivest BR, Coyer J & Tyler S (1999) The first known invasion of a free-living marine flatworm. Biological Invasions 1: 393-394

Rivier IK (1986) Composition and Distribution of the Winter Zooplankton Communities. Nauka Press, Leningrad 168 pp (in Russian)

Rivier IK (1993) Recent state of the Rybinsk reservoir zooplankton. In: Modern State of the Rybinsk Reservoir Ecosystem, pp 205-233. Nauka, Leningrad (in Russian)

Rivier IK (1998) The predatory Cladocera (Onychopoda: Podonidae, Polyphemidae, Cercopagidae) and Leptodorida of the world. Guides to the identification of the microinvertebrates of the continental waters of the world. 13. Leiden: Backhuys Publishing, 213 pp

Rizzi AJM (1992) *Crassostrea denticulata* (Born, 1791), een nieuwe oestersoort op de Nederlandse kust.Correspondentieblad Nederlandse Malacologische Vereniging 264: 928-932

Robba E (1987) The final occlusion of Tethys: its bearing on Mediterranean benthic mollusks. In: McKenzie KG (ed) Shallow Tethys, pp 405-426. Balkema, Rotterdam

Robert-Koch-Institute (2001) Ratgeber Infektionskrankheiten. www.rki.de/INFEKT/ RATGEBER/RAT.HTM (2001-09-03)

Roberts RJ & Shepherd CJ (1997) Handbook of trout and salmon diseases. Fishing News Books. 179 pp

Robertson PAW & Austin B (1994) Disease associated with cyprinids imported into the United Kingdom. International Symposium on Aquatic Animal Health, University of California School of Veterinary Medicine, Davis, Ca. p W-17.6

Robinson GR, Quinn JF & Stanton ML (1995) Invasibility of experimental habitat islands in a California winter annual grassland. Ecology 76: 786-794

Robinson JV & Edgemon MA (1988) An experimental evaluation of the effect of invasion history on community structure. Ecology 69: 1410-1417

Robinson JV & Dickerson JE (1984) Testing the invulnerability of laboratory island communities to invasion. Oecologia 61: 169-174

Rodgers SK & Cox EF (1999) Rate of spread of introduced rhodophytes *Kappaphycus alvarezii*, *Kappaphycus striatum*, and *Gracillaria salicornia* and their current distributions in Kane'ohe Bay, O'ahu, Hawai'i. Pacific Science 53: 232-241

Rodi AJ & Dauer DM (1996) Synonymy of *Marenzelleria viridis* (Verrill) and *Marenzelleria jonesi* Maciolek (Polychaeta: Spionidae). 24th Annual Benthic Ecology Meeting, Columbia, South Carolina, p 72

Rodin VE (1985) Spatial and functional structure of the red king crab populations. Izvestiya TINRO 110: 86-97 (in Russian)

Roditi HA, Caraco NF, Cole JJ & Strayer DL (1996) Filtration of Hudson River water by the zebra mussel (*Dreissena polymorpha*). Estuaries 19: 824-832

Rodman JE (1986) Introduction, establishment and replacement of sea-rockets (*Cakile*, Cruciferae) in Australia. Journal of Biogeography 13: 159-171

Rodriguez A (1987) Interaccion y cultivo de *Penaeus kerathurus* (Forskal, 1775) y *Penaeus japonicus* Bate, en el laboratorio y estanques de las Salinas de Cadiz. Cuadernos Marisqueros Publicación Técnica 8: 247-256

Röhner M, Bastrop R & Jürss K (1996) Colonization of Europe by two American genetic types or species of the *Marenzelleria* (Polychaeta: Spionidae). An electrophoretic analysis of allozymes. Marine Biology 127: 277-287

Rolik H & Rembiszewski JM (1987) Fishes and cyclostomes (Pisces et Cyclostomata). Fauna slodkowodna Polski, PWN, Warszawa, 314 pp (in Polish)

Rosecchi E, Poizat G & Crivelli AJ (1997) The introduction of freshwater fish and crayfish species in the Camargue: history, origins and changes in assemblages. Bulletin Francais de la Pêche de la Pisciculture 344-5: 221-232 (in French)

Rosell RS, Maguire CM & McCarthy TK (1999) First reported settlement of zebra mussels *Dreissena polymorpha* in the Erne system, Co Fermanagh, Northern Ireland. Biology and Environment, Proceedings of the Royal Irish Academy 98B: 191-193

Rosenfield A & Mann R (eds) (1992) Dispersal of Living Organisms into Aquatic Ecosystems. Maryland Sea Grant, College Park, MD

Rosenthal H (1976) Implications of transplantations to aquaculture and ecosystems, Paper presented to the FAO Technical Conference on Aquaculture. Kyoto, Japan, 26 May-2 June 1975. Rome, FAO, FIR:AQ,/Conf./76/E.67 (mimeo)

Rosenthal H (1980) Implications of transplantations to aquaculture and ecosystems. Marine Fisheries Review 5: 1-14

Rosenthal H, Gollasch S & Voigt M (eds) (2000) Final Report of the European Union Concerted Action "Testing Monitoring Systems for Risk Assessment of Harmful Introductions by Ships to European Waters". Contract MAS3-CT97-0111

Rosenthal H, McGlade J & Gollasch S (2001) The role of aquaculture in integrated coastal zone management. Bulletin of the Aquaculture Association of Canada 101(1): 5-10

Rosenzweig ML (1995) Species Diversity in Space and Time. Cambridge University Press, Cambridge, 458 pp

Rosenzweig ML (2001) The four questions: What does the introduction of exotic species do to diversity? Evolutionary Ecology Research 3: 361-367

Røsjorde HJ (1973) The red alga *Dasya baillouviana* (Gmel.) Mont., a new species in Norway. Blyttia 31: 169-173 (in Norwegian, English abstract)

Rossi R & Paesanti F (1992) Vongola verace: la situazione europea. Laguna 9: 24-29

Rosso A (1994) Segnalazione di *Electra tenella* (Hincks) (Bryozoa) lungo le coste sud-orientali della Sicilia. Bollettino Accademia Gioenia di Scienze Naturali, Catania 27 (346): 241-251

Roughgarden J (1986) Predicting invasions and rates of spread. In: Mooney HA & Drake JA (eds) Ecology of Biological Invasions of North America and Hawaii, pp 179-188. Springer Verlag, New York

Roy K, Jablonski D & Valentine JW (2001) Climate change, species range limits and body size in marine bivalves. Ecology Letters 4: 366-370

Roy S, Harris RP & Poulet SA (1989) Inefficient feeding by *Calanus helgolandicus* and *Temora longicornis* on *Coscinodiscus wailesii*: Quantitative estimation using chlorophyll-type pigments and effects on dissolved free amino acids. Marine Ecology Progress Series 52: 145-153

Rubinoff I (1968) Central America sea-level canal: possible biological effects. Science 161: 857-861

Rudnick DA, Halat KM & Resh VH (2000) Distribution, ecology and potential impacts of the Chinese mitten crab (*Eriocheir sinensis*) in San Francisco Bay. University of California, Berkeley, Water Resources Center, Contribution 26, 74 pp www.waterresources.ucr.edu

Rudolph K (1997) On the occurrence of the amphipod *Pontogammarus robustoides* Sars, 1894, in the region of Peene mouth. Natur und Museum 127: 306-312 (in German)

Rudolph K (2000) Alien malacostracan Crustacea in central Brandenburg. Actual state of the distribution. Neozoen 3: 10-11 (in German)

Rueness J & Rueness EK (2000) *Caulacanthus ustulatus* (Gigartinales, Rhodophyta) from Brittany (France) is an introduction from the Pacific Ocean Cryptogamie Algologie 21: 355-363

Rueness J (1977) The Norwegian algal flora. Universitetsforlaget, Oslo. 272 pp (in Norwegian)

Rueness J (1985) The Japanese seaweed – *Sargassum muticum*. Biological pollution of European coastal waters. Blyttia 43: 71-74 (in Norwegian)

Rueness J (1989) *Sargassum muticum* and other introduced Japanese macroalgae: Biological pollution of European coasts. Marine Pollution Bulletin 20: 173-176

Rueness J (1990) Dispersion of of the Japanese seaweed (*Sargassum muticum*) along the Skagerrak coast.] Blyttia 48: 19 (in Norwegian)

Rueness J (1998) Algae in colour. A field handbook on coastal macroalgae. Almater forlag, Oslo. 136 pp (in Norwegian)

Rueness J, Jacobsen T & Åsen PA (1990) Seaweed distribution along the Norwegian coast with special reference to *Ceramium shuttleworthianum* (Rhodophyta). Blyttia 48: 21-26 (in Norwegian, English abstract)

Ruggieri G (1967) The Miocene and later evolution of the Mediterranean Sea. In: Adams CG & Ager DV (eds) Aspects of Tethyan Biogeography. The Systematics Association Publishing, London

Ruiz GM (1998) Cross-cutting issues. In: Ridgeway RL, Gregg WP, Stinner RE & Brown AG (eds) Invasive Species Databases, Proceedings of a Workshop, pp 35-37. National Biological Information Infrastructure (NBII) USGS, Reston, Virginia

Ruiz GM & Hines AH (1997) The risk of nonindigenous species invasion in Prince William Sound associated with oil tanker traffic and ballast water management: pilot study. Smithsonian Environmental Research Centre, PFP 632.91.1

Ruiz GM, Carlton JT, Fofonoff P, Strayer D, Mills E, Eldridge L, Pauley G & Williams J (2001) Interpreting invasion patterns from ecological surveys. Report to United States Fish & Wildlife Service

Ruiz GM, Carlton JT, Grosholtz ED & Hines AH (1997) Global invasions of marine and estuarine habitats by non-indigenous species: mechanisms, extent and consequences. American Zoologist 37: 621-632

Ruiz GM, Fofonoff P, Hines AH & Grosholz ED (1999) Non-indigenous species as stressors in estuarine and marine communities: Assessing invasion impacts and interactions. Limnology and Oceanography 44: 950-972

Ruiz GM, Fofonoff PW, Carlton JT, Wonham MJ & Hines AH (2000) Invasion of coastal marine communities in North America: apparent patterns, processes, and biases. Annual Review of Ecology and Systematics 31: 481-531

Ruiz GM, Rawlings TK, Dobbs FC, Drake LA, Mullady T, Huq A & Colwell RR (2000) Global spread of microorganisms by ships: Ballast water discharged from vessels harbors a cocktail of potential pathogens. Nature 408: 49-50

Russell DJ (1992) The ecological invasion of Hawaiian reefs by two marine red algae, *Acanthophora spicifera* (Vahl) Boerg, and *Hypnea musciformis* (Wulfen) J. Ag., and their association with two native species, *Laurencia nidifica* J. Ag. and *Hypnea cervicornis* J. Ag. ICES Marine Science Symposia 194: 110-125

Russev B (1959) Contribution to the study of macrobenthos of the Bulgarian Danube. Comptes rendus de l'Academie bulgare des sciences 12(4): 345-348 (in German)

Russev B (1966) Zoobenthos of the Danube between 845 and 375 km of river flow. 1. Occurrence, distribution and ecology. Izviestija na Zoolgicheskija Institut i Muzej BAN 20: 55-132 (in German)

Rutherford ES, Rose KA, Mills EA, Forney JL, Mayer CM & Rudstam LG (1999) Individual-based model simulations of a zebra mussel (*Dreissena polymorpha*) induced energy shunt of walleye (*Stizostedion lucioperca*) and yellow perch (*Perca flavescens*) populations in Oneida lake, New York. Canadian Journal of Fisheries and Aquatic Sciences 56: 2148-2160

Ruusila V (1997) Canadian beaver is a more eager builder. In: Nummi P (ed) Nature of Finland, Mammals, p 95. Weilin & Göös, Porvoo (in Finnish)

Ryland JS (1967) Polyzoa. Oceanography and Marine Biology Annual Review 5: 343-369

Sabelli B (1969) Ritrovamenti malacologici a Pantellaria e nei banchi de pesca tunisini. Conchiglie 5: 8

Sacchi CF, Occhipinti Ambrogi A & Sconfietti R (1990) Les lagunes nord-adriatiques: un environnement conservateur ouvert aux nouveautés. Bullettin de la Società zoologique de France 114(3): 47-60

Safriel UN & Ritte U (1980) Criteria for the identification of potential colonizers. Biological Journal of the Linnean Society 13: 287-97

Safriel UN & Ritte U (1983) Universal correlates of colonizing ability. In: Swingland IR & Greenwood PJ (eds) The Ecology of Animal Movement, pp 215-239. Clarendon, Oxford, UK

Safronov GP (1993) Composition and ecology of the genus *Gammarus* Fabricius species of the southern part of East Siberia. PhD Thesis, Irkutsk: 23 (in Russian)

Safronov GP & Erbaeva EA (1998) Fauna of amphipods of Angara River and its reservoirs. In: Proceedings of International Conference of Environmental Problems of Large Rivers Basins. Toljatti: 242-243 (in Russian)

Salemaa H & Hietalahti V (1993) *Hemimysis anomala* G. O. Sars, (Crustacea: Mysidacea) - Immigration of a Pontocaspian mysid into the Baltic Sea. Annales Zoologici Fennici 30: 271-276

Salonen E & Mutenia A (1992) Stocking and changes in peled (*Coregonus peled* (Gmelin)) stocks, and fishery management in the Lokka and Porttipahta reservoirs, Northern Finland. Polskie Archiwum Hydrobiologii 39: 837-846

Sancholle M (1988) Présence de *Fucus spiralis* (Phaeophyceae) en Méditerranée occidentale. Cryptogamie, Algologie 9: 157-162

Sandlund OT, Schei PJ & Viken A (1999) Invasive Species and Biodiversity Management. Kluwer Academic Publishers, 431 pp

Sant N, Delegado O, Rodriguez-Prieto C & Ballesteros E (1996) The spreading of the introduced seaweed *Caulerpa taxifolia* (Vahl) C. Agardh in the Mediterranean Sea: Testing the boat transportation hypothesis. Botanica Marina 39: 427-430

Sarà M (1985) Ecological factors and their biogeographic consequences in the Mediterranean ecosystems. In: Moraitou-Apostolopoulou M & Kiortsis V (eds) Mediterranean Marine Ecosystems, pp 1-17. Plenum Press, New York

Sars GO (1896) Crustacea Caspia. Contributions to the knowledge of the carcinological fauna of the Caspian Sea. Amphipoda, Supplement. Bulletin de l'Academie imperiale des sciences de St-Petersburg 4: 421-489

Sars GO (1927) Notes on the Crustacean fauna of the Caspian Sea. In: A Tribute to Prof. N. M. Knipovich. VNIRO, Moscow (in Russian)

Sartoni G & Boddi S (1993) *Apoglossum gregarium* (Delesseriaeae, Rhodophyta), a new record for Mediterranean algal flora. Botanica Marina 36: 297-302

Sartoni G & Sarti M (1976) Sulla presenza a Livorno di *Aglaothamnion feldmanniae* L'Hardy-Halos. Informatore Botanico Italiano 8: 185

Satsmadjis J & Friligos N (1983) Red tide in Greek waters. Vie et Milieu 33: 111-117

Sauriau PG (1991) Spread of *Cyclope neritea* (Mollusca: Gastropoda) along the north-eastern Atlantic coasts in relation to oyster culture and to climatic fluctuations. Marine Biology 109: 299-309
Sauriau PG, Pichocki-Seyfried C, Walker P, de Montaudoüin X, Palud C & Héral M (1998) *Crepidula fornicata* L. (mollusque, gastéropode) en baie de Marennes-Oléron: cartographie des fonds par sonar à balayage latéral et estimation du stock. Oceanologia Acta 21: 353-362
Sauvageau C (1925) Sur la naturalisation en France d'une Floridée australasienne (*Asparagopsis armata* Harv.) et sur ses iodiques. Comptes Rendues Académie Sciences Paris 180
Savateeva EV (1985) Results of introduction of Baikalian amphipods and Ponto-Caspian mysids in Lake Ilmen. Proceedings of GosNIORKh 232: 3-15 (in Russian)
Savolainen R, Westman K & Pursiainen M (1997) Fecundity of Finnish noble crayfish *Astacus astacus* L., and signal crayfish, *Pacifastacus leniusculus*, in various natural habitats and in culture. Freshwater Crayfish 11: 319-338
Sax DF (2001) Latitudinal gradients and geographic ranges of exotic species: implications for biogeography. Journal of Biogeography 28: 139-150
Schaminée JHJ, Weeda EJ & Westhoff V (1995) The Vegetation of The Netherlands. Part 2. Plant communities of waters, swamps and wet heaths. Opulus Press, Uppsala & Leiden, 358 pp (in Dutch)
Schellenberg A (1937) Key and diagnoses of the kin forms of freshwater *Gammarus*, excluding species from the Baikal Lake and Australia. Zoologischer Anzeiger 117(11-12): 267-280 (in German)
Scheltema RS (1971) Dispersal of phytoplanktotrothic shipworm larvae (Bivalvia: Teredinidae) over long distances by ocean currents. Marine Biology 11: 5-11
Scherbina GH (2001) Self-acclimation of the Caspian polychaete *Hypania invalida* Stebb. in the Upper Volga basin. Zoological Journal 3: 278-284 (in Russian)
Scherbina GH, Arkhipova NP & Bakanov AI (1997) On the changes in the zoobenthos biodiversity in the Upper Volga and Gorki reservoirs. In: Problems of Biodiversity of Aquatic Organisms in Povolzh'ye Region. Zooplankton, Zoobenthos, pp 108-114. Institute of Ecology of the Volga Basin and Hydrobiological Society of the Russian Academy of Science Press, Tolyatti (in Russian)
Schiedek D (1997) *Marenzelleria* cf. *viridis* (Polychaeta: Spionidae) - ecophysiological adaptations to a life in the coastal waters of the Baltic Sea. Aquatic Ecology 31: 199-210
Schiffner V (1916) Studien über die Algen des adriatischen Meeres, Wissenschaftliche Meeresuntersuchgen Abteilung Helgoland 11: 129-198
Schlesch H (1932) Über die Einwanderung nordamerikanischer Meeresmollusken in Europa unter Berücksichtigung von *Petricola pholadiformis* Lam. und ihrer Verbreitung im dänischen Gebiet. Archiv für Molluskenkunde 64: 146-154
Schleswiger Nachrichten (1880) Hadersleben (Austern) 6.5.1880
Schleuter A & Schleuter M (1998) *Dendrocoelum romanodanubiale* (Turbellaria, Tricladida) und *Hemimysis anomala* (Crustacea: Mysidacea) zwei weitere Neozoen im Main. Lauterbornia 33: 125-127
Schleuter M, Schleuter A, Potel S & Banning M (1994) *Dikerogammarus haemobaphes* (Eichwald, 1841) (Gammaridae) from the Danube entered Main through Main-Danube canal. Lauterbornia 19: 155-159 (in German)
Schloesser DW & Nalepa TF (1994) Dramatic decline of unionid bivalves in offshore waters of western Lake Erie after infestation by the zebra mussel, *Dreissena polymorpha*. Canadian Journal of Fisheries and Aquatic Sciences 51: 2234-2242
Schloesser DW, bij de Vaate A & Zimmermen A (1994) A bibliography of *Dreissena polymorpha* in European and Russian waters: 1964-1993. Journal of Shellfish Research 13: 243-267
Schloesser DW, Smithee RD, Longton GD & Kovalak WP (1997) Zebra mussel induced mortality of unionids in firm substrata of Western Lake Erie and a habitat for survival. American Malacological Bulletin 14: 67-74
Schmidt RJ & Osenberg CW (1996) Detecting ecological impacts: concepts and applications in coastal habitats. Academic Press, London.
Schneider DW, Ellis CD & Cummings KS (1998) A transportation model assessment of the risk to native mussel communities from zebra mussel spread. Conservation Biology 12: 788-800
Schodduyn R (1931) Observations sur la flore et la faune des coquilles des huîtres. Bulletin de l'Institut Océanographique de Monaco 568: 1- 20
Schoener TW & Spiller DA (1995) Effect of predators and area invasion: An experiment with island spiders. Science 267: 1811-1813
Scholin CA & Anderson DM (1991) Population analysis of toxic and nontoxic *Alexandrium* species using ribosomial RNA signature sequences. In: Abstracts of Fifth International Conference on Toxic Marine Phytoplankton, p 113
Scholin CA, Hallegraeff GM & Anderson DM (1995) Molecular evolution of the *Alexandrium tamarense* 'species complex' (Dinophyceae): dispersal in the North American and West Pacific regions. Phycologia 34: 472-485
Schöll F (1990) The present occurrence of *Corophium curvispinum* Sars in the Rhine basin. Lauterbornia 5: 67-70 (in German)
Schöll F & Banning M (1996) First record of *Jaera istri* (Veuille) (Janiridae, Isopoda) in the Rhine. Lauterbornia 25: 61-62 (in German)
Schrenk AG (1848) Übersicht der Land- und Süsswassermollusken Livlands. Bulletin de la'Societe Imperiale des Naturalistes de Moscou 21
Schütz L (1961) Distribution and dispersal of the borer *Teredo navalis* L. and its entering of the Kiel Canal near Kiel. Kieler Meeresforschung 17: 228-236 (in German)
Schweimanns M (1993) Biodeteriation by wood destroying organisms along the Schleswig-Holstein coast (Mollusca: Pholadidae, Teredinidae; Crustacea: Isopoda). Schriften zur Malakozoologie, Cismar 6: 27-46 (in German)
SCOPE Newsletter (1998) Giant bloom of silica diatom algae in the North Sea in late summer of 1991. Scope Newsletter 33. scope.html at www.asi.fr (2001-11-13)
Scribner KT, Crane PA, Spearman WJ & Seeb LW (1998) DNA and allozyme markers provide concordant estimates of population differentiation: Analyses of U.S. and Canadian populations of Yukon River fall-run chum salmon (*Oncorhynchus keta*). Canadian Journal of Fisheries and Aquatic Sciences 55: 1748-1758
Sebestyen O (1937) Colonization of two new fauna-elements of Pontus-origin (*Dreissena polymorpha* Pall. and *Corophium curvispinum* G. O. Sars forma *devium Wundsch*) in Lake Balaton. Verhandlungen der Internationalen Vereinigung für Limnologie 8: 169-182
Seeb LW & Crane PA (1999) Allozymes and mitochondrial DNA discriminate Asian and North American populations of chum salmon in mixed-stock fisheries along the south coast of the Alaska Peninsula. Transactions of the American Fisheries Society 128: 88-103
Segerstråle S (1957) Baltic Sea. In: Hedgpeth JW (ed) Treatise on Marine Ecology and Paleoecology. Geological Society of America Memoirs 67: 751-800
Sei S, Licandro P, Zunini Sertorio T & Ferrari I (1999) Research on zooplankton in the Gulf of Rapallo. Chemistry and Ecology 16: 75-93
Sell P & Murrell G (1996) Flora of Great Britain and Ireland 5. Butomaceae-Orchidaceae. Cambridge University Press, Cambridge, 410 pp
Sellius, G. (1733) Historia naturalis teredinis seu xylophagi marini tubulo conchoidis speciatim Belgici. Trajecti ad Rehenum, [xxxiv]+356+[11] pp
Sembrat K (1953) Hydra. Popularne Monografie Zoologiczne 5, 79 pp (in Polish)
Sememoy VP (1974) On the oligochaete fauna (Oligochaeta) of the Rybinsk reservoir. Part II. Tubificidae. Information Bulletin of the Institute for Biology of Inland Waters 23: 37-39 (in Russian)
Semina GI & Zernova VV (1985) Generic evolutionary stages with reference to *Coscinodiscus* Ehr. (Bacillariophyta). In: Studies of Oceanic Phytoplankton, Moskva USSR Institute of Oceanology, 615 pp
Serbetis CD (1963) L'acclimatation de la *Meleagrina (Pinctada) margaritifera* (Lam.) en Grèce. Rapports et procès-verbaux des réunions, Commission internationale pour l'Exploration scientifique de la mer Méditerranée 17: 271-272
Settepassi F & Verdel V (1965) Continental Quaternary Mollusca of Lower Liri Valley (Southern Latium). Geologia Romana 4: 369-452
Shadrin NV (1985) Bioproductive features of water bodies in relation to their morphometry. In: Hydrobiology and Hydroparasitology of Zabaical Region. Novosibirsk, Nauka, pp 70-75 (in Russian)
Shadrin NV (1999) Ecosystem functioning and economics: interrelations on global and local scales. In: Pavlova EV & Shadrin NV (eds) Sevastopol Aquatory and Coast: Ecosystem Processes and Services for the Human Society. Sevastopol, Aquavita 1999, pp 10-24
Shadrin NV (2000a) Alien species in the Black Sea. Biologya morya 51: 25-31 (in Russian)
Shadrin NV (2000b) The species introducers in the Black Sea and in the Sea of Azov: causes and consequences. In: Exotic species in the European Seas of Russia. Apatity, 2000, pp 76 - 90 (in Russian)
Shadrin NV, Gubanova AD & Popova EV (1999) Long-term changes of *Acartia* (Copepoda) taxocene in the Sevastopol Bay. In: Pavlova EV & Shadrin NV (eds) Sevastopol Aquatory and Coast: Ecosystem Processes and Services for the Human Society. Sevastopol, Aquavita 1999, pp 159-167 (in Russian)
Sharonov IV (1971) The expansion of the area of some fishes with changes of the Volga regulation. In: Volga-I. Problems of Studying and Rational Use of Biological Resources of Water Bodies, pp 226-233. Kuybishev Publishing House Press, Kuybishev (in Russian)
Shcherbina GKh, Arkhipova NR & Bakanov AI (1997) On the changes in zoobenthos biological diversity of Upper Volga and Gorkovskoe reservoirs. In: Popchenko VI & Bychek EA (eds) Problems of biological diversity of aquatic organisms in Volga River basin. Institute of Ecology of Volga River Basin, Toljatti: 108-114 (in Russian)
Sherley G (ed) (2000) Invasive Species in the Pacific: A Technical Review and Draft Regional Strategy. South Pacific Regional Environment Programme (SPREP), Apia, Samoa, 190 pp
Sherr E & Sherr B (1988) Role of microbes in pelagic food webs: A revised concept. Limnology and Oceanography 33: 1225-1227
Shevtsova LV (1989) Importance of the Bug River *Dreissena* for decreasing organic matter and for the sedimentation of suspended substances. Hydrobiological Journal 25: 47-51
Shevtsova LV & Kharchenko TA (1981) The role of *Dreissena* in reworking of suspended organic matter in the North-Crimean Canal. Hydrobiological Journal 17: 33-36
Shevtsova LV, Zhdanova GA, Movchan VA & Primak AB (1986) Experimental interrelationship between *Dreissena* and planktic invertebrates. Hydrobiological Journal 22: 36-39
Shiganova TA (2000) Some results of study of *Mnemiopsis leidyi* biology in the Black Sea In: Volovik SP (ed) Ctenophore *Mnemiopsis leidyi* (A.Agassiz) in the Azov and Black Sea and consequences of its intrusion, pp 33-75. Rostov-on-Don
Shiganova T, Tarkan N, Dede A & Cebeci M (1995) Distribution of the ichthyo-jelly plankton *Mnemiopsis leidyi* (Agassiz, 1865) in the Marmara Sea (October 1992). Turkish Journal of Marine Science 1: 3-12
Shiganova TA, Bulgakova YuV, Sorokin PYu & Lukashev YuF (2000) Investigation of a new settler *Beroe ovata* in the Black Sea. Biology Bulletin 27: 202-209
Shigesada N & Kawasaki K (1997) Biological Invasions: Theory and Practice. Oxford University Press, Oxford, 205 pp
Shkorbatov GL, Karpevich AF & Antonov PI (1994) Ecological physiology. In: Starobogatov YI (ed) Freshwater Zebra Mussel *Dreissena polymorpha* (Pall.) (Bivalvia, Dreissenidae) Taxonomy, Ecology and Practical Use, pp 67-109. Nauka, Moscow (in Russian)

Shotts EB, Kleckner AL, Gratzek JB & Blue JL (1976) Bacterial flora of aquarium fishes and their shipping waters imported from Southeast Asia. Journal of the Fisheries Research Board of Canada 33: 732-735

Shulman GE (1972) Physiological-biochemical characteristics of annual cycles of fish. Pishcheprom Publishing House, Moscow. 368 pp (in Russian)

Shumway SE (1990) A review of the effects of algal blooms on shellfish and aquaculture. Journal of the World Aquaculture Society. 21(2): 65-104

Sidorovich V, Kruuk H & Macdonald DW (1999) Body size, and interactions between European and American mink (*Mustela lutreola* and *M. vison*) in Eastern Europe. Journal of Zoology 248: 521-527

Siegismund HR (1997) G-Stat, version 3.2, Genetical statistical programs for the analysis of population data. Department of Plant Ecology, University of Copenhagen, Denmark

Siekkinen J & Nummi P (1992) The captive breeding and release of waterfowl - a review. Suomen Riista 38: 45-55 (in Finnish, English summary)

Sikorski AV, Bick A & Bastrop R (in prep) Revision of the genus *Marenzelleria* (Spionidae, Polychaeta). Acta Zoologica

Sikorskii AV & Buzhinskaya GN (1998) The genus *Marenzelleria* (Polychaeta, Spionidae) in Seas of Russia. Zoologiceskij Zhurnal 77: 1111-1120 (in Russian)

Silina NI (1989) Species of the genus *Acartia* in the plankton of eastern part of the Gulf of Finland of the Baltic Sea. In: The investigations of water ecosystem. Proceedings of the Zoological Institute, Leningrad 205: 108-118 (in Russian)

Silva PC (1957) *Codium* in Scandinavian waters. Svensk Botanisk Tidskrift 51: 117-134

Simberloff D (1986) Introduced insects: A biogeographic and systamatic prespective. In: Mooney HA & Drake JA (eds) Ecology of Biological Invasions of North America and Hawaii, pp 3-24. Springer Verlag, New York

Simberloff D (1989) Which insect introductions succeed and which fail? In: Drake JA, Mooney HA, di Castri F, Groves RH, Kruger FJ, Rejmánek M & Williamson M (eds) Biological Invasions, pp 61-75. John Wiley & Sons, New York

Simberloff D (1996) Hybridization between native and introduced wildlife species. Wildlife Biology 2: 143-150

Simberloff D & Von Holle B (1994) Positive interactions of non-indigenous species: invasional meltdown? Biological Invasions 1: 21-32

Simon C, Gall EA & Deslandes E (2001) Expansion of the red alga *Grateloupia doryphora* along the coasts of Brittany (France). Hydrobiologia 443: 23-29

Sindermann CJ (1992) The introduction of marine species: What the future holds. In: DeVoe R (ed) Proceedings of the Conference & Workshop. Introductions & Transfers of Marine Species. Achieving a Balance between Economic Development and Resource Protection. Session VII, pp 143-148. South Carolina Sea Grant Consortium, USA

Sinitsyna OO & Protasov AA (1994) Structural and functional characteristics of benthic communities as criteria for the setting of thermal standards for cooling ponds. Hydrobiological Journal 30: 23-36

Sinner J, Forrest B, Taylor M, Dodgshun T, Brown S & Gibbs W (2000) Options for a national pest management strategy for the Asian kelp *Undaria pinnatifida*. Technical Report prepared for the New Zealand Ministry of Fisheries, 62 pp

Sipponen M & Hakkari L (1984) Brown trout (*Salmo trutta* m. *lacustris* (L.)) stockings as a compensation method in a polluted area in central Finland. EIFAC Technical Paper 42(1): 152-163

Sirenko BI & Piepenburg D (1994) Current knowledge on biodiversity and benthic zonation patterns of Eurasian Arctic shelf seas, with special reference to the Laptev Sea. Berichte zur Polarforschung 144: 69-77

Skalskaya IA (1994) On distribution of the Baikalian amphipod *Gmelinoides fasciatus* (Stebb.) in Rybinskoe Reservoir. Biology of Inland Waters Information Bulletin 96: 35-40 (in Russian)

Skalskaya IA (1996) Structure of population of the Baikalian amphipod *Gmelinoides fasciatus* (Stebb.) in Rybinskoe Reservoir. Biology of Inland Waters Information Bulletin 99: 29-35 (in Russian)

Skalskaya IA (1998) Establishment of the Baikalian amphipod in the Upper Volga basin. In: Proceedings of International Conference of Environmental Problems of Large Rivers Basins, Toljatti, pp 244-245 (in Russian)

Skerratt JH, Bowman JP, Nichols PD, Hallegraeff G & McMeekin TA (2000) Estuarine algicidal bacteria: their detection and effect on harmful algal species. 9th International Conference on Harmful Algal Blooms, Tasmania, Australia, 7-11 February 2000, p 57. University of Tasmania, Hobart (Abstract)

Skoldal HR & Dundas I (1991) The *Chrysochromulina polylepis* bloom in the Skagerrak and the Kattegat in May-June 1988: Environmental conditions, possible causes, and effects. ICES Cooperative Research Report, pp 1-59

Skora KE (1996) New and rare fish species from the Gulf of Gdansk. Zoologica Poloniae 41(suppl.): 113-130 (in Polish)

Skora KE & Stolarski J (1993) New fish species in the Gulf of Gdansk. Bulletin of the Sea Fisheries Institute, Gdynia 1(128): 83-84

Skora KE & Stolarski J (1995) Round goby - a fishy invader. WWF Baltic Bulletin 1: 46-47

Skora KE & Stolarski J (1996) *Neogobius melanostomus* (Pallas, 1811), a new immigrant species in the Baltic Sea. In: Styczynska-Jurewicz E (ed) Estuarine Ecosystems and Species. Proceedings of 2nd International Estuary Symposium, Gdansk, October 18-22, 1993: 101-108

Skora K, Olenin S & Gollasch S (1999) *Neogobius melanostomus*. In: Gollasch S, Minchin D, Rosenthal H & Voigt M (eds) Exotics Across the Ocean. Case Histories on Introduced Species, pp 69-73. Logos Verlag, Berlin

Skubinna JP, Coon TG & Batters TR (1995) Increased abundance and depth of submersed macrophytes in response to decreased turbidity in Saginaw Bay, Lake Huron. Journal of Great Lakes Research 21: 476-478

Skurdal J, Taugbøl T, Burba A, Edsman L, Söderbäck B, Styrishave B, Tuusti J & Westman K (1999) Crayfish introductions in the Nordic and Baltic countries. In: Gherardi F & Holdich DM (eds) Crayfish in Europe as Alien Species. How to Make the Best of a Bad Situation, pp 193-219 A.A. Balkema, Rotterdam, Brookfield

Skurdal J, Westman K & Bergan PI (eds) (1989) Crayfish culture in Europe. Workshop on Crayfish Culture, 16-19 November 1987, Trondheim, Norway, pp 39-48

Skvortsov AK (1995) On the systematics and nomenclature of the adventive species of g. *Epilobium* (Onagraceae) in Russia. Bulletin MOIP 100: 74-78 (in Russian)

Slatkin M (1993) Isolation by distance in equilibrium and non-equilibrium populations. Evolution 47: 264-279

Slobodkin LB (2001) The good, the bad and the reified. Evolutionary Ecology Research 3: 1-13

Slynko YV, Kiyashko VI & Yakovlev VN (2000) The list of cyclostomes and fish species in Volga river basin. In: Yakovlev VN (ed) Biodiversity of Volga River Basin, pp 38-46. YrGTU Press, Yaroslavl (in Russian)

Slynko YV, Kiyashko VI & Yakovlev VN (2001) Fish invaders in Upper Volga river basin. In: Kopylov AI (ed) Ecological Problems of Upper Volga, pp 84-86. YrGTU Press, Yaroslavl (in Russian)

Smayda TJ (1990) Novel and nuisance phytoplankton blooms in the sea: evidence for a global epidemic. In: Granéli E, Sundström B, Edler L & Anderson DM (eds) Toxic Marine Phytoplankton. Proceedings of the Fourth International Conference on Toxic Marine Phytoplankton (26-30 June 1989), pp 29-40. Elsevier, New York

Smit H, Van der Velde G, Smits RG & Coops H (1997) Ecosystem responses in the Rhine-Meuse Delta during two decades after enclosure and steps toward estuary restoration. Estuaries 20: 504-520

Smit HS, Bij de Vaate A, Reeders HH, Van Nes EH & Noordhuis R (1993) Colonisation, ecology and positive aspects of zebra mussels (*Dreissena polymorpha*) in the Netherlands. In: Nalepa TF & Schloesser DW (eds) Zebra Mussels: Biology, Impacts and Control, pp 55-77. Lewis Publishers

Smith CM & Walters LJ (1999) Fragmentation as a strategy for *Caulerpa* species: Fates of fragments and implications for management of an invasive weed. Marine Ecology PSZNI 20: 307-319

Smith D (1995) Chesapeake Bay ballast study. IUCN Species Survival Commission. Report 16

Smith G, Ruiz G, Frey M, Takata L, Altman S, Philips K, McCann L, Coss C, Lipski D, Huber T, Steves B & Chavez S (2001) Effectiveness of ballast water exchange. International Conference on Marine Bioinvasions, New Orleans. p 132 (Abstract)

Smith JMB (1996) Perspectives of an invasion of paradise: *Rhizophora stylosa* on Moorea, French Polynesia. Australian Geographical Studies 34(1)

Smith LD, Wonham MJ, McCann LD, Ruiz GM, Hines AH & Carlton JT (1999) Invasion pressure to a ballast-flooded estuary and an assessment of inoculant survival. Biological Invasions 1: 67-87

Smith TE & Kerr SR (1992) Introductions of species transported in ships' ballast waters: The risk to Canada's marine resources. Canadian Technical Report, Fisheries Aquatic Sciences 1867: 1-16

Smith VJ & Söderhäll K (1986) Crayfish pathology: an overview. Freshwater Crayfish 6: 199-211

Sneep J-W (1999) Control and eradication of non-native terrestrial vertebrates in the Netherlands. In: Workshop on the control and eradication of non-native terrestrial vertebrates, Malta, pp 51-57. Environmental Encounters, Council of Europe Publishing

Sneh A, Weissbrod T & Perath I (1975) Evidence for an ancient Egyptian frontier canal. American Scientist 63: 542-548

Snieszko SF (1973) Furunculosis of salmonidae. EIFAC Technical Paper 17 Suppl 2:157-62

Snovsky Z & Galil BS (1990) The occurrence of the American blue crab, *Callinectes sapidus* Rathbun, in the Sea of Galilee. Israel Journal of Aquaculture 42: 62-63

Snovsky Z & Shapiro J (1999) The fisheries and Aquaculture of Israel 1997 in figures. Ministry of Agriculture, Department of Fisheries, 42 pp

Sobol Z, Wladyslaw K & Bohdan W (1995) System for destruction of microorganisms occurring in ballast waters. Paper presented at IMO MEPC 38

Söderbäck B (1991) Interspecific dominance relationship and aggressive interactions in the freshwater crayfishes *Astacus astacus* (L.) and *Pacifastacus leniusculus* (Dana). Canadian Journal of Zoology 69: 1321-1325

Söderbäck B (1993) Population regulation in two co-occurring crayfish species. PhD Thesis, Acta Universitatis Upsaliensis, Uppsala University 434, 29 pp

Söderbäck B (1995) Replacement of the native crayfish *Astacus astacus* by the introduced species *Pacifastacus leniusculus* in a Swedish lake: possible causes and mechanisms. Freshwater Biology 33: 291-304

Sokolov AA (1964) Hydrography of the USSR, 534 pp. Gidrometeoizdat, Leningrad

Sokolov LI, Chepkin EA, Sokolov EL & Golovatiuk GJ (1989) New species in the Moskva river. Priroda 9: 80-81 (in Russian)

Sokolova NY, Izvekova EI, Lvova AA & Sakharova MI (1980a) Structure, distribution and seasonal dynamics of benthic densities and biomass. Trudy Vsesoiuznogo Gidrobiologicheskogo Obshchestva 23: 7-23 (in Russian)

Sokolova NY, Izvekova EI, Lvova AA & Sakharova MI (1980b) Features of benthos forming in small waterbodies using Uchinskoe Reservoir as an example. Trudy Vsesoiuznogo Gidrobiologicheskogo Obshchestva 23: 161-170 (in Russian)

Soletchnik P, Faury N, Razet D & Goulletquer P (1998) Hydrobiology of the Marennes-Oléron Bay. Seasonal indices and analysis of trends from 1978 to 1995. Hydrobiologia 386: 131-146

Solustri C & Morello E (2000) First record of *Anadara demiri* (Piani, 1981) (Bivalvia: Arcidae) in Italian Waters. I congresso Congiunto delle Societa Malacologiche del Mediterraneo, Genova 12-16 Novembre 2000

Sommer U (1995) An experimental test of the intermediate disturbance hypothesis using cultures of marine phytoplankton. Limnology and Oceanography 40: 1271-1277

Sonnenfeld P (1985) Models of upper Miocene evaporite genesis in the Mediterranean region. In: Stanley DJ & Wezel FC (eds) Geological Evolution of the Mediterranean Basin, pp 323-346. Springer Verlag, Heidelberg-New York

Sorbini L (1988) Biogeography and climatology of Pliocene and Messinian fossil fish of Eastern Central Italy. Bollettino del Museo civico di storia naturale di Verona 14: 1-85

Sordyl H, Bönsch R, Gercken J, Gosselck F, Kreuzberg M & Schulze H (1998) Dispersal and reproduction of the shipworm *Teredo navalis* L. in coastal waters of Mecklenburg-Western Pomerania. Deutsche Gewässerkundliche Mitteilungen 42: 142-149 (in German)

Southward AJ, Burton RS, Coles SL, Dando PR, DeFelice R, Hoover J, Parnell PE, Yamaguchi T & Newman WA (1998) Invasion of Hawaiian shores by an Atlantic barnacle. Marine Ecology Progress Series 165: 119-126

Southwood TRE (1977) Habitat, the templet for ecological strategies? Journal of Animal Ecology 46: 337-365

Sovinsky V (1904) Introduction to the study of fauna of the marine Ponto-Aralo-Caspian basin recognized as an independent zoogeographical province. Mémoires de la Societé des Naturalistes de Kieff, 18 (1-13): 487 pp; suppl. 1: 193 pp; suppl. 2: 197-216 (in French)

Spanier E & Galil BS (1991) Lessepsian migrations: a continous biogeographical process. Endeavour 16: 102-106

Specziár A, Tölg L & Bíró P (1997) Feeding strategy and growth of cyprinids in the littoral zone of Lake Balaton. Journal of Fish Biology 51:1109-1124

Spilinga C, Chiappafreddo U & Pirisinu Q (2000) *Dreissena polymorpha* (Pallas) al lago Trasimeno. Rivista Idrobiologia 39: 145-152

Sprung M (1991) "Costs" of reproduction: A study on metabolic requirements of the gonads and fecundity of the bivalve *Dreissena polymorpha*, Malacologia 33: 63-70

Stace CA (1991) New Flora of the British Isles. Cambridge University Press, Cambridge, 1226 pp

Stachowicz JJ, Whitlach RB & Osman RW (1999) Species diversity and invasion resistance in a marine ecosystem. Science 286: 1577- 1579

Staehr PA, Pedersen MF, Thomsen MS, Wernberg T & Krause-Jensen D (2000) Invasion of *Sargassum muticum* in Limfjorden (Denmark) and its possible impact on the indigenous macroalgal community. Marine Ecology Progress Series 207: 79-88

Stanczykowska A (1968) The filtration capacity of populations of *Dreissena polymorpha* Pall. in different lakes, as a factor affecting circulation of matter in the lake. Ekologia Polska B 14: 265-270

Stanczykowska A (1975) Ecosystem of the Mikolajskie lake. Regularities of the *Dreissena polymorpha* Pall. (Bivalvia) occurrence and its function in the lake. Polskie Archiwum Hydrobiologii 22: 73-78

Stanczykowska A (1977) Ecology of *Dreissena polymorpha* (Pall.) (Bivalvia) in lakes. Polskie Archiwum Hydrobiologii 24: 461-530

Stanczykowska A (1984) The effect of various phosphorus loadings on the occurrence of *Dreissena polymorpha* (Pall.). Limnologica (Berlin) 15: 535-539

Stanczykowska A (1997) Review of studies on *Dreissena polymorpha* (Pall.). Polskie Archiwum Hydrobiologii 44: 401-415

Stanczykowska A & Lewandowski K (1993) Effect of filtering activity of *Dreissena polymorpha* (Pall.) on the nutrient budget of the littoral of Lake Mikolajskie. Hydrobiologia 251: 73-79

Stanczykowska A & Lewandowski K (1993) Thirty years of studies of *Dreissena polymorpha* ecology in Mazurian Lakes of Northeastern Poland. In: Nalepa TF & Schloesser DW (eds) Zebra Mussels: Biology, Impacts, and Control, pp 3-33. Lewis Publishers, Ann Arbor, Michigan

Stanczykowska A & Stoczkowski R (1997) Are the changes in *Dreissena polymorpha* (Pall.) distribution in the Great Masurian Lakes related to trophic state? Polskie Archiwum Hydrobiologii 44: 417-429

Stanczykowska A, Lawacz W & Mattice J (1975) Use of field measurements of consumption and assimilation in evaluating of the role of *Dreissena polymorpha* (Pall.) in a lake ecosystem. Polskie Archiwum Hydrobiologii 22: 509-520

Stanczykowska A, Lewandowski K & Swierczynski M (1997) Summary of studies on *Dreissena polymorpha* (Pall.) conducted in the period 1993-95 in the Masurian and Pomeranian lakelands. Polskie Archiwum Hydrobiologii 44: 517-520

Standley LA (1983) *Carex kobomugi* Owhi, an adventive sedge new to New England. Rhodora 85: 265-269

Starobogatov YaI (1970) The Mollusk Fauna and Zoogeographic Distinguishing of the Earth's Continental Waters. Nauka Press, Leningrad, 372 pp (in Russian)

Starobogatov YI (ed) (1994) Taxonomy and paleontology. Freshwater Zebra Mussel *Dreissena polymorpha* (Pall.) (Bivalvia, Dreissenidae) Taxonomy, Ecology and Practical Use, pp 18-46. Nauka, Moscow (in Russian)

Starobogatov YaI & Andreeva SI (1994) Range and its history. In: Starobogatov YI (ed) Freshwater Zebra Mussel *Dreissena polymorpha* (Pall.) (Bivalvia, Dreissenidae) Taxonomy, Ecology and Practical Use, pp 47-56. Nauka, Moscow (in Russian)

Statzner B, Fievet E, Champagne JY, Morel R & Herouin E (2000) Crayfish as geomorphic agents and ecosystem engineers: biological behavior affects sand and gravel erosion in experimental streams. Limnology and Oceanography 45: 1030-1040

Stearns REC (1891) On accidentally introduced forms of animals. West American Scientist 7(57): 107-109

Steinitz W (1919) Memorial on the founding of a sea-station on the coast of Palestine for zoological investigations. Manuscript, 16 pp

Steinitz W (1929) Die Wanderung indopazifischer Arten ins Mittelmeer seit Beginn der Quartärperiode. Internationale Revue der gesamten Hydrobiologie und Hydrographie 22: 1-90

Steneck RS & Carlton JT (2001) Human alterations of marine communities: students beware! In: Bertness MD, Gaines SD & Hay ME (eds) Marine Community Ecology, pp 445-468. Sinauer Associates, Inc., Publishers, Sunderland, Massachusetts

Stergiou C, Christou N, Georgopoulos D, Zenetos A & Souvermetzoglou K (1997) Hellenic Seas: Physics, Chemistry, Biology & Fisheries. Oceanography and Marine Biology Annual review 35: 415-538

Steuer A (1936) Cumacea and stomatopoda von Alexandrien in Ägypten. Notizen des Deutsch-Italienischen Institutes für Meeresbiologie in Rovigno d'Istria 21: 1-19

Steuer A (1938) The fishery grounds near Alexandria, XVI Cumacea, Stomatopoda, Leptostraca. Hydrobiology and Fisheries Directorate Notes and Memoirs (Cairo) 26: 1-16

Stewart DJ, Woldemarian K, Dear G & Mochaba MF (1983) An outbreak of "Sekiten-byo" among cultured European eels, *Anguilla anguilla* L., in Scotland. Journal of Fish Diseases 6: 75-76

Stewart TW & Haynes JM (1994) Benthic macroinvertebrate communities of Southwestern Lake Ontario following invasion of *Dreissena*. Journal of Great Lakes Research 20: 479-493

Stewart TW, Gafford JC, Miner JG & Lowe RL (1999) *Dreissena*-shell habitat and antipredator behavior: combined effects on survivorship of snails co-occurring with molluscivorous fish. Journal of the North American Benthological Society 18: 274-283

Stewart TW, Miner JG & Lowe RL (1998a) Macroinvertebrate communities on hard substrates in western Lake Erie: Structuring effects of *Dreissena*. Journal of Great Lakes Research 24: 868-879

Stewart TW, Miner JG & Lowe RL (1998b) Quantifying mechanisms for zebra mussel effects on benthic macroinvertebrates: organic matter production and shell-generated habitat. Journal of the North-American Benthological Society 17: 81-94

Stigzelius J, Laine A, Rissanen J, Andersin A-B & Ilus E (1997) The introduction of the North American *Marenzelleria viridis* (Verrill 1873) into the Gulf of Finland and the Bothnian Sea. Annales Zoologici Fennici 34: 205-212

Stock JH (1974) The systematics of certain Ponto-Caspian Gammaridae (Crustacea, Amphipoda). Mitteilungen aus dem Hamburgischen Zoologischen Museum und Institut 70: 75-95

Stock JH (1993) Copepoda (Crustacea) associated with commercial and non-commercial Bivalvia in the East Scheldt, The Netherlands. Bijdragen tot de Dierkunde 63: 61-64

Stock JH (1995) *Gammarus* and *Chaetogammarus* (Crustacea, Amphipoda) from Macaronesia. Boletim do Museu Municipal do Funchal 45: 41-52

Stock JH, Mirzajani AR, Vonk R, Naderi S & Kiabi BH (1998) Limnic and brackish water Amphipoda (Crustacea) from Iran. Beaufortia 48: 173-234

Stoeckel, JA, Sparks RE, Blodgett KD, Whitney, SD & Raibley PT (1996) Interbasin dispersal of invading aquatic species. Illinois Natural History Survey Reports 341: 4-8

Stohlgren TJ, Binkley D, Chong GW, Kalkhan MA, Schell LD, Bull KA, Otsuki Y, Newman G, Bashkin M & Son Y (1999) Exotic plant species invade hot spots of native plant diversity. Ecological Monographs 69: 25-46

Straskraba M (1962) Amphipods of the Czechechoslovakia in the collection of Prof. Hrabe. I. Vestnik Ceskoslovenskoj Zoologiceskoj Spolecnosti 26(2): 117-145 (in German)

Strasser M (1999) *Mya arenaria* - an ancient invader of the North Sea coast. Helgoländer Meeresunterschungen 52: 309-324

Strayer DL (1999) Effects of alien species on freshwater mollusks in North America. Journal of the North American Benthological Society 18: 74-98

Strayer DL, Caraco NF, Cole JJ, Findlay S & Pace ML (1999) Transformation of freshwater ecosystems by bivalves - a case study of zebra mussels in the Hudson River. BioScience 49: 19-27

Strayer DL, Powell J, Ambrose P, Smith LC, Pace ML & Fischer DT (1996) Arrival, spread and early dynamics of a zebra mussel (*Dreissena polymorpha*) population in the Hudson River estuary. Canadian Journal of Fisheries and Aquatic Sciences 53: 1143-1149

Strayer DL, Smith LC & Hunter DC (1998) Effects of the zebra mussel (*Dreissena polymorpha*) invasion on the macrobenthos of the freshwater tidal Hudson River. Canadian Journal of Zoology 76: 419-425

Stuart MD & Chadderton WL (2001) Attempted eradication of the adventive Asian kelp, *Undaria pinnatifida*, in New Zealand. Phycologia 40 (4) Supplement: 20 (Abstract)

Stylinski CD & Allen EB (1999) Lack of native species recovery following severe exotic disturbance in southern Californian shrublands. Journal of Applied Ecology 36: 544-554

Subba Rao DV, Sprules WG, Locke A& Carlton JT 1994. Exotic phytoplankton from ships' ballast waters: risk of potential spread to mariculture sites on Canada's east coast. Canadian Data Report of Fisheries and Aquatic Science 937: 1-51

Subklew H-J (1963) Giftfarben oder Ultraschall zur Bekämpfung des Schiffsbewuchses. Schiffbautechnik (7) 381

Sumner FB (1909) On the occurrence of the littoral barnacle *Chthamalus stellatus* (Poli) at Woods Hole, Massachusetts Science n.s. 30: 373-374

Sumner FB, Osburn RC & Cole LJ (1913) A biological survey of the waters of Woods Hole and vicinity. Section 1 - Physical and zoological. Bulletin of the Bureau of Fisheries 31(1): 1-544

Sundet JH (1998) Bycatch of the king crab in ordinary fisheries. Fiskeriforskning. Rapport 1: 1-15 (in Norwegian)

Sundet JH (1999) Bycatch of the king crab in net and long-line fisheries in 1998. Fiskeriforskning. Rapport 2: 1-20 (in Norwegian)

Surkov SS (2001) Ecological aspects of the pink salmon acclimatization on the European North. In: Materiali VIII regional'noy nauchno-practicheskoy

conferentzii, pp 220-222. Izdatelstvo "Pravda Severa", Arkhangel'sk (in Russian)
Sutton CA, Murphy K, Martin RB & Hewitt C (1998) A Review and Evaluation of Ballast Water Sampling Protocols. CRIMP Technical Report 18
Svärdson G (1965) The American crayfish *Pacifastacus leniusculus* (Dana) introduced into Sweden. Report of the Institute of Freshwater Research, Drottningholm 46: 90-94
Svärdson G (1992) Ecological co-evolution of the parasitic fungus *Aphanomyces astaci* and its crayfish host. Finnish Fisheries Research 14: 135-143
Svärdson G (1995) The early history of signal crayfish introduction into Europe. Freshwater Crayfish 8: 68-77
Svärdson G, Fürst M & Fjälling A (1991) Population resilience of *Pacifastacus leniusculus* in Sweden. Finnish Fisheries Research 12: 165-177
Svetovidov AN (1964) Fishes of Black Sea. Fauna of the Soviet Union - Guide book. Opredelitel po faune SSSR, Akad. Nauk SSSR., Izdatelstvo Nauka, Moskva-Leningrad 86: 525 pp (in Russian)
Sykora KV (1990) History of the impact of man on the distribution of plant species. In: di Castri F, Hansen AJ & Debussche M (eds) Biological Invasions in Europe and the Mediterranean Basin, pp 37-50. Kluwer Academic Publishers, Dordrecht
Symstad AJ (2000) A test of the effects of functional group richness and composition on grassland invasibility. Ecology 81: 99-109
Szabo P (1994) Quality of effluent water from earthem fish ponds in Hungary. Journal of Applied Ichthyology 10: 326-334
Taiti S (1999) Terrestrial isopods from Midway Atoll (Crustacea: Oniscidea). Bishop Museum Occasional Paper 59: 37-38
Taiti S & Howarth FG (1996) Terrestrial isopods from the Hawaiian Islands (Isopoda: Oniscidea). Bishop Museum Occasional Paper 45: 59-71
Talley TS, Crooks JA & Levin LA (2001) Habitat utilization and alteration by the invasive burrowing isopod, *Sphaeroma quoyanum*, in California salt marshes. Marine Biology 138: 561-573
Talman, S, Bité JS, Campbell SJ, Holloway M, McArthur M, Ross DJ & Storey M (1999) Impacts of some introduced marine species found in Port Phillip Bay. In: Hewitt CL, Campbell ML, Thresher RE & Martin RB (eds) Marine Biological Invasions of Port Phillip Bay, Victoria. CRIMP Technical Report 20: 261-274. CSIRO Division of Marine Research, Hobart, Australia
Tangen K (1977) Blooms of *Gyrodinium aureolum* (Dinophyceae) in North European waters, accompanied by mortality in marine organisms. Sarsia 68: 1-7
Tarasov A (1996) Malacostraca of the middle and lower flows of Viatka River (Crustacea). Arthropoda Selecta 5 (3/4): 19-22
Taugbøl T & Skurdal J (1999) The future of native crayfish in Europe: How to make the best of a bad situation? In: Gherardi F & Holdich DM (eds) Crayfish in Europe as Alien Species. How to Make the Best of a Bad Situation, pp 271-279. A.A. Balkema, Rotterdam, Brookfield
Taylor AH & Rigby GR (2001) Suggested designs to facilitate improved management and treatment of ballast water on new and existing ships. Agriculture, Fisheries and Forestry, Australia, Ballast Water Research Series, Report 12
Taylor FJR, Fukuyo Y & Larsen J (1995) Taxonomy of harmful dinoflagellates. In: Hallegraeff GM, Anderson DN & Cembella AD (eds) Manual on Harmful Marine Microalgae, pp 283-317, UNESCO, France
Taylor JN, Courtenay WR & McCann JA (1984) Known impacts of exotic fishes in the continental United States. In: Courtenay WR & Stauffer JR (eds) Distribution, Biology and Management of Exotic Fishes, pp 322-373. John Hopkins University Press, Baltimore
Taylor MD & Bruce EJ (2000) Mid ocean ballast water exchange: shipboard trials of methods for verifying efficiency. Report prepared for Ministry of Fisheries. Cawthron Institute, Halifax, New Zealand, Report 524, 59 pp
Tegelström H & Sjöberg G (1995) Introduced Swedish Canada geese (*Branta canadensis*) have low levels of genetic variation as revealed by DNA fingerprinting. Journal of Evolutionary Biology 8: 195-207
Telesh IV, Alimov AF, Golubkov SM, Nikulina VN & Panov VE (1999) Response of aquatic communities to anthropogenic stress: a comparative study of Neva Bay and the eastern Gulf of Finland. Hydrobiologia 393: 95-105
Telesh IV, Bolshagin PV & Panov VE (2001) Quantitative estimation of the impact of alien species *Cercopagis pengoi* (Crustacea: Onychopoda) on the structure and functioning of plankton community in the Gulf of Finland, Baltic Sea. Doklady Biol Sci 377: 427-429
Telesh IV, Litvinchuk LF, Bolshagin PV, Krylov PI & Panov VE (2000) Peculiarities of biology of the Ponto-Caspian species *Cercopagis pengoi* (Crustacea: Onychopoda) in the Baltic Sea. In: Species Introducers in the European Seas in Russia, pp 130-151. Apatity (in Russian)
Templeton AR (1998) Nested clade analyses of phylogeographic data: testing hypothesis about gene flow and population history. Molecular Ecology 7: 381-397
Templeton AR (2001) Using phylogeographic analyses of gene trees to test species status and processes. Molecular Ecology 10: 779-791
Templeton AR, Routman E & Phillips CA (1995) Separating population structure from population history: a cladistic analysis of the geographical distribution of mitochondrial DNA haplotypes in the tiger salamander, *Ambystoma tigrinum*. Genetics 140: 767-782
Ten Winkel EN & Davids C (1982) Food selection by *Dreissena polymorpha* Pallas (Mollusca: Bivalvia). Freshwater Biology 12: 553-558
Thayer SA, Haas RC, Hunter RD & Kushler RH (1997) Zebra mussel (*Dreissena polymorpha*) effects on sediment, other zoobenthos, and the diet and growth of adult yellow perch (*Perca flavescens*) in pond enclosures. Canadian Journal of Fisheries and Aquatic Sciences 54: 1903-1915
The Astrakhan Reserve (1991) Agropromizdat Press, Moscow, 191 pp (in Russian)
The Kuybyshev Reservoir (1983) Nauka Press, Leningrad, 215 pp (in Russian)
The Royal Society (1983) Risk assessment: Report of a Royal Society Study Group. The Royal Society, London
The Rybinsk Reservoir and its Life (1972) Nauka Press, Leningrad, 360 pp (in Russian)
The Volga River and its Life (1978) Nauka Press, Leningrad, 352 pp (in Russian)
Thellier M, Echaubard M, Kloareg B, Laubier L & Pavé A (1997) Report to the French Minister for the Environment concerning the alga *Caulerpa taxifolia* in the Mediterranean. In: Dynamics of Invasive Marine Species: Application to the Expansion of *Caulerpa taxifolia* in the Mediterranean. Actes de Colloques Septembre 1997, Académie des Sciences, pp 345-352. Lavoiser Technique & Documentation, Paris
Thibaut T & Meinesz A (2001) Influence of *Posidonia oceanica* seagrass bed densities on settlement of the invasive introduced *Caulerpa taxifolia*. Phycologia 40(4) Supplement: 15 (Abstract)
Thibaut T, Meinesz A, Amade P, Charrier S, De Angelis K, Ierardi S, Mangialajo L, Melnick J & Vidal V (2001) *Elysia subornata* (Mollusca) a potential control agent of the alga *Caulerpa taxifolia* (Chlorophyta) in the Mediterranean Sea. Journal of the Marine Biological Association of the UK 81: 497-504
Thiebaut G, Rolland T, Robach F, Tremolieres M & Muller S (1997) Some consequences of the introduction of two macrophyte species, *Elodea canadensis* Michaux and *Elodea nuttallii* St. John, in continental aquatic ecosystems: Example of two areas in the North-East of France: Alsace Plain and Northern Vosges. Bulletin Francais de la Peche et de la Pisciculture 344-45: 441-452
Thiel H (1968) Die Einwanderung der Hydromeduse *Nemopsis bachei* L. Ag. aus dem ostamerikanischen Küstengebiet in die westeuropäischen Gewässer und in die Elbmündung. Abhandlungen des Naturwissen-schaftlichen Vereins in Hamburg N.F. 12: 81-94
Thienemann A (1950) Verbreitungsgeschichte der Süßwassertierwelt Europas. Versuch einer historischen Tiergeographie der europäischen Binnengewässer. Die Binnengewässer 18: 1-809
Thornton G (2000) Ballast water decontamination using heat as a biocide. Paper presented at Sea Australia Conference Sydney, Australia, 1-3 February
Thorson G (1961) Length of pelagic larval life in marine bottom invertebrates as related to larval transport by ocean currents. In: Sears M (ed) Oceanography, pp 455-474. American Association for the Advancement of Science, Publication 67
Thresher RE, Hewitt CL & Campbell ML (1999) Synthesis: Introduced and cryptogenic species. In: Hewitt CL, Campbell ML, Thresher RE & Martin RB (eds) Marine Biological Invasions of Port Phillip Bay, Victoria. CRIMP Technical Report 20 pp 283-295. CSIRO Division of Marine Research, Hobart, Australia
Thresher RE, Werner M, Høeg JT, Svane I, Glenner H, Murphy NE & Wittwer C (2000) Developing the options for managing marine pests: specifity trials on the parasitic castrator, *Sacculina carcini*, against the European crab, *Carcinus maenas*, and related species. Journal of Experimental Marine Biology and Ecology 254: 37-51
Throndsen J (1969) Flagellates of Norwegian coastal waters. Nytt magasin for botanikk 16: 161-216
Throndsen J (1990) *Olisthodiscus* and *Heterosigma*, - frustrating algae, especially for the specialists! Blyttia 48
Throndsen J (1996) Note on the taxonomy of *Heterosigma akashiwo* (Raphidophyceae). Phycologia 35: 367
Tilman D (1997) Community invasibility, recruitment limitation, and grassland biodiversity. Ecology 78: 81-92
Tilman D (1999) The ecological consequences of changes in biodiversity: a search for general principals. Ecology 80: 1455-1474
Timm V & Timm T (1993) The recent appearance of a Baikalian crustacean, *Gmelinoides fasciatus* (Stebbing, 1899) (Amphipoda, Gammaridae) in Lake Peipsi. Proceedings of the Estonian Academy of Sciences Biology 42 (2): 144-153
Timonov VV (1950) The main features of hydrological regime of the White Sea. In: In memoriam YuM Shokalsky. Part 2, Moscow-Leningrad, pp 206-236 (in Russian)
Tinenkova DC & Tarasova LI and Petrenko EL (2000). The settling of the introducer *Acartia clausi* in the Caspian Sea. In: Species introducers in the European Seas of Russia, pp 87-90. Murmansk (in Russian)
Tittizer T & Krebs F (eds) (1996) Ökosystemforschung: Der Rhein und seine Auen - eine Bilanz. Springer Verlag, Berlin, Heidelberg, New York
Tittizer T & Taxacher M (1997) First record of *Corbicula fluminea/fluminalis* (Müller 1774) (Corbiculidae, Mollusca) in the Danube. Lauterbornia 31: 103-107 (in German)
Tittizer T (1996) Vorkommen und Ausbreitung aquatischer Neozoen (Makrozoobenthos) in den Bundeswasserstrassen. In: Gebhardt H, Kinzelbach R & Schmidt-Fischer S (eds) Gebietsfremde Tierarten. Auswirkungen auf einheimische Arten, Lebensgemeinschaften und Biotopen, pp 49-86. Ecomed Verlagsgesellschaft, Landsberg
Tittizer T, Leuchs H & Banning M (1994) Macrozoobenthos of the Danube in the section Kelheim - Jochenstein (Danube - km 2414-2302). Limnologie actuel 2: 173-188 (in German)
Tittizer T, Schöll F, Banning M, Haybach A & Schleuter M (2000) Aquatische Neozoen im Makrozoobenthos der Binnenwasserstraßen Deutschlands. Lauterbornia 39: 1-72
Tognetto L, Bellato S, Moro I & Andreoli C (1995) Occurrence of *Ostreopsis ovata* (Dinophyceae) in the Tyrrhenian Sea during summer 1994. Botanica Marina 38: 291-295
Toivonen H & Meriläinen J (1980) Impact of the muskrat (*Ondatra zibethica* L.) on aquatic vegetation in small Finnish lakes. Developments in Hydrobiology 3: 131-138
Tom M & Galil BS (1990) The macrobentic associations of Haifa Bay, Mediterranean coast of Israel. Marine Ecology PSZNI 12: 75-86

Tomlin JR (1927) Report on the Mollusca (Amphineura, Gastropoda, Scaphopoda, Pelecypoda). Transactions of the Zoological Society of London 22: 291-320

Tømmerås BÅ (ed) (1994) Introductions of Foreign Organisms to Norway. Norsk Institutt for Naturforskning, Trondheim, 141 pp (in Norwegian)

Tong SM, Nygaard K, Bernard C, Vørs N & Patterson DJ (1998) Heterotrophic flagellates from the water column in Port Jackson, Sydney, Australia. European Journal of Protistology 34: 162-194

Tournier H (1972) Conditions d'acclimatation des crevettes *Penaeus kerathurus* et *Penaeus japonicus* dans les eaux du littoral languedocien. Science et Pêche 213: 1-13

Touw A & Rubers WV (1989) De Nederlandse bladmossen. Stichting Uitgeverij Koninklijke Nederlandse Natuurhistorische Vereniging, Utrecht, 532 pp

Tripos (1997) Ballastwateronderzoek. Voortgangsverslag 1997. Report 97. T0011

Trometer ES & Busch WDN (1999) Changes in age-0 fish growth and abundance following the introduction of zebra mussels *Dreissena polymorpha* in the Western basin of Lake Erie. North American Journal of Fisheries Management 19: 604-609

Trowbridge CD & Todd CD (1999a) The familiar is exotic: I. *Codium fragile* ssp. *atlanticum* on Scottish rocky intertidal shores. Botanical Journal of Scotland 50: 139-159

Trowbridge CD & Todd CD (1999b) The familiar is exotic: I. *Codium fragile* ssp. *tomentosoides* on Scottish rocky intertidal shores. Botanical Journal of Scotland 50: 161-179

Trowbridge CD (1995) Introduced versus native subspecies of *Codium fragile*: how distinctive is the invasive subspecies *tomentosoides*? Marine Biology 126: 193-204

Trowbridge CD (1998) Ecology of the green macroalga *Codium fragile* (Suringar) Hariot 1889: invasive and non-invasive subspecies. Oceanography and Marine Biology Annual Review 36: 1-64

Tsikhon-Lukashina EA, Reznichenko OG & Lukashova TA (1992) The feeding of ctenophora *Mnemiopsis* in coastal waters of the Black Sea. Oceanology 32: 724-729

Tsoi SCM, Lee SC, Wu WL & Morton BS (1991) Genetic variation in *Corbicula fluminea* (Bivalvia: Corbiculoidae) from Hong-Kong. Malacological Review 24: 25-34

Tugrul S & Salihoglu I (2000) Chemical oceanography of the Marmara Sea and the Turkish Strait System. In: Ozturk B, Kadioglu M & Ozturk H (eds) Marmara Sea 2000 Symposium, pp 327-346. TUDAV Publication, Istanbul (in Turkish)

Turgeon DD (ed) (1998) Common and scientific names of aquatic invertebrates from the United States and Canada: mollusks. American Fisheries Society Special Publication 26: 1-526

Türkay M (1989) Subtropische und tropische Elemente in der Fauna des ostlichen Mittelmeeres. Natur und Museum 119: 183-189

Turley CM (1999) The changing Mediterranean Sea - a sensitive ecosystem? Progress in Oceanography 44: 387-400

Turner RD (1966) A Survey and Illustrated Catalogue of the Teredinidae. Museum of Comparative Zoology, Harvard University, Cambridge, Massachusetts, 265 pp

Turner RD & Johnson, AC (1971) Biology of marine woodboring molluscs. In: Jones EGB & Eltringham SK (eds) Marine Borers, Fungi, and Fouling Organisms of Wood. OECD, Paris, 13: 259-301

Tutin TG, Burges NA, Chater AO, Edmondson JR, Heywood VH, Moore DM, Valentine DH, Walters SM & Webb DA (eds) (1993) Flora Europaea 1. 2nd ed. Psilotaceae to Plantanaceae. Cambridge University Press, Cambridge, 581 pp

Tutin TG, Heywood VH, Burges NA, Moore DM, Valentine DH, Walters SM & Webb DA (eds) (1968) Flora Europaea 2. Rosaceae to Umbelliferae. Cambridge University Press, Cambridge, 455 pp

Tutin TG, Heywood VH, Burges NA, Moore DM, Valentine DH, Walters SM & Webb DA (eds) (1972) Flora Europaea 3. Diapensiaceae to Myoporaceae. Cambridge University Press, Cambridge, 370 pp

Tutin TG, Heywood VH, Burges NA, Moore DM, Valentine DH, Walters SM & Webb DA (eds) (1976) Flora Europaea 4. Plantaginaceae to Compositae (and Rubiaceae). Cambridge University Press, Cambridge, 505 pp

Tutin TG, Heywood VH, Burges NA, Moore DM, Valentine DH, Walters SM & Webb DA (eds) (1980) Flora Europaea 5. Alismataceae to Orchidaceae (Monocotyledones). Cambridge University Press, Cambridge, 452 pp

Tyuryukanov AN, Fedorov VM & Timofeev-Ressovskiy NV (1996) Thoughts of Biosphere. RAEN Press, Moscow, 368 pp (in Russian)

Udekem d'Acoz C d' (1998) Kolonisatie van de Europees-Atlantische kusten door de borselkrab *Hemigrapsus penicillatus* (de Haan, 1835). De Strandvlo 18: 45-48

Udekem d'Acoz C d' (1999) Inventaire et distribution des crustacés décapodes de l'Atlantique nord-oriental, de la Méditerranée et des eaux continentales adjacentes au nord de 25°N. Patrimoines naturels 40: 1-383

Uitto A, Gorokhova E & Välipakka P (1999) Distribution of the non-indigenous *Cercopagis pengoi* in the coastal waters of the eastern Gulf of Finland. ICES Journal of Marine Science 56(Suppl.): 49-57

Ulbrich J (1930) Die Bisamratte. Heinrich, Dresden, 137 pp

Unestam T (1969a) Resistance to the crayfish plague in some American, Japanes and European crayfishes. Report of Institute Freshwater Research, Drottningholm 49: 202-209

Unestam T (1969b) Studies on *Aphanomyces astaci*, physiology and parasitic relationship to the crayfish. PhD Thesis, Uppsala Dissertations in Science 134, University of Uppsala, 22 pp

Unestam T (1972) On the host range and origin of the crayfish plague fungus. Report of the Institute of Freshwater Research Drottingholm 52: 192-198

Uriz M-J (1991) Possible influence of trawl fishery on recent expansion in the range of *Suberites tylobtusa* in the southeast Atlantic. Marine Pollution Bulletin 23: 309-315

Utter F & Ryman N (1993) Genetic markers and mixed stock fisheries. Fisheries 18: 11-21

Utting S & Spencer BE (1992) Introductions of marine bivalve molluscs into the United Kingdom for commercial culture - case histories. ICES Marine Science Symposia 194: 84-91

v. Siemaschko J (1847) Beitrag zur Kenntnis der Konchylien Russlands. Bulletin de la'Societe Imperiale des Naturalistes de Moscou 20: 93-131

Vadas, RL (1992) Littorinid grazing and algal patch dynamics. In: Grahame J, Mill PJ & Reid DG (eds) Proceedings of the Third International Symposium on Littorinid Biology, pp 197-209

Vadiya V & Shenuda S (1985) Role of the Suez Canal and flow from the Nile river in changing the salinity and fauna of the Mediterranean Sea. Journal of Ichtyology 25: 134-135

Väinölä R & Oulasvirta P (2001) The first record of *Maeotias marginata* (Cnidaria, Hydrozoa) from the Baltic Sea: a Pontocaspian invader. Sarsia 86: 401-404

Vaitonis G (1994) The formation of complexes of higher crustaceans in the waters of Lithuania and their meaning in biocenosis. PhD Thesis, Vilnius, 116 pp (in Russian)

Vallarino EA & Elias R (1997) The dynamics of an introduced *Balanus glandula* population in the southwestern Atlantic rocky shores. The consequences on the intertidal community. Marine Ecology PSZNI 18: 319-335

Valovirta I & Porkka M (1996) The distribution and abundance of *Dreissena polymorpha* (Pallas) in the eastern Gulf of Finland. Memoranda Societatis pro Fauna et Flora Fennica 72: 63-78

Van Banning P (1988) Management strategies to control diseases in the Dutch culture of edible oysters. American Fisheries Society Special Publication 18: 243-245

Van Benthem Jutting T (1937) De Portugeesche oester, *Ostrea angulata* Lam., in Zeeland. Basteria 2: 1-3

Van Benthem Jutting T (1943) Mollusca. C. Lamellibranchia. Fauna van Nederland 12: 1-477

Van Breemen PJ (1907) *Petricola pholadiformis* te Noordwijk en bij Den Helder in een veenbank. Tijdschrift Nederlandsche Dierkundige Vereeniging (2)10: 73

Van den Brink FWB & Van der Velde G (1998) Zoetwater-exoten in Nederland: aanwinst of verstoring? De Levende Natuur 99: 23-30

Van den Brink FWB, Paffen BGP, Oosterbroek FMJ & Van der Velde G (1993a) Immigration of *Echinogammarus* (Stebbing, 1899) (Crustacea: Amphipoda) into the Netherlands via the lower Rhine. Bulletin zoölogisch Museum Universiteit van Amsterdam 13: 167-170

Van den Brink FWB, Van der Velde G & Bij de Vaate A (1989) A note on immigration of *Corophium curvispinum* Sars, 1895 (Crustacea: Amphipoda) into the Netherlands via the River Rhine. Bulletin zoölogisch Museum Universiteit van Amsterdam 11: 211-213

Van den Brink FWB, Van der Velde G & Bij de Vaate A (1993a) Ecological aspects, explosive range extension and impact of a mass invader, *Corophium curvispinum* Sars, 1895 (Crustacea: Amphipoda), in the Lower Rhine (The Netherlands). Oecologia 93: 224-232

Van den Brink FWB, Van der Velde G & Cazemier WG (1990) The faunistic composition of the freshwater section of the River Rhine in The Netherlands: its present state and changes since 1900. Limnologie aktuell 1: 191-216

Van den Brink FWB, Van der Velde G, Buijse AD & Klink AG (1996) Biodiversity in the Lower Rhine and Meuse river-floodplains: its significance for ecological river management. Netherlands Journal of Aquatic Ecology 30: 129-149

van den Hoek C, Mann DG & Jahns HM (1995) Algae. An Introduction to Phycology. Cambridge University Press, Cambridge, 623 pp

Van der Meijden R (1996) Heukels' Flora van Nederland. 22th ed. Wolters-Noordhoff, Groningen, 678 pp

Van der Velde G, Paffen BGP, Van den Brink FWB, Bij de Vaate A & Jenner HA (1994) Decline of zebra mussel populations in the Rhine. Competition between two mass invaders (*Dreissena polymorpha* and *Corophium curvispinum*). Naturwissenschaften 81: 32-34

Van der Velde G, Rajagopal S, Kelleher B, Musko IB & Bij de Vaate (2000) Ecological impact of crustacean invaders. General considerations and examples from the Rhine River. In: von Vaupel Klein JC & Schram FR (eds) The biodiversity crisis and Crustacea. Proceedings of 4th International Crustacean Congress, Vol. 2, Brill, Leiden, Crustacean Issues 12: 3-33

Van der Velde G, Rajagopal S, Van den Brink FWB, Kelleher B, Paffen BGP, Kempers AJ & Bij de Vaate A (1998) Ecological impact of an exotic amphipod invasion in the River Rhine. In: Nienhuis PH, Leuven RSEW & Ragas AMJ (eds) New Concepts for Sustainable Management of River Basins, pp 159-169. Backhuys Publishers, Leiden

Van der Visch JA (1975) 'tLandt van Texsel. Den Burg

Van Dijk GM, Marteijn ECL & Schulte-Wülwer-Leidig A (1995) Ecological rehabilitation of the river Rhine: plans, progress and perspectives. Regulated Rivers: Research & Management 11: 377-388

van Oppen MJH, Draisma SGA, Olsen JL & Stam WT (1995) Multiple trans-Arctic passages in the red alga *Phycodrys rubens*: evidence from nuclear rDNA ITS sequences. Marine Biology 123: 179-188

Van Urk G & Marquenie JM (1989) Environmental behaviour of cadmium: who are at risk and why. Proceedings of the International Conference on Heavy Metals in the Environment 2: 456-459

Vanden Boeck A (1995) Etude des rôles potentiels des translocations par eaux de ballast de navires dans l'expansion géographique des efflorescences d'algues toxiques. Mémoire de licence en Géograpahie, Université Libre de Bruxelles

Vanderploeg HA, Johengen TH, Strickler JR, Liebig JR & Nalepa TF (1996) Zebra mussels may be promoting *Microcystis* blooms in Saginaw Bay and Lake Erie. Journal of the North-American Benthological Society 13: 181-182

Vanderploeg HA, Strickler JR, Liebig JR, Nalepa TF, Fahnenstiel GL, Gardner WS, Cavaletto JF, Fanslow DL & Johengen TH (1995) Do zebra mussels promote blue-green and metaphyton blooms in Saginaw Bay, and do these blooms affect the mussels? In: Proceedings of the 38th Conference of the International Association of the Great Lakes Research,1995, p 113

Vardala-Theodorou GE (1999) The occurrence of the Indo-Pacific molluscan species *Fulvia fragilis* (Forsskål, 1775) and *Bulla ampulla* L., 1758 in Elefsis Bay. Newsletter of the Hellenic Zoological Society 31: 10-11

Veltman CJ, Nee S & Crawley MJ (1990) Correlates of introduction success in exotic New Zealand birds. American Naturalist 147: 542-557

Verbitsky VB & Berezina NA (2002) The resistance of amphipod *Gmelinoides fasciatus* (Stebb.) to increasing of water temperature and salinity in different acclimate conditions. Russian Journal of General Biology (in press)

Verhecken A (1984) *Strombus decorus raybaudii* in de Middellandse Zee. Gloria Maris 23: 79-88

Verheyden C & Abbas A (1996) Impact du ragondin sur le milieu. In: Jouventin P, Micol T, Verheyden C & Guedon G (eds) Le Ragondin. Biologie et methods de limitation des populations, pp 44-54. Association de Coordination Technique Agricole, Paris

Verlaque M (1981) Contribution à la flore des algues marine de Méditerranée: espèces nouvelles pour la Méditerranée occidentale. Botanica Marina 24: 559-568

Verlaque M (1987) Contribution à l'étude du phytobenthos d'un écosystème photophile thermophile marine en Méditerranée occidentale. Étude structurale et dynamique du phytobenthos et analyse des relations faune-flore. Thèse de Sciences Naturelles, Université Aix-Marseille II, France, 389 pp

Verlaque M (1989) Contribution à la flore des algues marine de Méditerranée: espèces rares ou nouvelles pour les côtes françaises Botanica Marina 32: 101-113

Verlaque M (1994) Inventaire des plantes introduites en Méditerranée: origines et répercussions sur l'environnement et les activités humaines. Oceanologica Acta 17: 1-23

Verlaque M (2001) Checklist of the macroalgae of Thau Lagoon (Hérault, France), a hot spot of marine species introduction in Europe. Oceanologica Acta 24: 29-49

Verlaque M & Boudouresque CF (1991) *Stypopodium schimperi* (Buchinger ex Kützing) Verlaque & Boudouresque comb. nov. (Dictyotales, Fucophyceae), algue de Mer Rouge récemment apparue en Méditerranée. Cryptogamie, Algologie 12: 195-211

Verlaque M & Fritayre P (1994) Incidence de l'algue introduite *Caulerpa taxifolia* sur le phytobenthos de Méditerranée occidentale. 2 Les peuplements d'algues photophiles de l'infralittoral. In: Boudouresque CF, Meinesz A & Gravez V (eds) First International Workshop on *Caulerpa taxifolia*, pp 349-353. GIS Posidonie Publ., Marseille

Verlaque M & Riouall R (1989). Introduction de *Polysiphonia nigrescens* et d' *Antithamnion nipponicum* (Rhodophyta, Ceramiales) sur le littoral méditerranéen français. Cryptogamie, Algologie 10: 313-323

Verlaque M & Seridi (1991) *Antithamnion algeriensis* nov. sp. (Ceramiaceae, Rhodophyta) from Algeria (Mediterranean Sea) Botanica Marina 34: 153-160

Verlaque M, Boudouresque CF, Meinesz A & Graez V (2000) The *Caulerpa racemosa* complex (Caulerpaceae, Ulvophyceae) in the Mediterranean Sea. Botanica Marina 43: 49-68

Vermeij GJ (1991) When biotas meet: understanding biotic interchange. Science 253: 1099-1104

Vermeij GJ (1996) An agenda for invasion biology. Biological Conservation 78: 3-9

Verrill AE (1873) Report upon the invertebrate animals of Vineyard Sound and the adjacent waters, with an account of the physical characters of the region. Washington Rep. US Commission of Fish and Fisheries 1871-1872: 295-778

Vershinin NV (1967) Biology and distribution of *Gmelinoides fasciatus* Stebb. in Bratskoe Reservoir. Zoological Journal 46(7): 1024-1030 (in Russian)

Viejo RM (1997) The effects of colonization by *Sargassum muticum* on tidepool macroalgal assemblages. Journal of the Marine Biological Association of the United Kingdom 77: 325-340

Vigneaux E (1997) Introductions of freshwater decapod crustaceans into France. Can we speak of management? Bulletin Francais de la Pêche de la Pisciculture 344-5: 357-370

Vila M, Camp J, Garces E, Maso M & Delgado M (2001) High resolution spatio-temporal detection of potentially harmful dinoflagellates in confined waters of the NW Mediterranean. Journal of Plankton Research 23: 497-514

Vila M, Garcés E, Masó M & Camp J (2001) Is the distribution of the toxic dinoflagellate *Alexandrium catenella* expanding along the NW Mediterranean coast? Marine Ecology Progress Series 222:73-83

Villac MC, Persich G, Fernandes L, Paranhos R, Bonecker S, Garcia V, Odebrecht C, Tenenbaum D, Tristao ML, de Andrade S & Fadel A (2000) Ballast water exchange: testing the dilution method. Paper presented at 9th Conference on Harmful Algal Blooms, Hobart, Tasmania

Villac MC, Persich G, Fernandes L, Paranhos R, Dias C, Bonecker S, Garcia V, Odebrecht C, Tenenbaum D, Tristao ML, De Andrade S & Fadel A (2001) Ballast water exchange: testing the dilution method (Petroras, Brazil). In: Hallegraef GM, Blackburn SI, Bolch CJ & Lewis RJ (eds) Proceeding of 9th International Conference on Harmful Algal Blooms, pp 470-473. Intergovernmental Oceanographic Commission of UNESCO, Paris

Villari R (1988) Segnalazioni floristiche italiane: 565 *Halophila stipulacea* (Forsskål) Ascherson. Informatore Botanico Italiano 20: 672

Vincent T & Breton G (1999) Présence du crabe *Hemigrapsus penicillatus* (de Haan, 1835) dans les bassins du port du Havre (Normandie, France). Bulletin Trimestriel de la Société Géologique de Normandie et des Amis du Muséum du Havre 86: 19-23

Vinogradov ME (1990) Investigation of the pelagic ecosystem of the Black Sea (44th Cruise of the R/V Dmitriy Mendeleyev, 4 July-17 September 1989). Oceanology 30: 254-256

Vinogradov M, Vostokov S, Arashkevich E, Drits A, Musaeva E, Anokhina L & Shushkina E (2000) Peculiarity of the biology of exotic ctenophores and their role in the Black Sea ecosystem. In: Exotic Species in European Seas of Russia, pp 91-113. Apatity, Russia (in Russian)

Vinogradov ME, Shushkina EA & Nikolaeva GG (1993) The state of zoocoene in the open areas of the Black Sea during late summer 1992. Okeanologiya 33: 382-387

Vinogradov ME, Shushkina EA, Musaeva EI & Sorokin PYu (1989) Ctenophore *Mnemiopsis leidyi* (A. Agassiz) (Ctenophora: Lobata) - a new settler in the Black Sea. Oceanology 29: 293-298 (in Russian)

Vinogradov MYE, Shushkina EA, Musayeva EI & Sorokin PYU (1989) A newly acclimated species in the Black Sea: the ctenophore *Mnemiopsis leidyi* (Ctenophore: Lobata). Oceanology 29: 220-224

Vitousek PM (1990) Biological invasions and ecosystem processes: towards an integration of population biology and ecosystem studies. Oikos 57: 7-13

Vitousek PM, Mooney HA, Lubchenco J & Melillo JM (1997) Human domination of Earth's ecosystems. Science 277: 494-499

Vogt G (1999) Diseases of European freshwater crayfish, with particular emphasis on interspecific transmission of pathogens. In: Gherardi F & Holdich DM (eds) Crayfish in Europe as Alien Species. How to Make the Best of a Bad Situation, pp 87-103. A.A. Balkema, Rotterdam, Brookfield

Voigt M (2000) Treatment methods for ballast water. In: Maritime Conferences "Fresh water production and waste water treatment technologies for ships and islands" 15.-17.03.2000, Genoa, Italy. 11 pp

Volkov VV & Potina II (1977) Distribution, sizes and fecundity of *Gmelinoides fasciatus* (Stebbing), acclimatized in Gorkovskoe Reservoir. Fishery Studies of Inland Waters 21: 18-21 (in Russian)

Vollenweider RA, Munawar M & Stadelmann P (1974) A comparative review of phytoplankton and primary production in the Laurentian Great Lakes. Journal of Fisheries Research Board of Canada 31: 739-762

Volovik SP (ed) (2000) Ctenophore *Mnemiopsis leidyi* (A. Agassiz) in the Azov and Black Seas: Its Biology and Consequences of Its Intrusion. State Fisheries Commission of Russian Federation, Rostov-on-Don, 497 pp (in Russian)

Volovik SP, Dubinina VG & Semenov AD (1993) Hydrobiology and dynamics of fisheries in the Azov Sea. General Fisheries Council for the Mediterranean Studies and Reviews 64(2): 1-58

Vorobyeva NK, Rozhdestvenskaya AD & Sorokin AL (1985) Perspectives of fish-pond cultivation of trout and hoopid salmon in the White Sea In: Problems of study, rational exploitation and conservation of natural resources of the White Sea. Proceedings of the regional meeting, Archangel, p 221 (in Russian)

Vyushkova VP (1971) Discovery of *Comigerius maeoticus* (Polyphemidae, Cladocera) in the Volgograd zreservoir. Zoological Journal 50: 1875-1877 (in Russian)

Vyushkova VP & Gurova TV (1968) Discovery of the brackish-water crustacean *Calanipeda aquae-dulcis* Kritcz (Copepoda) in the Volgograd reservoir. Zoological Journal 47: 1726-1727 (in Russian)

Wade MJ, McKnight ML & Shaffer HB (1994) The effects of kin-structured colonization on nuclear and cytoplasmic genetic diversity. Evolution 48: 1114-1120

Waite T & Kazumi J (2001) Ballast Water Treatment Standards - Concepts and Issues. International Conference on Marine Bioinvasions, April 9-11, New Orleans, LA, USA

Waite TD, Kazumi J, Farmer L, Capo TR, Lane P, Hitchcock G, Smith SL & Smith SG (2001) Testing of Ballast Water Treatment Technologies at Large Scale. 1st International Ballast Water Treatment Research & Development Symposium of the Global Ballast Water Management Program (GloBallast, International Maritime Organization), London, UK March 26-30, 2001

Walford L & Wicklund R (1973) Contribution to a world-wide inventory of exotic marine and anadromous organisms. FAO Fisheries Technical Papers 121: 1-49

Wallentinus I (1999) Introduction and transfer of plants. In: Munro ALS, Uting SD & Wallentinus I (eds) Status of introductions of non-indigenous marine species to North Atlantic waters 1981-1990. ICES Cooperative Research Report 231: 1-43

Wallentinus I (1999) *Sargassum muticum*. In: Gollasch S, Minchin D, Rosenthal H & Voigt M (eds) Exotics Across the Ocean. Case Histories on Introduced Species, pp 21-30. Logos Verlag, Berlin

Wallentinus I (1999) *Undaria pinnatifida*. In: Gollasch S, Minchin D, Rosenthal H & Voigt M (eds) Exotics Across the Ocean. Case Histories on Introduced Species, pp 11-19. Logos Verlag, Berlin

Walne PR (1974) Culture of Bivalve Molluscs. 50 years experience at Conwy. Fishing News Books, West Byefleet. 173 pp

Waser PM & Elliot LF (1991) Dispersal and genetic structure in kangaroo rats. Evolution 45: 935-943

Waterstraat AK & Kohn J (1989) Contribution to the fauna of the Kummerov Lake; first record of the amphipod *Echinogammarus ischnus* Stebbing, 1899, in GDR. Archiv Freunde Naturgeschichte in Mecklenburgs 29: 93-106 (in German)

Wathern P (1984) Ecological Modelling in Impact Analysis. In: Roberts RD & Roberts TM (eds) Planning and Ecology, pp 80-98. Chapman and Hall, London

Wathern P (1988) Environmental Impact Assessment. Theory and Practice. Routledge, London

Watling L (1979) Zoogeographic affinities of northeastern North America gammaridean Amphipoda. Bulletin of the Biological Society of Washington 3: 256-282

Wawrzyniak-Wydrowska B & Gruszka P (2001) *Gammarus tigrinus* and *Pontogammarus robustoides* - new gammarid species in the river Odra/Oder estuary. Book of abstracts of Third International Symposium on: Functioning of Coastal Ecosystems in Various Geographical Regions, Gdynia, June 19-22, 2001, p 29

Weber E & D'Antonio CM (1999) Germination and growth responses of hybridizing *Carpobrotus* species (Aizoaceae) from coastal California to soil salinity. American Journal of Botany 86: 1257-1263

Weber M & Brandle R (1996) Some aspects of the extreme anoxia tolerance of the sweet flag, *Acorus calamus* L. Folia Geobotanica & Phytotaxonomica 31: 37-46

Wehrmann A, Herlyn M, Bungenstock F, Hertweck G & Millat H (2000) The distribution gap is closed. First record of naturally settled Pacific oysters *Crassostrea gigas* in the East Frisian Wadden Sea, North Sea. Senckenbergiana maritima 30: 153-160

Weidema IR (ed) (2000) Introduced Species in the Nordic Countries. Nordic Council of Ministers, Copenhagen. Nord 2000:13, 242 pp

Weis JS & Weis P (1996) The effects of using wood treated with chromated copper arsenate in shallow-water environments: a review. Estuaries 19: 306-310

Weitkamp DE & Katz M (1980) A review of dissolved gas supersaturation literature. Transactions of the American Fisheries Society 109: 659-702

Welcomme RL (1988) International Introductions of Inland Aquatic Species. FAO Fisheries Technical Paper 294, 318 pp

Welcomme RL (1991) International introductions of freshwater fish species into Europe. Finnish Fisheries Research 12:11-18

Welcomme RL (1992) A history of international introductions of inland aquatic species. ICES marine Science Symposia 194: 3-14

Wells RDS, De Winton MD & Clayton JS (1997) Successive macrophyte invasions within the submerged flora of Lake Tarawera, Central North Island, New Zealand. New Zealand Journal of Marine and Freshwater Research 31: 449-459

Wendling K (1993) Gewässergüte gestern und heute im rheinland-pfälzischen Rheinabschnitt. In: Ministerium für Umwelt Rheinland-Pfalz (ed) Die Biozönose des Rheins im Wandel. Lachs 2000? pp 79-87. Advanced Biology, Petersberg

Went AEJ (1980) Freshwater fishes. In: Keman RP, Mooney OV & Went AEJ (eds) The Introduction of Exotic Species: Advantages and Problems, pp 65-68. Royal Irish Academy, Dublin

Werner B (1948) Die amerikanische Pantoffelschnecke *Crepidula fornicata* L. im nordfriesischen Wattenmeer. Zoologisches Jahrbuch, Abteilung Systematik, Ökologie und Geographie der Tiere 77: 449-488

Westman K (1973) The population of the crayfish, *Astacus astacus* L. in Finland and the introduction of the American crayfish *Pacifastacus leniusculus* Dana. Freshwater Crayfish 1: 41-55

Westman K (1985) Effects of habitat modification on freshwater crayfish. In: Alabaster JS (ed) Habitat modification and freshwater fisheries. Proceedings of a symposium of the European Inland Fisheries Advisory Commission, Aarhus, Denmark, 23-25 May 1984: 245-255 (FAO), Butterworths, London

Westman K (1991) The crayfish fishery in Finland - its past, present and future. Finnish Fisheries Research 12: 187-216

Westman K (1992) Management of the noble crayfish *Astacus astacus* (L.) and the signal crayfish *Pacifastacus leniusculus* (Dana) in Finland. Finnish Fisheries Research 14: 39-51

Westman K (1995) The 6th Sture Abrahamsson Memorial Lecture. Introduction of alien crayfish in the development of crayfish fisheries: experience with signal crayfish (*Pacifastacus leniusculus* (Dana)) in Finland and the impact on the native noble crayfish (*Astacus astacus* (L.)). Freshwater Crayfish 10: 1-17

Westman K (2000) Comparison of the crayfish *Pacifastacus leniusculus* Dana, a species introduced into Finland, with the native species, *Astacus astacus* L., in allopatry and sympatry. PhD Thesis. Game and Fish Research, Helsinki, 50 +157 pp

Westman K & Manninen K (eds) (1996) Institutes, research workers and programmes related to research on crayfish in Europe. Kala- ja riistaraportteja 65, 82 pp

Westman K & Westman P (1992) Present status of crayfish management in Europe. Finnish Fisheries Research 14: 1-22

Westman K, Pursiainen M & Westman P (eds) (1990) Status of crayfish stocks, fisheries, diseases and culture in Europe. Report of the FAO European Inland Fisheries Advisory Commission (EIFAC) Working Party on Crayfish. Riista- ja kalatalouden tutkimuslaitos. Kalatutkimuksia - Fiskundersökningar 3: 1-206

Westman K, Savolainen R & Julkunen M (2002) Replacement of the native crayfish, *Astacus astacus* by the introduced species *Pacifastacus leniusculus* in a small, enclosed Finnish lake: a 30-year study. Ecography 25: 53-73

Wetsteyn LPMJ (2001) National report for The Netherlands. In: Report of the Working Group of Introductions and Transfers of Marine Organisms, Barcelona, Spain, 21-23 March 2001, pp 47-48. ICES CM 2001/ACME:08, Ref. E, F, Annex 3

White PS & Pickett STA (1985) Natural disturbance and patch dynamics: an introduction. In: Pickett STA & White PS (eds) The Ecology of Natural Disturbance and Patch Dynamics, pp 3-13. Academic Press, Sydney, Australia

Whittier TR & Kincaid TM (1999) Introduced fish in northeastern USA lakes: Regional extent, dominance, and effect on native species richness. Transactions of the American Fisheries Society 128: 769-783

WHO CEMP (1992) Environmental and Health Impact Assessment of Development Projects. A Handbook for Practitioners. Elsevier Applied Science, New York

Wiedemann AM & Pickart A (1996) The *Ammophila* problem on the Northwest Coast of North America. Landscape and Urban Planning 34: 287-299

Wiedenmann J, Baumstark A, Pillen TL, Meinesz A & Vogel W (2001) DNA fingerprints of *Caulerpa taxifolia* provide evidence for the introduction of an aquarium strain into the Mediterranean Sea and its close relationship to an Australian population. Marine Biology 138: 229-234

Wiik Ø & Nerland T (1972) Interesting finds of algae from Sunnmøre, NW Norway. Blyttia 30: 141-151

Wikander PB (1986) Suitability studies for extensive mariculture in Aust-Agder county. Norwegian Institute for Water Research, Report1898,159 pp (in Norwegian)

Wiktor J (1969) Biology of *Dreissena polymorpha* (Pall.) and its ecological effect in the Szczecin Firth. Studia i Materialy Morskiego Instytutu Rybackiego, Gdynia A2: 1-100 (in Polish)

Willan RC (1987) The mussel *Musculista senhousia* in Australia: another aggressive alien highlights the need for quarantine at ports. Bulletin of Marine Science 41: 475-89

Willan RC, Russell BC, Murfet NB, Moore KL, McEnnulty FR, Horner SK, Hewitt CL, Dally GM, Campbell ML & Bourke ST (2000) Outbreak of *Mytilopsis sallei* (Recluz, 1849) (Bivalvia: Dreissenidae) in Australia. Molluscan Research 20(2): 25-30

Williams AB (1984) Shrimps, Lobsters and Crabs of the Atlantic Coast of the Eastern United States, Maine to Florida. Smithsonian Institution Press, Washington D.C. 550 pp

Williams RJ, Griffiths FB, Van der Wal EJ & Kelly J (1988) Cargo vessel ballast water as a vector for the transport of non-indigenous marine species. Estuarine, Coastal and Shelf Science 26: 409-420

Williamson M (1989) Mathematical models of invasion. In: Drake JA, Mooney HA, di Castri F, Groves RH, Kruger FJ, Rejmánek M & Williamson M (eds) Biological Invasions, pp 329-350. John Wiley & Sons, New York

Williamson M (1994) Community response to transgenic plant release: Predictions from British experience of invasive plants and feral crop plants. Molecular Ecology 3: 75-79

Williamson M (1996) Biological Invasions. Chapman & Hall, London, 244 pp

Williamson M (1999) Invasions. Ecography 22: 5-12

Williamson MH & Brown KC (1986) The analysis and modelling of British invasions. Philosophical Transactions of the Royal Society of London B 314: 505-522

Williamson MH & Fitter A (1996) The characters of successful invaders. Biological Conservation 78: 163-170

Wisenden PA & Bailey RC (1995) Development of macroinvertebrate community structure associated with zebra mussel (*Dreissena polymorpha*) colonization of artificial substrates. Canadian Journal of Zoology 73: 1438-1443

Wisniewski R (1990) Shoals of *Dreissena polymorpha* as bio-processor of seston. Hydrobiologia 200/201: 451-458

Witkowski A (1989) Fishes introduced to Polish waters and their effect on environment. Przeglad Zoologiczny 33: 583-598 (in Polish)

Witkowski A (1996) Changes in the ichthyofauna of the Polish rivers: native and introduced species. Zoologica Poloniae 41(suppl.): 29-40 (in Polish)

Witt JDS, Hebert PDN & Morton WB (1997) *Echinogammarus ischnus*, another crustacean invader in the Laurentian Great Lakes basin. Canadian Journal of Fishery and Aquatic Sciences 54: 264-268

Wittman KJ (1995) Zur Einwanderung potamophiler Malacostraca in die obere Donau: *Limnomysis benedeni* (Mysidacea), *Corophium curvispinum* (Amphipoda) und *Atyaephyra desmarestii* (Decapoda). Lauterbornia 20: 77-86

Wohlenberg E (1935) Beobachtungen über das Seegras, *Zostera marina* L., und seine Erkrankung im nordfriesischen Wattenmeer. Nordelbingen 11: 1-19

Wolff WJ (1999) Exotic invaders of the meso-oligohaline zone of estuaries in The Netherlands: Why are there so many? Helgoländer Meeresuntersuchungen 52: 393-400

Wolnomiejski N (1970) The effects of *Dreissena polymorpha* Pall. aggregation on the differentiation of the benthonic macrofauna. Zecz nauk UMK 25: 31-39

Wonham MJ, Walton WC, Frese AM & Ruiz GM (1996) Transoceanic transport of ballast water: biological and physical dynamics of ballasted communities and the effectiveness of mid-ocean exchange. SERC Report

Wonham MJ, Walton WC, Ruiz GM, Frese AM & Galil BS (2001) Going to the source: role of the invasion pathway in determining potential invaders. Marine Ecology Progress Series 215: 1-12

Woo M, Smith C & Smith W (2000) Ecological interactions and impacts of invasive *Kappaphycus striatum* in Kane'ohe Bay, a tropical reef. In: Pederson J (ed) Marine Bioinvasions. Proceedings of the First National Conference, January 24-27, 1999, pp 186-192. Massachusetts Institute of Technology, Cambridge, MA

Woodward JB (1990) Ship ballast water designs to kill organisms before discharge. EPA Workshop on Zebra Mussels and other Introduced Aquatic Nuisance Species, Saginaw Valley State University, USA, 2 pp

Woomer PL, Muzira R, Bwamiki D, Mutetikka D, Amoding A & Bekunda MA (2000) Biological management of water hyacinth waste in Uganda. Biological Agriculture & Horticulture 17: 181-196

Wotton RS, Malmquist B, Muotka T & Larsson K (1998) Fecal pellets from a dense aggregation of suspension-feeders in a stream: an example of ecosystem engineering. Limnology and Oceanography 43: 719-725

Wundsch HH (1912) A new species of the genus *Corophium* Latr. from the Muggel Lake near Berlin. Zoologischer Anzeiger 39: 729-738 (in German)

Yablonskaya EA (1967) About nutrition of the Caspian Sea mollusks. Informacionnyi Buleten' instituta Biologii Vnytrennikh Vod 1: 47-50 (in Russian)

Yablonskaya EA (1969) Particulate matter as a food material for benthic organisms of the Caspian Sea. Trudy VNIRO 65:85-147 (in Russian)

Yacobi YZ, Zohary T, Kress N, Hecht A, Robarts RD, Waiser M, Wood AM & Li WKW (1995) Chlorophyll distribution throughout the southeastern Mediterranean in relation to the physical structure of the water mass. Journal of Marine Systems 6: 179-190

Yakovenko MYa (1995) The pink salmon as an object of fishery in the White Sea basin. In: Alimov AF (ed) Problems of study, rational exploitation and conservation of natural resources of the White Sea. Proceedings of the 6th regional meeting. St. Petersburg, pp 35-38 (in Russian)

Yakovleva AN (1975) Saratov Reservoir. Izvestiya GosNIORKh 102: 118-130 (in Russian)

Yilmaz A & Tugrul S (1998) The effect of cold- and warm-core eddies on the distribution and stoichiometry of dissolved nutrients in the northeastern Mediterranean Journal of Marine Systems 16: 253-268

Yoccoz NG, Nichols JD, & Boulinier T (2001) Monitoring of biological diversity in space and time. Trends in Ecology and Evolution 16: 446-453

Zaccaria M (1968) Studi geologici sulle isole del Dodecaneso (Mare Egeo) VIII. Fauna Calabriana dell'isola di Rodi. Rivista Italiana di Paleontologia e stratigrafia 74: 275-306

Zadoenko IN, Leis OA & Grigorjev VF (1985) Results and perspectives of acclimatization of the Baikalian amphipods in water bodies of the USSR. Proceedings of GosNIORKh 232: 30-34 (in Russian)

Zaika VE & Revkov NK (1994) Anatomy of gonads and spawning regime of ctenophore *Mnemiopsis* sp. in the Black Sea. Zoological Journal 73: 5-10 (in Russian)

Zaika VE & Sergeeva NG (1990) Morphology and development of ctenophore-colonizer *Mnemiopsis mccradyi* (Ctenophora, Lobata) in the Black Sea. Zoological Journal 69: 5-11 (in Russian)

Zaitsev Yu P (1993) Impact of eutrophication on the Black Sea fauna. Studies and Reviews 63: 59-86

Zaitsev YuP (1998) Marine hydrobiological investigations of National Academy of Science of Ukraine in the 1990s: Shelf and coastal water bodies of the Black Sea. Gidrobiologicheskij Zhurnal 34(6): 3-21 (in Russian)

Zaitsev Y & Ozturk B (eds) (2001) Exotic Species in the Aegean, Marmara, Black, Azov and Caspian Seas. Turkish Marine Research Foundation, Istanbul, 267 pp

Zaitsev Yu & Mamaev V (1997) Biological Diversity in the Black Sea: A Study of Change and Decline. New York, UN Publications. 208 pp

Zaitsev Yu & Öztürk B (eds) (2001) Exotic species in the Aegean, Marmara, Black, Azov and Caspian Seas. Turkish Marine Research Foundation, Istanbul, Turkey. 265 pp

Zakova Z, Palat M Kockova E & Toufar J (1994) Is it realistic to use water hyacinth for waste-water treatment and nutrient removal in central Europe? Water Science & Technology 30: 303-311

Zawisza J & Ciepielewski W (1973) Experimentally increased fish stock in the pond type lake Warniak. II. Changes of the autochtonous ichthyofauna due to the introduction of Carp (*Cyprinus carpio* L.) and bream (*Abramis brama* L.). Ekologia Polska 21: 424-444

Zenetos A (1994) *Scapharca demiri* (Piani, 1981): First finding in the North Aegean Sea. La Conchiglia 271: 37-38

Zenetos A (1997) STAR C/P 12.9.96. Hull fouling report NCMR Technical Report for Centrofin Management Inc.

Zenkevich LA (1963) Biology of the Seas of the USSR. Nauka, Moscow. 450 pp (in Russian)

Zettler ML (1996a) Successful establishment of the spionid polychaete *Marenzelleria viridis* (Verrill 1873), in the Darss-Zingst estuary (southern Baltic) and its influence on the indigenous macrozoobenthos. Archive of Fishery and Marine Research 43: 273-284

Zettler ML (1996b) Ökologische Untersuchungen am Neozoon *Marenzelleria viridis* (Verrill, 1873) (Polychaeta: Spionidae) in einem Küstengewässer der südlichen Ostsee. PhD Thesis, University of Rostock, 149 pp

Zettler ML (1997a) Population dynamics, growth and production of the neozoon *Marenzelleria* cf. *viridis* (Verrill, 1873) (Polychaeta: Spionidae) in the southern Baltic Sea. Aquatic Ecology 31: 177-186

Zettler ML (1997b) Bibliography on the genus *Marenzelleria* and its geographical distribution, principal topics and nomenclature. Aquatic Ecology 31: 233-258

Zettler ML (1998) On the distribution of Malacostraca (Crustacea) in freshwaters and offshore waters of Mecklenburg - Vorpommern. Lauterbornia 32: 49-65 (in German)

Zettler ML (1999) First record of *Dikerogammarus villosus* (Sovinski, 1894) and further record of *Gammarus varsoviensis* Jazdzewski, 1975 in Mecklenburg-Vorpommern (Crustacea: Amphipoda). Archiv der Freunde der Naturgeschichte Mecklenburgs 38: 231-233 (in German)

Zettler ML, Bönsch R & Gosselck F (2000) Distribution of macroinvertebrates in the Mecklenburg Bight (southern Baltic Sea) - recent and in a historic comparision. Meereswissenschaftliche Berichte 42: 1-144 (in German)

Zevina GB & Kuznetsova IA (1965) The part played by navigation in modifying the Caspian fauna. Oceanology 5: 101-108

Zhadin VI (1946) The travelling shellfish *Dreissena*. Priroda 5: 29-37 (in Russian)

Zhang F & Dickman M (1999) Mid-ocean exchange of container vessel ballast water. 1: seasonal factors affecting the transport of harmful diatoms and dinoflagellates. Marine Ecology Progress Series 176: 243-251

Zhuravel PA (1967) On *Dreissena bugensis* spread in man-made waterbodies. Gidrobiologicheskii Zhurnal 3(2): 87-90 (in Russian)

Zibrowius H (1974) *Oculina patagonica*, Scléractiniaire hermatypique introduit en Méditerranée. Helgoländer wissenschaftliche Meeresuntersuchungen 26: 153-173

Zibrowius H (1979) Serpulidae (Annelida Polychaeta) de l'océan Indien arrivés sur des coques de bateaux à Toulon (France). Rapports et procès-verbaux des réunions, Commission internationale pour l'Exploration scientifique de la mer Méditerranée 25/26 : 133-134

Zibrowius H (1991) Ongoing modification of the Mediterranean marine fauna and flora by the establishment of exotic species. Bulletin du Musée d'Histoire Naturelle de Marseilles 51: 83-107

Zibrowius H & Thorp CH (1989) A review of alien serpulid and spirorbid polychaetes in the British Isles. Cahiers de Biologie Marine 30: 270-285

Zmudzinski L (1957) Zoobentos of the Vistula Lagoon. Prace Morskiego Instytutu Rybackiego, Gdynia 9: 453-500 (in Polish)

Zmudzinski L (1990) Animal world of the Baltic Sea - Atlas of Macrofauna. Warszawa, 196 pp (in Polish)

Zmudzinski L (1999) *Cercopagis pengoi* (Cladocera) conquered the southern Baltic Sea. In: Baltic Coastal Zone. Institute of Biology and Environmental Protection. University of Education, Slupsk, 1998/1999 2: 95-96

Zmudzinski L, Chubarova-Solovjeva S, Dobrowolski Z, Gruszka P, Olenin S & Wolnomiejski N (1997) Expansion of the spionid polychaete *Marenzelleria viridis* in the southern part of the Baltic Sea. In: Andrushaitis A (ed) Proceedings of the 13th Symposium of the Baltic Marine Biologists. Riga, pp 127-129

Zolotarev V (1996) The Black Sea ecosystem changes related to the introduction of new mollusc species. Marine Ecology PSZNI 17: 227-236

Zonn IS (2000) Three Centuries at the Caspian (The Synchronism of Major Historical Events of XVIII–20 cc.), 72 pp. Moscow

Zouros E, Singh SM, Foltz DW & Mallet AL (1983) Post-settlement viability in the American Oyster (*Crassostrea virginica*): an overdominant phenotype. Genetical Research 41: 259-270

Zubchenko AV, Popov NG & Svenning MA (1998) Salmon rivers on the Kola Peninsula. Some results of acclimation of pink salmon (*Oncorhynchus gorbusha* (Walbaum). ICES C.M. 13: 11 pp

Zuljevic A, Thibaut T, Elloukal H & Meinesz (2001) Sea slug disperses the invasive *Caulerpa taxifolia*. Journal of the Marine Biological Association of the UK 81: 343-344

9 781402 008375

GPSR Compliance

The European Union's (EU) General Product Safety Regulation (GPSR) is a set of rules that requires consumer products to be safe and our obligations to ensure this.

If you have any concerns about our products, you can contact us on ProductSafety@springernature.com

In case Publisher is established outside the EU, the EU authorized representative is:

Springer Nature Customer Service Center GmbH
Europaplatz 3
69115 Heidelberg, Germany

Batch number: 10584520

Printed by Printforce, the Netherlands